P9-DMD-269

The Know-How of Face Transplantation

Maria Z. Siemionow
(Editor)

The Know-How of Face Transplantation

Editor
Maria Z. Siemionow, M.D., Ph.D., D.Sc
Department of Plastic Surgery
Cleveland Clinic
Cleveland, OH
USA
siemion@ccf.org

ISBN 978-0-85729-252-0 e-ISBN 978-0-85729-253-7
DOI 10.1007/978-0-85729-253-7
Springer London Dordrecht Heidelberg New York

A catalogue record for this book is available from the British Library

Library of Congress Control Number: 2011922529

Cover illustration: With permission from Cleveland Clinic Foundation and illustrator Mark Sabo, BFA

Cover design: eStudio Calamar, Figueres/Berlin

Printed on acid-free paper

Springer is part of Springer Science+Business Media (www.springer.com)

This book is dedicated to all organ donors and their families who are the silent heroes behind medical breakthroughs such as transplantation of the human face.

Preface

It is a great privilege to introduce the book *The Know-How of Face Transplantation* to those who are interested in innovations in plastic and reconstructive surgery as well as innovation in the transplantation field.

The idea to write *The Know-How of Face Transplantation* came early on, even before we had performed the first face transplant in the USA. The preparation process for face transplantation involved experimental studies, cadaver dissections, much legislative work and approval from different organizations including the Institutional Review Board (IRB), organ procurement organizations, coroner's office approvals, as well as approvals from different states' organ procurement organizations.

My intention was to share, with those interested in development of new programs for composite tissue allograft transplantation, our own experience and the experience of others in order to facilitate establishment of reconstructive transplantation programs in other US institutions, as well as in other countries worldwide.

Face transplantation has generated a lot of attention over the last six years, and it started in 2004 with the announcement of Cleveland Clinic granting the world's first IRB approval to proceed with human face transplantation. A lot of ethical, societal, as well as medical debate ensued after this approval was granted. The interest of the media nationally and internationally, as well as patient advocate and other groups, supported this breakthrough concept and procedure; however, many questions were raised regarding ethical issues, medical issues such as the need for lifelong immunosuppression, as well as issues of financial support for this novel procedure.

When I thought about the concept of creating a know-how manual for face transplantation, I did not want the reader to get the impression that this is a recipe which, when followed, will guarantee a 100% success rate. Based on the years of work and preparation and experience in developing the program of face transplantation at Cleveland Clinic, I realized that there are not only surgical and technical issues which need to be shared, but also issues related to experimental studies, anatomical cadaver dissections, ethics, legal approval and legislative issues, as well as societal, financial, and public relations concerns. This was a tremendous undertaking, and I have taken this work very personally due to the fact that a book on a breakthrough procedure which has been performed, for the first time, on only a few patients, brings a great responsibility to the Editor. In order to include the experience of all world experts who have performed face transplantation, I invited all surgeons who had participated, at the time of book production, in face transplantation programs, in their respective countries and institutions. These included Dr. Dubernard and Dr. Devauchelle, from Lyon, France, Dr. Lengele, from Belgium, Dr. Shuzhong, from China, Dr. Lantieri, from Paris, France, Dr. Cavadas, from Valencia, Spain, Dr. Pomahac and Dr. Pribaz, from Boston, as well as Dr. Butler, from London. I was hoping that all would contribute with a chapter sharing their experience. I also invited experts in transplantation ethics, infectious disease, rehabilitation, transplant immunology, media relation representatives, as well as organ procurement organization experts. I received overwhelming support from most of the centers; however, a few of the institutions' leaders decided to not participate in our educational journey. Therefore, we have included their experience in the review chapters summarizing the world experience with face transplantation.

The book has 72 contributing authors and 44 chapters which are divided into eight major sub-categories of topics outlined in the following order:

Part I Preclinical Aspects of Face Transplantation

There are 7 chapters in this part which discuss the issues of the face as a functional organ, the face as a sensory organ, and immunological aspects related to face transplantation. In addition, experimental studies in rodents, as well as large animal models including swine and primates, are discussed. Finally, the timeline and preparation for face transplantation in the cadaver model is presented.

Part II Clinical Aspects in Preparation for Face Transplantation

This part includes 8 chapters describing guidelines for technical aspects of face transplantation, anesthesia-related issues in face transplantation, as well as alternative approaches to face transplantation. In addition, ethical concerns, as well as psychological aspects of face transplantation, are thoroughly discussed, and physical therapy and rehabilitation, as well as prosthetic support relevant to face transplantation, are presented.

Part III Monitoring Aspects of Face Transplantation

This part includes 7 chapters and describes the important issues of how to monitor patients after facial transplantation, emphasizing details of immunological monitoring, pathological monitoring, and classification of facial graft rejection. In addition, the issues of brain plasticity, functional EEG, as well as sensory recovery and methods of assessment of cortical plasticity after face transplantation, are discussed.

Part IV Approval Process of Face Transplantation

Here, in 5 chapters, we outline the process of IRB approval, the ethical presentation of patients' informed consent, the legal and regulatory aspects of face donation and transplantation, the issue of death and end of life, as well as organ procurement organizations' approval process.

Part V Societal, Financial, and Public Relations Issues in Face Transplantation

This part summarizes, in 3 chapters, cultural, religious, and philosophical views on face transplantation, a comparative cost analysis of conventional reconstruction versus face transplantation, and finally, media-related aspects, viewed from a public relations perspective, on face transplantation.

Part VI World Experience with Face Transplantation

This part summarizes, in 7 chapters, the global experience with face transplantation: the facial allotransplantation experience in China, Cleveland Clinic's experience, the Spanish team's experience, as well as microsurgical aspects and sensory recovery following face transplantation. In addition, infectious issues related to face transplantation are outlined.

Part VII Future Directions in Face Transplantation

The 6 chapters in this part of the book discuss the military cases relevant to face transplantation, regenerative medicine approaches, the international registry of face transplantation, the aspects of concomitant face and upper extremity transplantation, immunosuppressive protocols for composite tissue transplantation, new cellular therapies, as well as novel aspects of tissue engineering in face transplantation.

Part VIII Current Status of Face Transplantation

This final part of the book summarizes, in one chapter, the technical and functional outcomes of the 13 face transplants performed thus far, between 2005 and 2010, by all institutions worldwide.

I hope that this book will help those who are planning to establish composite tissue allograft programs in their institutions and countries to understand that the approach to a novel procedure requires the cooperative effort of a team of multidisciplinary experts from different fields which are, quite often, far removed from the daily surgical activities of reconstructive surgeons. This book summarizes many of the issues which, as a surgeon, I had not considered when preparing for facial transplantation and which developed during the lengthy process of creating the face transplant program at Cleveland Clinic. I understand, from my interactions with contributing authors, that they have enjoyed the process of writing about a topic as new and undiscovered as face transplantation, a topic on which we do not yet have long-term patient outcomes to report and share.

It has been a privilege to work with so many of the field's experts in putting this book together. I hope that a careful process of preparation for face transplantation, as outlined in this book, will not be underestimated and justifies this procedure as a medical, ethical, and societal breakthrough. The final message which I want to convey is that the technical aspect of face transplantation is only one of many challenges, and the beginning of a fascinating journey of helping patients who have lost their faces, since "You need a face to face the world."

Contents

Contributors

George J. Agich, Ph.D. Department of Philosophy, Social Philosophy and Policy Center, Bowling Green State University, Bowling Green, OH, USA

Selman Altuntas, M.D. Department of Plastic Surgery, Cleveland Clinic, Cleveland, OH, USA

Medhat Askar, M.D., Ph.D. Allogen Laboratories, Transplant Center, Department of Surgery, Cleveland Clinic, Cleveland Clinic Lerner College of Medicine, Case Western Reserve University, Cleveland, OH, USA

Anthony Atala, M.D. Director, Wake Forest Institute for Regenerative Medicine, Wake Forest University School of Medicine, Medical Center Blvd., Winston-Salem, NC, USA

Robin Avery, M.D., FIDSA Department of Infectious Disease, Medicine Institute, Cleveland Clinic, Cleveland, OH, USA

Rolf N. Barth, M.D. Department of Surgery, University of Maryland School of Medicine, Baltimore, MD, USA

Stephen T. Bartlett, M.D. Department of Surgery, University of Maryland Medical Center, Baltimore, MD, USA

Wilma F. Bergfeld, M.D., FAAD Department of Dermatology and Pathology, Cleveland Clinic, Cleveland, OH, USA

Steven Bernard, M.D. Department of Plastic Surgery, Cleveland Clinic, Cleveland, OH, USA

Marino Blanes, M.D., Ph.D. Department of Infectious Diseases, Hospital Universitari La Fe de Valencia, Valencia, Spain

Gordon R. Bowen Lifebanc, Cleveland, OH, USA

Katrina A. Bramstedt, Ph.D. California Transplant Donor Network, Oakland, CA, USA

Grzegorz Brzezicki, M.D. Department of Plastic and Reconstructive Surgery, Cleveland Clinic, Cleveland, OH, USA

Federico Castro, M.D. Department of Plastic and Reconstructive Surgery, Anesthesiology Division, Clínica Cavadas, Valencia, Spain

Pedro C. Cavadas, M.D., Ph.D. Department of Plastic and Reconstructive Surgery, Clínica Cavadas, Valencia, Spain

Linda C. Cendales, M.D. Department of Plastic and Reconstructive Surgery, Division of Transplantation, Vascularized Composite Allotransplantation, Laboratory of Microsurgery, Emory University, Atlanta, GA, USA

Kathy L. Coffman, M.D., FAPM Department of Psychiatry and Psychology, Cleveland Clinic, Cleveland, OH, USA

Joanna Cwykiel, M.Sc. Department of Plastic Surgery, Cleveland Clinic, Cleveland, OH, USA

Jacek B. Cywinski, M.D. Department of General Anesthesiology, Cleveland Clinic Foundation, Anesthesiology Institute, Cleveland, OH, USA

Pamela L. Dixon, MOT, OTR/L Department of Physical Medicine and Rehabilitation, Cleveland Clinic, Cleveland, OH, USA

Risal Djohan, M.D. Department of Plastic Surgery, Cleveland Clinic, Cleveland, OH, USA

Mathieu Domalain, Ph.D. Department of Biomedical Engineering, Cleveland Clinic, Lerner Research Institute, Cleveland, OH, USA

D. John Doyle, M.D., Ph.D. Department of General Anesthesiology, Cleveland Clinic, Cleveland, OH, USA

Bijan Eghtesad, M.D. Department of HPB/Liver Transplant Surgery, Cleveland Clinic, Cleveland, OH, USA

Ann Marie Flores, P.T., Ph.D., CLT Department of Orthopaedics and Rehabilitation, Vanderbilt University School of Nursing, Vanderbilt University Medical Center, Nashville, TN, USA

John J. Fung, M.D., Ph.D. Department of General Surgery, Education Institute, Cleveland Clinic, Cleveland, OH, USA

James R. Gatherwright, M.D. Department of Plastic and Reconstructive Surgery, University Hospitals – Case Medical Center, Cleveland, OH, USA

Bahar Bassiri Gharb, M.D., FEBOPRAS Department of Plastic Surgery, Cleveland Clinic, Cleveland, OH, USA

Alexandra K. Glazier, J.D., M.P.H. New England Organ Bank, Inc, Waltham, MA, USA

Chad R. Gordon, D.O. Division of Plastic and Reconstructive Surgery, Department of Surgery, Massachusetts General Hospital, Boston, MA, USA

Lawrence J. Gottlieb, M.D., FACS Department of Surgery, Section of Plastic and Reconstructive Surgery, The University of Chicago Medical Center, Chicago, IL, USA

Robert G. Hale, D.D.S., COL US Army Dental Corps Craniomaxillofacial Research, U.S. Army Institute of Surgical Research, San Antonio, TX, USA

Charles Heald Community Outreach, LifeBanc, Cleveland, OH, USA

Michael L. Huband, D.D.S. Section of Maxillofacial Prosthetics, Department of Dentistry, Head and Neck Institute, Cleveland Clinic, Cleveland, OH, USA

Helen G. Hui-Chou, M.D. Division of Plastic and Reconstructive Surgery, The Johns Hopkins Hospital/University of Maryland School of Medicine, Baltimore, MD, USA

Javier Ibañez Mata Department of Plastic and Reconstructive Surgery, Clínica Cavadas, Valencia, Spain

Helen G. Hui-Chou, M.D. Division of Plastic and Reconstructive Surgery, The Johns Hopkins Hospital/University of Maryland School of Medicine, Baltimore, MD, USA

Luke S. Jones, B.S. Department of Surgery, University of Maryland School of Medicine, Baltimore, MD, USA

Jean Kanitakis, M.D. Department of Dermatology/Laboratory of Dermatopathology, Edouard Herriot Hospital Group, Lyon, France

Angela Kiska Corporate Communications, Cleveland Clinic, Cleveland, OH, USA

Aleksandra Klimczak, Ph.D. Department of Plastic Surgery, Cleveland Clinic, Cleveland, OH, USA and
Polish Academy of Sciences, Institute of Immunology and Experimental Therapy, Wroclaw, Poland

Yur-Ren Kuo, M.D., Ph.D., FACS Department of Plastic and Reconstructive Surgery, Chang Gung Memorial Hospital – Kaohsiung Medical Center, Kaohsiung, Taiwan

Krzysztof Kusza, M.D., Ph.D., D.Sc. Department of Anaesthesiology and Intensive Therapy, Nicolaus Copernicus University, Collegium Medicum, Poznan University of Medical Sciences, Bydgoszcz, Poland

Luis Landin, M.D. Department of Plastic and Reconstructive Surgery, Clínica Cavadas, Valencia, Spain

Daniel J. Lebovitz, M.D. Department of Pediatric Critical Care, Cleveland Clinic Lerner College of Medicine, Cleveland Clinic Children's Hospital, LifeBanc, Cleveland, OH, USA

Vernon W.-H. Lin, M.D., Ph.D. Department of Physical Medicine and Rehabilitation, Cleveland Clinic, Cleveland, OH, USA

Robert F. Lohman, M.D. Department of Orthopaedic and Plastic Surgery, Cleveland Clinic, Cleveland, OH, USA

Maria Madajka, Ph.D. Department of Plastic Surgery, Cleveland Clinic, Cleveland, OH, USA

Amanda Mendiola, M.D. Department of Surgery, Akron General Medical Center, Akron, OH, USA

Gerhard S. Mundinger, M.D. Division of Plastic and Reconstructive Surgery, Johns Hopkins Hospital/University of Maryland School of Medicine, Baltimore, MD, USA

Can Ozturk, M.D. Department of Dermatology and Plastic Surgery Institute, Cleveland Clinic, Cleveland, OH, USA

Frank Papay, M.D., FACS, FAAP Dermatology and Plastic Surgery Institute, Cleveland Clinic, Cleveland, OH, USA

Palmina Petruzzo, M.D. Department of Transplantation, Edouard Herriot Hospital, Lyon, France

Marc J. Popovich, M.D., FCCM Surgical Intensive Care Unit, Cleveland Clinic, Anesthesiology Institute, Cleveland, OH, USA

Antonio Rampazzo, M.D., FEBOPRAS Department of Plastic Surgery, Cleveland Clinic, Cleveland, OH, USA

Russell R. Reid, M.D., Ph.D. Department of Surgery/Plastic Surgery, University of Chicago, Chicago, IL, USA

Jose Rodrigo, M.D. Department of Plastic and Reconstructive Surgery, Clínica Cavadas, Valencia, Spain

Eduardo D. Rodriguez, M.D., D.D.S. Division of Plastic and Reconstructive Surgery, R Adams Cowley Shock Trauma Center, Johns Hopkins Hospital/ University of Maryland School of Medicine, Baltimore, MD, USA

Elliott H. Rose, M.D. Division of Plastic and Reconstructive Surgery, The Mount Sinai Medical Center, New York, NY, USA

Eileen M. Sheil Corporate Communications, Cleveland Clinic, Cleveland, OH, USA

Steven T. Shipley, DVM, DACLAM Veterinary Medicine, Comparative Medicine Program, University of Maryland School of Medicine, Baltimore, MD, USA

Maria Z. Siemionow, M.D., Ph.D., D.Sc. Department of Plastic Surgery, Cleveland Clinic, Cleveland, OH, USA

Vlodek Siemionow, Ph.D. Department of Plastic Surgery, Lerner Research Institute, Cleveland Clinic, Cleveland, OH, USA

Angela Sirigu, Ph.D. Center for Cognitive Neuroscience, CNRS, Bron, France

Erhan Sonmez, M.D., Ph.D. Department of Plastic and Reconstructive Surgery, Hacettepe University Hospital, Ankara, Turkey

Jason S. Stratton, M.D. Department of Pathology, Cleveland Clinic, Cleveland, OH, USA

Alessandro Thione, M.D., Ph.D. Department of Plastic and Reconstructive Surgery, Clínica Cavadas, Valencia, Spain

Claudia D. Vargas Department of Neurobiology, Institute of Biophysics Carlos Chagas Filho, Rio de Janeiro, Brazil

Tracy Wheeler Corporate Communications, Cleveland Clinic, Cleveland, OH, USA

Tsung-Lin Yang, M.D., Ph.D. Department of Otolaryngology, National Taiwan University Hospital, Taipei, Taiwan

James J. Yoo, M.D., Ph.D. Wake Forest University School of Medicine, Wake Forest Institute for Regenerative Medicine, Winston-Salem, NC, USA

Jose Maria Zarzalejos Andes, M.D. Department of Internal Medicine, Infectious Disease Unit, Hospital Universitario Gran Canaria Dr. Negrín, Las Palmas de Gran Canaria, Spain

Xiaoming Zhang, Ph.D. Medical Research Service, Louis Stokes Cleveland VA Medical Center, Cleveland, OH, USA

Fatih Zor, M.D. Department of Plastic Reconstructive and Aesthetic Surgery, Gulhane Military Medical Academy, Ankara, Turkey

Part I

Preclinical Aspects of Face Transplantation

1 Face as an Organ: The Functional Anatomy of the Face

Maria Z. Siemionow and Erhan Sonmez

Contents

M.Z. Siemionow (✉)
Department of Plastic Surgery, Cleveland Clinic, Cleveland, OH, USA
e-mail: siemiom@ccf.org

Abstract The role of the face in the daily interactions of a person through its expression of feelings, beauty, and identity is pertaining to life. Thus, severe facial trauma and disfigurement stemming from burns, tumor resection, and congenital and acquired malformation have deleterious effects on a person's life and expose a person to the stigmata of being different.

Face transplantation in humans, which has been performed worldwide, has raised the question of whether the face is just a "tissue" or if it is an "organ." This issue has been approached from different perspectives by different societies, agencies, and communities. We have summarized the anatomic, physiologic, and aesthetic functions of the human face in this chapter, and we propose that the face should be accepted as an organ. Additionally, face transplantation should be considered as an organ transplantation that enhances the quality of life to a degree comparable to that of solid organ transplantations.

1.1 Introduction

The face plays a central role in the daily interactions of a person through its expression of feelings, beauty, and identity. Consequently, severe facial trauma and disfigurement stemming from burns, tumor resection, and congenital and acquired malformation have deleterious effects on a person's life and expose a person to the stigmata of being different.[1,2] All of these conditions are difficult if not impossible to accurately reconstruct with autologous tissues despite the great advances in reconstructive surgery. The body contains no tissues possessing the texture, pliability, and complexity of the face. Therefore, the only option for restoring facial features in severely

M.Z. Siemionow (ed.), *The Know-How of Face Transplantation*,
DOI: 10.1007/978-0-85729-253-7_1,

disfigured patients remains transplantation of face from a human donor.[2] The concept of facial transplantation has become a reality with a total of ten cases worldwide at the time of this report since the first case that was reported from Lyon, France in 2005.[3,4]

Solid organ transplantations are essential for the continuation of life and have saved the lives of millions of people since the first kidney transplant, which was performed by Nobel laureate Joseph Murray of Brigham Women's Hospital in December 1954.[5] However, composite tissue allografts such as face, larynx, or hand, although certainly improving the life quality, are not essential for patient's survival. The risk of lifelong immunosuppression in patients receiving transplants which do not have a direct impact on their survival is the focus of debates regarding the use of composite tissue allotransplantation in daily practice. Although there has been a great improvement in quality and specificity of immunosuppressive drugs, their side effects are still of major concern.[2] Consequently, face transplantations in humans have raised the question of whether the face is just a "tissue" or an "organ." This issue has been approached from different perspectives by different societies, agencies, and communities. We have described the anatomic, physiologic, and aesthetic functions of the human face in this chapter.

1.2 Is the Face an "Organ" or a "Tissue"?

Based on the definitions of standard medical and general dictionaries, an "organ" (from Greek "organon," via Latin organum, "tool, implement") is a differentiated structure comprising tissues that perform a specialized function in an organism. On the other hand, "tissue" (ultimately from the Latin "texere," "to weave") is defined as an aggregate of similar cells, along with their intercellular substances, which comprise the materials that build structures in an organism.[6] Consequently, an organ comprises tissues from the conventional anatomic view. We want to examine this view from the perspective of the face and therefore want to raise the question of whether the face can be regarded as an organ that performs one or more specific functions or is it simply an aggregation of tissue with no discernible specific functions.

1.3 Anatomic and Physiologic Composition of the Face

The anatomic composition of face can be described starting from its most superficial to its deeper structures:

1.3.1 Skin

Based on our cadaver studies, we have confirmed that the surface area of the skin of the total face is 1,192 cm^2 with the scalp, and 675 cm^2 without the scalp.[7-10] It is composed of 3 functional layers: epidermis, dermis, and subcutis. Blood vessels and epidermal appendages such as hairs and glands are found in these layers[11] (Fig. 1.1).

The hair on the face is composed of hair shaft, root, and a shaft bulb at the base of the hair, similar to the hair all over the body. Root of the hair ends in the hair bulb, that lies in a sac-like pit in the skin called hair follicle. Hair follicles are lined with cells that synthesize the proteins needed for the growth of the hair. The oily coating of the hair shaft is secreted by a sebaceous gland which is associated with this follicle.[12]

Keratinocytes, fibroblasts, melonocytes, Langerhans cells, and the Merkel Cells are the five types of cells that make up the skin. Subcutis layer is composed of fat cells and the endothelial cells, and the superficial

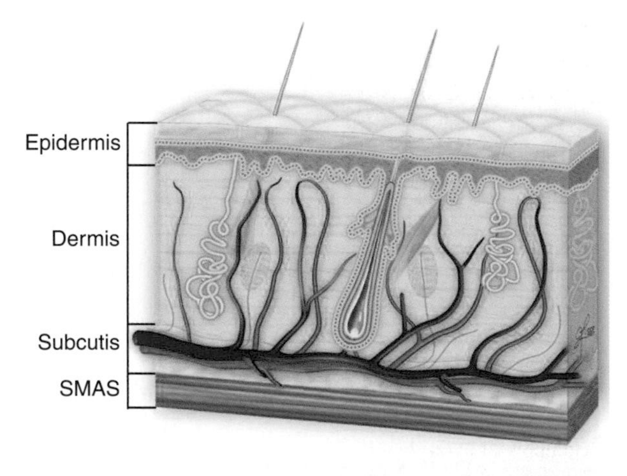

Fig. 1.1 Cross section of the skin (From Siemionow and Sonmez[2])

musculoaponeurotic system of the face is composed of striated muscle cells.[11]

The dermis (or Corium) is the layer of skin beneath the epidermis that contains appendages from the epidermis, such as hair follicles and sweat glands. The dermis is structurally divided into two areas: the "papillary region" is a superficial area adjacent to the epidermis which contains loose collagenous and elastic fibers, together with fibroblasts, mast cells, and macrophages. The deeper and thicker layer of the dermis is the "reticular region" which consists of dense, coarse bundles of collagenous fibers.[12]

The subcutis is the layer just beneath the dermis and is also known as the subcutaneous layer. It consists of a network of collagen, a layer of fatty areolar tissue that overlies the more densely structured fibrous fascia. The subcutaneous tissue serves as a "shock absorber" and insulation of heat of the body. Eccrine and apocrine glands are the sweat glands found all over the body, but eccrine glands predominate the skin of the face. They are particularly concentrated in the forehead skin.[11]

Sebaceous glands (halocrine glands) are found over the entire surface of the body except the palm, soles, and the dorsum of the feet. They are particularly concentrated in the skin of the face and scalp. The glands produce and secrete sebum which is a group of complex oils. The function of sebum is to lubricate and protect the skin against trauma and keep the moisture.[12]

The next deeper layer is called the superficial musculoaponeurotic system. This sheet is well developed in the scalp and face and includes the occipitofrontalis muscle, the tempoparietal fascia, the orbicularis oculi muscle, occipitofrontalis muscle, zygomatic muscles, levator labii superioris muscle, temporal branches of the facial nerve, superficial temporal vessels, and the auriculotemporal nerve. The superficial muscles of the face and their functions are summarized in Table 1.1 (Fig. 1.2).

The eyelids are composed of skin, subcutaneous tissue, orbicularis oculi muscle, submusculoareolar tissue, the fibrous layer consisting of the tarsus and the orbital septum, lid retractors of the upper and lower eyelids, retroseptal fat pads, and the conjunctiva from most superficial to the deeper consequently. The lids move through the action of the orbicularis oculi muscle and of the levator of the upper lid. The borders of the eyelids are lubricated by an oily secretion (called sebum) of the meibomian glands.[13]

The nose is composed of cartilaginous anterior portion and a bony posterior and superior portion. The

Table 1.1 Functions and innervations of the superficial muscles of the face

Superficial muscles of the face	Function	Innervation ("n" refers to nerve)
Frontalis	Pulls the eye brows	Temporal branch of facial n.
Auricularis posterior	May move the ears	Temporal branch of facial n.
Auricularis anterior	May move the ears	Temporal branch of facial n.
Auricularis superior	May move the ears	Temporal branch of facial n.
Orbicularis oculi	Squints the eyes	Temporal and Zygomatic branch of facial n.
Pyramidalis	Lowers glabella	Temporal branch of facial n.
Zygomaticus major	Pulls lip corners upward	Zygomatic and Buccal branch of facial n.
Zygomaticus minor	Pulls lip corners downward	Buccal branch of facial n.
Levator labii superioris	Pulls the upper lip upward	Buccal branch of facial n.
Levator labii superioris alaque nasi	Wrinkles nose	Buccal branch of facial n.
Levator anguli oris	Elevates lateral part of the lips	Zygomatic and Buccal branch of facial n.
Orbicularis oris	Purses the lips	Buccal and Mandibular branch of facial n.
Risorius	Retracts lip corners	Buccal branch of facial n.
Mentalis	Elevates the lower lip and skin	Mandibular branch of facial n.
Depressor anguli oris	Pulls the corners of the mouth down	Buccal and Mandibular branch of facial n.
Depressor labii inferioris	Pulls the lower lip down	Mandibular branch of facial n.
Platysma	Pulls the corners of the mouth down	Cervical branch of facial n.
Nasalis	Compresses the nostrils	Buccal branch of facial n.
Compressor naris	Constricts the nostrils	Buccal branch of facial n.
Depressor naris	Flares the alar parts of the nose	Buccal branch of facial n.

From Siemionow and Sonmez[2]

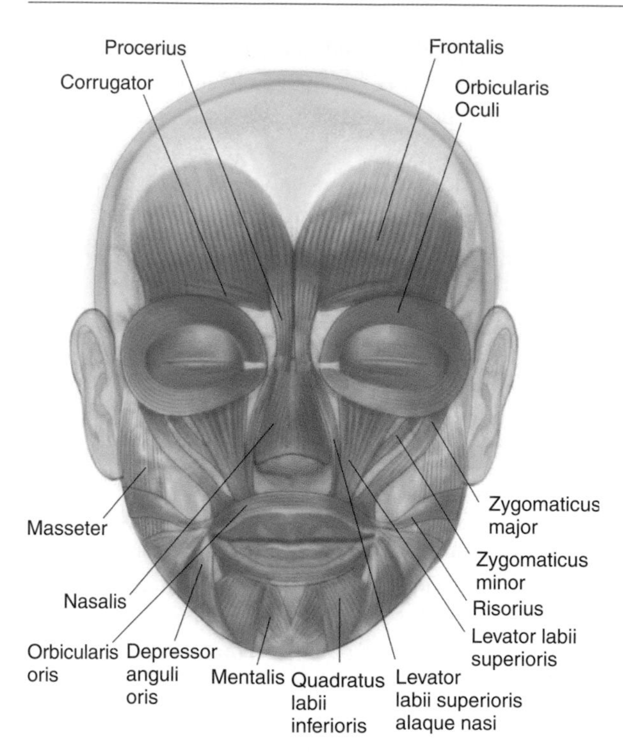

Fig. 1.2 Superficial muscles of the face (From Siemionow and Sonmez[2])

cartilaginous portion of the nose is made up of a paired set of cartilages which includes the greater alar (lower lateral), septal, lateral nasal (upper lateral), and lesser sesamoid cartilages. Paired nasal bones and the nasal processes of the maxillary and frontal bones form the bony framework of the nose. Posterosuperiorly, the bony nasal septum is composed of the perpendicular plate of the ethmoid.[2,14]

The lips are the soft, protruding, and movable parts of the face. The lower lip is usually somewhat larger than the upper one. The skin of the lips is very thin compared to the skin of the face and lacks sweat glands and sebaceous glands. The number of the melanocytes in the lip skin is very low, and because of this, the blood vessels appear through the skin of the lips and make the red coloring.[2]

The parotid gland, which is the largest of the major salivary glands, is located in the posterolateral side of the face, and in front of the external ear along the posterior border of the ramus of the mandible. It secretes saliva through Stensen's Duct into the oral cavity to facilitate mastication and swallowing. The paired submandibular glands (submaxillary glands) are located below the mandible on each sides of the jaw.[2]

In some parts of the face, the anatomic layers are condensed, forming the "retaining ligaments" which serve to anchor the skin of the face to the underlying bony structures. The zygomatic ligament is located in the cheek, anterior and superior to the parotid gland and posteroinferior to the malar eminence. The mandibular ligament is located on the jaw line and forms the anterior border of the jowl. The other two ligaments, the platysma-cutaneous and platysma-auricular ligaments, are aponeurotic condensations attaching the platysma to the underlying dermis. All of these ligaments support the facial structures and skin against gravitational pull.[15,16]

1.3.2 Vascularization of the Face

The arterial supply of the face relies on the terminal branches of the external carotid and internal carotid arteries. The superficial temporal artery and the internal maxillary artery supply the upper third and the deeper structures of the face, and the facial arteries supply the central and lower parts of the face. The ophthalmic artery, a collateral branch of the internal carotid artery, supplies the medial upper face and the periorbital area.

Most of the veins in the face run parallel to their corresponding arteries. These veins lack valves and therefore allow bidirectional blood flow. Because of this, wound infections of the perioral area and the upper lip have the potential to gain access to the cavernous sinus. Venous drainage of the face relies on the external, internal, and anterior jugular veins, which drain the superficial temporal vein; facial vein; and inferior labial and chin veins consequently[8,9,17] (Fig. 1.3).

1.3.3 Innervation of the Face

Innervation of the face is divided as sensory and motor. Sensory innervation to the face (and the rest of the head) is supplied by the sensory component of the trigeminal nerve (fifth Cranial Nerve). The trigeminal nerve divides into three major divisions which supply three major areas of the face.[3,10,17]

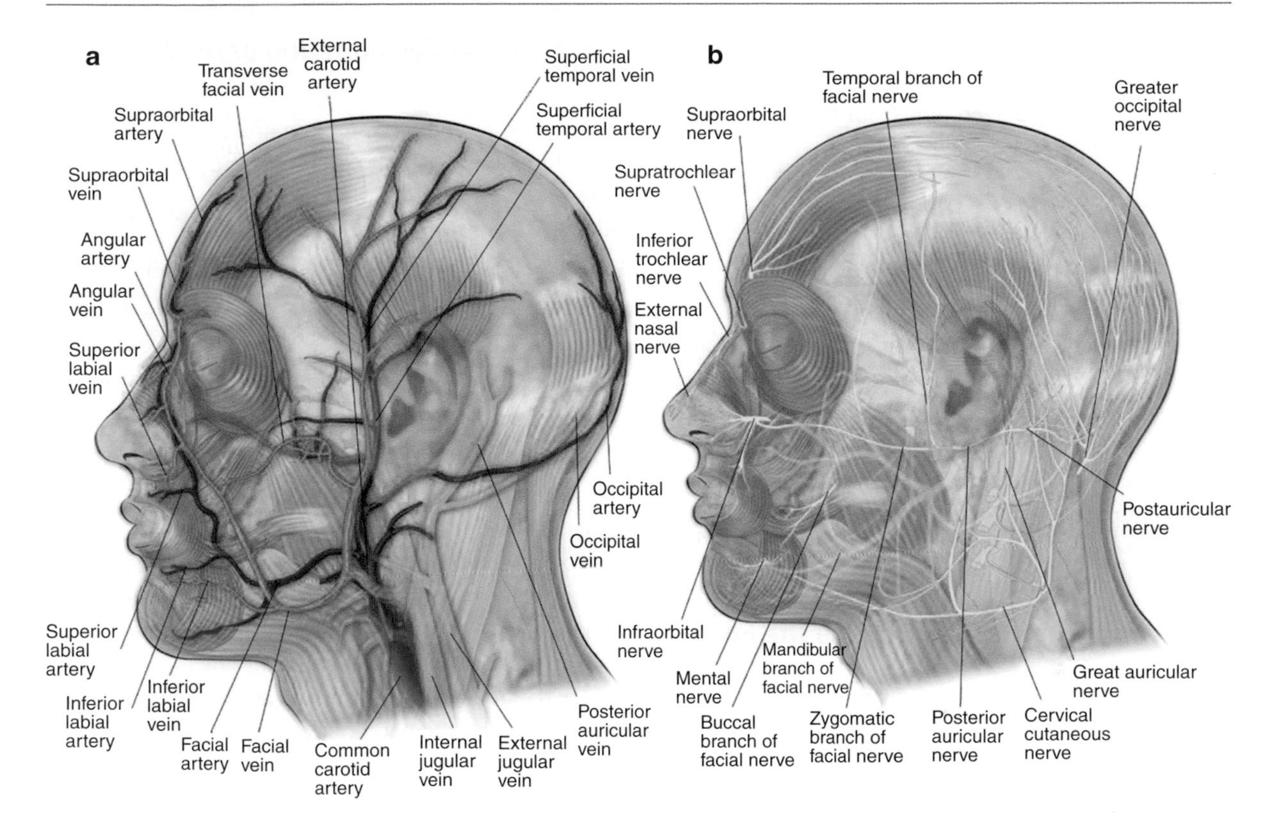

Fig. 1.3 Arteries, veins (**a**), and nerves (**b**) of the face (From Siemionow and Sonmez[2])

The ophthalmic division supplies the mucosa at the frontal sinus, the skin and conjunctiva, and skin over the forehead and scalp posterior to the region of the region of the lambdoid suture via the supraorbital branch. The maxillary division supplies the skin of the lower eyelid, cheek, nose, upper lip, and possibly the conjunctiva and skin over the maxilla via infraorbital branch. The mandibular division supplies the lower jaw, the lower lip, and the chin via mental branch.

All muscles of facial expression (plus platysma in the neck, the small muscles around the ear and the scalp muscles) are innervated by the facial nerve (seventh cranial nerve) via four branches. The temporal branch supplies the facial muscles superior to the zygomatic arc including the forehead muscles and the orbicularis oculi. The zygomatic branch innervates the muscles in the zygomatic, orbital, and infraorbital regions. The buccal branch supplies the buccinators and the muscles of the upper lip. The mandibular branch supplies the muscles of the lower lip and chin. The muscles of mastication are supplied by the mandibular division of the trigeminal nerve[3,4,10,17] (Fig. 1.3).

Sympathetic innervation of the face arises from the postganglionic cell bodies in the superior cervical ganglion that is located opposite the second and third cervical vertebra. Sympathetic preganglionic neurons that control the salivary glands are located in T1–T4 levels of the spinal cord. Stimulation of the sympathetic fibers leads to vasoconstriction of the vessels in the glands and cutaneous arteries of the face.[18]

Parasympathetic nerves are distributed to blood vessels in salivary glands and vessels of the nasal mucosa of the face. The major target of cranial parasympathetic pathways is the secretory glands associated with the secretion of tear (eye), saliva (mouth), and mucosa (nose). Stimulation of the parasympathetic system leads to vasodilatation and consequently secretion of watery fluid. Parasympathetic paraganglionic neurons that control the submaxillary and sublingual glands are located in the salivary nucleus (Cranial Nerve VII). Postganglionic fibers arise from postganglionic cell bodies in the submaxillary ganglion and course with the facial and trigeminal nerves to reach the glands. Parasympathetic preganglionic neurons that control the parotid glands are located in the inferior salivatory nucleus (Cranial Nerve IX), and its stimulation leads to secretion from the parotid gland via concomitant vasodilatation.[18]

1.4 Functions of the Face

Facial features arise from the subtle arrangement of many diverse tissues. The face is not simply a mask, but a functional, dynamic, and aesthetic organ. Functions of the face can be grouped as physiologic, expressive, and aesthetic functions. It also plays an important role in a person's identity.[2,19]

1.4.1 Physiologic Functions of the Face

The skin of the face serves as an anatomic barrier between the internal and external environment, providing bodily defense. The skin of the face is also a sensory organ containing a variety of nerve endings that react to warmth, cold, pressure, vibration, and tissue injury. The skin of the face regulates heat and controls evaporation, as is the case with the skin on other parts of the body, primarily through vasodilatation and vasoconstriction of the cutaneous blood vessels.[11]

The hair in the nose, ears, and around the eyes protects these sensitive areas from the infiltration of dust and other small particles. Eyebrows and eyelashes protect the eyes by decreasing the amount of light and particles that can enter the eyes.

Eyelids protect the ocular globes from mechanical injury and help to provide essential moisture for the conjunctiva and cornea.[20]

The nose warms and moisturizes the inspired air, removes bacteria and particulate debris, and conserves heat and moisture from expired air. It has an area of specialized cells which are responsible for the sense of smelling. The sense of smell plays a major role in the flavor of foods, and it is common for individuals who lose their sense of smell to report that food loses its taste although the food has only lost its aroma, but the taste (sweet, salty, sour, bitter) remains intact.

Nasal breathing (as opposed to mouth breathing) permits optimal pulmonary function.[21]

1.4.2 Expressive Function of the Face

The face is an organ of emotion apart from its physiologic functions. As an expressive organ, the face provides an effective and communicative presence to others. We constantly read facial expressions to understand what others are feeling. We can understand the happiness, anger, pain, sorrow, sadness, or even madness from the expressions on the face. Conscious and unconscious facial expressions are crucial in our encounters with others. We constantly perform a stream of facial movements when we communicate in person. We can say that approximately two thirds of our communication with others takes place via the nonverbal channels of the face.[22]

Consequently, the face can be accepted as the most powerful instrument of nonverbal communication, allowing us to express our thoughts and feelings and to decode the thoughts and feelings of others.[23]

There are numerous reports about the difficulties experienced by the people who are unable to use their faces to communicate effectively, whether through the absence of expression, or miscommunication resulting from altered expressions.[22-25]

1.4.3 Aesthetic Function of the Face

The face plays a critical role in physical attractiveness. It is perhaps the most important human art object. Attractive facial features in women may include a narrow facial shape; narrow, thin eyelids; and a slightly wider distance between the eyes; large eyes; a prominent zygomatic arc; thin eyebrows; a small nose; small chin; small jaw bones; and full lips. Male attractiveness includes prominent chin bones, large jaws, a prominent chin, thin lips, and thick eyebrows.[26]

A great deal of research has shown the robust effects of facial attractiveness on interpersonal perception. It has been observed that facially attractive people have social advantages. For instance, they are occupationally more successful, more popular, more assertive, and have more self-confidence. Social and developmental physiologists have proved that facial attractiveness produces a halo effect causing people to ascribe many positive qualities and characteristics to attractive persons.[27,28]

On the contrary, facial deformities from trauma, congenital disabilities, and post-surgical scars are perceived as dysfunctional and produce significant adverse consequences for the physiological and social functioning of affected persons. Functional facial deformities have been variously described as only those that impair respiration, eating, hearing, or speech; however, it has been documented in the literature that facial scars and cutaneous deformities can have a significant negative impact on social functionality.[25,29-32] Consequently, facial reconstructive goals include not only restoration of orifices for adequate respiration, vision, and alimentation, but also reestablishment of improved surface or contours meant to transform a deformity to an acceptable range of normal appearance. At that time "beauty" is not the goal, but rather the goal is to eliminate the negative stigma that arises from a facial configuration that lies outside of the range of normal.[33]

In a recent study, the authors confirmed that in the population, people do place a high value on a normal facial appearance that is delineated as distinct from a beautiful appearance, and normal appearance is accepted as an important function of the human face by the people. The subjects in that study ranked the face as "the most important body part to restore after an injury," followed by the hand, leg, arm, knee, and breast.[33]

1.4.4 Role of the Face in the Formation of Identity

The face plays a central role in the perception and formation of identity. Human beings recognize each other by first looking at the face. Our face develops from childhood into adulthood, then into middle age, and finally into senior years. Yet it retains features that were already prominent in childhood. Face help us understand who we are and where we come from with markers of genetic inheritance over many generations, providing evidence of parentage, ancestry, and racial identity.[34] These persisting features contribute to the relatively unchanging expression of the face and define its physiognomy. Disruption to one's facial appearance, particularly the inability to recognize oneself, represents the disruption of the body image and may constitute a major life crisis.[22,35]

The formation of identity function of the face cannot be totally transferred from donor to recipient without the transfer of all the underlying bony structures. The shape of the face is closely related to the shape of the underlying bones, a subject that has engendered debate with regard to facial identity transfer.[36]

1.5 Conclusion

Based on the functional anatomy of the face that is summarized in this chapter, we propose that the face should be accepted as an organ. Additionally, face transplantation should be considered as an organ transplantation that enhances the quality of life to a degree comparable to that of solid organ transplantations. The aim of this chapter is to emphasize multiple vital functions of face and to make awareness that without these functions, quality of life of severely disfigured patients is jeopardized.

References

1. Siemionow M, Papay F, Alam D, et al. Near-total human face transplantation for a severely disfigured patient in the USA. *Lancet*. 2009;374:203-209.
2. Siemionow M, Sonmez E. Face as an organ. *Ann Plast Surg*. 2008;61:345-352.
3. Devauchelle B, Badet L, Lengele B, et al. First human face allograft: early report. *Lancet*. 2006;368:203-209.
4. Dubernard JM, Lengele B, Morelon E, et al. Outcomes 18 months after the first human partial face transplantation. *N Engl J Med*. 2007;357:2451-2460.
5. Merrill JP, Murray JE, Harrison JH, Guild WR. Successful homotransplantation of the human kidney between identical twins. *J Am Med Assoc*. 1956;160:277-282.
6. Merriam-Webster Medical Dictionary. http://www.m-w.com/medical/. (Internet Communication).
7. Okie S. Facial transplantation: brave new face. *N Engl J Med*. 2006;354:889-894.
8. Siemionow M, Unal S, Agaoglu G, Sari A. A cadaver study in preparation for facial allograft transplantation in humans: part I. What are alternative sources for total facial defect coverage? *Plast Reconstr Surg*. 2006;117:864-872.
9. Siemionow M, Agaoglu G, Unal S. A cadaver study in preparation for facial allograft transplantation in humans: part II Mock facial transplantation. *Plast Reconstr Surg*. 2006;117:876-885.
10. Siemionow M, Papay F, Kulahci Y, et al. Coronal-posterior approach for face/scalp flap harvesting in preparation for face transplantation. *J Reconstr Microsurg*. 2006;22:399-405.
11. Greaves MW. Physiology of skin. *J Invest Dermatol*. 1976;67:66-69.

12. McDonald CJ. Structure and function of the skin. Are there differences between black and white skin? *Dermatol Clin.* 1988;6:343-347.
13. Burkat CN, Lemke BN. Anatomy of the orbit and its related structures. *Otolaryngol Clin North Am.* 2005;38:825-856.
14. Neskey D, Eloy JA, Casiano RR. Nasal, septal, and turbinate anatomy and embryology. *Otolaryngol Clin North Am.* 2009;42:193-205.
15. Thorne CH. Face lift. In: Thorne CH ed. *Grabb and Smith's Plastic Surgery.* Wolters Kluwer Health – Lippincott Williams & Wilkins, Philadelphia; 2007;498-508.
16. Furnas DW. The retaining ligaments of the cheek. *Plast Reconstr Surg.* 1989;83:11-16.
17. Petit F, Paraskevas A, Minns AB, Lee WP, Lantieri LA. Face transplantation: where do we stand? *Plast Reconstr Surg.* 2004;113:1429-1433.
18. Waxman SG. The autonomic nervous system. In: *Clinical Neuroanatomy.* New York: McGraw-Hill; 2010:248-263.
19. Hettiaratchy S, Butler PE. Face transplantation – fantasy or the future? *Lancet.* 2002;360:5-6.
20. Jelks GW, Jelks EB. The influence of orbital and eyelid anatomy on the palpebral aperture. *Clin Plast Surg.* 1991;18: 183-195.
21. Hornung DE. Nasal anatomy and the sense of smell. *Adv Otorhinolaryngol.* 2006;63:1-22.
22. Morris P, Bradley A, Doyal L, et al. Face transplantation: a review of the technical, immunological, psychological and clinical issues with recommendations for good practice. *Transplantation.* 2007;83:109-128.
23. Rumsey N. Psychological aspects of face transplantation: read the small print carefully. *Am J Bioeth.* 2004;4:22-25.
24. Morris PJ, Bradley JA, Doyal L, et al. Facial transplantation: a working party report from the Royal College of Surgeons of England. *Transplantation.* 2004;77:330-338.
25. Rumsey N, Harcourt D. Body image and disfigurement: issues and interventions. *Body Image.* 2004;1:83-97.
26. Honn M, Go G. The ideal of facial beauty: a review. *J Orofac Orthop.* 2007;68:6-16.
27. Bashour M. History and current concepts in the analysis of facial attractiveness. *Plast Reconstr Surg.* 2006;118: 741-756.
28. Langlois JH, Kalakanis L, Rubenstein AJ, Larson A, Hallam M, Smoot M. Maxims or myths of beauty? A meta-analytic and theoretical review. *Psychol Bull.* 2000;126(3): 390-423.
29. Brown BC, McKenna SP, Siddhi K, McGrouther DA, Bayat A. The hidden cost of skin scars: quality of life after skin scarring. *J Plast Reconstr Aesthet Surg.* 2008;61(9):1049-1058.
30. Rankin M, Borah GL. Perceived functional impact of abnormal facial appearance. *Plast Reconstr Surg.* 2003;111(7): 2140-2146.
31. Rumsey N, Clarke A, White P, Wyn-Williams M, Garlick W. Altered body image: appearance-related concerns of people with visible disfigurement. *J Adv Nurs.* 2004;48(5): 443-453.
32. Rumsey N, Harcourt D. Visible difference amongst children and adolescents: issues and interventions. *Dev Neurorehabil.* 2007;10(2):113-123.
33. Borah GL, Rankin MK. Appearance is a function of the face. *Plast Reconstr Surg.* 2010;125(3):873-878.
34. Gribbin J, Gribbin M. *Being Human.* London: Phoenix; 1995.
35. Bradbury E. Understanding the problems. In: Lansdown R, Rumsey N, Bradbury E, Carr A, Partridge J, eds. *Visibly Different: Coping with Disfigurement.* London: Butterworth-Heineman; 1997.
36. Clarke A, Butler PE. Facial transplantation: adding to the reconstructive options after severe facial injury and disease. *Expert Opin Biol Ther.* 2005;5(12):1539-1546.

The Face as a Sensory Organ

2

Maria Z. Siemionow, Bahar Bassiri Gharb,
and Antonio Rampazzo

Contents

M.Z. Siemionow (✉)
Department of Plastic Surgery, Cleveland Clinic,
Cleveland, OH, USA
e-mail: siemiom@ccf.org

Abstract The human face is a highly specialized organ which receives sensory information from the environment and transmits it to the cortex. The advent of facial transplantation has recently shown that excellent reconstruction of disfiguring defects can be achieved; thus, the expectations are now focused on functional recovery of the transplant. So far, restoration of the facial sensation has not received the same attention as the recovery of motor function. We describe the current knowledge of the sensory pathways of the human face and their respective functions, the available methods of sensory assessment, and the data on normal sensation. The topographical sensory anatomy of facial subunits is summarized, the trigemino-facial connections are illustrated, and the implications of these anatomical variations on facial allotransplantation are emphasized.

2.1 Introduction

Since 2005, 11 reports on face transplantation have confirmed that this procedure is technically and immunologically feasible. The goal of reconstructing severely disfiguring facial defects by coverage with similar tissues coming from human donors has been achieved. This opened the discussion on the best approach to achieve functional recovery of the transplanted face with restoration of fine facial movements and sensation. These two determinants of optimal functional recovery were restored differently for documented cases of face transplantation. In three patients, the facial nerve was repaired either directly (two patients)[1] or with interpositional nerve grafts (one patient),[2-4] whereas the sensory nerves were satisfactorily repaired only in one case.[1] These differences in the reconstructive approaches

M.Z. Siemionow (ed.), *The Know-How of Face Transplantation*,
DOI: 10.1007/978-0-85729-253-7_2,

to motor and sensory nerve repairs were mainly dictated by the extent of facial trauma before transplantation.

One of the fundamental functions of the human face is the ability to receive multimodal sensory information from the environment and to convey it to the cerebral cortex for integration and processing. The presence of normal sensation is important not only for the discrimination of touch, temperature, and pain, but also for initiation of vigilant or defense reactions. The presence of labial sensation helps in avoiding drooling while eating or drinking.[5] Stretching of the perioral skin contributes to the precise articulation in speech.[6] Interestingly, cutaneous stimulation increases the intensity of estimates of the olfactory system.[7] It has also been reported that facial skin cooling decreases the heart rate and increases blood pressure.[8] Finally, normal sensory pathways allow to draw pleasure and satisfaction when exposed to external stimuli.[9] It is clear that restoration of the above functions is expected and essential for the optimal outcomes following face transplantation.

To learn more about the importance of the face as a sensory organ, the aim of this chapter is to illustrate the complexity of the sensory pathways of the face and their specific functions, to review current methods of assessment of facial sensation, and to summarize the available data on normal sensation. Finally, the topographical sensory anatomy of facial subunits is summarized and the implications of sensory–motor communications on the mechanism of recovery of facial sensation after trauma and face allotransplantation are discussed.

2.2 Facial Skin Receptors and Their Function

Over 17,000 corpuscles have been reported in the human face, which contribute to several sensory functions.[10] For the discrimination of touch, four different types of receptors have been described in the hairy skin of the face and include Ruffini corpuscles, Meissner corpuscles, Merkel cell disks, and hair receptors (Fig. 2.1).

Ruffini corpuscles are especially sensitive to skin stretch, consist of axon terminals and surrounding Schwann cells that envelop tightly bundles of collagen fibrils and are associated with vellus hairs. They are innervated by the superficial portion of the dermal neural network.

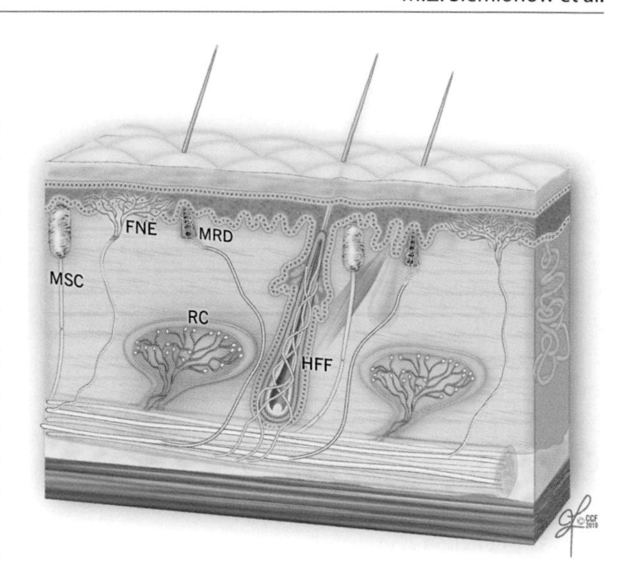

Fig. 2.1 The receptors of the human facial skin. *MSC* meissner corpuscle, *FNE* free nerve endings, *MRD* merkel disk, *RC* ruffini corpuscle, *HFF* hair follicle fiber (Reprinted with permission, Cleveland Clinic Center for Medical Art & Photography © 2010. All Rights Reserved)

Meissner corpuscles are more sensitive to stroking and fluttering of the skin and are localized in the dermal papillae. They are globular fluid-filled structures enclosing a stack of flattened epithelial cells. The terminal axons are entwined between the various layers of the corpuscles.

The *Merkel disk receptors* are formed by a small epithelial cell surrounding the nerve endings. Merkel receptors detect pressure applied on the skin and discriminate texture of objects. Two different types of Merkel cells have been described in facial skin.[11] The first type is localized in the dermis, on the external root sheath collar; it is not associated with nerve terminals and it is undifferentiated. The Merkel cells localized in the basal layer of the epidermis are associated with nerve terminals and have different granules within a single cell. An endocrine function has been attributed to them via regulation of the autonomic nerves.[12]

Interestingly, the *Pacini corpuscles*, which are well described in the fingertips and the palm of the hand where they are responsible for detection of vibrations, are absent in the skin of the human face.[12-14]

Hair follicle fibers work in a similar way to Meissner corpuscles, displaying a lower threshold for light stroking. They form a palisade of lanceolate terminals, which abut the external root sheath of the vellus hair in the region of the follicular neck. They derive from the deeper portion of dermal nerves.[15]

Beyond nerve fibers connected to mechanoreceptors and hair follicles, there are also free nerve endings with sensory function (temperature and pain). The intraepidermal nerve fibers terminate in different cellular layers, although the majority reach the stratum granulosum. The distribution of these endings is focal.[16,17] Kawakami et al.[16] reported that face presents with the highest distribution density of the free nerve endings. It was reported that beyond the sensory role, they may have also a trophic or immunoregulatory function.[17,18]

2.3 Physiology of Facial Sensation

2.3.1 Light Touch

The response of the face to the light touch is mediated by facial mechanoreceptors. These are associated with Aβ fibers (myelinated, 10–15 μm in diameter), divided into slowly and quickly adapting units. *Slowly adapting axons* initiate neural signals as soon as the skin is stimulated and continue to generate them as long as the cutaneous stimulus is sustained. *Ruffini complexes* and *Merkel cell disks* are the terminal corpuscles that are associated with these nerve fibers.[19]

Quickly adapting nerve fibers are activated only when new stimuli are applied. *Hair follicle fibers* and *Meissner corpuscles* are involved in transduction of these signals.[19,20] Innervation density for both quickly and slowly adapting fibers has been reported to increase from the upper face to the mid-face, followed by the lower face and the lip.[19]

Receptive field sizes of the facial skin afferents have been described to be similar in dimension to the receptive fields of the afferents innervating the vermilion (7–8 mm^2)[12] and the highest concentration of facial mechanoreceptors has been found at the corners of the mouth.[21]

2.3.2 Temperature and Pain

The perception of temperature changes and painful stimuli delivered to facial skin is not mediated by corpuscled receptors, but by small myelinated (Aδ) and unmyelinated fibers (C fibers). The *cold receptors* increase the firing rate with decreases in temperature, while *warmth receptors* increase the firing rate with increased temperature. In a study by Davies et al.,[22] raising of the temperature between 35°C and 40°C evoked a sensation of warming. In contrast, the cold receptors increased activity at lower temperatures ranging between 35°C and 15°C.

Nociceptors of the face are responsible for central transmission of painful stimuli and are activated by high threshold skin indentations (23 or 51 g),[23] as well as temperatures below 0°C[24] or above 47°C.[25]

2.4 The Ascending Pathways of Facial Sensation

The ascending pathways transmit the somatosensory information collected by the facial receptors, and conveyed by the peripheral axons of the *trigeminal sensory neurons* (along the peripheral pathways), to higher cortical centers for processing and integration.[26] These primary sensory neurons reside in the trigeminal ganglion (Gasser's ganglion or Ganglion Semilunaris) in the middle cranial fossa, from which afferent fibers pass into the mid-pons[27] (Fig. 2.2). The *second order sensory neurons* reside in trigeminal sensory nucleus. The sensory nucleus is divided into three subnuclei: the *principal sensory nucleus* is located in the pons and mediates *facial light touch* and *pressure* sensation; the *mesencephalic nucleus*, which receives *proprioceptive information* from the masticator muscles and the *nucleus of the spinal tract*, extends into the upper cervical cord (C2–C4) and is responsible for transmission of *facial pain* and *temperature* and secondarily *facial touch*.[28,29] After entering the pons, the pain and temperature fibers run caudally, forming the descending trigeminal tract and synapse with the second order neurons of the spinal nucleus. The axons of these neurons cross the midline and extend to the controlateral ventral posteromedial (VPM) nucleus of the thalamus, forming the *ventral trigeminothalamic tract* (or lemniscus).[30] Sensory fibers mediating light touch synapse in the principal sensory nucleus.[31] The secondary neuron axons ascend to the VPM nucleus of thalamus either contralaterally in the ventral trigeminothalamic tract (most) or ipsilaterally (*dorsal trigeminothalamic tract*)[29,30,32] (Fig. 2.2).

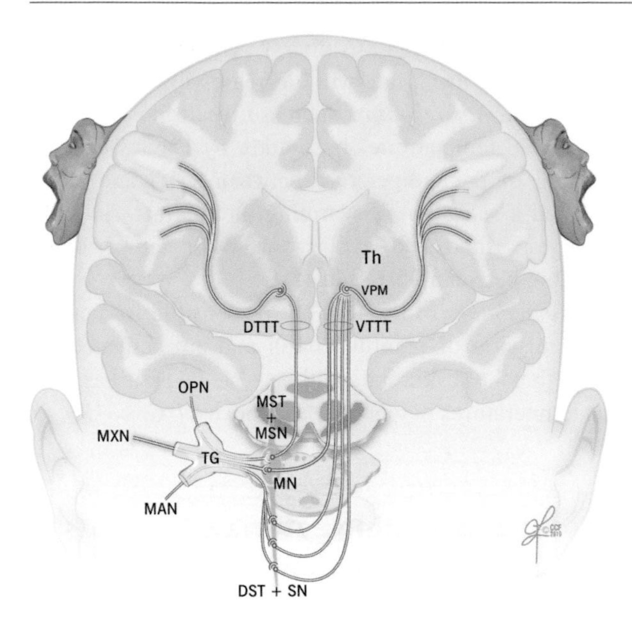

Fig. 2.2 Ascending pathways of facial sensation are shown from the peripheral branches of the trigeminal nerve which collect the sensory information and convey it to the Central Nervous system. *OPN* ophtalmic nerve, *MXN* maxillary nerve, *MAN* mandibular nerve, *TG* trigeminal ganglion, *MST+MSN* mesencephalic tract and nucleus, *MN* main sensory nucleus, *DST+SN* descending spinal tract and nucleus, *VTTT* ventral trigeminothalamic tract, *DTTT* dorsal trigeminothalamic tract, *Th* thalamus, *VPM* ventral posteromedial nucleus of thalamus (Reprinted with permission, Cleveland Clinic Center for Medical Art & Photography © 2010. All Rights Reserved)

From the ventroposteromedial thalamic nucleus, a third relay of fibers, the *thalamocortical tract* in the internal capsule, passes to the extensive face area of the main sensory neocortex (S-1 and S-2) of the post central gyrus (Brodman areas 3, 1, 2) and the upper bank of the sylvian fissure[32] (Fig. 2.2). Penfield reported that the representation of the facial structures was organized along the central sulcus, with the forehead in the superomedial region adjacent to the hand area, and the chin in the[33] inferolateral region.[33] The order of representation of the different facial subunits has been subject of debate. Tamura et al.[34] found, by using somatosensory-evoked magnetic fields, that topography of the areas representing intraoral structures along the central sulcus was the index finger, upper or lower lip, anterior or posterior tongue, and superior or inferior buccal mucosa, with a wide distribution, covering 30% of the S1 cortex. The skin-covered areas of the face were recently "relocated" between the thumb and the lip, which was in contrast to the original Penfield study.[35] Current studies in monkeys showed evidence for an upside-down representation;[36] however, studies by other investigators could not confirm either orientation.[37]

2.5 The Peripheral Pathways of Facial Sensation

The peripheral pathways, formed by the peripheral branches of the trigeminal sensory neurons, are responsible for conveying the sensory data from the facial skin to the Central Nervous System. The peripheral fibers of the primary sensory neurons exit the trigeminal ganglion organized into three trunks: the ophthalmic (V1), the maxillary (V2), and the mandibular (V3) nerves. The former two are purely sensory, whereas the latter is a mixed sensory and motor nerve. The trigeminal nerve collects sensibility of the full face except a small area around the mandibular angle and the auricular lobe, which is innervated by the great auricular nerve (C2–C3) (Fig. 2.3). The branches of these nerves supply sensation to the facial subunits of the upper face, mid-face, and lower face as summarized in Tables 2.1–2.3.

2.5.1 Trigemino-Facial Communications

When assessing the sensory recovery after facial trauma or transplantation, it is important to take into consideration the existence of direct connections between trigeminal nerve and facial nerve, which may play an important role in the mechanisms of facial sensory recovery.

The cutaneous branches of all three divisions of the trigeminal nerve and of the great auricular nerve show plexiform connections with the terminal rami of the facial nerve (Fig. 2.4). These connections can occur either in the proximal (auriculotemporal, great auricular) or distal (supraorbital, infraorbital, buccinators, mental) region of the facial nerve distribution. The auriculotemporal connections, to the upper division of the facial nerve, are the most consistent and sizable and represent the most constant pattern of the trigemino-facial communications.[38,39]

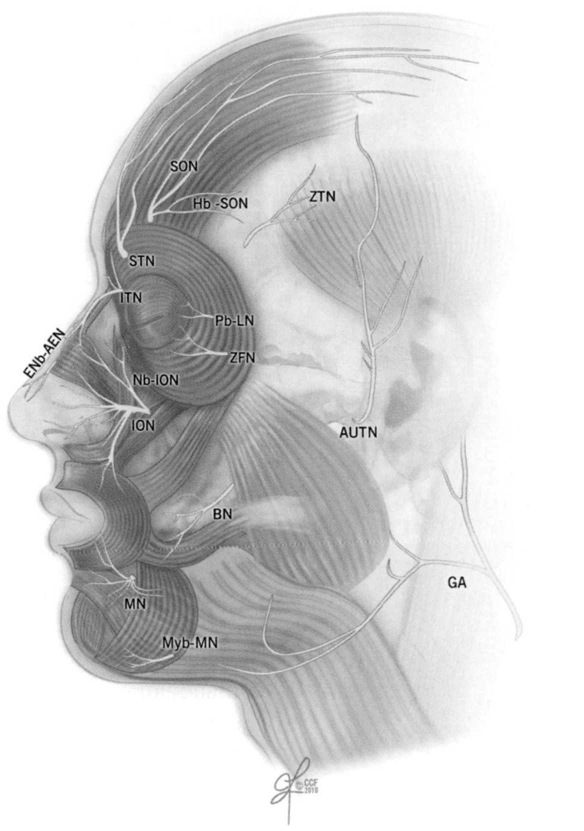

Fig. 2.3 Sensory innervation of the human face. *BN* buccal nerve, *ENb-AEN* external nasal branch-anterior ethmoidal nerve, *GA* great auricular nerve, *Hb-SON* horizontal branch supraorbital nerve, *ION* infraorbital nerve, *ITN* infratroclear nerve, *MN* mental nerve, mylohyoid branch-mental nerve, *Pb-LN* palpebral branch-lacrimal nerve, *SON* supraorbital nerve, *STN* supratroclear nerve, *ZFN* zygomaticofacial nerve, *ZTN* zygomaticotemporal nerve (Reprinted with permission, Cleveland Clinic Center for Medical Art & Photography © 2010. All Rights Reserved)

There is no common agreement on the function of the nerve fibers in the communicating rami. O'Connell and Huber reported that the relationship was of mere contiguity and after a joint journey, the sensory branches separated from the motor branches, each element finding the tissue it was destined to supply.[40,41] Others suggested that the trigeminal nerve fibers, in the communicating rami, conveyed proprioceptive information regarding the mimetic muscles, and the pseudomotor fibers to the integument or secretomotor fibers to the superficial part of the parotid and buccal mucosa glands.[42] Baumel implied[43] that sympathetic and parasympathetic fibers, not belonging to the trigeminal nerve, may also be the constituents of the communicating rami.

Finally, some investigators hypothesized presence of a sensory component in the facial nerve which explained the preservation of deep facial sensation after trigeminal neurectomy.[44,45] The connections between the facial nerve and the branches of the trigeminal and cervical nerves could also provide an additional motor supply to the superficial facial musculature.[41]

2.6 Evaluation of Facial Cutaneous Sensibility, Temperature, and Pain Thresholds

Current methods of sensibility testing evaluate the fiber–receptor complexes that mediate the perception of touch, temperature, and pain. *Pressure thresholds* (Semmes–Weinstein monofilament test) and *static two-point discrimination* (Disk-Criminator, Pressure-Specified Sensory Device) assess the function of slowly adapting fibers associated with Ruffini receptors and Merkel cell disks.[19,20] *Tactile discrimination* reflects the number of innervated sensory receptors. *Moving two-point discrimination* and *vibration stimuli* (Tuning fork) assess the function of quickly adapting nerve fibers, hair follicle fibers, and Meissner corpuscles.[19,20] Perception of two-point discrimination, vibration, and pressure threshold values improves from the lateral and posterior areas of the face to the midline, with the vermilion being the most sensitive area and the forehead being the least sensitive.[46,47] Table 2.4, summarizes the values of normal ranges for tactile discrimination of the human face.

Two-point discrimination and vibratory values in females are reported to be lower when compared to males, although these differences are not statistically significant. There are also no significant differences between left and right side of the human face.[46,48,49] Interestingly, all of the tests evidence higher values for smokers and subjects older than 45 years of age.[19,46]

The facial skin is relatively uniform in its sensitivity to warming. In response to *thermal stimuli*, the infraorbital region and nose are the most sensitive to warming, whereas other areas of the face do not differ

Table 2.1 Sensory innervation of the upper face including: forehead, temporal area, and upper eyelids

	Nerve	Origin	Exit foramen	Branches	Structures crossed	Areas innervated
Forehead	Supratroclear (V1)	Terminal branch of the frontal division of the ophthalmic nerve (V1)[64]	Upper medial corner of orbit[65,66]	Three or four branches,[67] connection to medial branch supraorbital nerve[68]	Corrugator and frontalis muscles[67]	Nasal root, glabella, midforehead[64,67,69]
	Supraorbital (V1)	Largest terminal branch of the frontal nerve[64]	Orbital margin, notch, or multiple foramens[70]	Superficial and deep branches	Frontalis muscle	Forehead skin from the orbital margin to the frontal hairline, parietofrontal scalp from midline to the superior temporal line up the level of the posterior edge of the helical rim of the ear[67]
	Zygomatic*o*-temporal (V2)	First branch of the zygomatic nerve	Zygomaticotemporal foramen	None	Temporalis muscle, deep and superficial temporal fascia	Lateral forehead
Upper eyelids	Supratroclear (V1)	Terminal branch of the frontal division of the ophthalmic nerve (V1)[64]	Upper medial corner of orbit[65,66]	Three or four branches,[67] connection to medial branch supraorbital nerve[68]	Corrugator and frontalis muscles[67]	Medial upper eyelid[64,67,69]
	Infratroclear (V1)	Largest branch of the nasociliary nerve (V1)	Anterior etmoidal foramen	Superior branches	Orbicularis Oculi	Skin of medial eyelids[64]
	Infratroclear (V1)	Largest branch of the nasociliary nerve (V1)	Anterior etmoidal foramen	Inferior branches	Orbicularis Oculi	Conjunctiva[64]
	Supraorbital (V1)	Largest terminal branch of the frontal nerve[64]	Orbital margin, notch, or multiple foramens[70]	Superficial and deep branches	Frontalis muscle	Central skin of upper eyelid
	Palpebral branch of the lacrimal nerve (V1)	Smallest branch of the ophthalmic nerve (V1)	None	None	Lacrimal gland, orbital septum	Lateral edge of the eyelid
Temporal area	Horizontal branch of the supraorbital nerve (V1)	Supraorbital nerve (V1)	None	None	Temporoparietal fascia	Skin with 1.5-cm radius at the border between the lateral forehead and temple[71]
	Zygomaticotemporal (V2)	First branch of the zygomatic nerve (V2)	Zygomaticotemporal foramen	Superficial (medial) division[67]	Temporalis muscle, deep and superficial temporal fascia	Lateral forehead and the anterior temporal areas of the skull[72,73]
	Auriculotemporal (V3)	Posterior division of the mandibular nerve (V3)	Between the bony external acoustic meatus and the articular eminence of the temporomandibular joint	Superficial temporal nerves	Lateral to neck of mandible, perforate superficial temporal fascia	Anterior temporal area[74]

Table 2.2 Sensory innervation of the mid-face including: lower eyelids, nose, upper lips, cheek, and preauricular area

	Nerve	Origin	Exit foramen	Branches	Structures crossed	Areas innervated
Lower eyelids, cheek and upper lips	Infraorbital (V1)	Terminal branch of the maxillary nerve (V2)[72]	Infraorbital foramen (single most cases, double or triple foramina reported)[75-77]	Inferior palpebral branches[78,79]	Orbicularis oculi muscle	Infraorbital area, from the lower eyelid margin to upper lip vermilion, from 2 cm lateral to the external canthus to 0.5 cm from the midline
	Zygomatic-facial (V2)	Second branch the zygomatic nerve, early branch of the Maxillary nerve (V2)[72]	Zygomaticofacial foramen	None	Orbicularis oculi muscle	Zygomatic area of cheek
Nose	Infratroclear (V1)	Largest branch of the nasociliary nerve (V1)	Anterior etmoidal foramen	Superior branches	Orbicularis oculi	Skin of the root of the nose
	Supratroclear (V1)	Terminal branch of the frontal division of the ophthalmic nerve (V1)[64]	Upper medial corner of orbit[65,66]	Three or four branches,[67] connection to medial branch supraorbital nerve[68]	Corrugator and frontalis muscles[67]	Nasal root, glabella[64,67,69]
	External nasal (V1)	Branch of anterior ethmoidal nerve (from the nasociliary division of the ophthalmic nerve-V1)	Nasal bone and the upper lateral cartilage	None	Superficial nasal musculoaponeurotic system	Nasal tip
	Nasal branches of the infraorbital nerve (V2)	Infraorbital nerve	Infraorbital foramen (single most cases, double, or triple foramina reported)[75-77]	Internal and external nasal branches	Orbicularis oculi muscle	Philtrum, nasal septum, and the vestibule of nose
Preauricular area	Auriculotemporal nerve (V3)	Posterior division of the mandibular nerve (V3)	Between the bony external acoustic meatus and the articular eminence of the temporomandibular joint	Anterior auricular nerve and the external acoustic meatus nerve	Lateral to neck of mandible, perforate superficial temporal fascia	Auricular anterior aspect, external acoustic meatus
	Great auricular nerve (C2–C3)	Superficial branch of the cervical plexus (C2–C3)	Erb's point	Anterior and posterior branches	Platisma muscle	Parotid gland and the lower preauricular region[80]

Table 2.3 Sensory innervation of the lower face including: lower lips and chin

	Nerve	Origin	Exit foramen	Branches	Structures crossed	Areas innervated
Lower eyelids, cheek and upper lips	Mental (V3)	Terminal branch of the inferior alveolar nerve (from the mandibular nerve) (V3)	Mental foramen[81,82]	Angular, medial inferior labial, lateral inferior labial, and mental branch	Depressor Anguli Oris muscle	Lower lip, vermilion, vestibular gengiva, and the skin of the chin
	Branch of the mylohyoid nerve (V3)	Terminal branch of the mandibular nerve (V3)[74,83]	None	None	Mylohyoid muscle	Submental skin

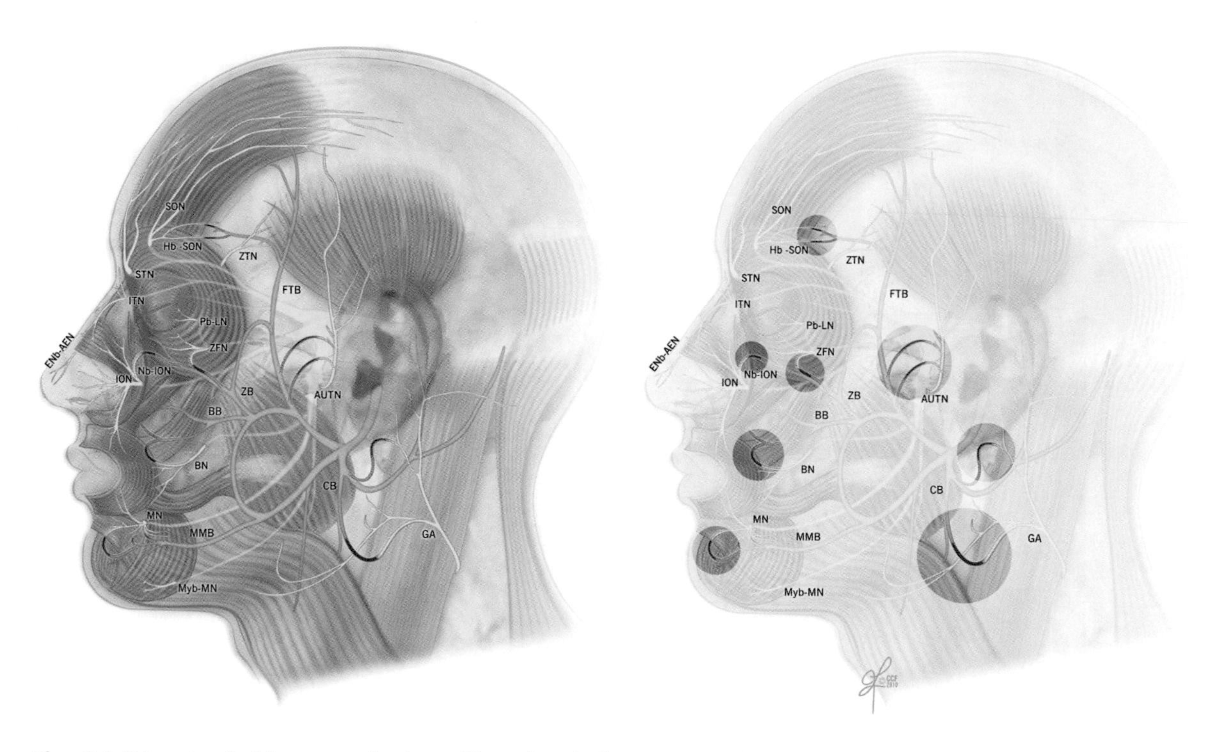

Fig. 2.4 Trigemino-facial communications. The trigeminal nerve and its branches are presented in yellow, the facial nerve in orange, and their communicating branches in blue. *BB* buccal branch of the facial nerve, *BN* buccal nerve, *CB* cervical branch of the facial nerve, *ENb-AEN*, external nasal branch-anterior ethmoidal nerve, *FTB* frontotemporal branch of the facial nerve, *GA* great auricular nerve, *Hb-SON* horizontal branch supraorbital nerve, *ION* infraorbital nerve, *ITN* infratroclear nerve, *MMB* marginal mandibular branch of the facial nerve, *MN* mental nerve, mylohyoid branch-mental nerve, *Pb-LN* palpebral branch-lacrimal nerve, *SON* supraorbital nerve, *STN* supratroclear nerve, *ZB* zygomatic branch of the facial nerve, *ZFN* zygomaticofacial nerve, *ZTN* zygomaticotemporal nerve (Reprinted with permission, Cleveland Clinic Center for Medical Art & Photography © 2010. All Rights Reserved)

significantly from one another.[50] There are not significant differences in sensitivity to cooling between oral mucosa and facial skin. Moreover, all extraoral sites are equally sensitive to cooling.[51]

In comparison to the above discussed neurosensory tests, the *pain detection threshold* represents the highest variability in data. Upper and lower labial areas are the most sensitive to pain stimuli, and the infraorbital areas are the least sensitive. In comparison, chin responses are in between these two areas.[52] The reasons for these spatial variations are undoubtedly numerous and most probably correlate with innervations density, epidermal thickness, and composition as well as with the receptors' depth.

Table 2.4 Physiologic tactile discrimination data for different areas of the human facial skin[19,20,46,48,49,52,84]

	Lateral forehead	Medial forehead	Zygoma	Cheek	Nasolabial skin	Paranasal	Superior labial skin	Upper vermilion	Upper lip mucosa	Inferior labial skin	Lower vermilion	Lower lip mucosa	Chin	Ear	Neck
Two-point discrimination (static)	13.4, 15.0	12.7, 13.0	10.58, 10.92	9.0, 13.1	7.4, 9.7	9.96, 10.38	3.25, 7.66	2.4, 7.0	2.4, 5.2	5.8, 7.04	3.0, 6.1	2.8, 4.6	5.4, 10.0	9.29, 13.2	13.7, 35.2
Two-point discrimination (dynamic)	11.0, 11.8	9.0, 11.1	9.33, 9.91	7.9, 10.0	6.6	9.63, 10.17	3.42, 3.8	4.2	6.1		4.5	3.0, 4.1	7.0, 5.1	12.9	11.5
Semmes-Weinstein (filament)	1.95	1.93		1.72, 1.84	1.75		1.75, 1.81	2.38	2.8	1.75, 1.77	2.26	2.91	1.95	1.79, 1.99	2.13
Vibratory threshold (voltage, mV)	22.0, 22.5	21.0, 22.8		17.0, 17.2	15		9.8	8.3	9.4		8.1	6.0, 9.3	11.0, 10.6	11.3	11.9
Pressure-specified sensory device 1-point static (g/mm)	0.88, 1.16	0.88, 0.95	0.71, 0.78		0.76, 0.80		0.77, 0.82	0.73, 0.77			0.69, 0.78		0.64, 0.82		
Pressure-specified sensory device 2-point static (g/mm)	31.11, 39.42	22.94, 33.38	20.58, 23.1		13.39, 13.53		11.08, 13.07	3.8, 3.98			3.27, 3.57		16.88, 17.18		
Pressure-specified sensory device 1-point moving (g/mm)	0.58, 0.71	0.56, 0.88	0.54, 0.58		0.54, 0.55		0.51, 0.57	0.49, 0.52			0.47, 0.54		0.52, 0.62		
Pressure-specified sensory device 2-point moving (g/mm)	10.16, 11.78	5.27, 8.23	5.2, 5.47		3.45, 3.72		2.66, 4.68	1.10, 1.15			1.13, 1.31		2.52, 4.28		

2.7 Conclusions

With 11 reported cases of facial allotransplantation, the technical challenges seem to be well addressed. The new challenges include achieving long-term survival under minimal immunosuppression and restoration of optimal sensory and motor functions after face transplantation. The motor function recovery after face transplantation is well documented and discussed, but the mechanisms of sensory recovery have not been adequately addressed. The presence of stable and diffuse connections between the facial and trigeminal nerves proves that the motor and sensory pathways of the human face are intrinsically interrelated; thus, every effort should be made to restore the continuity of both systems. The new era of facial reconstruction includes free tissue transfer and facial transplantation, so updating our knowledge on the sensory pathways of the face, including specific facial receptor systems, the ascending tracts, and the cortical responses to somatosensory stimulations, should add into our understanding of function and mechanisms of sensory restoration after application of modern reconstructive procedures. The facial skin presents the highest concentration of the sensory receptors in the entire body,[10] as confirmed by their fundamental role in collecting and transmitting external stimuli to the cerebral cortex for processing and integration. In contrast to the significant number of studies on composition and distribution of the sensory receptors within the hand, there are only few reported anatomical studies, performed principally on nonhuman primates, assessing the spectrum of receptors present in the facial skin. Only Munger and Halata[15] described the complex array of sensory receptors in the human face. There are, however, numerous micro-neurographic studies which confirmed the presence of four out of five types of tactile afferents which are known to innervate skin of the human hand as well, and excluded the presence of Pacini corpuscles.[12,14,21] For clinical evaluation of the sensation collected by facial receptors, different instruments and assessment devices have been described, but none of the tests have been accepted as the gold standard. Pressure-Specified Sensory Device (PSSD) is probably the most appropriate instrument for recording of human cutaneous pressure thresholds by measuring both the force and the distance at which one point can be distinguished from two points either static or moving.[47] Tests for thermal sensation are difficult to standardize since their accuracy and reliability have not been well determined.[53] Perception of the painful stimuli is usually not used in the clinical practice of sensory testing but has often been inferred from the necessity for local anesthesia when performing skin biopsies[2] or by pinprick tests. Based on previous reports, the presented summary of the range of normal values for discriminative thresholds may be useful for assessment of sensory recovery of facial sensation after free tissue transfers and in face transplant patients, since these tests are not routinely used in the current clinical practice. The reported higher tactile thresholds in patients older than 45 and in smokers should be taken into consideration during evaluation of sensibility of reconstructed or transplanted face especially if the donor age is different from the age of the recipient. During evaluation of sensory function return of a specific nerve branch, the areas to be tested should be scaled purely to the region of the repaired nerve distribution. For example, the middle third of the hemi-forehead should be tested for assessment of the supraorbital nerve, the central cheek and upper lip for the assessment of the infraorbital nerve, and the lower lip for the mental nerve. When testing thermal stimuli, the infraorbital region and nose should be considered as they were found to be the most sensitive to warm stimuli.[51] Upper and lower labial regions have been reported to be most sensitive to pain and if required can be used for assessment of painful stimuli while the infraorbital regions are the least sensitive[52] and should be avoided during sensory testing.

The distribution of the sensory nerves in the human face is usually described as a branching pattern from the main trunk to the distal rami. This approach is not helpful for sensory assessment from a reconstructive point of view, where focusing on the "complex" of peripheral nerves directly involved in the innervation of the reconstructed subunit would be more valuable. The summary offered in the Table 2.1 may be useful as a guide when deciding which is the most suitable area for sensory testing and which nerves should be considered for repair in composite unit transfers. Furthermore, the following anatomical features should be considered when repairing sensory nerves after trauma or during face transplantation. It is important to consider the fact that the supraorbital nerve exits the cranium through a bony foramen or notch, and multiple foramina have been reported. A high positioned supraorbital foramen with a long bony canal can be present in up to

24% of cases.[54] Moreover, the medial branch of the supraorbital nerve after exiting the foramen gives off multiple small rami and becomes superficial piercing the frontalis muscle. The infraorbital and mental nerves pass through long bony canals. They are already divided into their terminal branches at the level of foraminal exit, and they have a short distance to reach the skin. The presence of multiple foramina or rami, the long bony canals, and the short course of the main branches within soft tissues after exiting the cranium explains the reported difficulty in achieving direct repair of the sensory nerves in facial trauma and facial transplantation.[2] Osteotomy of the supraorbital canal, as proposed by Siemionow et al.,[55] or intraorbital division of the nerve and delivery of the proximal stump through the canal,[54] could be performed to lengthen the stump of the nerve available for repair. The infraorbital canal osteotomy and mandibular sagittal osteotomy are also useful to increase the length of these nerves during facial graft procurement.[55]

Beside the primitive reflex functions which aim to protect the individual from the noxious stimuli, the unconscious information conveyed by the trigeminal system assists in the fine tuning of the highly specialized facial functions. Livermore et al.[7] proved that the stimulation of the facial skin increases the perceived intensity of chemical stimuli applied simultaneously to the olfactory system. The stimulation of the trigeminal nerve has been shown to evoke systemic visceral reactions, when e.g., cooling of the face was associated with hypertension and bradycardia.[8] Finally, due to the absence of muscle spindles and tendon organs in the perioral muscle system,[56,57] the cutaneous receptors play an important role in the speech sensorimotor processes and adjustment of the articular motion.[6] The cutaneous and mucosal afferents discharge vigorously during labial contact and when stimulated by the air pressure, generated by speech sounds. Deformation or strain of the facial skin and mucosa associated with various phases of voluntary lip and jaw excursions provides proprioceptive information on facial movements.[58]

The second important aspect to consider is the response of the somatosensory cortex to the reconstruction of the sensory nerves. It has been proved that cortical reorganization following limb deafferentation involves reduction of the cortical representation in the motor and sensory cortices, with expansion of adjacent and controlateral areas.[59,60] Studies on the reorganization of the somatosensory cortex in patients undergoing hand transplantation showed the reversibility of this phenomenon. The patients showed activation of the primary somatosensory cortex, which started as early as 10 days and was observed[61] up to 2 years following transplantation.[62] Interestingly, Farne et el.[63] confirmed that the somatosensory perception of the transplanted hand was hampered when the ipsilateral face was simultaneously stimulated. This phenomenon disappeared 11 months after transplantation. During the remapping phase, when the transplanted hand reclaimed its original somatotopy, the face and the hand seemed to "compete" for the cortical representation and gave rise to a temporary overlapping area that received multiple conflicting inputs from two physically distant but cortically adjacent parts of the body. Equivalent modifications of the somatosensory cortex have not been studied in the face transplant patients. It would be interesting to evaluate if opposite changes can be detected. These findings are very important in the light of current attempt of simultaneous face and hand transplantation where potential competition for critical reeducation may take place jeopardizing functional outcome of one of the transplanted grafts.

In conclusion, we have illustrated the complexity of the sensory pathways of the human face and presented the role of facial sensation during interaction with the external environment. We believe that considering the present advancements in the field of facial transplantation, restoration of facial anatomy and function is crucial for the final outcome.

References

1. Dubernard JM, Lengele B, Morelon E, et al. Outcomes 18 months after the first human partial face transplantation. *N Engl J Med*. 2007;357:2451-2460.
2. Lantieri L, Meningaud JP, Grimbert P, et al. Repair of the lower and middle parts of the face by composite tissue allotransplantation in a patient with massive plexiform neurofibroma: a 1-year follow-up study. *Lancet*. 2008;372: 639-645.
3. Guo S, Han Y, Zhang X, et al. Human facial allotransplantation: a 2-year follow-up study. *Lancet*. 2008;372:631-638.
4. Siemionow M, Papay F, Alam D, et al. Near-total human face transplantation for a severely disfigured patient in the USA. *Lancet*. 2009;374:203-209.
5. Rogers SN, Lowe D, Patel M, Brown JS, Vaughan ED. Clinical function after primary surgery for oral and

oropharyngeal cancer: an 11-item examination. *Br J Oral Maxillofac Surg*. 2002;40:1-10.
6. Ito T, Gomi H. Cutaneous mechanoreceptors contribute to the generation of a cortical reflex in speech. *NeuroReport*. 2007;18:907-910.
7. Livermore A, Hummel T, Pauli E, Kobal G. Perception of olfactory and intranasal trigeminal stimuli following cutaneous electrical stimulation. *Experientia*. 1993;49:840-842.
8. LeBlanc J, Blais B, Barabe B, Cote J. Effects of temperature and wind on facial temperature, heart rate, and sensation. *J Appl Physiol*. 1976;40:127-131.
9. Loken LS, Wessberg J, Morrison I, McGlone F, Olausson H. Coding of pleasant touch by unmyelinated afferents in humans. *Nat Neurosci*. 2009;12:547-548.
10. Connor NP, Abbs JH. Orofacial proprioception: analyses of cutaneous mechanoreceptor population properties using artificial neural networks. *J Commun Disord*. 1998;31:535-542. 553.
11. Uchigasaki S, Suzuki H, Inoue K. Merkel cells in the vellus hair follicles of human facial skin: a study using confocal laser microscopy. *J Dermatol*. 2004;31:218-222.
12. Johansson RS, Trulsson M, Olsson KA, Westberg KG. Mechanoreceptor activity from the human face and oral mucosa. *Exp Brain Res*. 1988;72:204-208.
13. Nordin M, Thomander L. Intrafascicular multi-unit recordings from the human infra-orbital nerve. *Acta Physiol Scand*. 1989;135:139-148.
14. Bukowska M, Essick GK, Trulsson M. Functional properties of low-threshold mechanoreceptive afferents in the human labial mucosa. *Exp Brain Res*. 2010;201:59-64.
15. Munger BL, Halata Z. The sensorineural apparatus of the human eyelid. *Am J Anat*. 1984;170:181-204.
16. Kawakami T, Ishihara M, Mihara M. Distribution density of intraepidermal nerve fibers in normal human skin. *J Dermatol*. 2001;28:63-70.
17. Johansson O, Wang L, Hilliges M, Liang Y. Intraepidermal nerves in human skin: PGP 9.5 immunohistochemistry with special reference to the nerve density in skin from different body regions. *J Peripher Nerv Syst*. 1999;4:43-52.
18. Schulze E, Witt M, Fink T, Hofer A, Funk RH. Immunohistochemical detection of human skin nerve fibers. *Acta Histochem*. 1997;99:301-309.
19. Kesarwani A, Antonyshyn O, Mackinnon SE, Gruss JS, Novak C, Kelly L. Facial sensibility testing in the normal and posttraumatic population. *Ann Plast Surg*. 1989;22:416-425.
20. Fogaca WC, Sturtz GP, Surjan RC, Ferreira MC. Evaluation of cutaneous sensibility on infraorbital nerve area. *J Craniofac Surg*. 2005;16:953-956.
21. Nordin M, Hagbarth KE. Mechanoreceptive units in the human infra-orbital nerve. *Acta Physiol Scand*. 1989;135:149-161.
22. Davies SN, Goldsmith GE, Hellon RF, Mitchell D. Facial sensitivity to rates of temperature change: neurophysiological and psychophysical evidence from cats and humans. *J Physiol*. 1983;344:161-175.
23. Nordin M. Low-threshold mechanoreceptive and nociceptive units with unmyelinated (C) fibres in the human supraorbital nerve. *J Physiol*. 1990;426:229-240.
24. Chen CC, Rainville P, Bushnell MC. Noxious and innocuous cold discrimination in humans: evidence for separate afferent channels. *Pain*. 1996;68:33-43.
25. Bushnell MC, Taylor MB, Duncan GH, Dubner R. Discrimination of innocuous and noxious thermal stimuli applied to the face in human and monkey. *Somatosens Res*. 1983;1:119-129.
26. Gardner EP, Martin JH, Jessell TM. The bodily senses. In: Kandel ER, Schwartz JH, Jessell TM, eds. *Principles of Neural Science*. 4th ed. United States of America: McGraw-Hill; 2000:430-450.
27. Brannagan TH, Weimer LH. Cranial and peripheral nerve lesions. In: Rowland LP, Pedley TA, eds. *Merritt's Neurology*. 12th ed. Philadelphia: Lippincott Williams & Wilkins; 2010:506.
28. Laine FJ, Smoker WR. Anatomy of the cranial nerves. *Neuroimaging Clin North Am*. 1998;8:69-100.
29. Nemzek WR. The trigeminal nerve. *Top Magn Reson Imaging*. 1996;8:132-154.
30. Ropper AH, Samuels MA. Chapter 9: Other somatic sensation. In: Ropper AH, Samuels MA, eds. *Adams and Victor's Principles of Neurology*. 9th ed. New York: McGraw-Hill; 2009.
31. Eriksen K. Neurophysiology and the upper cervical subluxation. In: Eriksen K, Rochester RP, eds. *Orthospinology Procedures: An Evidenced-Based Approach to Spinal Care*. 1st ed. Philadelphia: Lippincott William & Wilkins; 2007: 183-207.
32. Terman GW, Bonica JJ. Spinal mechanisms and their modulation. In: Loeser JD, ed. *Bonica's Management of Pain*. 3rd ed. Philadelphia: Lippincott William & Wilkins; 2001:110-125.
33. Penfield W, Boldrey E. Somatic motor and sensory representation in the cerebral cortex of man as studied by electrical stimulation. *Brain*. 1937;60:389-433.
34. Tamura Y, Shibukawa Y, Shintani M, Kaneko Y, Ichinohe T. Oral structure representation in human somatosensory cortex. *Neuroimage*. 2008;43:128-135.
35. Nguyen BT, Inui K, Hoshiyama M, Nakata H, Kakigi R. Face representation in the human secondary somatosensory cortex. *Clin Neurophysiol*. 2005;116:1247-1253.
36. Servos P, Engel SA, Gati J, Menon R. fMRI evidence for an inverted face representation in human somatosensory cortex. *NeuroReport*. 1999;10:1393-1395.
37. Nguyen BT, Tran TD, Hoshiyama M, Inui K, Kakigi R. Face representation in the human primary somatosensory cortex. *Neurosci Res*. 2004;50:227-232.
38. Namking M, Boonruangsri P, Woraputtaporn W, Guldner FH. Communication between the facial and auriculotemporal nerves. *J Anat*. 1994;185(Pt 2):421-426.
39. Kwak HH, Park HD, Youn KH, et al. Branching patterns of the facial nerve and its communication with the auriculotemporal nerve. *Surg Radiol Anat*. 2004;26:494-500.
40. O'connell JE. The intraneural plexus and its significance. *J Anat*. 1936;70:468-497.
41. Huber E. Evolution of facial musculature and cutaneous field of trigeminus. *Q Rev Biol*. 1930;5:133-188.
42. Riessener D. Surgical procedure in tumors of parotid gland; preservation of facial nerve and prevention of postoperative fistulas. *AMA Arch Surg*. 1952;65:831-848.
43. Baumel JJ. Trigeminal-facial nerve communications. Their function in facial muscle innervation and reinnervation. *Arch Otolaryngol*. 1974;99:34-44.
44. Carmichael EA, Woolard HH. Some observations on the fifth and seventh cranial nerves. *Brain*. 1933;56:109-125.
45. Ley A, Guitart JM. Clinical observations on sensory effects of trigeminal dorsal root section. *J Neurol Neurosurg Psychiatry*. 1971;34:260-264.

46. Costas PD, Heatley G, Seckel BR. Normal sensation of the human face and neck. *Plast Reconstr Surg*. 1994;93: 1141-1145.
47. Dellon AL, Andonian E, DeJesus RA. Measuring sensibility of the trigeminal nerve. *Plast Reconstr Surg*. 2007;120: 1546-1550.
48. Chen CC, Essick GK, Kelly DG, Young MG, Nestor JM, Masse B. Gender-, side- and site-dependent variations in human perioral spatial resolution. *Arch Oral Biol*. 1995;40: 539-548.
49. Vriens JP, van der Glas HW. Extension of normal values on sensory function for facial areas using clinical tests on touch and two-point discrimination. *Int J Oral Maxillofac Surg*. 2009;38:1154-1158.
50. Green BG, Gelhard B. Perception of temperature on oral and facial skin. *Somatosens Res*. 1987;4:191-200.
51. Rath EM, Essick GK. Perioral somesthetic sensibility: do the skin of the lower face and the midface exhibit comparable sensitivity? *J Oral Maxillofac Surg*. 1990;48: 1181-1190.
52. Hung J, Samman N. Facial skin sensibility in a young healthy Chinese population. *Oral Surg Oral Med Oral Pathol Oral Radiol Endod*. 2009;107:776-781.
53. Kawano T, Kabasawa Y, Ashikawa S, Sato Y, Jinno S, Omura K. Accuracy and reliability of thermal threshold measurement in the chin using heat flux technique. *Oral Surg Oral Med Oral Pathol Oral Radiol Endod*. 2009;108:500-504.
54. Shimizu S, Osawa S, Utsuki S, Oka H, Fujii K. Course of the bony canal associated with high-positioned supraorbital foramina: an anatomic study to facilitate safe mobilization of the supraorbital nerve. *Minim Invasive Neurosurg*. 2008;51:119-123.
55. Siemionow M, Papay F, Kulahci Y, et al. Coronal-posterior approach for face/scalp flap harvesting in preparation for face transplantation. *J Reconstr Microsurg*. 2006;22:399-405.
56. Siemionow M, Agaoglu G, Unal S. A cadaver study in preparation for facial allograft transplantation in humans: part II. Mock facial transplantation. *Plast Reconstr Surg*. 2006;117: 876-885. Discussion 886-888.
57. Kubota K, Masegi T. Muscle spindle supply to the human jaw muscle. *J Dent Res*. 1977;56:901-909.
58. Trulsson M, Johansson RS. Orofacial mechanoreceptors in humans: encoding characteristics and responses during natural orofacial behaviors. *Behav Brain Res*. 2002;135:27-33.
59. Borsook D, Becerra L, Fishman S, et al. Acute plasticity in the human somatosensory cortex following amputation. *NeuroReport*. 1998;9:1013-1017.
60. Wall JT, Xu J, Wang X. Human brain plasticity: an emerging view of the multiple substrates and mechanisms that cause cortical changes and related sensory dysfunctions after injuries of sensory inputs from the body. *Brain Res Brain Res Rev*. 2002;39:181-215.
61. Neugroschl C, Denolin V, Schuind F, et al. Functional MRI activation of somatosensory and motor cortices in a hand-grafted patient with early clinical sensorimotor recovery. *Eur Radiol*. 2005;15:1806-1814.
62. Brenneis C, Loscher WN, Egger KE, et al. Cortical motor activation patterns following hand transplantation and replantation. *J Hand Surg Br*. 2005;30:530-533.
63. Farne A, Roy AC, Giraux P, Dubernard JM, Sirigu A. Face or hand, not both: perceptual correlates of reafferentation in a former amputee. *Curr Biol*. 2002;12:1342-1346.
64. Shankland WE. The trigeminal nerve. Part II: the ophthalmic division. *Cranio*. 2001;19:8-12.
65. Kimura K. Foramina and notches on the supraorbital margin in some racial groups. *Kaibogaku Zasshi*. 1977;52:203-209.
66. Webster RC, Gaunt JM, Hamdan US, Fuleihan NS, Giandello PR, Smith RC. Supraorbital and supratrochlear notches and foramina: anatomical variations and surgical relevance. *Laryngoscope*. 1986;96:311-315.
67. Knize DM. Transpalpebral approach to the corrugator supercilii and procerus muscles. *Plast Reconstr Surg*. 1995;95: 52-60. Discussion 61-62.
68. Malet T, Braun M, Fyad JP, George JL. Anatomic study of the distal supraorbital nerve. *Surg Radiol Anat*. 1997;19:377-384.
69. Andersen NB, Bovim G, Sjaastad O. The frontotemporal peripheral nerves. topographic variations of the supraorbital, supratrochlear and auriculotemporal nerves and their possible clinical significance. *Surg Radiol Anat*. 2001;23:97-104.
70. Beer GM, Putz R, Mager K, Schumacher M, Keil W. Variations of the frontal exit of the supraorbital nerve: an anatomic study. *Plast Reconstr Surg*. 1998;102:334-341.
71. Hwang K, Hwang JH, Cho HJ, Kim DJ, Chung IH. Horizontal branch of the supraorbital nerve and temporal branch of the facial nerve. *J Craniofac Surg*. 2005;16:647-649. Discussion 650.
72. Shankland WE II. The trigeminal nerve. Part III: the maxillary division. *Cranio*. 2001;19:78-83.
73. Hwang K, Suh MS, Lee SI, Chung IH. Zygomaticotemporal nerve passage in the orbit and temporal area. *J Craniofac Surg*. 2004;15:209-214.
74. Shankland WE II. The trigeminal nerve. Part IV: the mandibular division. *Cranio*. 2001;19:153-161.
75. Kazkayasi M, Ergin A, Ersoy M, Tekdemir I, Elhan A. Microscopic anatomy of the infraorbital canal, nerve, and foramen. *Otolaryngol Head Neck Surg*. 2003;129:692-697.
76. Canan S, Asim OM, Okan B, Ozek C, Alper M. Anatomic variations of the infraorbital foramen. *Ann Plast Surg*. 1999;43:613-617.
77. Aziz SR, Marchena JM, Puran A. Anatomic characteristics of the infraorbital foramen: a cadaver study. *J Oral Maxillofac Surg*. 2000;58:992-996.
78. Hu KS, Kwak HH, Song WC, et al. Branching patterns of the infraorbital nerve and topography within the infraorbital space. *J Craniofac Surg*. 2006;17:1111-1115.
79. Hwang K, Nam YS, Choi HG, Han SH, Hwang SH. Cutaneous innervation of lower eyelid. *J Craniofac Surg*. 2008;19:1675-1677.
80. Ginsberg LE, Eicher SA. Great auricular nerve: anatomy and imaging in a case of perineural tumor spread. *AJNR Am J Neuroradiol*. 2000;21:568-571.
81. Hwang K, Lee WJ, Song YB, Chung IH. Vulnerability of the inferior alveolar nerve and mental nerve during genioplasty: an anatomic study. *J Craniofac Surg*. 2005;16:10-14. Discussion 14.
82. Greenstein G, Tarnow D. The mental foramen and nerve: clinical and anatomical factors related to dental implant placement: a literature review. *J Periodontol*. 2006;77:1933-1943.
83. Hwang K, Han JY, Chung IH, Hwang SH. Cutaneous sensory branch of the mylohyoid nerve. *J Craniofac Surg*. 2005;16:343-345. Discussion 346.
84. Posnick JC, Zimbler AG, Grossman JA. Normal cutaneous sensibility of the face. *Plast Reconstr Surg*. 1990;86: 429-433. Discussion 434-435.

Immunological Aspects of Face Transplantation

3

Aleksandra Klimczak and Maria Z. Siemionow

Contents

M.Z. Siemionow (✉)
Department of Plastic Surgery, Cleveland Clinic,
Cleveland, OH, USA
e-mail: siemiom@ccf.org

Abstract Human facial transplantation is a form of composite tissue allotransplantation (CTA), and since November 2005, it has become a clinical reality. Face transplantation is still considered an experimental procedure in the clinic, and to date, 13 facial transplantations have been performed worldwide. We observe the progress in composite facial tissue allotransplantation, partial or full facial transplantation for severely disfigured patients. Facial CTA involves the transplantation of different type of tissues carrying different functions and immunologic characteristics. Immunogenicity of tissue components of the facial allograft and immunosuppressive strategies that reduce allogenic responses against the graft are discussed in this chapter.

Abbreviations

APC	antigen-presenting cell
ATG	anti-thymocyte immunoglobulin
CsA	cyclosporine A
CTA	composite tissue allotransplantation
CTLA	cytotoxic T lymphocyte–associated antigen
DDC	dermal dendritic cell
GVHD	graft vs. host disease
ICAM	intercellular adhesion molecule
IRHCTT	The International Registry on Hand and Composite Tissue Transplantation
LC	Langerhans cell
LFA	leukocyte function–associated antigen
mAb	monoclonal antibody
MHC	major histocompatibility complex
MMF	mycophenolate mofetil
NHP	non-human primates
PTLD	post-transplant lymphoproliferative disorder
RAPA	rapamycin
SALT	skin-associated lymphoid tissue

M.Z. Siemionow (ed.), *The Know-How of Face Transplantation*,
DOI: 10.1007/978-0-85729-253-7_3,

3.1 Introduction

The successful progress of composite tissue transplantation is an area of great promise in the field of plastic and reconstructive surgery. However, composite tissue transplantation, such as face or hand transplants, is still considered an experimental procedure. Applications of facial allograft transplantation predominantly improve the quality of life for severely disfigured patients by restoring anatomic, cosmetic, and functional integrities. To date, 13 facial transplantations have been performed worldwide (Table 3.1). Transplantation procedures performed in the clinic differ in the type of tissues which were transplanted to cover facial defect based on the severity of disfigurement. Thus, it is important to understand the immunological aspects of face transplantation since different tissue types may generate different immunological responses and affect graft acceptance and long-term survival.[1-6]

3.2 Face as a Composite Tissue Allograft

The facial composite tissue allograft (CTA) involves the transplantation of different type of tissues including skin, muscle, bone, lymph node, nerve, blood vessels, cartilage/soft tissue, salivary glands, and intraoral mucosa. In solid organ transplantation, allograft function is defined by the biochemical and physiologic properties of the transplanted organ. However, we have proposed that the face should be considered an organ, because face allotransplantation is essential for physical and social survival.[7] On the other hand, for the face allograft, as well as other CTA, the function and immunologic characteristic are more difficult to define because each individual component possesses unique properties that may affect the successful outcome of the allograft. In transplanting the face, we are transplanting histologically heterogeneous tissues, which represent different levels of immunogenicity and different types of immunologic responses.[8]

In this chapter, we focus on the immunogenicity of facial components of the allograft and immunosuppressive protocols that reduce allogenic responses against the graft and preserve normal function with limited adverse effects.

3.3 Immunogenicity of Facial Composite Tissue Allograft

3.3.1 *Immune System of the Skin*

The skin is the largest part of facial CTA and constitutes a highly immunogenic organ with physiologically active defense functions. Both epidermal and dermal

Table 3.1 World experience in face transplantations

Date	Location	Indication	Outcome – Survival
November 2005	Amiens, France	Dog bite	5.2 years - alive
April 2006	Xian, China	Bear attack	2.3 years – died
January 2007	Paris, France	Neurofibromatosis	4.0 years - alive
December 2008	Cleveland, USA	Shotgun trauma	2.1 years - alive
March 2009	Paris, France	Shotgun trauma	1.9 years - alive
April 2009 (concomitant bilateral hand transplant)	Paris, France	Third degree burns	2 months - died
April 2009	Boston, USA	Fall injury	1.8 year – alive
August 2009	Paris, France	Ballistic trauma	1.4 year - alive
August 2009	Valencia, Spain	Radiotherapy of aggressive tumor	1.4 year – alive
November 2009	Amiens, France	Explosion trauma	1.2 year - alive
January 2010	Seville, Spain	Neurofibromatosis	1.0 year - alive
April 2010	Barcelona, Spain	Shotgun trauma	10 months – alive
June 2010	Paris, France	Neurofibromatosis	7 months - alive

structures of the skin contain a set of different cell types having immunologic properties, and pro-inflammatory mediators that initiate and regulate immune response. The immunological microenvironment of the skin is known as skin-associated immune system (SALT). SALT is composed of diverse type of cells such as: antigen-presenting cells (APC), skin-seeking lymphocytes, keratinocytes and fibroblasts, dermal endothelial cells, and skin-draining lymph nodes.[9,10] These components of the skin are responsible for the relationship between the epidermal environment and the skin-draining lymph nodes, a specific immunologic subsystem that is of the utmost importance in the induction of immunity and tolerance.[10]

In the skin, professional APCs are represented by Langerhans cells (LC), localized in the epidermis, and by dermal dendritic cells (DDC) localized in the dermis.[11] The antigen-presenting cell functions of LC and DDC are accomplished by (1) expression of high levels of major histocompatibility complex (MHC) class I and MHC-II molecules, (2) expression of co-stimulatory molecules (CD80 and CD86), (3) ability to internalize and process antigen, (4) high migratory capacity, which allows them to transport antigen from periphery to draining lymph nodes, and (5) strong stimulatory function to allogenic T cells.[11,12]

LC and DDC adapt to the microenvironment and present differential phenotype according to their location. Functionally, it is probable that most of the LC and DDC remain inactive during the steady state; however, a low number become activated and carry self-antigen from the skin to the draining lymph nodes to present the antigen in a tolerogenic manner. In the presence of inflammatory conditions, skin-residing LC and DDC may have the ability to respond to the injury and mature into potent APC while still being able to maintain tolerance to self-antigen.

The immunogenic and tolerogenic functions of skin-resident APC to foreign stimuli constitute a major barrier to skin allotransplantation, since skin DC are essential for initiation of immune response and allograft rejection.[13] Depending on the state of skin maturity, DC may be tolerogenic or immunogenic and the manipulation of these properties may provide potential immunotherapies.[12]

Another important population of immunocompetent skin-resident cells localized in the dermis are T-lymphocytes represented by both CD4 [T-helper subpopulation (Th)] and CD8 [suppressor-cytotoxic subpopulation (Tc)].[14] Memory T cells, a Th1 subset, secrete pro-inflammatory cytokines (IL-2, IL-12, IFN-γ) and are responsible for initiation of cell-mediated immune response, whereas the Th2 subset producing IL-4, IL-5, IL-6, IL-10, and IL-13 cytokines is responsible for the humoral immune response.

Immune response in the skin may be also supported by keratinocytes, the main cellular component of the epidermis. Keratinocytes constitute a specific microenvironment and are a potential source of cytokines produced constitutively or upon induction of various stimuli. A variety of keratinocyte-derived cytokines play different functions. Pro-inflammatory cytokines IL-1, IL-6, IL-8, and TNF-α produced and secreted by keratinocytes have systemic effects on the immune system, influence keratinocyte proliferation and differentiation processes, and are a powerful attractant for inflammatory cells. Keratinocytes producing IL-7, IL-15 cytokines are considered a significant contributor in T-cell trafficking. Immunomodulatory cytokines IL-10, IL-12, IL-18 derived from keratinocytes are considered to be responsible for the systemic effect.[15] Moreover, upon stimulation by IFN-γ, keratinocytes express immunologically important surface antigens including MHC class II and intracellular adhesion molecule-1 (ICAM-1). These findings demonstrate that keratinocytes that become activated may act as an APC and are able to induce functional responses.[16]

3.3.2 Skin-Draining Lymph Nodes

When transplanting a face allograft, we should consider the presence of skin-draining lymph nodes. Lymph nodes are a source of immunocompetent T cells, B cells, and follicular dendritic cells. The presence of lymph nodes within transplanted tissues contributes to the induction of the recipient allo-immune response. After transplantation, recipient T cells migrate to the lymph nodes of the transplanted tissue where they undergo extensive proliferation and develop effector functions, and without proper immunosuppression, these effector mechanisms may lead to allograft rejection.[17] Dendritic cell populations of the skin-draining lymph nodes exist at different maturation stages, and they are capable of induction of primary or secondary immune responses against foreign antigens.[18]

3.3.3 Immunological Microenvironment of the Muscle

In face transplantation, muscle constitutes an important component to restore facial appearance, and the contribution of the muscle component in the facial allograft depends on the severity of injury. Muscle represents a tissue with specific immunological properties, and the immunoregulatory capacity of the muscle depends on positive and negative muscle-derived regulators. Studies have documented that under physiological conditions, mature muscle cells do not express MHC class I or class II molecules. Therefore, immune reactions triggered by, or directed against muscle cells, proceed along specific pathways. However, studies on cultured human myoblasts in the presence of IFN-γ, TNF-α, and IL-1β have demonstrated that these pro-inflammatory cytokines can induce expression of MHC class II molecules.[19] If muscle does express MHC class II molecules in vivo, they could hypothetically present not only viral or bacterial antigens but also muscle autoantigens or allo-antigens to CD4 T cells.[20] Moreover, skeletal muscle can express HLA-G "non-classical" MHC class I molecule, which has been characterized as a molecule that mediates immunotolerization.[21]

Under inflammatory conditions (e.g., inflammatory myopathies), muscle fibers express specific muscle-related non-classical co-stimulatory molecules ICOS-L, B7-H, B7-H2, members of the B7-family. They do not express the classical co-stimulatory molecules B7.1 and B7.2. As activated T cells present ICOS receptor, ICOS-L present on the muscle fibers is capable of interaction with T-cells' ICOS receptor[20] and triggers an immune response. Thus, under inflammatory conditions (IFN-γ, TNF-α), expressions of MHC class I and class II molecules and non-classical co-stimulatory molecules on the muscle fibers may play an active role in muscle–immune interactions, and muscle fibers may act as a non-professional APC.[22]

3.3.4 Nerve Immune Components

Proper innervation of the face allograft is responsible for appropriate sensory and motor functions of transplanted facial CTA. Peripheral nerves comprise neural and non-neural elements such as: (1) conducting axons, (2) insulating Schwann cells, and (3) surrounding connective tissue matrix. Immune components present in the nerve, such as fibroblasts, macrophages, mast cells, blood vessels, and fat, are localized in the internal and external epineurium.[23] Schwann cells represent a natural component of the nerve tissue and may act as immunomodulators by producing and secreting a variety of cytokines including pro-inflammatory cytokines IL-1, IL-6, and TNF-α and immunoregulatory cytokine TGF-β in a specific autocrine manner. Schwann cells may also synthesize other pro-inflammatory and immunoregulatory mediators such as prostaglandin E_2, thromboxane A_2, and leukotriene C_4, which may regulate the immune cascade in inflammatory conditions.[24] It was also reported that Schwann cells constitutively express MHC class I but not MHC class II molecules.[25] However, after nerve injury in immune-mediated disorders, in the presence of activated T-lymphocytes, Schwann cells released IFN-γ, and MHC class II molecules were also detected, suggesting that these Schwann cells may act as an APC and may contribute to the local immune response.[26,27]

The immunomodulatory function of Schwann cells is accomplished by production of erythropoietin which prevents axonal degeneration, reduces TNF-α production, Wallerian degeneration, and decreases pain-related behaviors after peripheral nerve injury.[28]

The peripheral nervous system is protected from the immune compartment by the blood–nerve barrier; however, activated T- lymphocytes and B-lymphocytes constantly patrol the peripheral nervous system, irrespective of their antigen specificity.[29] APCs in the peripheral nerve compartment are represented by macrophages, and their role as APCs was confirmed by expression of MHC class II molecules and co-stimulatory molecules B7-1 and B7-2, which are essential for effective antigen presentation to T cells, thereby modulating the local immune response.[29,30]

3.3.5 Bone and Bone Marrow

Bone constitutes an integral component of the face allograft especially when large facial defects such as the maxilla or mandible should be reconstructed. The antigenicity of the bone unit is considered to be low. In experimental models of CTA, such as limb or some face allograft models, vascularized bone containing hematolymphoid tissue with bone marrow cells was successfully transplanted. As we introduced in experimental models, the presence of viable donor hematopoietic cells within transplanted bone, under proper non-myeloablative conditions created by an immunosuppressive regimen, may play an immunomodulatory function, and may downregulate the host immune system to the allograft.[31-34] The donor-origin hematopoietic cells may be involved in tolerance induction. After CTA transplantation, donor bone marrow cells may migrate from transplanted tissues and colonize lymphoid and non-lymphoid organs of recipients. Engraftment of donor-origin cells into recipient lymphoid and non-lymphoid tissues is known as chimerism.[35] However, overrepresentation of donor hematopoietic cells within transplanted CTA without immunosuppression has the potential to attack the recipient immune system, leading to graft-versus-host disease (GVHD), which may be fatal.[36]

The immunomodulatory function of the bone marrow compartment may also be accomplished by bone marrow-derived dendritic cells. Depending on the maturation status, bone marrow-derived dendritic cells may act either as an APC or may lead to tolerance induction. After migration to T-cell areas of secondary lymphoid organs (e.g., draining lymph nodes), bone marrow-derived dendritic cells can both induce and regulate immune responses.[37]

3.3.6 Vessels and Immune Responses

Graft revascularization and blood supply is one of the major concerns for a successful outcome in a facial transplantation procedure.[6] Vessel endothelial cells, the main cellular component of the vessels, play a multifunctional role in the vascular system; their functions include regulation of thrombosis and thrombolysis, platelet adherence, modulation of vasomotor tone and blood flow, and regulation of immune and inflammatory responses.[8] An immune and inflammatory reaction is regulated by controlling leukocyte interaction with the blood vessels. Under inflammatory conditions, vessel endothelial cells may secrete pro-inflammatory cytokines IL-1, IL-6, IL-8, and subsequently activated endothelial cells induce expression of P-selectin, E-selectin, and cell adhesion molecules (ICAM-1 and ICAM-2, and VCAM-1), facilitating leukocyte extravasation into surrounding tissue.[38] Moreover, vessel endothelial cells express APC-related MHC class II molecules and co-stimulatory molecule CD40, leading to proliferation and differentiation of activated or memory T cells, but not naive T cells.[39]

3.3.7 Salivary Glands and Oral Mucosa

The immune function in the salivary glands is carried out by two complementary parts, which belong to the mucosa-associated lymphoid system. One is a secretory component, which acts as a glycoprotein receptor for immunoglobulins IgA and IgM, produced by acinar and ductal epithelial cells. The second part is the lymphoid tissue represented by lymphoid cells, either diffuse or organized in lymph nodes.[40]

Salivary glands are important effector sites in the mucosal immune network that possess lymphocyte populations, distinct from those in peripheral lymphoid sites, which regulate and mediate humoral and cellular immune responses, contributing to the protection of oral surfaces. The phenotypic studies identified unique mononuclear cell populations including T cells, B cells, and NK-cells.[41] These distinctions must be considered when designing effective immunotherapy in pathological processes occurring in mucosa-associated tissues.

Oral mucosa constitutes an integral part of facial composite tissue flap. Physiologically, oral mucosa is thought to be the most proximal extend of the mucosal immune system, recognizing and eliminating pathogens, while tolerating harmless commensals that are essential for maintaining immune homeostasis. The most important cellular population responsible for mucosal immune response are mucosal dendritic cells. The main population of mucosal dendritic cells are LC expressing the antigen-presenting molecule CD1a and activation and maturation markers.[42] Moreover, in oral mucosa, dermal and plasmacytoid dendritic cell pools have been identified. DDCs of oral mucosa origin contribute to the CD83+ mature dendritic cell pool in the lamina propria,[43] whereas plasmacytoid dendritic cells express a Toll-like receptor for the viral antigen.[44]

The oral mucosa also contains oral lymphoid foci representing effector sites for local immune response. Oral lymphoid foci are organized into lymphoid and myeloid elements, with lamina propria dendritic cells and CD4+ T cells, which include CD45RA and CD45RO subsets.[45]

3.4 Transplantation Tolerance

In the field of solid organ or CTA transplantation, the major problem is to induce and maintain antigen-specific transplantation tolerance, functionally defined as allograft survival with stable function in the absence of chronic immunosuppression.[46] However, in a clinical scenario, true tolerance is difficult to achieve due to heterogenicity of the transplanted tissues and donor–recipient immune status.[8,47] In clinic, an operationally tolerant state is possible to develop by different mechanisms generating a downregulated or unresponsive state to donor antigens inducing indefinite allograft survival, with minimal or no immunosuppression. These mechanisms of unresponsiveness to donor antigens may be accomplished by partial clonal deletion, clonal anergy, cytokine pattern alterations, and the presence of immunoregulatory cells.[48,49] As proposed by Sachs, in clinical practice, we should consider the downregulatory response as a tolerant state regardless of the mechanism.[50] This partially tolerant state with a minimal, non-toxic dose of maintenance immunosuppressive therapy, rather than absence of immunosuppression, is clinically relevant.

The concept of low dose of immunosuppression was introduced for the first time, by Calne, as a "prope" or "almost" tolerance. Donor-specific hypo-responsiveness is accomplished under a low dose of immunosuppressive therapy and leads to improved allograft survival, without acute or chronic rejection, and significantly reduces immunosuppression-related side effects.[51] A similar idea of minimal immunosuppression tolerance was referred to by Monaco, and this tolerant state was associated with immunoregulatory cell mechanisms under minimal doses of immunosuppressive regimens.[52]

The regulation of immune response and induction of transplantation tolerance can be achieved using approaches that induce peripheral and/or central tolerance to the allograft.

3.4.1 Central Tolerance in Clinical Transplantation

The thymus plays a key role in the maintenance of tolerance to self-antigens. In the physiological condition, central tolerance is accomplished by intrathymic clonal deletion of autoreactive T-lymphocytes in the process of negative selection. Negative selection occurs in the thymic medulla, when thymocytes recognize self-antigens and are destroyed to avoid antigen-reactive T lymphocytes reaching the periphery.[53]

In the field of solid organ transplants, central tolerance can be achieved by induction of donor-specific chimerism (co-existence of donor and recipient cells in the recipient compartments) which allows donor APCs to migrate into the recipient's thymus and induce negative selection of donor-reactive T cells.[54,55]

Successful donor-specific tolerance associated with the generation of donor-specific chimerism has been reported in kidney transplant recipients. Select patients with end-stage renal failure secondary to refractory multiple myeloma underwent kidney transplantation from the same living donor, and developed long-term kidney allograft acceptance without immunosuppression.[56-58]

Recently, a non-myeloablative conditioning protocol and induction of donor-specific chimerism were successfully used by the Massachusetts General Hospital transplantation team, in kidney transplant recipients.[59] Patients received a combination of bone marrow and kidney transplant from one-haplotype HLA-mismatched living related donors and except one patient who developed humoral rejection, all immunosuppressive therapy was discontinued between 9 and 14 months after transplantation, and patients demonstrated full tolerance to the kidney graft.

Interestingly, all of the kidney recipients developed only transient post-transplantation chimerism. This may suggest that the induction of central tolerance is dependent on central deletion of donor-reactive T cells, and peripheral mechanisms may be more applicable for long-term maintenance of transplantation tolerance. These studies are very promising and may lead to tolerance induction in transplant recipients; however, this method of tolerance induction is limited to living organ donors and is not applicable to face transplant recipients.

3.4.2 Peripheral Tolerance in Clinical Transplantation

After transplantation, circulating host memory T-lymphocytes become allo-reactive to donor antigens and are crucial in the initiation of the rejection response. To promote peripheral tolerance, the allo-reactive effector T-lymphocytes must be eliminated or inactivated and the regulatory mechanism must be enhanced.[60] Peripheral tolerance mechanisms are indeed active in extrathymic lymphoid tissues, and different strategies have been explored to accomplish peripheral tolerance to allo-antigens. Elimination of allo-reactive T-lymphocytes may be achieved by: (1) deletion of peripheral T cells using a lymphocyte-depletion protocol; (2) inhibition of T cell activation by blocking co-stimulatory signals; (3) interference with the effector or homing functions of activated T-lymphocytes by cytokine or chemokine alteration; (4) active suppression of effector T cells by antigen-specific regulatory T cells expressing CD4+/CD25+/FOXP3+ molecules.[61]

Only activated host T-lymphocytes are able to respond to allo-antigens and induce an immune response. For activation and differentiation of allo-reactive T cells into proliferating effector T cells, three distinct signals are necessary. The first step of activation of T cells (signal 1) is accomplished via TCR receptor present on host T cells through antigen recognition of donor-derived peptides presented in the context of MHC antigens expressed on the surface of donor APCs. The second step of T-cell activation requires delivery of a co-stimulatory signal (signal 2) through the binding of T-cell molecules, such as CD28, to their ligands, CD80 or CD86 molecules constitutively expressed on the activated APC.[62] If the activation process is incomplete, e.g., if TCR ligands are changed or in the absence of a co-stimulatory signal, T cells become unresponsive to proliferative signals and this state is referred to as anergy.[63] If both signals of activation, TCR-MHC interaction, and co-stimulatory signals are completed, the T cells are able to secrete IL-2 which interacts with its TCR to generate signal 3 of T-cell activation, and trigger T-cell proliferation and differentiation.[64]

Interruption of these signaling pathways of T-cell activation and differentiation is currently used in the clinic to prevent allograft rejection and may promote peripheral tolerance. In transplantation, peripheral tolerance to the allograft can be achieved using modern immunosuppressive regiments.

3.5 Immunosuppressive Strategies in Facial Transplantations

Based on clinical experience in solid organ transplants and composite tissue allografts such as hand and abdominal wall transplants, immunosuppressive protocols similar to those allografts have been applied to face transplantation.[65] Current immunosuppressive agents used for transplantation are given for three purposes: induction, maintenance, and treatment of rejection.

3.5.1 Induction Therapy

Induction therapies may promote regulatory mechanisms and the induction of peripheral transplantation tolerance. Most of the strategies for induction therapy applied in the clinic, in solid organ and CTA transplantation, are based on elimination or inactivation of allo-reactive T cells. Newer therapeutic approaches have been developed based on potentially allo-reactive cell depletion at the time of transplantation, when immune activation is most powerful. Cell-depleting strategies result in a significant reduction in circulating leukocytes capable of producing an allo-response at the time when the allograft is already susceptible to inflammatory damage following ischemia/reperfusion injury.

An immunomodulatory protocol that helps to reduce the immune response, extends organ survival, and diminishes systemic drug toxicity by specifically targeting T-cell subsets and modulating cytokine expression, may be accomplished by inhibition of activation signals 1, 2, or 3 of T cells (Fig. 3.1).

3.5.1.1 Inhibition of Signal 1

Depletion strategies by inhibition of activation signal 1 can be achieved by nonselective and selective T-cell depletion. The nonselective elimination of T cells (targeting all T cells, not only allo-reactive T cells) using

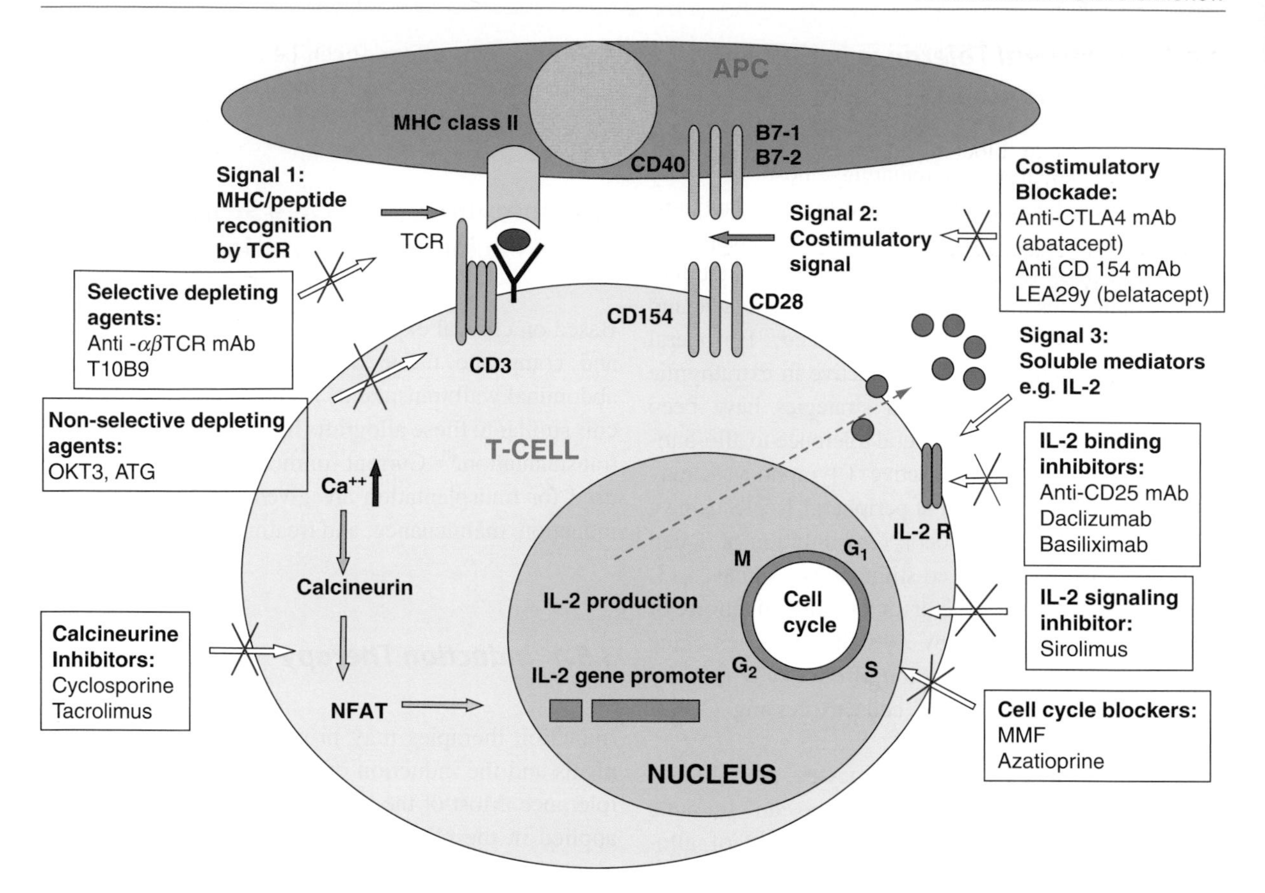

Fig. 3.1 T-cell activation and cellular targets of immunosuppressive drugs. *αβ-TCRmAb*, αβ–T-cell receptor monoclonal antibody; *APC*, antigen-presenting cells; *ATG*, antithymocyte globulin; *IL-2*, interleukin-2; *IL-2R*, interleukin-2 receptor; *mAb*, monoclonal antibody; *MHC*, major histocompatibility complex antigen; *MMF*, mycophenolate mofetil; *NAFT*, nuclear factor of activated T cells; *TCR*, T-cell receptor

polyclonal anti-thymocyte globulins (ATG), or monoclonal antibodies (mAb) such as humanized anti CD-52 mAb (Campath-1 H, alemtuzumab) or anti-CD3 mAb (muromonab-CD3) is frequently used in clinical transplantation as an induction therapy. ATGs produced by the immunization of horses or rabbits to human leukocytes is used as induction therapy in high immunological risk recipients, since they have broad T-cell specificity.[66,67]

The humanized anti CD-52 mAb known as Campath-1 H or alemtuzumab was successfully used for the first time by Calne et al. for kidney transplantation from deceased donors.[68] CD52 is expressed on T cells, B cells, monocytes, macrophages, natural killer cells (NK), and granulocytes.[69] ATG and anti CD-52 mAb are powerful lymphocyte-depleting agents capable of rapid and sustained depletion of circulating lymphocytes. The capacity of these agents induces profound and durable lymphopenia that can be associated with adverse effects such as immunodeficiency complications (e.g., viral infections CMV, EBV), or as reported following ATG treatment, the development of post-transplant lymphoproliferative disorders (PTLD).[70] Moreover, ATG and Campath-1 H may induce other adverse effects including thrombocytopenia, a cytokine-release syndrome or allergic response. However, the benefits of induction therapy with ATG or Campath-1 H outweigh the adverse effects, especially when induction therapy is supported with calcineurin inhibitors, cyclosporine A (CsA) or tacrolimus, or IL-2 signaling inhibitor sirolimus, agents used in maintenance of immunosuppression.[71]

Muromonab-CD3, mouse mAb binding CD3 component of TCR signal–transduction complex, has been used successfully for high-risk kidney transplant recipients.[72] However, because of side effects of muromonab-CD3, including a cytokine-release syndrome,[73] in CTA transplants, ATG is preferably used.

3.5.1.2 Inhibition of Signal 2

Transient inhibition of the T-cell co-stimulatory activation pathway by blocking CD154-CD40 or CD28-B7(CD80/CD86) interaction has been extensively studied in transplantation tolerance.[74,75]

CD40 is constitutively expressed on DC, B-lymphocytes, and macrophages – the cells with antigen-presenting function. CD154, the ligand of CD40, is induced in T cells after TCR–antigen interaction.[76] In a non-human primate (NHP) model, administration of humanized anti-CD154 mAb (clone hu5C8) promoted long-term skin allograft survival, but true tolerance was not achieved.[77] However, using a humanized anti-CD154 mAb, thromboembolic complications were observed and this agent was withdrawn from clinical trials.[78]

CD28 is constitutively expressed on CD4 T-lymphocytes and up to 50% of CD8 T-lymphocytes, and during the activation process, it binds to CD80 and CD86 molecules present on APCs. Cytotoxic T-lymphocyte-associated antigen-4 (CTLA-4) is induced on T cells after activation and binds the same ligands as CD28.[79] In NHP islet and kidney transplant models, inhibition of the co-stimulatory pathway CD28-B7(CD80/CD86) was accomplished by mAb CTLA4Ig (abatacept), which binds to the CD80 and CD86 receptors present on APCs.[80,81] A second-generation agent – LEA29Y (belatacept) – was recently introduced, and it had a higher affinity for CD80 and CD86 molecules than CTLA4Ig and was found to be more effective than CTLA4Ig when used in initial primate studies.[82] A selective co-stimulatory blocker, belatacept, was recently used in clinical trial in renal transplantation, as a primary maintenance immunosuppressant. Immunosuppressive protocol with belatacept significantly improved outcomes, and belatacept has appeared to preserve the renal function and reduce the rate of chronic allograft nephropathy compared to CsA-treated patients at 1 year after transplantation.[83,84]

However, experimental studies performed on NHP models showed that dual blockade of CD154-CD40 and CD28-B7(CD80/CD86) pathways acts synergistically and is more effective in preventing rejection and/or inducing tolerance.[74,85]

Recent experimental data suggested that co-stimulatory blockade may also have a positive effect on the expansion, survival, and function of Treg cells.[86,87]

3.5.1.3 Inhibition of Signal 3

The IL-2 receptor (called CD25), is upregulated after T-cell activation but is not expressed on most resting T cells. Activated T cells produce IL-2 that induces their proliferation via signaling through upregulated IL-2 receptors present on the surface of activated T cells (signal 3). IL-2 receptor antagonists such as daclizumab or basiliximab inhibit the IL-2-mediated proliferation and effector function of allo-reactive T cells. Dacilizumab or basiliximab is successfully used in kidney transplant recipients as an induction therapy.[88] These agents target activated T cells in the early post-transplantation period and do not cause significant lymphocyte depletion and have no significant side effects. An adverse result of IL-2 antagonist action may be associated with a deleterious effect on the Treg subset of CD4 T cells, which constitutively expresses CD25.[89]

3.5.1.4 Inhibition of Leukocyte Trafficking

Effective immune response to the allograft is accomplished when allorective T cells are able to migrate and infiltrate the graft. Cell migration is regulated by sphingosine-1-phosphate expressed on lymphocytes and DC. Sphingosine-1-phosphate modulator – fingolimod (FTY720) – inhibits lymphocyte trafficking to the allograft by lymphocyte sequestration in the lymphoid tissues such as lymph node or spleen. Clinical investigation with new immunomodulator FTY720 in kidney transplantation was discontinued because this drug generated a higher rate of side effects compared with standard immunosuppression.[90]

The inflammatory process induced initially by ischemia/reperfusion injury is associated with release of graft-derived chemokines and upregulation of cell adhesion molecules on vessel endothelial cells, including intercellular adhesion molecule-1 (ICAM-1), which mediate transmigration of activated T-lymphocytes. Leukocyte function-associated antigen-1 (LFA-1) has a multifaceted role in the immune response, including adhesion and trafficking of leukocytes, stabilizing the immune synapse of the MHC–TCR complex and providing co-stimulatory signals.[91] Clinical trials indicated that blockade of adhesion molecules such ICAM-1or LFA-1 could be another promising approach to immunosuppression in transplantation tolerance.[92]

3.5.2 Maintenance Immunosuppression

Maintenance immunosuppression is given to protect the allograft from an immune response of potentially persistent allo-reactive T cells. After the initial high dose of immunosuppressive regimen maintenance, immunosuppression is tapered safely within months after transplantation. However, only a small proportion of patients can completely withdraw from immunosuppressants or continue on a minimal non-toxic dose of immunosuppression.

Maintenance regimens mostly consisted of calcineurin inhibitors such as CsA and tacrolimus and/or antiproliferative agents such as azathioprine, mycophenolate mofetil (MMF), and rapamycin (sirolimus). These agents not only inhibit the immune response of effector T cells, but also affect the development and functional immunobiology of dendritic cells, thereby having an adverse effect on the interaction between APC and T cells.[93]

3.5.2.1 Calcineurin Inhibitors

The introduction of CsA into transplantation procedures remains a milestone in the current maintenance of immunosuppression. The immunosuppressive effect of CsA is accomplished by inhibition of Ca^{2+}-dependent TCR-mediated signal transduction leading to IL-2 production. CsA inhibits the intracellular calcium-calcineurin signaling pathway during the activation process of T cells and subsequently, IL-2 gene transcription, and ultimately inhibits IL-2 production and T-cell activation. Moreover, immunosuppressive effects of CsA on rodent and human DC have been reported.[94] CsA inhibits the antigen-presenting capacity of human peripheral blood DC subsets and human epidermal LC by downregulating surface co-stimulatory molecule expression,[95,96] and in this manner, inhibits signal 2 of the T-cell activation cascade. Thus, CsA inhibits DC-dependent production of IFN-γ, IL-2, and IL-4 by T cells and IL-6, IL-12p40, IL-12p70 by DC.[93]

Tacrolimus binds to the intracellular immunophilins FK506-binding proteins (FKBP), and the complex FK506 – FKBP blocks calcineurin, the same molecular target as CsA, inhibiting TCR-mediated signal transduction in T-lymphocytes.[97] However, direct effects of tacrolimus on DC have been also reported. Tacrolimus significantly reduces expression of MHC class II molecules and co-stimulatory molecules on mouse epidermal LC, and is more potent than other immunosuppressants in inhibiting DC cytokine production (IL-6, IL-12p40, IL-12p70).[98,99] Studies of the influence of tacrolimus on human DC displayed discrepant observations. One study reported that tacrolimus has no effect on co-stimulatory molecule expression; however, human DC displayed a reduced allo-stimulatory capacity and IL-12 production.[93] On the other hand, topical administration of tacrolimus in atopic dermatitis resulted in decreased expression of co-stimulatory molecules and suppresses cytokine production in epidermal DC.[100,101]

Thus, the immunosuppressive effect of calcineurin inhibitors acts primarily by direct immunosuppression of T-cell activation, and secondarily, via downregulation of the antigen-presenting function of DC.

Both CsA and tacrolimus have side effects, including nephrotoxicity and thrombotic microangiopathy. CsA is, more often than tacrolimus, a cause of hypertension or hyperlipidemia, but is rarely associated with post-transplantation diabetes, unlike tacrolimus. Many of the adverse effects of calcineurin inhibitors are dose dependent, and blood concentration monitoring is necessary to maintain allograft survival and minimize adverse effects.

3.5.2.2 Antiproliferative Drugs

Maintenance immunosuppression is also supported by purine nucleotide biosynthesis inhibitors such as azathioprine and mycophenolate mofetil (MMF) and by the mammalian target of rapamycin (RAPA) (mTOR) inhibitor – sirolimus.

Azathioprine was widely used for immunosuppression in organ transplant recipients for prophylaxis of acute rejection before calcineurin inhibitors were employed. This agent blocks de novo and salvage pathways of purine nucleotide biosynthesis and subsequently inhibits the proliferation of T and B cells. The immunosuppressive effect of azathioprine on the cells responsible for cellular and humoral response is enhanced by decreased allo-stimulatory capacity of epidermal LC on T cells.[102]

MMF is a nucleotide synthesis inhibitor and, in response to allogenic stimulation, inhibits both T- and B-cell proliferation.[103] Influences of MMF on DC subsets have been identified. MMF impairs the maturation and antigen-presenting function of epidermal LC.[104] Moreover, in combination with vitamin D_3, it could induce a tolerogenic phenotype of DC that promotes the frequency of CD4+/CD25+ regulatory T cells and promotes tolerance to the allografts.[105]

Side effects of MMF treatment are associated with gastrointestinal and hematological complications; however, its great efficacy in preventing acute rejection or chronic allograft dysfunction outweighs its adverse effects.

In transplant recipients, MMF usually is administered in combination with calcineurin inhibitors, and this strategy allows inhibition of different steps of T-cell activation while using lower dosages of these immunosuppressants, thereby reducing the development of side effects.

The antiproliferative capacity of sirolimus (rapamycin) is accomplished by inhibition of the activity of mTOR. This action results in inhibition of different biochemical pathways including: cytokine-induced proliferation, ribosome synthesis, and cell cycle progression into S phase, which are critical for T-cell activation.[106] In the experimental model, it was reported that sirolimus facilitates peripheral deletion of effector T cells by promoting apoptosis of allo-reactive T cells.[107] Moreover, in contrast to calcineurin inhibitors, sirolimus selectively expands the Treg population, which would be beneficial in inducing tolerance.[108] Sirolimus impairs human DC function by suppressing co-stimulatory molecule expression, inhibiting IL-12 and IL-10 production, and subsequently decreasing the allo-stimulatory capacity of T cells.[109]

Similar to other immunosuppressive drugs, sirolimus may induce nephrotoxicity, hyperlipidemia, and proteinuria and may impair wound healing; however, the advantages of using sirolimus make it a potent option.

3.5.3 Immunosuppressive Agents for Treatment of Rejection

Treatment of rejection episodes includes increases in the dosages of agents used for maintenance therapy. In the majority of cases, increased doses of calcineurin inhibitors, in combination with corticosteroid ointment, are sufficient for acute rejection reversal. In some cases, high doses of intravenous steroids are necessary. In steroid-resistant episodes, lymphodepletive agents such as ATG, Campath-1 H, or basiliximab should be considered to overcome the rejection episodes.

3.6 Immunosuppressive Protocols in Face Transplantation: Clinical Experience

Up-to-date information about the immunosuppressive regimen for the first four face transplant cases have been reported (Table 3.2). Information about the other seven cases of face transplants are based on the report in The International Registry on Hand and Composite Tissue Transplantation (IRHCTT) – (www.handregistry.com) and media releases, but details concerning immunosuppression are not reported.

Induction protocols include the lymphodepletive agent ATG, in three cases, and mAb against human IL-2 receptor, in one patient. In three of four transplants, the immunodepletive protocol was supported with tacrolimus, in addition to MMF and steroids – prednisone or methylprednisolone. Maintenance therapy consists of tacrolimus, MMF, and prednisone using dosages adjusted to the clinical presentation. In one patient, tacrolimus was replaced by sirolimus 11 months after facial transplantation due to nephrotoxicity of the calcineurin inhibitor.

The importance of developing donor-specific chimerism in CTA is debatable. Successful outcomes in kidney transplant recipients supported by donor bone marrow transplantation have encouraged the use of similar protocols with donor bone marrow cells for tolerance induction.[59] Bone marrow cells of donor-origin were included in the post-transplant therapeutic protocol for first face transplant recipient. Transient chimerism was detectable in the early post-transplant period.[2]

Table 3.2 Immunosuppressive protocols in face transplant recipients

Nr	Facial Transplantation (Location)	Transplanted Tissues	Induction Therapy	Maintenance Therapy	Treatment of Rejection
1	Amiens, France	Skin, muscle, nerves, artery and vein, oral mucosa	Thymoglobulin, Tacrolimus, MMF [1], Prednisone	Tacrolimus, MMF, Prednisone, Sirolimus (applied 11 months after transplantation)	Increased doses of Tacrolimus, MMF, Prednisone. Three boluses of Methylprednisolone
2	Xian, China	Skin, muscle, nerves, artery and vein, oral mucosa, parotid gland, nasal cartilage, zygoma	Anti- hIL2 receptor mAb, Tacrolimus, MMF, Methylprednisolone	Tacrolimus, MMF, Methylprednisolone switched to prednisone Anti- hIL2 receptor mAb	Increased doses of Tacrolimus, MMF, Methylprednisolone.
3	Paris, France	Skin, nerves, artery and vein, oral mucosa, nasal cartilage	Thymoglobulin, MMF, Prednisone	Tacrolimus, MMF, Prednisone	Three boluses of Prednisone, ATG
4	Cleveland, USA	Skin, muscle, nerves, artery and vein, oral mucosa, parotid glands, nasal cartilage, zygomas, maxilla	Thymoglobulin,MMF, Methylprednisolone	Tacrolimus, MMF, Prednisone	Single bolus of Methylprednisolone
5	Paris, France	Skin, muscle, nerves, artery and vein, oral mucosa, parotid glands, nasal cartilage, premaxilla	Thymoglobulin, MMF, Prednisone	Tacrolimus,MMF, Prednisone ECP[2]	Three boluses of Methylprednisolone
6	Paris, France (concomitant bilateral hand transplant)	Facial skin and scalp, muscle, nerves, artery and vein, oral mucosa, parotid glands, nasal and ear cartilage,	Thymoglobulin, MMF, Prednisone	Tacrolimus, MMF, Prednisone ECP	No rejection episode
7	Boston, USA	Skin, muscle, nerves, artery and vein, maxilla	Thymoglobulin,MMF, Prednisone	Tacrolimus, MMF (switched to mycophenolic acid) Prednisone	Three boluses of Methylprednisolone
8	Paris, France	Skin, muscle, nerves, artery and vein, nose, oral mucosa, maxilla and mandible	Thymoglobulin, MMF, Prednisone	Tacrolimus, MMF, Prednisone ECP	Three boluses of Methylprednisolone
9	Valencia, Spain	Skin, muscle, nerve, artery and vein, tongue, oral mucosa, and mandible	-	-	-
10	Amiens, France	Skin, muscle, nerve, artery and vein, and mandible	-	-	-
11	Seville, Spain	Skin, muscle, nerve, artery and vein, parotid glands, oral mucosa, partial nose	-	-	-
12	Barcelona, Spain	Skin, muscle, nerves, artery and vein, cheekbones, palate, nose, parotid glands, oral mucosa, maxilla and mandible	-	-	-
13	Paris, France	Skin, muscle, nerves, artery and vein, nose, parotid glands and ducts, oral mucosa,	-	-	-

Details concerning immunosuppressive protocols for patients 9 through 13 are still not reported and information about transplants are based on CTA registry (www.handregistry.com) and media releases

[1] Mycophenolate mofetil;

[2] ECP – Extracorporeal photophoresis

Acute rejection episodes occurred in all patients and were treated with increased dosages of the agents used for maintenance immunosuppression, and/or boluses of methylprednisolone[2,5] and, in one case, a bolus of prednisone was supported with lymphodepletive treatment using ATG.[4] All rescue therapies successfully reversed the rejection episodes.

3.7 Conclusion

Current experiences in face transplantation indicate that induction therapies which deplete or interfere with the activation and/or effector function of allo-reactive T cells are an effective means of inducing peripheral tolerance to the allograft. To maintain facial allograft survival and function, some degree of immunosuppression must be continued as long as the allograft is in place. Maintenance immunosuppression generally consists of calcineurin inhibitors, which inhibit alloantigen-dependent T-cell activation and can exert strong inhibitory effects on dendritic cell maturation and function. These agents offer potential avenues for the manipulation of dendritic cell – T-cell interactions to promote T-cell unresponsiveness to the allograft. Maintenance regimens may be enhanced by antiproliferative agents. To avoid the adverse side effects of lifelong immunosuppression, lower dosages of different immunosuppressive drugs, inhibiting different signals of T-cell activation, may be used. Finally, topical application of steroids (Clobetasol) or tacrolimus (Protopic) may play a role in maintenance protocols of face and other CTA transplants.

References

1. Devauchelle B, Badet L, Lengele B, et al. First human face allograft: early report. *Lancet*. 2006;368:203-209.
2. Dubernard JM, Lengele B, Morelon E, et al. Outcomes 18 months after the first human partial face transplantation. *N Engl J Med*. 2007;357:2451-2460.
3. Guo S, Han Y, Zhang X, et al. Human facial allotransplantation: a 2-year follow-up study. *Lancet*. 2008;372:631-638.
4. Lantieri L, Meningaud JP, Grimbert P, et al. Repair of the lower and middle parts of the face by composite tissue allotransplantation in a patient with massive plexiform neurofibroma: a 1-year follow-up study. *Lancet*. 2008;372: 639-645.
5. Siemionow M, Papay F, Alam D, et al. Near-total human face transplantation for a severely disfigured patient in the USA. *Lancet*. 2009;374:203-209.
6. Pomahac B, Lengele B, Ridgway EB, et al. Vascular considerations in composite midfacial allotransplantation. *Plast Reconstr Surg*. 2010;125:517-522.
7. Siemionow M, Sonmez E. Face as an organ. *Ann Plast Surg*. 2008;61:345-352.
8. Klimczak A, Siemionow M. Immunology of tissue transplantation. In: Siemionow M, Eisenmann-Klein M, eds. *Plastic and Reconstructive Surgery*. London: Springer; 2010.
9. Bos JD, Kapsenberg ML. The skin immune system: progress in cutaneous biology. *Immunol Today*. 1993;14:75-78.
10. Bos JD. *Skin Immune System (SIS)*. 2nd ed. Boca Raton, New York: CRC; 1997.
11. Mathers AR, Larregina AT. Professional antigen-presenting cells of the skin. *Immunol Res*. 2006;36:127-136.
12. Mutyambizi K, Berger CL, Edelson RL. The balance between immunity and tolerance: the role of Langerhans cells. *Cell Mol Life Sci*. 2009;66:831-840.
13. Rulifson IC, Szot GL, Palmer E, Bluestone JA. Inability to induce tolerance through direct antigen presentation. *Am J Transplant*. 2002;2:510-519.
14. Bos JD, Zonneveld I, Das PK, Krieg SR, van der Loos CM, Kapsenberg ML. The skin immune system (SIS): distribution and immunophenotype of lymphocyte subpopulations in normal human skin. *J Invest Dermatol*. 1987;88:569-573.
15. Grone A. Keratinocytes and cytokines. *Vet Immunol Immunopathol*. 2002;88:1-12.
16. Black AP, Ardern-Jones MR, Kasprowicz V, et al. Human keratinocyte induction of rapid effector function in antigen-specific memory CD4+ and CD8+ T cells. *Eur J Immunol*. 2007;37:1485-1493.
17. Wang J, Dong Y, Sun JZ, et al. Donor lymphoid organs are a major site of alloreactive T-cell priming following intestinal transplantation. *Am J Transplant*. 2006;6:2563-2571.
18. Henri S, Siret C, Machy P, Kissenpfennig A, Malissen B, Leserman L. Mature DC from skin and skin-draining LN retain the ability to acquire and efficiently present targeted antigen. *Eur J Immunol*. 2007;37:1184-1193.
19. Nagaraju K, Raben N, Merritt G, Loeffler L, Kirk K, Plotz P. A variety of cytokines and immunologically relevant surface molecules are expressed by normal human skeletal muscle cells under proinflammatory stimuli. *Clin Exp Immunol*. 1998;113:407-414.
20. Wiendl H, Hohlfeld R, Kieseier BC. Immunobiology of muscle: advances in understanding an immunological microenvironment. *Trends Immunol*. 2005;26:373-380.
21. Carosella ED, Moreau P, Aractingi S, Rouas-Freiss N. HLA-G: a shield against inflammatory aggression. *Trends Immunol*. 2001;22:553-555.
22. Wiendl H, Hohlfeld R, Kieseier BC. Muscle-derived positive and negative regulators of the immune response. *Curr Opin Rheumatol*. 2005;17:714-719.
23. Grant GA, Goodkin R, Kliot M. Evaluation and surgical management of peripheral nerve problems. *Neurosurgery*. 1999;44:825-839. Discussion 39-40.
24. Constable AL, Armati PJ, Toyka KV, Hartung HP. Production of prostanoids by Lewis rat Schwann cells in vitro. *Brain Res*. 1994;635:75-80.

25. Armati PJ, Pollard JD, Gatenby P. Rat and human Schwann cells in vitro can synthesize and express MHC molecules. *Muscle Nerve*. 1990;13:106-116.
26. Wohlleben G, Hartung HP, Gold R. Humoral and cellular immune functions of cytokine-treated Schwann cells. *Adv Exp Med Biol*. 1999;468:151-156.
27. Meyer zu Horste G, Hu W, Hartung HP, Lehmann HC, Kieseier BC. The immunocompetence of Schwann cells. *Muscle Nerve*. 2008;37:3-13.
28. Campana WM, Li X, Shubayev VI, Angert M, Cai K, Myers RR. Erythropoietin reduces Schwann cell TNF-alpha, Wallerian degeneration and pain-related behaviors after peripheral nerve injury. *Eur J Neurosci*. 2006;23: 617-626.
29. Kieseier BC, Hartung HP, Wiendl H. Immune circuitry in the peripheral nervous system. *Curr Opin Neurol*. 2006;19: 437-445.
30. Kiefer R, Kieseier BC, Stoll G, Hartung HP. The role of macrophages in immune-mediated damage to the peripheral nervous system. *Prog Neurobiol*. 2001;64:109-127.
31. Siemionow M, Izycki D, Ozer K, Ozmen S, Klimczak A. Role of thymus in operational tolerance induction in limb allograft transplant model. *Transplantation*. 2006;81:1568-1576.
32. Yazici I, Unal S, Siemionow M. Composite hemiface/calvaria transplantation model in rats. *Plast Reconstr Surg*. 2006;118:1321-1327.
33. Yazici I, Carnevale K, Klimczak A, Siemionow M. A new rat model of maxilla allotransplantation. *Ann Plast Surg*. 2007;58:338-344.
34. Kulahci Y, Siemionow M. A new composite hemiface/mandible/tongue transplantation model in rats. *Ann Plast Surg*. 2010;64:114-121.
35. Pree I, Pilat N, Wekerle T. Recent progress in tolerance induction through mixed chimerism. *Int Arch Allergy Immunol*. 2007;144:254-266.
36. Murase N, Starzl TE, Tanabe M, et al. Variable chimerism, graft-versus-host disease, and tolerance after different kinds of cell and whole organ transplantation from Lewis to brown Norway rats. *Transplantation*. 1995;60:158-171.
37. Barratt-Boyes SM, Thomson AW. Dendritic cells: tools and targets for transplant tolerance. *Am J Transplant*. 2005;5: 2807-2813.
38. Sumpio BE, Riley JT, Dardik A. Cells in focus: endothelial cell. *Int J Biochem Cell Biol*. 2002;34:1508-1512.
39. Ma W, Pober JS. Human endothelial cells effectively costimulate cytokine production by, but not differentiation of, naive CD4+ T cells. *J Immunol*. 1998;161:2158-2167.
40. Martinez-Madrigal F, Micheau C. Histology of the major salivary glands. *Am J Surg Pathol*. 1989;13:879-899.
41. O'Sullivan NL, Skandera CA, Montgomery PC. Lymphocyte lineages at mucosal effector sites: rat salivary glands. *J Immunol*. 2001;166:5522-5529.
42. Cutler CW, Jotwani R. Dendritic cells at the oral mucosal interface. *J Dent Res*. 2006;85:678-689.
43. Jotwani R, Cutler CW. Multiple dendritic cell (DC) subpopulations in human gingiva and association of mature DCs with CD4+ T-cells in situ. *J Dent Res*. 2003;82:736-741.
44. Liu YJ. IPC: professional type 1 interferon-producing cells and plasmacytoid dendritic cell precursors. *Annu Rev Immunol*. 2005;23:275-306.
45. Jotwani R, Palucka AK, Al-Quotub M, et al. Mature dendritic cells infiltrate the T cell-rich region of oral mucosa in chronic periodontitis: in situ, in vivo, and in vitro studies. *J Immunol*. 2001;167:4693-4700.
46. Monaco AP. Prospects and strategies for clinical tolerance. *Transplant Proc*. 2004;36:227-231.
47. Siemionow M, Klimczak A. Tolerance and future directions for composite tissue allograft transplants: part II. *Plast Reconstr Surg*. 2009;123:7e-17e.
48. Zheng XX, Sanchez-Fueyo A, Domenig C, Strom TB. The balance of deletion and regulation in allograft tolerance. *Immunol Rev*. 2003;196:75-84.
49. Stassen M, Schmitt E, Jonuleit H. Human CD(4+)CD(25+) regulatory T cells and infectious tolerance. *Transplantation*. 2004;77:S23-S25.
50. Sachs DH. Mixed chimerism as an approach to transplantation tolerance. *Clin Immunol*. 2000;95:S63-S68.
51. Calne RY. Prope tolerance–the future of organ transplantation from the laboratory to the clinic. *Int Immunopharmacol*. 2005;5:163-167.
52. Monaco AP. The beginning of clinical tolerance in solid organ allografts. *Exp Clin Transplant*. 2004;2:153-161.
53. Sprent J, Kishimoto H. The thymus and negative selection. *Immunol Rev*. 2002;185:126-135.
54. Starzl TE, Demetris AJ, Murase N, Ildstad S, Ricordi C, Trucco M. Cell migration, chimerism, and graft acceptance. *Lancet*. 1992;339:1579-1582.
55. Remuzzi G. Cellular basis of long-term organ transplant acceptance: pivotal role of intrathymic clonal deletion and thymic dependence of bone marrow microchimerism-associated tolerance. *Am J Kidney Dis*. 1998;31:197-212.
56. Spitzer TR, Delmonico F, Tolkoff-Rubin N, et al. Combined histocompatibility leukocyte antigen-matched donor bone marrow and renal transplantation for multiple myeloma with end stage renal disease: the induction of allograft tolerance through mixed lymphohematopoietic chimerism. *Transplantation*. 1999;68:480-484.
57. Buhler LH, Spitzer TR, Sykes M, et al. Induction of kidney allograft tolerance after transient lymphohematopoietic chimerism in patients with multiple myeloma and end-stage renal disease. *Transplantation*. 2002;74:1405-1409.
58. Fudaba Y, Spitzer TR, Shaffer J, et al. Myeloma responses and tolerance following combined kidney and nonmyeloablative marrow transplantation: in vivo and in vitro analyses. *Am J Transplant*. 2006;6:2121-2133.
59. Kawai T, Cosimi AB, Spitzer TR, et al. HLA-mismatched renal transplantation without maintenance immunosuppression. *N Engl J Med*. 2008;358:353-361.
60. Lechler RI, Garden OA, Turka LA. The complementary roles of deletion and regulation in transplantation tolerance. *Nat Rev Immunol*. 2003;3:147-158.
61. Golshayan D, Pascual M. Tolerance-inducing immunosuppressive strategies in clinical transplantation: an overview. *Drugs*. 2008;68:2113-2130.
62. Sayegh MH, Turka LA. The role of T-cell costimulatory activation pathways in transplant rejection. *N Engl J Med*. 1998;338:1813-1821.
63. Lechler R, Chai JG, Marelli-Berg F, Lombardi G. T-cell anergy and peripheral T-cell tolerance. *Philos Trans R Soc Lond B Biol Sci*. 2001;356:625-637.

64. Siemionow M, Klimczak A. Basics of immune responses in transplantation in preparation for application of composite tissue allografts in plastic and reconstructive surgery: part I. *Plast Reconstr Surg*. 2008;121:4e-12e.
65. Petruzzo P, Lanzetta M, Dubernard JM, et al. The international registry on hand and composite tissue transplantation. *Transplantation*. 2008;86:487-492.
66. Brennan DC, Flavin K, Lowell JA, et al. Leukocyte response to thymoglobulin or atgam for induction immunosuppression in a randomized, double-blind clinical trial in renal transplant recipients. *Transplant Proc*. 1999;31:16S-18S.
67. Brennan DC, Flavin K, Lowell JA, et al. A randomized, double-blinded comparison of thymoglobulin versus atgam for induction immunosuppressive therapy in adult renal transplant recipients. *Transplantation*. 1999;67:1011-1018.
68. Calne R, Friend P, Moffatt S, et al. Prope tolerance, perioperative campath 1 H, and low-dose cyclosporin monotherapy in renal allograft recipients. *Lancet*. 1998;351:1701-1702.
69. Watson CJ, Bradley JA, Friend PJ, et al. Alemtuzumab (CAMPATH 1 H) induction therapy in cadaveric kidney transplantation–efficacy and safety at five years. *Am J Transplant*. 2005;5:1347-1353.
70. Caillard S, Dharnidharka V, Agodoa L, Bohen E, Abbott K. Posttransplant lymphoproliferative disorders after renal transplantation in the United States in era of modern immunosuppression. *Transplantation*. 2005;80:1233-1243.
71. Pham PT, Lipshutz GS, Kawahji J, Singer JS, Pham PC. The evolving role of alemtuzumab (Campath-1 H) in renal transplantation. *Drug Des Devel Ther*. 2009;3:41-49.
72. Benfield MR, Tejani A, Harmon WE, et al. A randomized multicenter trial of OKT3 mAbs induction compared with intravenous cyclosporine in pediatric renal transplantation. *Pediatr Transplant*. 2005;9:282-292.
73. Bugelski PJ, Achuthanandam R, Capocasale RJ, Treacy G, Bouman-Thio E. Monoclonal antibody-induced cytokine-release syndrome. *Expert Rev Clin Immunol*. 2009;5:499-521.
74. Larsen CP, Elwood ET, Alexander DZ, et al. Long-term acceptance of skin and cardiac allografts after blocking CD40 and CD28 pathways. *Nature*. 1996;381:434-438.
75. Wekerle T, Sayegh MH, Ito H, et al. Anti-CD154 or CTLA4Ig obviates the need for thymic irradiation in a non-myeloablative conditioning regimen for the induction of mixed hematopoietic chimerism and tolerance. *Transplantation*. 1999;68:1348-1355.
76. Larsen CP, Pearson TC. The CD40 pathway in allograft rejection, acceptance, and tolerance. *Curr Opin Immunol*. 1997;9:641-647.
77. Elster EA, Xu H, Tadaki DK, et al. Treatment with the humanized CD154-specific monoclonal antibody, hu5C8, prevents acute rejection of primary skin allografts in nonhuman primates. *Transplantation*. 2001;72:1473-1478.
78. Kawai T, Andrews D, Colvin RB, Sachs DH, Cosimi AB. Thromboembolic complications after treatment with monoclonal antibody against CD40 ligand. *Nat Med*. 2000;6:114.
79. Alegre ML, Frauwirth KA, Thompson CB. T-cell regulation by CD28 and CTLA-4. *Nat Rev Immunol*. 2001;1:220-228.
80. Levisetti MG, Padrid PA, Szot GL, et al. Immunosuppressive effects of human CTLA4Ig in a non-human primate model of allogeneic pancreatic islet transplantation. *J Immunol*. 1997;159:5187-5191.
81. Kirk AD, Tadaki DK, Celniker A, et al. Induction therapy with monoclonal antibodies specific for CD80 and CD86 delays the onset of acute renal allograft rejection in non-human primates. *Transplantation*. 2001;72:377-384.
82. Larsen CP, Pearson TC, Adams AB, et al. Rational development of LEA29Y (belatacept), a high-affinity variant of CTLA4-Ig with potent immunosuppressive properties. *Am J Transplant*. 2005;5:443-453.
83. Vincenti F, Larsen C, Durrbach A, et al. Costimulation blockade with belatacept in renal transplantation. *N Engl J Med*. 2005;353:770-781.
84. Larsen CP, Knechtle SJ, Adams A, Pearson T, Kirk AD. A new look at blockade of T-cell costimulation: a therapeutic strategy for long-term maintenance immunosuppression. *Am J Transplant*. 2006;6:876-883.
85. Adams AB, Shirasugi N, Jones TR, et al. Development of a chimeric anti-CD40 monoclonal antibody that synergizes with LEA29Y to prolong islet allograft survival. *J Immunol*. 2005;174:542-550.
86. Tang Q, Henriksen KJ, Boden EK, et al. Cutting edge: CD28 controls peripheral homeostasis of CD4+CD25+ regulatory T cells. *J Immunol*. 2003;171:3348-3352.
87. Muller YD, Mai G, Morel P, et al. Anti-CD154 mAb and rapamycin induce T regulatory cell mediated tolerance in rat-to-mouse islet transplantation. *PLoS ONE*. 2010;5:e10352.
88. Vincenti F, de Andres A, Becker T, et al. Interleukin-2 receptor antagonist induction in modern immunosuppression regimens for renal transplant recipients. *Transpl Int*. 2006;19:446-457.
89. Baan CC, van der Mast BJ, Klepper M, et al. Differential effect of calcineurin inhibitors, anti-CD25 antibodies and rapamycin on the induction of FOXP3 in human T cells. *Transplantation*. 2005;80:110-117.
90. Tedesco-Silva H, Mourad G, Kahan BD, et al. FTY720, a novel immunomodulator: efficacy and safety results from the first phase 2A study in de novo renal transplantation. *Transplantation*. 2005;79:1553-1560.
91. Nicolls MR, Gill RG. LFA-1 (CD11a) as a therapeutic target. *Am J Transplant*. 2006;6:27-36.
92. Vincenti F, Mendez R, Pescovitz M, et al. A phase I/II randomized open-label multicenter trial of efalizumab, a humanized anti-CD11a, anti-LFA-1 in renal transplantation. *Am J Transplant*. 2007;7:1770-1777.
93. Abe M, Thomson AW. Influence of immunosuppressive drugs on dendritic cells. *Transpl Immunol*. 2003;11:357-365.
94. Lee JI, Ganster RW, Geller DA, Burckart GJ, Thomson AW, Lu L. Cyclosporine A inhibits the expression of costimulatory molecules on in vitro-generated dendritic cells: association with reduced nuclear translocation of nuclear factor kappa B. *Transplantation*. 1999;68:1255-1263.
95. Tajima K, Amakawa R, Ito T, Miyaji M, Takebayashi M, Fukuhara S. Immunomodulatory effects of cyclosporin A on human peripheral blood dendritic cell subsets. *Immunology*. 2003;108:321-328.
96. Teunissen MB, De Jager MH, Kapsenberg ML, Bos JD. Inhibitory effect of cyclosporin A on antigen and alloantigen presenting capacity of human epidermal Langerhans cells. *Br J Dermatol*. 1991;125:309-316.

97. Flanagan WM, Corthesy B, Bram RJ, Crabtree GR. Nuclear association of a T-cell transcription factor blocked by FK-506 and cyclosporin A. *Nature*. 1991;352:803-807.
98. Matsue H, Yang C, Matsue K, Edelbaum D, Mummert M, Takashima A. Contrasting impacts of immunosuppressive agents (rapamycin, FK506, cyclosporin A, and dexamethasone) on bidirectional dendritic cell-T cell interaction during antigen presentation. *J Immunol*. 2002;169:3555-3564.
99. Salgado CG, Nakamura K, Sugaya M, et al. Differential effects of cytokines and immunosuppressive drugs on CD40, B7-1, and B7-2 expression on purified epidermal Langerhans cells1. *J Invest Dermatol*. 1999;113:1021-1027.
100. Wollenberg A, Sharma S, von Bubnoff D, Geiger E, Haberstok J, Bieber T. Topical tacrolimus (FK506) leads to profound phenotypic and functional alterations of epidermal antigen-presenting dendritic cells in atopic dermatitis. *J Allergy Clin Immunol*. 2001;107:519-525.
101. Homey B, Assmann T, Vohr HW, et al. Topical FK506 suppresses cytokine and costimulatory molecule expression in epidermal and local draining lymph node cells during primary skin immune responses. *J Immunol*. 1998;160:5331-5340.
102. Liu HN, Wong CK. In vitro immunosuppressive effects of methotrexate and azathioprine on Langerhans cells. *Arch Dermatol Res*. 1997;289:94-97.
103. Fulton B, Markham A. Mycophenolate mofetil. A review of its pharmacodynamic and pharmacokinetic properties and clinical efficacy in renal transplantation. *Drugs*. 1996;51:278-298.
104. Mehling A, Grabbe S, Voskort M, Schwarz T, Luger TA, Beissert S. Mycophenolate mofetil impairs the maturation and function of murine dendritic cells. *J Immunol*. 2000;165:2374-2381.
105. Gregori S, Casorati M, Amuchastegui S, Smiroldo S, Davalli AM, Adorini L. Regulatory T cells induced by 1 alpha, 25-dihydroxyvitamin D3 and mycophenolate mofetil treatment mediate transplantation tolerance. *J Immunol*. 2001;167:1945-1953.
106. Sehgal SN. Rapamune (RAPA, rapamycin, sirolimus): mechanism of action immunosuppressive effect results from blockade of signal transduction and inhibition of cell cycle progression. *Clin Biochem*. 1998;31:335-340.
107. Wells AD, Li XC, Li Y, et al. Requirement for T-cell apoptosis in the induction of peripheral transplantation tolerance. *Nat Med*. 1999;5:1303-1307.
108. Segundo DS, Ruiz JC, Izquierdo M, et al. Calcineurin inhibitors, but not rapamycin, reduce percentages of CD4+CD25+FOXP3+ regulatory T cells in renal transplant recipients. *Transplantation*. 2006;82:550-557.
109. Monti P, Mercalli A, Leone BE, Valerio DC, Allavena P, Piemonti L. Rapamycin impairs antigen uptake of human dendritic cells. *Transplantation*. 2003;75:137-145.

4 Experimental Studies in Face Transplantation: Rodent Model

Maria Z. Siemionow and Fatih Zor

Contents

M.Z. Siemionow (✉)
Department of Plastic Surgery, Cleveland Clinic,
Cleveland, OH, USA
e-mail: siemiom@ccf.org

Abstract Rodents, especially rats, are the most frequently used animals in composite tissue allotransplantation (CTA) studies. There are several advantages of using rodents in experimental studies especially in transplantation studies. During the past 20 years of our research in the field of CTA transplants, we have designed and developed different craniofacial CTA models in rats testing different immunosuppressive protocols of tolerance induction. These models include full face/scalp transplants, hemiface transplants, composite hemiface/calvarium transplants, rat maxilla allotransplants, composite osteomusculocutaneous hemiface/mandible/tongue-flap transplants, and composite midface allotransplants. These models and the models that will be developed in the future will provide the scientific foundation for future success in CTA transplantation in the clinical setting.

Abbreviations

CsA	cyclosporine A
CTA	composite tissue allotransplantation
MEP	motor evoked potentials
MHC	major histocompatibility complex
SSEP	somatosensory evoked potentials

4.1 Introduction

The Face has an important role in identity and communication; hence, it defines a functionally and aesthetically important unit. Reconstruction of facial defects, especially after deep burns, tumor excision or trauma, remains a challenging task for plastic surgeons. Unfortunately, the aesthetic outcome of the currently available conventional reconstructive procedures

M.Z. Siemionow (ed.), *The Know-How of Face Transplantation*,
DOI: 10.1007/978-0-85729-253-7_4, © Springer-Verlag London Limited 2011

for facial reconstruction is not satisfactory and often the result is a tight, masklike face with a lack of facial expression and an unsatisfactory cosmetic appearance.[1] Composite tissue allotransplantation (CTA) has been recently introduced as a potential clinical treatment for complex reconstructive procedures.

Composite tissue allografts consist of heterogeneous tissues derived from different embryologic layers and include various tissues such as skin, fat, muscle, nerves, lymph nodes, bone, cartilage, ligaments, and bone marrow, with different antigenicity. Composite tissue allografts are considered to be more immunogenic than solid organ transplants. While cartilage, ligaments, and fat present low antigenicity, bone, muscles, nerves, and vessels present moderate antigenicity, and skin is the component that develops the most severe rejection because of the abundance of dendritic cells within the epidermis and dermis.[2,3] To study the mechanisms of CTA acceptance and rejection, different experimental models, strategies, and different immunosuppressive protocols have been used. Many facial transplantation models have been described in small and large laboratory animals.[4-13] In this chapter, we want to share our experience on facial transplantation models in rodents.

4.2 Rodent Models of Face Transplantation

Rodents, especially rats, are the most frequently used animals in CTA studies. There are several advantages of using rodents in experimental studies especially in transplantation studies. It is easier to perform a transplantation surgery in small animals. There is no additional person for the surgery such as an anesthesiologist and the surgeon can perform the whole procedure alone. Caging, preoperative, and postoperative care of small animals are easier than large ones. The surgical procedure can be performed easily in a short period. There are no special requirements for postoperative care. Another factor is finance and it is obvious that financially both the surgical and nonsurgical parts of experiments on rodents are less expensive. There are some disadvantages of rodent surgery, such as relative intolerance to blood loss, but all of them can be preventable by careful surgical technique. The dosage of anesthesia must be adjusted according to the surgical need and overdosages should be avoided. During surgery, care must be taken to minimize the blood loss. In case of transferring mucosal tissues to the recipient, special care is needed to prevent embedding of these mucosal tissues as they cause severe infection and subsequent death of the animal. Postoperatively, fluid resuscitation and keeping the animal warm is essential to a successful outcome.

During the past 20 years of our research in the field of CTA transplants, we have designed and developed different craniofacial CTA models in rats testing different immunosuppressive protocols of tolerance induction (Table 4.1). These models include full face/scalp transplants, hemiface transplants, composite hemiface/calvarium transplants, rat maxilla allotransplants, composite osteomusculocutaneous hemiface/

Table 4.1 Composite facial allotransplantaion models in rodents

Composite facial allotransplantaion model	Model	Author	Year	Composition of the allograft
Models including soft tissue only	Full face/scalp transplant model	Ulusal et al.	2003	Bilateral face and scalp. Mystacial pad, periorbital tissues are excluded
	Hemiface transplant model	Demir et al.	2004	Unilateral face and scalp
Models including bone	Composite hemiface/calvarium transplantation model	Yazici et al.	2006	Unilateral face and scalp including calvarial bone
	Maxilla allotransplantation model	Yazici et al.	2007	Maxillary bone
	Composite osteomusculocutaneous hemiface/mandible/tongue-flap model	Kulahci et al.	2010	Hemiface, mandibular bone with teeth, masseter muscle, and tongue
	Total osteocutaneous allotransplantation model	Altuntas et al.	2010	Hemiface, scalp, mystacial pad, premaxillay bone segment
Functional facial allotransplantation model	Composite midface transplant model with sensory and motor neuromuscular units	Zor et al.	2010	Mystacial pad, masseter muscle, premaxillary bone segment, facial and infraorbital nerves

mandible/tongue-flap transplants, and composite mid-face allotransplants.[8-13]

4.2.1 Soft Tissue Only Models

4.2.1.1 Full Face/Scalp Transplant Model

We have confirmed for the first time the feasibility of the total facial/scalp allograft transplantation across major histocompatibility complex (MHC) barriers in the rodent model. Transplants were performed between semi-allogeneic LBN ($RT1^{l+n}$) donors and Lewis ($RT1^{l}$) recipients.

In this model, the donor flap, based on the bilateral common carotid arteries and external jugular veins, including all facial skin, scalp, and bilateral ears is harvested. Flap marking is done by two circular markings. The first is passing 1 cm above the sternal notch, through the tips of the shoulders and continuing posteriorly, 1 cm caudal to the ear level. The second is passing caudal to the tip of the nose, continuing periorally 1 cm above the oral commissure and the lower lip. Periorbital structures are also marked. After the skin incisions are made, the superficial cervical portion of the platysma muscle and the loose areolar tissue are dissected in the anterior and lateral neck region. The external jugular vein and its anterior and posterior facial branches are identified bilaterally and these veins are prepared as used as the donor vein. The submandibular glandular branches draining to the anterior facial artery and vein are ligated and glands are removed. The sternocleidomastoid muscle is removed from its origin and insertion. The stylohyoid and omohyoid muscles and the greater horn of the hyoid bone are cut off and the common carotid artery and its main branches are exposed. The internal carotid artery and cervical branches of the external carotid artery are ligated. The posterior auricular branch of the facial nerve was identified and transected. The posterior auricular artery and vein are carefully dissected and preserved just anterior to the posterior auricular nerve. The ear is included in the flap with detachment of the external cartilaginous auricular canal from its bony insertion. The animal is then placed in the prone position. The flap is dissected in subgaleal plane toward the periorbital and perioral skin incisions. At the ear level, two of three branches of external carotid artery (posterior auricular and superficial temporal branches) are preserved. The third branch (internal maxillary branch) is ligated and transected. At the same level, the branches of the posterior facial vein were divided from the pterygoid and pharyngeal vein plexuses. The flap was thus completely freed from all surrounding tissues. Finally, the common carotid arteries and external jugular veins were divided bilaterally, creating the vascular pedicles of the donor.

In the recipient, using the same skin markings of the donor, a facial/scalp defect was created excising facial skin, scalp, and external ear structures. The facial nerves and muscles, and the perioral and the periorbital regions, are preserved to avoid functional deficits that could interfere with animal feeding, breathing, and eye closure (Fig. 4.1). Both common carotid arteries are used to vascularize the full facial/scalp flap. Arterial anastomoses are performed to the common carotid arteries (end-to-side) or external carotid arteries (end-to-end) of the recipients. Venous anastomoses are performed, connecting the external jugular and anterior facial veins (end-to-end).

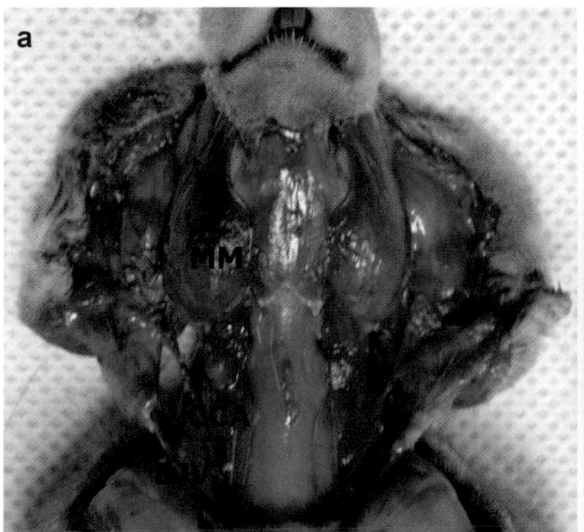

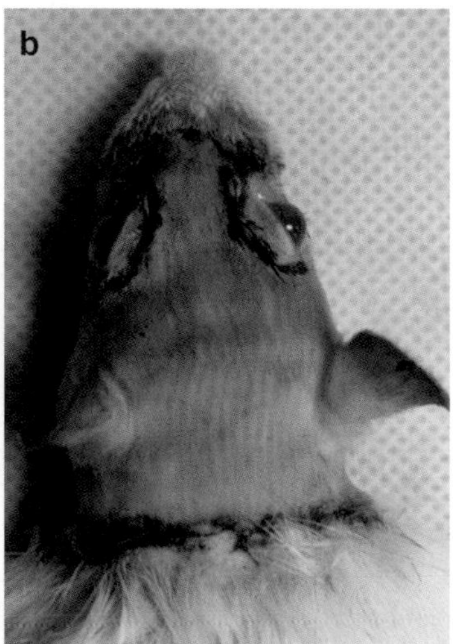

Fig. 4.1 Full face/scalp transplant model. (**a**) The full face/scalp flap is prepared depending on bilateral common carotid artery (*CCA*) and external jugular vein (*EJV*). The dissection plane is above the masseteric muscle (*MM*). (**b**) The flap is transferred to recipient animal. Note that the periorbital and periocular tissues are not included in the transplantation model

In this model, the recipient animal received CsA monotherapy at a dose of 16 mg/kg/day postoperatively, tapered to 2 mg/kg/day over 4 weeks, and maintained at this level during the follow-up period of over 200 days (Fig. 4.2).[14] Recently, to improve the survival of facial/scalp allograft recipients, we have introduced a new approach by modifying the arterial anastomoses in the recipient. The single (unilateral) common carotid artery of the recipient was used to vascularize the entire transplanted facial/scalp flap. Different modifications of the arterial anastomoses were performed. With these modifications, the postoperative mortality of the animals was significantly reduced by avoidance of complications associated with bilateral common carotid artery anastomoses.

In this model, facial/scalp allograft transplants were performed between fully allogeneic ACI (RT1a) donors and Lewis (RT11) recipient rats. The same CsA immunosuppressive protocol was used as in the previous model and resulted in over 180 days of facial/scalp allograft transplant survival.[15]

4.2.1.2 Hemiface Transplant Model

To further shorten surgery time and brain ischemia time, we have introduced a hemifacial allograft transplant model that is technically less challenging compared to the full facial/scalp model. This model was used to test induction of operational tolerance across MHC barriers. Hemifacial allograft transplants were performed between semi-allogeneic LBN

Fig. 4.2 Late postoperative result showing no rejection sign at postoperative 200 days

(RT1^{l+n}) and fully allogeneic ACI (RT1^{a}) donors and Lewis (RT1^{l}) recipients.

In this model, the same facial dissections were performed as described in full face/scalp transplant model but unilaterally.

After skin incision, dissection of the external carotid artery and its branches was performed from the midline neck approach. All branches of the external carotid artery in the neck, with the exception of the facial artery, facial vein, posterior auricular artery, and superficial temporal artery were ligated and divided. Facial and angular vessels were spared and left within the flap. Composite hemifacial/scalp flaps including the external ear and scalp, based on the common carotid artery and external jugular vein, were harvested from the donors. In the recipient, the hemifacial/scalp skin, including external ear was excised (Fig. 4.3). The arterial and venous anastomoses were performed to the common carotid artery (end-to-side) and to the external jugular vein (end-to-end), respectively. The same CsA monotherapy immunosuppressive protocol was used and 400 days survival was achieved for semi-allogeneic transplants and 330 days in the fully MHC mismatched hemifacial transplant recipients (Fig. 4.4).[8,16,17]

4.2.2 Soft Tissue and Bone Models

4.2.2.1 Composite Hemiface/Calvarium Transplantation Model

We introduced a new composite hemiface/calvarium transplantation model in the rat. The purpose of this composite tissue model was to extend application of the face/scalp transplantation model in the rat by incorporation of the vascularized calvarial bone, based on the same vascular pedicle, as a new treatment option for extensive craniomaxillofacial deformities with large bone defects. Composite hemiface/calvarium transplantations were performed across the MHC barrier between LBN and LEW rats.

In this model, dissection of the hemiface was performed as previously described by Demir et al.[16] Composite hemifacial/scalp flaps including the external ear and scalp, based on the common carotid artery and external jugular vein, were harvested from the donors. During flap dissection, after bone was reached,

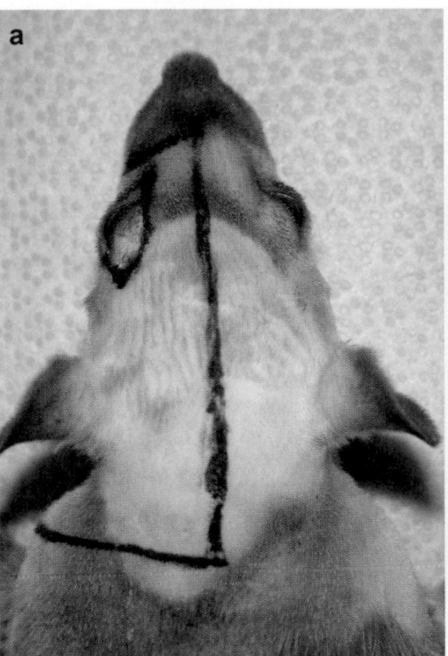

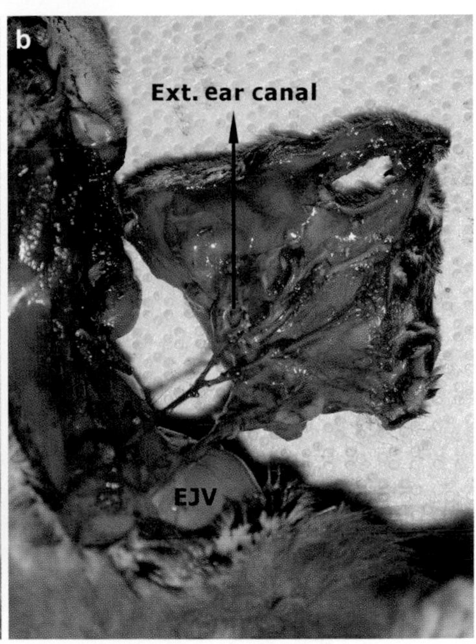

Fig. 4.3 Hemiface transplant model. (**a**) Flap markings are made similar to full face/scalp model but unilaterally. (**b**) Flap is elevated on unilateral common carotid artery (*CCA*) and external jugular vein (*EJV*). The external ear is also included in the flap

Fig. 4.4 Late postoperative result showing the flap and the recipient animal at 300 days posttransplant without any rejection sign

the line was traced up to the orbital roof and extended until the sagittal suture of the calvaria. The osteotomy surface for the frontal and temporoparietal area was cleared and dissection was conducted from the occipital area. Osteotomies were performed using fine-tipped scissors, and osteotomy borders included the sagittal suture at midline, supraorbital rim, orbital roof, zygomaticomaxillary junction, medial border of the temporal bone, and a segment of the occipital bone as the posterior border. The dura mater was left intact under the flap. Hemicalvarial bone and face grafts were dissected on the same pedicle of the common carotid artery and jugular vein and were transplanted to the deepithelized donor faces (Fig. 4.5). The arterial anastomosis was performed to the common carotid artery (end-to-side) and venous anastomosis was performed to the external jugular (end-to-end). The calvarial component of the composite flap was placed on the de-epithelialized surface of the donor rat face, above the facial musculature, and no bony fixation was performed.

All rats received tapered and continuous doses of CsA monotherapy. Evaluation methods included flap angiography, daily inspection, computed tomography (CT) scan, and bone histology. Flap angiography demonstrated an intact vascular supply to the bone. The average survival time was 154 days. No signs of rejection and no flap loss

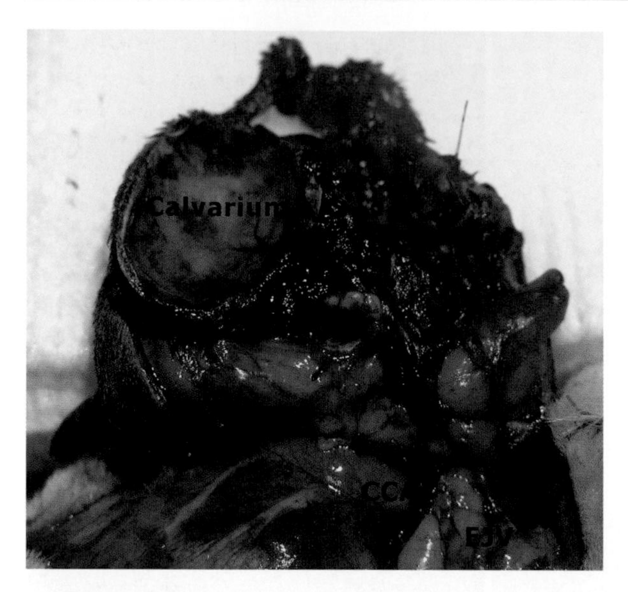

Fig. 4.5 The composite hemiface/calvarium flap is harvested on unilateral common carotid artery (*CCA*) and external jugular vein (*EJV*) similar to hemiface model but a piece of calvarial bone is included in the flap

were noted at 220 days posttransplantation. Bone histology at days 7, 30, 63, and 100 posttransplantation revealed viable bone at all time points, and CT scans taken at days 14, 30, and 100 revealed normal bones without resorption (Fig. 4.6). For extensive face deformities involving large bone- and soft-tissue defects, this new osteomusculocutaneous hemiface/calvarium flap model may serve to test new reconstructive options for coverage of multi-tissue defects in one surgical procedure.[9]

4.2.2.2 Maxilla Allotransplantation Model

We developed a rat model to test the effects of vascularized maxilla allotransplantation on composite maxillary substructures. Allograft maxilla transplantations were performed across the MHC barrier between ten LBN and ten LEW recipient rats under CsA monotherapy. Grafts were dissected along Le-Fort II osteotomy lines based on the common carotid artery and external jugular vein and transplanted to the anterior abdominal wall via microvascular anastomosis.

Briefly, following skin incision, the external carotid artery and its branches were exposed and the external carotid arteries were ligated and divided bilaterally. External jugular veins were identified in the supraclavicular area, and the anterior facial veins were ligated and divided bilaterally. Bilateral posterior facial veins were dissected up to the junction of the internal maxillary vein and transverse sinus. At this point, a

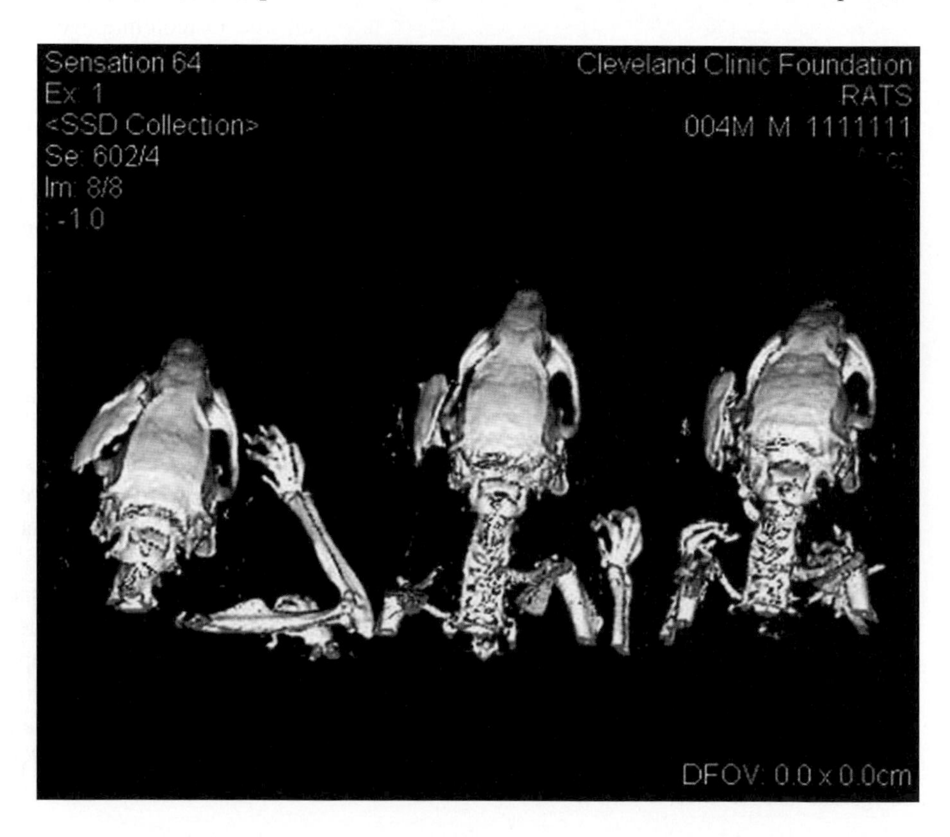

Fig. 4.6 Computed tomography (CT) scan showing the viable calvarial bone segment at 100 days posttransplant

tracheostomy was performed and a canula was inserted to secure the airway. Later, mandible and midline structures of the lower face was split and retracted bilaterally in order to expose whole palate. The dissection of maxilla was begun with a circumferential gingivobuccal and gingivolabial sulci incision. A supraperiosteal dissection was performed up to the frontal area, and the nasal cartilages. Laterally, the dissection was extended to the zygomatic arch and the infraorbital area. Following osteotomy to the zygomatic arch, lateral osteotomy lines are exposed. Bilateral osteomies were performed using fine-tipped scissors at the zygomatic processes of maxilla, the interocular plane anterior to the frontoparietal suture, from the orbital base to the temporal and occipital bones, and the atlantooccipital joint. After the osteotomies, the maxilla flap was gently detached from the brain, with particular care not to harm dural sinuses. A heterotopic transplantation was performed to the inguinal region of the recipient rat, using femoral artery and vein as recipient vessels. Only one vascular pedicle of the allograft was used for anastomoses and the other was ligated.

Allografts were examined by tomography, flow cytometry, angiography, and histology. Allograft survived up to 105 days without signs of rejection. A high level of donor-specific chimerism for T-cell and B-cell lineages was maintained. The incisors continued to grow; tooth buds, bone, cartilage, and mucosa remained intact (Fig. 4.7). Moderate inflammation of the nasal, oral mucosa, and keratinous metaplasia were noted histologically. We created a maxilla allotransplantation model that allows for studying immunologic responses and demonstrates potential clinical applications based on growth properties of the allograft. In the long-term surviving allograft recipients, over 105 days, there were no indications of flap loss, partial necrosis, or rejection. The incisors grew over 10 mm during the follow-up. Flow cytometry analysis of donor-specific chimerism in the peripheral blood was performed at day 105 posttransplant revealed 12.5% of CD4FITC/RT1n-Cy7 and 5.3% of CD8PE/RT1n-Cy7 T-cell subpopulations. Analysis of the B-cell population revealed 4.7% of CD45RAPE/RT1n-Cy7 donor-derived cells in the peripheral blood of maxilla recipients. Histologic evaluation revealed no signs of rejection and intact allograft structures, including teeth, tooth buds, teeth pulp, bone, cartilage, oral mucosa, nasal mucosa, and soft-palate musculature.[10]

4.2.2.3 Composite Osteomusculocutaneous Hemiface/Mandible/Tongue-Flap Model

We introduced a new model of composite osteomusculocutaneous hemiface/mandible/tongue allograft transplant, to extend application of the face/scalp transplantation model in the rat by incorporation of the vascularized mandible, masseter, and tongue, based on

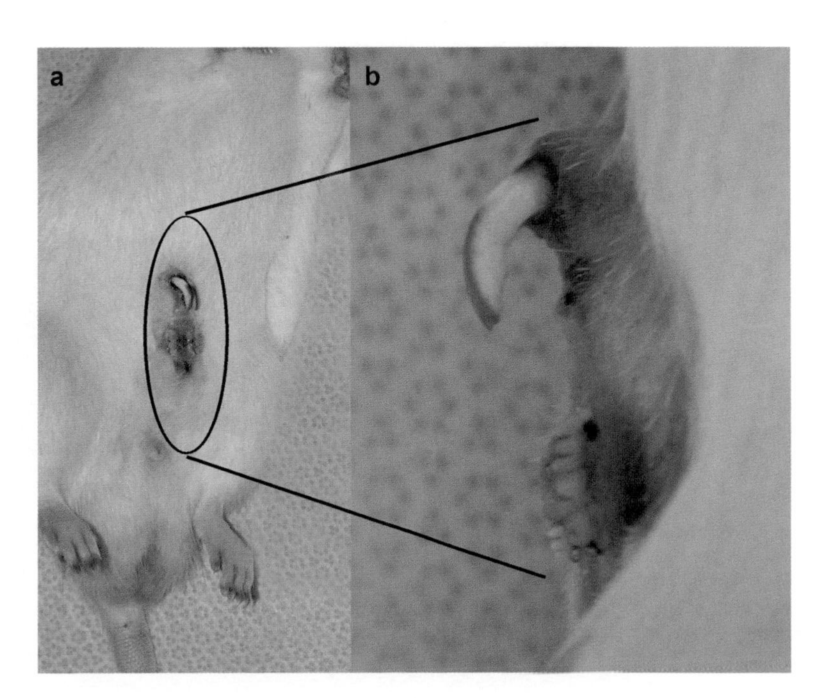

Fig. 4.7 Maxilla allotransplant model. (**a**) Heterotopic transplantation of macilla to the inguinal region of the recipient rat. (**b**) At the recipient site, the incisors continued to grow; teeth buds, bone, cartilage, and mucosa remained intact

the same vascular pedicle, as a new reconstructive option for extensive head and neck deformities with large soft- and bone-tissue defects. The feasibility of composite osteomusculocutaneous hemiface/mandible/tongue transplantations was tested in isotransplantation model between LEW rats and allotransplantation model performed across the MHC barrier between LBN donors and LEW recipients. Hemimandibular bone, masseter muscle, tongue, and hemifacial flaps were dissected on the same pedicle of the external carotid artery and jugular vein and were transplanted to the donor inguinal region (Fig. 4.8).

In the donor rat, a midline incision was performed and the submandibular gland was excised following ligation of glandular pedicle. Subplatysmal dissection was performed and external jugular vein and its two main branches are preserved. The sternocleidomuscle is excised and common carotid artery and its main branches are exposed. The posterior belly of the digastric muscle was excised, the omohyoid muscle was transected, and the greater horn of the hyoid bone was excised for better visualization of the external carotid artery and its branches. Internal carotid artery, superior thyroid artery, ascending pharyngeal artery, and ascending palatine artery were ligated and transected. Facial artery, superficial temporal artery, posterior auricular artery, lingual artery, and internal maxillary artery were preserved and included in the flap.

Before the dissections of the oral region, a tracheostomy was performed and a canula was inserted in the trachea in order to secure the airway. Mandible was split at midline and the tongue was included in the flap. In the perioral region, the facial artery and vein were and included in the flap after ligation of the superior and inferior labial branches. The facial flap was elevated toward the temporoparietal area and zygomatic arch, and the zygomatic arch is excised. At the level of ear, the external ear canal was incised and included in the flap. Side branches of the internal maxillary vein were ligated and the internal maxillary vein was spared. The capsule of the temporomandibular joint was opened the mandibular condyle was included in the flap. Upper gingivobuccal sulcus incision was performed and the flap was harvested depending on the common carotid artery and external jugular vein.

A heterotopic transplantation was performed to the inguinal region of the recipient rat. End-to-end anastomoses were performed between common carotid artery/femoral artery and external jugular vein/femoral vein.

All allogeneic transplant recipients received our established protocol of CsA monotherapy. Isograft controls survived indefinitely. All hemiface/mandible/tongue allotransplants survived over 100 days post-transplant (Fig. 4.9). Flap angiography demonstrated intact vascular supply to the bone. No signs of rejection

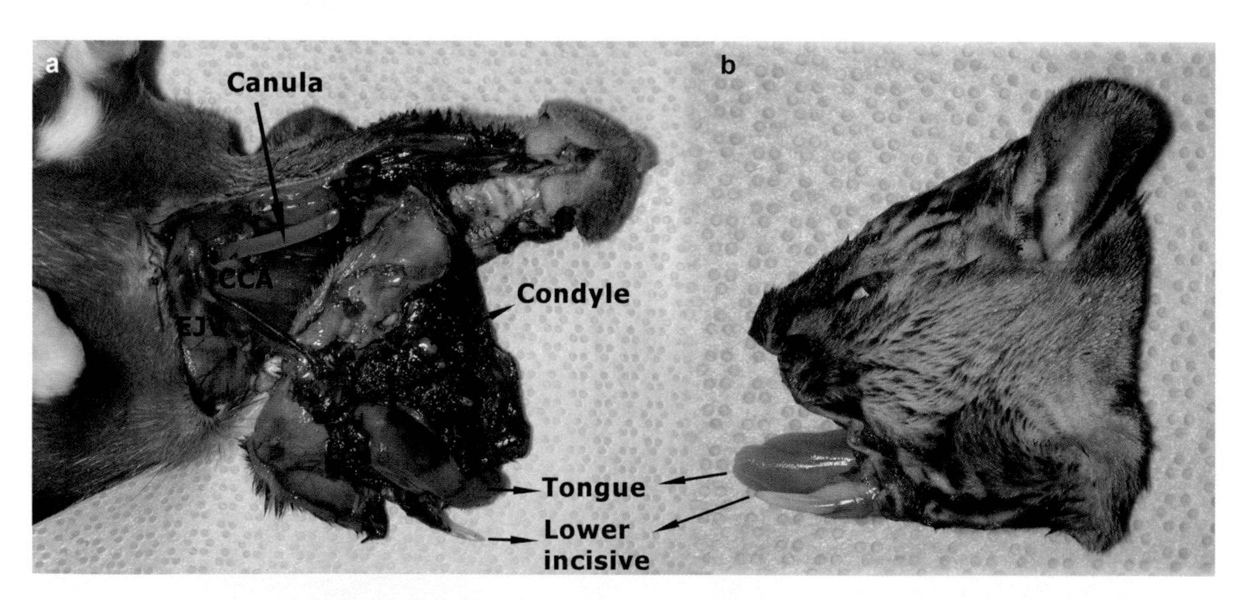

Fig. 4.8 Composite osteomusculocutaneous hemiface/mandible/tongue flap model. (**a**) The inner surface of the flap. (**b**) The outer surface of the flap. The flap is harvested on common carotid artery (*CCA*) and external jugular vein (*EJV*) and included hemimandible, tongue and masseter muscle. A tracheostomy canula is inserted for airway security

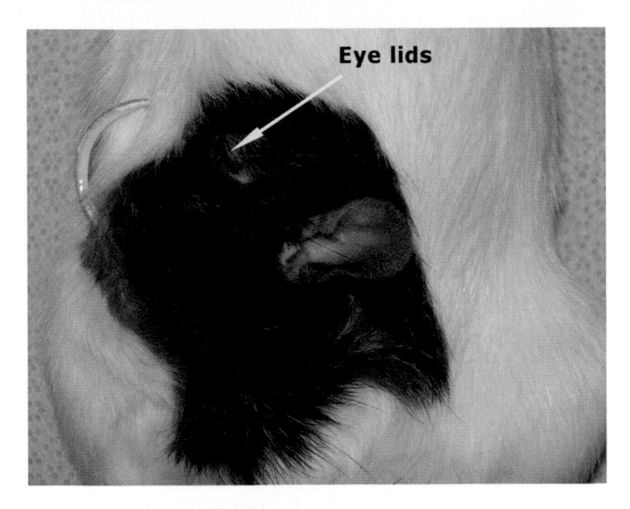

Fig. 4.9 Composite osteomusculocutaneous hemiface/mandible/tongue flap at 100 days posttransplant. The growing lower incisive teeth can be seen

and no flap loss were noted. CT scan and bone histology confirmed viability of bone components of the composite allografts. Viability of the tongue was confirmed by pink color, bleeding after puncture, and histology. H&E staining determined presence of viable bone marrow cells within the transplanted mandible. Donor-specific chimerism at day 100 posttransplant was evaluated by the presence of donor T cells (2.7% CD4/RT1n, 1.2% CD8/RT1n) and B cells (11.5% CD45RA/RT1n). Long-term allograft acceptance was accompanied by donor-specific chimerism supported by VBMT of the mandibular component. This model may serve as a new reconstructive option for coverage of extensive head and neck deformities involving large bone- and soft-tissue defects.[11,18-20]

4.2.2.4 Composite Midface Transplant Model with Sensory and Motor Neuromuscular Units

We developed a new rat model of composite midface allograft transplant with sensory and motor neuromuscular units by incorporation of vascularized premaxilla, and nose with infraorbital and facial nerves for evaluation of functional recovery of allotransplanted sensory and motor nerves following midface transplantation. Composite midface isotransplantations between LEW rats and midface allotransplants across the MHC barrier between LBN donors and LEW recipients were performed. Midfacial structures including nose, premaxillary bone segment, mystacial pad, masseter muscle, and lower lip were dissected on the same pedicle of the common carotid artery and external jugular vein (Fig. 4.10).

Following a midline neck incision, the submandibular gland and sternocleidomastoid muscle were excised and common carotid artery and external jugular vein are exposed. External jugular vein and its two main branches are dissected and preserved. The internal carotid artery, superior thyroid artery, ascending pharyngeal artery, ascending palatine

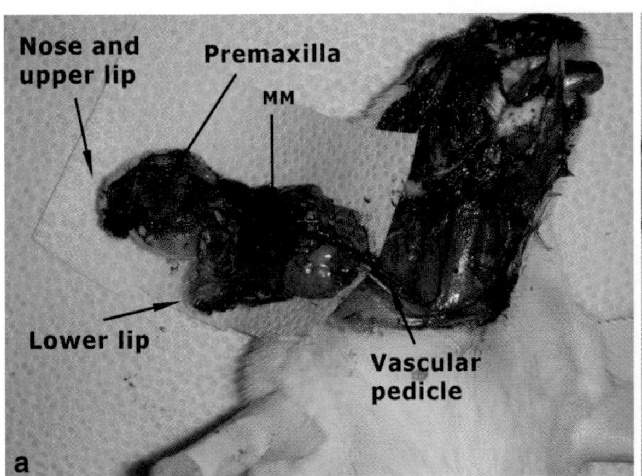

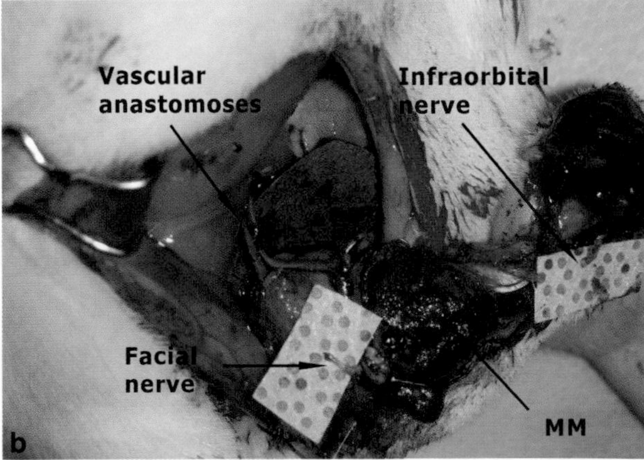

Fig. 4.10 Composite midface transplant model with sensory and motor neuromuscular units. (**a**) The vascular pedicle of the flap is composed of common carotid artery and external jugular vein. Premaxilla, nose, and both lips are included in the flap. The dissection plane was kept below the masseteric muscle (*MM*) in order to prevent iatrogenic injury to the facial nerve. (**b**) Vascular anastomoses are done and the flap perfusion is provided. Facial and infraorbital nerve is prepared for nerve coaptation

artery, superficial temporal artery, posterior auricular artery, lingual artery, and internal maxillary artery in the flap were ligated and transected. Of the branches of external carotid artery, only facial artery is preserved as the pedicle of the flap.

The surgical plane was below the masseter muscle in order to avoid iatrogenic damage to the branches of facial nerve during dissection. The facial nerve was transected at the stylomastoid foramen and the infraorbital nerve was transected at the level of infraorbital fissure and both nerves were included into the midface graft. The dissection was then carried around the nose, upper lip, right hemi lower lip, and right mystacial pad. Finally, periosteum was incised and premaxillary bone was transected transversely using a burr.

Composite midface flaps with sensory and motor units were transplanted to the donor inguinal region and vascular anastomoses were performed between pedicle of the flap and femoral vessels. Standard epineural neurorraphies were performed between the infraorbital nerve of donor and sapheneous nerve of the recipient and between facial nerve of donor and femoral nerve of recipient (Fig. 4.11). All allogeneic transplant recipients received CsA monotherapy.

Return of motor function was evaluated by observation of the return of movement to the mystacial pad. Sensory recovery was observed clinically by evasive behavior and defense reactions when the transplanted whiskers are pulled. Somatosensory evoked potentials (SSEP) and motor evoked potentials (MEP) were used to evaluate the sensory and motor recovery, respectively.

Successful flap transplantation was accomplished in all animals, with 100% flap survival rate over 100 days. Clinically, all grafts were pink and pliable during the entire observation period. The incisors continued to grow; tooth buds, bone, cartilage, and mucosa remained intact (Fig. 4.12). Motor recovery was observed at 21 days posttransplant and was confirmed by the movement of the mystical pad. Clinically evasive behavior and defense reactions were observed when transplanted whiskers were pulled. Computed tomography of composite nose flap showed persistence of the bony premaxilla. At 100 days posttransplant, SSEP and MEP tests revealed that sensory and motor recovery reached 67% of normal latency values for infraorbital nerve and 70% for facial nerve latency values. This model allows for single stage reconstruction of the central midface with functional recovery of the sensory and motor nerves.[12,21]

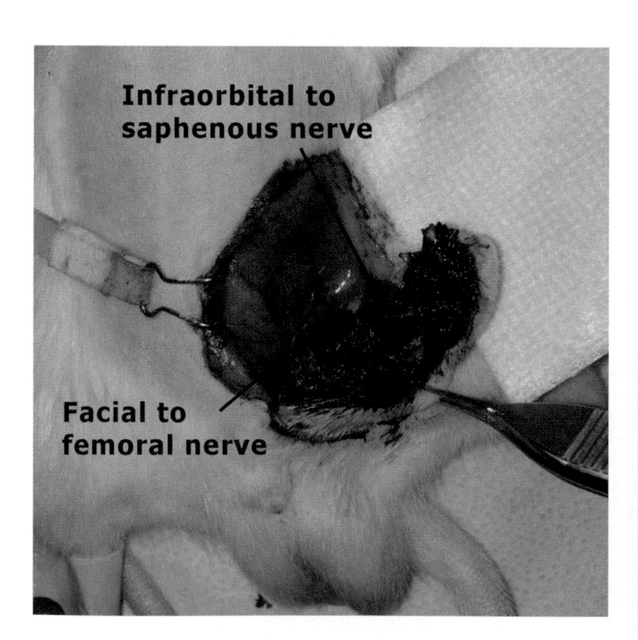

Fig. 4.11 Nerve coaptation was performed between facial and femoral nerves (motor unit) and infraorbital and saphenous nerves (sensory unit)

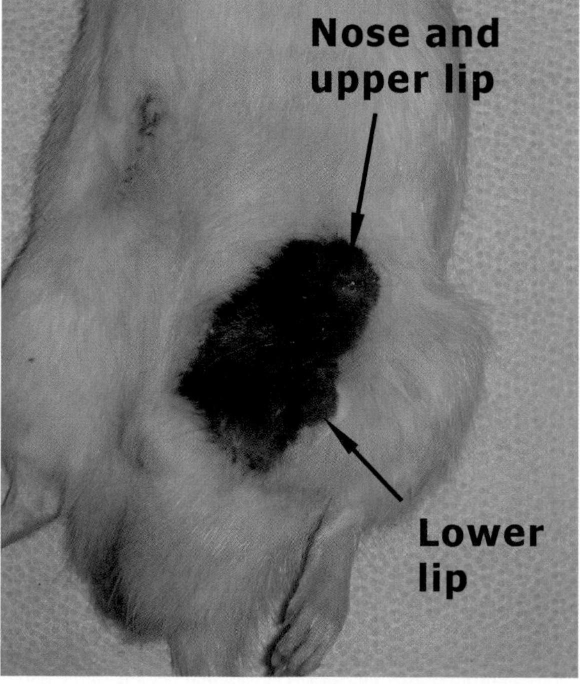

Fig. 4.12 Composite midface flap at 100 days posttransplant

4.2.3 Total Osteocutaneous Allotransplantation Model

The reconstruction of extensive composite tissue defects of the head and neck region is always challenging procedure for the reconstructive surgeon. Large composite tissue defects, sometimes involves the nose, anterior part of the maxilla, and whole hemiface usually occur after high-velocity gunshot injuries, cancer ablative surgeries, burns, or trauma. Recently, we developed a total osteocutaneous hemiface flap model and presented maintenance of donor-specific chimerism in a semi-allogeneic MHC-mismatched rats under low-dose cyclosporine A treatment. In this study, our aim was to extend application of the face/scalp transplantation model in the rat by incorporation of the vascularized nose, premaxilla, eyelids, and upper and lower lips, based on the same vascular pedicle, common carotid artery, and jugular vein, as a new reconstructive option for extensive head and neck deformities with large soft and bone tissue defects.

Ten transplantations were performed in two experimental groups. Group 1 (Isograft control) was performed between Lewis rats (*n* = 5) and Group 2 (allograft transplantation) was performed between Lewis–Brown Norway (LBN, $RT1^{l+n}$) donors and Lewis ($RT1^{l}$) recipients (*n* = 5).

Skin markings were made including whole hemiface, scalp, mystacial pad of the ipsilateral site, and the nose. Nose, premaxilla, and hemifacial flaps were dissected on the same pedicle of external carotid artery and jugular vein. Dissections were performed as described in composite midface allotransplantation model but the external ear, scalp, and periorbital structures were also included. Superficial temporal and facial artery that branches of external carotid artery was preserved and provided the arterial supply of the flap. External jugular vein and its two major branches are preserved for venous drainage of the flap (Fig. 4.13).

Composite flaps were transplanted to the inguinal regions of the recipient rat. Common carotid artery and external jugular vein of the graft were anastomosed with 10–0 sutures to the femoral artery and vein, respectively. Allograft recipients received cyclosporine A monotherapy. Cyclosporine was administered from 16 to 2 mg/kg within 4 weeks posttransplant and maintained at this level thereafter. Flap angiography and CT scan evaluated allograft viability. Flow cytometry analysis was performed to evaluate chimerism level in the peripheral blood of Lewis recipients during observation time.

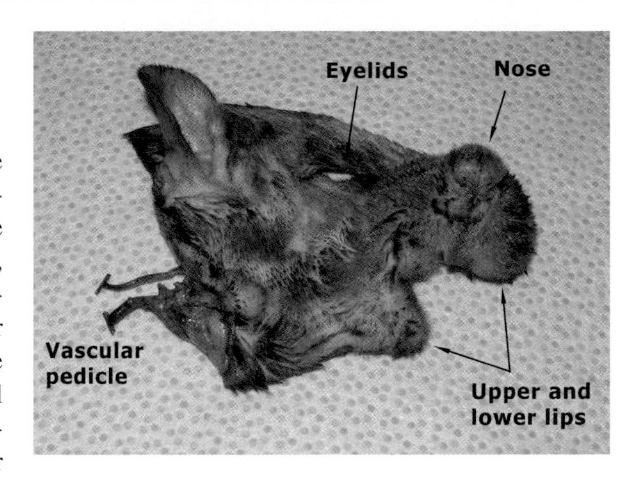

Fig. 4.13 Total hemiface allotransplant model. Whole hemifacial structures including external ear, nose, mystacial pad, eyelids with conjunctiva, and both lips are included in the flap

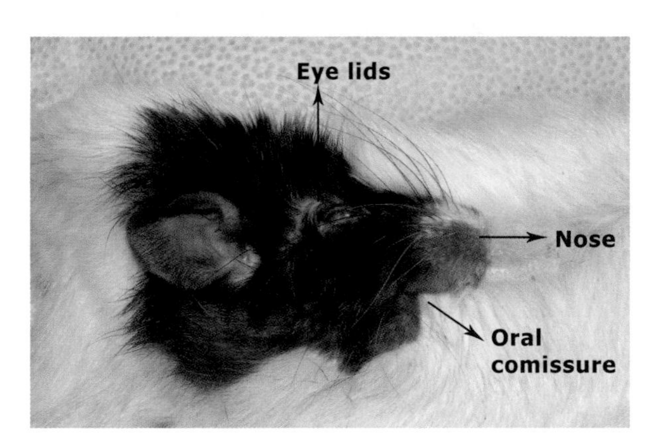

Fig. 4.14 Total hemiface/nose allotransplant at 100 days posttransplant. There is no sign of rejection

Isograft controls survived indefinitely. All total hemiface/nose allotransplants survived over 100 days posttransplant. There were no signs of rejection in the allotransplant (Fig. 4.14).

Histologic examination revealed intact nasal and oral mucosa, nasal septal cartilage, and tooth structures. Histology of the eyelid demonstrated integrity of the eyelid components of the flap.

CT scan demonstrated viability of bone components of the composite allografts at day 100 posttransplant.

Fluorescent immunostaining of donor showed MHC Class I (+) cells in recipient's skin, lymph node, and liver at 150 days postoperatively. Flow cytometry was used to show the presence of donor-origin cells ($RT1^{n}$) in the peripheral blood of nose transplant recipients and the chimerism level was found to be 1.5% at 100 days postoperatively.

In this study, we have introduced a new hemiface allotransplantation model including all the hemifacial soft tissues as well as the premaxillary bone segment. This model allows for the evaluation of all facial tissues following composite tissue allotransplantation. Moreover, in this allotransplantation model, long-term survival over 100 days was achieved.[22]

4.3 Conclusion

Since the first successful hand transplantation in France in 1998, CTA transplantation has opened a new era in the field of reconstructive surgery. So far, more than 50 CTA transplants have been reported and there have been 10 face transplantations performed worldwide.

It is obvious that CTA transplantation will improve patients' life quality, but this might be at the expense of decreasing the life expectancy of these patients. Currently, the main obstacle for CTA transplantation is the use of life-long immunosuppression therapy because of its well-known side effects, such as serious infections, organ toxicities, and malignancies. In addition, ethical, social, and psychological issues are raised when discussing face transplantation.

In order to overcome these obstacles there is no doubt that experimental studies are needed. Different experimental models allow us to evaluate the effects of various immunosuppression regimens and the fate of different tissues following composite tissue allotransplantation. A comparison of different composite facial allotransplantation models in rodents can be seen in Table 4.2.

These models and the models that will be developed in the future will provide the scientific foundation for future success in CTA transplantation in the clinical setting.

Table 4.2 Comparison of different composite facial allotransplantation models

Composite facial allotransplantaion model	Skin amount	Bone	Ear	Nose	Tongue	Conjunctiva and eyelids	Functional unit	Relative complexity	Transplantation site
Full face/scalp transplant model	++	–	+	–	–	–	–	++	Orthotopic
Hemiface transplant model	+	–	+	–	–	–	–	+	Orthotopic
Composite hemiface/calvarium transplantation model	+	Calvarium	+	–	–	–	–	+	Orthotopic
Maxilla allotransplantation model	–	Maxilla	–	–	–	–	–	++	Heterotopic
Composite osteomusculocutaneous hemiface/mandible/tongue-flap model	++	Hemimandible	+	–	+	+	–	++	Heterotopic
Total osteocutaneous allotransplantation model	++	Premaxilla	+	+	–	+	–	++	Heterotopic
Composite midface transplant model with sensory and motor neuromuscular units	+	Premaxilla	–	+	–	–	+	++	Heterotopic

References

1. Siemionow M, Agaoglu G. Allotransplantation of the face: how close are we? *Clin Plast Surg*. 2005;32:401-409.
2. Lee WP, Yaremchuk MJ, Pan YC, et al. Relative antigenicity of components of a vascularized limb allograft. *Plast Reconstr Surg*. 1991;87:401-411.
3. Cendales LC, Xu H, Bacher J, et al. Composite tissue allotransplantation: development of a preclinical model in nonhuman primates. *Transplantation*. 2005;80:1447-1454.
4. Shengwu Z, Qingfeng L, Hao J, et al. Developing a canine model of composite facial/scalp allograft transplantation. *Ann Plast Surg*. 2007;59:185-194.
5. Xudong Z, Shuzhong G, Yan H, Datai W, Yunzhi N, Linxi Z. A hemifacial transplantation model in rabbits. *Ann Plast Surg*. 2006;56:665-669.
6. Kuo YR, Shih HS, Lin CC, et al. Swine hemi-facial composite tissue allotransplantation: a model to study immune rejection. *J Surg Res*. 2009;153:268-273.
7. Silverman RP, Banks ND, Detolla LJ, et al. A heterotopic primate model for facial composite tissue transplantation. *Ann Plast Surg*. 2008;60:209-216.
8. Siemionow MZ, Demir Y, Sari A, Klimczak A. Facial tissue allograft transplantation. *Transplant Proc*. 2005;37:201-204.
9. Yazici I, Unal S, Siemionow M. Composite hemiface/calvaria transplantation model in rats. *Plast Reconstr Surg*. 2006;118:1321-1327.
10. Yazici I, Carnevale K, Klimczak A, Siemionow M. A new rat model of maxilla allotransplantation. *Ann Plast Surg*. 2007;58:338-344.
11. Kulahci Y, Klimczak A, Siemionow M. Long term survival of composite hemiface/mandible/tongue tissue allograft permitted by donor specific chimerism. *Plast Reconst Surg*. 2006;118(4 Suppl):34.
12. Zor F, Bozkurt M, Nair D, Siemionow M. A new composire nose allograft with sensory and motor reinnervation. 52nd Annual Meeting of the Ohio Valley Society of Plastic Surgeons. 15–17 May 2009, Indianapolis.
13. Ulusal BG, Ulusal AE, Ozmen S, Zins JE, Siemionow MZ. A new composite facial and scalp transplantation model in rats. *Plast Reconstr Surg*. 2003;112:1302-1311.
14. Siemionow M, Gozel-Ulusal B, Engin Ulusal A, Ozmen S, Izycki D, Zins JE. Functional tolerance following face transplantation in the rat. *Transplantation*. 2003;75:1607-1609.
15. Unal S, Agaoglu G, Zins J, Siemionow M. New surgical approach in facial transplantation extends survival of allograft recipients. *Ann Plast Surg*. 2005;55:297-303.
16. Demir Y, Ozmen S, Klimczak A, Mukherjee AL, Siemionow M. Tolerance induction in composite facial allograft transplantation in the rat model. *Plast Reconstr Surg*. 2004;114:1790-1801.
17. Siemionow M, Kulahci Y. Facial transplantation. *Sem Plast Surg*. 2007;21(4):259-268.
18. Siemionow M, Kulahci Y. Experimental models of composite tissue allograft transplants. *Sem Plast Surg*. 2007;21(4): 205-212.
19. Bozkurt M, Kulahci Y, Nasir S, Klimczak A, Siemionow M. Long term survival of composite hemiface/mandible/tongue tissue allograft permitted by donor specific chimerism. *Plast Reconstr Surg*. 2007;120(1 Suppl):82.
20. Kulahci Y, Siemionow M. A new composite hemiface/mandible/tongue transplantation model in rats. *Ann Plast Surg*. 2010;64:114-121.
21. Zor F, Bozkurt M, Nair D, Siemionow M. A new composite midface allotransplantation model with sensory and motor reinnervation. *Transpl Int*. 2010;23(6):649-656.
22. Altuntas SH, Zor F, Siemionow M. Total osteocutaneous hemiface allotransplantation model in rats. *Plast Reconstr Surg*. 2010;6S:117.

Experimental Studies in Face Transplantation: Swine Model

5

Yur-Ren Kuo

Contents

Y.-R. Kuo
Department of Plastic and Reconstructive Surgery,
Chang Gung Memorial Hospital – Kaohsiung
Medical Center, Kaohsiung, Taiwan
e-mail: t1207816@ms22.hinet.net, kuoyr@adm.cgmh.org.tw

Abstract Face composite tissue allotransplantation (CTA) was recently achieved in a human subject. However, the side effects of long-term immunosuppression and chronic rejection still need to be concerned. The goal of this chapter is to introduce swine hemifacial transplantation for preclinical studies. To design the hemifacial orthotopic transplant consisted of ear cartilage, auricular nerve, parotid gland and lymphoid tissue, muscle with surrounding hemifacial skin paddle, the vascular territories of the composite flap supplied by the superficial temporal artery, and its branches originating from the carotid artery were defined by angiography anatomic studies. The experimental results revealed that this model is suitable to investigate the new strategies for preclinical facial allotransplantation studies. Monitoring and modulation of early rejection in allo-skin and gland lymphoid tissue is a useful strategy to evaluate CTA survival.

Abbreviations

CsA cyclosporine A
CTA composite tissue allotransplantation
MHC major histocompatibility complex

5.1 Introduction

Composite tissue allotransplantation (CTA) (consisting of tissues such as skin, muscle, and bone) may serve as tremendous potential for composite defect following traumatic loss, tumor resection, or repair of congenital abnormalities.[1,2] Recently, important advances in human clinical trials was successfully performed.[3,4] Although not quite a routine yet, the practice of CTA is not rare.

M.Z. Siemionow (ed.), *The Know-How of Face Transplantation*,
DOI: 10.1007/978-0-85729-253-7_5,

Since 2005, nine face transplants have been performed in four countries: France, the USA, China, and Spain. These encouraging short-term outcomes, with the longest survivor approaching 5 years, have led to an increased interest in establishing face transplant programs worldwide.[5] Among many others such as donor source, ethics, psychology of recipient, and so on, immune rejection and its treatment continues to be the foremost among many big issues. As a matter of fact, application of immune suppress therapy is required. Despite its promising applications, the side effects of long-term immunosuppressive therapy and chronic rejection still need to be concerned.[2,6] Unlike many lifespan-prolonging solid organ transplants, CTA is an elective procedure for improving quality of life. Therefore, preclinical trials are needed to evaluate the long-term efficacy of new immunosuppressive strategies.

Preclinical animal models are essential for advancing CTA to clinical application. Investigations involving small animal models have comprehensively evaluated CTA rejection.[7,8] Although rodent models have shown predictable patterns of rejection, there exist fundamental differences between the immune system of humans and rats.[9,10] Therefore, rodent models may not be applicable in humans. Large animal models, especially swine and nonhuman primate, offer better characterization of the major histocompatibility complex (MHC), which is similar seen in humans, as compared to rodents.[11,12] Although large animal models are still different to human, however, large animal is necessary to be applied toward human clinical trial for surgeon's training and new immunosuppression protocol. Facial CTA, including total and hemifacial, have been performed previously in rodent models.[13,14] However, rare facial allotransplantation has been reported in a preclinical large animal study.[15,16] Therefore, this chapter intends to introduce the reproducibility of swine hemifacial transplantation for preclinical studies.[17]

5.2 Animal Model and Operative Technique

5.2.1 *Animals*

Outbred miniature swine (Lan-Yu strain and Hwa-Ban strain; age, 3 months; size, 12–20 kg) was included in the study. The miniature swine is an indigenous breed from Lau-Yu Islet, south-east of Taiwan. The inherited differences in donors and recipients from the original parental generation were identified. The study was conducted in accordance with *Guide for the Care and Use of Laboratory Animals* published by the National Institute of Health, USA. Experiments were conducted under the Institutional Animal Care and Use Committee (IACUC) protocol approved by the Chang Gung Memorial Hospital in Kaohsiung, Taiwan.

5.2.2 *Anesthesia and Surgical Anatomic Design*

The animals were premedicated with ketamine (10 mg/kg) and xylazine (1.5 mg/kg) intramuscular injection then placed in a supine position on the operating table and intubated. Anesthesia was maintained with Isoflurane inhalation throughout the procedure. The head and neck were shaved and painted with antiseptic iodine solution. The hemifacial flap was schematically marked (Fig. 5.1a). Upper and lower eyelids were not included in the flap. To design the hemifacial composite flap containing skin, muscle, ear cartilage, nerve, parotid gland, and surrounding tissue, the vascular territories of the composite flap supplied by the superficial temporal artery and its branches originating from the carotid artery were defined by angiography in preliminary anatomic studies (Fig. 5.1b). The model could be modified to include the facial nerve and innervated muscles together supplied by facial artery for investigation of the sensori-motor recover function. The model also could recruit a part of mandible or maxilla bone with surrounding muscles supplied by facial and maxillary artery as an osteomyocutaneous subunit model.

5.2.3 *Harvesting Swine Hemifacial Composite Tissue Allotransplant*

The skin was incised to the depth of the brachiocephalicus muscle in the anterior and posterior neck, to the depth of facial muscles in the facial region, and above the periosteal plane in the nasal and frontoparietal region. In the neck, dissection was continued superiorly above the sternomastoideus muscle to the level of angle of mandible, preserving the external jugular

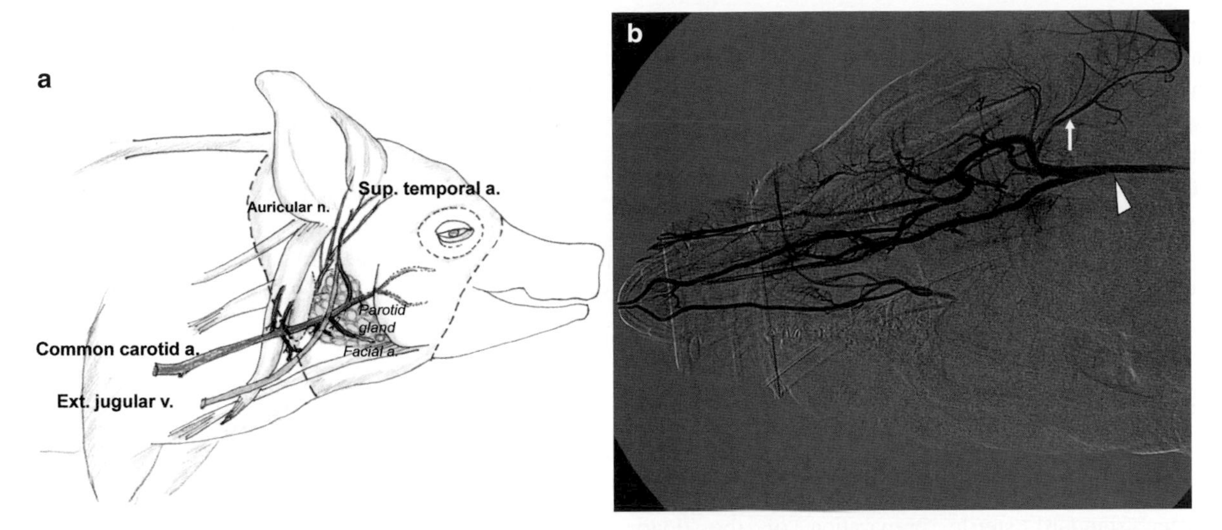

Fig. 5.1 (**a**) Schematic diagram of the orthotopic hemifacial composite tissue transplant model. The hemifacial flap tissue contained vascularized skin, lymphoid parotid gland, ear cartilage, part of muscle, auricular nerve, and surrounding soft tissue. (**b**) Angiography revealed vascular distribution of the hemifacial composite flap supplied by the superficial temporal artery (*arrow*) and its branches originating from common carotid artery (*arrowhead*)

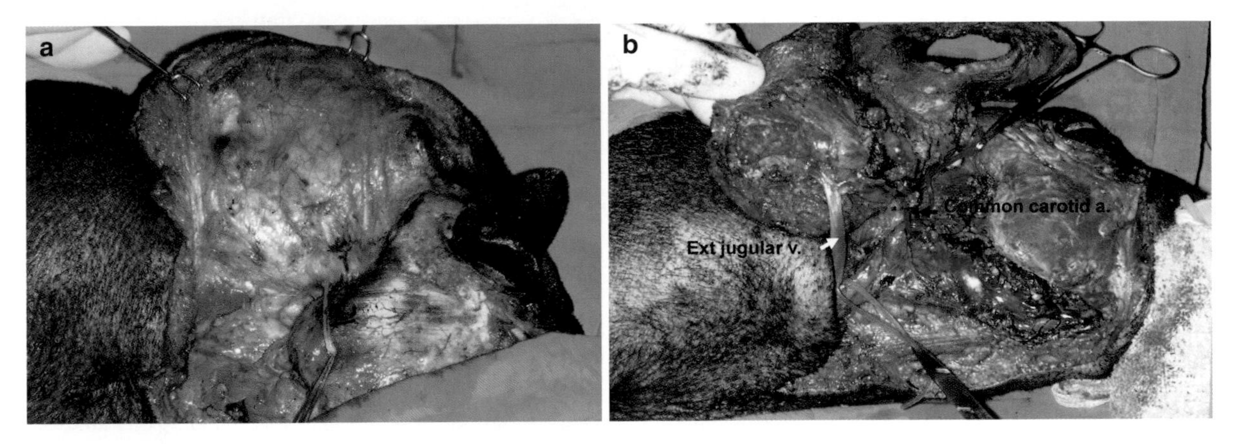

Fig. 5.2 (**a**) Intraoperative photo of hemifacial flap harvesting. Dissection was performed above the masseter muscle toward the ear. Facial artery and facial nerve (*arrow*) were identified and excluded from the flap. (**b**) The common carotid artery (*arrow*) and external jugular vein (*arrowhead*) were used as the vascular pedicle of the flap

vein. The submandibular gland was excised after ligation of the glandular branches of facial artery and vein. Facial artery and facial nerve were identified and excluded from the flap. Dissection was performed above the masseter muscle toward the ear (Fig. 5.2a). To preserve preauricular vascular structures, the parotid gland was included in the flap. In the retro-auricular region, the internal maxillary vein and the main trunk draining the pterygoid plexus were ligated and transected. The auricular nerve was preserved and included in the flap as a sensated flap for further sensory recover assessment. At the back of the neck, after transaction of platysma and levator auris longus muscles, the flap was elevated above the trapezius up to the posterior wall of the cartilaginous area of the external ear canal. The external ear canal was detached at the osteocartilaginous junction, and the external ear was kept within the flap.

The sternomastoideus muscle was detached, and bony ostectomy of jugular process in cervical spine and hyoid bony-cartilage part were performed. The dissection plane exposed the common carotid artery and its main branches, internal jugular vein, vagus and phrenic nerve, the external, and internal carotid arteries. The internal carotid artery, internal jugular vein, cranial thyroid artery, ascending pharyngeal artery, and lingual artery were ligated and transected. The common carotid artery and external jugular vein were dissected as the vascular pedicle of the flap (Fig. 5.2b). The flap revealed good circulation after harvest.

5.2.4 Preparation of the Donor Hemifacial Allotransplant

After standard sterile preparation of the donor swine, the hemifacial composite flap was harvested as described above. The common carotid artery and external jugular vein were divided to create the vascular pedicle of the flap. After dividing the vascular pedicle, a 24# catheter was inserted in the common carotid artery and heparinized normal saline solution (1 ml heparin 5,000 unit/ml in 500 ml 0.9% normal saline solution) with hydrostatic pressure was flushed into the allograft through the carotid artery until the venous outflow was clear. The hemifacial flap was put in a plastic bag and was preserved on the ice for cold preservation.

5.2.5 Preparation of the Recipient

The recipient animal was prepared in a similar fashion. Intravenous catheter was placed for intraoperative fluid management. This catheter was subsequently used for drawing blood samples and administering medicine postoperatively. A single lumen Hickman silicon catheter was inserted on the contralateral side of the external jugular vein under direct vision and tunneled in a posterior direction to exit high on the dorsal back. The incisions were closed in layers.

On the ipsilateral side of the recipient, the similar size full thickness of skin and subcutaneous tissue was trimmed. The external jugular vein was isolated anteriorly to the sternomastoideus muscle and prepared for venous anastomosis. Next, the sternomastoideus muscle was freed to expose the common carotid artery. Special attention was paid to keep vagus and phrenic nerves intact.

After preparation, the donor hemifacial flap was inset and sutured in the recipient. End-to-side anastomosis between common carotid artery of the recipient and donor vessel was performed under operating microscope magnification using 9-0 nylon sutures. Venous anastomosis was performed using standard end-to-end between the external jugular vein of the donor and the recipient. The external ear canal and flap skin was closed using 3-0 Vicryl and 3-0 Nylon. Immediate photography and 2 weeks posttransplant were revealed good circulation (Fig. 5.3a, b). The average time to complete the hemifacial transplant procedure (one donor to two recipients) was 10–11 h, and mean time of warm ischemia was 105 min.

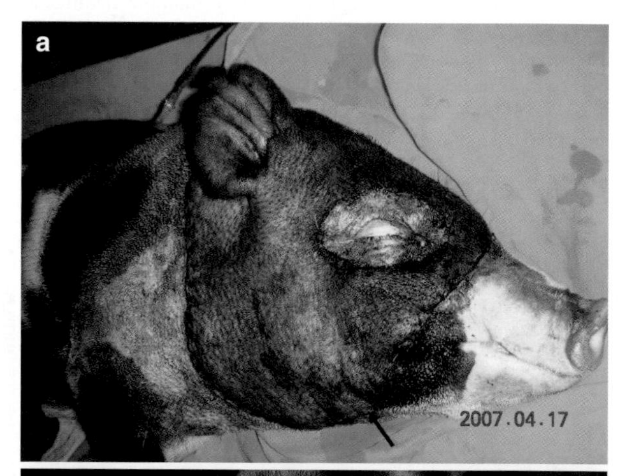

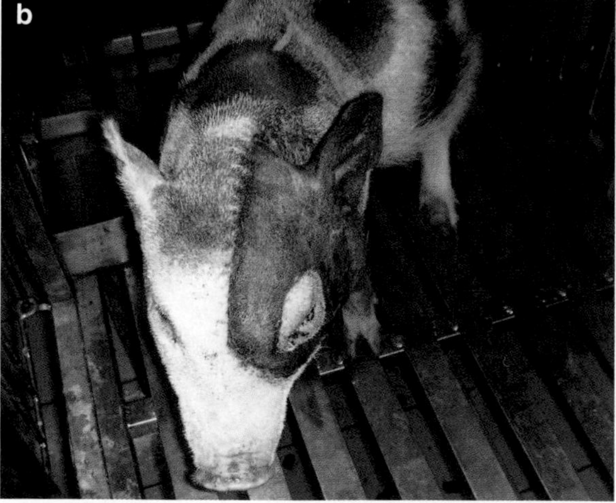

Fig. 5.3 (**a**) End-to-side anastomosis between common carotid artery of the recipient and donor vessel was performed. Venous anastomosis was performed using standard end-to-end between the external jugular vein of the donor and recipient. Immediate postoperative photograph of hemifacial transplants. (**b**) Two weeks postoperative photograph of hemifacial transplants

5.2.6 Postoperative Care

After the animal revived and was comfortably breathing, it was returned to its pen. The animal was monitored immediately after surgery in the recovery cage. Following recovery from surgery, the animal ambulated freely in its cage with no difficulty. Intravenous systemic antibiotics (Ampicillin or Augmentin) were given for 5–7 days. No anticoagulant drug was given postoperatively. The transplant was monitored on a daily basis for signs of rejection. The animal was also monitored for signs of distress, sepsis, or wound complications. All hemifacial allotransplants remained swollen for 2–3 weeks due to postoperative tissue-reactive edema and saliva gland hypersecretion.

5.2.7 Histological Evaluation of Graft Rejection

The transplanted face was observed daily for signs of rejection occurring in a reproducible sequence of epidermilysis, desquamation, eschar formation, and necrosis. Biopsy of donor skin, gland lymphoid tissue, and cartilage were obtained at specified predetermined time. At the clinically defined end point, animals were sacrificed. Histological evaluation of the graft biopsy for CTA rejection was referred to previously reported consensus scheme.[18,19] According to the severity of pathological changes, the rejection grades using a Banff classification were applied.[20]

5.3 Experimental Designs and Results

5.3.1 Experimental Study

The miniature swine were divided into three experimental groups. Group I ($n = 4$) received autologous hemifacial transplantation (Lan-Yu strain to Lan-Yu strain) as a normal control. Group II ($n = 4$) received hemifacial allotransplantation (Hwa-Ban strain to Lan-Yu strain) without treatment. Group III ($n = 3$) received cyclosporine A (CsA day 0~+28; 10 mg/kg for 2 weeks then 5 mg/kg for 2 weeks).

5.3.2 Result 1: Short-Term Immunosuppressant Prolonged Allograft Survival

All the transplants revealed swelling postoperatively and persisted for 2–3 weeks, especially the allotransplant as compared to that in autologous tissue transplantation. The autologous hemifacial transplant achieved 100% survival indefinitely until sacrifice 60 days posttransplant. In the control group, the results showed a progressive rejection of the allograft by day 7–28. The allograft with short-term CsA treatment revealed delayed the rejection between day 38 and day 49 postoperatively. This finding demonstrated that short-term CsA treatment could be significant prolongation of allograft survival as compared to the controls (Fig. 5.4).

5.3.3 Result 2: Histopathological Analysis of Allograft Biopsy

In control group, the histopathological analysis revealed severe rejection sign and abundant mononuclear infiltrations in lymphoid gland tissue and

Groups	CsA	Number	Survival days
Autologous	–	4	All survive until sacrifice
Control	–	5	28, 9, 7, 7, 21
CsA	+	3	36, 48, 38

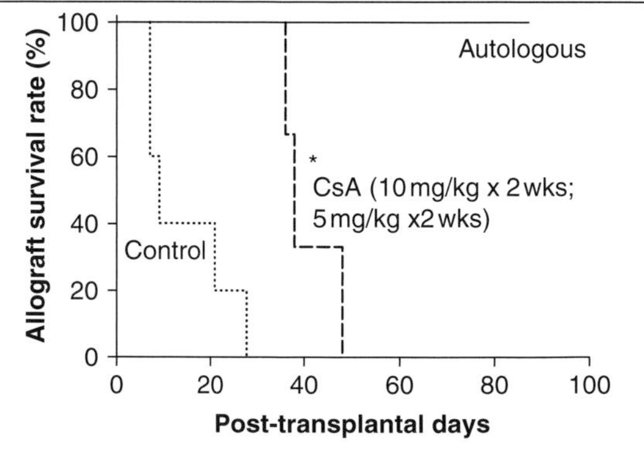

Fig. 5.4 Short-term immunosuppressant protocol prolonged allotransplant survival. The autologous transplant revealed 100% survival indefinitely 8 weeks postoperatively. The all-skin paddle in control group revealed progressive rejection of the allograft by day 7–28. The short-term cyclosporine A (CsA) treatment group showed only early mild rejection sign from day 38–49 postoperatively. This demonstrated that short-term CsA treatment significantly prolonged allotransplant survival as compared to the controls

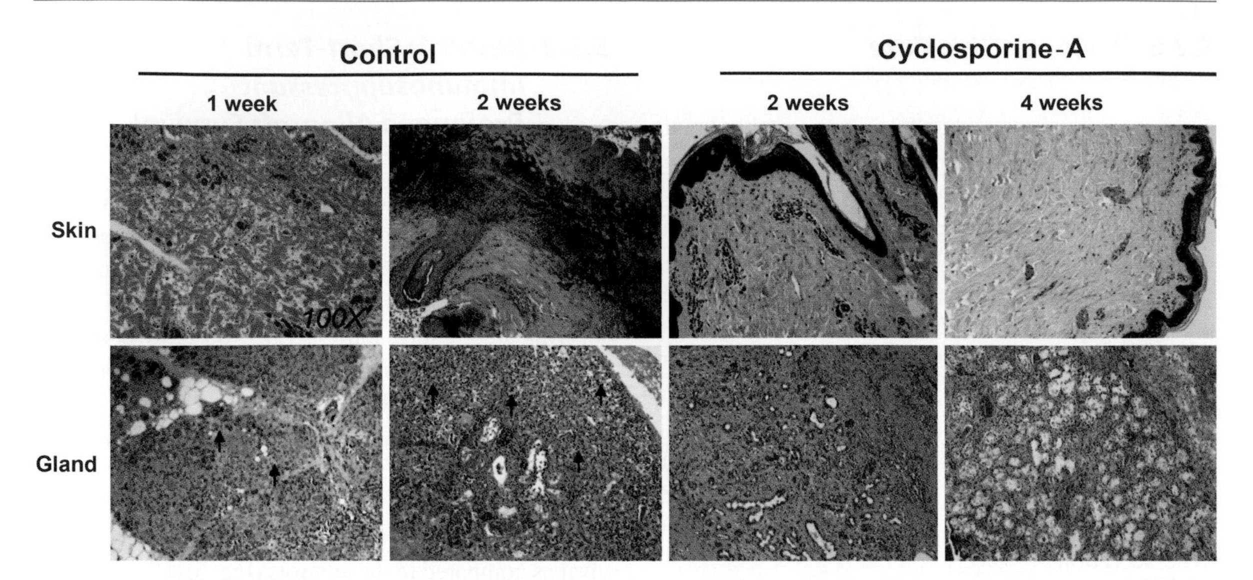

Fig. 5.5 Histopathological analysis of allotransplant tissue by using H&E staining. In control group, the histological examination revealed severe rejection sign and abundant mononuclear infiltrations in lymphoid gland tissue (grade III–IV) and allo-skin (grade III–IV), especial in the lymphoid gland tissue, at 2 weeks posttransplants. In contrast, the CsA treatment group showed only mild lymphocyte infiltration without significant rejection signs in 2 weeks and 4 weeks posttransplants (grade I–II). (Photo magnification is 100×)

allo-skin (grade III-IV), especially lymphoid tissue at 2 weeks posttransplant and sacrifice as compared to that in normal autologous lymphoid tissue and skin. In contrast, the CsA treatment group showed mild lymphocyte infiltration without significant rejection signs in 2 weeks (grade I) and 4 weeks (grade I–II) posttransplants (Fig. 5.5). However, there were no apparent differences in allo-cartilage between the control and CsA treatment group (grade 0). These analytical findings indicated that different antigenicities of the composite allografts tissues. Lymphoid gland tissue and allo-skin are both susceptible to early rejection.

5.4 Discussion

Facial allotransplantation in experimental rodents has been reported previously.[13] Clinical evidence indicates that small animal (rodent) model immunosuppression protocols are not consistently applicable because rodents tend to be more tolerant than humans to allograft transplantation.[9] However, establishment of a model for scientific research is not trivial. As a scientifically justified surgical model, it has to be reproducible and with a high success rate. It is always expensive and a lot of work to use large animals such as swine and primate as a model. Silverman and colleagues developed a heterotopic nonhuman primate facial CTA model including skin, masseter and a portion of pterygoid muscle, and mandible bone.[16] However, the results indicated this primate allotransplant model showed a big variation of CTA survival.

In this swine hemifacial allotransplantation model including the skin, lymphoid gland tissue, parts of the sternomastoideus and trapezius muscles, ear cartilage, and sensory auricular nerve was successfully developed. However, the surgical procedure needs experienced surgical team work to complete allotransplantation model.

In clinical observation, the autologous hemifacial transplant was 100% survival till sacrifice. However, autografts revealed swelling and saliva accumulation in the first 2 weeks postoperatively. The control group without treatment revealed progressive rejection by 1–4 weeks posttransplants. The short-term CsA treatment group showed only early mild rejection sign from 6 to 7 weeks posttransplant. This inference demonstrated that short-term immunosuppressant could significantly prolong allograft survival compared to that in controls. In contrast, rejection of the graft could be easily detected and monitored by inspection of the skin surface. This indicates our swine hemifacial model could be tested for further clinical trials.

Different antigenicities of the various tissues found within the CTA results in various rejections.[18,19] In this histopathological analysis, the control group revealed abundant lymphocyte infiltration in lymphoid gland tissue and allo-skin at 1 week posttransplant and sacrifice. In contrast, the CsA treatment group revealed less lymphocyte infiltration without significant rejection signs in 2–6 weeks posttransplant. However, there were no apparent differences in ear cartilage between the control and cyclosporine-A treatment group. These results demonstrated that modulation of early rejection in allo-skin and gland lymphoid tissue may be a key treatment strategy in CTA survival.

This experimental result warrants further preclinical studies of facial CTA in large animal models. However, some disadvantages were noted in this model. First, parotid saliva-pooling caused transplant swelling resulting in wound infection. The symptoms persisted for up to 2–3 weeks and debridement sometimes should be applied. This complication could be prevented by elongation of antibiotics, or saliva drainage by untightened suture over wound edge. Another shortcoming of this model is that functional outcome of facial animation could not be evaluated following CTA. However, assessment of innervation and sensation was beyond the scope of this chapter. Further modifications are needed to include facial innervated muscle and maxilla-mandible bone as a composite subunit for long-term function assessment.

In summary, this hemifacial transplant model is reproducible and warrants further preclinical immunological manipulation of the new strategies in large animal models.

References

1. Gorantla VS, Barker JH, Jones JW Jr, et al. Immunosuppressive agents in transplantation: mechanisms of action and current anti-rejection strategies. *Microsurgery*. 2000;20:420-429.
2. Tobin GR, Breidenbach WC III, Pidwell DJ, et al. Transplantation of the hand, face, and composite structures: evolution and current status. *Clin Plast Surg*. 2007;34:271-278.
3. Dubernard JM, Owen E, Herzberg G, et al. Human hand allograft: report on first 6 months. *Lancet*. 1999;353:1315-1320.
4. Devauchelle B, Badet L, Lengele B, et al. First human face allograft: early report. *Lancet*. 2006;368:203-209.
5. Siemionow M, Gordon CR. Overview of guidelines for establishing a face transplant program: a work in progress. *Am J Transplant*. 2010;10:1290-1296.
6. Hettiaratchy S, Randolph MA, Petit F, et al. Composite tissue allotransplantation-a new in plastic surgery? *Br J Plast Surg*. 2004;57:381-391.
7. Siemionow M, Oke R, Ozer K, et al. Induction of donor-specific tolerance in rat hind-limb allografts under antilymphocyte serum and cyclosporine A protocol. *J Hand Surg Am*. 2002;27:1095-1103.
8. Kuo YR, Huang CW, Goto S, et al. Alloantigen-pulsed host dendritic cells induce T-cell regulation and prolong allograft survival in a rat model of hindlimb Allotransplantation. *J Surg Res*. 2009;153:317-325.
9. Gunther E, Walter L. Comparative genomic aspects of rat, mouse and human MHC class I gene regions. *Cytogenet Cell Genet*. 2000;91:107-112.
10. Yuhki N, Beck T, Stephens RM, et al. Comparative genome organization of human, murine, and feline MHC class II region. *Genome Res*. 2003;13:1169-1179.
11. Chardon P, Renard C, Vaiman M. The major histocompatibility complex in swine. *Immunol Rev*. 1999;167:179-192.
12. Antczak DF. Structure and function of the major histocompatibility complex in domestic animals. *J Am Vet Med Assoc*. 1982;181:1030-1036.
13. Demir Y, Ozmen S, Klimczak A, et al. Tolerance induction in composite facial allograft transplantation in the rat model. *Plast Reconstr Surg*. 2004;114:1790-1801.
14. Ulusal BG, Ulusal AE, Ozmen S, et al. A new composite facial and scalp transplantation model in rats. *Plast Reconstr Surg*. 2003;112:1302-1311.
15. Shengwu Z, Qingfeng L, Hao J, et al. Developing a canine model of composite facial/scalp allograft transplantation. *Ann Plast Surg*. 2009;59:185-194.
16. Silverman RP, Banks ND, DeTolla LJ, et al. A heterotopic primate model for facial composite tissue transplantation. *Ann Plast Surg*. 2008;60:209-216.
17. Kuo YR, Shih HS, Lin CC, et al. Swine hemi-facial composite tissue allotransplantation: a model to study immune rejection. *J Surg Res*. 2009;153:268-273.
18. Kuo YR, Sacks JM, Wu WS, et al. Porcine heterotopic composite tissue allograft transplantation as a large animal model for preclinical study. *Chang Gung Med J*. 2006;29:268-274.
19. Lee WP, Yaremchuk MJ, Pan YC, et al. Relative antigenicity of components of a vascularized limb allograft. *Plast Reconstr Surg*. 1991;87:401-411.
20. Cendales LC, Kanitakis J, Schneebergerc S, et al. The Banff 2007 working classification of skin-containing composite tissue allograft pathology. *Am J Transplant*. 2008;8:1396-1400.

Experimental Studies in Face Transplantation: Primate Model

6

Eduardo D. Rodriguez, Gerhard S. Mundinger, Rolf N. Barth, Helen G. Hui-Chou, Steven T. Shipley, Luke S. Jones, and Stephen T. Bartlett

Contents

Abstract In offering optimal reconstruction for severe facial disfigurement, the advent of human face transplantation constitutes a landmark achievement in medicine and stands as a historical testament to the creativity, intelligence, ingenuity, and boldness of the human species. Facial allotransplantation has been modeled in rodents, canines, swine, and lagomorphs. However, human and rodent immune systems are dissimilar to a degree that precludes translation of tolerance induction protocols to humans. Nonhuman primates have long been used as translational models of human immunology and transplant immunobiology due to recent evolutionary divergence and shared major histocompatibility complex (MHC) II polymorphisms. We have developed a reproducible heterotopic model of nonhuman primate facial CTA permissive of long-term rejection-free survival. The purpose of this chapter is to share our experience in the development and maturation of this model, from surgical technique and immunosuppressive strategies, to experimental results and future directions.

Abbreviations

CTA	composite tissue allotransplantation
EBV	Epstein–Barr virus
GVHD	graft versus host disease
IHC	immunohistochemical staining
LCV	lymphocryptovirus
MHC	major histocompatibility complex
MLR	mixed lymphocyte reaction
MMF	mycophenolate mofetil
PTLD	posttransplant lymphoproliferative disorder
STR	short-tandem repeat
VBM	vascularized bone marrow

E.D. Rodriguez (✉)
Division of Plastic and Reconstructive Surgery, R Adams Cowley Shock Trauma Center, Johns Hopkins Hospital/University of Maryland School of Medicine, Baltimore, MD, USA
e-mail: erodriguez@umm.edu

M.Z. Siemionow (ed.), *The Know-How of Face Transplantation*,
DOI: 10.1007/978-0-85729-253-7_6,

6.1 Introduction

In offering optimal reconstruction for severe facial disfigurement, the advent of human face transplantation constitutes a landmark achievement in medicine and stands as a historical testament to the creativity, intelligence, ingenuity, and boldness of the human species. Clinical successes in face transplantation have been tempered by apprehension regarding the current necessity of costly, prolonged immunosuppressive therapy in otherwise healthy recipients, as well as the development of features consistent with chronic rejection in hand composite tissue allotransplants (CTAs).[1] Widespread application of face transplantation may hinge on the development of alternative methods of immunosuppression[2] and, perhaps, intrinsic graft composition.[3] Specifically, strategies promoting the development of tolerance with minimal requisite maintenance immunosuppression could minimize the risks of immunosuppression, lower the long-term costs of CTA, and, most importantly, extend allograft longevity.[4-6]

Despite recent success with a tolerance induction protocol in human renal transplantation,[7] no clinical tolerance induction protocols are currently available for human face transplantation due to the recognized immunological complexities of CTA relative to solid organ transplants.[8] Exploration of alternate immunosuppressive strategies in face transplantation is thus best undertaken in preclinical animal models to shield patients from the risks and consequences of novel experimental therapies.[9]

Facial allotransplantation has been modeled in rodents,[10-17] canines,[18] swine,[19] and lagomorphs.[20,21] Although tolerance induction has been reported in rodent face transplant models,[11] human and rodent immune systems are dissimilar to a degree that precludes translation of tolerance induction protocols to humans based on three of these studies.[8,22] Nonhuman primates have long been used as translational models of human immunology and transplant immunobiology due to recent evolutionary divergence and shared major histocompatibility complex (MHC) II polymorphisms.[23] Additionally, nonhuman primate anatomy closely resembles human anatomy. In the field of CTA, though nonhuman primate models of partial upper extremity[24-26] and hand[27,28] have been successfully developed, a single model of composite mandible transplantation was unable to achieve the reproducible technical success and prolonged graft survival requisite for progression to preclinical investigation.[29]

We have developed a reproducible heterotopic model of nonhuman primate facial CTA permissive of long-term rejection-free survival.[8,30,31] The purpose of this chapter is to share our experience in the development and maturation of this model, from surgical technique and immunosuppressive strategies, to experimental results and future directions.

6.2 Technique

Male and female cynomolgus macaques (*Macaca fascicularis*) weighing 3–8 kg are paired based on ABO blood group matching, and then mismatched using mixed lymphocyte reaction (MLR) assays. The day before transplantation, left internal jugular central lines are inserted into recipient monkeys under general anesthesia. Recipients are maintained in a jacket and tether system for 28 days to allow for continuous intravenous infusion of tacrolimus. Cefazolin at 22 mg/kg is administered twice daily for the full duration of the central line.

On the day of surgery, an oromandibular segment including masseter muscle, overlying skin, common carotid artery, and external and internal jugular veins, is harvested from the left jaw using loupe magnification (Figs. 6.1 and 6.2). A portion of vascularized mandible from the third molar to the condyle can additionally be included in the graft. The skin paddle extends from the oral commisure to the tragus, and from the inferior border of the mandible to the superior border of the zygoma. The facial, superior labial, superficial temporal, and transverse facial vessels are included in the graft. Immediately before division of the vascular pedicle, the donor monkey is administered 1,000 units of IV heparin. After pedicle division, the graft is flushed with approximately 100 mL of University of Wisconsin solution until the effluent is clear.

The graft is then transplanted heterotopically into the left lower abdominal wall of the recipient monkey (Fig. 6.3a). The common femoral artery and vein are dissected and used as recipient vessels. Immediately before beginning the vascular anastomoses, the

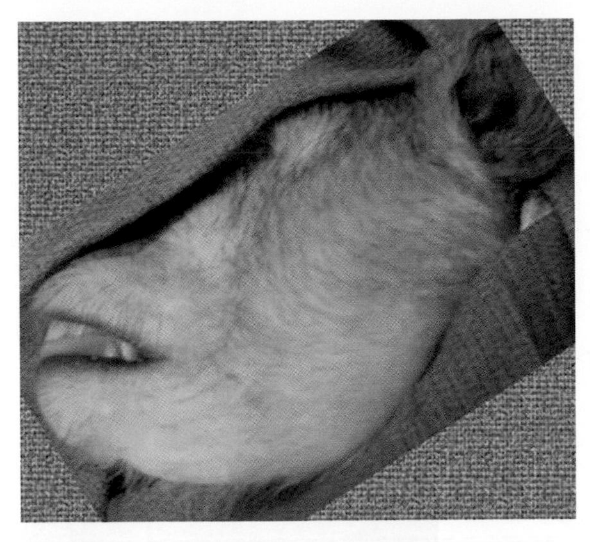

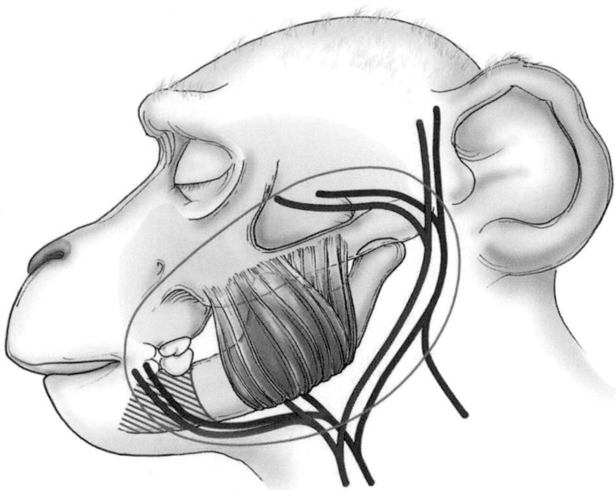

Fig. 6.1 (*Left*) *Macaca fascicularis* facial profile following positioning for donor surgery. (*Right*) Schematic of donor osteo-myocutaneous composite facial graft based on the common carotid artery and both jugular veins. Skin paddle (*Green circle*); Excluded hemimandibular bone (Mandibular cross-hatching) (Right image adapted from Barth[8] with permission)

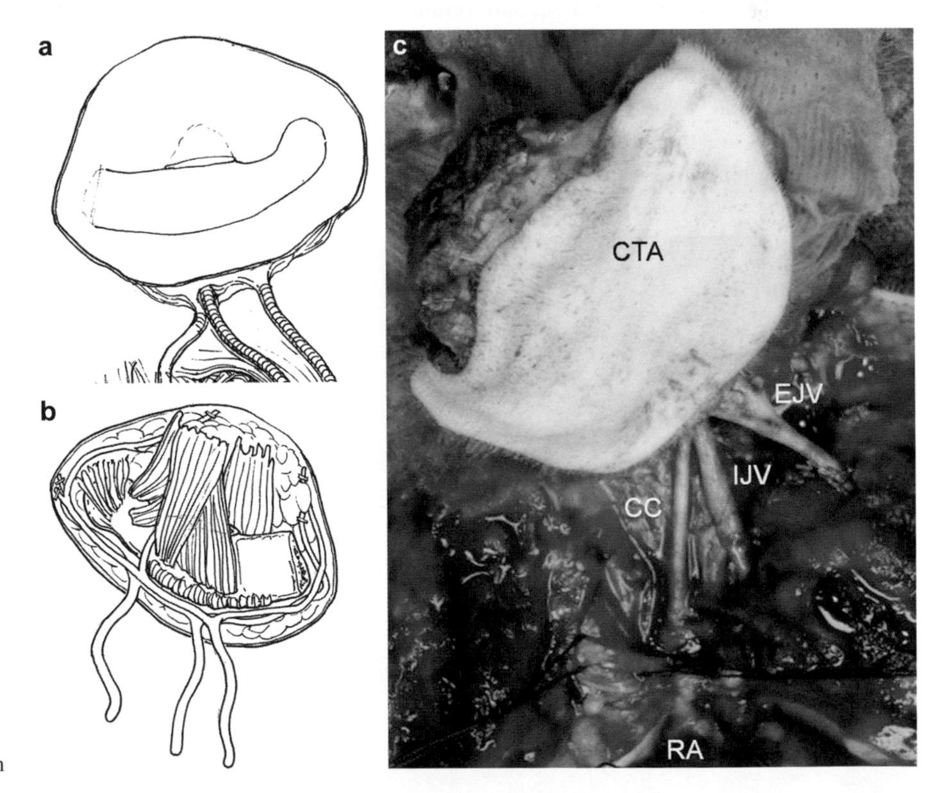

Fig. 6.2 (**a**) External and (**b**) internal schematic of facial subunit depicting bone, muscle, skin, and vessels. (**c**) Intraoperative photograph of the native graft in situ. *CTA* composite tissue allograft, *EJV* external jugular vein, *IJV* internal jugular vein, *CC* common carotid artery, *RA* right atrium

recipient receives 500 units of IV heparin. Three separate end-to-side anastomoses are then performed between the donor and recipient vessels with interrupted 9–0 nylon using an operative microscope. Anastomoses are performed between the donor common carotid and recipient femoral artery, donor external jugular vein and recipient femoral vein, and donor internal jugular vein and recipient femoral vein (Fig. 6.4). The arterial anastomosis is performed first to expeditiously establish graft reperfusion. We

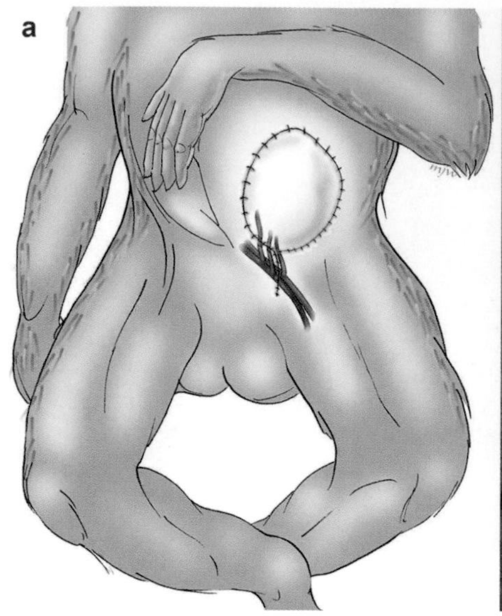

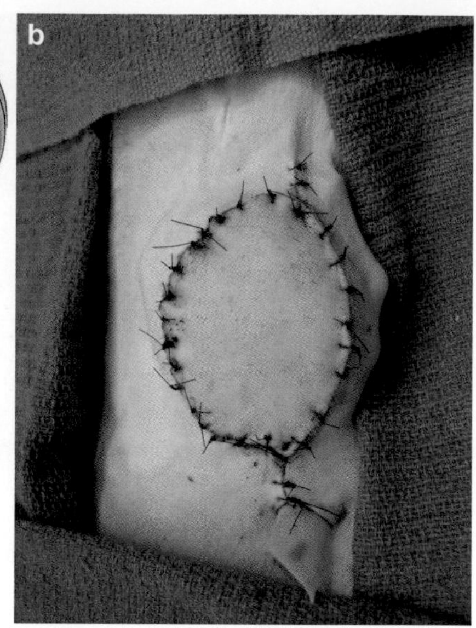

Fig. 6.3 (**a**) Schematic of composite facial subunit inset into the recipient left lower abdominal wall following completion of vascular anastomoses. (**b**) On-table photograph of transplanted graft at the conclusion of surgery

have found that inclusion of a second venous anastomosis greatly reduces postoperative graft swelling, as the external jugular vein accounts for, on average, 44% of total venous outflow from the graft.[31] Penrose drains are then sutured to the skin, and the flap is inset using interrupted stitches (Fig. 6.3b).

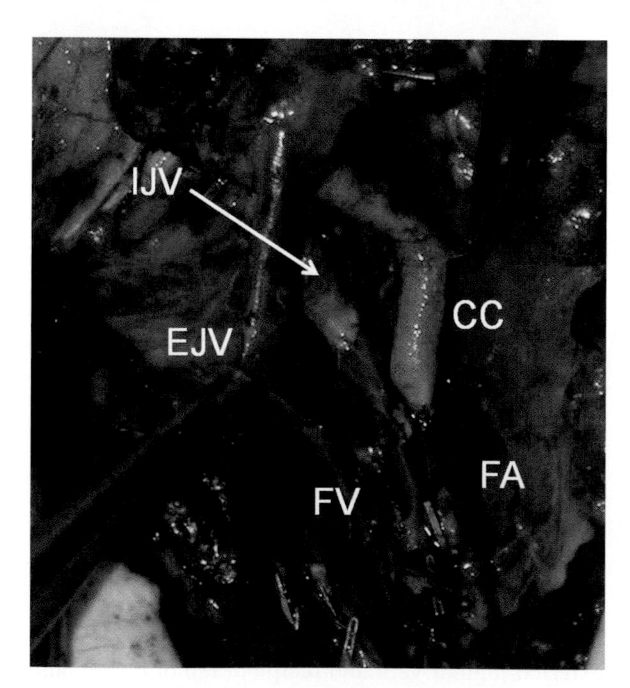

Fig. 6.4 Intraoperative photograph of arterial and dual venous end-to-side anastomoses between donor vessels and recipient femoral vessels. *EJV* external jugular vein, *IJV* internal jugular vein, *CC* common carotid artery, *FA* femoral artery, *FV* femoral vein

Drains are variably removed between post operative days 2–5. Postoperative analgesia is provided with buprenorphine at 0.02 mg/kg twice daily for the first 3 days postoperatively. Postoperatively for 28 days, recipients receive daily subcutaneous injections of dalteparin dosed at 35 units/kg and 20.25 mg of aspirin by mouth as thrombotic prophylaxis.

6.3 Immunosuppressive Strategies

To date we have performed a total of 19 transplants utilizing five immunosuppressive regimens (Table 6.1). All groups have included tacrolimus (FK506) immunosuppression. Tacrolimus administration began 1 day before surgery via continuous IV infusion. Continuous infusion was maintained for a total of 28 days before conversion to once daily intramuscular tacrolimus dosing. The goal level for the high-dose FK506 groups (Groups 1 and 2) was 30–50 ng/mL, and 20–30 ng/mL for the low-dose FK506 groups (Groups 3 and 4). Trough target levels after conversion to once daily dosing were 10–20 ng/mL. In Group 2, high-dose FK506 was converted to 5 mg/kg once daily rapamycin dosing after the 28-day period if continuous FK506 infusion. Average rapamycin level in this group was 11 ng/mL (range 7–18 ng/mL). In the third and fourth groups, antimetabolite therapy with oral mycophenolate mofetil

Table 6.1 Experimental surgical groups as defined by immunosuppressive regimen and inclusion of vascularized bone marrow (*VBM*). Tacrolimus (FK506); *MMF* mycophenolate mofetil

Group number	Immunosuppression	Bone and VBM	Goal FK506 level (ng/mL)
1	High FK506 ($n=6$)	Yes	30–50
2	High FK506→Rapamycin ($n=3$)	Yes	30–50
3	Low FK506/MMF ($n=4$)	Yes	20–30
4	Low FK506/MMF ($n=3$)	No	20–30
5	Low FK506/Anti-CD28 ($n=3$)	Yes	20–30

(MMF) dosed at 50 mg/kg per day was added to low-dose FK506. In Group 5, 28 days of anti-CD28 costimulatory blocking antibody (sc28AT) administered daily at 2 mg/kg IV was combined with low-dose tacrolimus therapy. In all experiments, rejection episodes (either clinical or histological) were not treated with additional therapies in order to study the natural resolution or progression of rejection occurring under the initial specific therapeutic regimen.

6.4 Graft Monitoring and Diagnostic Studies

All grafts are inspected daily and scheduled biopsies (4-mm punch or deep tissue core) are performed at scheduled intervals under propofol sedation. Blood is drawn weekly from the contralateral femoral vessels for complete blood counts, serum chemistry profiles, whole blood drug levels, and analysis by flow cytometry. Tissues are reviewed by a transplant pathologist with rejection classified according to Banff critera.[32] Short-tandem repeat (STR) genotypic analyses are used to determine whether cell populations of various origins are of donor or recipient origin. Mixed lymphocyte reactions (MLRs) are performed before transplantation and at scheduled intervals to assess the degree of recipient responsiveness to donor and third-party antigens. Flow cytometry is used to detect peripheral and central chimerism, evaluate regulatory T-cell populations, and quantify alloantibody production. With regard to imaging, selected animals undergo magnetic resonance imaging (MRI) to evaluate graft vasculature and transplanted tissue.

6.5 Results and Discussion by Immunosuppressive Regimen

In every treatment group, composite facial allografts all healed normally in the postoperative period as evidenced by normal skin appearance with hair growth (Fig. 6.5). Associations between clinical, histological, and MRI imaging rejection criteria were also demonstrable (Fig. 6.6). Consistent with patterns of rejection in other CTA models, skin has been the first target of acute rejection in our model, and may occur in the absence of rejection of other tissues. We have found that mild acute rejection episodes are not correlated with overall graft survival, in contrast to episodes of moderate and severe rejection.

We have previously reported our results with high-dose tacrolims monotherapy and the association of this immunosuppressive regimen with lymphocryptovirus (LCV) and posttransplant lymphoproliferative disorder (PTLD) (Group 1, Table 6.1).[8,31] Although recipients initially maintained on high-dose FK506 and then converted to rapamycin monotherapy (Group 2, Table 6.1) did not develop PTLD, conversion to rapamycin failed to prevent rejection, precluding prolonged graft survival, and did not promote the generation of chimerism. In comparing these groups, PTLD developed when grafts from LCV [the simian equivalent of Epstein–Barr virus (EBV)] positive donors were transplanted into LCV positive recipients.[8] Overall, MLR responses demonstrated mild sensitization to donor antigens, and mixed responsiveness to third-party antigens. We have not witnessed graft versus host disease (GVHD) in any of our transplants.

With combined low-dose tacrolimus and mycophenolate mofetil (MMF) therapy, we were able to achieve prolonged rejection-free survival in four monkeys without the development of PTLD (Group 3, Table 6.1).

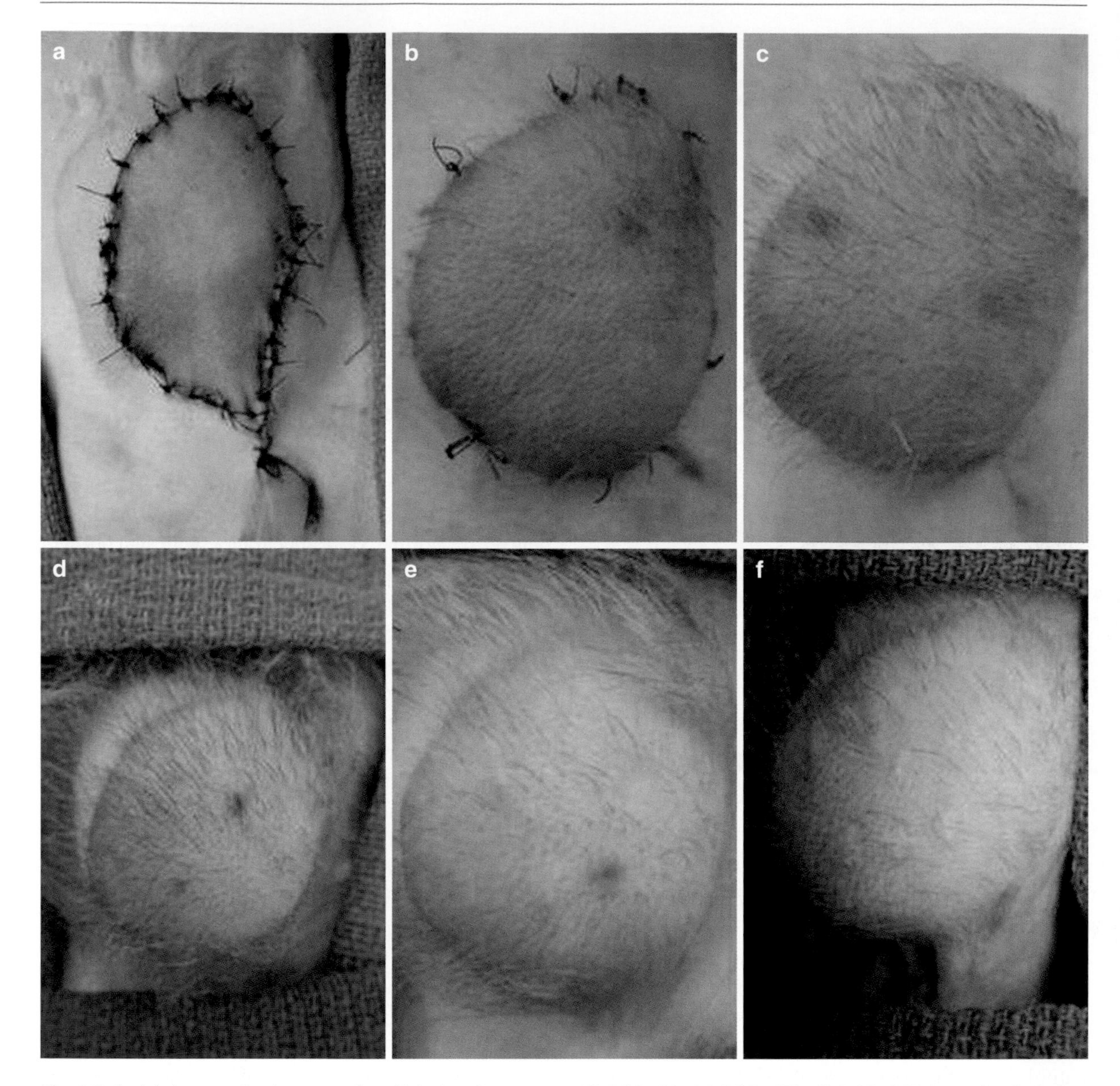

Fig. 6.5 Serial photographs demonstrating clinical graft appearance in a recipient maintained on low-dose FK506/MMF therapy (Group 3). (**a**) Immediate postoperative appearance. (**b**) Resolving erythema with hair growth on postoperative day (*POD*) 22. Note the punch-biopsy site in the upper right quadrant. (**c**) POD 44 with histological evidence of moderate rejection. (**d**) POD 98. (**e**) POD 158. (**f**) POD 275 immediately before withdrawal of immunosuppression. Note the pale, waxy graft appearance. This graft rejected acutely 18 days after withdrawal of immunosuppression. Post-necropsy histopathologic analysis revealed both acute and chronic rejection

Immunosuppression was weaned off in all four subjects, and all grafts subsequently succumbed to acute rejection after cessation of immunosuppression. Interestingly, some graft vessels demonstrated histopathologic signs of active transplant arteriopathy, neointimal proliferation, and graft fibrosis consistent with chronic rejection.[32,33] These changes were not evident in native recipient vessels. We have begun developing protocols for immunohistochemical (IHC) staining to determine the phenotype and activation status of immune cells involved in rejection responses.[34] Though acute rejection episodes have been well

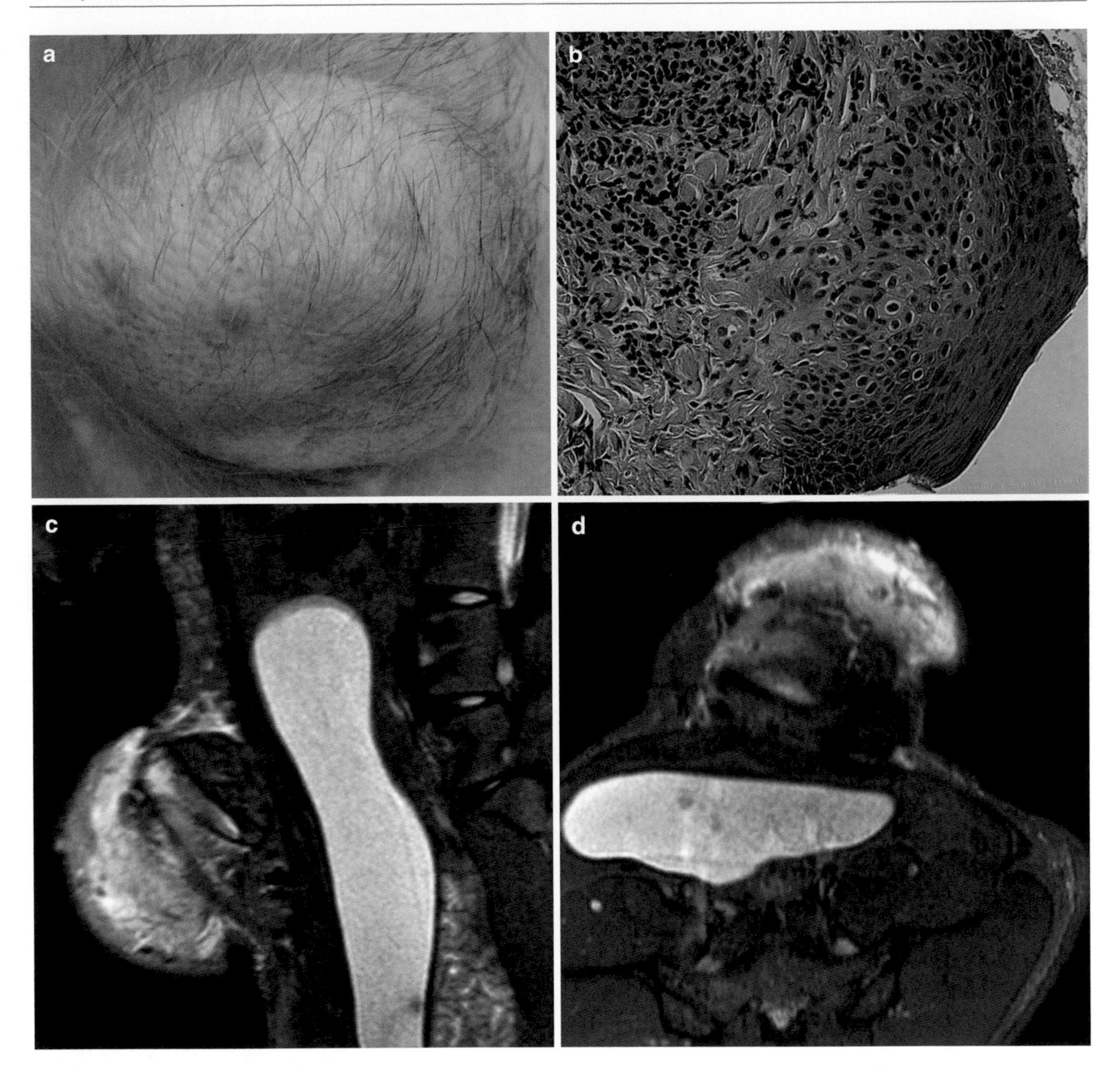

Fig. 6.6 Correlation of composite facial allograft clinical appearance, histology, and MRI imaging in the same subject on POD 176. (**a**) Clinical graft appearance. (**b**) Punch biopsy demonstrating moderate rejection (Banff grade II). (**c**) Sagittal and (**d**) axial short TI inversion recovery (*STIR*) sequence MRI images demonstrating graft soft tissue edema and inflammation relative to the recipient. The graft has good vascularity, and graft bone marrow is normal as compared to native iliac bone marrow

documented and characterized in human CTAs, the recognition of chronic rejection is an emerging phenomenon.[1,33,35] These findings have far-reaching implications for the investigation of methods to mitigate deleterious long-term graft changes throughout the field of CTA.

Having identified a viable long-term immunosuppressive strategy, we then explored the contribution of vascularized bone marrow (VMB) in promoting long-term rejection-free survival (Group 4, Table 6.1). We found that, in comparison to grafts under the same tacrolimus/MMF immunosuppression with bone (Group 3), grafts lacking VBM survived roughly half as long.[3] These data support a protocol combining vascularized bone marrow and simplified immunosuppressive therapy as a method to achieve prolonged graft survival without rejection episodes. Despite transient chimerism by flow cytometry, this strategy does not appear to translate to immunologic tolerance, as all grafts in both groups were ultimately lost to rejection after immunosuppression withdrawal.

Most recently, we have investigated the ability of costimulatory blockade provided by a selective CD28 antagonist (sc28AT) to promote long-term CTA survival via augmented regulatory cell populations (Group 5, Table 6.1).[36] The ability of costimulatory blockade therapy to provide effective immunosuppression and promote regulatory mechanisms facilitating long-term graft survival has been demonstrated in solid organ transplantation,[37] and suggested as a viable immunosuppressive strategy in CTA.[22] Current results from this group suggest that the addition of CD28 costimulatory blockade to low-dose tacrolimus therapy does not promote rejection-free CTA survival.[38]

6.6 Summary

In summary, we have achieved reproducible success using tacrolimus-based immunosuppression in a heterotopic nonhuman primate model of face transplantation through the addition of a second venous anastomosis to reduce postoperative graft swelling. This technical success has allowed for the nascent exploration of the unique immunobiology of different composite tissues and immunosuppressive/immunomodulatory strategies to achieve prolonged or permanent graft survival. Although tolerance to composite tissues can be readily induced in vitro and in rodent models by a variety of methods, it remains problematic to induce tolerance in large animal CTA models as our work demonstrates. The unique immunobiology of CTA may be secondary to the expression of a broad spectrum of tissue-specific antigens from skin, muscle, bone, mucosa, and nerve, leading to the development of split tolerance.[39]

Knowledge gained from these and future endeavors will ultimately benefit patients with severe facial deformities, allowing for safer, more durable, and more predictable allotransplantation of composite tissues.

Acknowledgments The authors would like to acknowledge Arthur Nam, Aruna Panda, Debra Kukuruga, Cinthia Drachenberg, Amir H. Dorafshar, and Theresa Alexander for their dedicated work on this project.

References

1. Breidenbach WC, Ravindra K, Blair B, et al. Update on the Louiville Hand Transplant Experience. *American Society for Reconstructive Microsurgery 2010 Annual Scientific Meetings* 2010.
2. Elster EA, Blair PJ, Kirk AD. Potential of costimulation-based therapies for composite tissue allotransplantation. *Microsurgery*. 2000;20:430-434.
3. Hui-Chou HG, Mundinger GS, Jones LS, et al. Vascularized bone marrow permits long-term rejection free survival of facial composite tissue allografts in non-human primates. *Am J Transplant*. 2010;10:91.
4. Siemionow M, Nasir S. Impact of donor bone marrow on survival of composite tissue allografts. *Ann Plast Surg*. 2008;60:455-462.
5. Yen EF, Hardinger K, Brennan DC, et al. Cost-effectiveness of extending Medicare coverage of immunosuppressive medications to the life of a kidney transplant. *Am J Transplant*. 2004;4:1703-1708.
6. Matas AJ, Humar A, Gillingham KJ, et al. Five preventable causes of kidney graft loss in the 1990s: a single-center analysis. *Kidney Int*. 2002;62:704-714.
7. Kawai T, Cosimi AB, Spitzer TR, et al. HLA-mismatched renal transplantation without maintenance immunosuppression. *N Engl J Med*. 2008;358:353-361.
8. Barth RN, Nam AJ, Stanwix MG, et al. Prolonged survival of composite facial allografts in non-human primates associated with posttransplant lymphoproliferative disorder. *Transplantation*. 2009;88:1242-1250.
9. Siemionow M, Klimczak A. Advances in the development of experimental composite tissue transplantation models. *Transpl Int*. 2010;23:2-13.
10. Landin L, Cavadas PC, Gonzalez E, et al. Functional outcome after facial allograft transplantation in rats. *J Plast Reconstr Aesthet Surg*. 2008;61:1034-1043.
11. Siemionow M, Agaoglu G. Face transplantation. In: Hewitt C, Lee W, Gordon C, eds. *Transplantation of Composite Tissue Allografts*. New York: Springer; 2008: 344-354.
12. Zor F, Bozkurt M, Nair D, Siemionow M. A new composite midface allotransplantation model with sensory and motor reinnervation. *Transpl Int*. 2010;23:649-656.
13. Kulahci Y, Siemionow M. A new composite hemiface/mandible/tongue transplantation model in rats. *Ann Plast Surg*. 2010;64:114-121.
14. Landin L, Cavadas PC, Gonzalez E, et al. Sensorimotor recovery after partial facial (mystacial pad) transplantation in rats. *Ann Plast Surg*. 2009;63:428-435.
15. Washington KM, Solari MG, Sacks JM, et al. A model for functional recovery and cortical reintegration after hemifacial composite tissue allotransplantation. *Plast Reconstr Surg*. 2009;123:26S-33S.
16. Yazici I, Unal S, Siemionow M. Composite hemiface/calvaria transplantation model in rats. *Plast Reconstr Surg*. 2006;118:1321-1327.
17. Ulusal AE, Ulusal BG, Hung LM, et al. Establishing a composite auricle allotransplantation model in rats: introduction to transplantation of facial subunits. *Plast Reconstr Surg*. 2005;116:811-817.

18. Shengwu Z, Qingfeng L, Hao J, et al. Developing a canine model of composite facial/scalp allograft transplantation. *Ann Plast Surg*. 2007;59:185-194.
19. Kuo YR, Shih HS, Lin CC, et al. Swine hemi-facial composite tissue allotransplantation: a model to study immune rejection. *J Surg Res*. 2009;153:268-273.
20. Xudong Z, Shuzhong G, Yan H, et al. A hemifacial transplantation model in rabbits. *Ann Plast Surg*. 2006;56: 665-669.
21. Rab M, Haslik W, Grunbeck M, et al. Free functional muscle transplantation for facial reanimation: experimental comparison between the one- and two-stage approach. *J Plast Reconstr Aesthet Surg*. 2006;59:797-806.
22. Siemionow M, Unal S. Strategies for tolerance induction in nonhuman primates. *Ann Plast Surg*. 2005;55:545-553.
23. Geluk A, Elferink DG, Slierendregt BL, et al. Evolutionary conservation of major histocompatibility complex-DR/ peptide/T cell interactions in primates. *J Exp Med*. 1993; 177:979-987.
24. Daniel RK, Egerszegi EP, Samulack DD, et al. Tissue transplants in primates for upper extremity reconstruction: a preliminary report. *J Hand Surg Am*. 1986;11:1-8.
25. Stevens HP, Hovius SE, Heeney JL, et al. Immunologic aspects and complications of composite tissue allografting for upper extremity reconstruction: a study in the rhesus monkey. *Transplant Proc*. 1991;23:623-625.
26. Cendales LC, Xu H, Bacher J, et al. Composite tissue allotransplantation: development of a preclinical model in nonhuman primates. *Transplantation*. 2005;80:1447-1454.
27. Stark GB, Swartz WM, Narayanan K, et al. Hand transplantation in baboons. *Transplant Proc*. 1987;19:3968-3971.
28. Stevens HP, Hovius SE, Vuzevski VD, et al. Immunological aspects of allogeneic partial hand transplantation in the rhesus monkey. *Transplant Proc*. 1990;22:2006-2008.
29. Gold ME, Randzio J, Kniha H, et al. Transplantation of vascularized composite mandibular allografts in young cynomolgus monkeys. *Ann Plast Surg*. 1991;26: 125-132.
30. Silverman RP, Banks ND, Detolla LJ, et al. A heterotopic primate model for facial composite tissue transplantation. *Ann Plast Surg*. 2008;60:209-216.
31. Barth RN, Bluebond-Langner R, Nam A, et al. Facial subunit composite tissue allografts in nonhuman primates: I. Technical and immunosuppressive requirements for prolonged graft survival. *Plast Reconstr Surg*. 2009;123:493-501.
32. Cendales LC, Kanitakis J, Schneeberger S, et al. The Banff 2007 working classification of skin-containing composite tissue allograft pathology. *Am J Transplant*. 2008;8:1396-1400.
33. Unadkat JV, Schneeberger S, Horibe EH, et al. Composite tissue vasculopathy and degeneration following multiple episodes of acute rejection in reconstructive transplantation. *Am J Transplant*. 2009;10:251-261.
34. Mundinger GS, Munivenkatappa R, Hui-Chou HG, et al. Chronic rejection in a non-human primate model of facial composite tissue allotransplantation: clinical, pathological, and immunohistochemical characterization. *23rd Annual International Congress of the Transplantation Society* 2010:#2633.
35. Unadkat JV, Haribe EK, Schneeberger S, et al. Systemic analysis of chronic rejection in composite tissue allotransplantation. *7th International Symposium on Composite Tissue Transplantation*. Seefeld/Tyrol, Austria 2007.
36. Vanhove B, Laflamme G, Coulon F, et al. Selective blockade of CD28 and not CTLA-4 with a single-chain Fv-alpha1-antitrypsin fusion antibody. *Blood*. 2003;102:564-570.
37. Poirier N, Azimzadeh AM, Zhang T, et al. Inducing CTLA-4-dependent immune regulation by selective CD28 blockade promotes regulatory T cells in organ transplantation. *Sci Transl Med*. 2010;2:17ra10.
38. Mundinger GS, Jones LS, Hui-Chou HG, et al. Costimulatory blockade does not promote survival of skin component in non-human-primate composite tissue allografts. *23rd Annual International Congress of the Transplantation Society* 2010:#2606.
39. Mathes DW, Randolph MA, Solari MG, et al. Split tolerance to a composite tissue allograft in a swine model. *Transplantation*. 2003;75:25-31.

Timeline and Evolution of Face Transplant Cadaver Models

7

Chad R. Gordon and Maria Z. Siemionow

Contents

Abstract The first worldwide IRB-approved protocol for facial transplantation was obtained by Dr. Siemionow and colleagues at the Cleveland Clinic in 2004. To date, the most complex near-total face transplant remains the one performed in December, 2008 at the Cleveland Clinic (Cleveland, OH). The purpose of this chapter is to describe in detail the strategic evolution of our team's tailored, cadaver dissection model in preparation for its first clinical application.

7.1 Introduction

The Cleveland Clinic's Department of Plastic Surgery (Cleveland, Ohio), under the guidance of Dr. Maria Siemionow, received the world's first IRB-approved protocol for face transplantation in 2004.[1] At this point in time, scarce literature had existed in regards to all preclinical and clinical aspects of facial composite tissue allotransplantation (CTA). More specifically, literature describing techniques for anatomical cadaver dissections necessary for both facial alloflap donor harvest (myocutaneous *and* osteocutaneous) and recipient alloflap insetting were nonexistent. With this in mind, the purpose of this chapter is to describe in detail the strategic evolution of our team's tailored, cadaver dissection model in preparation for its first clinical application. We will also chronologically examine the development of face transplant cadaver models around the world; following its first clinical application in 2005 by Dubernard et al. in Paris, France.[2]

7.2 The Very Beginning

Published contributions from our work over the last 6 years, in regards to cadaver dissections, have included landmark articles describing (1) autologous skin

C.R. Gordon (✉)
Division of Plastic and Reconstructive Surgery, Department of Surgery, Massachusetts General Hospital, Boston, MA, USA
e-mail: cgordon5@partners.org

M.Z. Siemionow (ed.), *The Know-How of Face Transplantation*,
DOI: 10.1007/978-0-85729-253-7_7,

coverage boundaries (i.e., free flaps) versus face transplantation, (2) optimal time estimates, recommended sequence of dissection, and pertinent anatomical structures, (3) issues surrounding facial appearance and identity transfer, (4) a coronal-posterior approach for pedicle length extension with donor harvest, and (5) definition and preservation of the vascular pedicle supplying the maxillary allograft. In fact, the very first article published on this subject appeared in the March 2006 issue of *Plastic & Reconstructive Surgery*; analyzing the hypothetical indication/limitations encountered with full facial burn patients requiring complete excision and free tissue transfer reconstruction[3-7] (Fig. 7.1).

Within this paper, the authors described unprecedented information gathered from this first-ever facial allotransplantation cadaver study. It revealed to the reader that all alternative autologous sources in the human body available for single flap transfer, including unilateral/bilateral free flaps such as radial forearm, anterolateral thigh, deep inferior epigastric perforator and scapular–parascapular, fail overwhelmingly in providing skin coverage for total face reconstruction measuring 675 sq cm. The author's conclusion was that for full face resurfacing, in the setting of burn injury and/or trauma, the most optimal aesthetic *and* functional skin match in regards to skin texture, pliability, and color could only be achieved with facial allotransplantation.[3]

Additionally, ten cadavers (donors and recipients) were dissected for evaluation of pertinent unknown variables such as total operative times for donor alloflap harvesting, pedicle lengths of neurovascular structures included in the graft, total operative time for graft inset, and total operative time for vascular anastomoses/nerve coaptations (Fig. 7.2). Interestingly, in

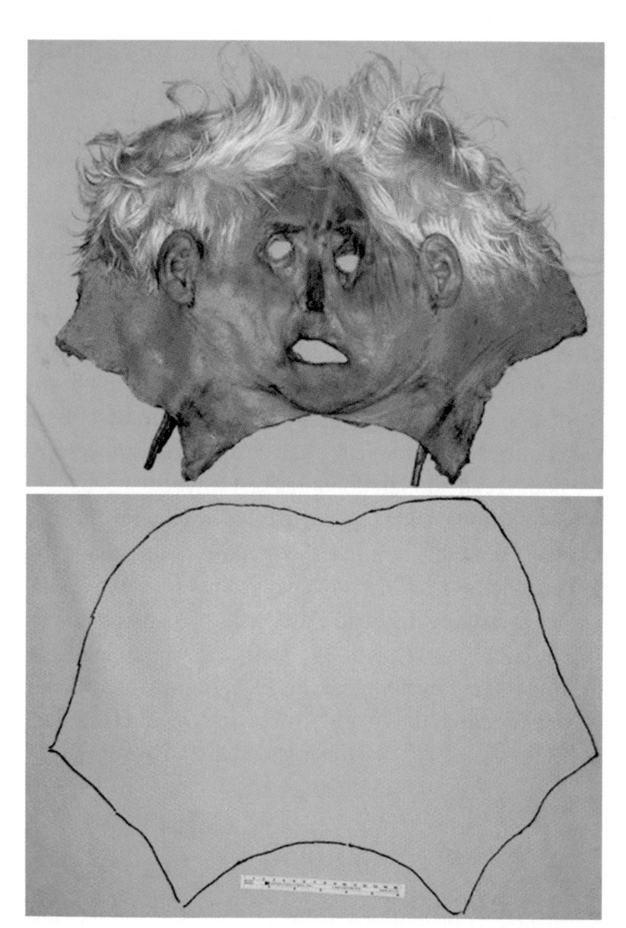

Fig. 7.1 (*Above*) Harvested total facial-scalp flap. (*Below*) Template of total facial-scalp flap for surface area measurements (From Siemionow et al.,[3] with permission)

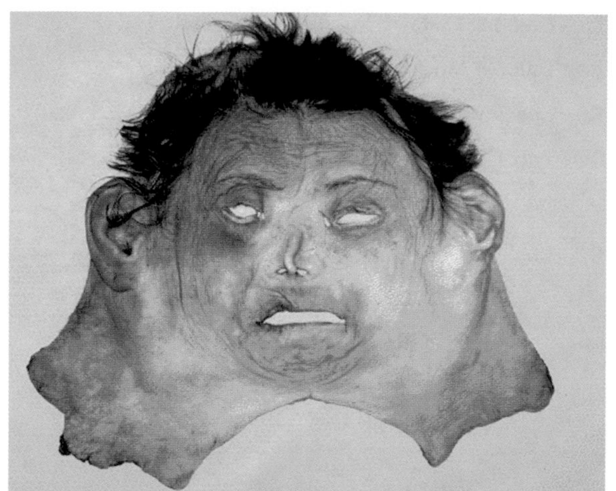

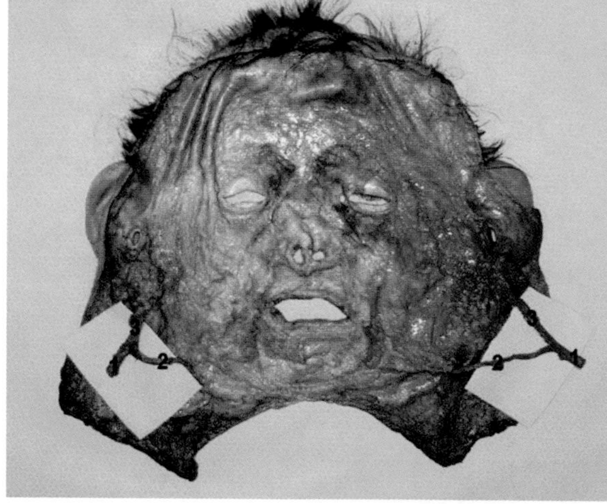

Fig. 7.2 Harvested total facial-scalp flap in the donor cadaver. (*Above*) The frontal surface area. (*Below*) The inverted surface area: *1*, the external carotid artery; *2*, the facial artery; *3*, the superficial temporal artery (From Siemionow et al.,[4] with permission)

this article, the authors reported a time necessary for mock cadaver face transplantation to be around 5 h and 20 min.[4] Although this described total myocutaneous facial alloflap transplant time, it turned out to be significantly less than the total, maxilla-containing face transplant operative time performed by the exact same institution 2 years later in 2008 (approximately 22 h). This prolonged operative time adequately demonstrates the complexity of face transplantation and how even the most prepared teams still find unexpected obstacles. It should be noted, however, that these cadaver mock transplants did not require meticulous hemostasis and microsurgical neurovascular repairs, thereby reducing total procurement times significantly.[8]

7.3 Evolving Cadaver Studies

That same year, approximately 4 months later, a second study involving a new generation of cadaver dissections was published by Siemionow and Agaoglu. It investigated the next evolving step, which was to prepare for the conflicting issues of "facial appearance" and potential "identity transfer." Once facial CTA began to be publicized and gain notoriety, some of the harshest critics were confident that all patients undergoing facial allotransplantation would lose his or her "facial identity," as opposed to others, who felt that the recipient's unaltered facial skeleton would determine soft-tissue draping of the alloflap, and that a close resemblance would remain following face transplantation. In this cadaver study, two served as mock recipients and eight as recipients. Eight facial flaps were placed alternatively on each of the two recipients, thereby allowing the authors to conclude that in fact the posttransplant appearance was a hybrid mixture of various features resembling *both* the donor and the recipient[5] (Fig. 7.3).

A fourth article describing cadaver studies soon followed 1 month later, in August 2006. This one described the development of a new surgical technique for facial alloflap harvest employing a coronal-posterior approach. Siemionow and coauthors reported multiple high-yield maneuvers allowing one the ability to extend neurovascular pedicle length for facial allograft inset, such as the supraorbital, infraorbital, and mental nerves; with the exact lengths measuring 3.5, 4.7, and 5.6 cm, respectively[6] (Figs. 7.4 and 7.5). Our team has since learned that patients with severe blast injury from close-range gunshots, for example, such as our first transplant patient, will have tremendous neural damage limiting available length for repair. Therefore, we have recommended planning ahead and harvesting extra donor nerve grafts (i.e., donor sural, spinal accessory, hypoglossal and/or vagus nerves) and veins (i.e., basilic, cephalic, and/or saphenous) just in case an interpositional nerve or vein graft is needed during the time of inset.[9]

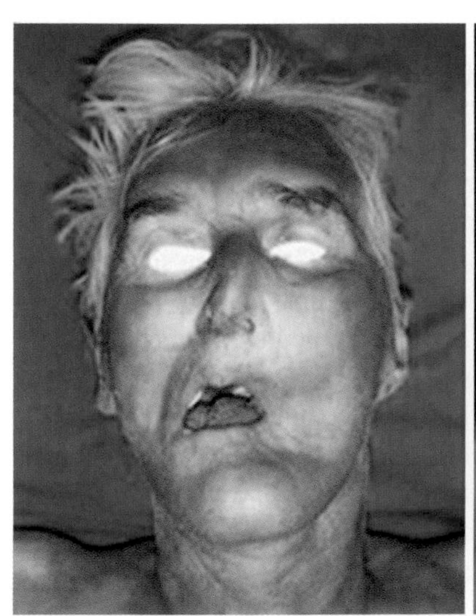

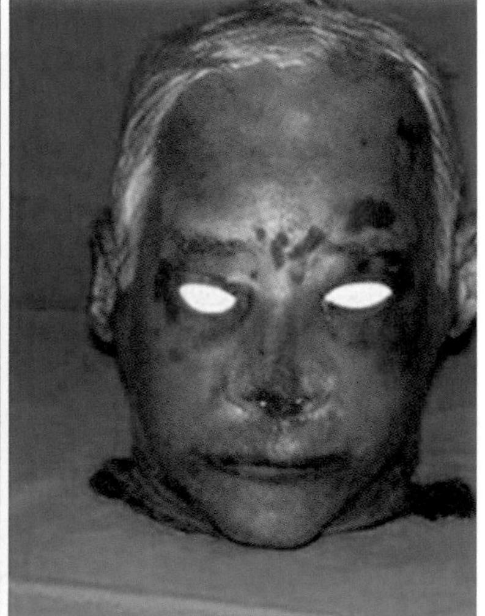

Fig. 7.3 Facial appearance of donor cadaver before facial flap harvesting (*left*). Frontal view of Styrofoam head model (*middle*). Appearance of Styrofoam head model after mounting of the harvested facial-scalp flap (*right*) (The eyes are covered according to HIPAA regulation) (From Siemionow et al.,[5] with permission)

Fig. 7.4 Appearance of the infraorbital nerve without osteotomy (From Siemionow et al.,[6] with permission)

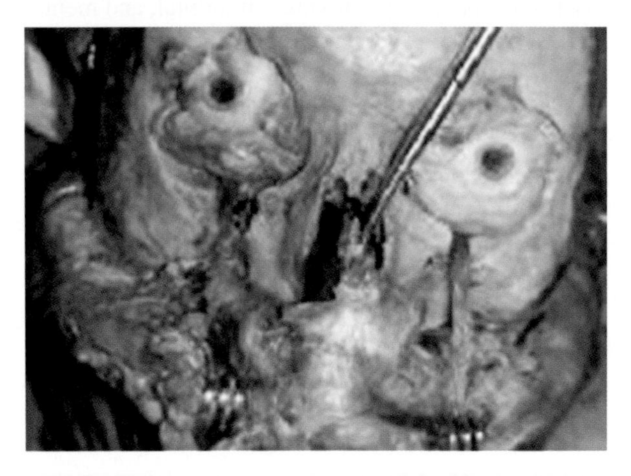

Fig. 7.5 Extended length of infraorbital nerve was achieved after osteotomizing the infraorbital rim and infraorbital groove (From Siemionow et al.,[6] with permission)

The first face transplant cadaver study published from an outside institution was written by Baccarani et al. at Duke University in November 2006.[10] In this paper, the authors summarized two new mock face transplant techniques. The first entailed a myocutanoeus alloflap raised via a dissection in the subgaleal, sub-SMAS, and subplatysmal planes. The second was dissected in a subperiosteal plane thereby allowing access for LeFort III-based osteotomies and harvest of an osteocutaneous alloflap.[11] Both flaps, however, were preserved on identical, bilateral pedicles consisting of the external carotid arteries and external jugular/facial veins. The value of this article was that it, for the first time, it identified the option of including an entire maxilla for the purpose of midface allotransplantation, similar to the ones later used by both Cleveland and Boston.[12]

In January 2007, a second cadaver study followed from Baccarani and colleagues.[13] In this study, four facial allografts were harvested as previously described in a subgaleal, sub-SMAS, and subplatysma plane. This allowed exchange of all four faces onto four different facial skeletons. By using photography and CT scans, the authors concluded that certain criteria such as (1) gender, (2) head/skeletal size, (3) soft-tissue features, and (4) skin color/texture were all important in preserving one's facial morphology posttransplant. This article highlighted the ongoing dilemma of aesthetically matching suitable recipients to appropriate donors. Unfortunately, in our experience, the great mismatch in terms of the limited number of consented face donors falls short in comparison to the amount of individuals desiring face transplantation. Therefore, preliminary screening tools such as the FACES score by Gordon et al., for optimizing limited resources and establishing an institutional registry in the future, have since been described.[14]

Later that same year, Wang et al. from China described a cadaver study comparing two facial flap-harvesting techniques for allotransplantation.[15] In an effort to shorten donor-graft harvest time and reduce warm-ischemia times, the authors transplanted six mock sub-SMAS myocutaneous alloflaps. However, each group varied in pedicle design including superficial temporal artery (STA), facial artery (FA), and external carotid artery (ECA). Total harvest times for STA and FA flaps were significantly less by almost 50% (113 min vs. 232 min, respectively). Therefore, the authors concluded that if allograft recovery was to be complicated by operative time and/or warm ischemia that the surgeon should attempt to avoid basing their dissection off of the ECA. We have since learned from various other studies that maxilla perfusion is dependent on the donor ECA pedicle design, and therefore, for those osteocutaneous alloflaps, avoiding the prolonged ECA donor dissection is unavoidable.[16,17]

Soon following, the first cadaver study article describing lower face allografts for transplantation was published by Meningaud and colleagues in July 2009.[16] Using 20 cadavers, the authors successfully demonstrated alloreconstruction of the lower two-thirds of one's face, and then followed this with its novel clinical application in January 2007. This alloflap design included perioral musculature, facial nerves (VII, V2, V3), and parotid glands, and was reported to be vascularized completely (on intraoperative inspection) just on a single facial artery pedicle. Graft harvest in the lab averaged 3 h, and then was prolonged to 4 h during the real transplant. All nerves, both sensory and motor, were repaired. Skin sensation was present at 4 months and voluntary motor function was observed at 5 months postoperatively. These timelines for sensorimotor recovery following face transplantation have remained relatively consistent, as reviewed by Gordon et al. in 2009.[12]

In July 2008, the first cadaver study describing LeFort II-based osteocutaneous face transplantation was written by Yazici et al.[7] Using six cadavers, in combination with MRI and angiography, the authors defined the vascular territories in the masticatory space along with the internal maxillary artery's anterior pterygomaxillary extension. This information is exceptionally important for the face transplant surgeons unfamiliar with maxillofacial surgery. Similarly designed vascular studies on LeFort III osteocutaneous alloflaps followed in June 2009 (Banks) and February 2010 (Pomahac).[17,18] Banks and colleagues are credited with making the distinction that bilateral vascular pedicles, including the internal maxillary arteries, are necessary for those facial alloflaps containing maxillary components.

7.4 Tailored Mock Cadaver Dissections

Since our protocol's inception, it quickly became obvious that each facial deformity differed in terms of size, depth, and extent.[14] Therefore, we soon realized the need for an individualized, tailored cadaver study for patients who became fully approved and cleared unanimously by the face transplant team. With this in mind, our goal was to design a cadaver model to the specific dimensions of our identified transplant candidate, in an effort to predict technical challenges and to avoid intraoperative delay (Figs. 7.6 and 7.7).

Three fresh cadavers were used in this study. A midface alloflap (skin, subcutaneous tissue, muscle, and bone) was harvested in all cadavers using a LeFort III basis, but each one involved varying amounts of maxilla. We found it advantageous to use a three-dimensional stereolithographic model during the dissection, allowing the recovery team full anticipation of the recipient's defect. Total harvest time averaged three hours [range = 2.5–3.5 h]. Flap dimensions were approximately 27 cm (wide) × 14 cm (tall) × 7 cm (deep). All three were preserved on external carotid artery and internal jugular/facial vein pedicles. The facial nerve was dissected within the parotid bilaterally and preserved with adequate length for transplantation. Each of the three flaps was placed upon the stereolithographic model in order to analyze cohesiveness (Fig. 7.8).

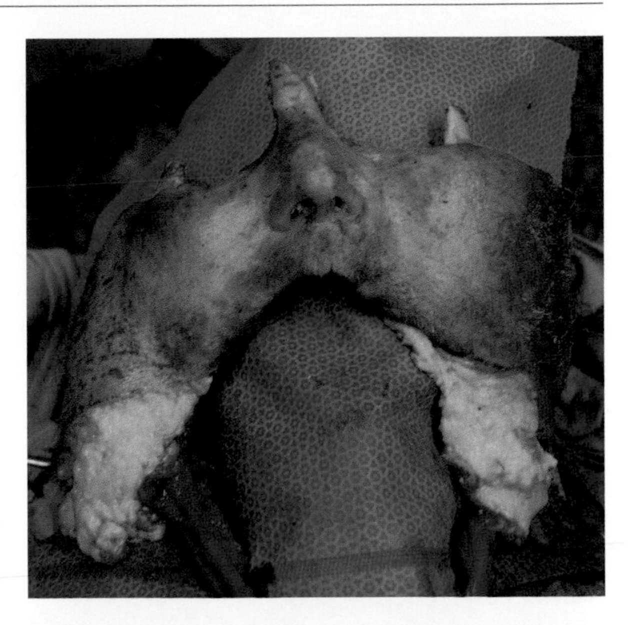

Fig. 7.6 Frontal view of mid-face alloflap customized to the specifics of our patient's reconstructive needs

Fig. 7.7 Posterior view of mid-face transplant graft

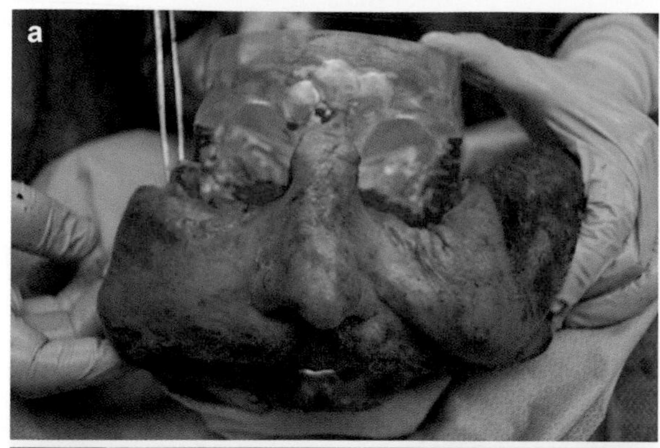

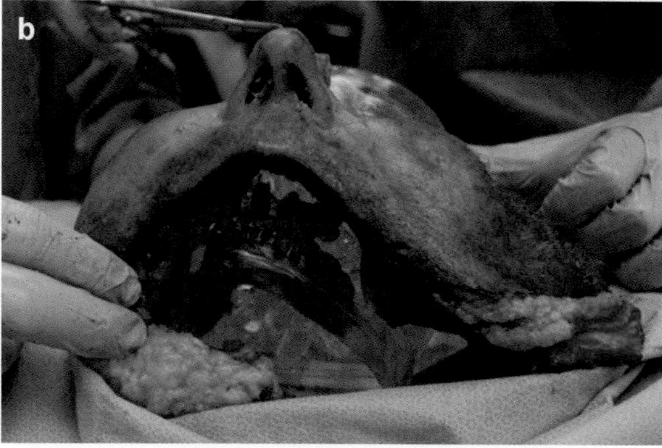

Fig. 7.8 Facial alloflap placed upon the three-dimensional model. (**a**) Frontal view. (**b**) Worm's eye view

7.5 Conclusions

In conclusion, a collaborative effort from countries such as USA, France, and China have contributed to the evolution of cadaver face transplant study. From the very first rat hindlimb allograft experiments showing successful tolerance with immunosuppression,[19] the clinical progression of facial CTA has remained encouraging and at this point has been well above what was originally expected.[20] In addition, based on the three anatomical dissections presented here, the authors introduce for the first time a new "tailored" technique for midface allotransplantation. This model, however, is only needed once a specific candidate has been chosen and confirmed unanimously by the team. In the meantime, we strongly suggest that all institutions preparing for face transplantation continue with generic mock face transplants.[21-23]

References

1. Okie S. Facial transplantation: brave new face. *N Engl J Med*. 2006;2:889-894.
2. Dubernard JM, Lengele B, Morelon E, et al. Outcomes 18 months after the first human partial face transplantation. *N Engl J Med*. 2007;357:2451.
3. Siemionow M, Unal S, Agaoglu G, Sari A. A cadaver study in preparation for facial allograft transplantation in humans: part I. What are alternative sources for total facial defect coverage? *Plast Reconstr Surg*. 2006;117:864-872.
4. Siemionow M, Agaoglu G, Unal S. A cadaver study in preparation for facial allograft transplantation in humans: part II. *Plast Reconstr Surg*. 2006;117:876-885.
5. Siemionow M, Agaoglu G. The issue of "facial appearance and identity transfer" after mock transplantation: a cadaver study in preparation for facial allograft transplantation in humans. *J Reconstr Microsurg*. 2006;22:329-334.
6. Siemionow M, Papay F, Kulahci Y, et al. Coronal-posterior approach for face/scalp flap harvesting in preparation for face transplantation. *J Reconstr Microsurg*. 2006;22: 399-405.
7. Yazici I, Cavusoglu T, Comert A, et al. Maxilla allograft for transplantation: an anatomical study. *Ann Plast Surg*. 2008;61:105-113.
8. Siemionow M, Papay F, Alam D, et al. First U.S. near-total human face transplantation – a paradigm shift for massive facial injuries. *Lancet*. 2009;374:203-209.
9. Siemionow M, Papay F, Djohan R, et al. First U.S. near-total human face transplantation – a paradigm shift for massive complex injuries. *Plast Reconstr Surg*. 2010;125: 111-122.
10. Baccarani A, Follmar KE, Baumeister SP, Marcus JR, Erdmann D, Levin LS. Technical and anatomical considerations of face harvest in face transplantation. *Ann Plast Surg*. 2006;57:483-488.
11. Follmar KE, Baccarani A, Das RR, et al. Osteocutaneous face transplantation. *J Plast Reconstr Aesthet Surg*. 2008; 61:518-524.
12. Gordon CR, Siemionow M, Papay F, et al. The world's experience with facial transplantation: what have we learned thus far? *Ann Plast Surg*. 2009;63:121-127.
13. Baccarani A, Follmar KE, Das RR, et al. A pilot study in sub-SMAS face transplantation: defining donor compatibility and assessing outcomes in a cadaver model. *Plast Reconstr Surg*. 2007;119:121-129.
14. Gordon CR, Siemionow M, Coffman K, et al. The Cleveland Clinic FACES score: a preliminary assessment tool for identifying the optimal face transplant candidate. *J Craniofac Surg*. 2009;20:1969-1974.
15. Wang HY, Li QF, Zheng SW, et al. Cadaveric comparison of two facial flap-harvesting techniques for alloplastic facial transplantation. *J Plast Reconstr Aesthet Surg*. 2007;60: 1175-1181.
16. Meningaud JP, Paraskevas A, Ingallina F, Bouhana E, Lantieri L. Face transplant graft procurement: a preclinical and clinical study. *Plast Reconstr Surg*. 2008;122: 1383-1389.

17. Banks ND, Hui-Chou HG, Tripathi S, et al. An anatomical study of external carotid artery vascular territories in face and midface flaps for transplantation. *Plast Reconstr Surg*. 2009;123:1677-1687.
18. Pomahac B, Lengele B, Ridgway EB, et al. Vascular considerations in composite midfacial allotransplantation. *Plast Reconstr Surg*. 2010;125:517-522.
19. Gordon CR, Nazzal J, Lee WPA, Siemionow M, Matthews MS, Hewitt CW. From experimental rat hindlimb to clinical face composite tissue allotransplantation: historical background and current status. *Microsurgery*. 2006;26:566-572.
20. Gordon CR, Siemionow M, Zins J. Composite tissue allotransplantation: a proposed classification system based on relative complexity. *Transplant Proc*. 2009;41:481-484.
21. Siemionow M, Gordon CR. Overview of guidelines for establishing a face transplant program: a work in progress. *Am J Transplant*. 2010;10:1290-1296.
22. Siemionow M, Zor F, Gordon CR. Face, upper extremity, and concomitant transplantation: potential concerns and challenges ahead. *Plast Reconstr Surg*. 2010;126:308-315.
23. Siemionow M, Gordon CR. IRB-based Recommendations for Medical Institutions Pursuing Protocol Approval for Face Transplantation. *Plast Reconstr Surg*. 2010;126(4):1232-9.

Part II

Clinical Aspects in Preparation for Face Transplantation

8 Guidelines and Technical Aspects in Face Composite Tissue Transplantation

Frank Papay

Contents

F. Papay
Dermatology and Plastic Surgery Institute,
Cleveland Clinic, Cleveland, OH, USA
e-mail: papayf@ccf.org

Abstract Face transplantation combines the surgical complexity of multiple subspecialties that require an organized systematic approach to the evaluation of potential face transplant candidates, donors and surgical team members. Once established, the face transplant team should comprise of the scientific, clinical and surgical expertise than can orchestrate a successful operation on a well chosen patient. Our efforts at the Cleveland Clinic's Department of Plastic Surgery first face transplantation procedure revealed guidelines and technical aspects in facial composite tissue transplantation that can guide other institutions in their efforts to expand composite tissue transplantation as not just a state-of-the-art surgery but rather the future standard of care for those debilitated patients that have exhausted all facets of facial reconstruction. This chapter is meant as only a guideline since each subsequent face transplantation patient will require a custom made surgical approach to both the donor face delivery and transplant procedure.

8.1 Introduction

Patients with large tissue defects resulting from trauma, tumor ablation, burns, or congenital defects number in the millions in the USA. A survey conducted by Langer and Vacanti in 1993 revealed that more than seven million people need tissue (skin, nerves, bone, cartilage, tendon, or ligaments) for some type of reconstruction each year.[1-5] This figure is considered to be more than double the number of solid organ transplants needed.

The vast majority of patients requiring these tissues are currently treated by reconstructive procedures that utilize autologous tissues and/or prosthetic alloplastic

M.Z. Siemionow (ed.), *The Know-How of Face Transplantation*,
DOI: 10.1007/978-0-85729-253-7_8,

materials. However, the best possible outcomes are achieved when these defects are repaired using native tissue, i.e., the same tissue lost to the trauma or disease. This is possible in cases of amputation due to trauma where the original tissue/body part is recovered from the accident scene and is reattached immediately. In such cases, good recovery of function and aesthetic appearance following reconstruction is expected if: (1) the amputated tissue is not damaged from the accident or handling, (2) the time elapsed between amputation and reattachment is short, and (3) the amputated tissue is cooled during the time it is ischemic (between amputation and reattachment). Unfortunately, in the majority of cases the original autogenous tissues are not available to be used for reconstructing these defects. This is because more often than not the above three criteria are not met or the original body part is destroyed (invaded by cancer, crushed by trauma beyond use, or severely burned) or did not exist in the first place (congenital defects). In the absence of the native body part/tissue (the majority of cases), surgeons must reconstruct these defects using autologous tissues from another anatomic site of the body and/or prosthetic materials. These reconstructive procedures consist of transferring one or combinations of several tissues from another part of the patients' own body to repair these defects. In the event the patients' own tissues are not sufficient to reconstruct a given defect, a variety of different prosthetic materials are custom made to mimic the form that is lost. Though these procedures have advanced a great deal over the years, they are still plagued by many drawbacks and their functional and aesthetic outcomes still do not come close to those achieved by procedures that use the native body part/tissue for reconstruction.

Limitations of currently used reconstructive procedures include poor functional and aesthetic outcomes; multiple surgical procedures to revise the definitive first surgery; prolonged rehabilitation resulting in patients not returning to work or normal life and becoming dependent on family members and the health care system for care; high costs of multiple surgeries/hospitalizations; donor site morbidity resulting from use of autologous tissues and postoperative complications associated with implanted prosthetic materials "foreign body" (infection, altered healing, rejection, etc.). One potential solution to this great need for native tissue is composite tissue allotransplantation (CTA). As solid organ transplantation revolutionized the treatment of terminal organ failure, CTA could fulfill the existing great need for native tissues to reconstruct large tissue defects.

Although in a few isolated clinical cases, tissues/structures (nerves, bone, joint, muscle, larynx, entire hands, and partial face) have been transplanted from donors,[4] CTA has not yet gained widespread clinical use. This can be attributed to one main reason: the risks posed by the immunosuppressive drugs required to prevent rejection are considered by many to be too high a price to pay for the benefits a patient would receive from one of these nonlife-threatening reconstructive procedures using a CTA.

Even though much progress has been made using autologous tissues and prosthetic materials, still the best functional and aesthetic outcomes are achieved when tissue(s) whose native form and function are most similar to the missing tissue(s) are used. Both functionally and aesthetically, the patient could return to work and lead a normal life in a very short time. The ability to use CTA in complex reconstructions eliminates most of the above listed limitations (poor functional and aesthetic outcomes, multiple revision surgeries, prolonged rehabilitation, high costs, donor site morbidity, and foreign body associated complications) and in doing so revolutionizes the field of reconstructive surgery.

At the time of this writing, ten face composite tissue allograft transplants have been performed throughout the world.[1-7] With the evolution of continued improvements in immunomodulation and refinements of transplantation surgical technique, the clinical applicability of face transplantation has arrived with the resolution of the ethical scrutiny of public and scientific debate. Currently there are a few academic medical institutions that are preparing efforts to perform partial or complete face allotransplantation. This chapter is intended to assist in such preparation by providing surgical guidelines in addition to specific technical details that have been learned from previous face transplants.

8.2 The Multidisciplinary Transplant Team

Face transplantation has inherent complexities that require an organized systematic multidisciplinary approach. Facial transplantation should therefore only be attempted at clinical institutions that possess the scientific, clinical, and surgical expertise that can be orchestrated in a specialized multidisciplinary team. Each institution should also establish and adhere to a strict clinical protocol that can be adapted to the face transplant recipient's particular deficits.

The face transplant surgical team should include individuals whose expertise involve the reconstruction of facial tissues, the understanding of the "micro-anatomy" of the skeletal and soft tissues of the head and neck in addition to those surgeons who have the experience of microvascular anastomosis, microscopic repair of nerves, and craniofacial/maxillofacial osteosynthesis.

Team members must truly act as a team with ego and personal ambitions put aside. A leader or conductor responsible for the orchestration of the surgical planning and outcome should be designated beforehand with all surgical team members agreeing to comply with the teams decisions both inside and outside the operative theater. Putting the patient's safety and surgical outcome above all personal incentives is a prerequisite of any team member.

8.3 Face Transplant Preoperative Anatomical Review

The head and neck contains an abundant blood supply that is accessible and often redundant through a combination of head and neck vascular arcades.[8] Facial reconstruction can be divided into three regions: (1) upper third (scalp and skull), (2) middle third (mid-face), and (3) lower third (mandibular). By considering the relevant anatomy and function of each facial subunit to be reconstructed, the ideal type of free-tissue allograft transfer can be designed containing various arrangements of soft tissue, fascia, and bone.[9-13] In addition, facial transplantation may also involve combinations of tissues requiring unique vascular considerations.

Access to donor and recipient vessels requires proper and careful preoperative radiological analysis and intraoperative dissection. Preoperative planning that details both the donor and recipient personal surgical and medical history may provide essential clues that will help in the planning of the surgical technical steps and allow certain critical operative periods in which the surgical transplantation procedure can be altered during the case or abandoned altogether.

In each particular facial transplantation case, the donor vessels are dissected beginning at the common carotid artery extending superiorly under the mandible to reach the carotid bifurcation and facial artery. In certain particular cases, exposure of the facial vessels can be aided by a mandibular osteotomy. If necessary, a preharvest donor and/or recipient surgical elective tracheostomy can be formed to allow full access to the neck and facial anatomy without being compromised technically by the endotracheal intubated airway applied through the nose or mouth.

Previous studies such as that by Takamatsu et al. discussed the selection of recipient vessels in head and neck microsurgical cases.[14] These authors recommend the superficial temporal vessels for upper third (scalp and skull) reconstruction, the facial and superficial temporal vessels for middle third (facial) defects, and the ipsilateral neck vasculature for lower third (mandibular) reconstruction. When the optimum vessels are not available, alternative recipients vessels may be identified such as: (1) adjacent small vessels in the area of first-choice vessels, (2) major neck vessels (i.e., external carotid and internal jugular vein), and (3) distant vessels (i.e., thyrocervical trunk). This alternative donor vascular selection may be true in facial transplantation candidates who have undergone previous microvascular reconstructive procedures obviating many first-choice vessels because of previous surgical scarring leading to lack of any identifiable vascular anatomy or simply because they primary selected vessels were previously ligated. Preoperative radiological imaging and intraoperative Doppler and/or color Doppler ultrasound may provide information on vessel patency that can save valuable time intraoperatively and prevent technical failures postoperatively.

Depending on which types of tissues are included in the facial allograft, the vascular supply to the flap may vary. Studies of the facial angiosomes suggest that multiple arteries would be necessary to adequately perfuse an entire panfacial flap; however, clinical and experimental data have suggested otherwise.[8] In the particular case at the Cleveland Clinic, the majority of the facial skin, glandular tissue, facial muscular, and a modified Lefort III anterior maxillary bony structure was supplied by the bilateral facial arteries alone. This case supports the concept of a periostial blood supply to the facial maxillary bone is enough to allow bony survival. Other case reports have also demonstrated successful perfusion of large segments of the facial soft tissues and scalp on a single vessel.[15] Meningaud also confirmed that complete revascularization of the soft tissues from the lower two-thirds of the face was possible by a single facial artery coaptation.[16] For flaps that intend to include soft tissues of the lateral cheek, ear, scalp, and forehead, inclusion of both the superficial temporal through the internal maxillary artery and facial artery branches would be necessary, with the external carotid as the source vessel.[8]

Although understanding of the perfusion of facial soft tissues has been aided by replantation in traumatic cases, there is a paucity of information available about the blood supply necessary to support composite facial flaps. Successful facial transplantation of a series of recent facial transplant cases continues to illustrate that the facial artery alone can adequately supply both the overlying soft tissues and bony elements of facial allografts that include the maxilla and zygoma.[7] These findings are further supported by anatomical studies showing that maxillary segments receive blood supply from the ascending palatine branch of the facial artery. This concept also shows that the blood supply to the bone is derived from periosteal rather then endosteal vessels.[17,18] Therefore during allograft harvest, it is critical that the periosteum remains attached to the maxilla for bony viability. In a case at the Cleveland Clinic, postoperative nasal endoscopy confirmed healthy nasal and sinus mucosa from the donor's allograft, which is further evidence of adequate tissue viability of the paranasal and sinus tissues based on perfusion by bilateral facial arteries.

Lastly, in cases of allografts that include the maxilla without overlying soft tissues, the facial artery is not the vessel of choice. Rather, such a flap would be dependent on endosteal supply from the bilateral internal maxillary arteries as source vessels because the periosteum is excluded and midline crossover at the palate is poor.[8,13,19] Regardless of the vessel chosen to supply a flap, if possible, the vessel should be dissected to a more proximal level where the caliber is greater and chances of technical failure are reduced. In the Cleveland Clinic case, the left and right donor facial artery supplied the flap, but it was dissected near the level of the external carotid origin where the anastomosis was performed.

Another technical consideration in face allograft transplantation is to perform osteotomies as one of the last technical steps in the flap harvest. Ligation of the distal external carotid artery before the zygoma osteotomies avoids potential hemorrhage from the 15 branches of the internal maxillary artery within the sphenopalatine fissure. This is critical to ensure that patients remain hemodynamically stable while solid organs are harvested by other transplant teams.

Another surgical consideration in face allograft harvesting is the adequate exposure of the lower orbital floor in preparation for a Lefort III osteotomy by performing bilateral orbital exonerations. The surgical removal of the orbital contents allows open exposure of the orbital floor, lateral and medial orbital walls that allow direct visualization for placement of a reciprocating saw, drill, or osteotomes for release of the maxilla. As described above, evidence that the periosteal blood supply to the maxilla is sufficient for maxillary bony survival allows the craniofacial surgeon to perform a "veneer" approach to facial bone harvesting of the maxilla. The term "veneer" is used since the anterior, medial, and lateral maxillary walls are harvested leaving behind the posterior maxillary wall so as not to disrupt the sphenopalatine (pterygopalatine) branches of the internal maxillary arteries. If disrupted, the author proposed that there may be uncontrolled posterior to the maxillary bleeding, which may induce an increased incidence of immediate postoperative morbidity. In addition, the "veneer" approach to maxillary bony harvesting allowed inset into the recipient's maxillary bed in an anterior–posterior position without the need to further remove scarred tissue near the middle cranial fossa skull base. Removal of tissue in this anatomic region can be technically difficult and prone with large vessels (carotid branches) and cranial nerve formina.

8.4 Donor Selection

Idealistically, the facial transplant donor and recipient are matched based on similar blood type, race, gender, bone size, age, and skin color/tone. After locating a suitable donor (long travel time must be accounted for with regards to total ischemia time), matching tests (lymphocytotoxic, HLA, B- and T-cell) are performed.

8.5 Preoperative Work-up

All candidates on the transplant list should undergo preoperative testing to analyze exact specifications necessary for allotransplant recovery. These specifications include

CT-scan (head/neck, chest, abdomen/pelvis), CT angiogram (vessel mapping, arterial and venous phase), MRI (to rule out preexisting soft-tissue abnormalities), Nerve sensory testing (Trigeminal nerve distribution), and EMG (Facial nerve function/status). Using these findings, an individualized protocol is then constructed for each face transplant patient and must be detailed accordingly.

In addition, each patient undergoes a complete dental and oropharyngeal examination to rule out any periodontal disease, impending dental abscesses or oropharyngeal carcinomas. All patients above 50 years are required to have preoperative upper/lower gastrointestinal endoscopy (to rule out undiagnosed polyps/carcinoma), and women above 40 years must be current with their mammography. A potential face transplant recipient in the perioperative period found to have an undiagnosed carcinoma (i.e., breast and gastrointestinal) would be severely detrimental due to the effects of chronic immunomodulation.

8.6 Donor Allograft Recovery

At the start of the facial allotransplant recovery, a phone-call algorithm is initiated by the face transplant coordinator. This process allows notification to the rest of the surgical team, anesthesia, tissue typing lab, public relations, and ICU nursing staff.

After all parties have been made aware, the face recovery team begins working in concert with all other involved transplant teams. Exact details, with regard to the organ/hand harvest sequence and the clamping of the aorta, need to be discussed preoperatively. In 2007, it was reported that 50% of the transplanted hands had been harvested in the operating room prior to solid organ retrieval (i.e., liver and kidneys), and that 50% had been recovered after the other organs.[20] The Louisville group recommends dissecting the hand last in an unstable patient (after all other organs have been dissected) and then retrieving the hand allotransplant prior to cross-clamping the aorta.[8,21,22] University of Wisconsin (UW) solution is used primarily for cold flush and facial/scalp preservation. The facial/scalp allotransplant is wrapped in moist sterile gauze for transport. The allotransplant should be kept on ice (≈ 4°C) in a secure, sterile container and then immediately transported to the hospital. While in route, it should be infused constantly with cold UW solution using either a facial or external carotid artery (ECA) cannula. Cold ischemia times for successful hand allotransplants have ranged from 30 min to 13 h.[20-29] Due to the paucity of data for face transplants the cold ischemia time is relatively unknown. At the end of the recovery operation, a facial prosthesis is placed on the donor in the event of an open-casket funeral. Obviously, this type of operation mandates a large, multifaceted two-team approach, with one team assigned to the recovery (or possibly two teams in the case of complete facial/scalp), with an additional team simultaneously preparing the recipient.[6,8,9]

In addition, as performed by the French teams, a sentinel vascularized myocutaneous alloflap should also be recovered and used for postoperative monitoring.[1-3,6] This can be performed by taking an additional aesthetic unit within the posterior facial neck region in which a biopsy can be performed in an inconspicuous region. In contrast to a hand transplant, frequent biopsies are aesthetically displeasing and should be performed in an as-needed basis. This sentinel flap, transplanted concomitantly to the facial allotransplant recipient, will allow frequent biopsies of histological value and, at the same time, will preserve the aesthetic value of the face transplant.

8.7 Face Transplant Procedure

A second team simultaneously prepares the recipient's head/neck in order to minimize total ischemia time. The recipient's head/neck region is dissected carefully, identifying (and labeling) viable bone, nerve, tendon and vessel endpoints, excising excess skin, and creating skin flaps for closure (Figs. 8.1–8.3). All pertinent endpoints are cut to predetermined lengths. Prior to implantation, all UW fluid is flushed and replaced with cool lactated ringers (LR) solution.

8.8 Donor Facial Bone Delivery and Recipient Osteosynthesis

As similar to an extremity replantation, a face transplant consisting of bone begins with osteosynthesis. Cranial osteosynthesis may involve the maxilla, orbital rim, mandible, zygomatic-maxillary complex, and/or cranial-vault, depending on the patient's extent of

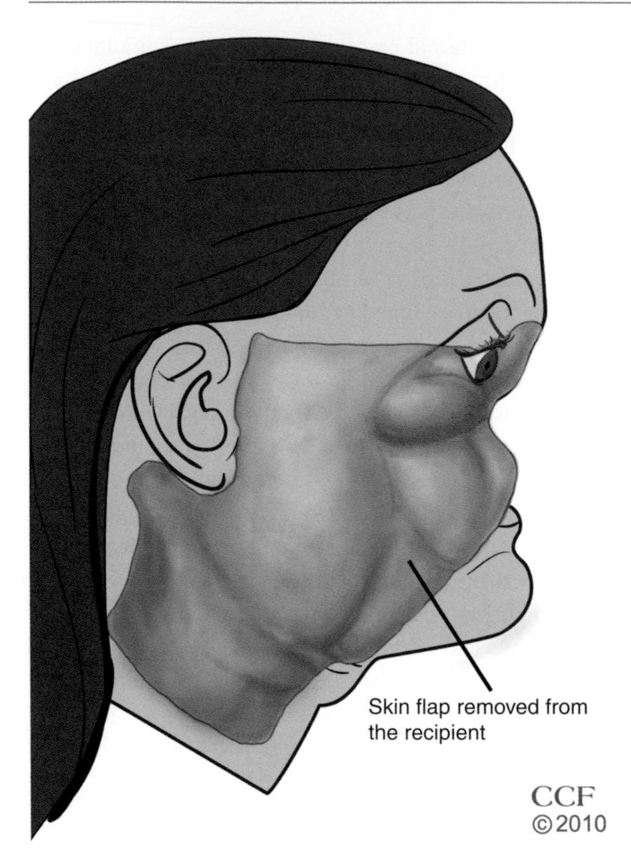

Fig. 8.1 Scarred skin removed from the recipient site

injury. The bone part is affixed with multiple plates and wires, as needed (Fig. 8.4). In certain cases, the width of the donor facial bones may be wider or narrower as compared to the recipient's. In the case of the Cleveland Clinic's first face transplant, the donor's zygomatic arch was anatomically wider and needed surgical modification to allow adequate apposition with the recipient's zygomatic arch. Careful attention is needed to allow the *X*-, *Y*-, and *Z*-axis to be accurately in position for facial width and projection in addition to appropriate dental occlusion. This can be somewhat challenging in light of poor dental hygiene and periodontal disease of the donor or recipient. In preparation for CTA facial transplantation, preoperative dental preparation and intraoperative dental extraction of the donor may be necessary to prevent any infectious complication after immunomodulation has been initiated for the recipient. If dental extraction is necessary for the donor's maxilla or mandible, preserving some viable dentition till the jaws are anatomically aligned will assist the surgeon in maxillary or mandibular positioning.

Attention to the sinus mucosa is also imperative (Fig. 8.5). If the sinus tracts do not line up (i.e., the frontal sinus of the recipient and frontal sinus duct of the donor), a potential sinus aeration and drainage obstruction could

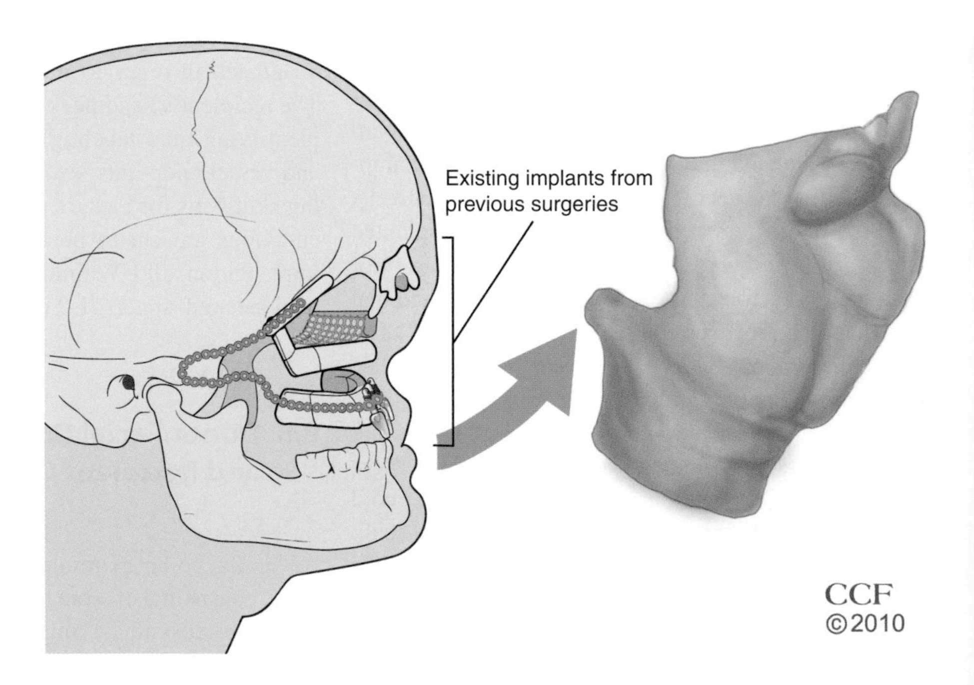

Fig. 8.2 Existing implants from the previous surgeries are shown

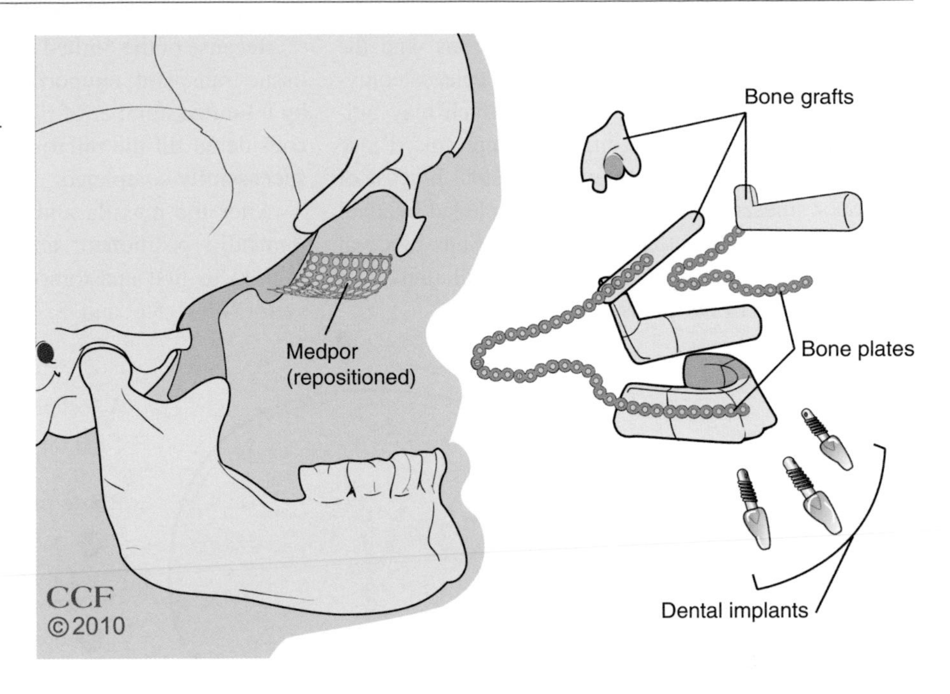

Fig. 8.3 Detailed view of previous bony reconstructions: nasal bone grafts and a free fibula flap with osteointegrated dental implants

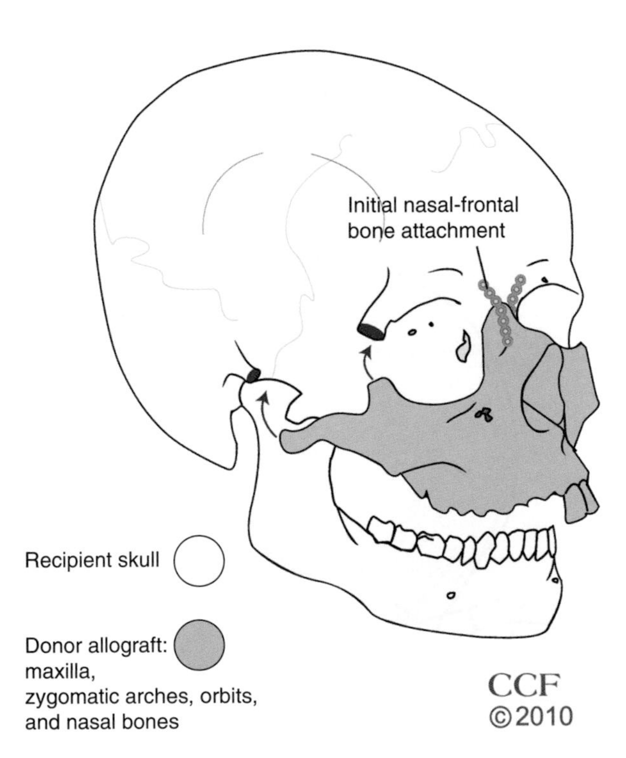

Fig. 8.4 Bone component of the facial allograft, including maxilla, zygomatic arches, orbits, and nasal bones is shown

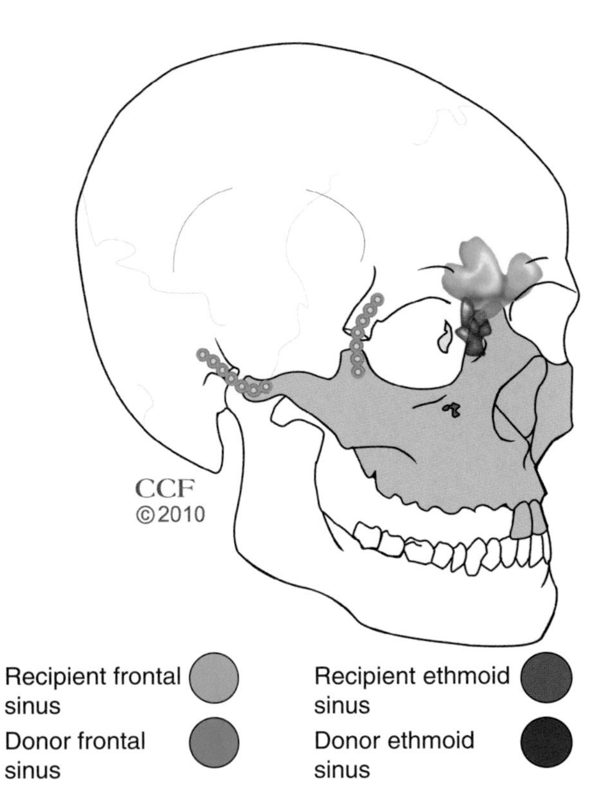

Fig. 8.5 Mucosa of the donor and recipient sinuses were removed to avoid postoperative infective complications

occur leading to mucoceles, sinus infections, and the infectious morbidities that may ensue (pyocele, osteomyelitis, meningitis, brain abscess, etc.), which may ultimately lead to sepsis in an immunocompromised face transplant patient. With this in mind, the sinus mucosa of the donor sinuses were stripped. This included bilateral donor sinus ethmoidectomies, maxillary sinus mucosal removal, and ensured patency of the frontal sinus duct (donor) into the frontal sinus (recipient).

Because of the limited ischemic time of a composite tissue transplant, temporizing the bony osteosynthesis by a limited number of plates and screws may also be considered till the microscopic anastomosis has been successfully completed.

After the maxilla and/or mandible has been anatomically positioned, an intraoperative microscope is used to first anastomose the arteries, followed by veins (Figs. 8.6 and 8.7). In bilateral facial artery,

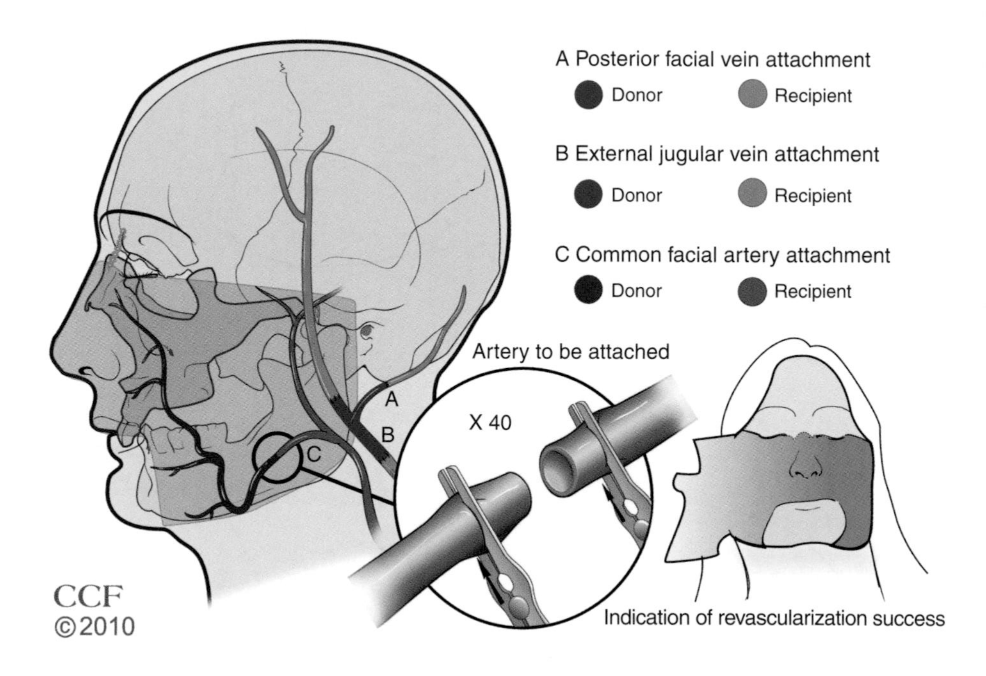

Fig. 8.6 Arterial and venous anastomosis are shown (*left side*)

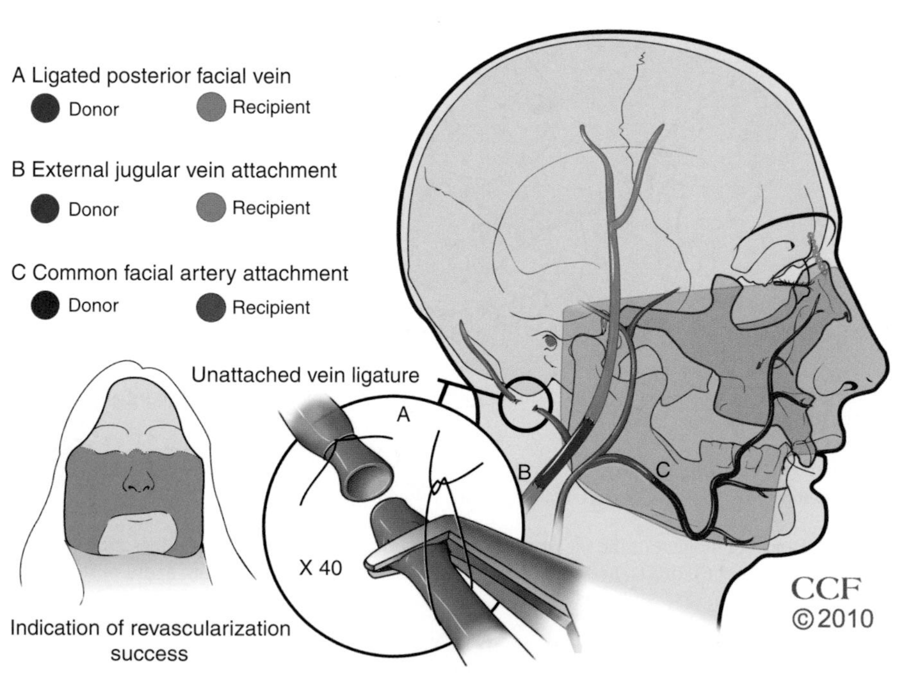

Fig. 8.7 Arterial and venous anastomosis are shown (*right side*): the posterior facial vein was ligated on the right side

anastomosis microsurgical clamp release of one side may be beneficial before the contralateral vessels are anastomosed to lessen the CTA transplant ischemic time, thereby allowing perfusion of the ipsilateral anastomosed facial tissue with some crossover to the contraleral side. This ischemic time is critical and should be monitored and announced throughout the case. Once perfusion of the CTA facial tissues has been completed and all remaining additional draining facial and neck veins are anastomosed, additional plates and screws can be used to finish the osteosynthesis of the maxilla and/or mandible (Fig. 8.8).

8.9 Donor to Recipient Motor and Sensory Nerve Attachment

After osteosynthesis and vascular perfusion has been secured, attention to other functional tissues can be addressed. Facial nerve and sensory nerve repairs are then initiated (Fig. 8.9). Donor nerve cable grafts can be utilized for such repairs. These donor grafts can be taken from the donor's hypoglossal, vagal, spinal accessory, or greater auricular nerves since they are in the relative surgical field and are adequate for cable graft

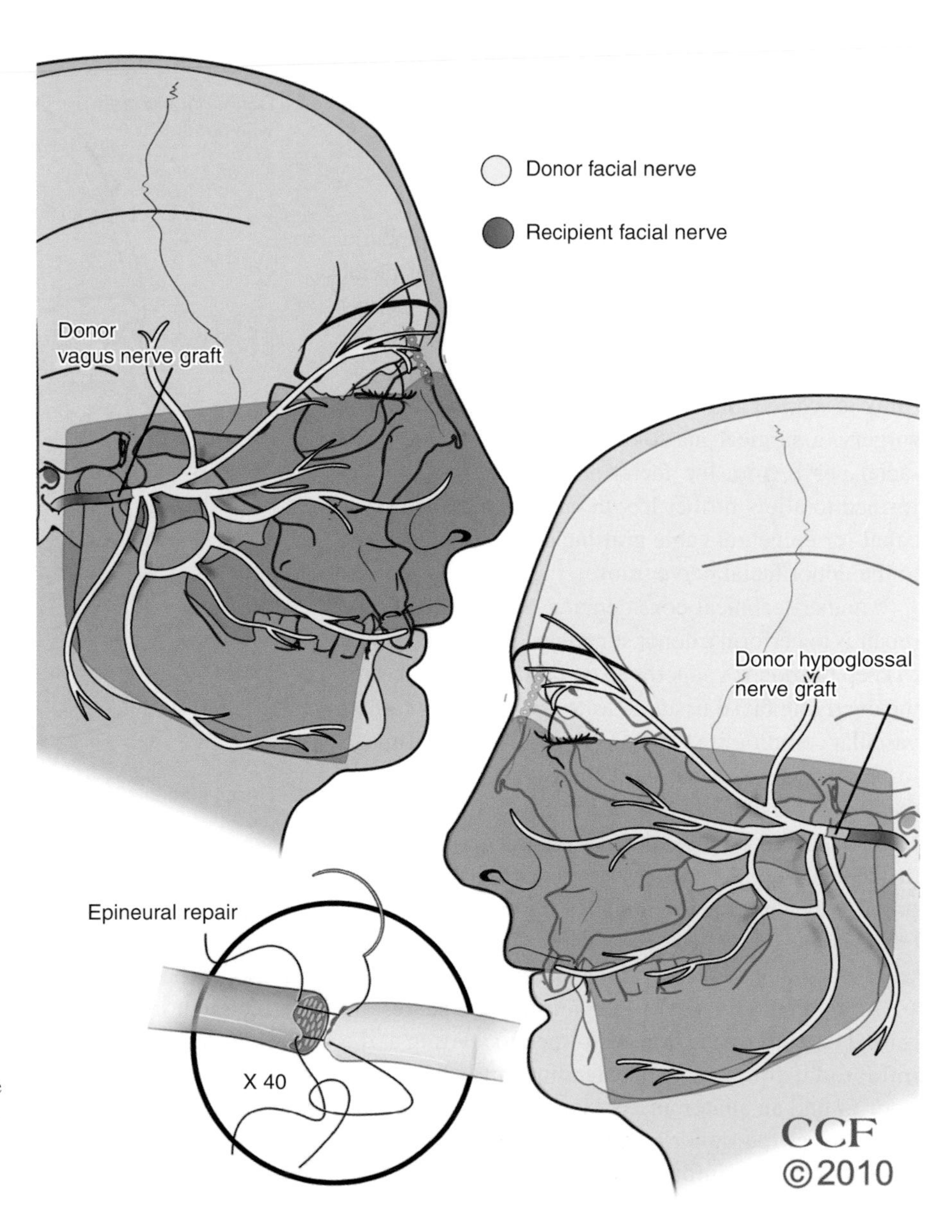

Fig. 8.8 Fixation of the bone component of the allograft is illustrated. The orbital floor was reconstructed with Medpor

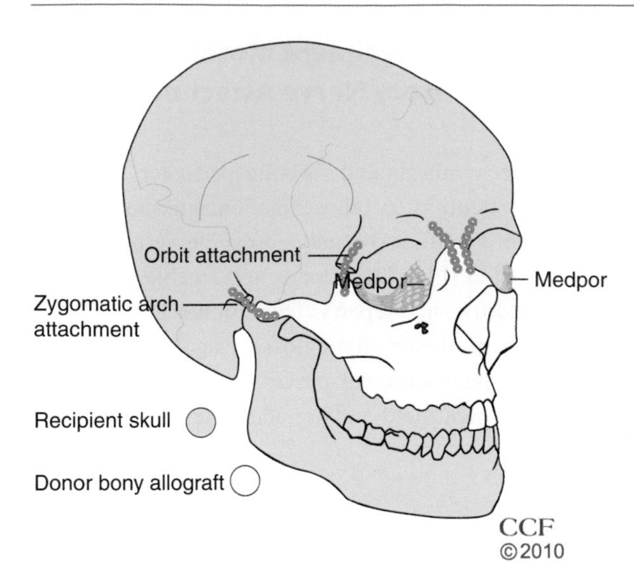

Fig. 8.9 Repair of the facial nerve required nerve grafts on both sides

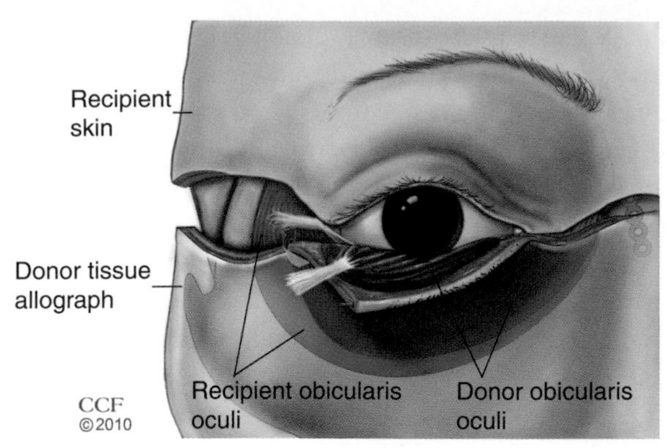

Fig. 8.10 Parotid gland was transferred with the flap to avoid damage to the facial nerve

diameter size and function. A microscopic epineural cable graft repair from the donor's functioning facial nerve trunk, mental nerves - inferior alveolar nerves V3, infraorbital nerves V2, and supraorbital nerves V1 can be performed if warranted. If the facial nerve trunk is scarred and damaged from previous trauma or surgery, a surgical mastoidectomy and release of the facial nerve from the facial nerve canal can be performed to allow further length outside the facial bony canal for epineural cable grafting of direct attachment to the donor facial nerve trunk.

Another technical consideration of the facial nerve repair is to perform a donor superficial parotidectomy or to keep the donor's superficial parotid gland attached to the overlying facial tissue so as to not allow any further vascular compromise to the overlying tissues or damage to the underlying facial nerve branches. In the case of the Cleveland Clinic's first face transplant, the surgical team elected to keep the donor's parotid tissue attached to the underlying facial nerve trunk (Fig. 8.10). This posed somewhat a problem in that the there appeared to be lower facial bulkiness and a wider facial gonial angle width due to the accessory donor parotid tissue.

To gain better access to the infraorbital nerves, an orbital osteotomy may be performed to starting at the infraorbital foramen and extending into the orbital floor to find an anatomically intact infraorbital nerve. In addition, the donor's infraorbital nerve may also be explored in a similar manner to allow an extended

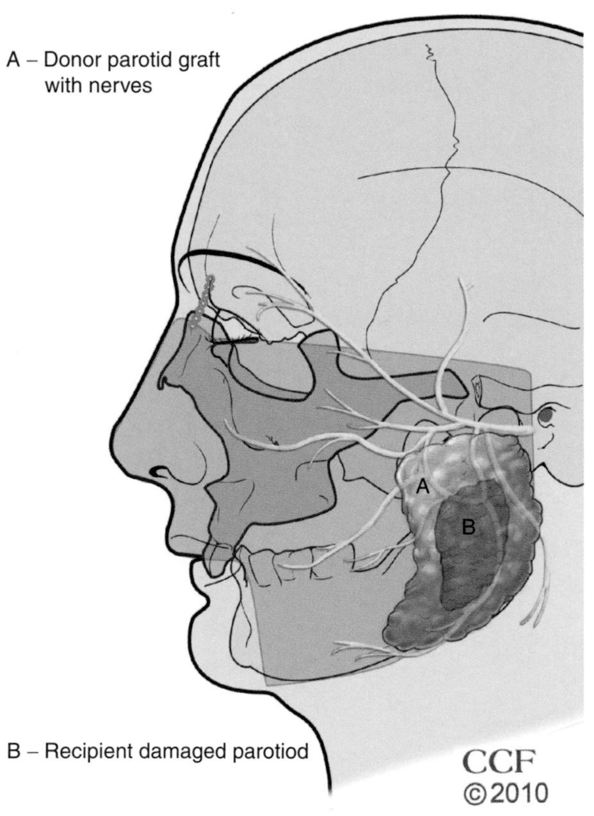

Fig. 8.11 Donor orbicularis oculi was fixed to the lateral canthus

release of the nerve for a tension-free attachment to the recipient infraorbital nerve bed.

Skin incisions are designed to assure a tension-free closure of donor/recipient skin edges, and if possible, lie within aesthetic unit junctions or within aesthetic acceptable areas (i.e., nasolabial folds) (Figs. 8.11–8.13). If incomplete skin coverage is obtained, a full-thickness

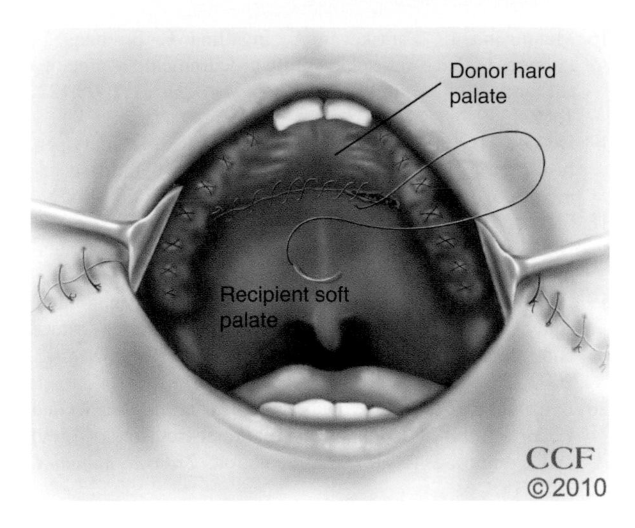

Fig. 8.12 Donor hard palate was repaired to recipient's soft palate

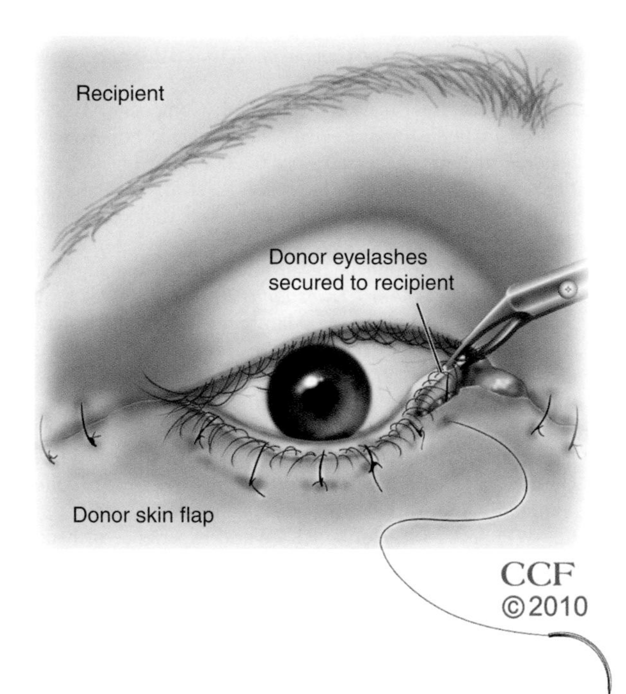

Fig. 8.13 Donor eyelashes were secured to recipient's

skin graft (FTSG) is used for closure. The transplanted face is then dressed carefully to allow physiotherapy to commence postoperatively. The main objective at this time is to minimize unwanted motion restriction, and to allow a patent airway and oral cavity for nutrition (or a PEG tube may be used if necessary).

8.10 Technical Failure of the Facial Transplantation

As with any microsurgical procedure, there is the possibility of clotting of the arteries or veins that have been anastomosed within the immediate postoperative period. If this happened, it would be apparent within hours. If rapid diagnosis of the problem were made, the anastomosis might be salvageable by reexploration and reanastomosis of the vessels. If that salvage surgery failed, the transplant would have to be removed. This is unlikely to occur after the second day following the transplant. If it does happen, it is classified as a technical failure and is quite distinct from immunologic rejection. Acute rejection of the transplant would be apparent generally within days or weeks and, unless reversed by medications, would lead to necrosis of the transplant tissue.

In the event of either a technical failure or acute rejection, the transplant would have to be removed. Because previous skin grafts would have been removed before the transplantation, the patient would have to have further skin grafts or microvascular place flaps of their own tissue to replace the failed rejected tissue, assuming that there were sufficient healthy donor skin sites. In this event, there is the possibility that there would be even more scarring than there was originally. The risk of free-tissue transfer failure for technical reasons in experienced units is considered to be less than 5%.

To the reconstructive surgeon, facial transplantation would constitute a major breakthrough in restoration of a quality of life to those whose faces have been destroyed by accident or tumor. It is therefore worthy of study.

The microsurgical skills and anatomic knowledge required for this procedure are already well established and well known. However, at present, this is not only a question of technical achievement. The immunosuppression needed, the psychological impact on the recipient and on the donor family, and the ethical concerns are the issues that must be considered. The need for lifetime immunosuppression carries considerable long-term risks that appear to outweigh any premature attempt to open the gates to facial transplantation. It remains unclear how acceptable and valid consent can be obtained from potential recipients, given the uncertainties about the risks and benefits that accompany the highly experimental character of the procedure.

References

1. Devauchelle B, Badet L, Lengele B, et al. First human face allograft: early report. *Lancet*. 2006;368:203-209.
2. Dubernard JM, Devauchelle B. Face transplantation. *Lancet*. 2008;372:603-604.
3. Dubernard JM, Lengele B, Morelon E, et al. Outcomes 18 months after the first human partial face transplantation. *N Engl J Med*. 2007;357:2451-2460.
4. Kanitakis J, Badet L, Petruzzo P, et al. Clinicopathologic monitoring of the skin and oral mucosa of the first human face allograft: report on the first eight months. *Transplantation*. 2006;82:1610-1615.
5. Lantieri L, Meningaud JP, Grimbert P, et al. Repair of the lower and middle parts of the face by composite tissue allotransplantation in a patient with massive plexiform neurofibroma: a 1-year follow-up study. *Lancet*. 2008;372:639-645.
6. Lengele BG, Testelin S, Dakpe S, et al. "Facial graft": about the first facial allotransplantation of composite tissues [in French]. *Ann Chir Plast Esthét*. 2007;52:475-484.
7. Siemionow M, Papay F, Alam D, et al. Near-total human face transplantation for a severely disfigured patient in the USA. *Lancet*. 2009;374:203-209.
8. Houseman ND, Taylor GI, Pan WR. The angiosomes of the head and neck: anatomic study and clinical applications. *Plast Reconstr Surg*. 2000;105:2287-2313.
9. Cordeiro PG, Disa JJ. Challenges in midface reconstruction. *Semin Surg Oncol*. 2000;19:218-225.
10. Cordeiro PG, Santamaria E. A classification system and algorithm for reconstruction of maxillectomy and midfacial defects. *Plast Reconstr Surg*. 2000;105:2331-2346. Discussion 2347-2348.
11. Parrett BM, Pomahac B, Orgill DP, Pribaz JJ. The role of free-tissue transfer for head and neck burn reconstruction. *Plast Reconstr Surg*. 2007;120:1871-1878.
12. Pribaz JJ, Weiss DD, Mulliken JB, Eriksson E. Prelaminated free flap reconstruction of complex central facial defects. *Plast Reconstr Surg*. 1999;104:357-365.
13. Yazici I, Cavusoglu T, Comert A, et al. Maxilla allograft for transplantation: an anatomical study. *Ann Plast Surg*. 2008;61:105-513.
14. Takamatsu A, Harashina T, Inoue T. Selection of appropriate recipient vessels in difficult, microsurgical head and neck reconstruction. *J Reconstr Microsurg*. 1996;12:499-507. Discussion 508-513.
15. Wilhelmi BJ, Kang RH, Movassaghi K, Ganchi PA, Lee WP. First successful replantation of face and scalp with single-artery repair: model for face and scalp transplantation. *Ann Plast Surg*. 2003;50:535-540.
16. Meningaud JP, Paraskevas A, Ingallina F, Bouhana E, Lantieri L. Face transplant graft procurement: a preclinical and clinical study. *Plast Reconstr Surg*. 2008;122: 1383-1389.
17. Dodson TB, Bays RA, Neuenschwander MC. Maxillary perfusion during Le Fort I osteotomy after ligation of the descending palatine artery. *J Oral Maxillofac Surg*. 1997;55:51-55.
18. Siebert JW, Angrigiani C, McCarthy JG, Longaker MT. Blood supply of the Le Fort I maxillary segment: an anatomic study. *Plast Reconstr Surg*. 1997;100:843-851.
19. Banks ND, Hui-Chou HG, Tripathi S, et al. An anatomical study of external carotid artery vascular territories in face and midface flaps for transplantation. *Plast Reconstr Surg*. 2009;123:1677-1687.
20. Lanzetta M, Petruzzo P, Dubernard JM, et al. The Second Report (1998–2006) of the International Registry of Hand and Composite Tissue Transplantation. *Transpl Immunol*. 2007;18:1-6.
21. Nation's fourth hand transplant performed at Jewish Hospital Hand Care Center by Kleinert Kutz and University of Louisville Surgeons. 2008, July 12.Hand Transplant www.handtransplant.com/news_release/071208.html July 12, 2008.
22. Gorantla VS, Breidenbach WC. Hand transplantation: the Louisville experience. In: Hewitt CW, Lee WPA, Gordon CR, eds. *Transplantation of Composite Tissue Allografts*. New York: Springer; 2008:215-233.
23. International Registry of Hand and Composite Tissue Transplantation. Hand Registry home page. http://www.handregistry.com. Accessed April 6, 2008.
24. Lanzetta M, Petruzzo P, Margreiter R, et al. The international registry on hand and composite tissue transplantation. *Transplantation*. 2005;79(9):1210-1214.
25. Dubernard JM. Composite tissue allografts: a challenge for transplantologists. *Am J Transplant*. 2005;5:1580-1581.
26. Schneeberger S, Ninkovic M, Margreiter R. Hand transplantation: the Innsbruck experience. In: Hewitt CW, Lee WPA, Gordon CR, eds. *Transplantation of Composite Tissue Allografts*. New York: Springer; 2008:193-208.
27. Petruzzo P, Morelon E, Kanitakis J, et al. Hand transplantation: the Lyon experience. In: Hewitt CW, Lee WPA, Gordon CR, eds. *Transplantation of Composite Tissue Allografts*. New York: Springer; 2008:209-214.
28. Lee WP, Sacks JM, Horibe EK. World experience of hand transplantation independent assessment. In: Hewitt CW, Lee WPA, Gordon CR, eds. *Transplantation of Composite Tissue Allografts*. New York: Springer; 2008.
29. Ploplys E, Mathes D. Survey on current attitudes towards composite tissue transplantation in north america among hand surgeons. *Plast Reconstr Surg*. 2008;121(6S):92.

9 Anesthetic Care for Face Transplantation

Jacek B. Cywinski, D. John Doyle, and Krzysztof Kusza

Contents

Abstract This chapter will address the current state of clinical experience and scientific knowledge in the perioperative management of the face transplant recipient and donor. Areas of controversy will be discussed, and practical approaches to the preoperative, intraoperative and postoperative care of the face transplant recipient will be presented. Several key points should be emphasized: airway management can be a special challenge and often involves tracheostomies in both the donor and the recipient; fluid management is especially important in ensuring good graft perfusion in both the donor and the recipient; and there may be periods where muscle relaxation should be avoided to facilitate nerve identification using electrical stimulation. A critically important consideration is that the anesthesia and surgical teams jointly discuss their clinical plans.

Abbreviations

ICU	intensive care unit
PICC	peripherally inserted central catheter
TA-GVHD	transfusion-associated graft-versus-host disease
TEG	thromboelastograph®
TIVA	total intravenous anesthesia
TRALI	transfusion-related associated lung injury

9.1 Introduction

Only a few face transplants have been done in the world to date, and each procedure has been unique with respect to recipient indications and the nature and volume of the transplanted tissue. Consequently, little

J.B. Cywinski (✉)
Department of General Anesthesiology, Cleveland Clinic Foundation, Anesthesiology Institute, Cleveland, OH, USA
e-mail: cywinsj@ccf.org

M.Z. Siemionow (ed.), *The Know-How of Face Transplantation*,
DOI: 10.1007/978-0-85729-253-7_9, © Springer-Verlag London Limited 2011

information has been published regarding the anesthetic management of face transplant recipients. This chapter will address anesthetic considerations for this extraordinarily complex procedure, which includes perioperative management considerations for both the donor and the recipient.

9.2 Management of the Donor During Face Tissue and Multiorgan Harvest

The anesthetic and ICU care of the multiorgan donor has been well described in the literature.[1-4] The key goals in such a setting are (1) ensuring that the brain-death documentation is correct and complete, (2) ensuring adequate perfusion of all tissues of interest, and (3) providing adequate muscle relaxation where needed. Of these goals, that of providing adequate tissue perfusion is often the most difficult given the physiological changes frequently found in brain-dead patients. In the cardiac system, such changes commonly include disturbances of cardiac rate and rhythm, myocardial ischemia and impairment, increased pulmonary artery pressures, and systemic hypotension requiring support with pressors. Electrolyte, endocrine, hematologic, and pulmonary changes can similarly complicate anesthetic management.

Although principles similar to conventional organ procurement apply when consideration is given to the procurement of a composite facial graft, one important difference needs to be considered. Due to the surgical complexity involved, harvesting of the facial graft should ordinarily always be performed before other organs are retrieved; this can sometimes take a significant amount of time. Careful anesthetic management during this long procurement period can have a significant impact on the quality of the graft.

Surgical planning for the harvest, including estimation of the amount of composite tissue needed and the various muscles, vessels, and nerves to be procured, is planned preoperatively with the goal of optimally matching it to the recipient's needs. In some instances, a practice "mock harvest" is performed in a cadaver lab to finalize surgical strategy. To avoid interference with the surgical field, a donor tracheostomy is usually performed first. Since dissection of the nerves must be carried out precisely to avoid nerve damage, a nerve stimulator is often used by the surgical team to identify the facial nerves. For this reason, muscle relaxants should be avoided until dissection of the nerves is complete. During that time, if needed, a motionless surgical field can be provided using a volatile anesthetics or with total intravenous anesthesia (TIVA) involving an intravenous infusion of an opioid (i.e., remifentanil). In such a case, however, careful attention must be paid to avoiding hypotension, as this potentially compromises perfusion of the harvested graft. Not infrequently circulatory function must be supported with intravascular volume expansion and vasoactive drugs. Consider, however, that although maintenance of adequate perfusion pressure is very important, high doses of vasoconstrictors can make it difficult to identify small facial blood vessels during the surgery and may also compromise tissue perfusion through vasoconstriction. Consequently, intravascular volume expansion may be better suited as a first step to restore blood pressure. Finally, since the face tissue procurement period may be quite lengthy, it is important to maintain adequate hemodynamics in the donor to avoid potential ischemia to any other organs that are to be eventually procured.

9.3 Preoperative Assessment of the Recipient

The anesthetic evaluation of candidates for face transplant is an integral part of the pretransplant process. For the most part, their preoperative anesthetic evaluation is similar to the evaluation of any surgical patient, with the major goals of identifying potential anesthetic problems and optimizing clinical conditions which might adversely impact on postoperative outcome. The patient evaluation must be tailored to each individual patient. Although detailed criteria for approval do not yet exist, a general consensus exists that the patient must be sufficiently fit to undergo very prolonged anesthesia and surgery in order to be considered for face transplantation. Such patients should be free from any comorbidity which could significantly impact on the postoperative course in a negative way.

Anesthetic evaluation starts with detailed review of the medical history and physical examination findings.

A review of comorbid conditions will help determine the patient's surgical risk as well as help tailor the pre-transplant testing. Particular attention needs to be paid to the details of face injury and past reconstructive efforts, especially because they may greatly influence airway management plans. Detailed evaluation of cardiopulmonary status is required; in our program, we modified our preoperative evaluation guidelines for liver transplant candidates for this purpose (see Table 9.1). Although these guidelines appear to be exceptionally comprehensive in terms of cardiac testing, the risk of proceeding in the presence of an important undetected significant cardiac comorbidity (coronary disease, valvular disorder) is simply unacceptable. Evaluation of respiratory function focuses on assessment of pulmonary reserve; if the patient has a history of any lung disease, detailed evaluation is mandatory. A careful review of the clinical findings and radiological images will help identify any underlying disease processes.

Understanding the details of the patient's airway anatomy is critical since it guides intraoperative airway management. Some patients may present with a tracheostomy (with or without tracheostomy tube) while some may be able to breathe through natural upper airway passages despite possible extensive scaring from trauma or previous reconstructive surgery. A review of head and neck computed tomography scans may be helpful in delineating anatomical structures and developing a plan for airway management. In some cases, flexible fiber-optic examination of the airway may be very helpful to reveal further details.

The importance of communication with the surgical team about the surgical plan cannot be overestimated, and planning for airway management is an especially important component of the discussion. All candidates for face transplantation have significant facial disfigurement; in some cases, this disfigurement compromises the upper airway. Mask ventilation after the induction of general anesthesia may be very difficult in such instances. In cases where a tracheostomy is not present preoperatively, patients will often require oral awake fiber-optic intubation. In such cases, following successful intubation, the surgical team will usually proceed with tracheostomy so that the endotracheal tube is out of the surgical field.

Table 9.1 Cardiac workup for face transplant candidates

1. **2D transthoracic echo with RVSP estimation**: ALL patients undergoing evaluation for face transplantation.
2. **Stress test – Dobutamine Stress Echo (DSE)**: • ALL patients 50 years and older. • Patients younger than 50 who have one or more of the following risk factors: – History of DM > one year – Hyperlipidemia – Strong family history of CAD – History of angina – ECG changes indicating prior MI – Unexplained shortness of breath. Any stress-induced ischemia or equivocal findings on the stress test will warrant a cardiology consult and left heart catheterization. If the DSE is non-diagnostic due to an inability to achieve > 85% MPHR, the patient will be referred to a cardiologist to direct further workup to rule out significant CAD (preferably by left heart catheterization).
3. **Patients with known CAD (i.e., previous MI, PCI, CABG)**: ALL patients with established CAD will be referred to a cardiologist for further workup (preferably by left heart catheterization).
4. **Valvular disease of the heart** Patients with valvular stenosis or regurgitation graded > 2+ will be referred to a cardiologist. If the valve lesion does not require immediate surgical intervention, the cardiologist will determine the appropriate follow-up with diagnostic tests (echo) and cardiology clinic appointments.

Modified from Cleveland Clinic Orthotopic Liver Transplant Assessment Protocol

CABG Coronary artery bypass grafting, *CAD* Coronary artery disease, *DM* Diabetes mellitus, *DSE* Dobutamine Stress Echo, *MI* Myocardial infarction, *MPHR* Maximum predicted heart rate, *PCI* Percutaneous coronary intervention, *RVSP* Right ventricular systolic pressure

9.4 Intraoperative Management

9.4.1 *Monitoring*

In patients with normal cardiopulmonary status, the need for advanced anesthetic monitors is diminished. In most cases, standard ASA monitors[5] in conjunction with an arterial line and central venous access will be sufficient. In addition to the monitoring of central venous pressure changes through the use of a central line, the patient's volume status can be ascertained by monitoring urine output, and by looking at systolic pressure variation and related indices.[6] As emphasized earlier, maintenance of adequate intravascular volume is important in both the donor and the recipient in order that graft perfusion be adequate at all times. The use of electroencephalographic depth of anesthesia monitoring (e.g., BIS monitoring) may be complicated by proximity of the sensors to the surgical field.

9.4.2 *Airway Management*

Airway management in the donor usually involves performing a tracheostomy on the patient to avoid interference with harvested tissues. A wire-reinforced endotracheal tube is often used in this setting. Airway management in the recipient will necessarily be influenced by the nature of the facial injuries; in many cases, a tracheostomy will be present and airway management merely involves the insertion of an appropriate endotracheal tube into the tracheal stoma. In recipient patients who do not have a tracheostomy, one common approach involves awake of fiber-optic oral intubation, followed by a tracheostomy procedure. Again, a wire-reinforced endotracheal tube is often used.

9.4.3 *Vascular Access*

As with all surgery, vascular access is important in the face transplant patient. Although the procedures can be very lengthy in duration, massive blood loss is usually not encountered. One or two peripheral intravenous catheters and a central line are usually adequate. Where a central line is planned, discussion with the surgical team as to the proposed location is imperative, as a subclavian site may be preferable to the use of the internal jugular site in order to avoid impinging on the surgical field. In some instances, placement of a central venous catheter in the neck (internal jugular vein) or in the upper chest (subclavian vein) will be inadvisable due to planned vascular anastomoses and concern of venous outflow from the graft. Although less preferable due to concerns about infection, the femoral vein may be used as an access site in cases where central venous access is deemed to be essential. A peripherally inserted central catheter (PICC line) is another choice that will be helpful in some cases. The reader is also reminded of the importance of using strict sterile technique with central line insertion to avoid infections.[7] Finally, many anesthesiologists will want to use ultrasound guidance to assist in central line insertion.[8]

9.4.4 *Choice of the Anesthetic (Volatile Agent vs. TIVA)*

Little data is available to guide the clinician regarding the choice of anesthetic technique; the likelihood is that both volatile agents as well as total intravenous anesthesia (TIVA) will be adequate as long as attention is paid to underlying principles. Given the long duration of the surgery, avoidance of nitrous oxide with its adverse effects following prolonged administration[9] would seem to be prudent. Where muscle relaxation must be avoided, as in portions of the surgery where nerve testing is required, the use of a remifentanil infusion can be helpful in reducing movement. Unfortunately, remifentanil use is often associated with hypotension; this may make remifentanil unsuitable in some cases, at least in large doses. Another consideration is that the use of volatile anesthetic agents is believed to protect against intraoperative awareness better than the use of total intravenous anesthesia. Given that the electroencephalographic monitoring of depth of anesthesia may sometimes be difficult in these patients, the use of volatile anesthetic agent like isoflurane may be advantageous; also, the depth of anesthesia achieved with volatile anesthetics directly correlates with the index of peripheral perfusion.

The choice of anesthetic technique for the transplant of composite tissue grafts may play an important role in influencing outcome. In experimental models, certain anesthetics have been shown to improve blood flow in the microcirculation of free flaps as well as in the flow through micro- and macro-vascular anastomoses.[10] Although almost entirely eliminated from modern clinical practice, the effect of halothane on free flap microcirculation has been studied extensively in animal models, especially its effect on post capillary venules.[11] Other halogenated volatile anesthetics have microcirculatory effects similar to halothane under normovolemic conditions.[12] The postcapillary venule is the primary blood reservoir, receiving all the formed blood elements passing through the capillary network and for which the postcapillary venule is the anatomically designed drainage site. Exposure to a volatile anesthetic may improve free flap survival by reducing the number of leucocytes flowing through microcirculation, decreasing the number of leucocytes adhering to the endothelium of vessel wall, as well as by decreasing the number of leucocytes migrating outside the vessel. These accumulated leucocytes can potentially cause tissue injury via the release of proinflammatory mediators (released via degranulation) during graft reperfusion. Impaired drainage of blood from the microcirculation due to endothelial edema and the presence of leucocytes adhering to the endothelial wall of venules can cause decreased capillary blood flow in the territory draining blood into affected venules. This is paralleled by a significant decrease in tissue oxygenation. It appears that other halogenated volatile anesthetics (i.e., sevoflurane) offer a similar protective effect on microcirculation, although the effect may be smaller.[11,13,14]

Propofol infusions in experimental models of free musculocutaneous flaps have demonstrated to cause quite significant increases in endothelial edema, with a decrease in the number of rolling leucocytes and lymphocytes passing through postcapillary venules. At the same time, the number of leucocytes and lymphocytes adhering to the endothelial wall and migrating outside the postcapillary venule was significantly increased. Leucocyte adhesion is a clear manifestation of their increased chemotactic activity and reduced microcirculatory blood flow. Leucocytes contain a wide variety of proinflammatory mediators which can be released in an uncontrolled fashion and potentiate the activity of free oxygen radicals upon graft reperfusion. For these reasons, prolonged propofol infusions may be less than ideal as a choice for patients undergoing composite tissue grafts. Also, in experimental models, propofol demonstrated immunomodulatory properties resulting in inhibition of lymphocyte activity.[15,16]

The effect of opioids on the microcirculation of free flaps is less defined; however, based on animal model experiments, they decrease the diameter of the arterioles in skeletal muscle, which decreases blood flow in the free flap, as manifested by increased velocities of the blood flow measured with Doppler flow meter.[11,17]

9.4.5 Fluid Management

Fluid management in face transplantation is not substantially different from fluid management in any other long surgical procedure involving microvascular free flaps. Maintenance of adequate graft perfusion is paramount. The adequacy of fluid management is judged by assessing the patient's urine output and changes in central venous pressure, as well as by noting variations in systolic pressure as a consequence of positive pressure ventilation. Colloid administration may be especially helpful in maintaining intravascular volume, but concerns about impaired clotting must be born in mind.[18] Fluid management in these patients may also involve the administration of blood and blood products such as plasma or platelets, depending on the clinical circumstances. Some surgeons request that the patient's hemoglobin be kept around 10 g/dl with a view to producing optimal rheological properties in the blood.

9.4.6 Transfusion of Blood Products

General guidelines for the transfusion of blood components are available in the published literature; transfusion practices in the face transplant recipient need to be highly individualized and guided by the clinical situation.[19-21] Red blood cells are indicated in symptomatic, anemic patients to restore oxygen-carrying capacity. There is a general consensus that transfusion of red blood cells is not indicated in relatively healthy, asymptomatic individuals until the hemoglobin level drops below 6 g/dL. However, in the face transplant recipients, the threshold for transfusion of red cells may be slightly higher to provide adequate oxygen-carrying capacity to the graft. Excessive increases in hematocrit can adversely affect the rheologic properties of blood and compromise microcirculation. Since microvascular bleeding and hematoma formation is a potentially devastating postoperative complication, the coagulation status of the recipient needs to be monitored closely, especially in cases with large blood loss (> one blood volume). Whole blood clotting analysis, as assessed with the Thromboelastograph® (TEG) and Sonoclot®, provides a dynamic picture of the entire clotting process and may be best suited to assist decision making about fresh frozen plasma and platelet transfusions. Although evidence-based data supporting administration of fresh frozen plasma in patients with INR values <2.0 are lacking,[20] this may be appropriate where surgical bleeding from impaired coagulation could be disastrous. Surgical patients with microvascular bleeding usually require a platelet transfusion if the platelet count is less than 50×10^9/L and rarely require therapy if it is greater than 100×10^9/L.[19] Transfusion of the blood products is not risk free, and it may cause acute and delayed complications, in addition to infection, fever, urticaria, and hemolysis.[21] Transfusion-Related Acute Lung Injury (TRALI), Transfusion-Associated Graft-versus-Host Disease (TA-GVHD), and immunomodulation are other potential complications that should be borne in mind.

9.4.7 Pressors

While the use of pressors such as phenylephrine is in common clinical use in anesthesia, their use is often discouraged in cases where microvascular surgery is involved. This is because of concerns about graft perfusion. In addition, use of vasoconstricting vasopressors can make it difficult for the surgeon to identify and handle vascular structures in the graft procurement phase of the operation. Such considerations emphasize the importance of maintaining an adequate intravascular volume at all times.

9.4.8 Muscle Relaxants

Muscle relaxants may be needed at various phases of both the donor and recipient operations. The central consideration in their use concerns the fact that both the donor and recipient operations may at times require use of a nerve stimulator to help identify various nerves. The use of a shorter acting muscle relaxant like rocuronium is preferable to a longer acting drug like pancuronium for this reason. Obviously, it is important to communicate with the surgical team on this matter.

9.4.9 Postoperative Sedation

Postoperatively, the face transplant recipient will be ventilated through a tracheostomy tube in an intensive care setting. Standard ventilator settings based on the patient's particulars will generally be used. During this period, postoperative sedation will be necessary and can be achieved by a variety of means. An infusion of propofol in conjunction with an opiate such as fentanyl is one means of achieving the needed sedation. This infusion will generally be titrated to a clinical endpoint, such as the Ramsey Sedation Scale (Table 9.2) or the Modified Observer's Assessment of Alertness/Sedation Scale (Table 9.3).

Table 9.2 Ramsey sedation scale

Score	Responsiveness
1	Patient is anxious and agitated or restless, or both
2	Patient is cooperative, oriented and tranquil
3	Patient responds to commands only
4	Patient exhibits brisk response to light glabellar tap or loud auditory stimulus
5	Patient exhibits a sluggish response to light glabellar tap or loud auditory stimulus
6	Patient exhibits no response

Source: Ramsay et al.[22]

Table 9.3 Modified observer's assessment of alertness/sedation scale

Responsiveness	Score
Agitated	6
Responds readily to name spoken in normal tone (alert)	5
Lethargic response to name spoken in normal tone	4
Responds only after name is called loudly and/or repeatedly	3
Responds only after mild prodding or shaking	2
Does not respond to mild prodding or shaking	1
Does not respond to deep stimulus	0

Source: Cohen et al.[23]

9.5 Anesthetic Team

Face transplant operations are usually carried out in an academic medical center where an anesthetic team approach is employed for anesthesia delivery. This team usually involves an attending anesthesiologist in conjunction with a resident or CRNA. Given the long duration of the procedure, multiple teams may be involved. With the transfer of patient care from one team to the other, it is imperative that good communication and good documentation be carried out.

9.6 Providing Anesthetic Care to Face Transplant Recipients for Subsequent Surgical Procedures

As with all transplant patients, subsequent procedures are sometimes necessary. In the case of face transplant patients, this will commonly be for revisions pertaining to the original transplant surgery, although such patients may also require surgery unrelated to the original transplant. A special issue in either situation concerns airway management in cases where the composite graft involves maxillary or mandibular structures, since the forces associated with direct laryngoscopy for intubation purposes could conceivably cause damage to incompletely healed bony structures. Consequently, in cases where the tracheostomy has healed, over two options must be considered. One option is to reopen the tracheostomy under local anesthesia and use the new tracheostomy site for anesthesia. Another option, and the one we prefer, is to perform awake fiber-optic intubation.

9.7 Conclusions

Although many aspects of anesthesia for face transplantation surgery are straightforward, anesthesia for this procedure involves a number of special issues that are important for obtaining a successful surgical outcome. Airway management can be a special challenge which often involves tracheostomies in both the donor and the recipient. Fluid management is especially important in ensuring good graft perfusion in both the donor and the recipient. The long duration of the procedure necessarily entails more than one anesthesia team. In addition, there may be periods where muscle relaxation should be avoided to facilitate nerve identification using electrical stimulation. An especially important consideration is that the anesthesia and surgical teams discuss their mutual plans so that they are mutually compatible.

References

1. Hevesi ZG, Lopukhin SY, Angelini G, et al. Supportive care after brain death for the donor candidate. *Int Anesthesiol Clin*. 2006;44:21-34.
2. Venkateswaran RV, Patchell VB, Wilson IC, et al. Early donor management increases the retrieval rate of lungs for transplantation. *Ann Thorac Surg*. 2008;85:278-286.
3. Shah VR. Aggressive management of multiorgan donor. *Transplant Proc*. 2008;40:1087-1090.
4. DuBose J, Salim A. Aggressive organ donor management protocol. *J Intensive Care Med*. 2008;23:367-375.
5. American Society of Anesthesiologists. Standards for basic anesthetic monitoring. American Society of Anesthesiologists. 2005. http://www.asahq.org/publicationsAndServices/standards/02.pdf. Accessed May 14, 2010.
6. Marik PE, Cavallazzi R, Vasu T, et al. Dynamic changes in arterial waveform derived variables and fluid responsiveness in mechanically ventilated patients: a systematic review of the literature. *Crit Care Med*. 2009;37:2642-2647.
7. Maki DG, Kluger DM, Crnich CJ. The risk of bloodstream infection in adults with different intravascular devices: a systematic review of 200 published prospective studies. *Mayo Clin Proc*. 2006;81:1159-1171.
8. Gann M Jr, Sardi A. Improved results using ultrasound guidance for central venous access. *Am Surg*. 2003;69:1104-1107.
9. Renard D, Dutray A, Remy A, et al. Subacute combined degeneration of the spinal cord caused by nitrous oxide anaesthesia. *Neurol Sci*. 2009;30:75-76.
10. Adams J, Charlton P. Anaesthesia for microvascular free tissue transfer. *Br J Anaesth CEPD Rev*. 2003;3:33-37.
11. Kusza K, Siemionow M, Nalbantoglu U, et al. Microcirculatory response to halothane and isoflurane anesthesia. *Ann Plast Surg*. 1999;43:57-66.
12. Sigurdsson GH, Banic A, Wheatley AM, et al. Effects of halothane and isoflurane anaesthesia on microcirculatory blood flow in musculocutaneous flaps. *Br J Anaesth*. 1994;73:826-832.
13. Hagau N, Longrois D. Anesthesia for free vascularized tissue transfer. *Microsurgery*. 2009;29:161-167.
14. Liu X, Peter FW, Barker JH, et al. Leukocyte-endothelium interaction in arterioles after ischemia and reperfusion. *J Surg Res*. 1999;87:77-84.
15. Kusza K, Blaszyk M, Siemionow M, et al. Alteration in peripheral microcirculatory haemodynamics of muscle flaps during propofol infusion anaesthesia. *Anaesthesiol Intens Ther*. 2002;34:187-193.

16. Holzmann A, Schmidt H, Gebhardt MM, et al. Propofol-induced alterations in the microcirculation of hamster striated muscle. *Br J Anaesth*. 1995;75:452-456.
17. Brookes ZL, Brown NJ, Reilly CS. The dose-dependent effects of fentanyl on rat skeletal muscle microcirculation in vivo. *Anesth Analg*. 2003;96:456-462.
18. Marx G, Schuerholz T. Fluid-induced coagulopathy: does the type of fluid make a difference? *Crit Care*. 2010;14:118.
19. American Society of Anesthesiologists Task Force on Perioperative Blood Transfusion and Adjuvant Therapies. American Society of Anesthesiologists Task Force on Perioperative Blood Transfusion and Adjuvant Therapies. Practice guidelines for perioperative blood transfusion and adjuvant therapies. *Anesthesiology*. 2006;105:198-208.
20. Triulzi DJ. The art of plasma transfusion therapy. *Transfusion*. 2006;46(8):1268-1270.
21. American Society of Anesthesiologists Committee on Transfusion Medicine. Questions and Answers About Blood Management, 4th Edn. American Society of Anesthesiologists. 2008. http://www.asahq.org/publicationsAndServices/transfusion.pdf. Accessed May 14, 2010.
22. Ramsay MA, Savege TM, Simpson BR, Goodwin R. Controlled sedation with alphaxalone-alphadolone. *Br Med J*. 1974;2:656-659.
23. Cohen LB, DeLegge MH, Aisenberg J, et al. AGA Institute review of endoscopic sedation. *Gastroenterology*. 2007;133:675-701.

Alternative Approaches to Face Transplantation: Microsurgical Approach

10

Elliott H. Rose

Contents

E.H. Rose
Division of Plastic and Reconstructive Surgery,
The Mount Sinai Medical Center,
New York, NY, USA
e-mail: erose@facemakernyc.com

Abstract History was made in the USA in December 2008 when news of the ground-breaking face transplant in a 45-year-old woman was performed by Dr. Maria Siemionow and her team at the Cleveland Clinic. With the flurry of public interest engendered by early transplant success, the debate has been framed by the bio-ethics community regarding the autonomy of the recipient (i.e., informed consent), unpredictable effect on those receiving the transplant (life-long immunosuppression and exposure to medical sequelae), and the influence on society as a whole (economic/cost factors). In view of the high stakes of CTA face transplant surgery (not to mention the large amount of visibility in the press), *traditional methods of facial restoration should not be overlooked.* Considering the significant morbidity inherent to life-long immunosuppression in the transplant patient, each transplant candidate should be carefully evaluated for conservative options utilizing autogenous tissue to address facial deformity in the decision-making process. The following chapter addresses the salient issues regarding the debate between conventional reconstructive options and face transplantation with a particular focus on promulgating the beneficial aspects of traditional reconstruction in select patients.

Abbreviations

CMV cytomegalovirus
CTA composite tissue allotransplantation
LT left
RT right

M.Z. Siemionow (ed.), *The Know-How of Face Transplantation*,
DOI: 10.1007/978-0-85729-253-7_10,

10.1 World Face Transplant Experience

History was made in the USA in December 2008 when news of the ground breaking face transplant in a 45-year-old woman was performed by Dr. Maria Siemionow and her team at the Cleveland Clinic. This, the most complex of the composite face transplants reported to date, entailed a Composite Le Fort III midfacial skeletal transplant including soft tissue, total nose, lower eyelids, upper lip, bilateral zygomas, maxillae, parotid glands, and upper dentition.[1] At the time of the surgery, this was the *fourth known* successful composite facial transplant performed. Since that landmark surgery, an additional seven facial transplants have been performed including a "total" face transplant performed by Spanish surgeons in March 2010 on a man injured in a shooting accident.

An extensive review of the world's first seven face transplants was published in November 2009 (not including the two face transplant reported in Spain in January 2010 and April 2010 respectively).[2] Seventy-one percent were male (five patients) and 29% female (two patients). Etiologies were traumatic in 86% (gunshot wounds, animal attacks, fall, and burns) and non-traumatic in 14% (neurofibromatosis). Two deaths occurred (29%) – a Chinese patient at 2 years post transplant who discontinued his immunosuppression regimen on the advice of a "witch doctor"[3] and a French patient who developed overwhelming sepsis 2 months following a triple transplantation (face and two hands).[4] At least one episode of acute graft rejection occurred in all patients (2–3 in most), reversed by increased doses of immunosuppressant therapy and corticosteroids.[1,5-7] Significant post-transplant complications included acute renal failure, thrombocytopenia, hemolytic anemia, thrombotic microangiopathy, thrombocytopenia, delirium, CMV viremia, steroid-induced diabetes, and fatal sepsis. The RCS Working Party report estimates of a 10% chance of acute rejection within the first year and 30–50% chance of chronic rejection in the 2–5-year-period following face transplantation.[8]

Functional and aesthetic outcomes were generally quite promising. On average, sensory recovery was noted at 3–6 months. Motor function tended to appear at 9–12 months with demonstrated degrees of mimetic facial movement, symmetrical smiling, and lip occlusion.[2] The first face transplant, performed on November 27, 2005 by Dubenard in Amiens, France, reported that she was "satisfied" with her appearance and was more "comfortable" in public.[5] Additional refinement surgeries such as scar revisions, graft contouring, cartilage grafting, tissue rearrangement of soft tissue redundancy, lid tightening have helped to achieve optimal aesthetic results.

10.2 Bioethical Issues: Cosmetic vs. Functional?

With the flurry of public interest engendered by early transplant success, the debate has been framed by the bio-ethics community regarding the autonomy of the recipient (i.e., informed consent), unpredictable effect on those receiving the transplant (life-long immunosuppression and exposure to medical sequelae), and the influence on society as a whole (economic/cost factors).[9] Critics, of course, consign face transplants as *cosmetic in nature* and argue that the experimental character of the procedure makes it impossible for the transplant team to provide an objective informed consent because of the lopsided information portrayed in the press.[10,11] Characterized as an "external transplant" (unlike solid organ transplants), a face transplant is always "visible" and, as a multi-functioning organ, requires aggressive rehabilitation to integrate allograft function via cortical reorganization to achieve a modicum of success in daily activities. Proponents, by contrast, contend that the benefits of the new treatment outweigh the risks. In fact, 71% of facially disfigured patients were still willing to undergo the procedure even after being informed that the possibility of acute rejection was 50% with the first year.[12] Sieimionow has asserted that the face is "unmatched in our relationship to our daily social interactions and well-being."[13] Missing facial parts preclude vital daily functions such as smiling, drinking, eating, and speaking.[14] White argues that face transplants are not "immoral" because it both restores critical organ function and indisputably restores human appearance.[15]

The Cleveland Clinic has developed an objective grading scheme for identifying the optimal face transplant candidate. The FACES Score (range 10–60) grades prospective face transplant candidates in 5 categories and stratifies patients within the institution according to complexity.[16] Patients with evidence of poor medical compliance, unsatisfactory psychological evaluation, end-stage disease, or significant co-morbidity were excluded as candidates. Transplant surgeons must be extremely selective in candidate selection and work in conjunction with psychiatry and bio-ethic colleagues.

10.3 Life-Long Immunosuppression

In the 1980s, the introduction of cyclosporine revolutionized the field of transplant surgery. In three decades, the scope has evolved from experimental rat hindlimb surgery to composite tissue allotransplantation (CTA).[17] Following the success of hand transplantation and other allografts including the trachea, peripheral nerve, flexor tendon apparatus, vascularized knee, larynx and abdominal wall, partial and full face transplants have become a reality. For the most part, the immunosuppression regimens of the first 4 reviewed face transplants have mimicked those of their solid organ counterparts. Immunosuppression was often induced with anti-thymocyte globulin followed by various cocktails of Tacrolimus, mycophenolate mofetil (MMF), and prednisone with or without injection of donor bone marrow hematopoietic cells to induce chimerism.[18] The face transplant recipients, such as solid organ transplants, are saddled with the burden of *life-long immunosuppression* but with the caveat of a "non-life-threatening injury" and the onus of high-risk medical sequelae such as melanoma, CMV, lymphoma, infiltrative pneumonias, diabetes, avascular necrosis, opportunistic infections, kidney failure, etc. However, as newer options for reduced immunosuppressive therapy become available, the horizon for face transplant indications will be greatly expanded.[19] In view of the high stakes of CTA face transplant surgery (not to mention the large amount of visibility in the press), *traditional methods of facial restoration should not be overlooked.* Considering the significant morbidity inherent to life-long immunosuppression in the transplant patient, each transplant candidate should be carefully evaluated for conservative options utilizing autogenous tissue to address facial deformity in the decision-making process.

10.4 Face Transplant vs. Autogenous Reconstruction?

In the author's opinion, CTA face transplant may be the *procedure of choice* for extensive central/ mid facial defects (nose, peri-oral, and peri-ocular particularly with associated architectural deficits), although very elegant reconstructions of central facial features have been achieved by Menick and others.[20,21] In the author's hands, very acceptable facial restoration can be achieved with the application of *multistaged, pre-patterned, pre-sculpted microsurgical flap transfers* for most near total facial defects.[22,23] Flap design mimicking "aesthetic subunits" hides the scars at the junction of facial planes[24,25] (Fig. 10.1). Extensive intra-operative flap sculpting immediately restores the contours and planes of facial geometry and precludes the need for additional flap contouring at a later stages[26] (Fig. 10.2). Others have advocated "super thin" microvascular free flaps for resurfacing of large contour

Fig. 10.1 Aesthetic subunits of the face

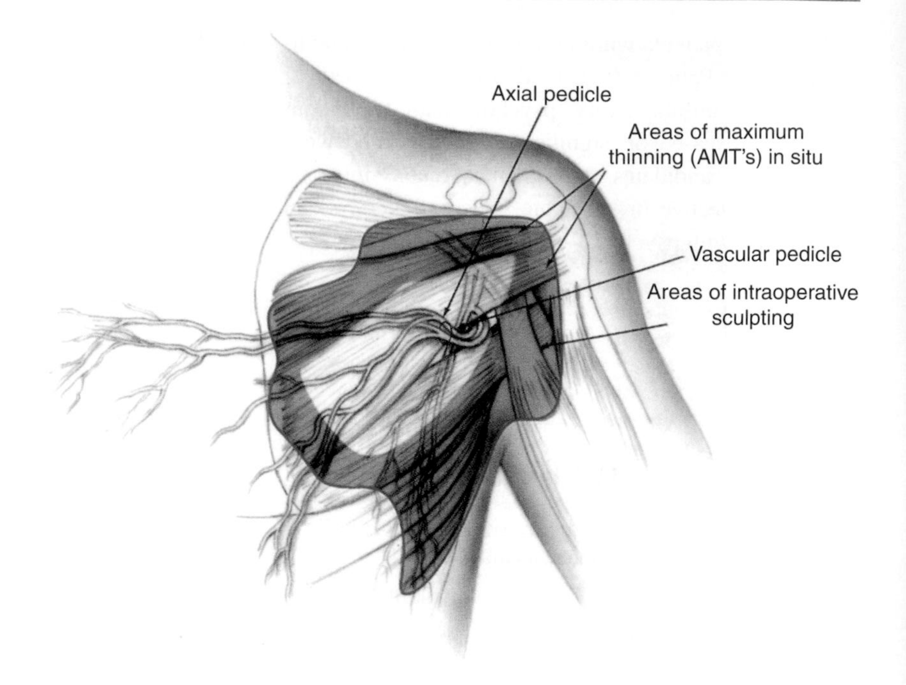

Fig. 10.2 Design of *pre-patterned, pre-sculpted* autogenous free flap transfers for inset into facial defects like "pieces of a jig-saw puzzle" (Reprinted from Rose[33])

sensitive areas.[27-29] The soft skin texture provided by the composite flap transfer has the *look and feel of normal facial skin* and provides the "palette" for camouflage make-up.[30] Most importantly, in the employment of autogenous free transfers, there is *no need for lifelong immunosuppression* and the attendant morbidity (including death). Albeit, the multi-stage nature of the autogenous reconstructions is more time-consuming, but the outcomes are relatively permanent, medical treatment has a finite length, and the underlying "fear" of late rejection (face "falling off") is eliminated. Economic impact is *limited* to the initial stages of treatment negating the need for the annual expenses for immunosuppressant therapy (see chapter 29).

10.5 Steps in Multistaged Aesthetic Autogenous Facial Restoration

1. Rebuild facial architecture with cartilage/ bone grafting
2. Establish deep structural support of facial foundation with fascia lata slings
3. Staged segmental replacement of "aesthetic" facial units with *pre-patterned microsurgical or pedicled tissue transfers*
4. Aggressive *intra-operative sculpting*
5. Seams hidden at junction of facial planes
6. Secondary debulking/suction assisted lipectomy to achieve facial contour
7. Refinement surgery for definition of facial features
8. Laser resurfacing
9. Cosmetic camouflage

10.6 Clinical Examples

10.6.1 Clinical Case Example 1

A 12-year boy was referred from Ireland to the unit at Mount Sinai Medical Center for facial reconstruction after sustaining a near total facial burn as a toddler when his Halloween costume caught fire. Eleven prior surgeries for eyelid ectropion repair, chin and lip

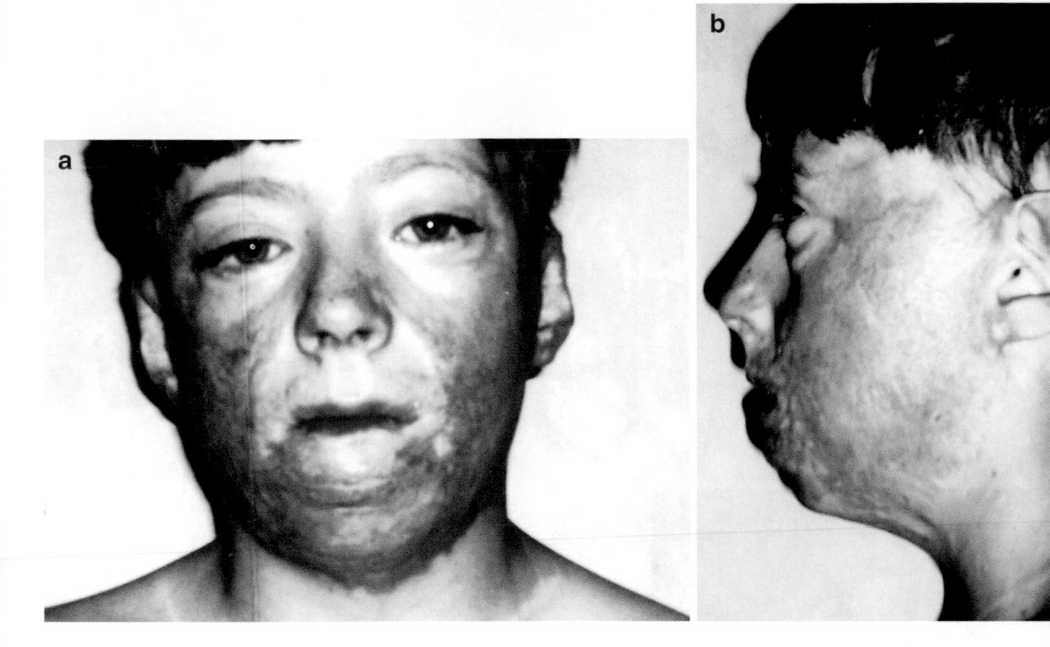

Fig. 10.3 Case 1. 12-years-old Irish boy with near total facial flame burns. (**a**) Pre-op frontal. Dense keloid scars over both cheeks, temporal areas, lower lip, chin, neck and jawline. (**b**) Profile. Chin was markedly retrusive with dense bands of contracting scar extended obliquely across the cervico-mental crease to the sternal notch (Reprinted from Rose[30] with permission)

correction were marginally successful. On exam, the dense keloid scars enveloped almost the entire face, distorting facial planes and creating a "mask-like facies." The deforming scar extended over both cheeks, temporal areas, lower lip, chin, neck, and jawline (Fig. 10.3a). Of note, the central face (nose and upper/lower lips) was relatively uninvolved. On profile, the chin was markedly retrusive with dense bands of contracting scar extended obliquely across the cervico-mental crease to the sternal notch (Fig. 10.3b). Traction from the neck scar caused evagination of the lower lip with exposure of lower dentition and alveolar ridge. First-stage reconstruction entailed insertion of bimalar fascia lata slings for lower lip suspension (Fig. 10.4) and patterned microvascular free radial forearm flap to the chin/neck subunit (Fig. 10.5). This was followed by *sequential* patterned scapular flaps to the RT (Fig. 10.6) and LT cheeks/hemi-face (Fig. 10.7), respectively, and simultaneous placement of fascia lata slings from the malar arches to the lateral lip modioli for buttressing of the deep facial foundation and lateral lip support

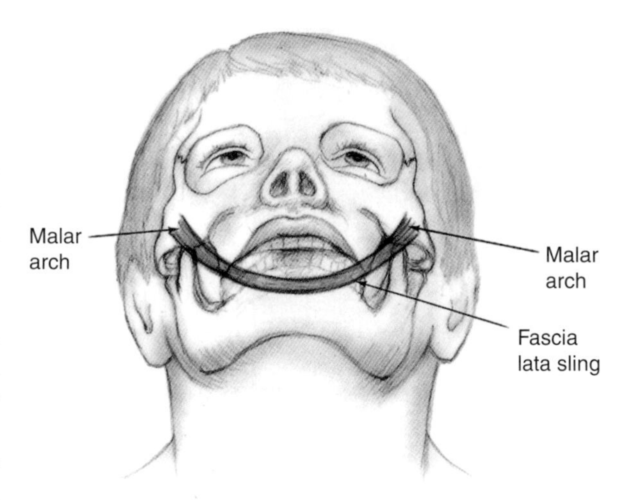

Fig. 10.4 Case 1. Bimalar fascia lata sling for lower lip/chin support (Reprinted from Rose[33])

(Fig. 10.8). Additional "refinement" procedures included debulking and contouring of the cheek and neck flaps, SAL, canthoplasty OU, insertion of Porex chin implant, multiple scar revisions, dermal

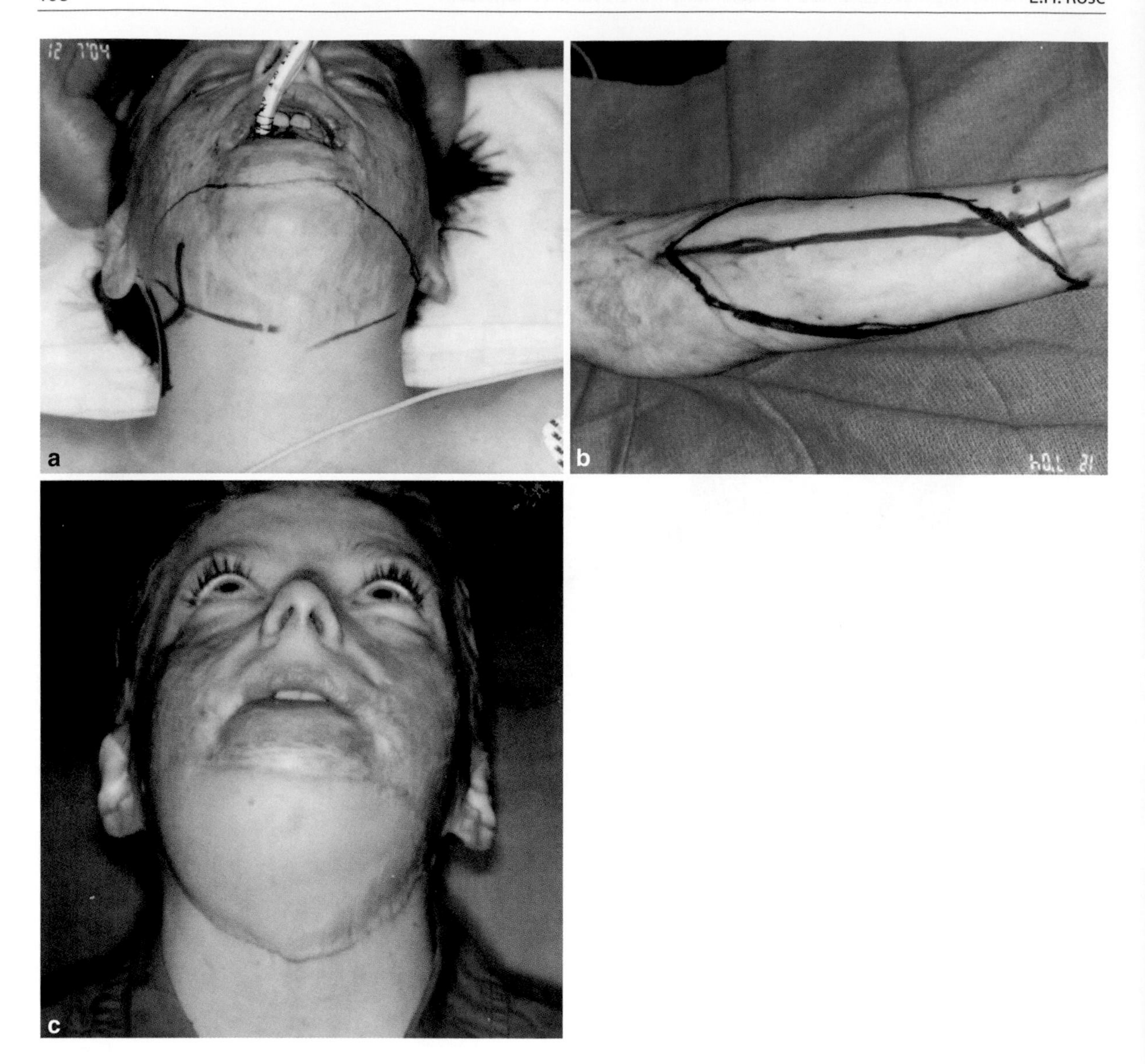

Fig. 10.5 Case 1. (**a**) Intra-operative keloid resection of aesthetic neck unit. (**b**) Design of patterned radial forearm flap. (**c**) Intermediate, after inset of neck flap (Reprinted from Rose[30] with permission)

placation of the nasolabial creases, and laser resurfacing of the facial scars (Fig. 10.9). After 1 year, the final surgery facial contours are restored with sculpted soft tissue conforming to facial geometry (Fig. 10.10a). Seams are hidden at the junction of the aesthetic subunits. On profile, cervico-mental angle is acute and well defined with good chin shape and projection (Fig. 10.10b). The patient has reintegrated with his peers and actively plays on his school soccer team.

Total length of treatment was 2½ years. Total number of surgeries = 5. Total hospital cost was $84,517.

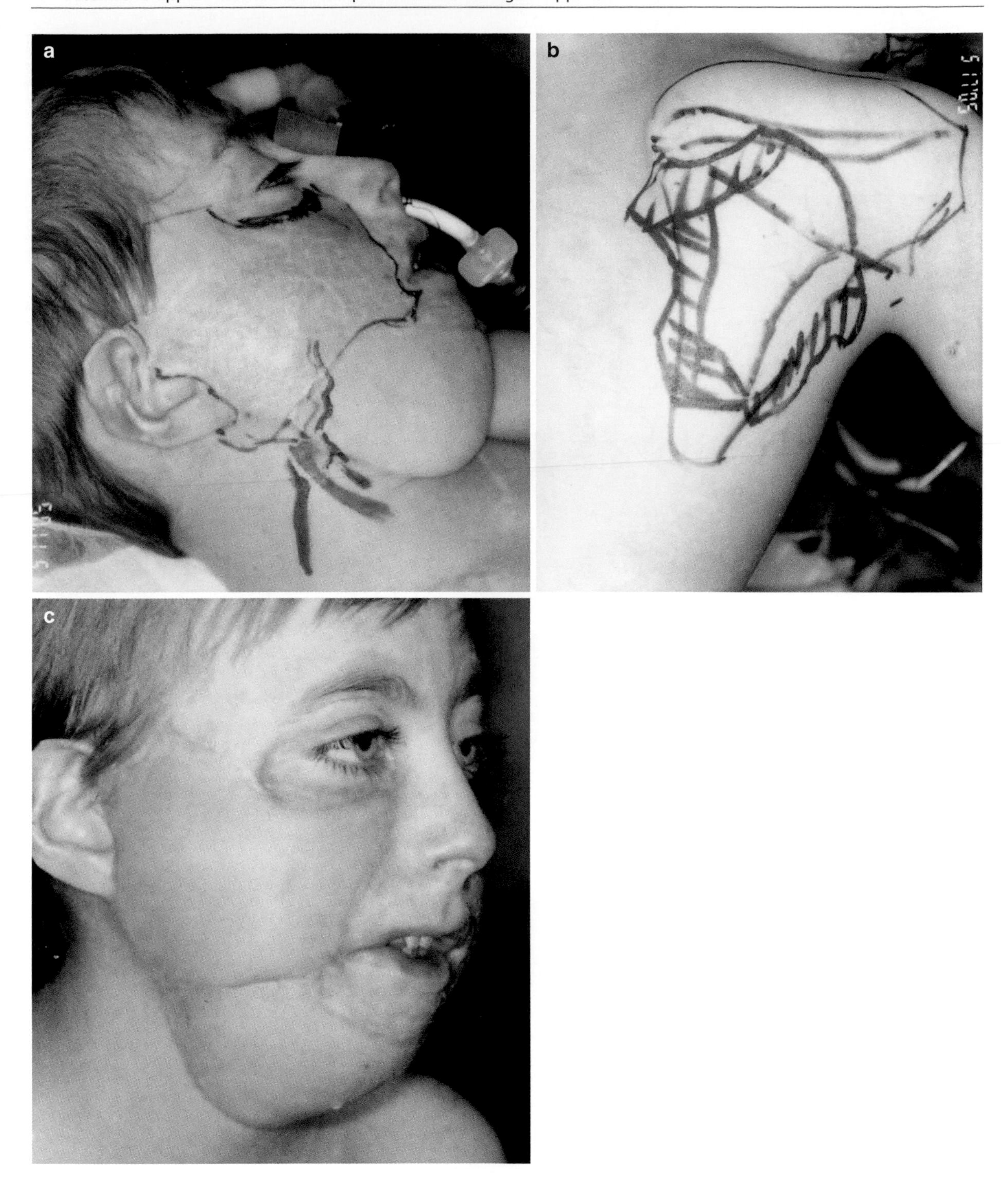

Fig. 10.6 Case 1. (**a**) Keloid excision RT cheek unit. (**b**) Design of pre-patterned scapular flap. Note intra-operative sculpting. (**c**) Intermediate, after inset of RT cheek flap (Reprinted from Rose[30] with permission)

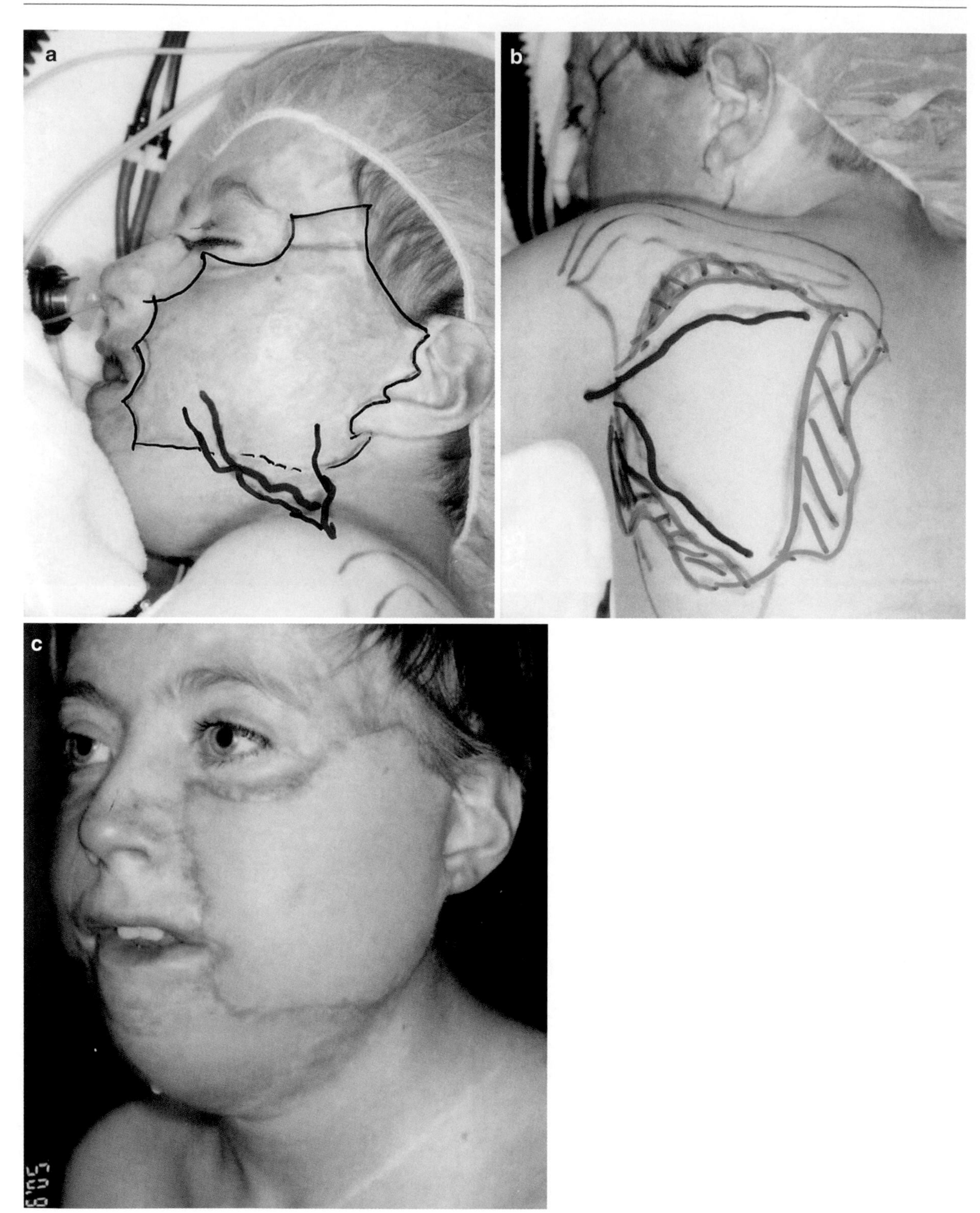

Fig. 10.7 Case 1. (**a**) Keloid excision of LT cheek unit. (**b**) Design of pre-patterned, pre-sculpted scapular flap. (**c**) Intermediate, after inset of LT cheek flap

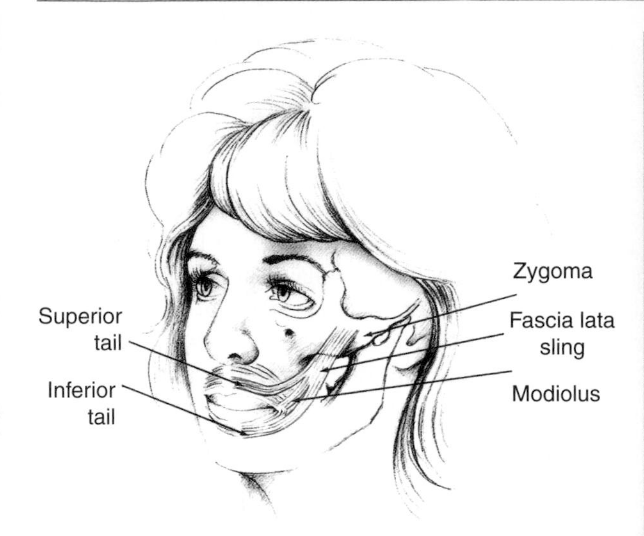

Fig. 10.8 Graphic of fascia lata sling for lateral lip suspension and support of deep facial foundation (Reprinted from Rose[33])

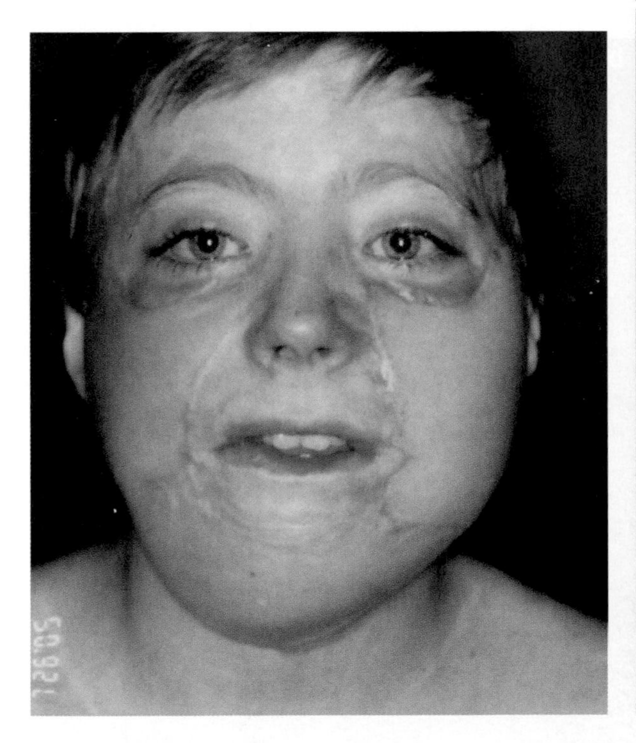

Fig. 10.9 Case 1. Intermediate frontal view prior to additional "refinement" procedures included debulking and contouring of the cheek and neck flaps, SAL, canthoplasty OU, insertion of Porex chin implant, multiple scar revisions, dermal placation of the nasolabial creases, and laser resurfacing of the facial scars

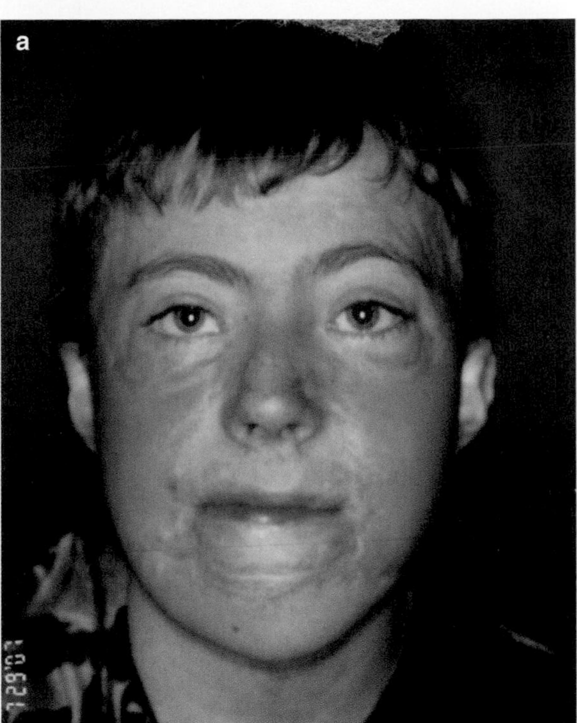

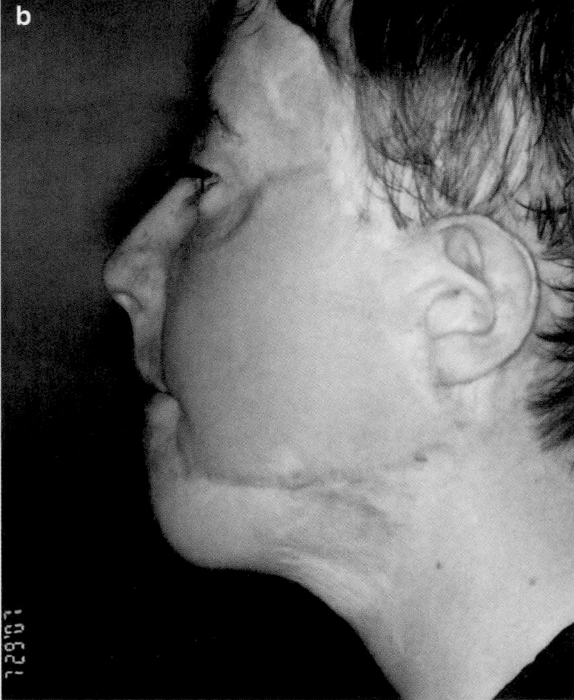

Fig. 10.10 Case 1. Post operative, at 1 year after last surgery. (**a**) Frontal view. Facial contours are restored with sculpted soft tissue conforming to facial geometry. Seams are hidden at the junction of the aesthetic subunits. (**b**) Lateral view. Cervicomental angle is acute and well-defined with good chin shape and projection (Reprinted from Rose[30] with permission)

10.6.2 Clinical Case Example 2

A 10-year-old girl sustained 80% TBSA burns in a crib fire in her native Columbia. She was abandoned by her biological parents and subsequently adopted by medical foster parents in Tampa, Florida. Prior to her transfer to Mount Sinai for facial restoration, she underwent 10+ prior reconstructive surgeries (Z-plasties, tissue expansion, etc) with little success. On exam, grotesque facial scarring was observed with distortion of facial planes, ocular displacement of the LT peri-ocular adnexae, nasal collapse, and significant lip contraction (Fig. 10.11a). On profile, the chin was markedly retrusive with dense plaques of keloid scar in the hemi-facial and peri-oral planes (Fig. 10.11b). Significant lower lip ectropion created exposure of the lower dentition. Projection of the nasal bridge and tip was deficient as well as large patches of scalp alopecia and absence of the LT ear. Initial stage of reconstruction entailed wide excision of keloid of the LT hemi-face and scalp, insertion of a fascia lata sling for lateral lip suspension and support of the deep facial foundation, and resurfacing with a patterned, sculpted microvascular free scapular flap tailored to the defect (Fig. 10.12). Second stage followed with a mirror image patterned, sculpted microvascular free scapular flap to the RT hemi-face and placement of a fascia lata sling to the RT lateral commissure (Fig. 10.13). Peri-ocular reconstruction entailed re-alignment of the medial canthal ligament by transnasal wire fixation and re-suspension of the lateral canthal ligament by wire fixation to the lateral orbital rim. Both upper and lower lids were resurfaced with a single sheet graft to the orbital subunit with a slit for the ciliary aperture (Fig. 10.14). Total nasal reconstruction included architectural modification of the nasal tip with conchal cartilage grafts and external resurfacing with a patterned, pedicled forehead flap (Fig. 10.15). The divided base of the nasal pedicle was "piggy-backed" to the lower eyelid for ectropion repair prior to permanent inset (Fig. 10.16). Nostril patency was re-established with full thickness skin grafts wrapped around a nasal stent. Additional "refinement"

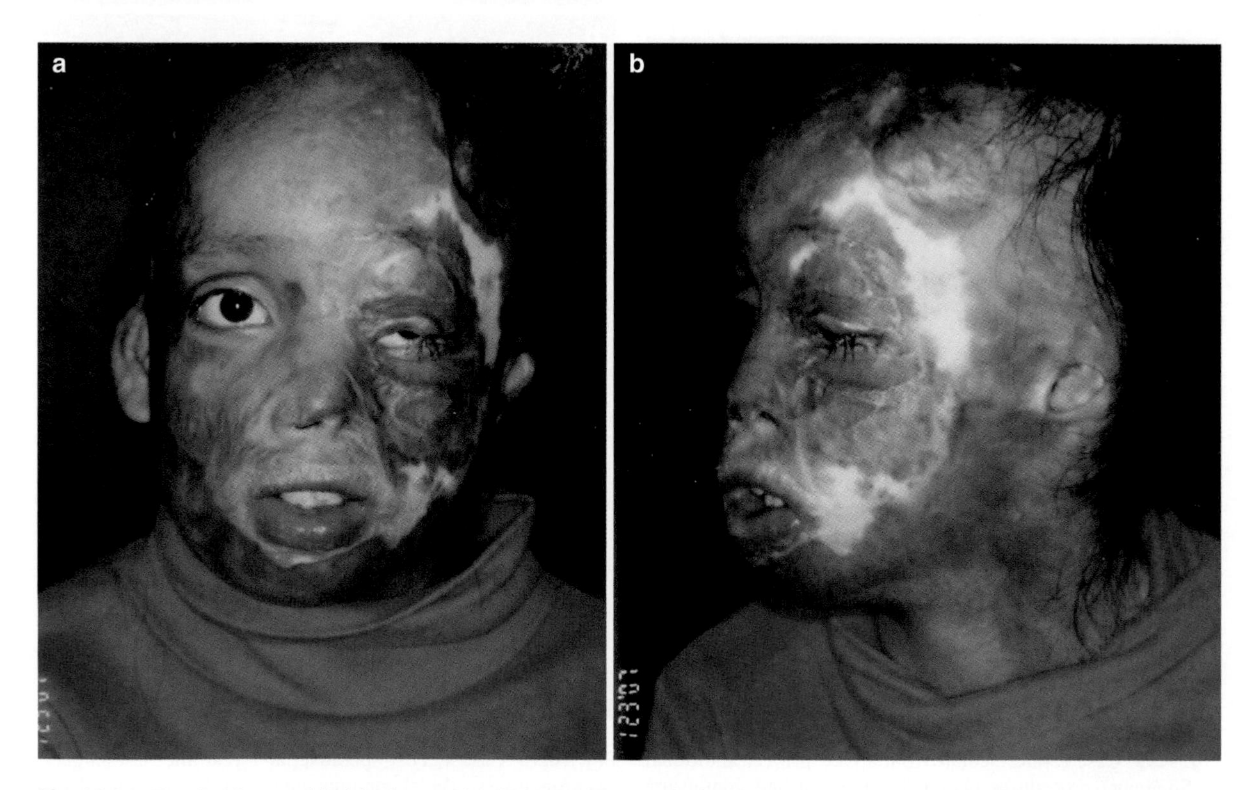

Fig. 10.11 Case 2. 10-year-old girl who sustained 80% TBSA burns in a crib fire as an infant. 10+ prior surgeries (Z-plasty, tissue expansion, etc.) with little success. (**a**) Frontal view. Grotesque facial scarring with distortion of facial planes, ocular displacement of the LT peri-ocular adnexae, nasal collapse and significant lip contraction. (**b**) Profile. Chin markedly retrusive with dense plaques of keloid scar in the hemi-facial and peri-oral planes. Projection of the nasal bridge and tip is deficient as well as large patches of scalp alopecia and absence of the LT ear

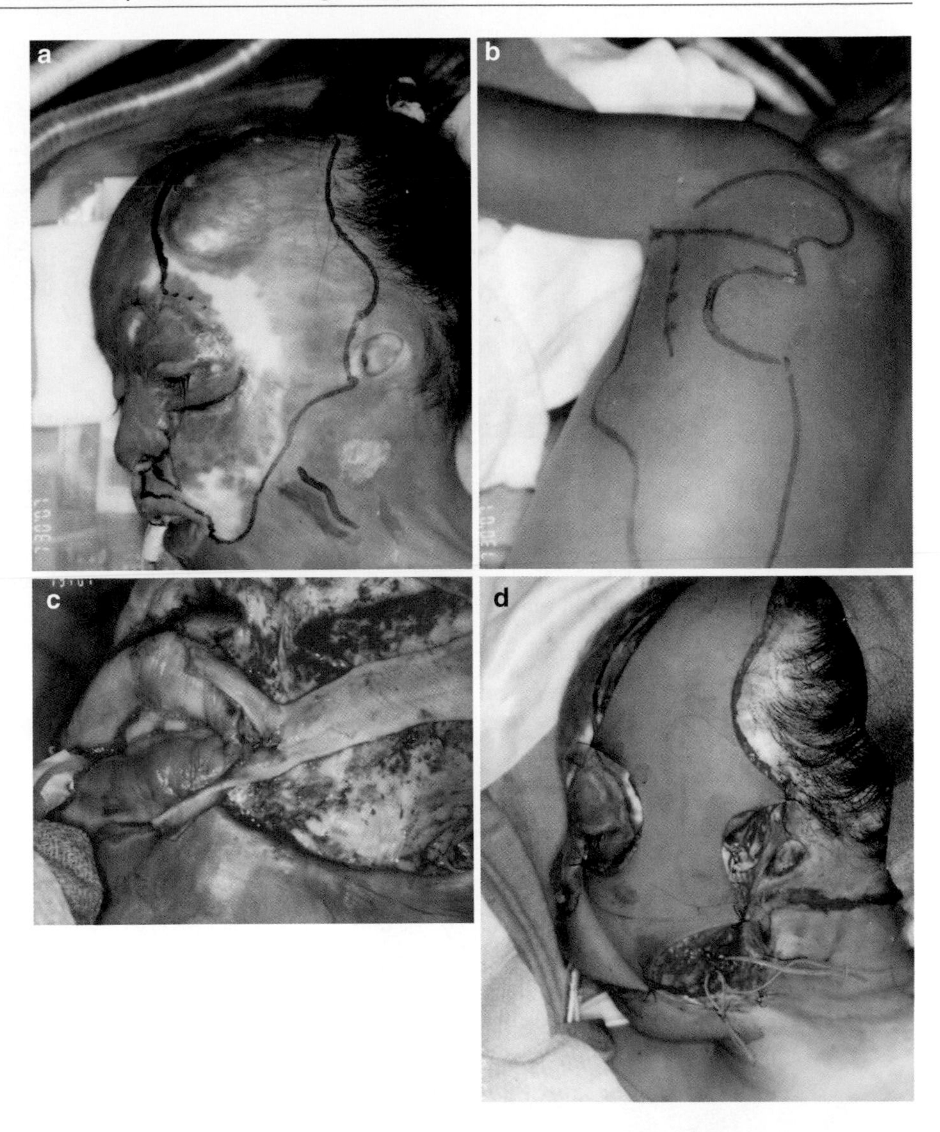

Fig. 10.12 Case 2. First stage free flap transfer (**a**) Keloid excision of LT hemi-face. (**b**) Design of patterned scapular flap. (**c**) Inset of fascia lata sling to lateral lip. (**d**) Intra-operative transfer of patterned free flap

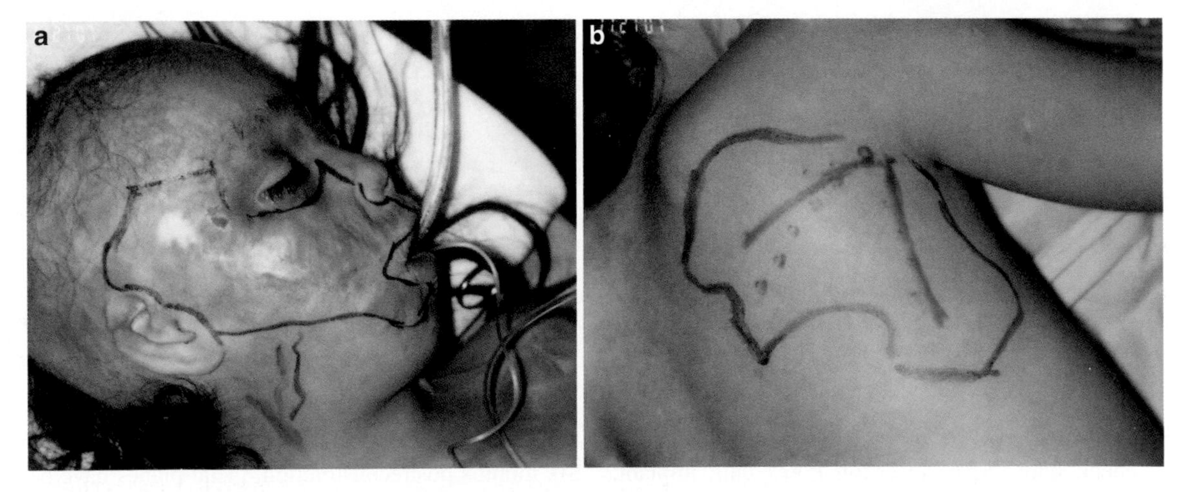

Fig. 10.13 Case 2. Second stage free flap transfer. (**a**) Keloid excision RT hemi-face. (**b**) Design of patterned scapular flap

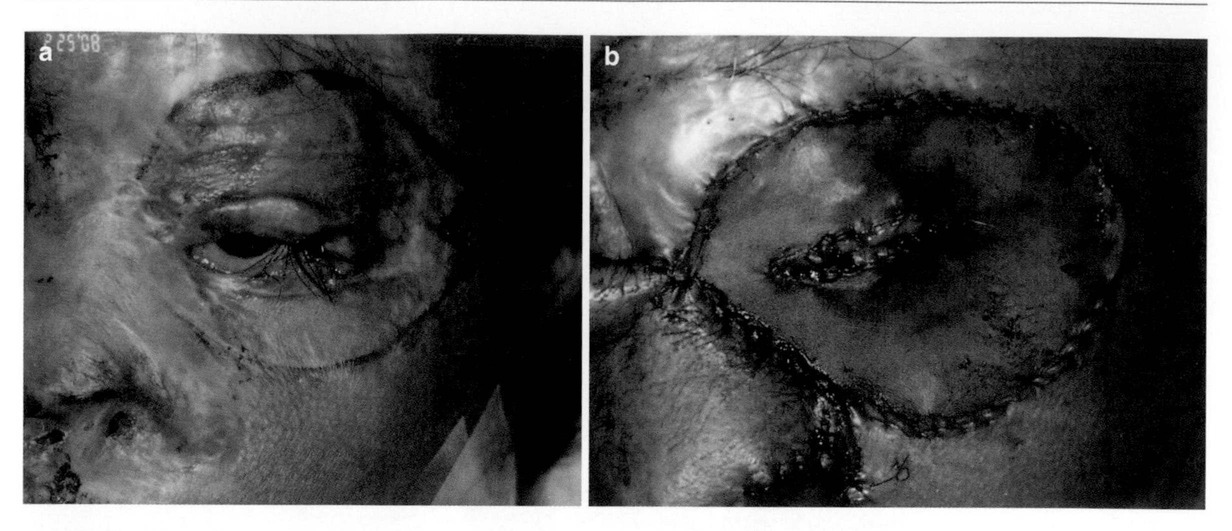

Fig. 10.14 Case 2. Peri-ocular reconstruction. (**a**) Pattern of excision. (**b**) Both upper and lower lids resurfaced with a single sheet graft to the orbital subunit with a slit for the ciliary aperture. Medial canthal ligament re-aligned by transnasal wire fixation and lateral canthal ligament re-suspended to the lateral orbital rim

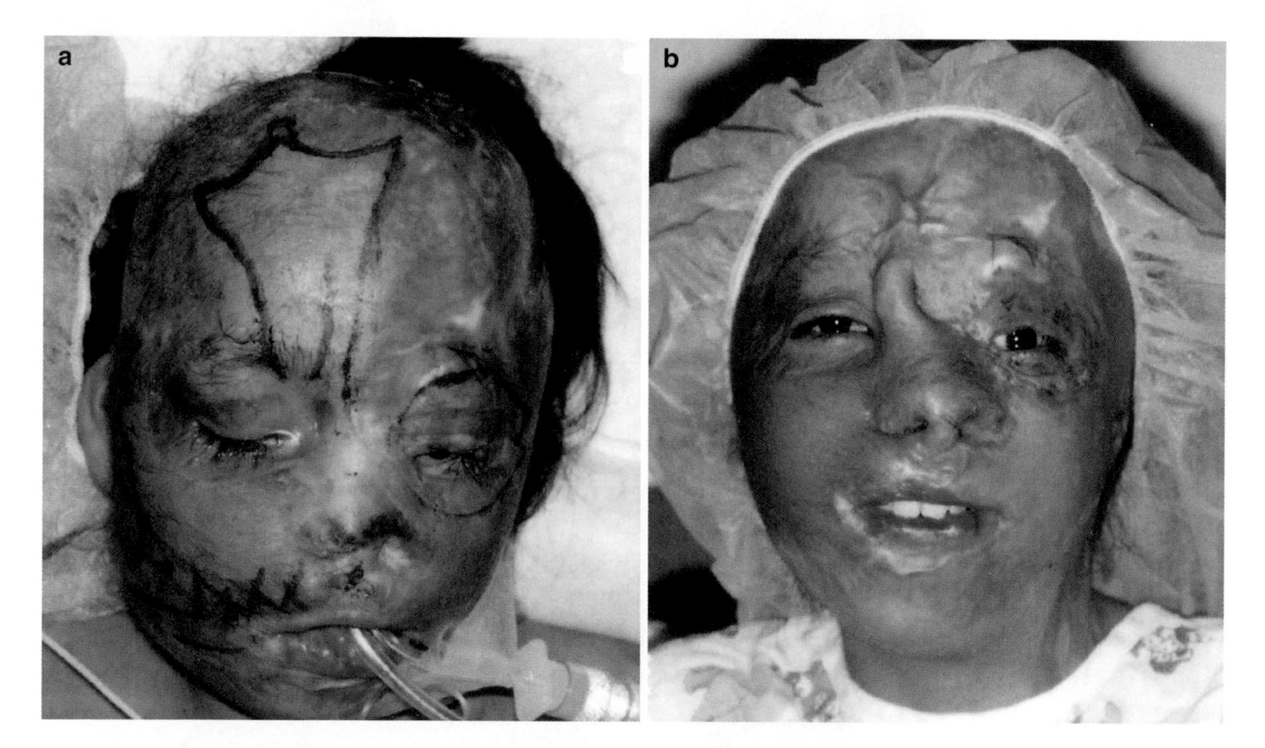

Fig. 10.15 Case 2. Total nasal reconstruction. (**a**) Design of patterned pedicled forehead flap. (**b**) Architectural modification of the nasal tip with conchal cartilage grafts and external resurfacing with a patterned, pedicled forehead flap

procedures included debulking/contouring of the nasal and cheek flaps, SAL, insertion Porex chin implant, levator advancement OS, dermal strip grafts for upper lip augmentation, nostril thinning and repositioning, multiple scar revisions, and laser resurfacing (Fig. 10.17). Six months postreconstruction, facial planes have been restored with soft, textured surfaces. Facial structures (nose, lips, eyes) are balanced and complementary

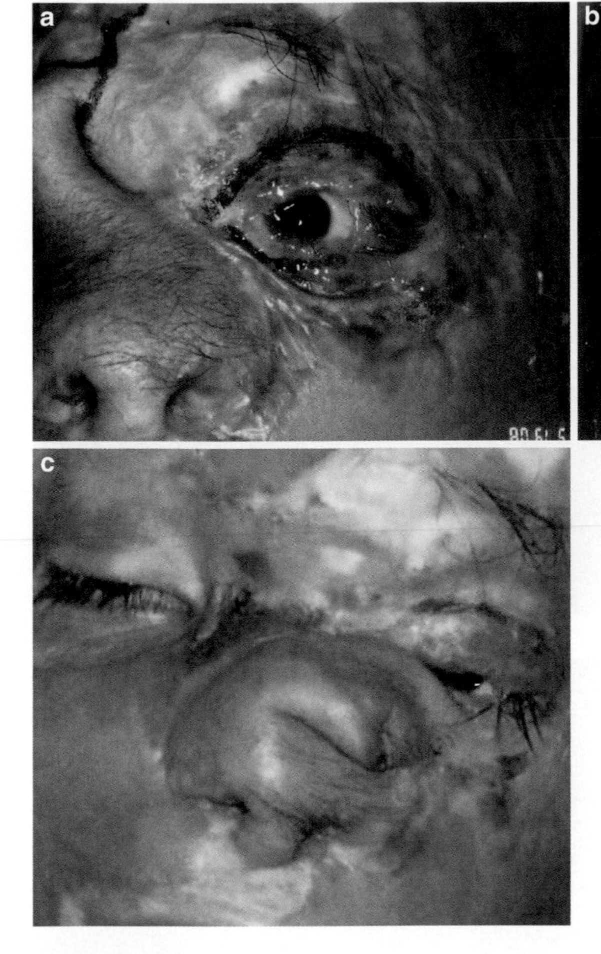

Fig. 10.16 Case 2. Lower lid ectropion reconstruction. (**a**) LT lower lid ectropion. (**b**, **c**) Divided base of the nasal pedicle "piggy-backed" to the lower eyelid for ectropion repair prior to permanent inset. Cartilage graft placed in lower lid for architectural support

(Fig. 10.18a). Animation is complete, and smile is symmetrical (Fig. 10.18b). Eyelid adnexal structures are realigned, and nostril shape is restored. On profile, nasal, lip, and chin projection are proportional (Fig. 10.19).

Total length of treatment was 3½ years. Total number of surgeries = 9. Total hospital cost was $325,900.

10.7 Advantage/Disadvantages of Autogenous Facial Restoration

10.7.1 Advantages of Autogenous Facial Restoration

1. No necessity for life-long immunosuppression
2. No psychological identification with donor
3. No fear of rejection or graft vs. host reaction (face "falling off")
4. Avoidance of other medical problems associated with CTA – melanoma, lymphoma, CMV, infiltrative pneumonias, diabetes mellitus, kidney failure, etc
5. Treatment is finished *for all intent and purposes* at the conclusion of the multistage process
6. Cost is limited to initial stages of treatment

10.7.2 Disadvantages of Autogenous Facial Restoration

1. Multiple stages required to achieve success
2. Multiple hospitalizations and anesthetic exposures
3. Great challenges to achieve "perfection" in the central portion of the face (nose, peri-ocular, and peri-oral)

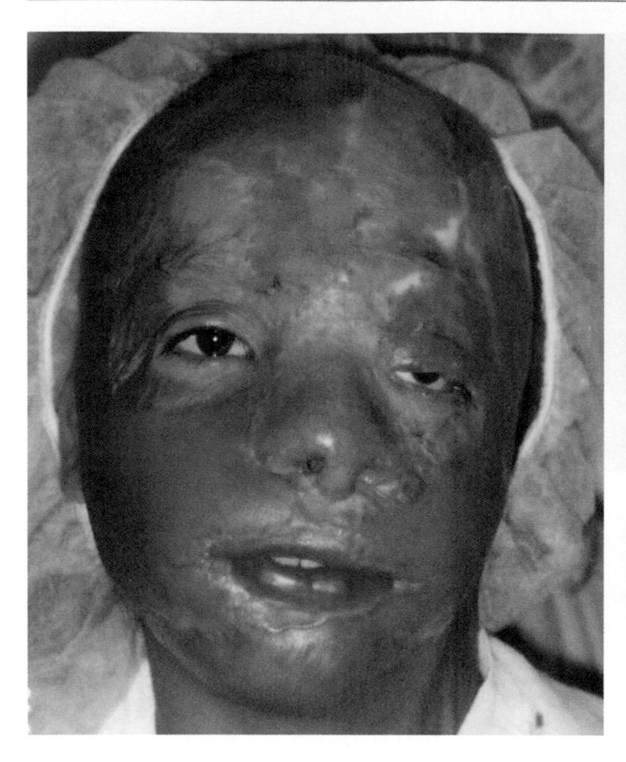

Fig. 10.17 Case 2. Intermediate stage, prior to additional "refinement" procedures included debulking/contouring of the nasal and cheek flaps, SAL, insertion Porex chin implant, levator advancement OS, dermal strip grafts for upper lip augmentation, nostril thinning and repositioning, multiple scar revisions, and laser resurfacing

4. Limited donor sites in extensive burn victims
5. Psychological "burn out" of multiple surgeries (both surgeon and patient).
6. Economic impact of multiple surgeries
7. Need for multiple insurance authorizations

10.8 Be "Fair and Balanced" in Selection Process

The author personally sees a "bright future" for CTA face transplants, particularly as newer immunosuppression technology emerges to achieve tolerance to the composite facial tissue by the existence of chimerism in the recipient after transplantation (the presence of two different cell lines within the patient). Unfortunately, to date, there has been no evidence of chimerism in any of the transplant recipients.[5,7] However, experience in solid organ transplants has shown that in the pediatric population, a few have developed sustainable chimerism and tolerance, allowing for withdrawal of immunosuppression.[31,32]

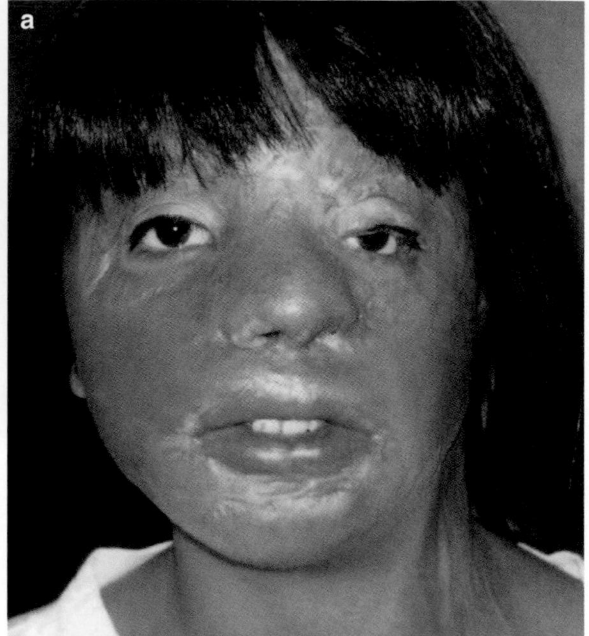

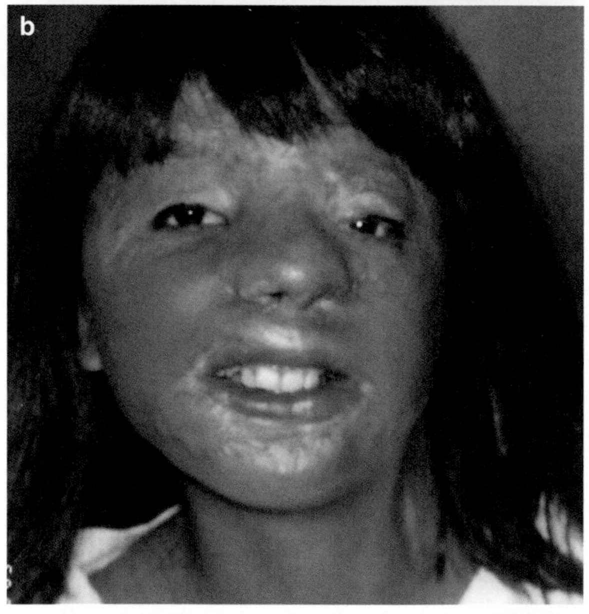

Fig. 10.18 Case 2. Postoperative at 6 months. Frontal view. (**a**) Facial planes have been restored with soft, textured surfaces. Facial structures (nose, lips, eyes) are balanced and complementary. (**b**) Animation is complete, and smile is symmetrical. Eyelid adnexal structures are re-aligned, and nostril shape is restored

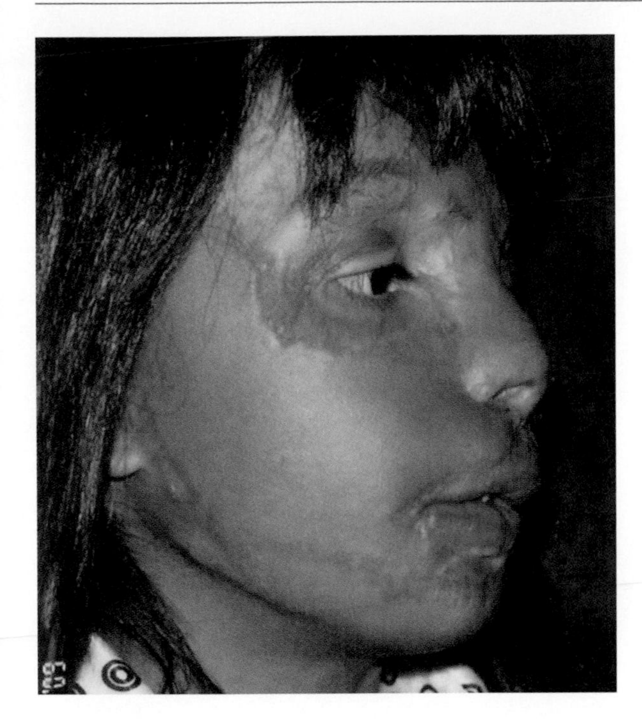

Fig. 10.19 Case 2. Postoperative at 6 months. Profile view. Nasal, lip, and chin projection are proportional

10.9 Conclusion

In the zeal for innovation, the reconstructive surgeon is advised not to neglect more traditional methods of autogenous facial transfers, particularly in more peripheral facial defects (cheeks, forehead, neck). *Prepatterned, sculpted flaps* are remarkably successful in achieving aesthetic excellence in many complex facial deformities.[23,26,30] In the author's opinion, CTA may be the *procedure of choice* for extensive central/ mid facial defects affecting nasal architecture and lip/palatal alignment, particularly with significant skeletal abnormalities (as exemplified in Dr. Siemionow's case). The ability to "customize" the bony and soft tissue elements in harvesting allograft tissue and transfer as a "single unit" in a one-stage operation is metaphorically replacing the missing "piece of the puzzle" in a precise and accurate way (unlike the need to "mold" the autogenous flap to fit the defect). Given these skills, it is critically important at this early juncture in CTA development that judicious decisions be made by the face transplant "pioneers" in terms of patient selection and safety, so as not to jeopardize the minions of transplant surgeons that will follow.

References

1. Siemienow M, Papay F, Alam D, et al. First U.S. near-total human face transplantation – a paradigm shift for massive facial injuries. *Lancet*. 2009;374:203-209.
2. Gordon CR, Siemienow M, Papay F. The world's experience with facial transplantation: what have we learned thus far? *Ann Plast Surg*. 2009;63:572-578.
3. Chinese face transplant Li Guoxing dies. Available at http://www.news.com.au/story/0,27574,24829166–23109,00.html Accessed on 20 Dec 2008.
4. Face-and-hands transplant patient dies. Available at: http://www.msnbc.msn.com/id/31367511/ Accessed on 15 Jun 2009.
5. Dubernard JM, Lengele B, Morelon E, et al. Outcomes 18 months after the first human partial face transplantation. *N Engl J Med*. 2007;357:2451-2460.
6. Guo S, Han Y, Ahang X, et al. Human facial allotransplantation: a 2-year follow up study. *Lancet*. 2008;372:631-638.
7. Lantieri I, Meningaud JP, Grimbert P, et al. Repair of the lower and middle parts of the face by composite tissue allotransplantation in a patient with massive plexiform neurofibroma: a 1 year follow-up study. *Lancet*. 2008;372:639-645.
8. Morris P, Bradley A, Doyal L, et al. Facial transplantation: Working Party report: 2003. Available at http://www.reseng.ac.uk/reseng/content/publications/docs/facial_transplantation.html, 2007. Accessed on Nov 2006.
9. Rumsey N. Psychlogical aspects of face transplantation; read the small print carefully. *Am J Bioethics*. 2004;4:22-25.
10. Morreim EH. About face: downplaying the role of the press in facial transplantation research. *Am J Bioeth*. 2004; 4:27-29.
11. Barker JH, Brown CS, Cunningham M, et al. Ethical considerations in human facial transplantation. *Ann Plast Surg*. 2008;60:103-109.
12. Barker JH, Furr A, Cunningham M, et al. Investigation of risk acceptance in facial transplantation. *Plast Reconstr Surg*. 2006;118:663-670.
13. Siemienow M, Sonmez E. Face as an organ. *Ann Plast Surg*. 2008;61:345-352.
14. Lengele BG. Current concepts and future challenges in facial transplantation. *Clin Plast Surg*. 2007;36:507-521.
15. White BE, Brassington I. Facial allograft transplants: where's the catch? *J Med Ethics*. 2008;34:723-726.
16. Gordon CR, Siemienow M, Coffman K, et al. The Cleveland Clinic FACES score: a preliminary assessment tool for identifying the optimal face transplant candidate. *J Craniofac Surg*. 2009;20:214-220.
17. Gordon CR, Nazzal J, Lozano-Calderan SA, et al. From experimental rat hindlimb to clinical face composite tissue allotransplantation: historical background and current status. *Microsurgery*. 2006;26:566-572.
18. Hui-Chou HG, Nam AJ, Rodriguez ED. Clinical composite tissue allotransplantation: a review of the first four global experiences and future implications. *Plast Reconstr Surg*. 2010;125:538-546.
19. Madani H, Hettiaratchy S, Clarke A, et al. Immunosuppression in an emerging field of plastic reconstructive surgery: composite tissue allotransplantation. *J Plast Reconstr Aesthet Surg*. 2008;61:245-249.

20. Menick FJ. Defects of the nose, lips, and cheek: rebuilding the composite defect. *Plast Reconstr Surg*. 2007;120: 887-898.
21. Burget G, Walton R. Optimal use of microvascular free flaps, cartilage grafts, and a paramedian forehead flap for aesthetic reconstruction of the nose and adjacent facial units. *Plast Reconstr Surg*. 2007;120:1228.
22. Rose EH, Norris MS. The versatile temporoparietal fascial flap: adaptability to a variety of composite defects. *Plast Reconstr Surg*. 1990;85:224-232.
23. Rose EH. Prepatterned microsurgical tissue transfers for replacement of aesthetic facial subunits. In: Rose EH, ed. *Aesthetic Facial Restoration*. Philadelphia: Lipincott-Raven; 1998.
24. Gonzalez-Ulloa M. Restoration of the face covering by means of selected skin in regional aesthetic units. *Br J Plast Surg*. 1956;9:212-221.
25. Menick FJ. Artistry in aesthetic surgery. Aesthetic perception and the subunit principle. *Clin Plast Surg*. 1987;14:723-735.
26. Rose EH. Aesthetic restoration of the severely disfigured face in burn victims: a comprehensive strategy. *Plast Reconstr Surg*. 1995;96:1573-1585.
27. Hyakusoku H, Pennington DG, Gao JH. Microvascular augmentation of the super thin occipito-cervico-dorsal flap. *Br J Plast Surg*. 1994;47:465-469.
28. Chin T, Ogawa R, Murakami M, et al. An anatomical study and clinical cases of "super-thin flaps" with transverse cervical perforators. *Br J Plast Surg*. 2005;58:550-555.
29. Gao JH, Ogawa R, Hyakusoku H, et al. Reconstruction of the face and neck scar contractures using staged transfer of expanded "super thin flaps. *Burns*. 2007;33:760-763.
30. Rose EH. Prepatterned, sculpted free flaps for facial burns. In: Hyakusoku, ed. *Chapter 44: Color Atlas of Burn Reconstructive Surgery*. Berlin, Heidelberg: Springer; 2010.
31. Tzakis AG, Reyes J, Zeevi A, et al. Early tolerance in pediatric liver allograft recipients. *J Pedatr Surg*. 1994;29:754-756.
32. Mineo D, Ricordi C. Chimerism and liver transplant tolerance. *J Hepatol*. 2008;49:478-480.
33. Rose EH. *Aesthetic Facial Restoration*. Philadelphia: Lipincott-Raven; 1998.

Multi-staged Autologous Reconstruction of the Face

11

Lawrence J. Gottlieb and Russell R. Reid

Contents

Abstract In 1948, Sir Archibald McIndoe, when referring to reconstruction of the burned face, stated: "The aim is to produce a face which in sum is symmetrical in its separate parts, of good color and texture and freely mobile so that expression of mood is possible in all its infinite variety". Three different cases are presented at varying levels of age and deformity. The techniques and plans designed to correct these deformities have been described in order to demonstrate the intense level of planning and technical understanding necessary for successful execution. We have demonstrated three different complex patients that have been treated successfully with current techniques. While not readily available for direct comparison, the level of deformity requires intense speculation and definitive treatment. While the debate regarding autologous versus interventional procedures will continue, these important issues deserve ongoing discussion in the light of informed debate.

Abbreviations

CTA Composite tissue allotransplantation
DIEP Deep inferior epigastric perforator

11.1 Introduction

The first documentation of autologous facial reconstruction dates back to the ancient writings of Susruta before the sixth century BCE. As described in a letter to the editor of *Gentleman's Magazine* in 1794, the ancient Indian method of nasal reconstruction required at least two stages: insetting and dividing.[1] The beginning of multistaged autologous reconstruction of the

L.J. Gottlieb (✉)
Department of Surgery, Section of Plastic and Reconstructive Surgery, The University of Chicago Medical Center, Chicago, IL, USA
e-mail: lgottlie@surgery.bsd.uchicago.edu

M.Z. Siemionow (ed.), *The Know-How of Face Transplantation*,
DOI: 10.1007/978-0-85729-253-7_11,

face in the Western world is generally thought to date back to the fifteenth century with the use of delayed skin flaps from the arm by the Branca family of Sicily; this method was refined and widely popularized by Gaspere Tagliacozzi at the end of the sixteenth century.

A number of multistaged pedicled flaps were described in the nineteenth century, including Sabattini's cross lip flap in 1837, Mutter's epaulette flap in 1841, and Halsted's "waltzed" flap for neck burn contracture in 1896. Although one stage island flaps were described by Gersuny in 1887 and Monks in 1898, the multistaged tubed-pedicle flap became popular after independent descriptions by Filitov and Gillies in 1917.[2] In the same year, Esser disparaged these techniques for lip reconstruction stating, "The methods in general use did not satisfy me, as the results were not sufficient in an aesthetic and functional way. The use of the pedicle flap of skin from the arm or wandering flap from the breast is generally besides its disagreeable technic for patients, decidedly disfiguring. The color, paleness, hairlessness, flaccidness and other particulars or qualities differ so much from the skin of the face, especially in the neighborhood of the nose, that such a technical successful plastic only succeeds in closing the defect, but does not construct a proper lip."[3] In 1934, he describes the advantages of axial pattern flaps of the head and neck using direct palpation to find the arteries.[4] However, it is not until the 1950s that the initial description of a musculocutaneous flap is described for head and neck reconstruction.[5] In 1963, Bakamjian describes a single stage reconstruction of a palate with a sternocleidomastoid flap.[6] Two years later, he describes the versatile multistaged deltopectoral flap that revolutionized head and neck cancer reconstruction.[7] In 1967, Fujino, in studying the circulation of the skin, describes the importance of perforator vessels for flap viability[8] – a concept that seemed to go unnoticed for two ensuing decades.

The next major advance in pedicle flap development for head and neck reconstruction was Ariyan's description of the pectoralis myocutaneous flap.[9] Meanwhile, the 1960s and 1970s witnessed the birth of clinically relevant reconstructive microsurgery for head and neck reconstruction throughout the world. In 1964, Nakayama et al. reported a series of microsurgical free-tissue transfers using vascularized intestinal segments to the neck for esophageal reconstruction using 3–4-mm vessels following cancer resections.[10] McLean and Buncke introduced clinical microvascular surgery of the *head and neck* to the Western world with the report of a successful transfer of omentum to the scalp in 1972.[11] Two years later, Taylor and colleagues, from Australia, introduced the concept of free vascularized bone flaps to reconstruct head and neck cancer defects.[12] Presently, an improved understanding of anatomy and clinically relevant knowledge of tissue circulation, along with the development of new tissue transfer techniques, have brought reconstructive surgery to new heights of current accomplishments with autologous reconstructive techniques.

Although the newer autologous tissue transfer techniques have allowed reconstructive surgeons to move large portions of the body to a new location in one stage, much of Esser's critique, related to tubed-pedicle flaps, hold true for more modern transfer techniques of distant tissue. Even after reconstructive surgeons have advanced from a perspective of being satisfied with amorphous "blobs" of tissue surviving transfer to an era of refinement, it is rare to have an aesthetically and functionally acceptable result without secondary procedures. Moreover, despite Esser's apt comments, there are diseases unique to the modern era (gunshot wounds, high-speed motor-vehicle collisions, and radiotherapy) that force every reconstructive surgeon's hand to ascend the reconstructive elevator and resort to distant tissue that require multiple revisions.

11.2 Functional Units of the Face

The function of the face is to look normal or "human" in both repose and animation and to express the spectrum of emotions allowing for successful interpersonal interaction. In 1948, Sir Archibald McIndoe, referring to reconstruction of the burned face, stated "The aim is to produce a face which in sum is symmetrical in its separate parts, of good color and texture and freely mobile so that expression of mood is possible in all its infinite variety".[13] Influenced by McIndoe's results, Dr. Mario Gonzalez-Ulloa described the aesthetic units of the face delineating lines of demarcation of different segments.[14,15] Most surgeons think of these lines as mere guides to the placement of seams or suture lines, so they can be hidden in natural skin lines or shadows. Although this concept has been helpful to many reconstructive surgeons, the true importance of Gonzalez-Ulloa's work is that each of these aesthetic units has unique skin texture, color, thickness, and histologic structure.[16] Burget and Menick have gone on to further subdivide the aesthetic units of the nose and lips to subunits.[17,18] They emphasize not only the skin cover, but also the importance of structure and lining in reconstructing the semblance of a nose.[19]

Likewise, when we think of a functional face, we should not only think of the skin covering but we must think of all the components beneath the skin that give the face its three-dimensional human characteristics, and allow for communication and expression of emotions. As such, functional units of the face are three-dimensional considerations of aesthetic units. Each individual component of facial units, including its underlying musculature and/or structural support, has specialized functions.

The forehead and brow skin with its underlying frontalis muscles contributes to facial expression, and the frontal bone contributes to the protection of the eyes and brain. The eyelids with its fine delicate structures protect the eyes and produce lubrication to maintain a continuously moist cornea. When irritated, it has the ability to increase tear production as well as a drainage system to manage excess fluid. The amazingly coordinated blink reflex (primarily of the upper lid) protects the globe from foreign bodies and helps spread the tears and other secretions on the eye surface to keep it moist. The eyelashes serve to heighten the protection of the eye from dust and foreign debris, as well as from perspiration. Since the upper and lower eyelids have different structures and functions, they should each be considered individual functional units. The nose and the midface (the platform of the nose) have been considered the "keystone" of the face and have significant aesthetic and social importance. Additionally, its complex lining warms and humidifies inhaled air, aerates the paranasal sinuses, and delivers air to the cribiform plate, so smell can be sensed by the olfactory nerves. The lips allow for articulating certain sounds, maintenance of oral competence during eating, drinking, sucking, and speaking, as well as expression of emotions with the ability to smile and kiss. In concert with motion of the mandible, they allow access to the mouth not only for food but also for oral and dental hygiene. Its mucosal lining keeps the inner surface of lips moist and probably has immune functions as well. Similar to the eyelids, the upper and lower lips, although working in concert with each other, should be considered separate functional units. The cheeks contain skin, fat, salivary glands, motor and sensory nerves, muscles of mastication, muscles of facial expression, and mucosal lining. The maxillary bones provide structural support of the cheek while the mandible maintains its inferior border. These osseous structures not only support the soft tissues of the face, but also set the critical dimensions (facial height, width, and projection) via vertical and horizontal buttresses. The external ears are important for a normal facial appearance but are otherwise vestigial structures that define the lateral borders of the cheeks and provide a structure to hang jewelry from and keep eyeglasses in place. The chin serves as support for the lips and tongue as well as the lower border of the facial outline. The underlying skeleton of the face not only gives it structural support and protection but also supports the teeth and contains mucosally lined aerated sinuses that make the skull lighter and helps the voice resonate.

Reconstruction of facial units should not only be directed to restore a skin surface similar to that which was lost, with seams hidden by natural skin folds, shadows, and hair, but also to restore the components beneath the skin. The mobile structures of the face, notably the upper eyelids and lips, are the most challenging to reconstruct not only because no true autologous equivalent of these specialized tissues exist, but also because these tissues need to work appropriately in repose and animation, both mechanically and aesthetically, since function and aesthetics are inextricably linked in these functional units. Furthermore, the presence of scars and gravity will invariably have deforming effects on mobile structures. Finally, to further complicate functional unit reconstruction, mobile functional units may require fully operational adjacent structures. Consider perfectly reconstructed lips with no functional cheek muscles!

Although current autologous reconstructive techniques can be used to successfully reconstruct most individual functional units, they frequently fall short of rebuilding these parts to act harmoniously with the rest of the face when large segments of multiple contiguous functional units are destroyed.

11.3 Goals

When reconstructive surgeons are faced with a facial defect or deformity, whether from tumor, trauma, burn, infection, or congenital etiology, goals and priorities must be determined. In some instances, such as palliation, the goal is to close a wound in the safest, most expeditious way with the least morbidity. In others, the initial aim may be to close the wound with planning on reconstructing the defect later. Lastly, the goal may be to rebuild a part with the intent to set initial building blocks for a delayed multistaged reconstruction.

If the goal is merely to close a wound, then one should use the sequential thought process of the wound closure ladder, choosing the simplest, least morbid method available. However, if the goal is to reconstruct a part (or functional unit), then the "best" method should be undertaken to obtain the desired aesthetic and mechanical outcome.[20] It is rare that a complex deformity can be satisfactorily reconstructed using autologous tissue in one stage without at least small refinements later on. Defining the desired aesthetic and mechanical outcome requires discretion on the part of the surgeon and patient, often balancing risks versus benefits of each potential reconstructive option.

In treating patients who suffer from severely disfiguring diseases or injuries, the minimal goal should be to restore the patient's positive sense of self and ability to successfully interact with family, friends, and the rest of society. In many patients this just requires small adjustments to a distorted facial feature; in others it means a major reconstructive effort. Unfortunately, in some the limitations of our technical abilities preclude the accomplishment of this goal with autologous tissue.

When dealing with a partial loss of a facial unit or structure, adjacent "like" tissue can frequently be used to provide an ideal replacement. When the defect is too large to close with "like" tissue, then regional or distant tissue is required. Invariably, if distant skin is used for reconstruction, the result will be lacking: Skin from different parts of the body and even from different parts of the face, head, and neck have different qualities of color, texture, composition, pliability, elasticity, and attachments to deeper tissue, whether they be subcutaneous fat or muscle fibers. For example, although the skin above the clavicles tends to have a better color match to skin of the face as compared to skin from distant sites, its other characteristics (e.g., texture, composition, pliability, elasticity, and attachment to deeper tissues) frequently make it less than an ideal replacement. An exception to this is, perhaps, the likeness of the forehead skin to that of the nasal tip. However, this may require future flap thinning to achieve a desirable result and tends to have a more favorable color match in lighter skin individuals.

Whatever tissue is used to reconstruct a part, one should try to hide scars along the borders of the aesthetic units of the face or within (or parallel to) natural skin lines. In addition, although the human face is not perfectly symmetrical, obtaining relative symmetry and avoiding distortion are essential to reconstruction. The idealized goal in facial reconstruction is for a casual observer to not be able to notice the scars (or the reconstruction) in normal social interactions.

11.4 Case Examples

Describing all the various techniques and advances described since Tagliacozzi for achieving these goals is beyond the scope of this chapter. Rather some illustrative case examples of multistaged autologous reconstruction will be presented.

Case 1 (Fig. 11.1a–g) – This 60-year-old woman presented with an infected, indolent basal cell carcinoma on the right side of her face. After initial extirpation, wound closure was delayed until bacteriologic control of the wound and negative pathologic margins could be assured. Wound closure was accomplished with a latissimus dorsi musculocutaneous flap. The muscle portion of the flap (covered by a skin graft) was used to close the scalp, and the skin portion was used to close the lateral portion of the face. Multiple thinning procedures were then performed to provide a normal contour.

Case 2 (Fig. 11.2a–k) – This patient is a 40-year-old male who sustained a gunshot wound to his face. The resulting three-dimensional defect is missing lining, structural support, and cover as well as muscle and nerves. After an initial debridement, the wound was redebrided including a completion parotidectomy with tagging of the proximal portion of the facial nerve. The first stage of reconstruction provided not

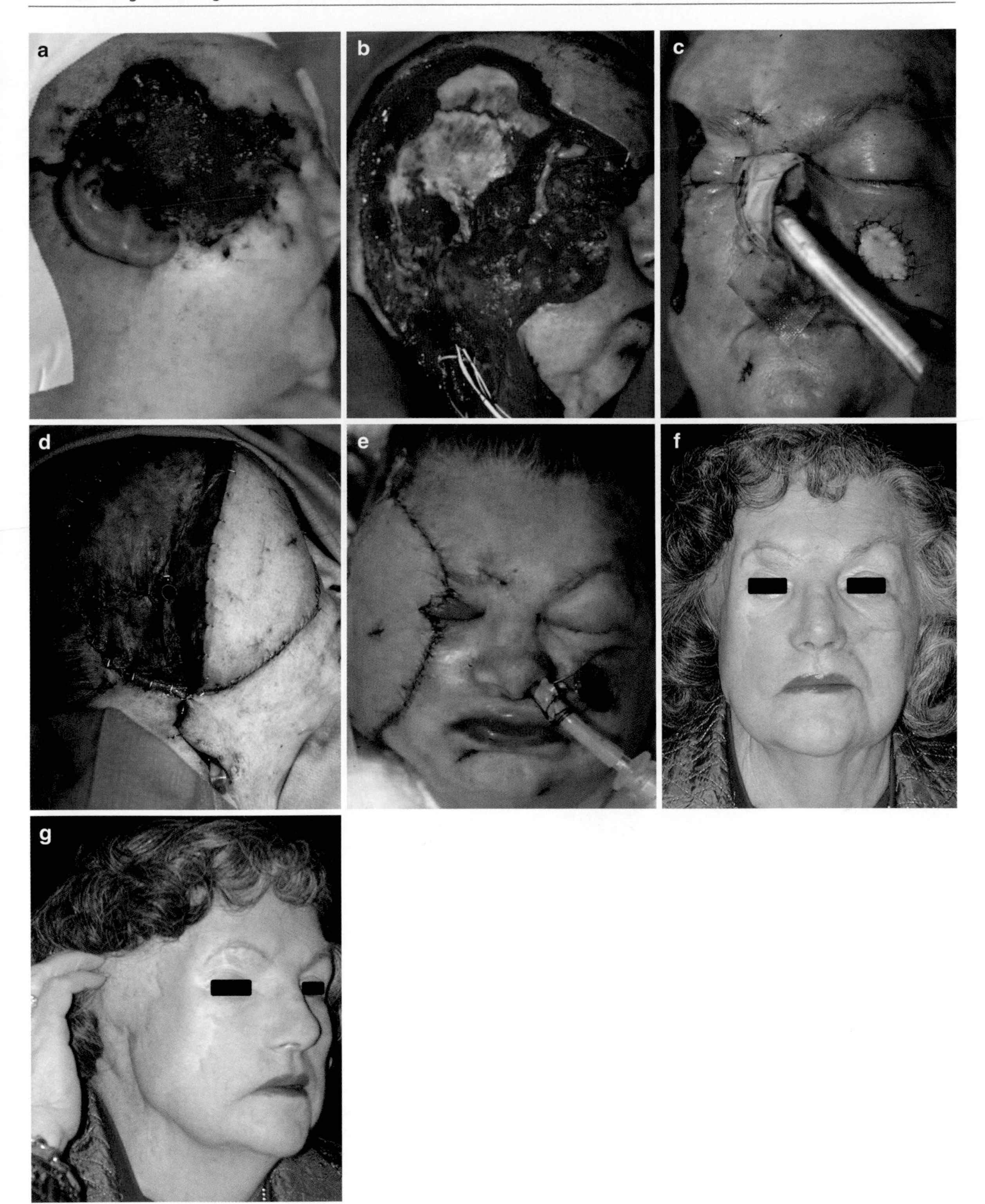

Fig. 11.1 (**a**) A 60-year-old female with infected basal cell carcinoma invading side of face and scalp. (**b**, **c**) After extirpation and obtaining bacteriologic balance of wound. Resection included part of brow, lateral eyelid skin, temporalis muscle, parotid and facial nerve leaving exposed skull and lateral orbital rim. (**d**, **e**) Latissimus dorsi musculocutaneous flap used for closure with posterior (scalp) area closed with muscle portion of flap (covered with a skin graft) and the anterior portion of the wound closed with the skin portion of the flap. (**f**, **g**) Two years after multiple thinning procedures and peri-orbital refinements

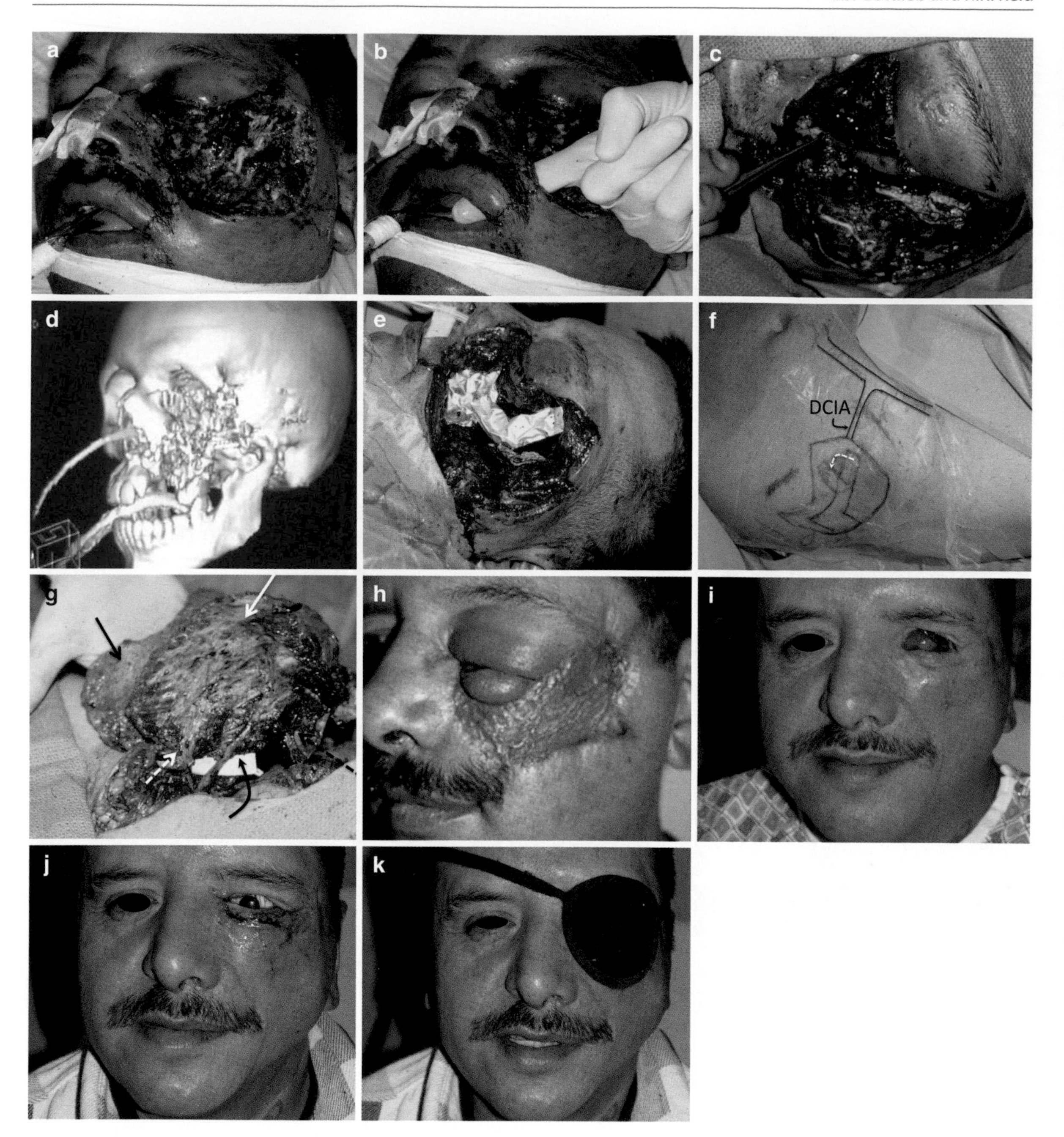

Fig. 11.2 (**a**, **b**) A 40-year-old male with gunshot wound to face with loss of maxilla, orbital floor, eye, lining, and overlying soft tissue. (**c**) Avulsion of lower eyelid. (**d**) Three-dimensional CT scan showing bony injury. (**e**) Tin foil template of anticipated bony reconstruction. (**f**) Design of Iiiac crest osteomusculocutaneous flap based on the deep circumflex iliac artery (*DCIA*). White dashed arrow depicts position of bone after osteotomy. (**g**) Undersurface of flap. Internal oblique muscle (*solid white arrow*) based on the ascending branch (*dashed white arrow*) of DCIA (*curved black arrow*) used for lining. Skin and subcutaneous tissue (*black arrow*) and iliac crest (*dashed black arrow*). (**h**) Marginal circulation of skin paddle and bulkiness of flap led to removal of skin and subcutaneous tissue and closure with a skin graft. (**i**) After tissue expansion to replace skin graft and facial animation with a neurotized segmental latissimus muscle free flap. (**j**) With orbital prosthesis after recent composite graft to lower eyelid. (**k**) Despite multiple operations on left lower eyelid over a 6-year period, the patient was never satisfied enough with the result to go in public without his eye patch

only lining and cover to close the wound but also provided structural support to maintain 3-dimensional contour. The next stage was to expand the lower cheek/upper neck tissue and provide color-matched skin with scars that were hidden along the borders of the aesthetic unit. The last stage of cheek reconstruction was to provide support and animation to the corner of the mouth with a neurotized segmental latissimus dorsi functional muscle transfer. The lower eyelid was dealt with conventional techniques of local flaps and composite grafts.

Case 3 (Fig. 11.3a–t) – This 8-year-old boy sustained a massive facial injury from a pit bull attack. His entire right cheek, right lower eyelid skin, nose, left cheek, 20% of left upper lip, 10% of left lower lip, entire left upper and lower eyelids, left forehead, and both ears were missing. After initial stabilization and establishment of an airway, the patient was taken to the OR for exploration, evaluation, debridement, salvage repair, and provision of physiologic cover. During that initial procedure, the left eye was covered with conjunctival flaps, the right parotid duct was repaired (the left was not found), and avulsed small branches of the right facial nerve were inserted directly into muscle. The left facial nerve was not found. In addition, his left oral commissure was repaired and associated lacerations of surrounding skin were closed. Temporary physiologic closure was

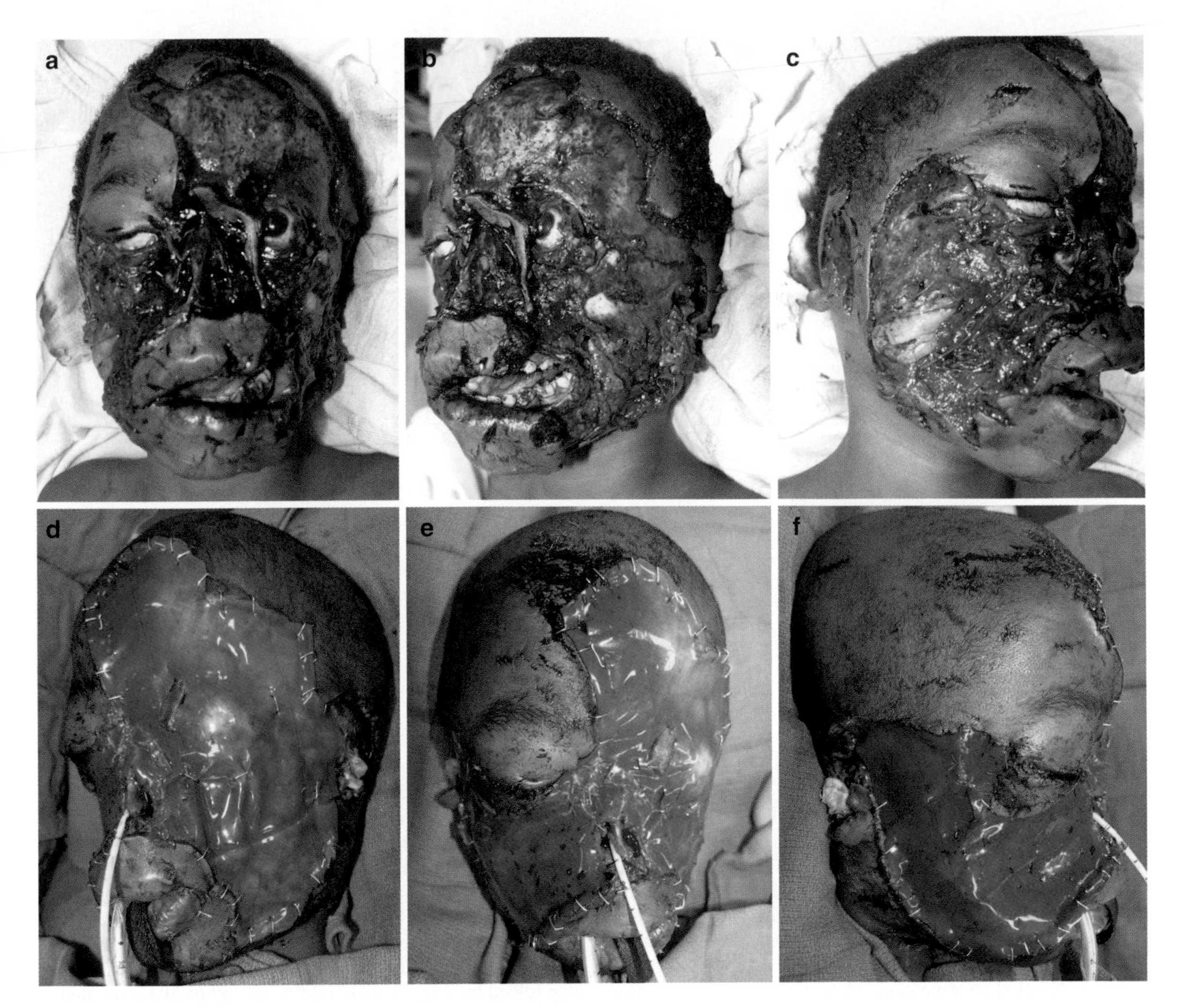

Fig. 11.3 (**a–c**) Facial injuries of an 8-year-old boy after being mauled by a pit bull. At presentation he was alert, awake, and talking. (**d–f**) Temporary physiologic coverage with Integra™ Bilayer Matrix Wound Dressing

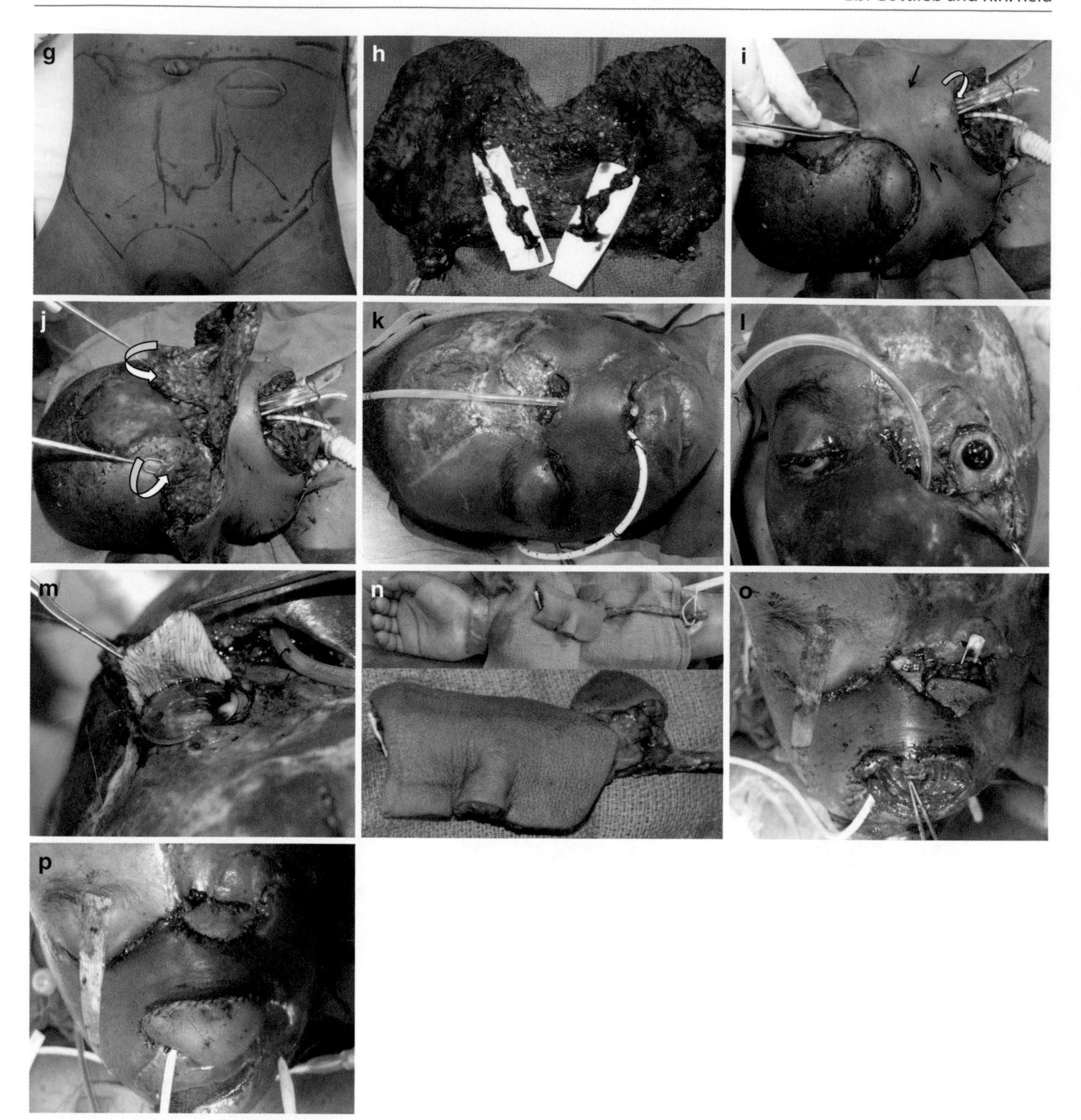

Fig. 11.3 (continued) (**g**) Markings of flap on lower abdomen. (**h**) Undersurface of double pedicle DIEP flap. (**i**) Flap transferred with microvascular anastomosis to bilateral facial arteries and veins. A portion of the flap flipped under center portion of flap for nasal lining (*curved white arrow*). Flap suspended with multiple bone anchoring sutures (*black arrows*). (**j**) Sub-scarpa fat flaps folded under main flap (*curved white arrows*) to provide fullness in malar areas and simultaneously thin the upper portion of flap. (**k**) Nasal lining flap died as did a small area medial to left eye. Nasal skin projection temporarily maintained with a Foley catheter balloon. Feeding tube in right nostril. (**l**) Retraction of upper left side of flap demonstrating loss of conjunctival flaps' complete absence of left upper and lower eyelids. (**m**) Insetting of mucoperiosteal hard palate graft for eyelid lining. (**n**) Double paddle radial forearm flap. The larger paddle will be used to resurface nasal lining and nasal tip, the smaller flap on a proximal perforator to resurface wound breakdown of upper portion of flap. (**o**) Nasal lining inset, small paddle passed to upper portion of flap. Cranial bone inserted. Microvascular anastomosis to DIEP pedicle of abdominal flap. (**p**) Radial forearm flap folded back to create nasal tip. Second paddle inset medial to left eye

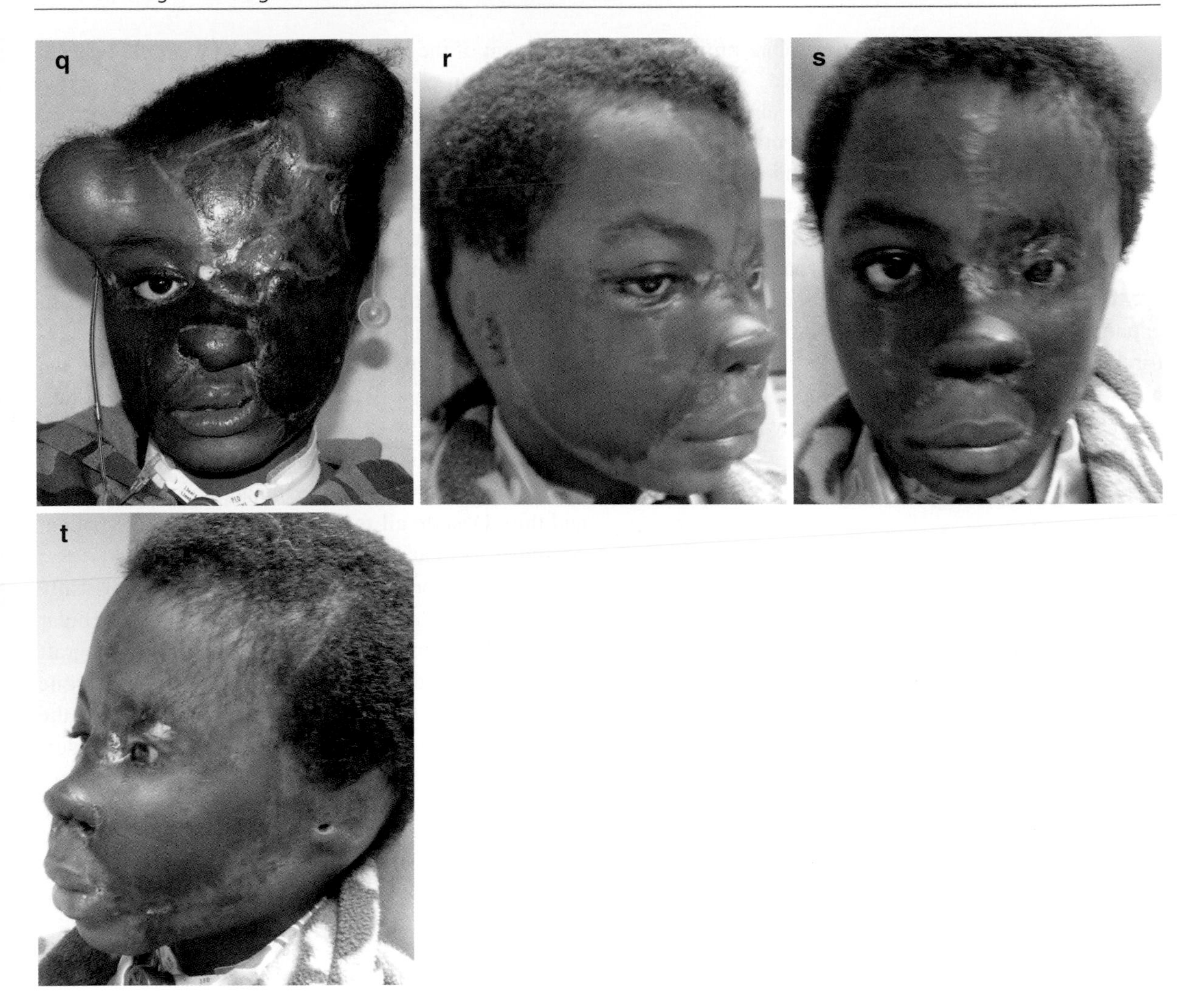

Fig. 11.3 (continued) (**q**) Ten weeks after injury with tissue expanders to facilitate forehead reconstruction. (**r–t**) One year after injury after 18 operations including two free flaps

obtained with a collagen-glycosaminoglycan biodegradable matrix (INTEGRA™ Bilayer Matrix Wound Dressing, Integra LifeSciences, Plainsboro, NJ.). A double pedicle deep inferior epigastric perforator (DIEP) lower abdominal free flap was transferred 12 days after the injury. It was decided not to include the left forehead defect so as not to preclude closure of the abdominal donor site; therefore, the left side of the forehead was closed with a skin graft. The abdominal flap was suspended to bone using multiple bone-anchoring (Mitek™, DePuy Mitek, Inc.) sutures. Nasal lining was provided with a turned-in portion of the inferior portion of the abdominal flap, and the left eye was totally covered with the flap. Temporary nasal projection was accomplished with cadaver cartilage grafts. The superior portion of the abdominal flap was thinned by dissecting the subscarpa's fat layer, which was then turned under the flap to provide more bulk in the malar area. Postoperatively, the nasal lining portion failed and the area just medial to the left eye broke down. After debridement of the failed nasal lining, the outside nasal skin height was maintained with an intranasal Foley catheter. As expected, the loss of both upper and lower left eyelids was the most significant challenge. The left eyelid conjunctival flaps had separated, and a full thickness hard palate mucosal graft was used for conjunctival replacement. The thinned abdominal flap was used for eyelid skin and to revascularize the mucosal graft. One month later, autologous costochondral cartilage grafts were inserted between the mucosal graft and the outside skin. At a subsequent operation, a remnant of levator was found which was mobilized and secured to the neo-tarsus. Secondary nasal reconstruction was performed using a double paddle radial forearm flap that was able to provide nasal lining, nasal tip, and skin for the breakdown just medial to the left eye. Tissue expanders were used to reconstruct the left side of the forehead. Multiple small revisions of the eyelids, nose, and lips were performed every 3 or 4 weeks for the first year, alternating sites to allow for swelling to settle.

11.5 Case Analysis

Although, ideally, reconstruction of functional units of the face would restore the three-dimensional loss of tissue within the anatomic boundaries of the aesthetic units, not all defects are amenable to this. When the majority of the unit is present, then it should be preserved. If the majority of the unit is not present, the reconstructive plan should incorporate putting seams and junctions along the borders of the newly established unit. Each of these cases demonstrates different approaches individualized for each patient predicated on the defect, anatomy, and goals.

In case 1, the initial goal was to close the wound. The latissimus dorsi flap was chosen because of its large surface area and the fact that muscle resurfaces the scalp well; in addition, it was anticipated that residual hair could camouflage the defect. Although the skin and subcutaneous tissue component were clearly too bulky to aesthetically resurface the lateral face, subsequent multistaged thinning was planned realizing that normal contour was the key to success and any color discrepancy could be corrected with makeup in this 60-year-old woman.

In case 2, reasonably symmetrical cheek projection with the iliac crest flap could be restored during the initial operation, planning on resurfacing the entire cheek unit during subsequent procedures. Addressing the entire cheek unit with expansion of adjacent skin not only converted the patch scar to linear ones along the natural borders of the cheek but also allowed for placement of the segmental latissimus muscle beneath an intact skin cover. An asymmetrical but acceptable smile was achieved. Loss of the eye could only be restored with a prosthetic eye and the reconstruction of the lower eyelid was minimally acceptable. The biggest challenge was establishing the appropriate level of a platform for his ocular prosthesis and obtaining a stable lower eyelid that was symmetrical with the contralateral one. This reconstruction required a total of 21 operations spanning 6 years including one tightening procedure and one debulking procedure of the latissimus flap as well as three operations to treat complications of the iliac crest flap donor site. Although the functional unit of his upper cheek was restored, the lower portion (overlying the mandible) was too tight and thin. Despite all reconstructive efforts, the patient continues to wear an eye patch.

In case 3, the initial goal was to physiologically close the wound as expeditiously as possible. Acellular bilayer matrix was chosen over allograft or xenograft because it was thought that removal (of the silicone layer) would be the least likely method to disturb the wound with the residual facial nerve sitting on the surface. The typical algorithm after revascularization of the acellular bilayer matrix would have been to replace the temporary silicone layer with a skin graft, subsequently to be replaced by flaps at a later date. Recognizing that skin grafting this defect would result in a severely distorted face, it was decided to skip this step and to try providing a more human-looking face in a shorter time frame. Therefore, 2 weeks after the injury, the major reconstructive component, a double pedicle lower abdominal free flap, was transferred with plans for subsequent refining operations. Unlike typical burn reconstruction, where skin grafts can satisfactorily resurface an entire face, the three-dimensional loss of multiple facial functional units, including muscles, nerves and subcutaneous tissue (in addition to the overlying skin cover), required reconstruction with distant tissue that could be sculpted to provide reasonably normal three-dimensional contours to this destroyed face. This patient required 18 operations in the first year, including 2 free flaps and 16 smaller procedures. His left eyelids did not function normally, and he was unable to smile. Although at 1 year post-injury his face looked and functioned far better than when he initially presented, it was still significantly disfigured, precluding patient's integration back to a normal classroom setting. At 14 months post-injury, he spontaneously regained some animation to the right side of his lips (where the avulsed facial nerve was reimplanted into muscle for direct neurotization), and at 16 months he underwent a facial reanimation procedure on the left side. He has subsequently undergone an additional dozen operations.

11.6 Discussion

Skin resurfacing of the cheek is generally straightforward. Small defects can usually be closed with adjacent tissue rearrangements or local/regional flaps. Large skin defects of entire cheek units can be successfully resurfaced with thick skin grafts. When subcutaneous tissue is missing as well, regional flaps, expanded flaps, prefabricated flaps, or distant flaps, whether pedicled or free, is usually preferable to skin grafts. When distant skin (either as a graft or flap) is used to resurface both sides of the face, a reasonably successful outcome can be achieved. When a unilateral cheek is resurfaced using distant skin, whether as a graft or flap, the result is usually less than satisfactory due to discrepancies of color, texture, composition, pliability, and elasticity of the skin when compared to the normal side. These differences are accentuated when only part of an aesthetic or functional unit is replaced. Multidimensional loss of cheek tissues is more of a challenge to reconstruct satisfactorily as normal contour tends to be the most important element in success of most cases. Despite accomplishing acceptable results in these patients' main defect, the difficulties in obtaining symmetrical eyelid reconstruction detract from an ideal outcome.

Multistaged autologous tissue reconstruction does not truly reconstruct a missing or deformed part or unit but makes a semblance of the part or unit. Considering Gillies' principle of replacing like-with-like, it is hard to believe that multistaged autologous tissue reconstruction of functional units of the face will ever look and function as well as reconstructive transplantation with age-, gender-, color-, and size-matched composite tissue allograft (CTA) replacing exactly what was lost. In addition, donor site scars and morbidity that are inherent with autologous reconstruction must be considered. The balance of course is between the risks and benefits.

11.7 Conclusions

Advances in autologous reconstructive concepts, principles, and techniques allow us to close a wound of almost any size. Functional and aesthetically acceptable reconstruction is more of a challenge and usually requires a multistaged approach. Most individual parts or functional units can be reconstructed in an acceptable way, with perhaps, the exception of total loss of the upper eyelid or total loss of the upper lip in a child or female patient especially when associated with an adjoining unit loss or dysfunction. However, when no autologous donor tissue is available, as in a >90% burn victim, when all autologous donor sites have been used (and failed to reconstruct successfully), when multiple functional units (especially if contiguous) are missing or deformed, or when both upper and lower lips or eyelids are missing, we fall short in accomplishing our goal of making a functional human face capable of restoring the patient's sense of self and ability to successfully interact with society. In these circumstances, CTA becomes a potential option, which should be considered for providing patients with an optimal reconstructive outcome.

References

1. B.L. Letter to the Editor. *Gentleman's Magazine*, London, October, 1794, p. 891; Reprinted in Silvergirl's Surgery. In: McDowell F., ed. *Plastic Surgery*. Austin, TX: Slivergirl, Inc.; 1977, p. 47.
2. Wallace AF. History of plastic surgery. *J Roy Soc Med*. 1978;71:834-838.
3. Esser JF. Studies in plastic surgery of the face: I. Use of skin from the neck to replace face defects. II. Plastic operations about the mouth. III The epidermal inlay. *Ann Surg*. 1917;65: 297-315.
4. Esser JF. Preservation of innervation and circulatory supply in plastic restoration of upper lip. *Ann Surg*. 1934;99:101-111.
5. Owens N. A compound neck pedicle designed for repair of massive facial defects: formation, developments and application. *Plast Reconstr Surg*. 1955;15:369.
6. Bakamjian VY. A technique for primary reconstruction of the palate after radical maxillectomy for cancer. *Plast Reconstr Surg*. 1963;31:103.
7. Bakamjian VY. A two stage method for pharyngo-esophageal reconstruction with a pectoral skin flap. *Plast Reconstr Surg*. 1965;36:173.

8. Fujino T. Contribution of the axial and perforator vasculature in circulation of flaps. *Plast Reconstr Surg*. 1967;39(2): 125-137.
9. Ariyan S. The pectoralis major mycutaneous flap: a versatile flap for reconstruction in head and neck. *Plast Reconstr Surg*. 1979;63:73.
10. Nakayama K, Yamamoto K, Tamiya T, et al. Experience with free autografts of the bowel with a new venous anastomosis apparatus. *Surgery*. 1964;55:796-802.
11. McLean BH, Buncke HJ Jr. Autotransplantation of omentum to a large scalp defect with microsurgical revascularization. *Plast Reconstr Surg*. 1972;49:268.
12. Taylor GI, Miller GD, Ham FJ. The free vascularized bone graft. A clinical extension of microvascular techniques. *Plast Reconstr Surg*. 1975;55:533-544.
13. Donelan MB. Discussion of Aesthetic restoration of the severely disfigured face in burn victims: a comprehensive strategy. *Plast Reconstr Surg*. 1995;96:1586-1587.
14. Gonzalez-Ulloa M. Restoration of the face covering by means of selected skin in regional aesthetic units. *Br J Plast Surg*. 1956;9(3):212-221.
15. Gonzalez-Ulloa M. Regional aesthetic units of the face. *Plast Reconstr Surg*. 1987;79(3):489-490 [Letter].
16. Gonzalez-Ulloa M, Castillo A, Stevens E. Preliminary study of the total restoration of the facial skin. *Plast Reconstr Surg*. 1954;13:151-161.
17. Burget GC, Menick FJ. The subunit principle in nasal reconstruction. *Plast Reconstr Surg*. 1985;76:239-247.
18. Burget GC, Menick FJ. Aesthetic restoration of one-half the upper lip. *Plast Reconstr Surg*. 1986;78(5):583-593.
19. Burget GC, Menick FJ. Nasal reconstruction: seeking a fourth dimension. *Plast Reconstr Surg*. 1986;78(2): 145-157.
20. Gottlieb LJ, Krieger LM. Editorial: from the reconstructive ladder to the reconstructive elevator. *Plast Reconstr Surg*. 1994;93(7):1503-1504.

12 Ethical Aspects of Face Transplantation

George J. Agich

Contents

G.J. Agich
Department of Philosophy, Social Philosophy and Policy Center, Bowling Green State University, Bowling Green, OH, USA
e-mail: agichg@bgsu.edu

Abstract To date, ten face transplants have been performed around the world with two deaths neither of which was directly caused by the transplant. So, while this procedure is still innovative, it is clearly technically feasible, though it is fraught not only with risk, but ethical concerns. In the end, whether face transplantation will become a regular reconstructive surgical alternative for severe facial deformity will depend not only on scientific and technical issues, but the degree to which the clinical ethical problems are identified and addressed. In this chapter, I will focus on the concrete ways that these issues need to be addressed in developing and implementing face transplantation protocols.

12.1 Introduction

To date, ten face transplants have been performed around the world with two deaths neither of which was directly attributed to the transplant itself. So, while this procedure is still innovative, it is clearly technically feasible, though it is fraught not only with risk, but ethical concerns. It is thus ethically important that it be treated as investigational and conducted under the external review of IRB or research ethics committee oversight. Even though the initial challenges of the surgical procedures, including graft procurement, have been overcome in achieving outcomes that initially appear to be satisfactory, the long-term advantages and disadvantages of this procedure are unclear, so it is best to regard face transplantation as an innovative procedure that is not yet mature.[1] Therefore, face transplantation will continue to be a subject rife with ethical controversy.

The ethics of face transplantation became a lively issue in bioethics when it emerged as a topic in the

M.Z. Siemionow (ed.), *The Know-How of Face Transplantation*,
DOI: 10.1007/978-0-85729-253-7_12,

public media in December 2002.[2-10] At the winter meeting of the British Association of Plastic Surgeons, Peter Butler announced his intention to perform the first procedure. Media reports subsequently featured the procedure in a series of provocative articles that sensationalized the surgery. In these reports, science fiction and film scenarios were far more prominent than clinical or scientific sources, so it is no wonder that the initial bioethics reaction to face transplantation was one of caution or skepticism, laced with what bioethicists have termed a "yuk" reaction meriting a skeptical and cautious response.[11,12] The initial reaction of formal review committees in the UK in December 2003[13] and France in February 2004[14] similarly urged caution and stressed the need for further research before the procedure should be undertaken. This work has proceeded.[1,15-31] As face transplantation cases have been reported, the bioethics reaction has moderated to the point where bioethical discussion is far more balanced in reflecting on the actual ethical issues associated with the procedure.[32-44] Like any example of advanced medical care, the financial and resource costs are significant and raise considerations of equity and social justice. Although procuring facial tissue for use in transplantation raises concerns, they are comparable to the ethical questions regarding organ procurement that have been raised in the past. Since these considerations are not unique to face transplantation, they are omitted for the purposes of this paper, which, instead, concentrates on the clinical ethical questions that arise in face transplantation.

By the phrase *clinical ethical questions*, I mean those ethical questions that are unavoidable in developing and implementing face transplantation protocols. These questions are eminently practical and have direct implications for the specific procedures that have to be followed and processes built into the transplant protocol. Many of the theoretical ethical questions and concerns have been discussed widely in the bioethics literature such as informed consent and risk/benefit considerations.[35,36,45-51] Rather than repeat these points, I focus this chapter on the concrete ways that these issues need to be addressed in developing and implementing face transplantation protocols.

The University of Louisville team that pioneered hand transplantation has published extensively on the ethics of face transplantation.[33,38,40,52-55] Their guidelines have been published repeatedly, and they rely on the framework that Francis Moore articulated for addressing ethical questions in clinical trials of new drugs and innovative procedures in surgery.[56-58] The first of these criteria for assessing the ethics of any surgical innovation, according to Moore, is that the scientific background in support of the innovation must be fully adequate before it is ethical to undertake it.[57,58]

This was the one of the initial points of concern raised by the UK review committee in 2003.[59] Since that time, publication of research performing face transplants in animal models,[29,60-65] improved understanding of the immunosuppressant requirements for composite tissue allografts,[30,66] computer and cadaveric work demonstrating that the result of a full or partial face transplant will be a composite of the donor face and the recipient,[19,60,67-69] and reported results of previous surgeries have done much to fortify the scientific basis for the procedure. Even though there is a more robust scientific basis than at the time when it was first proposed, face transplantation is by no means ready for prime time, and for that reason it must be regarded as experimental and subject to the oversight of institutional review committees. Since the risks associated with the surgical procedures involved in face transplantation are not more significant than conventional reconstructive procedures using the patient's own tissue, a central practical ethical question in face transplantation is the construction of an appropriately skilled team to undertake the surgical procedures involved and to provide follow-up care. This point cannot be over-stressed, because the literature on face transplantation often addresses general ethical principles and concerns without drawing implications for the practical construction of an ethically sound protocol. The extensive specific competencies that are required must not just exist as a general knowledge in the field, but must be possessed by the relevant team members. Ethically, we can require in addition that support personnel must include not only the usual and relevantly qualified and experienced clinical support services, but also psychological or clinical ethics support, which are often overlooked.

This is a general point that might be taken ethically as having been settled by reports of successful face transplantation procedures undertaken around the world. However, the ethical requirement about the adequacy of the scientific background implies that the specific surgical and other support teams involved must be fully competent and capable of performing the procedures involved. In part, this is what Moore meant by the criterion of field strength of the team undertaking the innovation.[58] The implication is that it is not ethical to undertake an innovative surgical procedure such as face transplantation unless all members of the team possess the relevant skill and experience in all of the techniques involved, not just surgical but all the supportive services as well.

Before the procedure could be ethically undertaken at any particular institution, the team must be able to function effectively as a team, since on-the-job development of experience in coordinating the complex services required during such a complicated procedure would be ethically unacceptable. In addition to all of the well-discussed ethical considerations involving face transplantation, such as informed consent or risk and benefit considerations, there are several lesser discussed clinical ethical concerns about face transplantation that a transplant team must address: the psychological impact of the procedure, including the issues associated with patient consent, publicity, and the media interest in face transplantation, and the post-surgical risks including those associated with the long-term reliance on immunosuppressive medications. These concerns and other practical challenges associated with developing and implementing clinical protocols for face transplantation are ethically based on the clinical needs of patients with severe facial deformities rather than philosophical issues about the nature of the face in the constitution of personal or social identity, or the already mentioned immature and incomplete status of the scientific background.[35]

12.2 Psychological Aspects

As Siemionow has argued, the face is best approached as an organ[24] and, as the French National Consultative Ethics Committee stressed, the face is an organ of expressivity.[14] In this sense, the face is uniquely unlike other organs. As an organ of expressivity, the face is uniquely identified with the patient both in terms of the individual's self-identity and social identity. Thus, in approaching face transplantation, one must be mindful of the effects of the projected reconstruction for achieving such expressivity. Transplantation of the skin envelope would not transfer the donor visage, but would result in a "face" comprised of features like skin color of the donor tissue and the recipient's own underlying facial structures and shape. The procedure seems to meld the donor and recipient faces, so the critical objective is to achieve a result that is capable of communicating to and with others the patient's feelings and thoughts through facial movements. The outcome needs to be acceptable by the patients as their own face, albeit a newly reconstituted one.

The early ethics literature on face transplantation uncritically accepted the perspective of science fiction and film, which seriously distorted the therapeutic purpose of the procedure and seriously marginalized the important symbolic, social, and psychological significance of the face. Reconstruction of severe facial deformity is not a cosmetic procedure undertaken for vanity; neither is it a facial identity swap (as in the film *Face/Off* starring John Travolta and Nicholas Cage). Instead, it represents an effort to restore the important functional organ of the human face to patients who have suffered severe facial deformity. For this reason, the psychological assessment of the impact of face transplantation must begin with a psychological assessment of the status of the patient suffering from the facial deformity.

It is well recognized that the adverse psychological impact that patients with severe facial deformities experience in interacting with others in public spaces can significantly impair subjective quality of life and create stresses that challenge normal coping abilities. As Agich and Siemionow have argued, diminished quality of life is the stark effect of their deformities, because the rest of us cannot accept them as they are.[35] It is reasonable that what they want is to recover a facial appearance that will not elicit revulsion or avoidance; therefore, assessment of patient expectations is ethically critical to justifying this surgery. Coffman et al. argue that psychological assessment of face transplantation surgery must also take into account the body image, mood changes, quality of life, self-esteem, and social reintegration of recipients.[70] We concur that this is an important and much under-stressed ethical consideration.

Early reports seem to indicate that while face transplantation appears to decrease depression and improve quality of life and social reintegration, it is unclear whether it alters anxiety or significantly improves self-esteem.[70,71] Therefore, it is important to differentiate these factors. Developing scales for measuring psychological distress and social reintegration of transplant recipients is an important need both for screening the transplant candidates and for assessing the success of face transplantation. Arguably, improving psychological assessment is ethically important, especially as the experience with face transplantation adequately addresses scientific and technical issues. Unlike solid organ transplantation, graft loss will not portend the patient's demise, but rather the need for reconstructive procedures, and the disappointment and need to again confront a social world that is inhospitable to their deformity. Heretofore, ethical analysis has focused on the scientific background and the standard ethical considerations such as informed consent for assessing the cost and benefit of the procedure for particular patients. Therefore, psychological assessment of the quality of life of patients with successful face transplants, along with psychological support and counseling for patients who experienced graft loss or complications, should be a *sine qua non* of any ethically sound face transplantation program.

Some of the early ethical reservations about face transplantation centered on the question whether patients suffering from severe facial deformities would be able to provide a fully informed consent, because of the concern that their decision making would be impaired by the experience of loss and alienation associated with their deformities.[72-74] Hence, they might be willing to pursue even an unrealistically optimistic outcome without fully considering the inherent risks of the surgery. Having a period of time in which to experience and attempt to adjust to their deformity and the challenges associated with both reconstructive surgeries, and the experiences of living with deformity, is an ideal preparation for a realistic assessment by the patient of the procedures and risks.

The ethical standards for informed consent in innovative medical contexts go beyond the extensive informed consent forms, and extend to an assessment of the patient's understanding of the intervention. Relying on James Drane's sliding scale model of assessing competence for informed consent,[75] face transplantation surgery is clearly subject to the highest standard of consent, which involves the requirement that the candidate for the surgery exhibit both a critical and reflexive understanding of the risks and benefits of the surgery. For this reason, the initial candidates for this procedure would have to have failed prior reconstructive attempts since one of the risks of face transplantation is the loss of the graft, which would necessitate additional reconstructive procedures. It is hard to imagine the psychological devastation that a patient would experience at having lost, for a second time, a functioning face due to rejection only to have to deal with reconstructive surgeries in an attempt to repair the graft loss. The prior experience of the rigors of reconstructive surgery thus strengthens the ability of the patient to give a full informed consent.

The significantly diminished quality of life that patients with severe facial deformity experience is an ethically important component in the assessment of candidates for this procedure. The early criticism that because face transplantation is not lifesaving, its risks are not ethically justified, fails to acknowledge that other solid organ transplants, such as living donor kidney transplants for patients who are functioning on dialysis, are justified based on cost-benefit considerations and overall improvement in quality of life, not saving life. Similarly, other well-accepted reconstructive procedures, such as surgery for cleft lip and palate, are ethically accepted, because they improve the child's quality of life and increase the potential for social interaction and normal development. As in these cases, the face transplant patient's own psychological need and desire for improving the ability to function socially is ethically compelling. Therefore, screening of candidates for face transplantation surgery will need both careful clinical ethical assessment of their decision making and psychological assessment of the degree to which their ability to function socially and their subjective quality of life are impaired. This is a practical requirement, and not just a theoretical consideration, that justifies the surgery. The clinical ethical and psychological assessment of candidates for face transplantation surgery is thus an ethically critical component in the transplant process. The patient and family should receive counseling about the risks and challenges of entering the group of first patients to undergo a procedure that has drawn so much media and public attention.

12.3 The Publicity and Privacy Dilemma

The issue of publicity is an important ethical question for face transplantation that is seldom addressed in the literature.[35,76] Given media interest, face transplantation is a hot button topic from a public relations perspective. It garners the host institution significant publicity. The institution's and surgical team's interest in publicity, however, can ethically complicate protocol development. It is illustrative that there are contrasting approaches to this question. The University of Louisville has widely published, and communicated in the media, its commitment to developing a face transplant program over many years. In addition to the solid scientific and clinical work done at that institution in composite tissue transplantation and the ethical background of face transplantation, the limelight was the place in which protocol development occurred. This was a deliberate choice. The justification for this approach is that publicity and transparency is one of the ethical requirements in Moore's approach to surgical innovation. This is not the place to offer a detailed alternative interpretation of Moore's criterion of publicity, but simply to point out that publicity in surgical innovations has two components. The first concerns technical and scientific aspects of an innovative procedure. Few commentators will deny that it is ethically required that these be shared with professional colleagues to benefit from peer review and criticism, which is what Moore himself stressed. The second, however, concerns public announcement of the institutional commitment to developing a face transplantation program, which the Louisville team merges with the first requirement saying that "Open Display and Public and Professional Discussion and Evaluation are key components of these ethical guidelines."[39] This is ethically problematic, because the public's reaction to and interest in face transplantation should not be a decisive factor in the development of the protocol and should not motivate the team. The demands of publicity and media management can detract from scrupulous attention to the myriad details associated with sound protocol development and implementation. The potentially corrosive effect that media attention can have on protocol development, team preparation, the integrity of the IRB review process, and patient and family confidentiality is unfortunately often overlooked in the ethical discussions of face transplantation. In fact, publicity can create conflicts of interest for the institution and team. It can create pressures that should not be allowed to influence the pace of protocol development, but once released, these pressures are difficult to control. In addition, publicity surrounding face transplantation, like other high publicity innovations, must be addressed in light of patient and family confidentiality and privacy.[76,77] Specific efforts must be made not only within the transplant team and ancillary support services, but the institution at large, to guard the confidentiality of all patients and families receiving innovative interventions like face transplantation.

Receiving a face transplantation should not mean that the patient must forgo confidentiality. It is unethical to expect face transplantation patients to be willing to undergo the kind of media scrutiny that has occurred heretofore even if it is in the interest of the institution and team. Managing the media and publicity is an ethically important requirement in order to protect important patient/family values of confidentiality and privacy. While it is difficult, if not impossible, to prevent attention post discharge, it is troubling ethically that institutional and team interest in self-promotion is not more carefully addressed in order to avoid conflict with ethical obligations to patients' privacy and confidentiality. This has implications that the transplant team must address within the institution. Cooperation from institutional leaders and media relations personnel within the institution must be achieved, and education sessions for all those involved, even peripherally, in the transplant must be conducted.

12.4 Risks of Immunosuppression and Rejection

Among the well-recognized medical risks of face transplantation are transplant rejection or failure and the risks associated with life-long immunosuppression such as increased risk for infection, metabolic disturbances (diabetes), and development of malignancies (lymphoma).[30] Because rejection of facial tissue is not immediately life-threatening, face transplantation is different from other solid organ transplantation. Nonetheless, ethical commentators have been concerned about the devastating effect of such a "loss of face" posttransplant on the psychological well-being of the recipient. Since the treatment option following rejection involves skin re-grafting from the patient's own body, an important selection criterion for patients for this procedure must be that they have intact tissue that can be used for such reconstructive surgery in the event of graft rejection.

Although the bioethics community has questioned whether improving quality of life can justify the risks of immunosuppression for face transplantation,[78] it is clear that this criticism does not seriously consider the alternative for these patients, which involve multiple burdensome reconstructive procedures and a life of social isolation.[72] Risk acceptance must thus be viewed in this context. Of course, some commentators have argued that the risks of immunosuppression are not justified for the potential quality of life improvement for the person with severe facial deformity. Underlying the ethical worries about the risks of immunosuppression is the tacit acceptance that undertaking transplantation is legitimate only in the case of lifesaving. It is further assumed that the ethical justification of solid organ transplantation is ethically justified on the basis of achieving a lifesaving outcome for the transplant recipient. This common view is, however, a significant oversimplification of the ethics of organ transplantation, which is justifiably performed for quality of life improvement as well.[79] The growth of living kidney donation for individuals for whom dialysis remains feasible contradicts this assumption as does the fact that solid organ transplants that would be immediately lifesaving for an individual who is terminally ill are deemed unethical, because the recipient will not benefit significantly. The transplant would only save the life of the patient short-term. Thus, the ethical criticism of face transplantation, because the immunosuppressant and other risks are too great given that the procedure is not lifesaving fail simply because transplant patients who are too sick to benefit significantly are justifiably denied transplants based on quality of life considerations even when the procedure would be immediately lifesaving. Critics also worry that the loss of the graft would be devastating and would necessitate additional reconstructive procedures,[47,78] but it would not be life-threatening as some have claimed.[80] Nevertheless, the potential for graft rejection means that candidates for face transplantation must have appropriate tissue-sites for re-grafting should the allograft be rejected. On the positive side, if the graft were rejected, immunosuppression would be stopped and its attendant risks would be eliminated. It will remain an open question whether face transplantation should be accepted as a non-experimental alternative for severe facial deformities until long-term quality-of-life data are available.

12.5 Conclusion

Reports of the international face transplantation experience have shown that although the challenges for this procedure are substantial, they are not insurmountable. However, because the procedures are likely to be performed episodically in many centers around the world, it is ethically imperative that full sharing of data associated with the procedures be made available for the profession to continue to build the scientific background for this procedure. The development of a registry or other mechanism to share such data will become even more important as the number of cases increase.

Some of the ethical opposition to face transplantation trials reflects the conviction that people suffering from severe facial deformities should simply endure their condition. These critics devalue the patient's own distress, so they conclude that the risks of the procedure outweigh the benefits. The insensitivity of society toward these deformed individuals has also distorted the ethical discussion of this procedure and this is likely to continue.[35] Until "normal" society is able to look beyond the public face of deformity to the face that reveals the patient's inner worth and dignity, the need for reconstructive surgery will remain. However, whether face transplantation will become a regular reconstructive surgical alternative for severe facial deformity will depend not only on the scientific and technical improvements, but the degree to which the clinical ethical problems identified above are sensitively addressed.[81]

References

1. Gordon CR, Siemionow M, Papay F, et al. The world's experience with facial transplantation: what have we learned thus far? *Ann Plast Surg*. 2009;63(5):572-578.
2. Anonymous. In your face: if the idea of a hand transplant seems hard to take, just wait. *New Sci*. 1998;160(2154):3.
3. Altman LK. New direction for transplants raises hopes and questions. *New York Times*. May 2, 1999;1:46–47.
4. Hettiaratchy S, Butler PE. Face transplantation – fantasy or the future? *Lancet*. 2002;360(9326):5-6.
5. Betterhumans Staff. Face transplant planned for Irish girl. Betterhumans.com. March 6, 2003.
6. Randerson J. Face transplants 'still too risky'. *New Sci*. 2003;180(2422):10.
7. American Society of Plastic Surgeons. Science fiction of face transplants may be closer than you think: reconstructive plastic surgeons perform miracle transplants now. American

Society of Plastic Surgeons. 2004. http://www.plasticsurgery.org/Media/Press_Releases/Science_Fiction_of_Face_Transplants_May_Be_Closer_than_You_Think.html. Accessed June 3, 2010.
8. Clark J. Face transplants technically possible, but "very hazardous". *CMAJ*. 2004;170(3):323.
9. Kirby R. Face switching procedures could be the next big thing. *The Salt Lake Tribune*. January 22, 2004.
10. LaFee S. Trading faces: proposed transplants promise new outlook but at unseen cost. *San Diego Union-Tribune*. February 18, 2004.
11. Baylis F. A face is not just like a hand: pace Barker. *Am J Bioeth*. 2004;4(3):30-32.
12. Caplan A, Katz D. About face. *Hastings Cent Rep*. 2003; 33(1):8.
13. Working Party on Facial Transplantation of the Royal College of Surgeons of England. Facial Transplantation: Working Party Report. London, UK: Royal College of Surgeons of England;2003.
14. Comité Consultatif National d' Ethique pour les sciences de la vie et de la santé. L' allotransplantation de tissu composite (ATC) au niveau de la face (Greffe totale ou partielle d' un visage). Paris: Comité Consultatif National d'Ethique pour les sciences de la vie et de la santé, Opinion 82. 2004.
15. Gordon CR, Nazzal J, Lozano-Calderan SA, et al. From experimental rat hindlimb to clinical face composite tissue allotransplantation: historical background and current status. *Microsurgery*. 2006;26(8):566-572.
16. Gordon CR, Siemionow M, Coffman K, et al. The Cleveland Clinic FACES Score: a preliminary assessment tool for identifying the optimal face transplant candidate. *J Craniofac Surg*. 2009;20(6):1969-1974.
17. Knobloch K, Vogt PM. Plastic surgeons' tradition in transplantation medicine in light of composite tissue allotransplantation. *J Am Coll Surg*. 2009;209(5):674-675.
18. Pomahac B, Aflaki P, Chandraker A, Pribaz JJ. Facial transplantation and immunosuppressed patients: a new frontier in reconstructive surgery. *Transplantation*. 2008;85(12):1693-1697.
19. Pomahac B, Aflaki P, Nelson C, Balas B. Evaluation of appearance transfer and persistence in central face transplantation: a computer simulation analysis. *J Plast Reconstr Aesthet Surg*. 2010;63(5):733-738.
20. Pomahac B, Lengele B, Ridgway EB, et al. Vascular considerations in composite midfacial allotransplantation. *Plast Reconstr Surg*. 2010;125(2):517-522.
21. Siemionow M, Papay F, Kulahci Y, et al. Coronal-posterior approach for face/scalp flap harvesting in preparation for face transplantation. *J Reconstr Microsurg*. 2006;22(6): 399-405.
22. Siemionow M, Agaoglu G. The issue of "facial appearance and identity transfer" after mock transplantation: a cadaver study in preparation for facial allograft transplantation in humans. *J Reconstr Microsurg*. 2006;22(5):329-334.
23. Siemionow M, Agaoglu G, Unal S. A cadaver study in preparation for facial allograft transplantation in humans: part II Mock facial transplantation. *Plast Reconstr Surg*. 2006; 117(3):876-885.
24. Siemionow M, Sonmez E. Face as an organ. *Ann Plast Surg*. 2008;61(3):345-352.
25. Siemionow M, Papay F, Alam D, et al. Near-total human face transplantation for a severely disfigured patient in the USA. *Lancet*. 2009;374(9685):203-209.
26. Siemionow MZ, Papay F, Djohan R. First U.S. near-total human face transplantation: a paradigm shift for massive complex injuries. *Plast Reconstr Surg*. 2010;125(1):111-122.
27. Thorburn G, Hettiaratchy S, Ashcroft R, Butler PE. Patient selection for facial transplantation III: ethical considerations. *Int J Surg*. 2004;2(2):118-119.
28. Toure G, Meningaud JP, Bertrand JC, Hervé C. Facial transplantation: a comprehensive review of the literature. *J Oral Maxillofac Surg*. 2006;64(5):789-793.
29. Unal S, Agaoglu G, Zins J, Siemionow M. New surgical approach in facial transplantation extends survival of allograft recipients. *Ann Plast Surg*. 2005;55(3):297-303.
30. Vasilic D, Alloway RR, Barker JH, et al. Risk assessment of immunosuppressive therapy in facial transplantation. *Plast Reconstr Surg*. 2007;120(3):657-668.
31. Yi C, Guo S. Facial transplantation: lessons so far. *Lancet*. 2009;374(9685):177-178.
32. Goering S. Facing the consequences of facial transplantation: individual choices, social effects. *Am J Bioeth*. 2004;4(3):37-39.
33. Wiggins OP, Barker JH, Martinez S, et al. On the ethics of facial transplantation research. *Am J Bioeth*. 2004;4(3): 1-12.
34. White RJ. Ethics of the face transplant: focus on the positive. *New York Times*. December 11, 2005.
35. Agich GJ, Siemionow M. Until they have faces: the ethics of facial allograft transplantation. *J Med Ethics*. 2005;31(12): 707-709.
36. Clark PA. Face transplantation: part II – an ethical perspective. *Med Sci Monit*. 2005;11(2):RA41-RA47.
37. Ad-El DD. On the ethics of composite tissue allotransplantation (facial transplantation). *Plast Reconstr Surg*. 2007; 119(2):747.
38. Barker JH, Furr A, McGuire S, Cunningham M, Wiggins O, Banis JC Jr. On the ethics of composite tissue allotransplantation (facial transplantation). *Plast Reconstr Surg*. 2007; 119(5):1621-1622.
39. Brown CS, Gander B, Cunningham M, et al. Ethical considerations in face transplantation. *Int J Surg*. 2007;5(5):353-364.
40. Barker JH, Brown CS, Cunningham M, et al. Ethical considerations in human facial tissue allotransplantation. *Ann Plast Surg*. 2008;60(1):103-109.
41. Grayling AC. Face transplantation and living a flourishing life. *Lancet*. 2008;371:707-708.
42. Johnson SE, Corsten MJ. Facial transplantation in a new era: what are the ethical implications? *Curr Opin Otolaryngol Head Neck Surg*. 2009;17(4):274-278.
43. O'Neill H, Godden D. Ethical issues of facial transplantation. *Br J Oral Maxillofac Surg*. 2009;47(6):443-445.
44. Alexander AJ, Alam DS, Gullane PJ, Lengele BG, Adamson PA. Arguing the ethics of facial transplantation. *Arch Facial Plast Surg*. 2010;12(1):60-63.
45. Martin J. Face transplantation: psychological, social, ethical and legal considerations. *Rev Méd Suisse*. 2005;1(8):583-584.
46. Trachtman H. Facing the truth: a response to "On the ethics of facial transplantation research" by Wiggins et al. *Am J Bioeth*. 2004;4(3):W33-W34.

47. Strong C. Should we be putting a good face on facial transplantation? *Am J Bioeth.* 2004;4(3):13-14.
48. Petit F, Paraskevas A, Lantieri L. A surgeons' perspective on the ethics of face transplantation. *Am J Bioeth.* 2004;4(3):14-16.
49. Butler PE, Clarke A, Ashcroft RE. Face transplantation: when and for whom? *Am J Bioeth.* 2004;4(3):16-17.
50. Peled ZM, Pribaz JJ. Face transplantation: the view from Harvard Medical School. *S Med J.* 2006;99(4):414-416.
51. Walton RL, Levin LS. Face transplantation: the view from Duke University and the University Of Chicago. *South Med J.* 2006;99(4):417-418.
52. Banis JC, Barker JH, Cunningham M, et al. Response to selected commentaries on the AJOB target article on the ethics of facial transplantation research. *Am J Bioeth.* 2004; 4(3):W23-W31.
53. Barker JH, Furr A, Cunningham M, et al. Investigation of risk acceptance in facial transplantation. *Plast Reconstr Surg.* 2006;118(3):663-670.
54. Barker JH, Furr LA, McGuire S, et al. Patient expectations in facial transplantation. *Ann Plast Surg.* 2008;61(1):68-72.
55. Barker JH, Stamos N, Furr A. Research and events leading to facial transplantation. *Clin Plast Surg.* 2007;34(2):233-250. ix.
56. Moore FD. Therapeutic innovation: ethical boundaries in the initial clinical trials of new drugs and surgical procedures. *CA Cancer J Clin.* 1970;20(4):212-227.
57. Moore FD. The desperate case: CARE (costs, applicability, research, ethics). *JAMA.* 1989;261(10):1483-1484.
58. Moore FD. Therapeutic innovation: ethical boundaries in the initial clinical trials of new drugs and surgical procedures. *Daedalus.* 1969;98:502-522.
59. The English position: position paper of the Royal College of Surgeons of England. *South Med J.* 2006;99(4):431.
60. Barth RN, Nam AJ, Stanwix MG, et al. Prolonged survival of composite facial allografts in non-human primates associated with posttransplant lymphoproliferative disorder. *Transplantation.* 2009;88(11):1242-1250.
61. Landin L, Cavadas PC, Gonzalez E, Caballero-Hidalgo A, Rodriguez-Perez JC. Sensorimotor recovery after partial facial (mystacial pad) transplantation in rats. *Ann Plast Surg.* 2009;63(4):428-435.
62. Morris P, Bradley A, Doyal L, et al. Face transplantation: a review of the technical, immunological, psychological and clinical issues with recommendations for good practice. *Transplantation.* 2007;83(2):109-128.
63. Siemionow M, Klimczak A. Advances in the development of experimental composite tissue transplantation models. *Transpl Int.* 2010;23(1):2-13.
64. Ulusal BG, Ulusal AE, Ozmen S, Zins JE, Siemionow MZ. A new composite facial and scalp transplantation model in rats. *Plast Reconstr Surg.* 2003;112(5):1302-1311.
65. Washington KM, Solari MG, Sacks JM, et al. A model for functional recovery and cortical reintegration after hemifacial composite tissue allotransplantation. *Plast Reconstr Surg.* 2009;123(2 Suppl):26S-33S.
66. Whitaker IS, Duggan EM, Alloway RR, et al. Composite tissue allotransplantation: a review of relevant immunological issues for plastic surgeons. *J Plast Reconstr Aesthet Surg.* 2008;61(5):481-492.
67. Belanger M, Harris PG, Nikolis A, Danino AM. Comparative analysis between scientific and the media communication following facial transplantation. *Transplant Proc.* 2009;41(2): 485-488.
68. Siemionow M, Ulusal B, Engine UA, Ozmen S, Izycki D, Zins JE. Functional tolerance following face transplantation in the rat. *Transplantation.* 2003;75(9):1607-1609.
69. Silverman RP, Banks ND, Detolla LI, et al. A heterotopic primate model for facial composite tissue transplantation. *Ann Plast Surg.* 2008;60(2):209-216.
70. Coffman KL, Gordon C, Siemionow M. Psychological outcomes with face transplantation: overview and case report. *Curr Opin Organ Transplant.* 2010;15(2):236-240.
71. Brill SE, Clarke A, Veale DM, Butler PE. Psychological management and body image issues in facial transplantation. *Body Image.* 2006;3(1):1-15.
72. Hurlburt M. Facial transplantation: understanding the interests of patients and hurdles to informed consent. *Med Sci Monit.* 2007;13(8):RA147-RA153.
73. Renshaw A, Clarke A, Diver AJ, Ashcroft RE, Butler PE. Informed consent for facial transplantation. *Transpl Int.* 2006;19(11):861-867.
74. Rumsey N. Psychological aspects of face transplantation: read the small print carefully. *Am J Bioeth.* 2004;4(3): 22-25.
75. Drane JF. Competency to give an informed consent. *JAMA.* 1984;252(7):925-927.
76. Morreim EH. About face: downplaying the role of the press in facial transplantation research. *Am J Bioeth.* 2004;4(3): 27-29.
77. Morreim EH. High-profile research and the media: the case of the AbioCor artificial heart. *Hastings Cent Rep.* 2004; 34(1):11-24.
78. Maschke KJ, Trump E. Facial transplantation research: a need for additional deliberation. *Am J Bioeth.* 2004;4(3): 33-35.
79. Agich GJ. Extension of organ transplantation: some ethical considerations. *Mt Sinai J Med.* 2003;70(3):141-147.
80. Allen W. Louisville doctors studying face transplants weigh ethical issues. *The Courier-Journal.* November 16, 2003.
81. Powell T. Face transplant: real and imagined ethical challenges. *J Law Med Ethics.* 2006;34(1):111-115.

Psychological Aspects of Face Transplantation

13

Kathy L. Coffman, Chad R. Gordon, and Maria Z. Siemionow

Contents

K.L. Coffman (✉)
Department of Psychiatry and Psychology, Cleveland Clinic, Cleveland, OH, USA
e-mail: coffmak@ccf.org

Abstract The first face transplantation in France in 2005 started a new era, raising many ethical and psychological issues only speculated about before this event. Roughly 10% of the US population has some form of facial disfigurement that severely compromises the ability to lead a normal life. Face transplantation appears to decrease depression and verbal abuse that patients experience in public, improve quality of life and societal reintegration, though it may not alter anxiety, self-esteem, or sexual functioning. Furthermore, there is a critical need for modification of existing rating scales to allow effective assessment of face transplant candidates before and after transplantation. More systematic data should be collected to further examine whether the long-term physical and psychological outcomes of facial transplantation outweigh the risks of ongoing immunosuppression in a surgery that is not life saving, but may be life enhancing. Unlike solid organ transplants which are internal, and therefore "invisible," transplantation of composite tissues, such as the face and hand, presents a very "visible" difference. Facial transplantation presents a new challenge for preoperative and post-transplant care in a vulnerable population.

Abbreviations

CAPS-1	clinician–administered PTSD scale
CMV	cytomegalovirus
MMPI-PTSD	Minnesota multiphasic personality inventory PTSD scale
PASTAS	physical appearance state and trait anxiety scale
PDS	Posttraumatic stress diagnostic scale
PTSD	Posttraumatic stress disorder

M.Z. Siemionow (ed.), *The Know-How of Face Transplantation*,
DOI: 10.1007/978-0-85729-253-7_13, © Springer-Verlag London Limited 2011

13.1 Introduction

The first face transplantation in France in 2005 started a new era, raising many ethical and psychological issues only speculated about before this event.[1]

The face can be viewed as an organ with numerous functions including:

- Communication
- Consumption of food
- Expression of emotion, affection, and sexuality
- Perception (gustatory, olfactory, tactile, vision),[2] and
- Conveying social information such as age, ethnicity, gender identity, and biological sex.[3]

Face transplantation is seen as a last resort after traditional reconstructive techniques have failed to restore function, rather than for cosmesis alone.

Goals include

- Regaining movement and sensation of the facial structures, and
- Restoring a more normal facial appearance.

The issue of appearance transfer has been investigated through cadaver study[4] and computer simulation.[5]

13.2 Prevalence of Facial Disfigurement and Facial Transplantation

Roughly 10% of the US population has some form of facial disfigurement that severely compromises the ability to lead a normal life.[6]

Suggestions have been made to use more positive terminology, such as visible distinction or visible difference rather than disfigurement. The support group Changing Faces puts the number of people in the UK with a visible difference at 400,000 in 2001.[7]

The etiologies of visible difference include acquired (disease and trauma) and congenital conditions.[8,9]

To date, there have been 9 transplants in 4 countries, including France (4), China (1), USA (2), and Spain (2) since 2005. The indications for face transplant have included: 2 for attacks by animals (bear, dog), 2 due to burn injuries, 1 congenital disorder (neurofibromatosis), 3 with close range gunshot wounds, and 1 after treatment for an aggressive tumor. Of these composite tissue allotransplantations, 55.5%, 5/9 were osseomyocutaneous.

13.3 Psychological Co-morbidity and Facial Disfigurement

Psychological co-morbidity may vary depending on the cause and duration of the facial disfigurement. Adaptation to facial surgery in adulthood may vary across a broad spectrum.[10]

Generally, patients presenting with facial trauma in urban centers are due to assaults in unemployed, unmarried young males in their 30s related to substance abuse, high levels of anxiety, depression, hostility, and poor impulse control.[11]

About 23% of these patients will have significant posttraumatic stress disorder symptoms 1-year after the injury related to pain severity, level of stress in the year prior to injury, poor social supports, and prior history of trauma.[12]

Other factors that predispose to PTSD after facial injury include older age, female gender, and pain from the injury.[13]

Facial surgery patients need three types of support: emotional, informational, and practical.[14]

In burn patients, various factors predicted better adjustment including: less avoidant coping, lower functional disability (for men), more recreational activities, more problem solving (for women), and more social support.[15]

Among those with facial disfigurement due to head and neck cancer, there are generally low levels of depression, high levels of life happiness, and positive feelings of well-being, though women show more depression and less happiness. However, for women with head and neck cancer, social support seemed to buffer the impact of disfigurement.[16] These patients may not report lower quality of life than normal populations.[17]

A conceptual framework for coping with disfigurement suggested that dysfunction at the time of discharge may lead to either denial or obsession with the defect, depression, non-compliance with follow-up care, and social isolation. Dropkin concluded that coping effectively preoperatively predicted coping well postoperatively. Reintegration of body image was indicated by reduced anxiety, attending to self-care, and resocialization.[18,19]

There may be more psychological disturbance in those with facial disfigurement than the general population as measured on the General Health Questionnaire and the Hospital Anxiety and Depression Scale.[20] There may be more addiction, anxiety, altered body image, depression, marital problems, social avoidance and social anxiety, PTSD, and poorer quality of life in those with various disfiguring conditions.[21,22] However, contrary to expectations, adjustment may not be predicted by extent, severity, or type of facial disfigurement.[23-26]

The impact of facial disfigurement may vary with the patient's developmental stage in life.[27] In infants with congenital facial disfigurement, bonding with parents may be affected, particularly if the difference affects facial expression[28,29] and language development. In children, craniofacial conditions may result in behavioral problems (aggression, hyperactivity, learning disorders, oppositional defiance, or social inhibition), and psychological symptoms including anxiety and depression that continue into adulthood.[30,31]

Teasing about facial differences may be experienced at any age, but generally is experienced in the 4–12-year-old cohort. Adults with craniofacial conditions may experience discrimination, have interpersonal problems and marry later, and may have panic attacks.[32-36] They may also have more difficulty leaving familiar surroundings during transitions involving changing schools, jobs, or neighborhoods, and developing new coping strategies for interacting with strangers and establishing new friends.[37] Rumsey and Harcourt provide a more thorough treatment of developmental issues in children with visible differences and their families.[9]

13.4 Comparing Face Transplantation with Solid Organ Transplantation

In comparing facial transplantation with solid organ transplantation, there are similarities and differences. The differences include:

- Face transplant, like pancreatic transplant, has not been shown to improve survival, but is done to enhance quality of life.[38]
- Higher mortality than some solid organ transplants – 25%, 2/8 at 2 years post-transplant.
- Timing of rejection is later in facial transplantation generally occurring between days 18 and 120.
- Longer hospital stay up to 6 months, much longer than most solid organ transplants other than small bowel transplant recipients.[39]
- There is an increased emphasis on informed consent for an experimental procedure that is not life saving, but hopefully life-enhancing.[40]
- There is a demanding Speech therapy regimen, so patients must be motivated.
- Other issues include long-standing tracheotomy care, and PEG tube feeding.
- Chronic pain disorders and potential addiction from the injury and procedures.
- There must be a rescue plan in case the face transplant fails; the recipient must have enough skin available to do another flap to cover the facial structures.
- There is an increased focus on societal reintegration after surgery.
- Media training is needed for recipients and also tight security postoperatively due to the intense interest of the media and public.

Duration of surgery for facial transplantation has ranged from 15 to 22 h, with as little as 500 cc blood loss. Facial sensation may return within 3–6 months, with motor function recovering by 1 year after the transplant.

Immunosuppression for face transplantation is similar to standard immunosuppression for solid organs. For facial transplantation, a target level of 12–15 ng/ml for tacrolimus is used for the first 3 months and 10–12 ng/ml thereafter, in combination with MMF and Prednisone. Weekly biopsies are done on the skin and oral mucosa for 1 month, then biweekly for 2 months, then monthly. Speech therapy is daily for the first 6 weeks, including static and dynamic exercises, gentle massage, and sensory re-education.

Infections transmitted to face transplant recipients so far from the donor include: CMV in two recipients and *T. pallidum* in one recipient, peri-oral herpes in one patient, and molluscum contagiosum in one patient.[39]

Only one recipient has returned to work thus far.[41]

The ethical issues in face transplant are paramount, have been addressed elsewhere at length, and were considered at our institution 5 years before the first face transplant was done.[42,43]

Patients must be educated about the potential risks inherent with transplantation, including infection, rejection, length of hospital stay and recuperation, surgical risks, and risk of cancers with long-term immunosuppression.[44]

Infection occurs in over 80% of heart, liver, or kidney transplant, though generally are not severe, yet 40% of mortality after transplantation are attributed to infectious causes.

Certain risks are difficult to quantitate for composite tissue grafts, such as face transplantation, for example: chronic rejection, the possibility of graft-versus-host disease, or the impact of metabolic side-effects such as hypertension or hypercholesteremia with Cyclosporine A, diabetes or neurological side-effects with tacrolimus, or osteonecrosis, cardiovascular risks, cataract or glaucoma with corticosteroids, or gastrointestinal side-effects and leucopenia from mycophenolate mofetil.

The experience with immunosuppression is not yet sufficient to know whether minimizing protocols, with gradual steroid withdrawal, and low levels of calcineurin inhibitors will be possible in face transplant recipients. There is some evidence with other grafts that m-Tor inhibitors may prevent chronic rejection. The risks of non-adherence to immunosuppression with grafts that are not life sustaining may be higher than with other organs. With composite tissue grafts, the skin is the primary target for rejection, and generally muscle and bone are spared. With face transplant rejection, mild rejection is seen only on biopsy, though when rejection progresses, this is readily apparent as the face appears sunburned. Topical tacrolimus may be used, but the efficacy has not yet been proven in facial transplantation.

The risks of cancers postoperatively with hand transplant were estimated by extrapolation from kidney data, and thought to be about 3%, with one third of these being skin cancers, some of which are preventable with good sun screen prophylaxis. The risk of cancers in face transplant recipients may be comparable. However, as Schuind et al. noted, these estimates are made based on patients many of whom are older than 40 with multiple medical co-morbidities and having been on dialysis.[38]

Fatigue due to CMV may compromise quality of life if this is transmitted from the donor graft. CMV may become resistant to current antiviral drugs.[45] Since face transplant is intended to improve quality of life, not as a life-saving procedure, it may be prudent to reconsider whether CMV in the donor should be a contraindication if the recipient is CMV negative, as in hand transplantation.[38]

A similar consideration could be made for Epstein–Barr virus if discordant in recipient and donor, in light of the risks of post-transplant lymphoproliferative disorder (PTLD). The incidence of non-Hodgkins lymphoma is estimated at 0.3–0.4% in the first year after transplant with solid organ transplantation and 0.06–0.09% per year thereafter. Recently, cases of PTLD have been seen years from the original transplantation. Kaposi's sarcoma has also been seen in solid organ transplant recipients, but is generally treatable by switching from calcineurin inhibitor to Sirolimus, which inhibits mTOR, and has anti-cancer properties. There has not been an increase in other common types of cancer seen among transplant recipients, such as breast, colon, lung, and prostate cancers.[46,47]

Preoperative assessment of face transplant candidates may be hampered as many patients have severe speech impediments that impair communication, if they lack mid-face structures such as the maxilla and upper and lower incisors, palate, nose, and lips. Surgical attachment of an artificial palate or using an obturator to close the gap in the palate can markedly reduce the intelligibility of speech. Writing boards may help, but may be difficult to use post-transplant with visual impairment and tremor due calcineurin inhibitors. A reading machine can be used for teaching about transplantation for patients that are legally blind, but retain some vision. Other aids can be helpful; a talking watch, and alarm stick that can be set to chime at the times medications are due. Visual impairment may result in some mistakes in adherence to the immunosuppression medication regimen, and may be considered a relative contraindication to transplant. Total blindness is generally considered an absolute contraindication for face transplantation due to difficulties with postoperative rehabilitation.

13.5 Patient Selection in Face Transplantation

The timing of evaluation for facial transplantation must allow

- Time for grieving losses and coming to grips with the injuries sustained
- Treatment of posttraumatic stress disorder and any depression, as well as rehabilitation.

Goals of psychiatric evaluation are as follows

- Select motivated patients
- Discuss options besides face transplant
- Discuss risks and benefits of transplantation
- Discuss the success rate and rescue procedures
- Provide education about immunosuppression regimen
- Recognize need for smoking or substance abuse rehabilitation
- Identify psychiatric disorders requiring treatment for better outcomes

In order to establish a registry of prospective face transplant candidates, a rating scale was developed, the Cleveland Clinic FACES score which is analogous to the MELD score for liver transplant candidates.[48]

13.6 Psychiatric Contraindications to Face Transplant Surgery

- Active bulimia nervosa
- Active psychotic disorder
- Severe personality disorders
- Active substance abuse or dependence
- Non-adherence to the medical regimen
- Mental retardation without adequate social support
- Suicide attempts or psychiatric admission within the past year

13.7 Psychiatric Evaluation of Face Transplant Surgery Candidates

Many predictions were made before the first facial transplant occurred anticipating what personality traits and behavior would typify the successful candidate.

Some of these predictions were accurate, such as the need for high levels of self-esteem based on factors other than physical appearance. In addition, the belief that disfigurement does not preclude happiness and a good quality of life is important. Taking an active approach to the comments made by the public about the patient's disfigurement is good preparation for handling the intense media attention and comments by the public after a face transplant.[49] Avoidant strategies can decrease anxiety, but may preclude or delay the positive coping and self-care needed for successful face transplantation.

However, some predictions were unrealistic. Patients who believe others judge them based on appearance are accurately perceiving reality.[50] Studies have shown that opinions are formed within minutes of an introduction, and a great deal of this assessment is based on appearance, involving encoding social information in the amygdala and posterior cingulate cortex.[51]

Key to patient selection is the distinction between assertive coping strategies in handling the injury and social encounters, and long-term avoidant strategies. Lazarus described the dilemma as the conflict between "protection of the self" versus "presentation of the self."[52] Avoidant strategies may be used temporarily for some months to decrease anxiety and allow recovery; however, long-term passivity predicts poor adjustment after craniofacial injury.[53]

Avoidant strategies include

- Social withdrawal
- Covering the injuries continually with makeup, masks, or hats
- Not talking about the extent of the injuries
- Not mourning the losses due to the injuries
- Not touching or looking at the facial injuries in the mirror
- Not confronting the functional losses (eating, drinking, speech, vision)
- Excessive and repeated verbal denial that the injury occurred

Assertive coping strategies include

- Taking the initiative in social interactions
- Calmly confronting negative reactions from others
- Educating others about facial disfigurement
- Use of social skills (firm handshake, good eye contact, smiling, and nodding)

Callahan describes the paradox that the bodily part that is injured is also the tool needed for reintegration of the sense of self.[54]

Candidates may have some anxiety, depression, and social anxiety, especially if prior reconstructive surgeries have failed. Patients may have minor residual symptoms of posttraumatic stress disorder that need to be addressed to help patients tolerate further interventions without a severe exacerbation, and to assist sleep. Depression or anxiety that compromises functioning should be treated prior to listing for transplantation.

Patients may have undergone multiple surgical procedures in attempts to ameliorate disfigurement, and this is not necessarily a contraindication to facial transplantation. This may limit options for rescue procedures due to loss of skin suitable for grafting. However, the belief that only transplantation will bring happiness can be problematic.

Girotto et al. noted many chronic sequelae after complex facial fractures, and these symptoms may also be seen in face transplant candidates with facial disfigurement.[55]

- Anosmia or change in olfactory and gustatory sensation
- Painful dentition
- Diplopia or decreased vision
- Epiphora (uncontrollable tearing)
- Chronic headache
- Mastication problems or drooling
- Facial numbness or pain
- Shifting orofacial structures or a
- Chronic pain disorder related to the initial injury and/or subsequent reconstructive surgeries requiring large amounts of opioids for pain management.

Unfortunately, creating composite structures such as nose, eyelids, and lips is beyond the scope of surgical interventions at this time, though some envision applications of selective tissue engineering in vitro for craniofacial regeneration.[56]

The lack of confidence in social situations may not be an absolute contraindication to face transplant. Social confidence may vary based on the time since injury, and with the type of social situation. In addition, patients with facial disfigurement may perceive reactions from the public ranging from avoidance, fear, revulsion or staring, to physical or verbal abuse.[57-59]

We must be cautious about raising false hopes in potential candidates, and continue to provide psychological support to those that are not deemed to be suitable candidates for face transplantation.[44]

13.8 Psychological Tasks in Adjusting to Face Transplantation

Aside from function, the face is intimately connected with our identity and sense of individuality. As Attar observed in the ancient Persian poem, "You can never see your own face, only a reflection, not the face itself."[60] More recent authors have imagined that "wearing another person's face may raise complex issues of identity."[44] Wearing another person's face like a mask restores the person's ability to move in society inconspicuously, without comments and questions from others about their visual difference. Symbolic interaction theory hypothesizes that people form identity and self-esteem through interpreting how others behave toward them.[61,62]

One psychological task for every organ transplant recipient involves incorporation of the organ. In the 1970s, Muslin theorized that the transplant recipient may go through several steps to incorporate the organ including

1. Perceiving the organ as a foreign object
2. Perceiving the organ and donor as transitional objects
3. Perceiving the organ as a personal belonging
4. Letting go off the donor as a transitional object
5. Integrating the organ into the recipient's self-schema

D.W. Winnicott's transitional model describes the psychological process in pre-verbal childhood, whereby the child adopts a transitional object for comfort when the parent is absent. Recipients may either idealize the donor as a protective parental or god-like rescuing figure, identify with the donor in a twin-ship relationship, or project negative introjects onto the donor as an alter ego, or may be viewed as a persecutor by patients with conflicted family relationships. Patients may employ this defense against fears, and communicate with the donor through magical thinking via thought transference. Recipients with borderline personality structure or childlike fantasies may use transitional objects to cope with anxiety. Goetzmann theorized if the recipient continues to use the donor or organ as transitional objects, this preoccupation may delay social and professional reintegration.[63]

The first partial face transplant recipient in France has confirmed some of these ideas in her interviews, indicating that incorporating the face of her donor has been challenging. She is grieving both the death of her donor and the loss of her former appearance. She stated, "I used to think of her every day and 'talk' to her." She voiced the thought, "I'd like to meet her family it would help me get used to her face. I accept it as if it was my face, but when I talk about it I say 'the' face or 'her' face. She's there somewhere. It will never be my face. I mean, it's my face without being mine. The lips are different. I had a smaller nose, a bit turned up at the end." She thought if she could watch the film of the donor's face being removed and grafted onto her own face, "I'm sure it will be tough to watch but it would close something for me. May be that way I can say goodbye to her." She has expressed identification with the donor as well, calling her "a twin sister" since reports indicated that the donor had committed suicide. She had expected to look more like she had prior to the injury. She has stated that she felt stronger since the transplant, but expressed guilt that she was given so much after having done a "stupid thing." She also observed after being kissed on the cheeks by a clerk that recognized her that others no longer thought of her "as a victim of the plague."[64,65]

13.9 Psychometric Testing in Face Transplant Candidates

Previous face transplant teams have not quantitatively investigated body image, mood changes, perception of teasing, quality of life, self-esteem, or social reintegration.

We have identified a significant void in rating scales and instruments specific for psychiatric assessment and applicability to facial transplantation. Several rating scales were modified specifically for facial transplantation, such as the Perception of Teasing-FACES and the Physical Appearance State and Trait Anxiety Scale (PASTAS).

In view of the etiologies of facial disfigurement, including congenital, trauma, and cancer, it is likely that the incidence of posttraumatic stress disorder will be high in face transplant candidates. Since our patient was 4 years post-trauma, her level of PTSD symptoms was low, though there were minor exacerbations during periods of high stress during the post-transplant period requiring an increase in escitalopram dosage.

However, in anticipation of future candidates, a review of PTSD instruments may prove useful. Generally, a trade-off must be made between the best instrument and the most practical and time-efficient instrument clinically.

For initial screening for the presence PTSD, the 10 item Trauma Screening Questionnaire may be superior to several other screening measures including the PTSD Checklist,[66] the Posttraumatic Stress Diagnostic Scale,[67] the Davidson Trauma Scale[68] and the 4 item SPAN,[69] and the BPTSD-6.[70]

For screening purposes, documenting severity of symptoms and tracking all the DSM-IV-based criteria in an efficient way, the self-rated Posttraumatic Stress Diagnostic Scale (PDS) may suffice. This 49-item scale can be administered in 10–15 min, correlates with the Beck Depression Inventory and State-Trait Anxiety Inventory, and has good reliability and validity.[67]

Another measure that assesses DSM-IV criteria for PTSD, which can be used for tracking changes in symptom severity, is the Davidson Trauma Scale, a 17-item scale. This rating scale has good test-retest reliability, shows a high correlation with other PTSD measures, and is not confounded by extroversion/introversion personality traits.[68]

A more thorough and detailed interview is the Clinician–Administered PTSD Scale (CAPS-1), but this takes about 45 min to administer, though it provides a multi-dimensional view of the severity of PTSD, corresponds to established DSM-IV diagnostic criteria, delineates both current and lifetime diagnostic time frames for those with history of multiple traumatic events, with high sensitivity and specificity as well as being a reliable and valid instrument.[71]

Other instruments frequently used with PTSD patients include: the Impact of Event Scale, the Mississippi scale, and the Minnesota Multiphasic Personality Inventory PTSD Scale (MMPI-PTSD) – all of which may be used for screening for baseline symptoms, but none are diagnostic measures or useful for measuring treatment outcomes.

13.10 Case Presentation

This 45-year-old separated female was referred for an evaluation for facial transplantation 4 years after a gunshot wound to the face perpetrated by her husband. The patient was legally blind, but read with a magnifying glass. Transplantation was first mentioned in 2004 prior to reconstructive surgery. She had chronic pain from shifting facial tissues, multiple screws and plates in her forehead, and a malpositioned right eye prosthesis. Pain was rated as 6–7/10. She understood there were options including a customized facial silicone prosthesis, but viewed this as artificial like breast implants, and preferred a real human face. Compliance with medications was good. The idea of donor tissue did not arouse guilt or superstitions as she had a donor card. She had good social supports. Coping was through humor and assertive behavior, for example, attending the Pittsburgh School for the Blind for mobility training and to learn plumbing. She actively approached others that made comments about her appearance and engaged them, despite having had strangers scream and run away or call her names in public.

Her main functional complaints included: eating through a straw, as food fell out of her mouth embarrassing her. She could not drink from a cup. She had no sense of smell, which affected cooking and detection of spoiled foods or gas leaks. Her expectations of the face transplant were as follows: to be able to wear her glasses, to be able to smell, eat food normally, and to look normal.

13.10.1 Previous Psychiatric History

Past psychiatric history included treatment for Major depression, and PTSD. She had Alcohol abuse ceased in 2004 at the time of the trauma. She was taking escitalopram 10 mg daily, zolpidem 5 mg at bedtime, and lorazepam 0.5 mg TID. She had a previous trial of sertraline. She had no history of inpatient psychiatric admission or any attempts to harm herself or others.

There was alcoholism in both parents. There was no family history of depression, bipolar disorder, or schizophrenia. The patient had worked in a painting and wallpapering business with her husband, and had co-owned a restaurant/bar where the assault occurred. She had numerous hobbies. She was raised Christian, but was not active in a church. Mental status exam was not remarkable, as she was alert and oriented, without any hallucinations or delusions. She noted some problems with seeing an image of her husband and the flash from the 12-gauge shotgun if lights were turned on unexpectedly. Cognition was intact, though she showed mild distractibility on serial sevens after the third sum, but caught her mistake and finished the task. Insight and judgment were excellent.[72]

13.10.2 Findings: Psychological Outcomes

More information will become available over the next few years regarding the psychological outcomes in face transplant recipients. So far, however, no other team has attempted to look in a quantitative way at psychological outcomes. Appearance self-rating jumped from 3/10 after her injury to 7/10 within 6 weeks of transplantation, and is now 7–8/10. STAI state anxiety score stayed constant as did Rosenberg self-esteem score. Her self-esteem was well developed prior to the injury and to a large extent was not dependent on her appearance, but on pride in her work activities.

The score on the PASTAS-State (Physical Appearance State and Trait Anxiety Scale: State) rating scale rose from 15 at 2 weeks after transplant to 23 by 10 weeks postoperatively due to steroid-related weight gain, but fell markedly to 6 at 11 months postoperatively reflecting markedly less concern about body image. This was mirrored by the decreased score from week 4 to week 10 on the Body Self-Relations Questionnaire: Appearance Evaluation subscale.

In our case, the patient's Beck Depression Inventory score declined from 16 to 6 by 3 months post-face transplant while on escitalopram. On 12/7/09, the BDI score was 14 reflecting CMV infection and other issues at home. At last evaluation, March 16, 2010, the patient did not report any symptoms of depression on escitalopram 40 mg daily.

After transplantation, the scores on the Social Environment domain on the PAIS-SR (Psychosocial Adjustment to Illness Scale-Self-Rated) quality of life rating scale showed a steady and significant decrease reflecting societal reintegration (Fig. 13.1).

The slight rise at 15 months was due to readmission to the hospital for infections with *C. dificile* and aeromonas diarrhea, and cholecystectomy. Overall, Quality of life scores on the PAIS-SR have continued to improve since the transplant. On the PAIS-SR, the patient rated changes in her appearance that made her less attractive before transplant as "extremely," after transplant rated this as "a little bit."

After the transplant, she took cues about her appearance from others, especially her daughter as predicted by symbolic interaction theory. Interestingly, mother and daughter found similarities between the new face and family traits reaffirming family ties.

PAIS-SR Psychological distress rose at 3 months post-transplant and then fell markedly over the next 4 months until CMV infection caused extreme fatigue (Fig. 13.2). Now that she is receiving a new medication and fatigue is lifting, the psychological distress is improving again. Although the SF-36 and WHOQOL-BREF were utilized, the PAIS-SR was more useful in reflecting social reintegration and

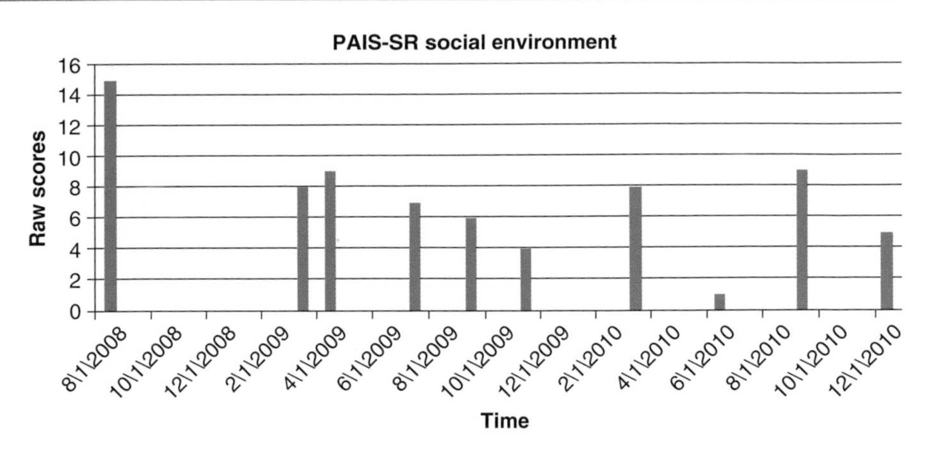

Fig. 13.1 Social reintegration after December 2008 face transplant. The rating scale was repeated after the patient started socializing 3 months after facial transplantation. A lower score indicates better functioning

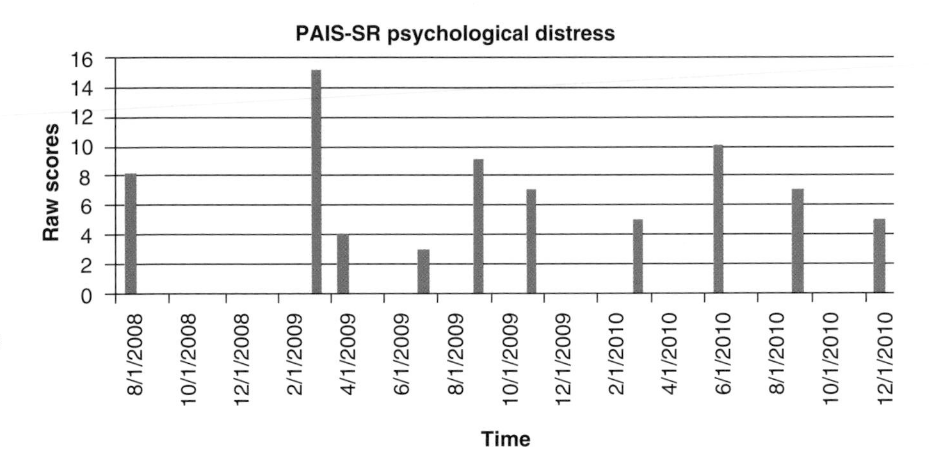

Fig. 13.2 Psychological distress: a lower score indicates better functioning. The rating scale was repeated starting in March after she started socializing

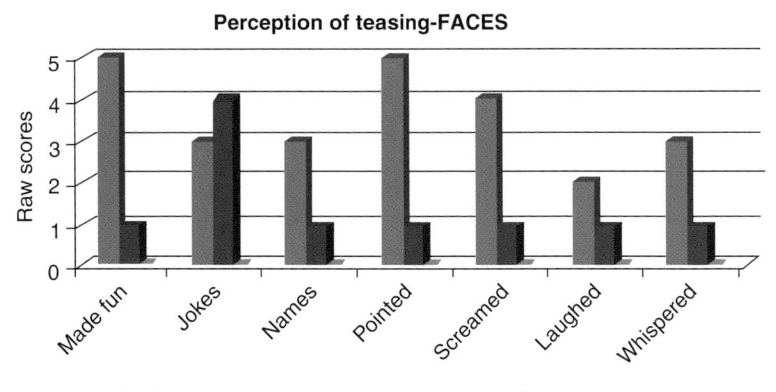

Fig. 13.3 Teasing before and after face transplant. This rating scale reflects the decreased perception by the recipient of negative reactions from the public after the face transplant

psychological distress, and other domains such as sexual functioning and attitudes toward health care. Scores on the FACES- Perception of Teasing Scale fell dramatically from before she saw her new face to the present, reflecting perceptions of lower levels of teasing and verbal abuse. She is also less bothered by the type of teasing now. She rated jokes after the face transplant as happening "often," but she was "not upset." Jokes before transplant were often derogatory, rather than humorous (Fig. 13.3). She often receives positive affirmations from the public now, and rarely did before the face transplant. Unfortunately, until there is a series of face transplant patients that have been studied regarding incidence of depression and anxiety pre- and postoperatively, there can be no comparison to the solid organ transplant recipients.

13.11 Summary

Face transplantation appears to decrease depression and verbal abuse patients experience in public, improve quality of life, and societal reintegration, though it may not alter anxiety, self-esteem, or sexual functioning. The PAIS-SR may have advantages over the SF-36 and WHOQOL-BREF rating scales for measuring psychological distress and social reintegration. Furthermore, there is a critical need for modification of existing rating scales to allow effective assessment of face transplant candidates before and after transplantation. More systematic data should be collected to further examine whether the long-term physical and psychological outcomes of facial transplantation outweigh the risks of ongoing immunosuppression in a surgery that is not life saving, but may be life enhancing. Unlike solid organ transplants which are internal, and therefore "invisible," transplantation of composite tissues, such as the face and hand, presents a very "visible" difference. Facial transplantation presents a new challenge for preoperative and post-transplant care in a vulnerable population.

References

1. Dubernard JM, Lengelé B, Morelon E, et al. Outcomes 18 months after the first human partial face transplantation. *N Engl J Med*. 2007;357:2451-2560.
2. Siemionow M, Sonmez E. Face as an organ. *Ann Plast Surg*. 2008;61:345-352.
3. Furr LA, Wiggins O, Cunningham M, et al. Psychosocial implications of disfigurement and the future of human face transplantation. *Plast Reconstr Surg*. 2007;120:559-565.
4. Siemionow M, Agaoglu G. The issue of "facial appearance and identity transfer" after mock transplantation: a cadaver study in preparation for facial allograft transplantation in humans. *J Reconstr Microsurg*. 2006;22:329-334.
5. Pomahac B, Aflaki P, Nelson C, et al. Evaluation of appearance transfer and persistence in central face transplantation: a computer simulation analysis. *J Plast Reconstr Aesthet Surg*. 2010;63:733-738.
6. Valente SM. Visual disfigurement and depression. *Plast Surg Nurs*. 2004;24:140-146.
7. Changing Faces. *Facing Disfigurement with Confidence*. London: Changing Faces; 2001.
8. Thompson A, Kent G. Adjusting to disfigurement: process involved in dealing with being visibly different. *Clin Psychol Rev*. 2001;21:663-692.
9. Rumsey N, Harcourt D. Body image and disfigurement: issues and interventions. *Body Image*. 2004;1:83-97.
10. Furness P, Garrud P, Faulder A, Swift J. Coming to terms: a grounded theory of adaptation to facial surgery in adulthood. *J Health Psychol*. 2006;11:454-466.
11. Glynn SM. The psychosocial characteristics and needs of patients presenting with orofacial injury. *Oral Maxillofac Surg Clin North Am*. 2010;22:209-215.
12. Glynn SM, Shetty V, Elliot-Brown K, Leathers R, Belin TR, Wang J. Chronic posttraumatic stress disorder after facial injury: a 1-year prospective cohort study. *J Trauma*. 2007;62:410-418.
13. Bisson JI, Shepherd JP, Dhutia M. Psychological sequela of facial trauma. *J Trauma*. 1997;43:496-500.
14. Furness PJ. Exploring supportive care needs and experiences of facial surgery patients. *Br J Nurs*. 2005;14:641-645.
15. Brown B, Roberts J, Browne G, Byrne C, Love B, Streiner D. Gender differences in variables associated with psychosocial adjustment to burn injury. *Res Nurs Health*. 1988;11:23-30.
16. Katz MR, Irish JC, Devins GM, Rodin GM, Gullane PJ. Psychosocial adjustment in head and neck cancer: the impact of disfigurement, gender and social support. *Head Neck*. 2003;25:103-112.
17. Vickery LE, Latchford G, Hewison J, Bellew M, Feber T. The impact of head and neck cancer and facial disfigurement on the quality of life of patients and their partners. *Head Neck*. 2003;25:289-296.
18. Dropkin MJ. Coping with disfigurement and dysfunction after head and neck cancer surgery: a conceptual framework. *Semin Oncol Nurs*. 1989;5:213-219.
19. Dropkin MJ. Body image and quality of life after head and neck cancer surgery. *Cancer Pract*. 1999;7:309-313.
20. Newell R, Marks I. Phobic nature of social difficulty in facially disfigured people. *Br J Psychiatry*. 2000;176:177-181.
21. Rumsey N, Clarke A, Musa M. Altered body image: the psychological needs of patients. *Br J Community Nurs*. 2002;7:563-566.
22. Levine E, Degutis L, Pruzinsky T, Shin J, Persing JA. Quality of life and facial trauma: psychological and body image effects. *Ann Plast Surg*. 2005;54:502-510.
23. Rumsey N. Body image and congenital conditions with visible differences. In: Cash TF, Pruzinsky T, eds. *Body Image: A Handbook of Theory, Research, and Clinical Practice*. New York: Guilford; 2002:226-233.
24. Rumsey N. Optimizing body image in disfiguring congenital conditions. In: Cash TF, Pruzinsky T, eds. *Body Image: A Handbook of Theory, Research, and Clinical Practice*. New York: Guilford; 2002:431-439.
25. Bradbury E. *Counselling People with Disfigurement*. Leicester: British Psychological Society; 1996.
26. Malt U, Ugland O. A long-term psychosocial follow-up study of burned adults: review of the literature. *Burns*. 1980;6:190-197.
27. Desousa A. Psychological issues in oral and maxillofacial reconstructive surgery. *Br J Oral Maxillofac Surg*. 2008; 46:661-664.
28. Walters E. Problems faced by children and families living with visible differences. In: Lansdown R, Rumey N, Bradbury E, Carr T, Partridge J, eds. *Visible Difference: Coping with Disfigurement*. Oxford: Butterworth-Heinemann; 1997:112-120.
29. Speltz M, Endriga M, Mason C. Early predictors of attachment in infants with cleft lip and/or palate. *Child Dev*. 1997;68:12-25.
30. Richman LC. Behavior and achievement of cleft palate children. *Cleft Palate J*. 1976;13:4-10.
31. Robinson E, Rumsey M, Partridge J. An evaluation of the impact of social interaction skills training for facially disfigured people. *Br J Plast Surg*. 1996;49:281-289.

32. McWilliams BJ, Paradise LP. Educational, occupational, and marital status of cleft palate adults. *Cleft Palate J*. 1973;10:223-229.
33. Ramstad T, Ottem E, Shaw WC. Psychosocial adjustment in Norwegian adults who had undergone standardised treatment of complete cleft lip and palate. II. Self- reported problems and concerns with appearance. *Scand J Plast Reconstr Surg Hand Surg*. 1995;29:329-336.
34. Rumsey N, Clarke A, White P, Wyn-Williams M, Garlick W. Altered body image: appearance-related concerns of people with visible disfigurement. *J Adv Nurs*. 2004;48:443-453.
35. Sarwer DB, Bartlett SP, Whitaker LA, Paige KT, Pertschuk MJ, Wadden TA. Adult psychological functioning of individuals born with craniofacial anomalies. *Plast Reconstr Surg*. 1999; 103:412-418.
36. Pope AW, Ward J. Factors associated with peer social competence in preadolescents with craniofacial anomalies. *J Pediatr Psychol*. 1997;22:455-469.
37. Bradbury E. Understanding the problems. In: Lansdown R, Rumey N, Bradbury E, Carr T, Partridge J, eds. *Visibly Different: Coping with Disfigurement*. Oxford: Butterworth-Heinemann; 1997:180-193.
38. Schuind F, Abramowicz D, Schneeberger S. Hand transplantation: the state-of-the-art. *J Hand Surg Eur*. 2007;32:2-17.
39. Gordon CR, Siemionow M, Papay F, et al. The world's experience with facial transplantation: what have we learned thus far? *Ann Plast Surg*. 2009;63:572-578.
40. Renshaw A, Clarke A, Diver AJ, et al. Informed consent for facial transplantation. *Transpl Int*. 2006;19:861-867.
41. Lantieri L, Meningaud JP, Grimbert P, et al. Repair of the lower and middle parts of the face by composite tissue allotransplantation in a patient with massive plexiform neurofibroma: a 1-year follow-up study. *Lancet*. 2008;372: 639-645.
42. Brown CS, Gander B, Cunningham M, et al. Ethical considerations in face transplantation. *Int J Surg*. 2007;5:353-364.
43. Siemionow M, Bramstedt K, Kodish E. Ethical issues in face transplantation. *Curr Opin Organ Transplant*. 2007;12: 193-197.
44. Morris P, Bradley A, Doyal L, et al. Face transplantation: a review of the technical, immunological, psychological and clinical issues with recommendations for good practice. *Transplantation*. 2007;27(83):109-128.
45. Torres-Madriz G, Boucher HW. Immunocompromised hosts: perspectives in the treatment and prophylaxis of cytomegalovirus disease in solid-organ transplant recipients. *Clin Infect Dis*. 2008;47:702-711.
46. First MR, Peddi VR. Malignancies complicating organ transplantation. *Transplant Proc*. 1998;30:2768-2770.
47. Penn I. Posttransplant malignancies. *Transplant Proc*. 1999; 31:1260-1262.
48. Gordon CR, Siemionow M, Coffman K, et al. The Cleveland clinic FACES score: a preliminary assessment tool for identifying the optimal face transplant candidate. *J Craniofac Surg*. 2009;20:1969-1974.
49. Clarke A, Butler PEM. Patient selection for facial transplantation II: psychological consideration. *Int J Surg*. 2004;2: 116-118.
50. Farrer R. Psychological considerations in face transplantation. *Int J Surg*. 2004;2:77-78.
51. Schiller D, Freeman JB, Mitchell JP, et al. A neural mechanism of first impressions. *Nat Neurosci*. 2009;12:508-514.
52. Lazarus R. Coping theory and research: past, present and future. *Psychosom Med*. 1993;55:234-247.
53. Horowitz MJ. Stress-response syndromes: a review of posttraumatic and adjustment disorders. *Hosp Community Psychiatry*. 1986;37:241-249.
54. Calahan C. Facial disfigurement and sense of self in head and neck cancer. *Soc Work Health Care*. 2004;40:73-87.
55. Girotto JA, MacKenzie E, Fowler C, Redett R, Robertson B, Manson PN. Long-term physical impairment and functional outcomes after complex facial fractures. *Plast Reconstr Surg*. 2001;108:312-327.
56. Scheller EL, Krebsbach PH. Gene therapy: design and prospects for craniofacial regeneration. *J Dent Res*. 2009;88:585-596.
57. Houston V, Bull R. Do people avoid sitting next to someone who is facially disfigured? *Eur J Soc Psychol*. 1994;24: 279-284.
58. Macgregor FC. Facial disfigurement: problems and management of social interaction and implications for mental health. *Aesthet Plast Surg*. 1990;14:249-257.
59. Lansdown R, Rumsey N, Bradbury E, Carr T, Partridge J. *Visibly Different: Coping with Disfigurement*. Oxford: Butterworth Heinemann; 1997.
60. Attar. "Looking for Your own face." *Persian Poets*. In: Washington, P, ed. New York/Toronto: AA Knopf Inc;2000:59.
61. Cunningham M, Barbee A, Philhower C. Dimensions of facial physical attractiveness: the intersection of biology and culture. In: Rhodes G, Zebrowitz L, eds. *Advances in Visual Cognition*, vol. 1. Stamford, CT: Conn: JAI/Ablex; 2002.
62. Cash TF. Body image and plastic surgery. In: Sarwer DB, Pruzinsky T, Cash TF, et al., eds. *Psychological Aspects of Reconstructive and Cosmetic Surgery: Clinical, Empirical, and Ethical Perspective*. Philadelphia: Lippincott Williams &Wilkins; 2006:37-59.
63. Goetzmann L. "Is it me, or isn't it?" – Transplanted organs and their donors as transitional objects. *Am J Psychoanal*. 2004;64:279-289.
64. http://www.telegraph.co.uk/news/worldnews/europe/france/3367041/Face-transplant-woman-struggles-with-identity.html
65. Times Online from the Sunday Times January 17, 2010 www.timesonline.co.uk/tol/life_and_style/.../article 6987682.ece
66. Blanchard EB, Jones-Alexander J, Buckley TC, Forneris CA. Psychometric properties of the PTSD checklist (PCL). *Behav Res Ther*. 1996;34:669-673.
67. Foa EB, Cashman L, Jaycox L, et al. The validation of a self-report measure of posttraumatic stress disorder: the posttraumatic diagnostic scale. *Psychol Assess*. 1997;9:445-451.
68. Davidson JR, Book SW, Colket JT, et al. Assessment of a new self-rating scale for post-traumatic stress disorder. *Psychol Med*. 1997;27:153-160.
69. Meltzer-Brody S, Churchill E, Davison JRT. Derivation of the SPAN, a brief diagnostic screening test for post-traumatic stress disorder. *Psychiatry Res*. 1999;88:63-70.
70. Fullerton CS, Ursano RJ, Epstein RS, et al. Measurement of posttraumatic stress disorder in community samples. *Nord J Psychiatry*. 2000;54:5-12.
71. Dudley DB, Weathers FW, Nagy LM, et al. The development of a clinician administered PTSD scale. *J Trauma Stress*. 1995;8:75-91.
72. Coffman KL, Gordon C, Siemionow M. Psychological outcomes with face transplantation; overview and case report. *Curr Opin Organ Transplant*. 2010;15(2):236-240.

Physical Medicine and Rehabilitation after Face Transplantation

14

Pamela L. Dixon, Xiaoming Zhang, Mathieu Domalain, Ann Marie Flores, and Vernon W.-H. Lin

Contents

Abstract Our face is the first thing that others notice and remember about us. Our faces, however, serve more than aesthetic importance and value. The function the face plays in basic every day activities cannot be overstated. The face plays an important role in functional needs such as speech, communicative competence, eye protection, and emotional expressiveness. However, the ability to use these functions following face transplantation may not be complete. Therefore, Physical medicine and rehabilitation is a necessary aspect of the rehabilitation process following this procedure. It is necessary for both the patient and transplant team to realize the importance of PT, OT, and SLP and their role in recovery. Face transplantation offers a new area for the Physical Medicine and Rehabilitation specialist to explore and provide an invaluable service.

Abbreviations

ADL	Activities of daily living
EMG	Electromyography
FDI	Facial disability index
FGS	Facial Grading Scale
FIM	Functional Independence Measure
IADL	Independent activities of daily living
ICF	International Classification of Functioning Disability and Health

V.W.-H. Lin (✉)
Department of Physical Medicine and Rehabilitation, Cleveland Clinic, Cleveland, OH, USA
e-mail: linv@ccf.org

M.Z. Siemionow (ed.), *The Know-How of Face Transplantation*,
DOI: 10.1007/978-0-85729-253-7_14,

OT	Occupational therapy
PT	Physical therapy
SLP	Speech language pathology
WHO	World Health Organization

God has given you one face, and you make yourselves another.

— William Shakespeare

Our face is the first thing that others notice and remember about us. Our face, however, serves more than aesthetic importance and value. The function the face plays in basic every day activities cannot be overstated. The face plays an important role in functional needs such as speech, communication, eye protection, and emotional expressiveness. The latter function bears significant social and psychological importance, because two-thirds of our communication takes place through nonverbal facial expressions.[1] The following chapter is a comprehensive critical review of the literature that has been developed for potential technologies and procedures that can be considered in facial transplant rehabilitation, but does not fully reflect what any one facility has followed up to date. Psychosocial function in facial rehabilitation is essential to ensure optimal success with recovery. For the purpose of this chapter, we will discuss psychosocial function as it correlates to and reinforces physical function in rehabilitation. A more in-depth review and discussion of psychosocial function and the facial transplant patient will be covered in another chapter of this book.

14.1 General Principles of Rehabilitation

Physical medicine and rehabilitation is the medical specialty concerned with restoring and maintaining the highest possible level of function, independence, and quality of life. This relatively new specialty has evolved into a number of subspecialties to meet the needs of patients from all age groups with diverse primary medical issues including neurological, cardiac, pulmonary, amputees, sports injury, orthopedic, and pain related. In 2001, the World Health Organization (WHO) published the International Classification of Functioning, Disability and Health guidelines (ICF).[2] Impairment is defined as a loss or abnormality in body structure or physiological function (including mental functions). Activity is defined as the execution of a task or action by an individual, whereas activity limitations are difficulties an individual may have in executing activities. Participation restrictions are problems an individual may experience while engaging in life situations. The presence of a participation restriction is determined by comparing an individual's participation to that of an individual without disability in the same culture or society.

To rehabilitate means "to enable." Rehabilitation is the process of helping an individual achieve the highest level of independence and quality of life possible. Facial transplant rehabilitation may focus on motor and sensory function restoration, speech, and swallowing, Therapeutic activities that improve independence in activities of daily living, community reintegration, and alleviation of psychosocial barriers are also paramount. The overall goal of facial transplant rehabilitation is to restore facial function and movement in order for the patient to confidently participate in meaningful activities, thus increasing physical, emotional, social, and spiritual independence (see Fig. 14.1).

To date, ten face transplants have been performed in four countries: France, the United States (US), China, and Spain. Facial transplantation is becoming a more feasible solution for disfigured patients.[3-7] Following surgery, involvement of skilled rehabilitative services is integral to achieving optimal outcome of recovery. Through skilled rehabilitation, interventions are implemented to regain facial movements involved with

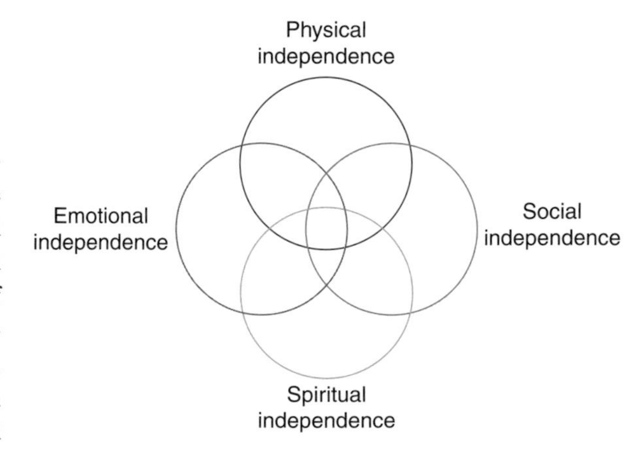

Fig. 14.1 Goals of skilled rehabilitation for facial transplant patient

expression and interpersonal communication, as well as to eliminate or diminish unwanted patterns of movement.[8,9] Skilled intervention also focuses on the functional independence of the patient including mobility, endurance, and activities of daily living.

14.2 Rehabilitation Team

Rehabilitation efforts for patients with facial transplant require a comprehensive treatment approach to address their unique and often multidimensional problems. The interdisciplinary team consists of practitioners from different medical specialities/disciplines who work together for a common patient, with shared goals and responsibility. In any team, communication and collaboration are vital to ensure the best possible treatment plan for the patient. The interdisciplinary rehabilitation team brings different perspectives and areas of skilled expertise with a unified outcome in mind for the patient. Thus, the team becomes stronger and is able to bring forth a multitude of solutions for rehabilitating a patient that can never be achieved by one discipline alone.[10,11]

Surgeons and physicians lead the coordination of the interdisciplinary team members (see Fig. 14.2). The ideal facial rehabilitation team includes physiatrists (rehabilitation medicine specialists), nurses, physical therapists, occupational therapists, speech-language pathologists, recreational therapists, nutritionists, social workers, psychologists, orthotists, and prosthetists among others. Chaplains, vocational counselors, home care agencies, support groups, and educational outreach programs serve as important additional resources that can aid patients in facial rehabilitation. At the epicenter of the rehabilitation process, the patient and the patient's family members serve as key team members.

Depending on the facility (or country) where a facial transplantation occurs, the role each specific discipline may take in a patient's rehabilitation varies. In some instances, physical therapy (PT), or Physiotherapy,

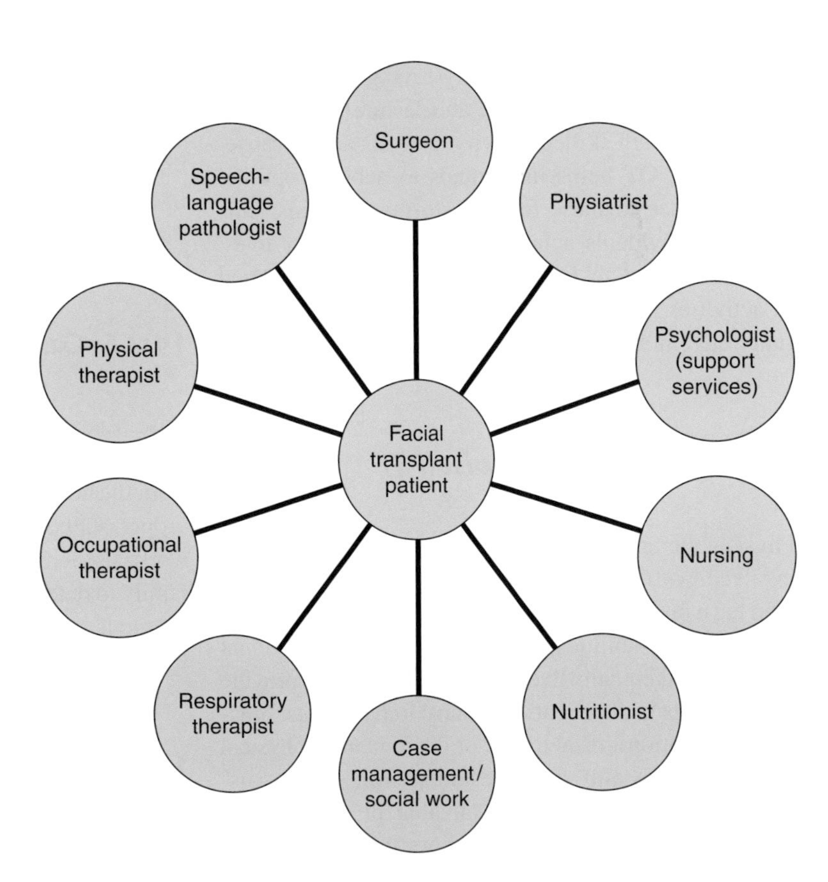

Fig. 14.2 Interdisciplinary team members in facial rehabilitation

may take the lead with facial movement and training whereas in other instances, speech-language pathology (SLP) may take the primary role. Similarly, Speech-Language Pathology (SLP) may take the lead on swallowing or olfactory training in one facility (or country), whereas in another facility (or country), occupational therapy (OT) may take primary responsibility for those performance areas. The factors that may preclude a discipline from working on a certain function include skill set, experience, availability, and philosophy of the facility (or country) where transplant occurs. Regardless of the role each therapy discipline may take, the ultimate focus of the team remains solely on the rehabilitation needs of the facial transplant patient.

14.2.1 Occupational Therapist (OT)

Occupational therapy is a science-driven, evidence-based profession that enables people to live life to its fullest by promoting health and preventing (or live better) with illness, injury, or disability.[12] Occupational therapy is classified within three occupational performance areas: Activities of Daily Living (ADL), work and productive performance, and leisure/play performance. Through skilled activity analysis and purposeful activity, OT helps individuals to achieve optimal independence in these three performance areas, thus helping individuals achieve fulfillment in their lives. Areas of focus in OT include, but are not limited to activities such as bathing, dressing, feeding skills, functional mobility, and home management.

14.2.2 Physical Therapist (PT)

Physical therapy is an evidence-based health profession involved with diagnosis and treatment of individuals who have medical problems or health-related conditions that limit their abilities to move and perform functional activities in their daily lives.[13] Physical therapists provide services to patients/clients who have impairments, functional limitations, disabilities, or decrement in physical function and health status resulting from injury, disease, or other causes. This valuable profession also addresses risk factors and behaviors that may impede optimal functioning and plays a role in providing prevention, promoting health, wellness, and fitness.[14] Areas of focus include but are not limited to muscle strength, motor function, balance, and cardiovascular strength.

14.2.3 Speech-Language Pathologist (SLP)

Speech-language pathologists work with the full range of communication and related disorders.[15] They evaluate, diagnose, and treat speech, language, voice, resonance, cognition, and feeding/swallowing disorders across the human lifespan due to many potential etiologies. These include laryngeal, pharyngeal, and oro-facial anomalies, developmental disabilities, neurological diseases, and genetic disorders.

14.2.4 Respiratory Therapist

Respiratory therapists provide patient care services and education regarding the management of respiratory symptoms and treatments aimed at optimizing respiratory function and oxygenation. The respiratory therapist serves a vital role in the rehabilitation team in that they work with the facial transplant patient to maintain the functions of the respiratory system.

14.2.5 Case-Management/Social Worker

Case-Management serves the primary role of ensuring the patient has the most appropriate plan for discharge from the hospital. This involves ensuring patient has the proper support and resources upon discharge. Social work also collaborates with the patient and patient's family to determine needs. Decision making toward discharge is also based on feedback and recommendations from other members of the interdisciplinary team.

14.2.6 Physiatrist

The physiatrist (or rehabilitation medicine specialist) is a neuromuscular and musculoskeletal physician trained to assess functional disability, biomechanics, and human movement. Physiatrists are also experts in understanding and incorporating their knowledge of medical co-morbidities such as cardiac, pulmonary, and rheumatologic disease into optimal patient care. The ability to appropriately prescribe medications, modalities, orthotics/prosthetics, and assistive devices that accounts for competing medical and psychosocial issues is key to the success of rehabilitation after facial transplantation. The physiatrist serves as the vital link between the clinical team and the interdisciplinary team to coordinate rehabilitation services for the patient. They prepare a comprehensive plan for treatment and follow the care plan determined in order to increase function, reduce pain, and minimize activity limitation due to disability.

14.3 Clinical Evaluation

14.3.1 Physiatric Assessment

The goal of a rehabilitation physician (physiatrist) is to restore people with an activity limitation to the fullest possible physical, mental, social, and economic independence. Thus, a physiatric evaluation has to be globally focused, and not limited to one organ system. A rehabilitation evaluation is also interdisciplinary, incorporating feedback, opinions, and findings from the various specialities/disciplines involved in the care of the patient. The valuable information for the evaluation comes from the combined efforts of the physiatrists, other physicians, rehabilitation team members, consulting services, and above all from the patient. A rehabilitation evaluation may be in an inpatient setting, or in an outpatient setting, before or after the facial transplant procedure.

14.3.1.1 History

A patient history should include the chief complaint, history of present illness, medical history, surgical history, facial procedure history, functional history, and a list of medications. The functional history should address communication, eating, grooming, bathing, toileting, dressing, transfers, and mobility. Functional history should also emphasize facial functions, such as eye movement, oral function, facial expression, and communication. A review of systems should include the following: constitutional symptoms, head and neck, respiratory, cardiovascular, gastrointestinal, genitourinary, neurological, and musculoskeletal symptoms. Patient profiles should include: (1) Personal history that includes psychological and psychiatric history, life-style-related history, diet, alcohol, and drug history; (2) Social history that includes family and home environment; and (3) Vocational history that includes education, training, work history, and finances. As part of the psychosocial history, the patient's adjustment to disability, decision-making capacity, and medical compliance history need to be reviewed and appropriately recorded. Psychosocial history also includes the degree of motivation, realistic expectation, potential for psychological regression, perceived body-image adaptation and anticipated comfort with donated facial transplant.[16,17]

14.3.1.2 Physical Examination

Physical examination, performed by a physiatrist or a rehabilitation specialist, should seek physical findings to support and formulate the diagnosis further. This will also help to define activity limitations and participation restrictions that emanate from the impairment. The physical examination also helps the physiatrist identify any remaining physical, psychosocial, and intellectual strengths which will help to re-establish functional independence. A physical examination includes the following areas: vital signs and general appearance, integumentary (with emphasis on the face) and lymphatics, head (observe for major deformities or surgical scars), eyes (visual acuity and eye movement), ears, nose, mouth and throat, neck, chest, heart and peripheral vascular system, abdomen, genitourinary system, and musculoskeletal system. A neurological examination should include a mental status and cognitive evaluation (orientation, attention, recall, general fund of information, calculations, proverbs, similarities, and judgment), speech and language function (listening, reading, speaking, and writing), cranial nerves (I. Olfactory, II. Optic, III/IV/VI. Oculomotor [and Trochlear and Abducens], V. Trigeminal, VII. Facial, VIII. Vestibular, IX. Glossopharyngeal, X. Vagus, XI. Accessory, XII. Hypoglossal), motor function, reflexes, coordination, sensation, and perception. Motor and sensory functions of the facial nerve, facial expression, facial symmetry, oral function, and olfaction should also be emphasized. Facial nerve assessments include eyelid closure, forehead wrinkling, facial grimace, pouting, movements of the cheek, symmetry of the face at rest and during movements, and ability to retract the angles of the mouth and flatten cheek. The strength of orbicularis oculi and oris muscles in their sphincter action can be tested by an attempt to overcome their action with the examining fingers. The strength of the buccinator muscle can be tested by asking patients to inflate a balloon.

14.3.1.3 Functional Evaluation

In addition to a history and physical examination, a functional examination is an important part of the clinical evaluation. This includes an assessment of function related to activities of daily living: eating, grooming, bathing, toileting, dressing, transfers, and mobility. Several self-care scales have been developed and applied in rehabilitation settings. The Barthel Index is a weighted scale for measuring basic Activities of Daily Living (ADLs) in people with chronic disability.[18] A person with a maximum score of 100 points is defined as continent, able to feed and dress himself or herself, walk at least a block, and climb and descend stairs. The Functional Independence Measure (FIM) evolved from a task force of the American Congress of Rehabilitation Medicine and the American Academy of Physical Medicine and Rehabilitation. It has been used to document the severity of disability as well as the outcomes of rehabilitation treatment as part of a uniform data system. The FIM consists of 18 items organized under 6 categories, including self-care (e.g., eating, grooming, bathing, upper body dressing, lower body dressing, and toileting); sphincter control (i.e., bowel and bladder management); mobility (e.g., transfers for toilet, tub, or shower, as well as bed, chair, and wheelchair); locomotion (e.g., walking, wheelchair, and stairs); communication, including comprehension and expression; and social cognition (e.g., social interaction, problem solving, and memory).[19-26]

In addition, there are two other scales that have been used for facial rehabilitation: The Facial Grading Scale (FGS) and the Facial Disability Index (FDI). The FGS was developed to evaluate and monitor facial function after facial nerve insult.[27] It evaluates facial impairment in three areas: (1) resting posture of the eye, the nasolabial (cheek) fold, and the corner of the mouth; (2) voluntary movement for five expressions in five regions of the face, forehead wrinkles, eye closure, open mouth smile, snarl, and pucker; and (3) synkinesis. Another potential assessment tool that may be used for evaluating facial function in post-facial transplant patients is the Facial Disability Index (FDI).[28] The FDI is a disease-specific, self-report instrument for the assessment of the disabilities of patients with facial nerve disorders, and scored as two subscales, namely physical and social subscales. Both of these assessment tools will need further evaluation and testing before they can be adopted as standard tools for use in patients with facial transplantation.

14.3.1.4 Formulation of Rehabilitation Plan

In addition to the history, physical examination, and functional evaluation, a physiatrist should be able to integrate other relevant clinical information into the assessment. This clinical information may include the patient's medical condition, peri-operative recovery status, imaging and laboratory studies, electrophysiological studies, medication history, and immunological status. The physiatrist then utilizes his clinical skills to further integrate inputs from other interdisciplinary rehabilitation clinicians to develop a comprehensive interdisciplinary rehabilitation plan. The interdisciplinary plan is based on both short- and long-term goals which will then be reviewed in an interdisciplinary setting at regular intervals, typically on a weekly basis.

14.3.2 Neurophysiological Evaluation of the Facial Nerves

14.3.2.1 Neurophysiology of Facial Nerve

The facial nerve leaves the lateral aspect of the brain stem at the lower border of the pons, and together with the intermedius and the VIII cranial nerves, it enters the internal auditory meatus and passes through it. It then enters the facial canal, which, in its first part, is directed laterally. At the geniculate ganglion, it makes a sharp bend in a posterolateral direction, proceeds caudally, and leaves the skull through the stylomastoid foramen. The facial canal lies close to the tympanic cavity and in its dorsolateral and caudal course is separated from it only by a thin bone lamella. On leaving the stylomastoid foramen, the facial nerve pierces the parotic gland and splits into several branches, which spread out in a fan-like manner to reach the facial muscles. It also gives off motor fibers to the stapedius muscle in the middle ear and to the stylohyoid and posterior belly of the digastric muscles and to the platysma.[29,30]

14.3.2.2 Facial Nerve Conduction Studies

Facial nerve conduction studies may be performed prior to the facial transplant to ensure normal function of the donor nerves compared with the recipient's nerves. These studies may also be performed periodically after a facial transplant to determine the degree of neuroregeneration and facial muscle recovery. Electrical stimuli are delivered to the facial nerve through surface electrodes placed near the stylomastoid foramen, that is, just below and anterior to the mastoid bone. Surface recording electrodes are placed over the orbicularis oculi, nasalis, orbicularis oris, or mentalis muscles. The reference electrode is placed either over the nasal bone or the same muscle of the opposite side. The ground electrode is often placed on the lip, chin, or wrist. Normal latencies range from 2.5 to 5.0 ms, with a mean of 3.2 ms.[31]

Nerve conduction of the lower and upper motorneuronal pathways of the facial nerve provides information regarding the potency of the nerve. Lower motorneuronal analysis of the facial nerve, as it emerges from behind the stylomastoid foramen, is routinely conducted using electrical nerve stimulation. Site of stimulation can be localized antero-inferior to the mastoid bone. Surface electrodes record activity within the zygomatic (upper) or the mandibular (lower) branches as contraction of frontalis, buccinator, orbicularis oculi, orbicularis oris, or nasalis is measured. Latencies range between 2.5 and 5 ms (see Fig. 14.3a – S1).

To address pathologies that may be more proximal to the site of facial nerve exiting from the stylomastoid foramen, lower motorneuronal nerve conduction can be conducted using Transcranial Magnetic Stimulation (TMS). TMS is a noninvasive, safe and, unlike electrical stimulation, painless method of assessing the integrity of neuronal pathways.[32] It is based on the principle of electromagnetic induction wherein rapidly alternating current passing through a coil of wire generates a magnetic field, which with sufficient strength and transient application can induce electrical currents in the brain to depolarize underlying neurons. For clinical neurophysiology, TMS offers an important advantage compared to electrical stimulation as relatively deeper neuronal structures can be targeted since the induced current passes through skull and intervening tissues with little attenuation and without causing pain.

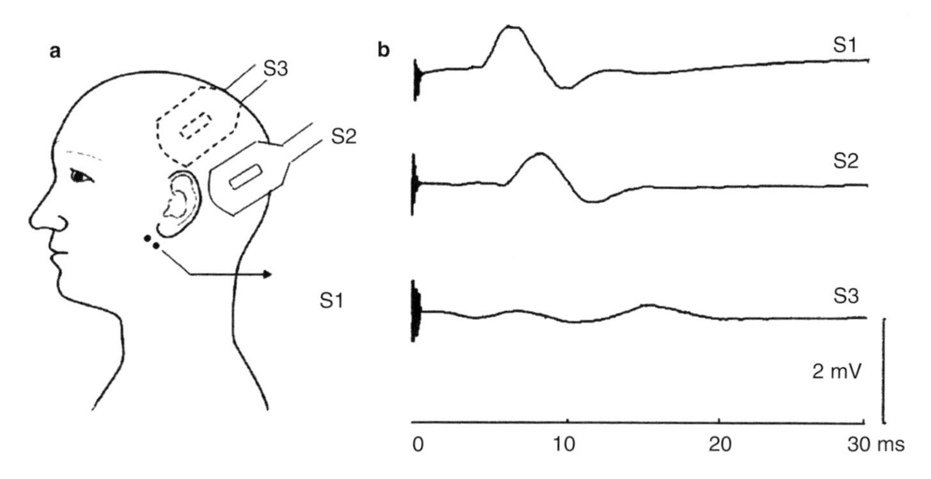

Fig. 14.3 Nerve conduction in the upper and lower (proximal and distal) facial nerve pathways. (**a**) Represents location of stimulation: S1 refers to the placement of electrical stimuli (monophasic, single pulse, 0.2 ms) to target distal facial nerve as it exits the stylomastoid foramen; S2 displays the site of stimulation of proximal facial nerve (behind ipsilateral ear) using a racetrack-shaped TMS coil at an intensity of 50% of stimulator output and S3 indicates stimulation of face region of contralateral motor cortex (right fronto-parietal) using TMS at 73% of stimulator output (**b**) displays evoked responses from target muscle (left frontalis), recorded using surface electrodes, following stimulation at sites S1, S2, and S3

In Fig. 14.3a (S2), we demonstrate a method of testing conduction in proximal facial nerve. A racetrack-shaped TMS coil (with greater specificity compared to a circular coil) has been placed behind the ear in the parieto-temporal region. The direction of the coil is in line with the assumed path of the nerve. At a relative low intensity (about 50% of the stimulator output; MagPro R30, Dantec, Denmark), a nerve conduction response can be generated at slightly longer latencies compared to S1 (approximately 6 ms or greater) with comparable amplitude.

Lastly, TMS can also be used to study the upper motorneuronal path of the facial nerve (motor cortical region dedicated to facial nerve). Motor-evoked potentials can be generated from the contralateral "face" area of the primary motor cortex. The center of the TMS coil is placed approximately 2–3 cm anterior to the vertex and about 6–8 cm lateral to the mid-sagittal plane (see Fig. 14.3a – S3) in the frontal region (precentral gyrus). Motor-evoked potentials can be generated at low intensities (40–60% of machine output). These potentials have longer latencies compared to S1 and S2 and smaller amplitudes (see Fig. 14.3a – S3).

14.3.2.3 Electromyography of Facial Muscle

Electromyography (EMG) is valuable in the assessment of signs of nerve degeneration and regeneration.[33-36] Approximately 10–14 days after the facial transplant surgery or facial nerve injury, typical spontaneous activities (positive sharp waves and fibrillation potentials) will be seen because of Wallerian degeneration. Depending on the type of nerve repair, regeneration potentials can be detected 4–6 months postoperatively. EMG should be used to monitor facial transplant facial nerve recovery. If regeneration potentials are not detected 9–12 months postoperatively, the nerve transplant/repair has failed.

14.3.3 Physical, Occupational Therapy, and Speech-Language Pathology Assessment

The ultimate objective of PTs, OTs, and SLPs is to return the individual patient to the maximal level of independence through the attainment or re-attainment of optimal function. Functional activity, therefore, is the heart and soul of all rehabilitation efforts. Clinical decision making involves a series of sequential steps that therapists utilize to plan an effective treatment that meets the needs and goals of the patient as well as the team. These steps include (1) patient assessment of function and dysfunction; (2) analysis and interpretation of the assessment data; (3) establishment of long-term and short-term goals; (4) development and implementation of treatment intervention; and (5) monitoring and reassessment of progress toward goals.[37]

14.3.4 Assessment of Function/ Dysfunction

A therapy assessment begins with obtaining information about the patient, including both subjective and objective history. The therapist gathers thorough information regarding patient's prior functional level, demographics, available social support, patient's primary complaint, previous medical history, course of present illness/injury, and most importantly, patient's own personal goals and expectations. The therapist obtains this information with the intention of gaining a full understanding of the patient, and their life, in order to determine the best plan of care possible.

A therapy assessment is a dynamic, evolving process in which a therapist determines the elements to be tested based on the patient's condition and information gained from individual's history. As the therapist completes the assessment, the attention is on both listening and observing how the patient responds to functional skill testing. Although PT, OT, and SLP generally follow the same assessment process detailed above, their focus of assessment varies based on their individual discipline, clinical knowledge, and training.

14.3.5 Analysis and Interpretation of Assessment Data

Using clinical reasoning, knowledge, and skills, a therapist analyzes the data and patient performance results. Through analysis of the assessment, the therapist identifies the patient's strengths and problem areas needing intervention. The therapist weighs the patient's strengths and deficits along with their prior level of function, roles, and available social support to determine intervention needed for a safe discharge. Through working with the patient and the interdisciplinary team, the therapist determines the functional skills that should be prioritized in treatment.

14.3.6 Establishment of Short-Term and Long-Term Goals

The therapist incorporates the results of the assessment with patient's goals to determine short-term and long-term goals. Long-term goals set the anticipated optimal performance level of patient at the end of rehabilitation process in order to reach maximal independence. Long-term goals should include (1) the patient's level of independence and safety as well as the amount of supervision or assistance required for an activity and (2) adaptive equipment or method necessary to perform an activity safely and adequately.[37] Short-term goals concentrate on the skills a patient needs to achieve the long-term goals and achieve the desired outcome of the rehabilitation plan. Throughout a patient's progress, the therapist continually reassesses long-term and short-term goals, as well as the skilled intervention provided.

14.3.7 Treatment Plan and Intervention

Once the therapist sets goals, the next step is to determine what skilled and functional activities need to be part of the treatment plan in order to achieve goals. The determination of the treatment plan includes treatment principles, precautions and protocols based on illness/injury, as well as the anticipated frequency of treatment needed to achieve goals.

Treatment intervention provided by the therapist needs to be skilled, purposeful, and functional. Intervention should correlate to the goals set and the anticipated discharge outcome for the patient. Special attention during therapy intervention needs to be centered on safety levels of the patient and level of assistance needed, including physical level of assistance, frequency of assistance or cues (verbal or tactile), and utilization of adaptive equipment or methods. All three therapy disciplines work closely together in order to reinforce education, safety, and function. For the ultimate benefit and outcomes of the patient, education and rehabilitation plans should reinforce, and at times even overlap, to provide consistency for the patient.

14.3.8 Reassessment of Treatment Plans and Outcome

Throughout provision of skilled intervention, the therapist continually reassesses patient progress and response to treatment. Utilizing clinical knowledge and reasoning, progress toward treatment goals should be analyzed at each session. As patient progress is continually reassessed, decisions will need to be made on whether new goals are needed, the timeframe for anticipated progress and recovery, and ultimately the most appropriate and safest discharge plan for the patient (see Fig. 14.4).

14.4 Continuum of Care (Fig. 14.5)

14.4.1 Acute Phase (1–30 Days Following Surgery)

Rehabilitation (OT, PT, SLP) should begin immediately after surgery (Table 14.1). During the initial part of the acute phase, the patient may be in the intensive care unit following surgery, prior to being transferred to nursing unit. Initial rehabilitation in the acute phase with patients is geared toward helping an individual achieve medical stability, recovery, and healing. This includes basic activities such as respiratory function, ability to eat/drink, and sleeping comfortably. As the patient is able to tolerate increased activity, it is important to view the person not as a patient, but as a "whole person" who fills social roles in the environment and community. These roles are vitally important to the patient and most likely one of their ultimate goals for recovery.[38]

Patient education begins from the very first contact with the patient and the family members. It is essential to continue patient and family education throughout the rehabilitation process. An important part of the patient education is an understanding of postoperative

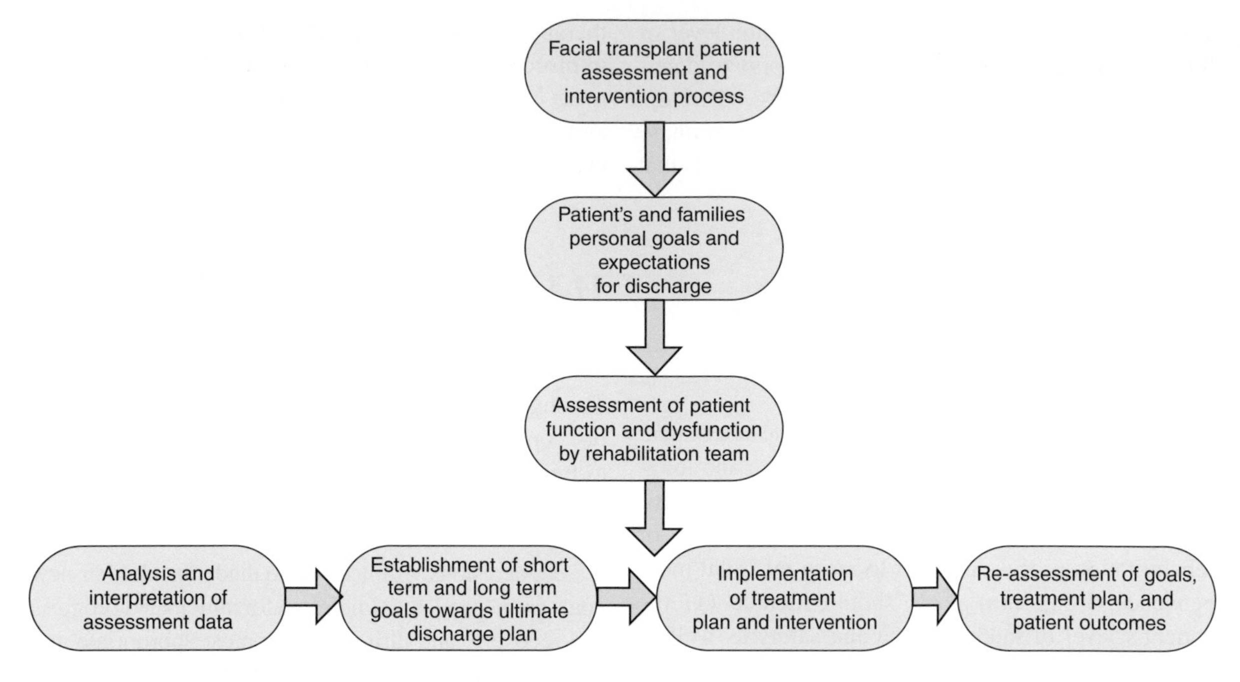

Fig. 14.4 Facial transplant patient assessment and intervention

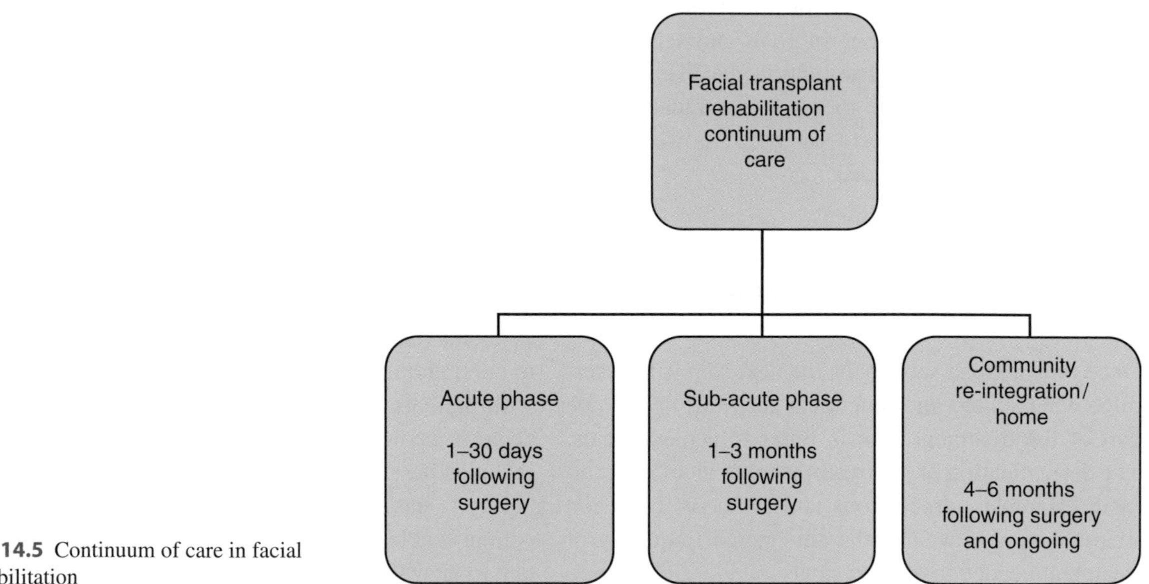

Fig. 14.5 Continuum of care in facial rehabilitation

precautions and general exercise precautions that will need to be adhered to in daily functional activities. The earlier this information is presented, the better the patient and family members will know what to expect. These precautions should be reviewed and reinforced as often as needed. The therapist should provide education to patient on rehabilitation milestones to be aware of expected physical, biomechanical, and sensory changes. The rehabilitation team members must stress the importance of patient compliance with precautions and prescribed exercise program throughout a patient's progress toward goals. As the patient begins to feel stronger and more confident, compliance with precautions can be more challenging, requiring so frequent prompting and education. During the acute phase, patient education should be constant, consistent, and

Table 14.1 Facial rehabilitation process table

Phase	Focus of intervention
Acute phase (1–30 days following surgery)	• Patient education-post-op precautions and activity precautions, physical recovery milestones • Positioning • Achieving medical stability and healing • Respiratory function • Eating/drinking ability • Increasing activity tolerance and endurance • Basic sensory function • Mobility and transfers • Simple self-care/activities of daily living • Cognitive and communicative functioning • Psychosocial coping strategies
Sub-acute phase (1–3 months following surgery)	• Patient education-individualized home exercise program, precautions with increased activity, adaptive equipment/adapted methods • Increase levels of activity and training • Mobility/physical activity (treadmill, bicycle) • Activities of daily living (ADLs) • Instrumental activities of daily living (IADLs) • Static and dynamic facial exercises • Advance diet as tolerated • Olfactory re-education • Visual processing • Preparation for community-re-entry/discharge-safety and wellness • Psychosocial coping strategies
Community re-entry/home (4–6 months following surgery and ongoing)	• Patient education-independence with individualized home exercise program, post-discharge activity precautions, adaptive equipment/adapted methods • Community reintegration • Instrumental ADLs • Ongoing therapy services as necessary to address facial movement/expression, sensation training, visual processing, olfaction, speech, swallowing • Psychosocial coping strategies • Social support groups and community resources

reinforced among all team members to ensure patient and family understanding and adherence.

After most surgeries, active range of motion as tolerated is generally indicated. In the case of facial transplant, active range of motion and flexibility exercises of the involved area should be restricted to preserve tissue integrity and promote healing. Depending on the extent of reconstruction, movement may be restricted for at least 6–8 weeks after surgery, if not longer, to prevent transplant failure, wound dehiscence, and infection. The patient will also be instructed on positioning techniques to help maintain comfort in various positions including supine, sitting, standing, transfers, and mobility. Position and movement of head and neck during functional activities will also be discussed with the patient, as one's face, head, and neck functions are interrelated. In the acute phase of care, the patient may still require physical assistance and/or adaptive equipment for transfers and mobility. Self-care skills such as how to protect their face and participate in activities such as washing face, combing hair, scratching face, or even blowing nose become a focus.

A thorough assessment of the patient's overall level of independence and function will drive the course of therapy. Assessment should include active and passive range of motion of the cervical spine, temperomandibular joints, upper and lower extremities, muscle strength and flexibility, endurance, sensory function of face, ADL performance, and specifically observations of face movement and function. The muscle groups and the functions involved in facial expression should be a point of observation for the therapist; extreme caution should be taken in accordance with surgeon's recommendations and approval. The range of motion and skin sensation in the head and neck region should also be checked.[39] Additionally, when the patient is physically ready and has physician approval, aerobic activity such as walking on treadmill or bicycling should become part of exercise program to increase patient's strength and endurance. Client-centered practice is critical in this phase of therapy. Taking into account patient's values, goals, and needs is imperative to ensure patient's motivation and progress toward recovery.

14.4.2 Sub-acute Phase (1–3 Months Following Surgery)

As the patient enters this phase of rehabilitation, they have begun initial stages of healing and are now able to participate in increased activity and training. Static and dynamic facial exercises, and focus on movements such as lip suspension and mouth occlusion, become a larger part of daily treatment. Higher level functional activities related to ADLs and mobility can be implemented into treatment sessions. Involvement can begin with interventional techniques such as visual processing, olfactory training, and preparation for community re-entry and discharge to home.

In this phase, the patient should become more independent with their home exercise program and require less demonstration or cueing. The patient's personal goals for future functioning in social, family, and work roles should drive therapy intervention. Once a decision has been made on a post-discharge plan, the therapy team will focus skilled care on activities the individual needs to be independent in their environment and society.

14.4.3 Community Reintegration/Home (4–6 Months Following Surgery and Ongoing)

As the patient progresses toward independence and discharge from the hospital, education and demonstration of independence with home exercise program are critical. Therapists work to ensure that the patient demonstrates the most independent level possible with functional performance including safety, mobility, activities of daily living, visual processing, safe olfactory sensation training, speech, and swallowing. If any areas of concern continue to exist, therapists will ensure that patient is instructed with adaptive and safe techniques in the above areas to function in their environment.

Therapists can assist in educating patients with a focus on strategies to complete functional activities with wellness and health in mind. OT, PT, and SLP are an integral component to ensuring patient has a sense of accomplishment with their level of independence and function. The meaning, "satisfaction" and "sense of accomplishment" that we feel with our daily functions and roles are tied to our overall psychological function. As the patient prepares for discharge, the therapy team will work on community re-entry including independence in Instrumental Activities of Daily Living (IADLs) such as home-making activities and community reintegration including going shopping at the drugstore or going to the bank to complete money management tasks.

For a facial transplant patient, community reintegration might be the most challenging encounter as patients need to deal with their identity, acceptance of body image, functioning in social roles, and overall quality of life. The emotional and psychological aspect of fulfilling social roles and adjusting to social situations can be overwhelming and daunting. If a patient does not have adequate support from relationships in their lives, it is essential that they take advantage of social support groups and outside resources that can assist them with coping strategies.

During this phase, the patient is discharged from the hospital and starts to re-enter and reintegrate into life and society. Through this process, patients become their own therapist and drive their own recovery. Meanwhile, the interactions with family, friends, and the rest of the community may be difficult at first. The patient may have been getting more used to the hospital environment, especially as their hospital stay may have lasted weeks, if not months. Treatment and education before discharge will consist of preparing the patient and family for the difficulties anticipated after discharge, as well as strategies for coping.

As the patient begins navigating the community and their home environment, he or she should be encouraged to participate in recommended support groups or resources for further assistance. Depending on the individual patient's progress, they may require additional therapy services through home care or outpatient setting.

Once out in the community, the patient will encounter new challenges that will bring up new questions and areas of function that may need to be addressed. Thus, the need for therapy may be ongoing following discharge to home, which can range from a few weeks to a few months.

14.5 Major Therapy Interventions

The interventions that are commonly used in facial neuromuscular rehabilitation can be categorized as follows: patient education, facial muscle therapy (muscle relaxation or stimulation, mirror exercise, facial muscle training, and EMG biofeedback), speech and swallowing training, olfactory sensation and smell training, activities of daily living, vision training, and psychosocial function (see Fig. 14.6).

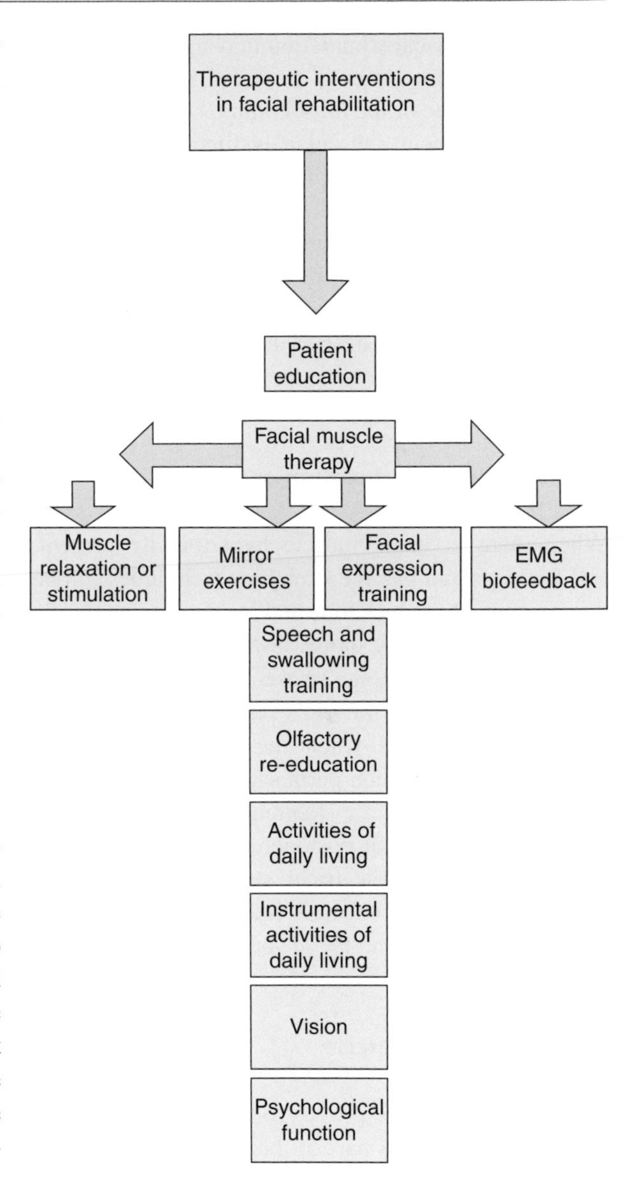

Fig. 14.6 Therapeutic interventions in facial rehabilitation

14.5.1 Patient Education

Patient education is the most important factor for successful facial rehabilitation. In-depth patient education throughout the continuum of care, including before and after the facial transplant procedure, is needed to ensure that a patient has a clear understanding and direction with routine functional activities. It is the interdisciplinary team's responsibility to ensure that all points of education are covered in treatment. The team must ensure that the information provided to the patient and family is consistent and reinforces precautions and goals of a facial procedure.

14.5.2 Facial Muscle Therapy

Neuromuscular dysfunction of the facial musculature after facial transplant may fall into the following four categories: (1) loss of muscle strength, (2) loss of isolated motor control, (3) muscle hypertonicity, or (4) synkinesis (defined as "an unwanted movement accompanying a desired motion"). Facial motor activity may begin to return between 2 and 3 months after the facial transplant, which usually is after partial restoration of the sensory function of the facial nerve. In the case of loss of muscle strength, possibly due to lack of nerve regeneration, it may result in extended facial palsy. Treatment should include sensory stimulation techniques and isolated strengthening exercise to the affected muscles. For cases that show loss of isolated motor control, surface EMG or mirror training can be used to train the movement of individual muscle. In late phase of neuroregeneration (greater than 6 months post-facial transplant), when patients begin to develop muscle hypertonicity or facial asymmetry, mirror feedback, passive stretching, and massage can be used to reduce muscle hypertonicity. Overly aggressive and prolonged electrical stimulation may aggravate hypertonicity, and thus should be used in moderation. Synkinesis is one of the most common as well as most difficult facial neuromuscular dysfunction to resolve. Great patience will be required to unlink the desired motion from the undesired one. This is done by making the slow, small amplitude symmetric desired motion while preventing or releasing the undesired synkinetic contraction. Once this is accomplished, the speed and force of the desired movement are gradually increased while suppressing the synkinesis.

14.5.2.1 Muscle Relaxation or Stimulation

When there is facial muscle hypertonicity present, self-massage and passive stretching techniques can be demonstrated to reduce tone and regain normal muscle length. The patient is shown how to use the opposite hand to produce massage strokes that go in the opposite direction of the muscle's pull. Stimulation techniques can be used when the facial muscles are hypotonic, with surgeon permission at the appropriate time in recovery phase (community re-entry/home phase) and along with postoperative precautions and protocol in mind.[9] Electrical stimulation should be employed carefully because inappropriate use may increase synkinesis and facial spasms.

14.5.2.2 Mirror Exercise

Mirror exercises have proven to be effective methods of training patient, as part of their exercise programs. In front of a mirror, a patient performs facial movement exercises involving the isolation of each muscle or muscle group on either side of the face. Patient practices the facial motion instructed, while watching movement in the mirror, so there is instant visual feedback. Variations can occur with these exercises, such as changes in frequency, amount of pressure, speed, and direction (unilateral, bilateral, or both sides).

14.5.2.3 Facial Muscle Training (Facial Expression)

Our facial expressions are porthole to our emotions and feelings. Through the expressions on a person's face, we see the flow of both nonverbal and verbal communication.[40] In facial muscle training, the therapist (often primarily the SLP) can help retrain the facial transplant patient in the appropriate use of facial muscle expressions in daily and social interactions. There is minimal information available regarding facial retraining. In one study of facial training, Beurskens and Heymans evaluated facial retraining which involved emotional expression exercises, synkinesis exercises, and other techniques.[41] They found that patients' self-assessments and measurements of mouth movements were improved through treatment. Other methods of training facial expression can involve having patient practice facial expressions seen in pictures and then practice in front of a mirror. Potential future methods include the use of motion capture and force sensors which can be used to quantitatively assess training outcomes.

14.5.2.4 EMG Biofeedback

EMG biofeedback using surface electrodes is one of the most effective facial neuromuscular re-education tools available. It provides extremely sensitive real-time visual and/or auditory feedback information to the patient while attempting to relax hypertonic muscles, increase force of targeted muscle contraction, prevent undesired muscle activity in adjacent areas, or decrease and eliminate synkinesis patterns. When combined with specific mirror exercises and a daily home exercise program done between treatment sessions, optimal results can be obtained.[42-44] Biofeedback has the added advantage of providing encouragement to the patient and family that the nerve is indeed recovering – a vital factor when dealing with the slow nature of facial nerve recovery.

14.5.3 Speech Training and Swallowing Training

When a person looks at someone's face (another person's), they often focus on the features such as the lips or nose, commenting on size, shape, and structure. Our lips and nose serve many other purposes besides just aesthetic beauty. It is very difficult for one to function in everyday activities without the functional use of lips and nose. The lips play an important role in swallowing, which determines whether someone is able to adequately accept liquid or solid food for proper nutrition. The lips also play a role in facial expression, such as smiling, frowning, pleasure, or anger. When a patient loses his/her ability to smile, it can be very difficult to participate in relationships and society. Rehabilitation of the ability to smile evolves from the re-education of muscles and motor control. The zygomaticus and risorius muscles are pointed out, and the patient can practice smiling, often referring back to mirror exercises. Rehabilitation of speech and swallowing after facial reconstructive surgery has not been previously well described in the literature.[45] What we can extract from our review is that swallowing problems experienced post-facial transplant surgery are dependent on a combination of several factors, including degree of initial trauma/injury, extent and type of reconstruction needed, and rate and speed of recovery with skilled intervention. As future rehabilitation protocols are developed, methods utilized for treating after face paralysis can be adapted for facial transplant patients.

14.5.4 Olfactory Re-education

The role and effects of the olfactory system on performance and daily life is more instrumental than we realize. The olfactory system (sensation of smell) plays a vital part in our general safety and participation in daily activities. Our sense of smell is conveyed by receptors that lie deep within the nasal cavity. Smell is important for detection of noxious and pleasant odors and is associated with the pleasure of taste. With connections to the limbic system, smell also has a role in our memories and our emotions, thus influencing meaning and satisfaction in our lives. Occupational therapy (and speech therapy as an adjunct) plays an important role in helping patients after facial transplantation to regain olfactory sensation. This is important because it helps patients re-establish the ability to smell, experience pleasure and displeasure with smells in their environment. Occupational therapists can evaluate an individual's ability to smell and once the olfactory nerve regeneration is complete, can assist with helping that individual to relearn important smells. This is done by incorporating familiar scents as well as scents to rebuild safety awareness in treatment sessions.

Therapy involving peripheral nerve reattachment for hands has demonstrated the benefits of sensory and motor re-education; however, reattachment of the olfactory nerve is a new area of research.[36] One method for the assessment and intervention of olfactory sensation involves exposing the patient to various familiar and natural odors within similarly colored bottles ensuring that the patient cannot identify the odor by bottle type. Odors that can be used include coffee, almond, chocolate, lemon oil, and peppermint. Vinegar, ammonia, or other irritating chemical odors should be used with caution as they tend to be irritating to patients. Occupational therapists should assess whether the patient is able to detect and identify odors as well as if they can smell through only one or both nostrils.[44] As the olfactory nerve continues to regenerate, OTs can continue to work with the patient and expose them to various types of odors. Speech-language pathologists can assist with training the patient to use the muscles of the mouth to create a negative pressure and increase nasal airflow with attempting to smell.[37] Facial transplant patients can also become an integral part of their own treatment as they independently work on olfactory re-education activities as part of their individualized home program.

14.5.5 Activities of Daily Living

Activities of Daily Living (ADLs) are part of our everyday lives. Some activities are so natural to us that we complete them without even realizing what is involved in the activity. ADLs are activities that are oriented toward taking care of one's own body and inherent to survival and well being. ADLs include but are not limited to activities such as bathing, dressing, eating, feeding, personal hygiene, and grooming. Instrumental Activities of Daily Living (IADLs) are activities that support daily life within the home and community that often require more complex steps and skills. IADLs include, but are not limited to caregiving of others such as children and pets, communication management and mobility, financial management, meal preparation shopping, home establishment/management.[46]

With facial transplantation, OTs work to help the patient achieve optimal independence with ADLs and

IADLs. After a facial transplant procedure, a patient must learn how to touch their face and clean their face with the appropriate amount of pressure. In collaboration with SLP, OT helps the patient to be independent with eating and drinking using proper facial movements and adaptive equipment, adapted strategies as needed. Occupational therapists can help an individual gain the ability to participate in daily activities such as dressing, bathing, and home management tasks. An OT might suggest adaptive equipment for both hospital stay (bedside commode or built up feeding utensils) or at home (adaptive equipment for bathroom).

14.5.6 Vision

Our vision is our guide to the world around us. We depend on our vision for feedback on how we approach our everyday activities, as well as for the decisions one makes as how to anticipate and adapt to the environment around us. Visual processing skills influence so many important functions including motor planning, postural control, cognition, perception, and safety. Visual processing occurs through two modes: focal vision/attentive vision and ambient, peripheral, or pre-attentive vision. Focal vision allows us the ability and attention needed for perception and discrimination of an object. Ambient vision works in connection with proprioceptive, kinesthetic, tactile, and vestibular systems.[47]

Occupational therapists play a role in helping a patient after facial transplant deal with their visual abilities to function and adapt if necessary. Patients who undergo facial transplant surgery may have a variety of visual deficits depending on origin and extent of facial disfigurement. Occupational therapists are not responsible for conducting extensive eye examinations as they are more the role of an ophthalmologist or vision specialist. Occupational therapists use skilled interventions to identify visual processing issues a patient may be experiencing and plan treatment activities to help a patient function. Strategies that OT utilize for adaptation for vision include alteration of the environment, ensuring consistency of environment, using adapting techniques to complete a task, utilizing variations of lighting, and use of adaptive equipment to aid with vision. Adaptive equipment used most commonly may include a magnifying glass, screen reader, and Braille or other type of labeling using sensory (materials to help patient with identification) or color coding (bright or darker colors for discrimination between objects).[48]

14.5.7 Psychosocial Function

A person's interests, values, and goals, as well as their overall psychological, cognitive, and emotional capabilities have an impact on how they react to the environment around them. When a person faces an injury, disability, or traumatic event, these challenges can be intensified and increased. With facial transplant patients, literature has reported that psychosocial capacities are important for successful recovery. The therapist and the patient work together to assess patient's level of function (physical and psychosocial) within the environment and to identify areas needing change. Once areas needing change are identified, intervention is implemented toward the goal of maximizing patient's strengths to overcome any potential deficits.[49] Intervention may include participation in new or existing meaningful activities and roles, development of coping strategies through education and practice, and providing skill training relevant to daily functional life. The entire rehabilitation team has a critical role in assisting facial transplant patients in their ability to develop and implement effective coping strategies to deal with challenges they may encounter.

14.6 Case Presentations

14.6.1 Case 1

14.6.1.1 The Patient

The patient is a 38-year-old female who suffered trauma to her face after a dog chewed her face while she was unconscious. The severe dog bite affected all of the soft tissues, amputating her distal nose, upper and lower lips, the whole chin, and adjacent parts of right and left cheeks. The patient underwent an operation for facial transplant at the University Hospital in Amiens, France.[50-53]

14.6.1.2 Prior to Facial Transplantation Surgery

Due to the extent of injury to facial structures, the patient's ability for functional activities such as communication (verbal and nonverbal), eating, drinking, and olfaction sensation was significantly impacted. Patient's routine participation in daily Activities of Daily Living and social engagement were also affected. Prior to facial transplant, the goal was to prepare the patient physically, mentally, and emotionally for the upcoming surgery. Physiotherapy (physical therapy) was initiated pre-operatively for scar management and prevention of muscle atrophy. Physiotherapy was provided with the goal of preserving facial function, including facial movements involved with facial movement and facial expression. The hope was that through implementation of therapy and preservation of function, the patient would have an increased likelihood of success for recovery following facial transplantation. Since trauma to facial function was extensive, physiotherapist's ability to work on functional activities such as speech, feeding, and olfaction was limited. Patient education throughout pre-operative phase remained a focus of treatment intervention.

14.6.1.3 Following Facial Transplantation Surgery

Following facial transplant, the patient was noted to have mild edema but in general good medical condition. Physiotherapy was initiated 48 h after surgery with a plan of care requiring twice a day therapy. Physiotherapy intervention was focused on sensory re-education, passive and active facial exercises including restoration of lip movement and mouth occlusion. Patient education was a vital component of the treatment plan, including patient's understanding of postoperative precautions, physiotherapy protocols, and psychosocial function.

14.6.1.4 Functional Outcomes

Functionally, the patient responded well to physiotherapy intervention. Following surgery, the patient began to participate in sensory proprioception activities including mobilization of tongue to experience new facial features (lip and cheeks). The patient demonstrated adequate ability for eating and drinking 4 days postoperative, and this continued to be an area for skilled intervention as physiotherapy continued. Physiotherapy intervention also focused on symmetric facial expression such as ability to smile, as well as quality and function of speech.

Following the initial surgery, the patient underwent a second surgery which resulted in positive physical and psychological outcomes. The patient continued to receive physiotherapy to address facial movement and function. Overall, the patient demonstrated quicker recovery with sensory function than motor function. Sensory discrimination quickly recovered postoperatively while heat and cold sensation steadily recovered over a 6-month period of time post surgery. Facial movements, such as motion of the nose, chin, lips, muscles involved with smiling, mastication, and phonation, continued to improve progressively over the next 6–18 months. Ultimately, the patient experienced a successful reintegration into her social environment and community.

14.6.2 Case 2

14.6.2.1 Introduction

The following case study was an unprecedented clinical experience and should be viewed as multiple clinicians' effort to respond to the challenge of speech, swallowing, and physical rehabilitation after facial transplant surgery. Relevant treatment issues include: (a) no pre-injury speech/voice profile available for use as a reference target, (b) care strategies were based on application of general stimulation principles via trial and error, (c) all strategies were supported and guided by the surgeons to both maximize the surgical benefits and avoid disruption of tissue healing, (d) coordinated, interdisciplinary interaction was perceived as the ideal care model.

14.6.2.2 The Patient

The patient was a 45-year-old female who had a history of severe mid-face trauma due to a gunshot wound 4 years previously. In December 2008, the patient underwent near-total face transplantation at the Cleveland Clinic, where 80% of face was replaced with a tailored composite tissue allograft.

14.6.2.3 Intensive Care Unit (ICU) Stay

Initial PT and OT evaluations for facial transplant were initiated 7 days postoperative surgery. At this time, the patient required "total assistance" for mobility and ADLs. Initially, PT and OT provided concurrent therapy sessions due to the patient's decreased activity tolerance and schedule limitations from medical tests and procedures. Her plan of care in the ICU was for 5 days a week of therapy to address mobility and ADLs. Facial movement exercises were not completed at this time due to the patient's initial healing period.

14.6.2.4 Nursing Floor

Therapy continued once patient transferred to the regular nursing floor at 22 days postoperative surgery. Once the patient was settled on regular nursing floor, PT and OT re-evaluated the patient at that time and set new goals to include increasing activity tolerance with mobility, time management within the hospital schedule to include a set time to rest during the day, increasing independence with ADLs while maintaining postoperative surgery precautions, and increasing smell recognition. In addition, the patient identified a goal of increasing independence with precautions, therapy exercise handouts, and medication management by regaining the ability to read while in the hospital. The patient used a screen reader at home to enlarge text. A similar screen reader was provided similar to the patient during her hospital stay. Physical therapy worked with patient on strength, mobility, and endurance. Focus was on independence with home exercise program. The patient had access to stationary bike and treadmill in an adjacent room. Through work with PT, she was independent with mobility and with her individualized exercise program.

14.6.2.5 Pre-obturator Treatment

Speech-language pathology (SLP) was first consulted following patient's transfer from the intensive care unit to the regular nursing floor in the acute hospital. Therapeutic assessment was initiated to determine baseline speech function and to begin facial reanimation training and tissue stimulation. Patient was seen 1–2 times per day, 5 days per week for 15–45 min sessions during the acute hospital stay.

Given the significant edema of the donor tissue, along with the ongoing healing of the suture lines, utmost care was taken not to be overly aggressive with the oral-motor exercises and digital stimulation. The patient had been receiving tube feeding through a Percutaneous Endoscopic Gastrostomy (PEG) prior to the transplant surgery, and oral intake trials were not initiated until after the palatal obturator was fitted. She also previously had a #6 Shiley tracheostomy tube which was capped allowing voicing with excellent quality and a strong cough. The primary issue affecting speech intelligibility was the palatal defect resulting in significant hypernasality. When the nares were occluded with finger pinching, nasality was reduced, but absence of upper lip movement affected all anterior labial speech and non-speech sounds to some degree. Patient demonstrated the ability to compensate for this by articulating the lower lip up against the transplanted middle incisors.

Pre-obturator treatment sessions consisted of traditional oral-motor exercises including slow and repetitive facial movements as well as gentle digital stretching, tapping of the muscle, and massaging across the entire face in a very deliberate manner. Hands-on techniques were used to stimulate sensory pathways and blood flow with a secondary goal of decreasing headache pain.

Patient was given 11 basic exercises to work on independently and with her nurses 2–3 times daily for 10 min. These included jaw range of motion movements, labial retraction and pursing, scrunching of the nose and whole face, eye brow raising and eye lid lowering along with production of exaggerated labial speech and non-speech movements (e.g., "wow," "woo"). Because patient was legally blind, all written material had to be in large (greater than 1 in. height) letters for her to read; she required a partner to assist her in all exercise work early on. This also made mirror feedback or visual cues for motor movements very challenging. Tactile cues were imperative as patient was encouraged to touch the therapist's face as movements were demonstrated to her prior to her own duplication attempt. Occupational therapy provided the patient with a reading magnifier which was helpful in improving overall communication abilities by allowing her to read the written directions/words and to view diagrams and drawings. Collaboratively, SLP and OT also began integrating functional facial gesturing into treatment including kissing, winking, blinking, and blowing.

The patient was noted to be working too diligently and "overdoing" her exercises which resulted in use of her chest, neck, and hands during her performance. Speech-language pathology worked on isolating only the facial muscles without excessively straining the body. In addition, SLP began introducing more functional "sucking" and "blowing" activities. These included using an incentive spirometer with the large mouthpiece to improve the labial seal, and also straws of varying sizes to "suck up" tissue paper and blow cotton balls across the table. Light pinching of the nares was still required at this time to avoid loss of air through the nose.

Over the first 4 weeks of therapy, the patient maintained an optimistic, motivated attitude along with a strong spirit. Response variability did occur for multiple reasons including periodic biopsies and sinus irrigations, medication, pain, general fatigue, and other medical and social issues. Humor and light-hearted conversation was frequently injected into the treatment sessions to offset any monotony or depression. Facial massage was at times the only treatment provided as it subjectively eased tension and pain.

14.6.2.6 Post Obturator Treatment

At week 4 of treatment, dentistry provided the patient with her initial palatal obturator which did not include teeth. The patient reported improved listening intelligibility, initiation of a liquid diet, and ingestion of oral medications, immediately after having this obturator placed. Oral phase swallowing therapy was initiated, assisting patient with functional drinking from a cup and straw. There was no clinical evidence of pharyngeal dysphagia and no indication of need for further instrumental swallowing studies.

Many adjustments of the prosthesis were made to close off the palatal defect completely as well as the addition and modification of teeth to the plate to improve dental occlusion for mastication and to refine speech clarity. The patient was now allowed to eat soft solid foods and tolerated them well.

As care shifted to a more acute rehabilitation level, treatment was provided 3–5 times per week but with more aggressive digital stimulation to further facilitate facial sensation and motor movement. Facial muscle re-education/massage now involved intra-oral stimulation using finger pressure points and an iced laryngeal mirror for tapping and rubbing. The patient was discouraged at this time due to her inability to feel the cold stimulus, but did report sensing deeper finger pressure and stretching especially nearer the suture lines at the ears. Facial expression exercises were initiated and included exaggerated smiling, frowning, surprise, fear, pain, and anger.

14.6.2.7 Post-discharge from Acute Care

Upon discharge from the hospital, OT re-evaluated the patient again. Plan of care continued at 5 days per week, and new goals were set to include increasing independence with the IADLs of cooking, learning stress management and relaxation techniques, and continuing olfactory re-education to address safety concerns once patient returned to her home. Occupational therapist worked with the patient to complete community reintegration activities by accompanying the patient as she performed functional mobility in the local community area for the first time since her surgery. The goal of community re-entry activities was to help prepare patient for what she would experience once in the home environment. The OT was able to provide education and instruction to patient on safety techniques, as well as recommendations for function of various tasks. Upon discharge from OT services, the patient was independent with ADLs and modified independent with IADLs. The patient was also able to correctly identify the smell of coffee and had strategies to continue to re-educate herself on smells while at home. As the patient had demonstrated independence with mobility, strength, and her exercise/fitness program at discharge from nursing floor, PT services were discharged at that time.

After 18 weeks of acute care and acute rehabilitation level speech therapy at the Cleveland Clinic, the patient was discharged to home with plans for ongoing speech therapy in her local area. With learned compensatory patterns, reduced rate of speech, and slightly over-exaggerated articulation, speech intelligibility was considered very good, but with persistent sound distortions. The patient was eating and drinking unassisted without difficulty. Ultimately, the patient was successfully reintroduced to her home and community environment.

14.7 Discussion

Rehabilitation is a vital component of any facial transplantation surgery. The interdisciplinary team, including PT, OT, and SLP, is an integral part of the team contributing to the success of rehabilitation. Through client-centered care, the rehabilitation team works together to ensure that the patient's physical, psychological, and emotional needs are met.

It is important for an individual considering face transplant surgery to be aware of the potential need for therapy services postoperatively. Therapy being involved pre-operatively and having a full awareness of patient's abilities and deficits prior to surgery can assist in ensuring a better postoperative plan of care. This therapy can start while the patient is still in the ICU or even prior to admission to the hospital. Speech-language pathology involvement in pre-operative surgery can help to prepare for the challenge of how to rehabilitate speech and swallowing ability as well as how to undergo facial reanimation training for transplant patients.

It is important for the surgeon to inform the therapy team regarding peri-operative care, anticipated outcomes, and pending procedures. Examples of this useful information include any additional surgeries that the patient is expected to undergo either while in the hospital or upon discharge, a timeline for when the patient can get the incisions wet, and return to regular face washing, a time for when the patient may begin eating again, and if the patient can have his or her head in a gravity-dependent position.

If at all possible, continuity of care should occur throughout the rehabilitation process ideally with the same clinician from pre-operative, acute care, to acute rehabilitation, home, and outpatient therapy. Since facial transplantation is such a new frontier, it is imperative that members of the rehabilitation team gain a better understanding of how patients proceed throughout the continuum of care. This is how we will gain a true understanding of the proper protocols and procedures needed to ensure most successful rehabilitation outcomes. Successful rehabilitation results in improved patient satisfaction, self-esteem, and quality of life.[5]

Facial neuromuscular rehabilitation is effective for achieving optimal function of recovery. Rehabilitation can occur across the health care continuum from onset of injury/illness to pre-surgery to post surgery and then community re-entry. Each patient displays unique injury and movement patterns, so each individual needs an individually tailored treatment and exercise program. The strategies mentioned in this chapter have only begun to break the surface of potential interventions that can help a facial transplant patient optimize their function and satisfaction in life.

There is much that needs to be studied in the area of facial rehabilitation. Research, identification, and implementation of new treatment modalities and techniques will be needed as more facial transplants are completed across the globe. Sequence of facial movements and static/dynamic facial exercises, and how these movements affect nerve growth in facial transplant patients need to be further investigated. Use and timing of modalities, such as electrical stimulation, should be further considered.

As facial transplants become more prevalent, members of the rehabilitation team will need to further identify the most appropriate tools and techniques to utilize in facial transplant rehabilitation. As most therapists do not regularly work with patients with facial muscle and nerve injuries, and have little specialized training in facial muscle therapy, this will need to be an area of further education and training to ensure therapists are empowered to provide patients with quality and competent care.

14.8 Conclusion

Facial transplantation is an exciting new frontier in medicine and rehabilitation. It has added new hope to individuals with illness/injuries such as trauma, burns, wounds, cancer, or congenital developmental effects. For members of the rehabilitation team, it has sparked an exciting new area of rehabilitation, where we as clinicians are about to embark into unforeseen areas. Most importantly, it gives the clinician a valuable opportunity to make a vital difference in the life of a patient who deserves to have a second chance at experiencing quality of life and functional independence.

Acknowledgments The authors wish to thank Angela Broadnax OTR/L, Brian Hedman SLP, Ela Plow Ph.D., PT, and Vinoth Ranganathan MSE, M.B.A. for their editorial comments and assistance.

References

1. Siemionow M, Kulahai Y. Facial transplantation. *Semin Plast Surg*. 2007;21:259-268.
2. World Health Organization. *International Classification of Functioning, Disability and Health*. Geneva: WHO; 2001.
3. Siemionow M, Gordon CR. Overview of guidelines for establishing a face transplant program: a work in progress. *Am J Transplant*. 2010;10:1290-1296.
4. Chenggang Y, Yan H, Xudong Z, et al. Some issues in facial transplantation. *Am J Transplant*. 2008;8:2169-2172.
5. Toure G, Meningaud JP, Bertrand JC, et al. Facial transplantation: a comprehensive review of the literature. *J Oral Maxillofac Surg*. 2006;64:789-793.
6. Morris P, Bradley A, Doyal L, et al. Face transplantation: a review of the technical, immunological, psychological and clinical issues with recommendations for good practice. *Transplantation*. 2007;83:109-128.
7. Barker JH, Stamos N, Furr A, et al. Research and events leading to facial transplantation. *Clin Plast Surg*. 2007; 34:233-250. ix.
8. Diels H. New concepts in nonsurgical facial nerve rehabilitation. In: Myers E, Bluestone C, eds. *Advances in Otolaryngology-Head and Neck Surgery*. Chicago: Mosby-Year Book; 1995:289-315.
9. Diels HJ. Facial paralysis: is there a role for a therapist? *Facial Plast Surg*. 2000;16:361-364.
10. Westberg J, Jason H. *Collaborative Clinical Education: The Foundation of Effective Healthcare*. New York: Springer; 1993.
11. Baldwin D. *The Role of Interdisciplinary Education & Teamwork in Primary Care and Health Care Reform ed.* Rockville: BoHP, Health Resources and Services Administration; 1994.
12. Stav WB. Occupational therapy and older drivers: research, education, and practice. *Gerontol Geriatr Educ*. 2008;29: 336-350.
13. Dean E. Physical therapy in the 21st century (Part I): toward practice informed by epidemiology and the crisis of lifestyle conditions. *Physiother Theory Pract*. 2009;25:330-353.
14. American Physical Therapy Association. *Guide to Physical Therapist Practice*, 2nd ed. Alexandria, VI: American Physical Therapy Association. *Phys Ther*. 2001;81:9-746.
15. Schneider SL, Sataloff RT. Voice therapy for the professional voice. *Otolaryngol Clin North Am*. 2007;40:1133-1149. ix.
16. Gordon CR, Siemionow M, Coffman K, et al. The Cleveland Clinic FACES Score: a preliminary assessment tool for identifying the optimal face transplant candidate. *J Craniofac Surg*. 2009;20:1969-1974.
17. Brill SE, Clarke A, Veale DM, et al. Psychological management and body image issues in facial transplantation. *Body Image*. 2006;3:1-15.
18. Cabanero-Martinez MJ, Cabrero-Garcia J, Richart-Martinez M, et al. The Spanish versions of the Barthel index (BI) and the Katz index (KI) of activities of daily living (ADL): a structured review. *Arch Gerontol Geriatr*. 2009;49:e77-e84.
19. Granger CV, Hamilton BB. The Uniform Data System for Medical Rehabilitation report of first admissions for 1992. *Am J Phys Med Rehabil*. 1994;73:51-55.
20. Heinemann AW, Kirk P, Hastie BA, et al. Relationships between disability measures and nursing effort during medical rehabilitation for patients with traumatic brain and spinal cord injury. *Arch Phys Med Rehabil*. 1997;78: 143-149.
21. Khan F, Pallant JF, Brand C, et al. Effectiveness of rehabilitation intervention in persons with multiple sclerosis: a randomised controlled trial. *J Neurol Neurosurg Psychiatry*. 2008;79:1230-1235.
22. Segal ME, Gillard M, Schall R. Telephone and in-person proxy agreement between stroke patients and caregivers for the functional independence measure. *Am J Phys Med Rehabil*. 1996;75:208-212.
23. Choo B, Umraw N, Gomez M, et al. The utility of the functional independence measure (FIM) in discharge planning for burn patients. *Burns*. 2006;32:20-23.
24. D'Aquila MA, Smith T, Organ D, et al. Validation of a lateropulsion scale for patients recovering from stroke. *Clin Rehabil*. 2004;18:102-109.
25. Mylotte JM, Graham R, Kahler L, et al. Impact of nosocomial infection on length of stay and functional improvement among patients admitted to an acute rehabilitation unit. *Infect Control Hosp Epidemiol*. 2001;22:83-87.
26. Oczkowski WJ, Barreca S. The functional independence measure: its use to identify rehabilitation needs in stroke survivors. *Arch Phys Med Rehabil*. 1993;74:1291-1294.
27. Davis RE, Telischi FF. Traumatic facial nerve injuries: review of diagnosis and treatment. *J Craniomaxillofac Trauma*. 1995;1:30-41.
28. VanSwearingen JM, Brach JS. The Facial Disability Index: reliability and validity of a disability assessment instrument for disorders of the facial neuromuscular system. *Phys Ther*. 1996;76:1288-1298; Discussion 1298-1300.
29. Berry H. Traumatic peripheral nerve lesions. In: *Clinical Electromyography*. 2nd ed. Boston: Butterworth-Heinemann; 1993:323-390.
30. Valls-Sole J. Neurophysiological assessment of trigeminal nerve reflexes in disorders of central and peripheral nervous system. *Clin Neurophysiol*. 2005;116:2255-2265.
31. Ongerboer de Visser BW, Cruccu G. Neurophysiologic examination of the trigeminal, facial, hypoglossal, and spinal accessory nerves in cranial neuropathies and brain stem disorders. In: Brown WF, Bolton CF, eds. *Clinical Electromyography*. 2nd ed. Boston: Butterworth-Heinemann; 1993:61-92.
32. Kobayashi M, Pascual-Leone A. Transcranial magnetic stimulation in neurology. *Lancet Neurol*. 2003;2:145-156.
33. Sittel C, Guntinas-Lichius O, Streppel M, et al. Variability of repeated facial nerve electroneurography in healthy subjects. *Laryngoscope*. 1998;108:1177-1180.
34. Sittel C, Stennert E. Prognostic value of electromyography in acute peripheral facial nerve palsy. *Otol Neurotol*. 2001; 22:100-104.
35. Jungehulsing M, Guntinas-Lichius O, Stennert E. Rehabilitation in chronic facial paralysis. 1 [in German]. *HNO*. 2001;49:418-426.

36. Stitik TP, Foye PM, Nadler SF. Electromyography in craniomaxillofacial trauma. *J Craniomaxillofac Trauma*. 1999;5: 39-46.
37. O'Sullivan SB, Schmitz JT. *Physical Rehabilitation: Assessment and Treatment*. 2nd ed. Philadelphia: F. A. Davis; 1988.
38. Christiansen C. Occupational therapy: intervention for life performance. In: Christiansen C, Baum C, eds. *Occupational Therapy: Overcoming Human Performance Deficits*. Thorofare: Slack; 1991.
39. Siemionow M, Hivelin M. Face transplantation: clinical application of the concept. *Pol Przegl Chir*. 2008;80:571-578.
40. Byrne PJ. Importance of facial expression in facial nerve rehabilitation. *Curr Opin Otolaryngol Head Neck Surg*. 2004;12:332-335.
41. Beurskens C, Heymans P. Positive effects of mime therapy on sequelae of facial paralysis: stiffness, lip mobility, and social and physical aspects of facial disability. *Otol Neurotol*. 2003;24:677-681.
42. Piza-Katzer H, Estermann D. Cognitive re-education and early functional mobilisation in hand therapy after bilateral hand transplantation and heterotopic hand replantation – two case reports. *Acta Neurochir Suppl*. 2007;100:169-171.
43. Risberg-Berlin B, Ylitalo R, Finizia C. Screening and rehabilitation of olfaction after total laryngectomy in Swedish patients: results from an intervention study using the nasal airflow-inducing maneuver. *Arch Otolaryngol Head Neck Surg*. 2006;132:301-306.
44. Williams P, Beth E. *Occupational Therapy: Practice Skills for Physical Dysfunction*. London: Elsevier; 2001.
45. Clayton NA, Ledgard JP, Haertsch PA, et al. Rehabilitation of speech and swallowing after burns reconstructive surgery of the lips and nose. *J Burn Care Res*. 2009;30:1039-1045.
46. American Occupational Therapy Association. Occupational therapy practice framework: domain and process. *Am J Occup Ther*. 2002;56:609-639.
47. Zoltan B. *Vision, Perception, & Cognition A Manual for Treatment of the Neurologically Impaired Adult*. Thorofare: Slack; 1996.
48. Hopkins HL, Smith HD. *Willard and Spackman's Occupational Therapy*. Philadelphia: Lippincott Williams & Wilkins; 1993.
49. Bruce MA, Borg B. *Psychosocial Occupational Therapy Frames of Reference for Intervention*. Thorofare: Slack; 1993.
50. Devauchelle B, Badet L, Lengele B, et al. First human face allograft: early report. *Lancet*. 2006;368:203-209.
51. Dubernard JM, Lengele B, Morelon E, et al. Outcomes 18 months after the first human partial face transplantation. *N Engl J Med*. 2007;357:2451-2460.
52. Devauchelle B, Lengele BG, Moure C, et al. Mediatisation and "facial graft" or "to whom belongs the facial graft"? *Ann Chir Plast Esthét*. 2007;52:528-530.
53. Lengele BG, Testelin S, Dakpe S, et al. "Facial graft": about the first facial allotransplantation of composite tissues. *Ann Chir Plast Esthét*. 2007;52:475-484.

15 Prosthetic Support in Face Transplantation

Michael L. Huband

Contents

M.L. Huband
Section of Maxillofacial Prosthetics, Department of Dentistry, Head and Neck Institute, Cleveland Clinic, Cleveland, OH, USA
e-mail: hubandm@ccf.org

Abstract After face transplantation, the patient may still present with defects which were not corrected or correctable through surgery. This may include loss of teeth and bone, malocclusion and arch size discrepancies, communication between the oral and nasal cavities and sinuses, difficulties with speech and swallowing, palatopharyngeal incompetence, palatopharyngeal insufficiency, functional or anatomic deficits of the tongue, narrowing of the nostrils, and loss of anatomical structures such as eyes, ears, or nose. To correct these defects and improve function and esthetics, a variety of prostheses and appliances are available. These are fabricated by dentists who specialize in Prosthodontics and subspecialize in Maxillofacial Prosthetics. Although the face transplant patient will present clinicians with new challenges, established techniques and sound prosthodontic and prosthetic principles will provide the foundation for the delivery of prosthetic support and treatment of this patient population. For optimal results, the Maxillofacial Prosthodontist should be a member of the transplant team and consulted during the pre-surgical, surgical, and post-surgical phases of face transplantation.

15.1 Prosthodontics and Maxillofacial Prosthetics

Prosthodontics is one of the nine recognized specialties in dentistry. This specialty is concerned with the replacement of lost teeth and associated oral structures through the fabrication of crowns and bridges, and complete and partial dentures. Maxillofacial Prosthetics is a sub-specialty of Prosthodontics which is concerned

M.Z. Siemionow (ed.), *The Know-How of Face Transplantation*,
DOI: 10.1007/978-0-85729-253-7_15, © Springer-Verlag London Limited 2011

with the replacement of lost teeth and associated oral, head and neck, and facial structures through the fabrication of intra-oral and extra-oral appliances and prostheses. In essence, it may be considered the sub-specialty of non-surgical or prosthetic reconstruction of the oral cavity, head and neck, and face.

Maxillofacial Prosthodontists are trained in the management and rehabilitation of patients with oral and head and neck deformities resulting from congenital or acquired defects. They are able to fabricate intra-oral appliances and prostheses to assist with restoring mastication, speech, swallowing, and esthetics. Additionally extra-oral, or facial prostheses may be constructed to enhance esthetics and replace missing structures including the eyes, ears, and nose. Although the face transplant patient will present clinicians with new challenges, established techniques and sound prosthodontic and prosthetic principles will provide the foundation for the delivery of prosthetic support and treatment of this patient population.

15.2 Prosthetic Support Before and During Transplantation

Ideally, the recipient and the donor should undergo a thorough visual and radiographic dental evaluation prior to face transplantation. This is to identify and remove current and possible future sources of infection. Established dental guidelines for other solid tissue transplant patients should be followed.

As part of the pre-surgical evaluation, an analysis of the occlusal and skeletal relationships should be preformed. From this, surgical splints may be fabricated to assist the surgical team in establishment of proper position of the teeth, mandible, and maxilla. During surgery, the Maxillofacial Prosthodontist should be consulted to help determine where surgical margins, which impact on the design and success of prosthesis, should be placed. For example, if an orbit cannot be restored with native or donor tissue and a prosthesis is planned for, are there adequate dimensions to the defect to permit insertion of an ocular or facial prosthesis? If a portion of an ear needs to be replaced prosthetically, a more esthetic result is often achieved when the remaining auricular tissue is removed and the entire ear is restored with prosthesis.

15.3 Prosthetic Support After Transplantation

After face transplantation, the patient may still present with defects which were not corrected or correctable through surgery. This may include loss of teeth and bone, malocclusion and arch size discrepancies, communication between the oral cavity and nasal cavities and sinuses, problems with speech and swallowing, palatopharyngeal incompetence, palatopharyngeal insufficiency, functional or anatomic deficits of the tongue, narrowing of the nostrils, loss of anatomical structures such as eyes, ears, or nose. To correct these defects and improve function and esthetics, a variety of prostheses and appliances are available, including complete and partial dentures, obturators, palatal lift appliances, pharyngeal obturators, palatal augmentation prostheses, nasal conformers, and facial and ocular prostheses.

15.3.1 Removable and Fixed Prostheses

Missing teeth may be restored with conventional dental prostheses such as fixed bridges, removable partial dentures, complete dentures, and dental implants. Often these types of dental prostheses are combined with other prostheses including obturators and palatal augmentation prostheses. In doing so, combination-prostheses replace missing teeth while enhancing other functions such as speech and swallowing.

15.3.2 Occlusion and Arch Size Discrepancies

When the maxilla and/or mandible are part of the transplanted tissue, there is a chance that the donor bone and teeth will not align or occlude with the native teeth and bony structures (Fig. 15.1). The resulting malocclusion and arch size discrepancies may be treated with occlusal guards or through the modification of other types of prostheses. A common practice in this situation is to place prosthetic teeth in the proper position for esthetics and add additional material or a second row of teeth to provide for a stable occlusion and function (Fig. 15.2).

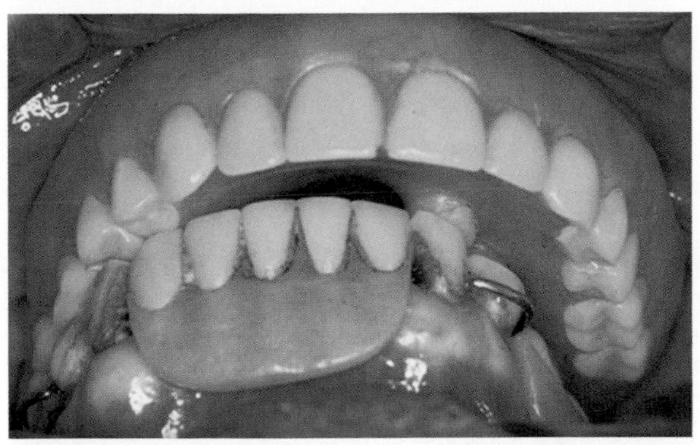

Fig. 15.1 Arch discrepancy. Teeth not touching on right side

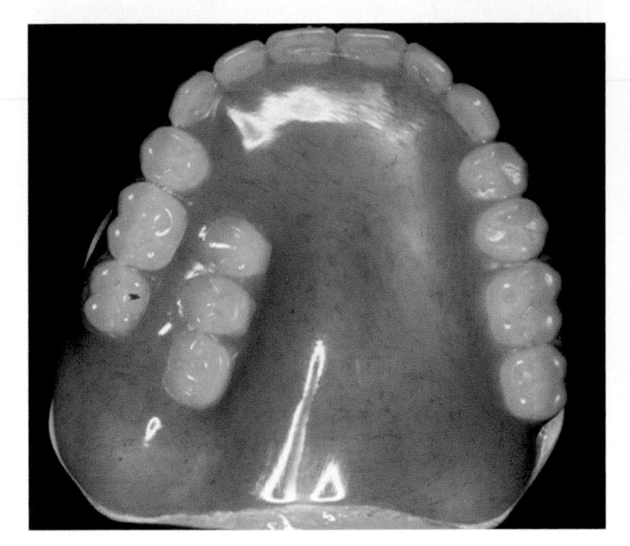

Fig. 15.2 Second row of teeth added to provide for occlusion

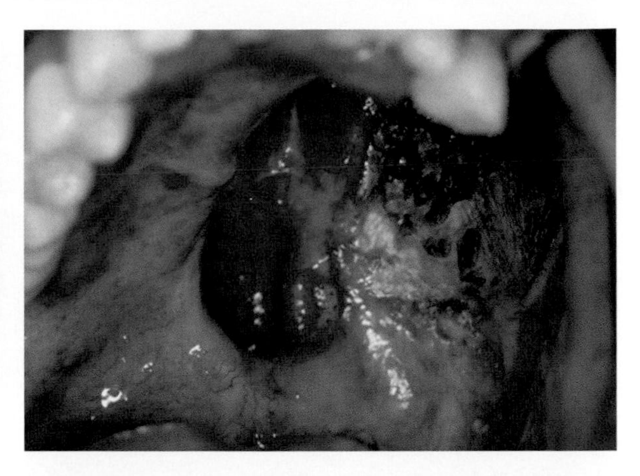

Fig. 15.3 Palatal defect with communication

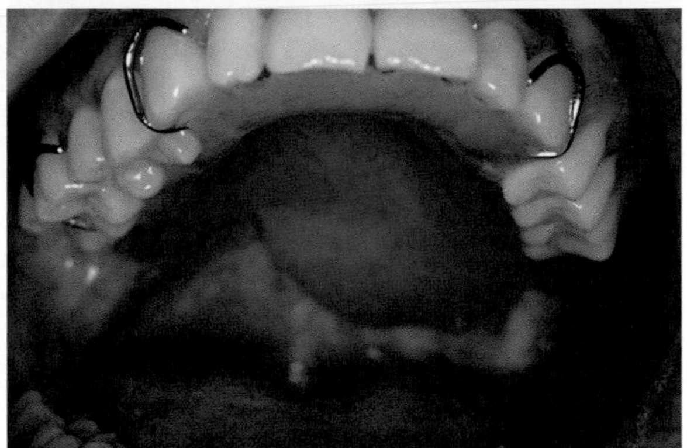

Fig. 15.4 Obturator prosthesis in place

15.3.3 Obturators

Perforations of the hard palate create a communication between anatomical compartments which allows for the oral cavity, maxillary sinus, nasal cavity, and nasopharynx to become a confluent space (Fig. 15.3). This creates disabilities in speech and swallowing. Air, liquids, and food may escape from the oral cavity to exit from the nose. This makes taking in adequate nutrition through the oral cavity difficult if not impossible. The patient's speech becomes hypernasal and unintelligible due to the inability to impound air in the oral cavity. An obturator prosthesis helps to restore the integrity between the oral and nasal compartments by providing a seal to facilitate normal swallowing and speech (Figs. 15.4 and 15.5).[1] A common treatment sequence is a surgical obturator during the initial phase of healing, followed by an interim obturator, then placement of a definitive obturator.[2]

A surgical obturator is placed at the time of surgery or within few days after surgery, and it is used to replace the missing components of the maxilla. It is left in place for several days and then replaced by an interim obturator, which is used during the intermediate phase of healing. After healing is complete, the patient is then fitted with a definitive obturator.

Surgical obturator prostheses are often secured in place by sutures, wires, or screws. Interim obturators are usually held in place by clasping the teeth, much like an orthodontic retainer, or added to a new or existing complete denture. Definitive obturators may be retained by a complete denture, clasping the remaining

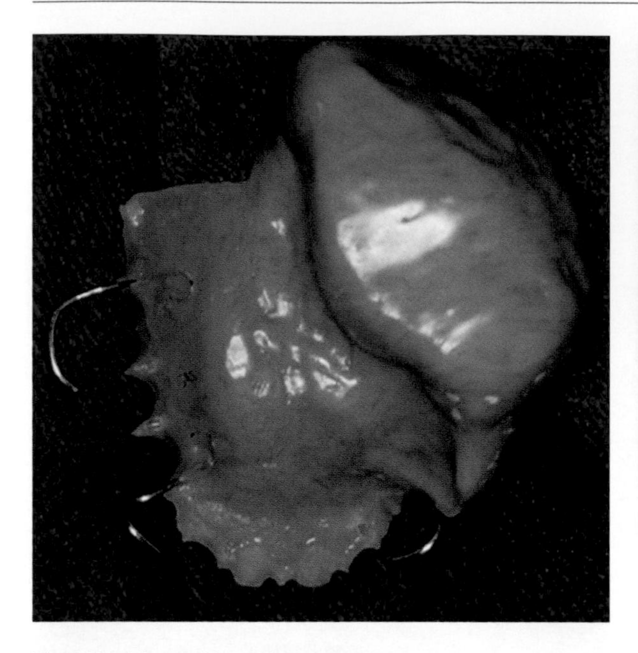

Fig. 15.5 Tissue side of obturator

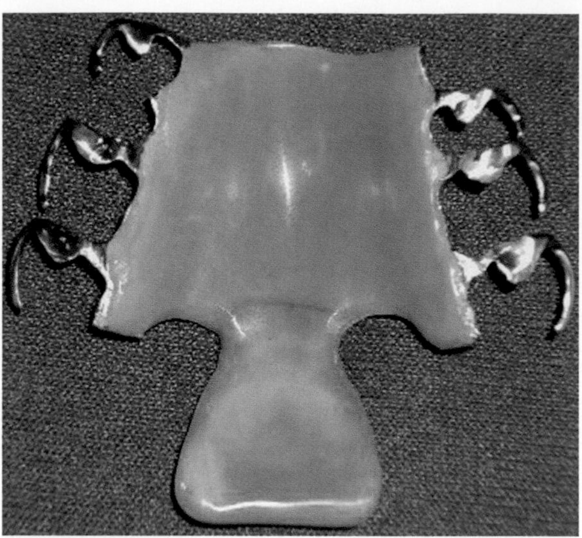

Fig. 15.6 Palatal lift appliance. Note extension to reposition the soft palate

dentition, or dental implants. When an obturator prosthesis is part of a complete denture, the use of a denture adhesive or dental implants is required in most situations to retain the prosthesis.

15.3.4 Palatal Lift Appliances

A palatal lift appliance is used to mechanically reposition and elevate the soft palate when it is functionally impaired to achieve the closed position during speech and swallowing.[2] Commonly, teeth or dental implants are required for anchorage, as the force to close the soft plate must be less than the retentive properties of the prosthesis. When teeth are present, they are clasped similar to an orthodontic appliance (Fig. 15.6). Dental implants may provide anchorage using retentive attachments. For this type of prosthesis to function, the soft palate must be mobile and have adequate extension and bulk for closure during speech and swallowing.

15.3.5 Pharyngeal Obturators

When normal velopharyngeal function is interrupted, from a surgical or congenital defect, a communication is created between the oral and nasal cavities resulting in hypernasal and unintelligible speech and regurgitation of food and liquids into the nasal cavity.[2] Since there is a tissue deficit, repositioning the soft palate will not close the communication. Therefore, a pharyngeal obturator is used to fill the communication, thus producing a seal between the oral and nasal cavities, and sinuses during speech and swallowing. The older terminology for this type of prosthesis is a speech bulb.

15.3.6 Palatal Augmentation Prostheses

During the oral phase of swallowing, the tongue comes into contact with the hard palate and is braced against the roof of the mouth. This allows the tongue to act as plunger forcing the bolus of food posterior. When the tongue can no longer adequately move to contact the hard palate, a palatal augmentation prosthesis can be used to create an artificial palate which is positioned to bring the level of the palate down to contact the tongue during function.

Similarly, during speech, the tongue contacts the anterior portion of the palate or the teeth during "T" and "TH" sounds, and contacts the posterior area of the palate when forming "K" and "G" sounds. To construct a palatal augmentation prosthesis, a functional impression or palatogram is generated while the patient repeats a series of words (Fig. 15.7). From the

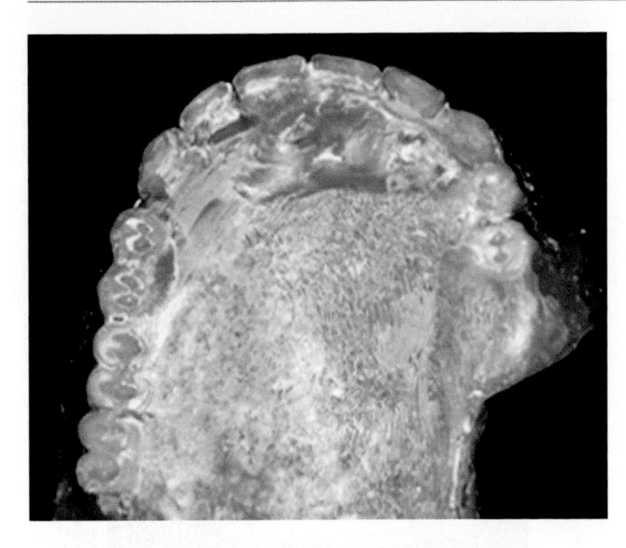

Fig. 15.7 Palatogram recorded on denture base

impression or palatogram, the desired contours of the prosthesis are generated and reproduced in the prosthesis.[1] For dentate patients, the prosthesis is retained by placing clasps on the teeth in a manner similar to an orthodontic retainer. If the patient is edentulous, it may be added to a new or existing denture (Fig. 15.8).

A quick screening to determine if your patient would benefit from this type of prosthesis includes:

- Is the tongue tethered or does it deviate during protrusion?
- Is the problem with moving a bolus during the oral phase of swallowing?
- Is the patient unable to produce a crisp and clear sound during the production of "T," "TH," "K," and "G" sounds? An easy test is to have them repeat aloud "Go get Gary because kit kat is tip top"

If the answer to any of these questions is "Yes," then this type of prosthesis may be appropriate.

15.3.7 Nasal Conformers

A conformer is a device used to maintain a passageway, opening, or body cavity. Nasal conformers are employed to preserve or increase the opening diameter of the nostrils and may be made from stock medical supplies, such as a piece of tubing, or custom fabricated in acrylic resin from an impression of the patient's nares. Nasal conformers are hollow to allow for nasal breathing. These devices support soft-tissues following reconstructive surgery and help to reduce narrowing from scar tissue formation during healing. They may be secured with sutures or removable for daily cleaning and adjustments. It is common practice to periodically add to the outer diameter of the nasal conformer to allow for it to function as a tissue expander (Figs. 15.9 and 15.10).

15.3.8 Facial and Ocular Prostheses

Facial and ocular prostheses are individually and custom fabricated for the patient. This service has an overlap with

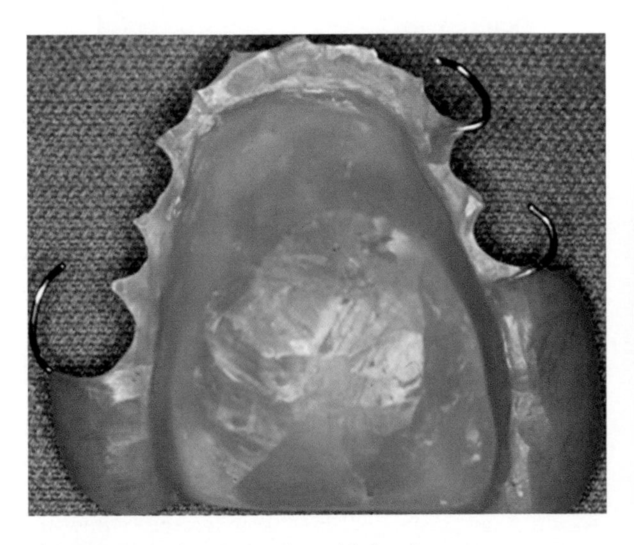

Fig. 15.8 Palatal augmentation added to denture

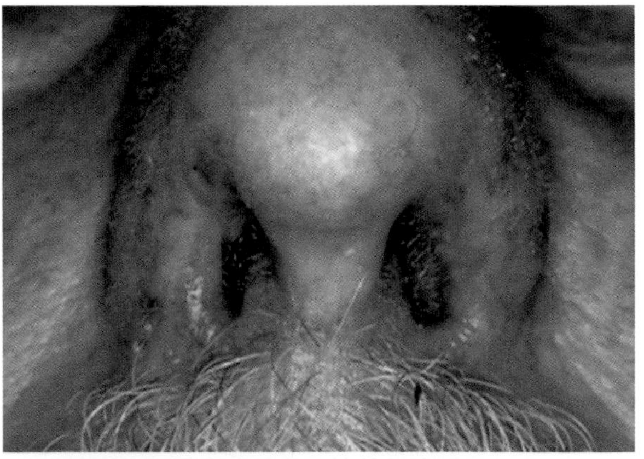

Fig. 15.9 Constricted nares

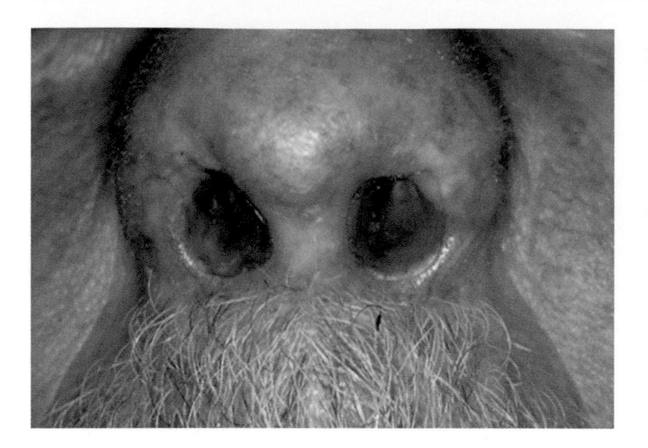

Fig. 15.10 Conformers in place. Note increase opening

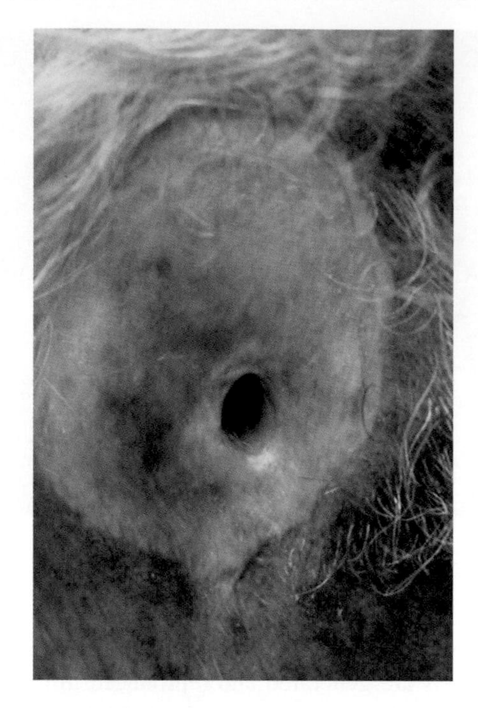

Fig. 15.11 Partial loss of the ear

other providers as facial and ocular prostheses may be created by a Maxillofacial Prosthodontist, Prosthodontist, Dental Technician, Anaplastologist, or Ocularist. These prostheses are retained with adhesive or craniofacial implants. The advantages of prosthetic replacement include: It is a reversible procedure with reduced surgery and associated surgical risks, and may offer a more predictable esthetic result. Disadvantages include: The prosthesis may become detached from the patient at an inopportune time, concerns by the patient that the prosthesis will fall off, difficulty in applying adhesive and proper placement of the prosthesis, skin irritation and adhesive allergies, and the need for remaking the prosthesis every 3 months to 2 years.

Fabrication of the prosthesis begins by creating a wax or clay sculpting on a cast of the patient. The sculpting is then viewed on the patient to determine what modifications need to be performed in order to achieve the correct anatomical contours and optimal esthetics. After refining the sculpting, a mold is made to process the prosthesis. Most facial prostheses are processed in a silicon material. Pigments are used for intrinsic and extrinsic coloration to match the prosthesis with the skin tones of the patient (Figs. 15.11 and 15.12). Ocular prostheses, which replace only the eye, are processed in acrylic resin and retained by the patient's eyelids. Orbital prostheses contain an ocular prosthesis and replace the eyelids and structures surrounding the eye.

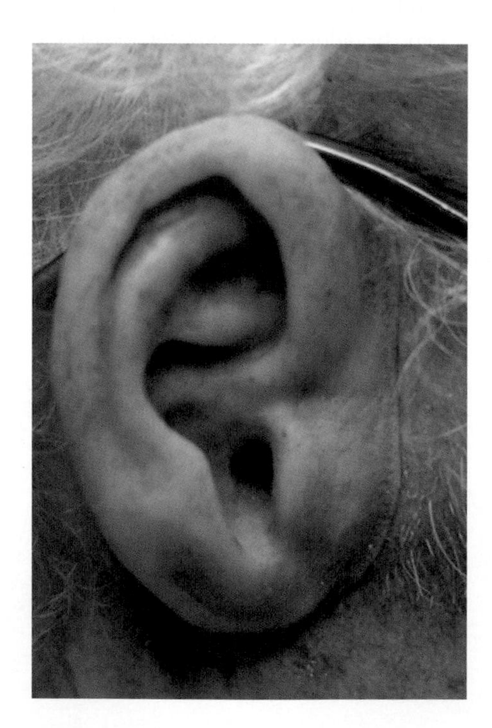

Fig. 15.12 Ear prosthesis in place

15.4 Conclusion and Recommendations

Although the face transplant patient will present clinicians with new challenges, established techniques and sound prosthodontic principles will provide the foundation for the delivery of prosthetic support and treatment of this patient population. Many problems which are not correctable with surgery or require correction while awaiting further surgery may be treated with various appliances and prostheses provided under the care and direction of a Maxillofacial Prosthodontist. For optimal results, the Maxillofacial Prosthodontist should be a member of the transplant team and consulted during the pre-surgical, surgical, and post-surgical phases of face transplantation.

References

1. Taylor TD. *Clinical Maxillofacial Prosthetics*. Chicago: Quintessence; 2000.
2. American Academy of Maxillofacial Prosthetics. *Coding and Reimbursement Manual*. New Orleans: AAMP; 1997.

Part III

Monitoring Aspects of Face Transplantation

Immunological Monitoring

16

Medhat Askar

Contents

Abstract Facial transplants involve vascularized allografts which are prone to various types of antibody- and cell-mediated rejection. The requirement for multiple non-immunological layers of donor suitability in this context dictates careful assessment of which immunological criteria to consider. Although the clinical practice of facial transplants is fairly recent, experience from other organ and tissue transplants has shown that pre-transplant immunological risk assessment and post-transplant monitoring are critical to maximize graft survival. The real challenge is to correlate short- and long-term outcomes of this procedure with individual tests to determine their clinical relevance in this unique setting. An overview of the landscape of modern immunological testing methods will be presented with emphasis on the clinical applicability of these methods. Both routine and novel testing methods will be reviewed. Relevant studies to assess the clinical significance of different methods in other solid organ and hematopoietic stem cell transplants in humans and from animal models of composite tissue allografts (CTA) will be highlighted.

Abbreviations

AMR	antibody-mediated rejection
ATP	adenosine tri-phosphate
CDC XM	cytotoxic crossmatch
CTA	composite tissue allotransplantation
DSA	donor-specific HLA antibodies
ELISA	enzyme-linked Immunosorbent Assay
ELISPOT	enzyme-linked immunospot
FCXM	flow cytometric crossmatch
GvHD	graft-versus-host disease
HLA	human leukocyte antigen
HPC	human progenitor cells

M. Askar
Allogen Laboratories, Transplant Center,
Department of Surgery, Cleveland Clinic, Cleveland Clinic Lerner College of Medicine, Case Western Reserve University, Cleveland, OH, USA
e-mail: askarm@ccf.org

M.Z. Siemionow (ed.), *The Know-How of Face Transplantation*,
DOI: 10.1007/978-0-85729-253-7_16, © Springer-Verlag London Limited 2011

MICA	major histocompatibility complex class I related chain A
NP	not published
PBL	peripheral blood lymphocytes
PCR	polymerase chain reactions
PHA	phytohemagglutinin
PICC	peripherally inserted central catheter
PRA	panel reactive antibodies
SSOP	sequence-specific oligonucleotide probes
SSP	sequence-specific primer
TNF	tumor necrosis factor

16.1 Introduction

Face transplant is the most recent frontier of transplantation with a unique set of immunological challenges. While face transplants have proven successful, the long-term immunological consequences remain unknown. A specific histocompatibility testing protocol to assess the immunological risk in a prospective face transplant candidate and her/his compatibility with a given donor is still to be established. An additional layer of complexity in face transplantation is imposed by the heterogeneity of the tissues transplanted which raises the concern of how a specimen from one tissue such as a mucous membrane is representative of a pathological process taking place in other tissues such as the skin. Therefore, histologic surveillance may be less informative since immunological responses within the various facial allograft components may be different at the same time and simply may overestimate or underestimate the overall severity of a rejection episode or the response to treatment. For these reasons, post-transplant noninvasive immunological monitoring using a panel of serologic and cellular assays and potentially other novel testing methods such as gene and protein arrays is critical to optimize long-term graft survival. This chapter is not intended to be a comprehensive technical review of histocompatibility testing methods but rather to focus on the clinical relevance of these methods in other transplant contexts. An overview of the landscape of modern testing methods will be presented and for further detailed discussion of such methods, suggested references will be provided for each method. The conceptual framework of designing appropriate protocols for post-transplant immune monitoring of face transplant recipients is predicated by lessons from other solid organs, vascularized composite tissue and hematopoietic stem cell transplants in humans and from animal models of composite tissue allografts (CTA).[1-3] In developing such a protocol, it is critical to maintain the delicate balance between maximizing access to donors, minimizing the potential for serious immunological consequences, and avoiding excessive stringency that may preclude all candidates.

16.2 Pre-transplant Assessment

Strategies for post-transplant monitoring depend at least in-part on pre-transplant risk factors such as preformed donor-specific HLA antibodies (DSA). Pre-transplant assessment includes HLA typing, HLA antibody screening, and crossmatch testing. Experience with all other allografts (with the possible exception of liver transplants) indicates that DSA, at least at high levels, should be avoided to reduce the risk of hyperacute rejection. In the nine reported face transplants performed to date, no hyperacute rejections occurred; however, there are no grounds to suggest that the potential does not exist. Thus, it is prudent to avoid face transplants across positive crossmatch or high levels of donor HLA-specific antibodies.

16.2.1 HLA Typing

Historically, HLA typing was performed using serological methods, but these methods are less likely to play a significant role in the assessment of face transplant candidates. In recent years, most laboratories adopted molecular methods for HLA typing. Molecular methods include sequence-specific primer PCR (SSP), sequence-specific oligonucleotide probes (SSOP), and direct DNA sequencing, and are described elsewhere.[4] In all these methods, DNA is isolated from the subject to be typed and amplified using standard polymerase chain reactions (PCR). The sequence of the amplification products is determined, and the HLA type is assigned by comparison to published HLA allele sequences.

HLA typing of recipient/donor pairs allows the degree of HLA matching to be assessed and to exclude donors with HLA antigens to which the recipient has corresponding preformed DSA. The degree of HLA matching was published in four out of the nine reported face transplants and is shown in Table 16.1.[5] The best match reported was the first French case with only one DR mismatch between the donor and the recipient. Since multiple layers of non-HLA matching requirements such as gender and skin color already restrict which donors will be considered for any given recipient, the addition of a stringent match for HLA antigens may be too restrictive. Although the overall experience with face allografts is quite limited, the data suggest that HLA mismatching does not affect short-term outcome. Further support for this notion comes from reports of successful hand transplants across multiple HLA mismatches between the donor and recipient.[6,7] The impact of HLA mismatching on long-term outcomes in this context remains to be determined. If matching is to be pursued, the renal transplant literature indicates that HLA-DRB1 matching be given priority. In kidney transplantation, zero HLA-DRB1-mismatched grafts confer a statistically significant survival advantage compared to one and two HLA-DRB1-mismatched grafts.[8] In hematopoietic progenitor cell (HPC) transplants, HLA-DRB1 matching is critical even when HPC source is umbilical cord blood where HLA matching criteria are less restrictive than marrow- or peripheral blood-derived HPC.[9,10]

Table 16.1 Donor-recipient HLA mismatch and recipient sensitization

	First French case	Chinese case	Second French case	Cleveland case
HLA-A, -B, DRB1 mismatches	1/6	3/6	3/6	4/6
PRA %	NP	99	NP	Class I=67[a] Class II=29

The number of donor-recipient HLA-mismatched antigens and the degree of recipient sensitization in four face transplant recipients (no further published information is available)

[a]PRA in the Cleveland case was calculated PRA (cPRA) based on antibodies identified using single antigen Luminex beads. No donor-specific HLA antibodies were detected at the time of transplantation

NP not published

16.2.2 HLA Antibody Screening and Characterization

Allosensitization to HLA and non-HLA antigens results from exposure to these antigens during pregnancy, blood transfusion, and/or a prior transplant. The deleterious effect of DSA in solid organ transplants has been appreciated for over four decades.[11] More recently, it has been suggested that allosensitization may lead to failure of engraftment in HPC transplants.[12,13] The allosensitization status has been reported in two of the nine reported cases of face transplant. The Chinese case was reported as having panel reactive antibodies (PRA) of 99% (Table 16.1). The Cleveland case had a calculated panel reactive antibodies (cPRA) of 67% class I and 29% class II based on antibodies identified using single antigen Luminex beads. PRA calculation was based on antigen frequency in a historical cohort of close to 6,000 deceased donors tested in Allogen Laboratories at the Cleveland Clinic. In this case, the patient had no DSA against any of the mismatched donor HLA antigens. In seven of the nine reported cases, at least one episode of acute graft rejection was encountered.[5] Of those, five patients have been reversed successfully, with two mortalities that may have been rejection-related including the Chinese case. However, in the latter case, the patient was reported noncompliant with his immunosuppression. Methods of HLA antibody detection and characterization are described elsewhere.[14] Categorically, these methods are either cell-based or solid phase–based assays. In cell-based assays, recipient sera are tested against a panel of donor lymphocytes with different HLA types, ideally, representing common HLA types. In solid phase–based assays, sera are tested against affinity-purified or recombinant HLA antigens attached to plastic plates in ELISA format or to plastic beads in flow (Fig. 16.1a) or Luminex format.[15] In general, solid phase assays have superior sensitivity and specificity compared to cell-based assays. Importantly, cell-based assays do not distinguish HLA from non-HLA antibodies; whereas, solid phase–based assays detect only HLA antibodies.

The application of single antigen solid phase assays allows a sensitive and specific identification of unacceptable antigens to compute cPRA, a more predictive estimate of the probability of finding a donor for which the patient has no DSA based on the antigen frequencies in the donor population.[16]

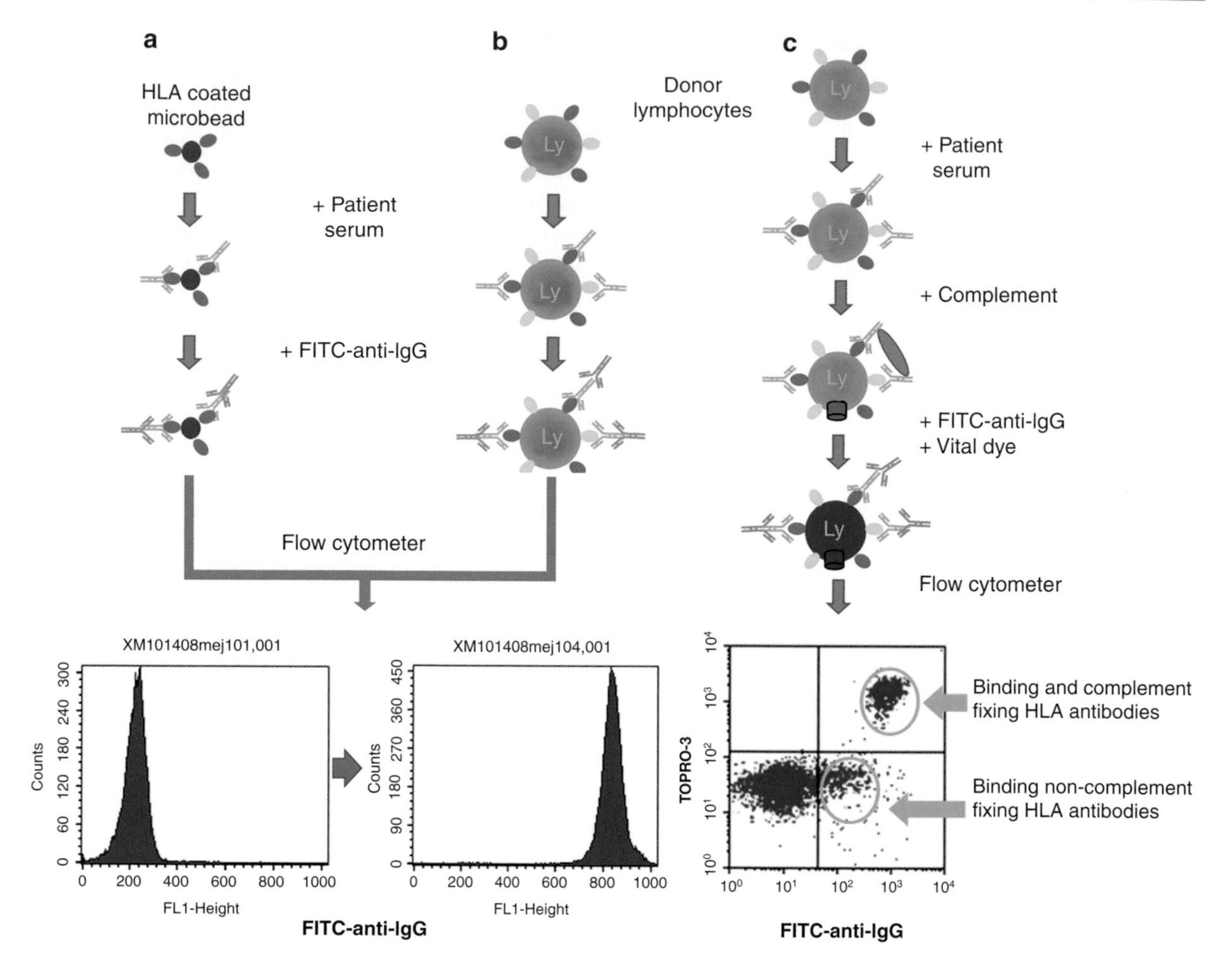

Fig. 16.1 Schematic representation of the resemblance and differences between examples of flow cytometry–based HLA antibody detection and crossmatch assays. (**a**) In the flow bead assay, HLA-coated microbeads are incubated with the patient serum followed by addition of FITC-labeled anti-human IgG. (**b**) Flow cytometry crossmatch (*FCXM*) is similar to A except that instead of beads, donor lymphocytes (Ly) are incubated with patient serum followed by addition of FITC-labeled anti-human IgG. (**c**) Cytotoxic FCXM is similar to B except that prior to the addition of anti-human IgG, complement is added, followed by addition of FITC-labeled anti-human IgG and a vital stain (e.g., TOPRO-3). Complement-fixing antibodies will lead to cell lysis and the vital dye uptake by complement-lysed cells (upper right quadrant on the scatter plot). Non-complement-fixing antibodies will be identified by FITC-labeled anti-human IgG (lower right quadrant). Both methods A and B do not distinguish complement-fixing versus non-complement-fixing antibodies. Both B and C (cell-based assays) do not distinguish HLA from Non-HLA antibodies versus A (solid phase–based assays) detects only HLA antibodies

16.2.3 Complement-Fixing Antibodies

Recent reports have suggested that complement-fixing DSA may pose a particular immunologic risk of graft loss in kidney and heart transplant recipients.[17,18] Antibody-mediated complement activation and deposition of various split components including C4d contribute to allograft damage even in the absence of assembly of the terminal lytic complex (C5–C9) and cell lysis.[19] Therefore, deposition of C4d along the peritubular capillaries of renal biopsies has been incorporated in Banff criteria for diagnosis of antibody-mediated rejection (AMR).[20] Detection of these antibodies relies on determining their ability to fix C4d onto Luminex beads and has been described elsewhere.[21] At this point, there is not enough information regarding the role of complement-fixing versus non-complement-fixing antibodies in AMR in face transplants. Nevertheless, C4d has been proposed as among other pieces of information that should be gathered in order to define AMR in CTA in the Banff 2007 Working Classification of Skin-Containing CTA Pathology.[22]

16.2.4 *Crossmatch, the Real and the Virtual*

16.2.4.1 Clinical Relevance and Classical Methods

The clinical relevance of crossmatching in kidney transplantation has been recognized over four decades.[23] A positive cytotoxic crossmatch (CDC XM) at the time of transplant is a major risk factor for development of hyperacute rejection and primary nonfunction. The more sensitive flow cytometric crossmatch (FCXM) identifies patients at risk for antibody-mediated rejection and graft loss. A positive FCXM in prospective kidney transplant recipients with a negative CDC XM was associated with a higher incidence of rejection, a higher risk of vascular rejection, and a worse graft survival.[24,25] None of the published face transplant experiences were performed across a positive crossmatch. Similarly, in hand transplantation, having a donor with negative crossmatch was always pursued.[26,27]

Crossmatch methods and our understanding of the strength and limitations of each method have steadily evolved. Most importantly, we have learned that a positive crossmatch result must be interpreted in the context of mismatched donor HLA antigens and recipient HLA antibodies. Various crossmatch methods are described elsewhere.[28,29] These methods could be complement dependent, flow cytometry based, or solid phase based. In complement-dependent crossmatch, recipient sera are incubated with donor cells and cell death indicates a positive crossmatch result. Flow cytometry–based assays (Fig. 16.1b) rely on detection of cell-bound anti-donor antibodies in the recipient serum using fluorescent-labeled anti-human antibody and a flow cytometer. It is noteworthy that standard crossmatch methods (cell-based assays) do not distinguish HLA from Non-HLA antibodies; whereas, solid phase–based assays detect only HLA antibodies.

16.2.4.2 Novel Crossmatching Methods

Flow cytometric crossmatching detects low level and non-complement-fixing antibodies but does not typically distinguish complement-fixing from non-complement-fixing antibodies. Recently, a modified version of the flow cytometric crossmatch (Fig. 16.1c) has been reported. The modified assay combines the ability to detect low level antibodies with the ability to distinguish between complement-fixing and non-complement-fixing antibodies.[30] The clinical relevance of this assay in terms of correlation with graft outcomes has not been established yet.

Solid phase–based crossmatch assays are currently available in ELISA and Luminex format and rely on solubilization of donor cell membranes to release donor HLA molecules. In the Luminex version, the lysate prepared with donor lymphocytes is incubated with capture beads to enable binding of the solubilized donor HLA molecules onto the beads. The mixture is then washed and incubated with recipient serum. Finally, fluorescent-labeled anti-human immunoglobulin is added and fluorescence is detected by Luminex.[31]

16.2.4.3 Virtual Crossmatching

Virtual crossmatching has been long contemplated, but it was only during the latter half of the last decade, that it was implemented in thoracic transplants in an effort to circumvent the limitation of short ischemia time that precludes performing a prospective crossmatch on donors from geographically distant sites.[32,33] In this algorithm, crossmatch results can be predicted based on the donor-mismatched antigens and recipient DSA. A virtual crossmatch involving a donor whose mismatched antigens do not correspond to HLA antibodies in a prospective recipient would predict a negative result. Currently, this concept has become practiced in all solid organs, particularly in highly sensitized patients, and has been proposed for HPC transplants.[34,35] Virtual crossmatching could expedite informed decision making by the face transplant team regarding accepting a given donor particularly in a highly sensitized patient by eliminating the wait time for a prospective crossmatch to be performed. It can also potentially increase facial allograft donor pool by sharing facial allografts across large geographic areas since no pre-procurement crossmatch testing is necessary provided that the logistics and regulations are conducive.

16.3 Post-transplant Monitoring

The objectives of post-transplant monitoring are to detect various types of rejection early enough before they become irreversible, to adjust the level of immunosuppression, and potentially to identify a subset of recipients who might be appropriate candidates for immunosuppression minimization or even complete withdrawal if they demonstrate features of allograft tolerance.

16.3.1 Classical Immune Monitoring Assays

16.3.1.1 HLA Antibody Monitoring

The interest in post-transplant monitoring of solid organ transplant recipients to detect DSA has been steadily growing. Several reports have shown a strong association between the presence of DSA post-transplant and acute and chronic allograft injury and graft loss in kidney, heart, and lung transplants [36-38] Interestingly, the appearance of posttransplantation antibodies directed against donor HLA-A, -B, -Cw, -DR, and -DQ mismatches was reported to precede kidney allograft loss and to be strongly predictive of transplant failure.[39] Indeed, as shown in Fig. 16.2, in the Cleveland case, an increase in both class I (A23) and class II (DQ9) DSA was detected on post-transplant day 41 and preceded a biopsy diagnosis of subclinical rejection of the graft mucosa (Banff III/IV)

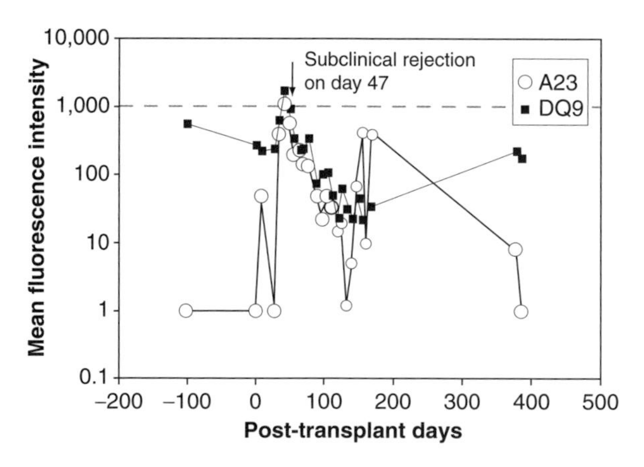

Fig. 16.2 Post-transplant temporal kinetics of DSA levels against mismatched class I (HLA-A23) and class II (DQ9) donor HLA antigens using single antigen Luminex beads (positive cutoff = 1,000). The DSA peak occurred on day 41 and preceded an episode of subclinical rejection on day 47 (indicated by the *arrow*)

without any clinical evidence of skin rejection (Banff 0/IV) on day 47.[40] Antirejection therapy was initiated and rejection was reversed accompanied by DSA reduction as of day 48 and confirmed by normal biopsy on day 50. Granted, this is just a single observation; but nevertheless, it underscores the potential relevance of DSA monitoring in guiding appropriate therapeutic interventions.

16.3.1.2 Non-HLA Antibody Monitoring

There is a long history of non-HLA antibodies being produced after renal and cardiac transplantation. Non-HLA antibodies may occur as alloantibodies or autoantibodies. The described antigenic targets for non-HLA antibodies include Major Histocompatibility Complex Class I–related Chain A (MICA), other endothelial antigens, various minor histocompatibility antigens, vascular receptors, adhesion molecules, and intermediate filaments. Acute and chronic kidney allograft rejections have been reported in HLA-identical sibling transplants underscoring the relevance of an immune response against non-HLA targets.[41,42] A synergism between HLA and non-HLA antibodies including autoantibodies has been suggested to be significantly associated with development of chronic rejection after lung transplantation.[43,44] Currently, screening for non-HLA antibodies is not routinely performed pre-transplant or monitored post-transplant in solid organ or HPC transplants. It has been reported that non-HLA antibodies such as MICA antibodies may be present at the time of rejection in the absence of HLA antibodies.[45,46] Post-transplant testing for non-HLA antibodies may be helpful in instances where no HLA antibodies are detected in the context of biopsy diagnosed rejection, particularly since pre-transplant screening for such antibodies is not routinely performed. Until more data become available regarding the presence of HLA antibodies at the time of rejection episodes, our ability to discern a potential role for non-HLA antibodies in facial allograft rejection remains limited.

16.3.1.3 Monitoring T-Cell Functions

Monitoring of T cell functions may prove critical in guiding individualized immunosuppression based on alloreactive and anti-infectious responses. Assays that monitor T cells can be antigen-specific or nonspecific.

Antigen-Nonspecific Assays

The ImmuKnow assay (Cylex Inc., Columbia, MD) applies a nonspecific approach by measuring intracellular ATP production of CD4+ T helper (Th) cells by luminescence after overnight phytohemagglutinin (PHA) stimulation of whole blood. A multicenter, cross-sectional, cohort analysis, including 127 kidney, liver, pancreas, and simultaneous kidney/pancreas transplant recipients, has shown that responses fell categorically into strong (ATP≥525 ng/mL), moderate (226–524), and low (≤225) categories. These zones broadly were associated with risk of rejection (high values) or infections due to over immunosuppression (low values).[47] In the Cleveland case, ImmuKnow assay was monitored every 1–2 weeks. The highest ATP value of 455 ng/mL (>60% higher than the previous reading) was observed on post-transplant day 48 and coincided with the episode of subclinical rejection (Fig. 16.3). The nadir ATP values were observed between days 125 and 132 (72 ng/mL) and coincided with an episode of pseudomonas bronchitis. The condition was resolved by discontinuation of MMF and valganciclovir and administration of piperacillin-tazobactam. However, during that time, the patient had 1.82 K/μL WBC, 77% neutrophils, and CD4+ cell count of 26/mm^3 which arguably would have been sufficient to support the diagnosis of infection and may have been responsible at least in-part for such low ATP values. Further, the patient had more serious infection episodes such as an episode of *Pseudomonas* bloodstream infection on day 62 related to peripherally inserted central catheter (PICC)-line infection when ImmunoKnow value was not very low (255 ng/mL). In our experience as well as others, a single ATP level has a limited informative value compared to trends observed with serial longitudinal monitoring that are more meaningful. In addition, ATP values should be interpreted in the context of other clinical, laboratory, and pathology findings; particularly, values within the first 1 or 2 months post-transplant are unstable and must be interpreted with caution. It has also been suggested that ImmuKnow results need to be interpreted with caution in patients receiving Thymoglobulin induction therapy prior to kidney transplantation. In this group, low ATP levels identified patients at increased risk for infection; however, high ATP values failed to correlate with rejection.[48]

Other antigen nonspecific assays assessing general T-cell activation markers such as serum soluble CD30 (sCD30) have been developed. Soluble CD30 is a transmembrane glycoprotein member of the tumor necrosis factor (TNF) superfamily expressed on T cells (including alloantigen-activated CD4 and CD8 cells), among other populations.[49] After activation, sCD30 is cleaved from the surface of activated CD30+ cells and can be detected in the serum of most normal individuals. However, elevated post-transplant serum sCD30 levels have been strongly correlated with acute kidney allograft rejection.[50,51]

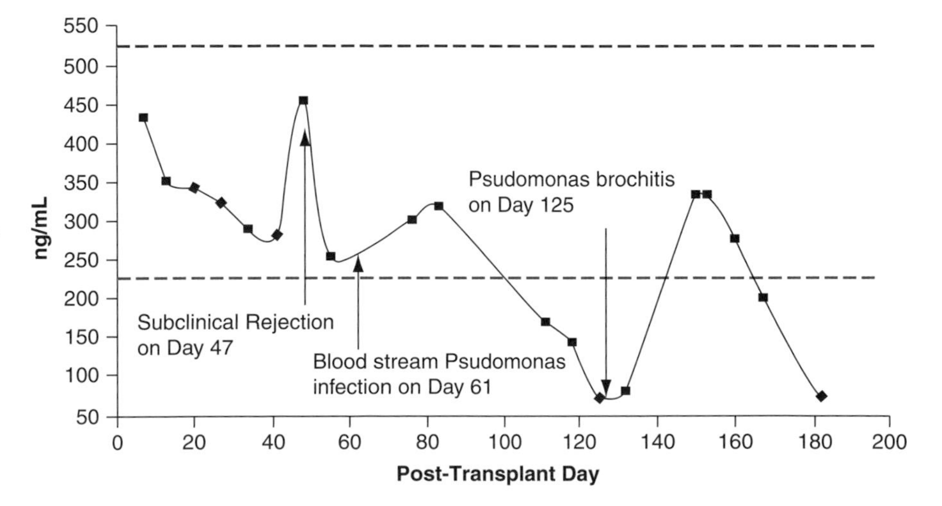

Fig. 16.3 Monitoring of ATP levels (ng/mL) measured by ImmuKnow assay in the first 180 days post-transplant. ATP values coinciding with rejection (day 47) and infection episodes (days 61 and 125) are indicated by the *arrows*. The *blue* (225) and *red* (525) *dashed lines* indicate the lower and upper limits of moderate response, respectively

Antigen Specific Assays

A more methodologically involved antigen-specific assay is based on cytokine enzyme-linked immunospot (ELISPOT) assay. In principle, this assay is capable of quantifying cytokine secretion by individual, antigen-reactive T cells within a population of peripheral blood lymphocytes (PBLs). A significant expansion of IFN-g-producing donor-reactive memory PBLs was reported to be detectable at 4–6 months post-transplant in kidney transplant recipients who had experienced an acute rejection episode compared with those with a stable post-transplant course.[52]

16.3.1.4 Donor Lymphoid Chimerism

In the first face transplant, donor hematopoietic progenitor cells were infused on postoperative days 4 and 11 as a tolerogenic measure.[53] In at least four other facial transplants, a vascularized bone marrow component was included.[5] Animal models suggest that vascularized bone marrow transplantation has unique characteristics. It potentially reduces the risk of delayed engraftment, engraftment failure, graft-versus-host disease, and aids in inducing immunologic tolerance.[54]

In the first French case, post-transplant chimerism was monitored frequently on whole blood, CD3+ cells, CD56+ cells, total and purified CD34+ bone marrow cells by real-time quantitative polymerase chain reaction (RQ-PCR) with the use of TaqMan technology as described elsewhere.[55] Of the many assessments for microchimerism, only one suggested that microchimerism was present 2 months after transplantation (0.1%) among the CD34+-enriched population of bone marrow cells.[53] Interestingly, in five patients who received combined bone marrow and kidney transplants from HLA single-haplotype-mismatched living related donors as a tolerogenic measure, transient chimerism developed in all recipients.[56] In four out of five recipients, it was possible to discontinue all immunosuppressive therapy 9–14 months after the transplantation, and renal function remained stable for 2.0–5.3 years since transplantation suggesting that initial post-transplant transient chimerism may be associated with development of allograft tolerance. However, post-transplant mixed donor chimerism could be a double-edged sword. Acute graft-versus-host disease (GvHD), a common complication of allogeneic hematopoietic progenitor cell transplantation, has been recognized as a rare but potentially fatal complication of solid organ transplants particularly of those organs with a high content of passenger leukocytes such as the liver.[57] GvHD occurs when immunocompetent donor T cells from the graft recognize disparate alloantigens of host cells. Detection of a high percentage of donor lymphoid chimerism has been reported to be a useful adjunct test to support the diagnosis of GvHD.[58] To date, no cases of GvHD were reported in the limited number of face transplants; however, the risk potentially exists and testing for donor lymphoid chimerism on the earliest appearance of clinical manifestations suggestive of GvHD could be life saving.[59]

16.3.2 Novel Approaches for Immune Monitoring

Recent advances in many high-throughput 'omic techniques, such as genomics, metabolomics, antibiomics, peptidomics, and proteomics, have been adapted to develop novel biomarkers for acute rejection, chronic rejection, and operational tolerance. A comparison between MicroRNA (miRNA) expression profile of acute rejection and the controls identified 20 miRNA differently expressed in acute rejection after renal transplantation.[60] Recently, Proteomic and metabolomic strategies have been proposed for frequent noninvasive measurements in tissue fluids, allowing for serial post-transplant monitoring of allografts. A common limitation of these approaches thus far has been the lack of reproducibility of identified signature profiles. Due to the high cost of these technology platforms and the complexity of data analysis, they are unlikely to be used for routine monitoring in the clinic anytime soon. Nevertheless, the use of these approaches holds the promise of providing rapid and global views regarding the profiles of different disease states and potentially identifying important diagnostic markers.

16.4 Future Directions

Until histocompatibility practices in transplanted facial allografts are systematically documented and long-term outcomes are correlated to specific tests, the clinical relevance of any given test and the impact on short- and long-term outcomes remain unclear. Moving forward, it will be critical that teams document their immunological testing protocols, correlate clinical outcomes to specific tests performed, and share these experiences through peer-reviewed publications.

It is certainly reasonable that many of the predictors of facial allograft dysfunction and loss will be shared with other organ or tissue transplants. Taken together, post-face transplant immune monitoring will most likely rely on a panel of tests performed on multiple sample sources and collected serially. As the number of face transplants grows and immunological data are collected systematically and correlated with post-transplant outcome, experience will determine the constellation of assays that will be reliable predictors of short- and long-term graft function and survival.

Acknowledgments I would like to thank Drs. Robin Avery, Bijan Eghtesad, Titte Srinivas, and Maria Siemionow from the Cleveland Clinic Transplant Center, Drs. Howard Gebel and Robert Bray from the Department of Pathology at Emory University, and Garnett Smith from the Cleveland Clinic Lerner College of Medicine for providing clinical follow-up data and insightful discussion.

References

1. Carter V. My approach to cardiothoracic transplantation and the role of the histocompatibility and immunogenetics laboratory in a rapidly developing field. *J Clin Pathol.* 2010;63:189-193.
2. Sage D. My approach to the immunogenetics of haematopoietic stem cell transplant matching. *J Clin Pathol.* 2010;63:194-198.
3. Siemionow M, Klimczak A. Advances in the development of experimental composite tissue transplantation models. *Transpl Int.* 2010;23:2-13.
4. Baxter-Lowe LA, Hurley CK. Advancement and clinical implications of HLA typing in allogeneic hematopoietic stem cell transplantation. *Cancer Treat Res.* 2009;144:1-24.
5. Gordon CR, Siemionow M, Papay F, et al. The world's experience with facial transplantation: what have we learned thus far? *Ann Plast Surg.* 2009;63:572-578.
6. Jones JW, Gruber SA, Barker JH, Breidenbach WC. Successful hand transplantation. One-year follow-up. Louisville Hand transplant team. *N Engl J Med.* 2000;343(7):468-473.
7. Gordon CR, Siemionow M. Requirements for the development of a hand transplantation program. *Ann Plast Surg.* 2009;63:262-273.
8. Opelz G. New immunosuppressants and HLA matching. *Transplant Proc.* 2001;33(1-2):467-468.
9. Lee SJ, Klein J, Haagenson M, et al. High-resolution donor-recipient HLA matching contributes to the success of unrelated donor marrow transplantation. *Blood.* 2007;110:4576-4583.
10. Kamani N, Spellman S, Hurley CK, et al. State of the art review: HLA matching and outcome of unrelated donor umbilical cord blood transplants. *Biol Blood Marrow Transplant.* 2008;14:1-6.
11. Kissmeyer-Nielsen F, Olsen S, Petersen VP, Fjeldborg O. Hyperacute rejection of kidney allografts, associated with pre-existing humoral antibodies against donor cells. *Lancet.* 1966;2:662-665.
12. Ciurea SO, de Lima M, Cano P, et al. High risk of graft failure in patients with anti-HLA antibodies undergoing haploidentical stem-cell transplantation. *Transplantation.* 2009;88:1019-1024.
13. Spellman S, Bray R, Rosen-Bronson S, et al. The detection of donor-directed, HLA-specific alloantibodies in recipients of unrelated hematopoietic cell transplantation is predictive of graft failure. *Blood.* 2010;115:2704-2708.
14. Gebel HM, Moussa O, Eckels DD, Bray RA. Donor-reactive HLA antibodies in renal allograft recipients: considerations, complications, and conundrums. *Hum Immunol.* 2009;70:610-617.
15. El-Awar N, Lee J, Terasaki PI. HLA antibody identification with single antigen beads compared to conventional methods. *Hum Immunol.* 2005;66(9):989-997.
16. Cecka JM. Calculated PRA (CPRA): the new measure of sensitization for transplant candidates. *Am J Transplant.* 2010;10:26-29.
17. Wahrmann M, Bartel G, Exner M, et al. Clinical relevance of preformed C4d-fixing and non-C4d-fixing HLA single antigen reactivity in renal allograft recipients. *Transpl Int.* 2009;22:982-989.
18. Rose ML, Smith JD. Clinical relevance of complement-fixing antibodies in cardiac transplantation. *Hum Immunol.* 2009;70:605-609.
19. Baldwin WM III, Qian Z, Wasowska B, Sanfilippo F. Complement causes allograft injury by cell activation rather than lysis. *Transplantation.* 1999;67:1498-1499.
20. Solez K, Colvin RB, Racusen LC, et al. Banff 07 classification of renal allograft pathology: updates and future directions. *Am J Transplant.* 2008;8:753-760.
21. Smith JD, Hamour IM, Banner NR, Rose ML. C4d fixing, luminex binding antibodies – a new tool for prediction of graft failure after heart transplantation. *Am J Transplant.* 2007;7:2809-2815.
22. Cendales LC, Kanitakis J, Schneeberger S, et al. The Banff 2007 working classification of skin-containing composite tissue allograft pathology. *Am J Transplant.* 2008;8:1396-1400.
23. Patel R, Terasaki PI. Significance of the positive crossmatch test in kidney transplantation. *N Engl J Med.* 1969;280:735-739.
24. Ilham MA, Winkler S, Coates E, Rizzello A, Rees TJ, Asderakis A. Clinical significance of a positive flow crossmatch on the outcomes of cadaveric renal transplants. *Transplant Proc.* 2008;40:1839-1843.
25. Graff RJ, Xiao H, Schnitzler MA, et al. The role of positive flow cytometry crossmatch in late renal allograft loss. *Hum Immunol.* 2009;70:502-505.
26. Zheng XF, Pei GX, Qiu YR, Zhu LJ, Gu LQ. Serial monitoring of immunological parameters following human hand transplant. *Clin Transplant.* 2004;18:119-123.

27. Brandacher G, Ninkovic M, Piza-Katzer H, et al. The Innsbruck hand transplant program: update at 8 years after the first transplant. *Transplant Proc*. 2009;41:491-494.
28. Gebel HM, Lebeck LK. Crossmatch procedures used in organ transplantation. *Clin Lab Med*. 1991;11:603-620.
29. Gebel HM, Bray RA, Nickerson P. Pre-transplant assessment of donor-reactive, HLA-specific antibodies in renal transplantation: contraindication vs. risk. *Am J Transplant*. 2003;3:1488-1500.
30. Saw CL, Bray RA, Gebel HM. Cytotoxicity and antibody binding by flow cytometry: a single assay to simultaneously assess two parameters. *Cytom B Clin Cytom*. 2008;74: 287-294.
31. Billen EV, Voorter CE, Christiaans MH, van den Berg-Loonen EM. Luminex donor-specific crossmatches. *Tissue Antigens*. 2008;71:507-513.
32. Delmonico FL, Fuller A, Cosimi AB, et al. New approaches to donor crossmatching and successful transplantation of highly sensitized patients. *Transplantation*. 1983;36:629-633.
33. Appel JZ 3rd, Hartwig MG, Cantu E 3rd, Palmer SM, Reinsmoen NL, Davis RD. Role of flow cytometry to define unacceptable HLA antigens in lung transplant recipients with HLA-specific antibodies. *Transplantation*. 2006;81:1049-1057.
34. Amico P, Honger G, Steiger J, Schaub S. Utility of the virtual crossmatch in solid organ transplantation. *Curr Opin Organ Transplant*. 2009;14:656-661.
35. Gutman JA, McKinney SK, Pereira S, et al. Prospective monitoring for alloimmunization in cord blood transplantation: "virtual crossmatch" can be used to demonstrate donor-directed antibodies. *Transplantation*. 2009;87:415-418.
36. Einecke G, Sis B, Reeve J, et al. Antibody-mediated microcirculation injury is the major cause of late kidney transplant failure. *Am J Transplant*. 2009;9:2520-2531.
37. Morales-Buenrostro LE, Castro R, Terasaki PI. A single human leukocyte antigen-antibody test after heart or lung transplantation is predictive of survival. *Transplantation*. 2008;85:478-481.
38. Girnita AL, McCurry KR, Zeevi A. Increased lung allograft failure in patients with HLA-specific antibody. *Clin Transpl*. 2007:231-239.
39. Worthington JE, Martin S, Al-Husseini DM, Dyer PA, Johnson RW. Posttransplantation production of donor HLA-specific antibodies as a predictor of renal transplant outcome. *Transplantation*. 2003;75:1034-1040.
40. Siemionow M, Papay F, Alam D, et al. Near-total human face transplantation for a severely disfigured patient in the USA. *Lancet*. 2009;18(374):203-209.
41. Ahern AT, Artruc SB, DellaPelle P, et al. Hyperacute rejection of HLA-AB-identical renal allografts associated with B lymphocyte and endothelial reactive antibodies. *Transplantation*. 1982;33:103-106.
42. Opelz G. Non-HLA transplantation immunity revealed by lymphocytotoxic antibodies. *Lancet*. 2005;365:1570-1576.
43. Angaswamy N, Saini D, Ramachandran S, et al. Development of antibodies to human leukocyte antigen precedes development of antibodies to major histocompatibility class I-related chain A and are significantly associated with development of chronic rejection after human lung transplantation. *Hum Immunol*. 2010;71:560-565.
44. Nath DS, Basha HI, Mohanakumar T. Antihuman leukocyte antigen antibody-induced autoimmunity: role in chronic rejection. *Curr Opin Organ Transplant*. 2010;15:16-20.
45. Sumitran-Karuppan S, Tyden G, Reinholt F, Berg U, Moller E. Hyperacute rejections of two consecutive renal allografts and early loss of the third transplant caused by non-HLA antibodies specific for endothelial cells. *Transpl Immunol*. 1997;5:321-327.
46. Magro CM, Klinger DM, Adams PW, et al. Evidence that humoral allograft rejection in lung transplant patients is not histocompatibility antigen-related. *Am J Transplant*. 2003;3: 1264-1272.
47. Kowalski R, Post D, Schneider MC, et al. Immune cell function testing: an adjunct to therapeutic drug monitoring in transplant patient management. *Clin Transplant*. 2003;17:77-88.
48. Serban G, Whittaker V, Fan J, et al. Significance of immune cell function monitoring in renal transplantation after thymoglobulin induction therapy. *Hum Immunol*. 2009;70: 882-890.
49. Smith CA, Gruss HJ, Davis T, et al. CD30 antigen, a marker for Hodgkin's lymphoma, is a receptor whose ligand defines an emerging family of cytokines with homology to TNF. *Cell*. 1993;73:1349-1360.
50. Pelzl S, Opelz G, Daniel V, Wiesel M, Susal C. Evaluation of posttransplantation soluble CD30 for diagnosis of acute renal allograft rejection. *Transplantation*. 2003;75:421-423.
51. Wang D, Wu GJ, Wu WZ, et al. Pre- and post-transplant monitoring of soluble CD30 levels as predictor of acute renal allograft rejection. *Transpl Immunol*. 2007;17:278-282.
52. Gebauer BS, Hricik DE, Atallah A, et al. Evolution of the enzyme-linked immunosorbent spot assay for post-transplant alloreactivity as a potentially useful immune monitoring tool. *Am J Transplant*. 2002;2:857-866.
53. Dubernard JM, Lengele B, Morelon E, et al. Outcomes 18 months after the first human partial face transplantation. *N Engl J Med*. 2007;357:2451-2460.
54. Gordon CR, Tai CY, Suzuki H, et al. Review of vascularized bone marrow transplantation: current status and future clinical applications. *Microsurgery*. 2007;27:348-353.
55. Alizadeh M, Bernard M, Danic B, et al. Quantitative assessment of hematopoietic chimerism after bone marrow transplantation by real-time quantitative polymerase chain reaction. *Blood*. 2002;99:4618-4625.
56. Kawai T, Cosimi AB, Spitzer TR, et al. HLA-mismatched renal transplantation without maintenance immunosuppression. *N Engl J Med*. 2008;358:353-361.
57. Smith DM, Agura E, Netto G, et al. Liver transplant-associated graft-versus-host disease. *Transplantation*. 2003;75:118-126.
58. Domiati-Saad R, Klintmalm GB, Netto G, Agura ED, Chinnakotla S, Smith DM. Acute graft versus host disease after liver transplantation: patterns of lymphocyte chimerism. *Am J Transplant*. 2005;5:2968-2973.
59. Stotler CJ, Eghtesad B, Hsi E, Silver B. Rapid resolution of GVHD after orthotopic liver transplantation in a patient treated with alefacept. *Blood*. 2009;113(21):5365-5366.
60. Dai Y, Sui W, Lan H, Yan Q, Huang H, Huang Y. Comprehensive analysis of microRNA expression patterns in renal biopsies of lupus nephritis patients. *Rheumatol Int*. 2009;29:749-754.

17 A Single Institution's Experience With Diagnosing Acute Cellular Rejection in Facial Allotransplantation

Jason S. Stratton and Wilma F. Bergfeld

Contents

Abstract Recent advances in immunosuppression and surgical techniques have progressed to make face transplants possible. These composite tissue grafts consist of skin, subcutaneous tissue, muscle, nerve, and bone. Accurate clinical and histologic rejection surveillance is vital to preserve the function of the graft. In December 2008, the first near-total face transplant was performed. Reviewing the pathology from this case reveals that the clinical impression and skin histology showed good correlation. However, the mucosal biopsies showed histologic signs of acute cellular rejection that far exceeded that of the skin biopsies. This discrepancy made it difficult for the pathology team to decide with certainty whether these changes truly represented acute cellular rejection.

Abbreviations

ACR	Acute cellular rejection
CCF	Cleveland Clinic Foundation
CMV	Cytomegalovirus
FA	Facial allotransplantation
H&E	Hematoxylin & Eosin
PAS	Periodic Acid-Schiff Stain
TUNEL	Terminal deoxynucleotidyl transferase dUTP nick end labeling

17.1 Introduction

A new class of transplants has emerged with the advent of improved microsurgical and immunosuppression techniques. This class is termed composite tissue allografts. They are composed of aggregates of tissues such as muscle, bone, nerve, skin, mucosa, cartilage,

J.S. Stratton (✉)
Department of Pathology,
Cleveland Clinic, Cleveland, OH, USA
e-mail: strattj@ccf.org

M.Z. Siemionow (ed.), *The Know-How of Face Transplantation*,
DOI: 10.1007/978-0-85729-253-7_17, © Springer-Verlag London Limited 2011

and even teeth. Examples of these allotransplants include: arm, hand, abdomen, scalp, and face. In December of 2008, the Cleveland Clinic Foundation (CCF) performed the fourth of nine face transplants performed to date. The facial graft included skin from the lower eyelids to below the mandible, buccal mucosa, bone, and cartilage from nose and zygomatic arches, and the two upper front teeth.[1] The pathology team was directly responsible for the histologic detection of acute cellular rejection (ACR), chronic rejection, infection, and various other histologic changes. Given the limited worldwide experience in face allotransplantation (FA) pathology and the variety of confounding factors, such as infection and medication effect, the diagnosis of ACR is difficult and needs to be made in a context of close clinical consultation and with a discussion of the possible differential diagnoses.

17.2 Methods

The system of rejection surveillance adopted at the CCF was based on previously published cases and on the suggestion in the Banff classification for composite tissue allograft pathology.[2] Paired mucosal and skin biopsies were formalin fixed and paraffin embedded. Hematoxylin & eosin (H&E) and periodic acid-Schiff stain (PAS) stains were obtained, and immunohistochemical studies were performed for CD3, CD8, CD20, CD68, CD30, FoXP3, K167, HMb45, CD1a, S100, Factor XIIIa, CD31, and CD34. On an intermittent basis, C4d stain was performed to look for antibody-mediated rejection and a TUNEL assay was performed to confirm the presence of apoptotic keratinocytes (Fig. 17.1). The H&E and PAS stains were obtained 1 day prior to the immunostains, and grading was entirely based on these stains. Grading was performed using a slightly modified version of the Banff classification that was discussed with our clinical staff (Table 17.1). Under this classification, grade I ACR represented a mild lymphocytic perivascular infiltrate. Grade II represented a moderate lymphocytic infiltrate and/or epidermal changes with at most spongiosis of the overlying epidermis. Grade III contained epidermal damage in the form of clusters of two or more keratinocytes, and grade IV contained epidermal necrosis.

The H&E is the most important stain when evaluating FA biopsies. That said the PAS stain proved valuable with its ability to detect fungal organisms and

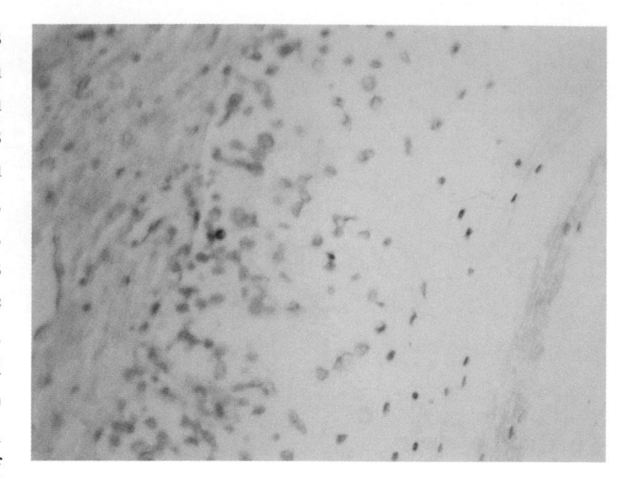

Fig. 17.1 TUNEL assay showing apoptotic keratinocyte

highlight basement membranes and apoptotic cells. The immunohistochemical studies were rarely contributory, and we frequently found ourselves questioning their utility. A brief examination of the immunostains showed CD1a, S100, HMB45, CD31, CD34, CD68, and Factor XIIIa, staining that is qualitatively and quantitatively similar to native skin. The epidermal proliferative fraction was identified by staining to ki-67 and was quantitatively similar to native skin. Seventy to ninety percent of lymphocytes were CD3 positive T-cells. CD4:8 ratio was consistently between 1.5:1 and 3:1. CD20 only identified rare aggregates of B-cells. CD30 identified exceedingly rare activated lymphocytes. Staining for regulatory T-cells with FoxP3 identified 1–15% of the total lymphocytes. Perhaps, in the future, a more simplified approach, such as CD3, CD20, Foxp3, and unstained slides for further studies might prove more appropriate and cost-effective. C4d was not contributory, in our experience, but could be useful if vessel endothelial damage is detected by H&E stain. The TUNEL (Terminal deoxynucleotidyl transferase dUTP nick end labeling) is a technique that detects apoptosis by highlighting DNA fragmentation. In our experience, it proved effective at detecting apoptotic keratinocytes that were readily identifiable by the H&E and PAS stains, and thus was not useful. Early pathological studies on FA rightfully included a broad range of testing designed to detect any number of unexpected findings. As the literature builds on the pathology of FA, a more focused approach toward testing might be warranted, with increased focus on getting the correct diagnosis to the clinician as quickly as possible.

Table 17.1 Modified BANFF classification used for grading on biopsies[2]

Grade	Clinical features	Histologic features
I	Predominantly normal skin and mucosa	Mild to moderate lymphocytic perivascular infiltrate in the superficial to middle dermis
II	Slightly scaly erythematous macules on skin, erythema of mucosa	Moderate lymphocytic perivascular infiltrate in the superficial to middle dermis Mild epidermal interface changes with or without spongiosis without keratinocyte death
III	Scattered or confluent lichenoid erythematous scaly papules and plaques on skin, plaques on mucosa	Moderate to severe lymphocytic perivascular infiltrate filling the dermis Interface inflammation with at least clusters of at least two apoptotic keratinocytes
IV	Confluent erythematous scaly plaques on skin, plaques, and ulceration of mucosa	Grade III with epidermal necrosis

17.3 Results and Interpretation

If rejection is to be thought of as a graft wide process, it would be expected that both the mucosa and skin biopsies would show similar histologic changes. Instead, at the CCF, the mucosal biopsies showed more frequent signs of ACR. These changes are only histologic, lacking any clinical suspicion or symptom of ACR and also lacking concurrent histologic changes in the skin.

The skin biopsies had excellent clinical correlation. Nineteen percent of biopsies had histologic signs of ACR. Each of the biopsies that showed histologic signs of rejection fell within 2 episodes that had clinical symptoms suspicious for rejection: a perinasal papule and whole graft erythema. In the skin biopsies, histologic symptoms of ACR progressed similarly to that described in the BANFF classification. In the skin biopsies, we saw collections of perivascular CD3 positive lymphocytes that expand to fill the dermis. These lymphocytes eventually involved the epidermis with spongiosis, then causing keratinocytes death (Fig. 17.2). When follicular units were present, they were involved to the degree of the overlying epidermis.

The mucosal biopsies differed from the skin biopsies in that they had poor clinical correlation. Seventy six percent of mucosal biopsies had histologic signs of ACR; the majority of these biopsies did not have concurrent clinical symptoms. Possible explanations for this lack of correlation include the difficulty in detecting erythema on mucosa, essentially making this inflammation a subclinical finding, and the possibility that the inflammation represents pathology different from ACR. Histologic examination of the mucosal biopsies also reveals a pattern of inflammation that is different from the skin. This pattern is predominantly interface with little submucosal or perivascular inflammation (Fig. 17.3). These mucosal biopsies contain multiple foci of interface mucositis that progresses from spongiosis to focal keratinocyte apoptosis. Interestingly, the only mucosal biopsies to have significant perivascular inflammation were during the episode of whole graft erythema (Fig. 17.4) possibly suggesting that this interface inflammation is non-specific and could represent drug effect, infection, or a mechanism other than ACR.

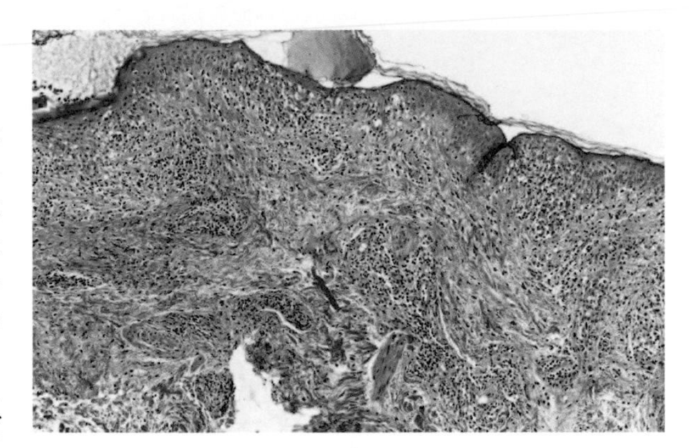

Fig. 17.2 Grade III ACR of the skin

To look into this further, we discussed our findings with the clinical team. After discovering that there was indeed no clinical finding, we sent one of our mucosal slides out for expert consultation and they returned with a diagnosis of ACR grade III. We then considered infectious causes, ruled out CMV and fungal sources. Given the patient's medications, including immune modulators, we did not feel it was possible to totally

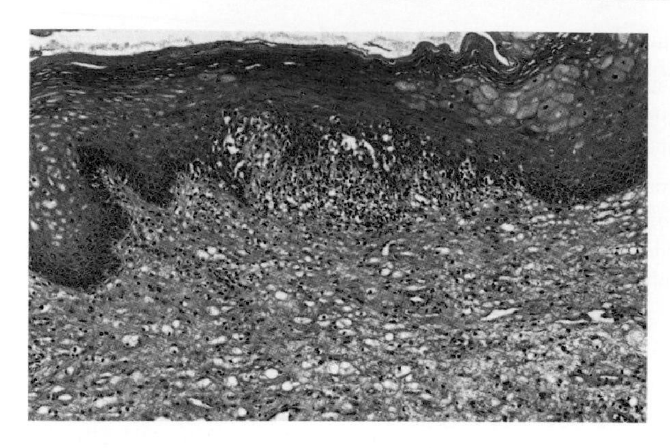

Fig. 17.3 Grade III ACR of the mucosa with apoptotic keratinocytes

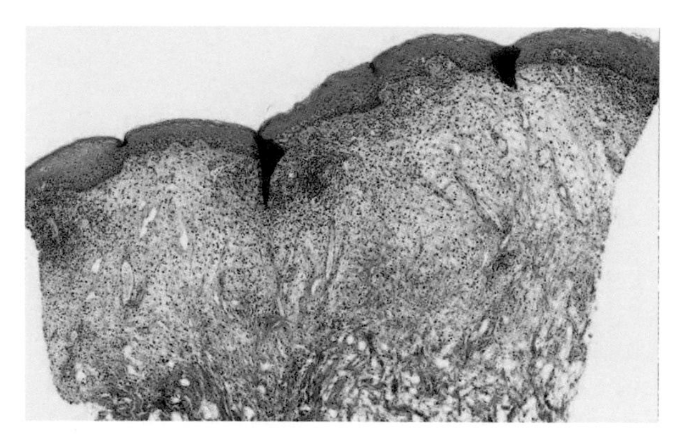

Fig. 17.4 Grade III of the mucosa in an episode of clinically suspected rejection

exclude medication side effects. So, we discussed our finding again with the clinical team. The decision was made to diagnose these mucosal biopsies with only interface inflammation as acute cellular rejection grade II unless there was a cluster of at least two apoptotic keratinocytes, which would then be grade III. The clinical team, for its part, decided not to treat if rejection was not suspected clinically.

17.4 Discussion

The difficulty that we faced in diagnosing ACR, in our case of FA, predominately involved frequent mucosal interface inflammation. One possible etiology for the inflammation was medication effect. The immunosuppressants used for the first 6 months were mycophenolate and tacrolimus. After 6 months, mycophenolate was exchanged for sirolimus. Mycophenolate has been shown to induce graft-versus-host type injury to the large bowel through its inhibition of purine biosynthesis.[3] It, more recently, has been shown to damage the mucosal squamous esophagus.[4] Sirolimus has been shown to cause oral ulcers without associated skin pathology.[5] Because of these side effects, we did not feel we were able to completely exclude medication side effect as a possible cause of the mucosal inflammation.

If this mucosal inflammation truly represents ACR, it does not seem to be specifically discussed in the BANFF classification. Some authors have noted that the mucosa does show more signs of acute cellular rejection than the skin.[6] When the BANFF classification was developed, only one case of FA had a significant number of mucosal biopsies. This is in contrast to skin biopsies from 28 hand transplants, 9 abdominal walls transplants, and 1 knee transplant.[2]

A possible explanation to the difference in presentation between skin and mucosa is the theory of split tolerance. Split tolerance states that different tissues have varying tolerance toward ACR.[7-9] In animal models, for example, when organ skin allotransplants are performed, the skin is more prone to rejection than the solid organ. Likewise, it has been suggested that bone, muscle, nerve, and adnexa have increased tolerance to ACR. Some authors have postulated that the mucosa's lack of adnexal structures makes it more susceptible to ACR.[6] Another hypothesis is that a larger population of semi-mature dendrite cells could promote graft tolerance by expression of CD40, CD80, CD86, and lack of expression of IL-1, IL-6, and IL-12. They have also been hypothesized to increase populations of CD4+, CD25+ regulatory T-cells secrete IL-10 and can promote graft tolerance.[10] Each of these possible mechanisms remains an area for future research and collaboration.

FA remains a new field and each new case brings new knowledge, but also new unanswered questions. In pathology, we endeavor to create a report that is clinically useful to the clinicians. This is difficult when the histologic picture is not specific. It is hopeful, now that new cases are being performed and more mucosal biopsies are available, a better understanding will be gained on the way ACR affects the mucosa, skin, and the other tissues of these complicated composite grafts.

References

1. Siemionow MZ, Papay F, Alam D, et al. Near-total human face transplantation for a severely disfigured patient in the USA. *Lancet*. 2009;374:203-209.
2. Cendales LC, Kanitakis J, Schneeberger S, et al. The Banff 2007 working classification of skin-containing composite tissue allograft pathology. *Am J Transplant*. 2008;8:1396-1400.
3. Papadimitriou JC, Cangro CB, Lustberg A, et al. Histologic features of mycophenolate mofetil-related colitis: a graft-versus-host disease-like pattern. *Int J Surg Pathol*. 2003;11:295-302.
4. Nguyen T, Park JY, Scudiere JR, et al. Mycophenolic acid (cellcept and myofortic) induced injury of the upper GI tract. *Am J Surg Pathol*. 2009;33:1355-1363.
5. Fricain JC, Cellerie K, Sibaud V, et al. Oral ulcers in kidney allograft recipients treated with sirolimus. *Ann Dermatol Vénéréol*. 2008;135:737-741.
6. Kanitakis J, Badet L, Petruzzo P, et al. Clinicopathologic monitoring of the skin and oral mucosa of the first human face allograft: report on the first eight months. *Transplantation*. 2006;82:1610-1615.
7. Swearingen B, Ravindra K, Xu H, et al. Science of composite tissue allotransplantation. *Transplantation*. 2008;86: 627-635.
8. Mathes DW, Randolph MA, Solari MG, et al. Split tolerance to a composite tissue allograft in a swine model. *Transplantation*. 2003;75:25-31.
9. Lee WP, Yaremchuk MJ, Pan YC, et al. Relative antigenicity of components of a vascularized limb allograft. *Plast Reconstr Surg*. 1991;87:401-411.
10. Fu BM, He XS, Yu S, et al. Tolerogenic semimature dendritic cells induce effector T-cell hyporesponsiveness by the activation of antigen-specific CD4+ CD25+ T-regulatory cells. *Exp Clin Transplant*. 2009;7:149-156.

18 Classification of Face Rejection: Banff classification for CTA

Jean Kanitakis and Linda C. Cendales

Contents

Abstract Facial allotransplantation (FA) has recently emerged as a new viable option for reconstruction of severe facial tissue defects that are not amenable to conventional reconstructive techniques. FA falls within the spectrum of Composite Tissue Allografts (CTA), and as such may undergo allograft rejection. The experience obtained so far from the limited number of FA shows that the recipients develop, in the first post-graft months, signs of (skin) rejection that can be reversed with adjustment of the immunosuppressive treatment. The severity of skin rejection can be assessed with a pathological score that was proposed during the 2007 Banff meeting in La Coruna, Spain (Banff CTA-07) and classifies rejection in five grades (0–IV) according to the severity of pathological changes in the skin. There are still several questions that remain so far unanswered regarding rejection in FA, including namely the role of skin-infiltrating cells and the possibility of development of chronic rejection.

Abbreviations

AMR Antibody-mediated rejection
CTA Composite tissue allografts
FA Facial allotransplantation
GVHD Graft versus host disease
HES Hematoxylin-eosin-saffron
PAS Periodic Acid Schiff stain

18.1 Introduction

Facial allotransplantation (FA) has recently emerged as a new viable option for reconstruction of severe facial tissue defects (secondary mostly to traumatic injuries and burns) that are not amenable to conventional

J. Kanitakis (✉)
Department of Dermatology/Laboratory of Dermatopathology, Edouard Herriot Hospital Group, Lyon, France
e-mail: jean.kanitakis@univ-lyon1.fr

M.Z. Siemionow (ed.), *The Know-How of Face Transplantation*,
DOI: 10.1007/978-0-85729-253-7_18, © Springer-Verlag London Limited 2011

reconstructive techniques using autologous tissues. FA falls within the wider spectrum of Composite Tissue Allografts (CTA), i.e., allografts containing embryologically heterogeneous tissues such as skin, nerves, vessels, bones (including bone marrow), tendons, and muscles. Human FA provides various combinations of skin, muscle, and/or bone. After the first (partial) mid-face transplantation performed in France (Amiens/Lyon),[1,2] ten additional facial allografts have been performed worldwide in humans in France (Paris and Amiens/Lyon), China (Xia), USA (Cleveland and Boston), and Spain (Valencia and Barcelona).[3-9] Therefore, although the present results are promising, human FA remains a challenging, and for some, controversial procedure[10] that is still in its experimental stage.[11] In view of the small number of FA performed so far, and the fact that the latest ones have not yet been published in detail in the medical literature, only sparse data are available on mid- and long-term functional, esthetic and immunologic outcomes, so that the lessons that have so far been learned[6] should be viewed as preliminary at best.

18.2 Rejection in Facial Allotransplantation

Similarly to other types of allografts (including CTA), FA elicits a strong allo-immune response; therefore, their recipients need to receive life-long immunosuppression, following induction, in order to prevent allograft rejection. Despite this, the experience gained so far shows that CTA, including FA, regularly undergo episodes of graft rejection, namely, in the early post-graft period. These manifest clinically by cutaneous changes including pink or erythematous macules that may gradually progress (in the case of hand allografts) to red infiltrated, scaly lichenoid papules with or without edema and nail changes in more advanced cases.[12-14]

In the case of FA recipients, rejection has manifested clinically as early as from day 18 post-graft with diffuse redness of the allografted facial skin, and less frequently with edema and congestion.[3-5,15] In the case of two FA recipients followed in Lyon, who received a sentinel vascularized skin flap of donor's origin, erythematous macules and/or redness developed concomitantly with clinical lesions of the face.[2,9,15] Rejection episodes can in most cases be reversed with adequate adjustment of the immunosuppressive treatment.

Early detection of rejection in FA is crucial in order to treat the recipient precociously so as to stop the development of persistent rejection. Experience obtained from previous CTA (mainly hands/forearms) showed that clinico-pathological evaluation of the skin is the most efficient way to detect allograft rejection. Skin biopsies obtained from CTA showed that pathological changes during allograft rejection vary in intensity depending on the severity of rejection. They affect initially the dermis and may spread to the epidermis and hypodermis at more advanced stages. Dermal changes consist mainly in an inflammatory cell infiltration with T-cells (including CD3+, CD4+, CD8+, TIA-1+ cytotoxic cells and FoxP3+ T-regulatory cells), CD68+ monocytic cells and more rarely eosinophils. This infiltrate initially forms perivascular cuffs and nodules, and later spreads to the interstitial dermis, the epidermis and hypodermis. Epidermal/adnexal changes include mainly keratinocyte necrosis/apoptosis, inflammatory cell exocytosis, more rarely spongiosis, acanthosis, papillomatosis, and ortho-hyperkeratosis. During very severe rejection episodes, the infiltrate may extend to the hypodermis, forming perivascular and periadnexal nodules. On the basis of the intensity of these changes, four scoring systems were initially proposed to assess the severity of CTA rejection.[16-19] At the Ninth Banff Conference on Allograft Pathology in La Coruna, Spain, a symposium on CTA rejection was held (26 June 2007) and proposed a working classification (Banff CTA-07) for the categorization of CTA rejection.[20] This classification was derived from a consensus discussion session attended by most senior authors of the afore-mentioned published classification systems. It was based on findings of skin rejection, since deeper tissues have not yet been, with few exceptions, sufficiently studied in human CTA. In one such case, where several tissues were studied during persistent rejection of a hand allograft due to non-adherence to the immunosuppressive treatment, the skin was found to be the most severely affected tissue,[12] thus confirming previous findings obtained in animal models.[21] It seemed therefore relevant to rely on skin findings for diagnosing rejection of skin-containing CTA. Additionally, the skin can be examined clinically and can be easily biopsied.

The Banff CTA-07 symposium considered that the skin specimen necessary for evaluation of possible rejection should be obtained with a 4-mm (or larger) punch taken from the most erythematous and/or indurated (but apparently viable) area of involved skin. The structures required to constitute an adequate sample are the epidermis and its adnexa, dermis, subcutaneous tissue, and vessels. The recommendations for slide preparation are hematoxylin and eosin (H&E) and periodic acid Schiff (PAS) stains. Immunohistochemical labelings are recommended « as needed », based on HES findings and/or for research purposes; these include (but are not limited to) CD3, CD4, CD8, CD19, CD20, and CD68, as well as HLA-DR, CMV, and C4d.[20]

18.3 Pathologic Classification of Rejection in Face Transplantation: The Banff CTA-07 Score

The Banff CTA-07 classification of CTA rejection, established in order to score rejection of skin-containing CTA, is a tiered system that comprises the following five grades (0–IV) of severity[20]:

- Grade 0 (no rejection): No or rare inflammatory dermal infiltrates (some degree of perivascular inflammatory infiltrate can be found in biopsies of normal-looking skin, especially on the face) (Fig. 18.1);
- Grade I (mild rejection): Mild (lymphocytic) perivascular infiltration – no involvement of the overlying epidermis (Fig. 18.2);
- Grade II (moderate rejection): Moderate-to-dense perivascular inflammation (mainly lymphocytic), with or without mild epidermal and/or adnexal involvement (limited to spongiosis and exocytosis) – no epidermal necrosis or apoptosis (Fig. 18.3);
- Grade III (severe rejection): Dense dermal inflammation and epidermal involvement with epithelial apoptosis and/or necrosis, interface dermatitis (Fig. 18.4);
- Grade IV (necrotizing acute rejection): Frank necrosis of epidermis or other skin structures.

The scoring systems established for grading rejection of skin-containing CTA, including the Banff CTA-07, have been used to assess rejection in the few cases of FA performed worldwide.[3-5,9,15] Biopsies from FA have been taken from the allografted facial skin, the allografted oral mucosa, or the donor full-thickness skin placed as a sentinel skin graft on the recipient's skin, serving as donor site for biopsies in order to spare the face.[9,15] The established Banff CTA-07 as well as other

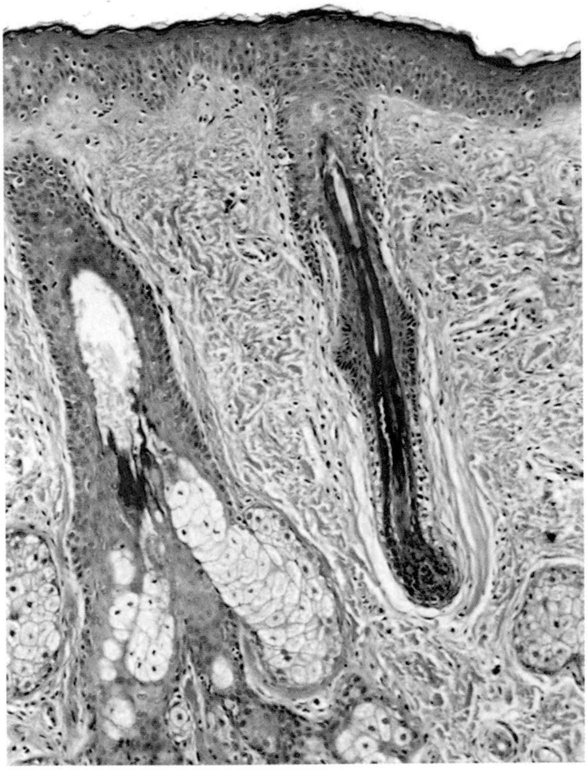

Fig. 18.1 Normal-looking skin from the chin of a face allograft recipient during the first days post-graft (grade 0 rejection) (HES stain)

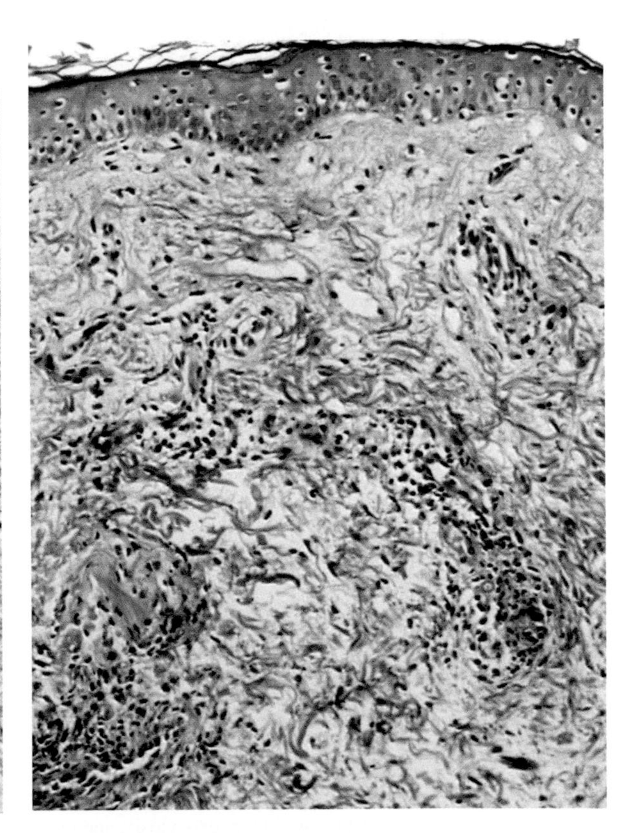

Fig. 18.2 Grade I (mild) rejection (sentinel skin graft of a face allograft recipient): a mild perivascular lymphocytic infiltration is seen in the upper dermis (HES stain)

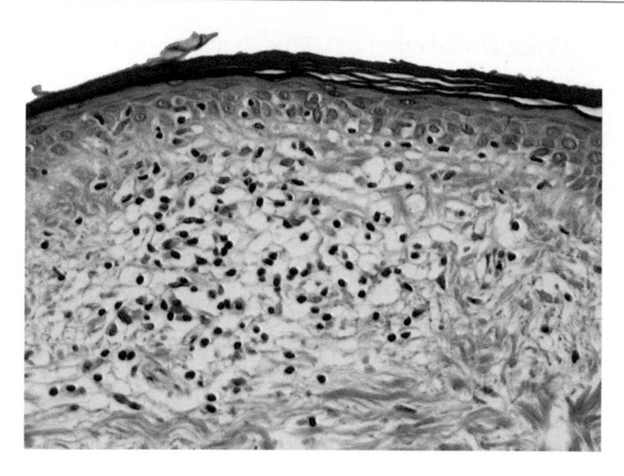

Fig. 18.3 Grade II (moderate) rejection (sentinel skin graft of a face allograft recipient): a moderate lymphocytic infiltration is seen in the upper edematous dermis, giving rise to exocytosis in the epidermis (HES stain)

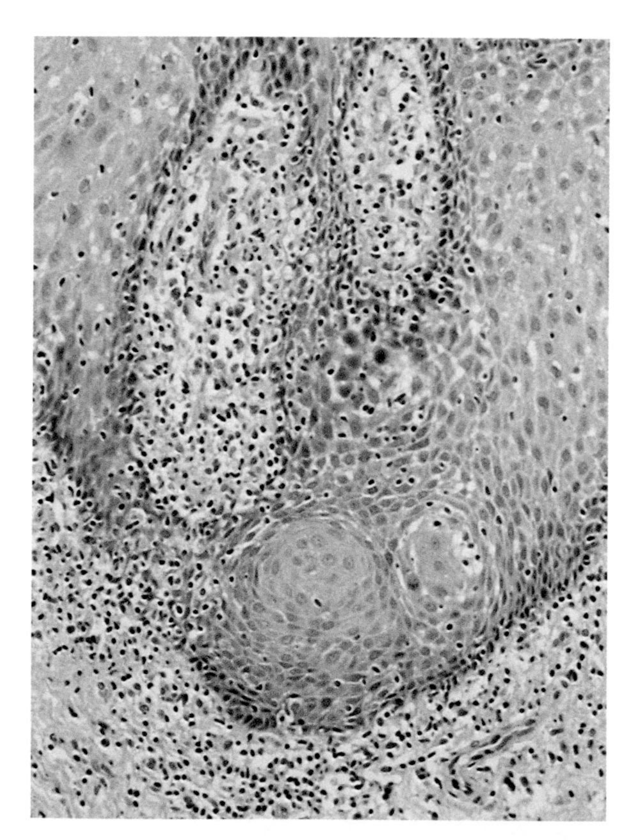

Fig. 18.4 Grade III (severe) rejection (oral mucosa of a face allograft recipient): a dense lymphocytic infiltration is seen in the upper corium, associated to significant exocytosis, and basal cell vacuolization (HES stain)

systems proposed for CTA proved applicable for the assessment of FA rejection since they were based on changes observed in the skin, despite the fact that the latter may show some microscopic differences according to the anatomical location (e.g., face vs. hands) and is also slightly different from the oral mucosa. In the three patients where details on rejection were given, rejection grades varied from 0 to III. Cells infiltrating the skin included mainly CD3+, CD4+, CD8+, TiA-1+, and Fox-P3+ T-cells (Fig. 18.5).[3,15] In the cases where skin and oral mucosa biopsies were taken concomitantly, the latter showed more severe changes compared with the former.[4,5,15] The explanation of this finding is presently unknown; it could be related to a different distribution and density of antigenic structures/cells, such as endothelial and dendritic cells. Furthermore, when bilateral (right and left) mucosa biopsies were taken concomitantly, discrepancies were occasionally noted between the two sites as to rejection grade,[15] suggesting a patchy infiltration pattern also observed and well described in solid organ transplants.

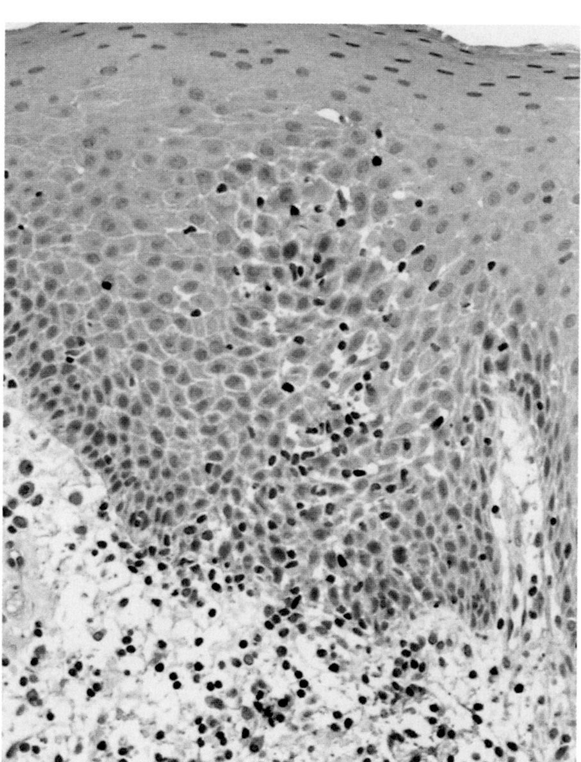

Fig. 18.5 FoxP3+ T-regulatory cells in the oral mucosa (corium and epithelium) of a mucosa of a face allograft recipient (immunoperoxidase revealed with diaminobenzidin)

Complementary pathologic studies that have been investigated in order to get further insight into the mechanisms of rejection and ultimately diagnose it more specifically include:

- Immunophenotyping of the cell infiltrate in the skin. As mentioned above, this is T-cell predominant, made of T-helper and cytotoxic T-cells (CD3+, CD4+, CD8+, TIA-1+) and FoxP3+ T-reg cells:[15,22] Whether the composition of the infiltrate changes with time after transplantation, and overall whether the cellular markers studied correlate with the severity of rejection remains to be established;
- Detection of C4d on skin biopsies. C4d is a complement degradation product deposited on endothelial cells of tissues during antibody-mediated rejection (AMR) in several allotransplants such as the kidney. The results obtained on CTA have so far been inconclusive: endothelial C4d deposition during CTA rejection in the skin has been reported in some studies,[22,23] but not in others, including namely FA.[24] The existence of AMR in CTA (supported by the presence of donor-specific antibodies) has not been convincingly shown so far.

It should be reminded here that the pathological changes seen in the skin (and mucosa) of CTA during rejection are not specific, but can mimic a variety of inflammatory and tumoral dermatoses, including allergic contact dermatitis, (pseudo)lymphomas, insect bites, lichen planus, drug eruptions, dermatophytoses, viral rashes, and GVHD, to name but a few.[25] Furthermore, as with other organ transplants, CTA rejection can coincide with other diseases (e.g., superficial fungal infection of the skin and/or oral mucosa). All these pitfalls should be known by the pathologists involved in interpretation of the slides so as not to overdiagnose rejection. Ancillary techniques, such as histochemical stains (PAS) or study of clonality of the lymphocytic infiltrate, may provide useful clues as to the correct diagnosis. In all cases, clinicopathologic correlation, entailing close collaboration between the pathologist and the transplant physicians, is mandatory for establishing the diagnosis of rejection.

18.4 Future Directions

As pointed out above, pathological data regarding human FA and vascularized CTA in general are still very sparse; therefore, there are several issues in this domain that remain to be further studied and answered:

- Which is the optimal tissue for assessing (globally) allograft rejection? This is not a purely theoretic question, since the decision to adjust/increase the immunosuppressive treatment usually relies on the result of pathological examination. Indeed, as stated previously, microscopic examination of tissues seems to be presently the most efficient indicator of CTA rejection. Both in experimental (animal) and human studies, the skin appeared as the most antigenic tissue among those contained in a CTA. After the advent of FA, it was noted that oral mucosa shows more severe pathological changes than the skin during rejection. The obvious advantage of oral mucosa versus skin is that post-biopsy scars on the former are not visible, contrasting with those of the facial skin that are most often visible and cosmetically undesired. Possible rejection of underlying deeper tissues (muscles, bone) in CTA (including FA) has not been studied so far pathologically, partly because biopsies of these tissues are (obviously) more difficult to obtain. The correlation between skin, mucosal, and underlying tissue rejection in CTA remains to be further studied, as this will allow to define which tissue reflects best, from a clinically relevant point of view, allograft rejection.
- What is the functional role of cells infiltrating the skin regarding rejection/tolerance? The Banff CTA-07 score does not take into account the immunophenotype of skin-infiltrating cells; however, it seems likely that phenotypically different cells (e.g., TIA-1+ cytotoxic vs. FoxP3+ T-regulatory cells) play different, possibly opposing, roles in the local immunological process. Better knowledge of the function of these cells in the infiltrate of CTA will probably lead to amendments of the Banff CTA rejection score, in order to take into account the composition of the cell infiltrate.
- Will chronic rejection develop in FA recipients? So far, chronic rejection (namely vasculopathy) has not been observed in FA nor in other CTA, with the exception of a hand-transplant patient who lost his graft because of arteriopathy.[26] Of note, this patient was receiving reduced immunosuppressive treatment, a fact that may have contributed to development of rejection, as happened with the first

hand-transplant patient following treatment discontinuation.[12] Thus, the possibility that chronic rejection will develop in FA in the long term cannot be excluded. The Banff CTA 07 does not presently include features of chronic rejection (such as myointimal vascular proliferation or fibrosis) but – as with other transplants – the scoring system will evolve, if necessary, as more clinical and experimental data becomes available. In solid organ (kidney) transplantation, higher incidences of acute rejection episodes are associated with higher rates of chronic rejection, and when considering the relatively high incidence of acute rejection observed in FA (as in hand-transplant) recipients, it may appear surprising that chronic rejection has not yet been seen more frequently. This could be due to several factors: (1) relatively short post-transplant follow-up (5 years in the case of FA), (2) low incidence of associated risk factors (such as hypertension and dyslipidemia) commonly seen in solid organ – but not CTA-transplant recipients, (3) a lower vascular susceptibility of CTA to the toxic effects of immunosuppressive drugs, and (4) early identification and reversal of acute rejection in FA transplants allowed by the possibility to rapidly diagnose skin rejection. Conversely, the relatively high frequency of diagnosis of rejection in CTA may be due, at least in part, to the fact that skin changes can be readily observed in CTA (as compared with other inner organ transplants). Whether chronic rejection will develop in the long term in the skin or other tissues in FA still remains to be observed. As noted above, the Banff CTA-07 will evolve as these data becomes available. Hopefully, the numerous questions that remain as yet unanswered in the field of human FA will be settled when larger clinicopathologic experience will be obtained in the future.

References

1. Devauchelle B, Badet L, Lengelé B, et al. First human face allograft: early report. *Lancet*. 2006;368:203-209.
2. Dubernard J, Lengelé B, Morelon E, et al. Outcomes 18 months after the first human partial face transplantation. *N Engl J Med*. 2007;357:2451-2460.
3. Guo S, Han Y, Zhang X, et al. Human facial allotransplantation: a 2-year follow-up study. *Lancet*. 2008;372:631-638.
4. Gordon CR, Siemionow M, Papay F, et al. The world's experience with facial transplantation: what have we learned thus far? *Ann Plast Surg*. 2009;63:572-578.
5. Eaton L. Spanish doctors carry out first transplantation of a full face. *BMJ*. 2010;340:c2303.
6. Lantieri L, Meningaud JP, Grimbert P, et al. Repair of the lower and middle parts of the face by composite tissue allotransplantation in a patient with massive plexiform neurofibroma: a 1-year follow-up study. *Lancet*. 2008;372:639-645.
7. Pomahac B, Pribaz J, Eriksson E, et al. Restoration of facial form and function after severe disfigurement from burn injury by a composite facial allograft. *Am J Transplant*. 2011. doi: 10.1111/j.1600-6143.2010.03368.x. [Epub ahead of print]
8. Siemionow MZ, Papay F, Alam D, et al. Near-total human face transplantation for a severely disfigured patient in the USA. *Lancet*. 2009;374:203-209.
9. Morelon E, Testelin S, Petruzzo P, et al. New partial face allograft transplantation: report on first three months. Presented at: the XXIIIth International Congress of the Transplantation Society; August 15–19, 2010; Vancouver, British Columbia, Canada.
10. Strong C. An ongoing issue concerning facial transplantation. *Am J Transplant*. 2010;10:1115-1116.
11. Siemionow M, Gordon C. Overview of guidelines for establishing a face transplant program: a work in progress. *Am J Transplant*. 2010;10:1290-1296.
12. Kanitakis J, Jullien D, Petruzzo P, et al. Clinicopathologic features of graft rejection of the first human hand allograft. *Transplantation*. 2003;76:688-693.
13. Schneeberger S, Gorantla V, van Riet R, et al. Atypical acute rejection after hand transplantation. *Am J Transplant*. 2008; 8:688-696.
14. Cendales L, Breidenbach W. Hand transplantation. *Hand Clin*. 2001;17:449-510.
15. Kanitakis J, Badet L, Petruzzo P, et al. Clinicopathological monitoring of the skin and oral mucosa of the first human face allograft. Report on the first eight months. *Transplantation*. 2006;82:1610-1615.
16. Bejarano PA, Levi D, Nassiri M, et al. The pathology of full-thickness cadaver skin transplant for large abdominal defects. *Am J Surg Pathol*. 2004;28:670-675.
17. Schneeberger S, Kreczy A, Brandacher G, Steurer W, Margreiter R. Steroid- and ATG-resistant rejection after double forearm transplantation responds to Campath-1H. *Am J Transplant*. 2004;4:1372-1374.
18. Kanitakis J, Petruzzo P, Jullien D, et al. Pathological score for the evaluation of allograft rejection in human hand (composite tissue) allotransplantation. *Eur J Dermatol*. 2005;15:235-238.
19. Cendales L, Kirk A, Moresi M, et al. Composite tissue allotransplantation: classification of clinical acute skin rejection. *Transplantation*. 2006;81:418-422.
20. Cendales L, Kanitakis J, Schneeberger S, et al. The Banff 2007 working classification of skin-containing composite tissue allograft pathology. *Am J Transplant*. 2008;8:1396-1400.
21. Lee W, Yaremchuk M, Pan Y, Randolph MA, Tan CM, Weiland AJ. Relative antigenicity of components of a vascularized limb allograft. *Plast Reconstr Surg*. 1991;87:401-411.
22. Hautz T, Zelger B, Grahammer J, et al. Molecular markers and targeted therapy of skin rejection in composite tissue allotransplantation. *Am J Transplant*. 2010;10:1200-1209.

23. Landin L, Cavadas P, Ibanez I, et al. CD3+ mediated rejection and C4d deposition in two composite tissue (bilateral hand) allograft recipients after induction with alemtuzumab. *Transplantation*. 2009;87:776-781.
24. Kanitakis J, McGregor B, Badet L, et al. Absence of C4d deposition in human composite tissue (hands and face) allograft biopsies: an immunoperoxidase study. *Transplantation*. 2007;84:265-267.
25. Kanitakis J. The challenge of dermatopathological diagnosis of rejection of composite tissue allografts: a review. *J Cutan Pathol*. 2008;35:738-744.
26. Breidenbach W, Ravindra K, Blair B, Burns C, et al. Transplant arteriopathy in clinical hand transplantation. Presented at: the 9th Meeting of the International Society of Hand and Composite Tissue Allotransplantation; September 11–12, 2009; Valencia, Spain.

Brain Plasticity After Hand and Face Allotransplantation

19

Claudia D. Vargas and Angela Sirigu

Contents

A. Sirigu (✉)
Center for Cognitive Neuroscience, CNRS, Bron, France
e-mail: sirigu@isc.cnrs.fr

Abstract The traumatic amputation of a hand is devastating because instantly dispossesses an individual from extremely well-developed upper limb sensory functions as well as the capacity to perform precision movements. Likewise, the face can be considered as a sophisticated organ of expressivity and communication, carrying important symbolic, social, and psychological significance. Thus, severe hand or facial traumatic loss can be lifelong impairing and strongly dysfunctional. Recent advances in the domain of transplantation are endowing severely deformed and/or functionally impaired patients with the possibility of receiving composite tissue allografts (CTA). The hand and face allograft are examples of CTA transplantation that contain skin, subcutaneous tissues, muscles, vessels, and nerves. Changes in the cortical motor representations induced by traumatic hand amputation have been shown to be overturned after hand allograft. Based on principles of plasticity underlying hand amputation and allograft, we will herein discuss hypotheses and set predictions regarding cortical changes after limb and face allograft.

Abbreviations

EMG	Electromyography
FMRI	Functional Magnetic Resonance Imaging
TMS	Transcranial Magnetic Stimulation

19.1 Introduction

It was long believed that the synaptic networks, and consequently, the functional organization of the brain were hard wired from birth and could not change during

M.Z. Siemionow (ed.), *The Know-How of Face Transplantation*,
DOI: 10.1007/978-0-85729-253-7_19, © Springer-Verlag London Limited 2011

adult life. This view was first challenged by Donald Hebb, who suggested more than 50 years ago[1] that synapses were continuously remodeled by experience. The term "plasticity" was coined to refer to the brain's capacity for such changes, occurring as a response of immediate or longer-lasting body peripheral modifications. Seminal experiments performed subsequently in animal models and humans indicated that the cortical representation of body parts is continuously modulated in response to activity, behavior, and skill acquisition.[2-7]

19.2 How Changes in the Body Periphery Affect Cortical Maps

Among the brain regions shown to undergo plastic modifications, it is now well established that the primary sensory (S1) and motor (M1) cortical regions are highly influenced by changes occurring at the body's periphery. Evidences from human and animal models show that when deprived of their afferent sensory input and/or its motor effectors, S1 and M1 undergo major plastic modifications.[3,4,8-18] We herein will focus on two major changes in the cortical sensorimotor representations induced by peripheral modification: those induced by traumatic hand amputation[11,12,15] and those occurring after hand allograft.[19,20]

19.3 Cortical Plasticity After Hand Amputation

In humans, the effects of traumatic upper limb amputation have been extensively investigated.[21,22] Curiously, patients often report a global feeling that the missing body part is still present. This feeling is frequently associated with specific sensory and kinesthetic sensations and pain in the missing limb. Many patients further describe that the phantom limb can be moved voluntarily.[21,22]

From the sensory perspective, hand amputation represents an acute deafferentation injury with immediate and long-standing influence on the corresponding representational areas in S1 as well as in adjacent cortical territories.[13] Using non-invasive neuromagnetic imaging techniques to determine cortical reorganization in humans,[14,15,23-25] S1 was shown to undergo a massive reorganization after hand amputation, with the territory corresponding originally to the hand now responding to stimulation of the face. Early studies from Ramachandran et al.[26] in forearm amputees offered a perceptual correlate of these topographical changes by showing that face stimulation evoked precisely localized referred sensations in phantom digits. From the motor side, functional investigation of human M1 reorganization after amputation has demonstrated that instead of becoming inactive, the hand area is activated during proximal limb movements,[12,19,27] the cortical stimulation of this region evoking contraction of proximal upper limb muscles.[11,16,28,29] In addition, face and forearm motor representations which surround the representation of the missing hand have also been shown to expand into the de-efferented cortex,[29,30] with the expansion of lip movements into the former hand area correlating positively with the amount of phantom limb pain.[31]

This sensorimotor reorganization has been understood as an invasion of the adjacent arm and face representations into the deafferented hand area,[20] reflecting a local competition for neural resources.[32] In fact, studies employing TMS paired-pulse protocols have shown less intracortical inhibition in the region corresponding to the amputated limb when compared with the intact limb's region in M1,[33,34] suggesting that modulation of inhibitory cortical circuits might play a fundamental role in representational changes that follow amputation.

19.4 Plastic Changes Occurring After Hand Allograft

Transplantation to replace the amputated body part offers the opportunity to study how new grafted muscles are processed in M1 and their effects on the long-term cortical changes provoked by the amputation. Recent functional resonance imaging (fMRI) results from transplanted patients indicated that a reversal of the long-standing amputation-induced reorganization was possible, with the hand allograft overthrowing the long-standing amputation-induced reorganization in M1.[19,35] However, cortical reorganization was mostly documented for movement involving extrinsic musculature (present before the transplant) and therefore it was not clear whether and how the newly intrinsic transplanted muscles reacquired a functional status in M1. Using transcranial magnetic stimulation (TMS) in former amputees who received double hand allograft, we observed the gradual reappearance of intrinsic hand muscles representation in the patient's M1, with distinct time courses observed for left and right hand muscle representations[20] (Fig. 19.1). Although it was not possible to precisely define how the coarser peripheral reconnection and the intense rehabilitation training interacted in determining the degree and extent of functional gain after hand allograft, it was concluded that the process of motor cortical plasticity extends to the recognition of newly transplanted muscles in order to build novel limb motor synergies, this plasticity being closely tied to motor recovery.

Among the factors that could block the emergence of plastic changes associated to functional skill reacquisition after a hand allograft is the lack of precise reconnection on the periphery. Previous studies in human unilateral upper limb replant recipients suggested that sensory reinnervation often remains incomplete even after many years.[36] However, fMRI results indicate that the restoration of afferent input (albeit incomplete) leads to activation in the region corresponding to the hand representation in S1.[19,37] Likewise, behavioral results gathered in one of the bilateral hand-grafted patients[38] and in upper limb amputees whose arm nerves were redirected to chest muscles[39] indicated that peripheral and central sensory pathways remain viable even after prolonged periods of amputation-induced disuse, and that somatosensory circuits of the human brain readily reintegrate peripheral information pending its availability.

From the motor side, it was shown in monkeys with amputated segments that de-efferented motoneurons preserve their functional efficacy by aberrantly innervating more proximal muscles.[40] If muscle contact is prevented during peripheral nerve regeneration, motor neurons tend to preferentially reinnervate the skin.[41] In mice, the specificity and percent of motor plate's reinnervation is severely degraded after a peripheral nerve cut.[42] Accordingly, clinical evaluation performed longitudinally in transplanted patients indicated that hand representation could remain dysfunctional as long as reinnervation of the hand is deficient.[35,36]

TMS results obtained in bilateral hand allograft patients[20] indicated that newly transplanted intrinsic muscles, which are extremely important for fine and skillful hand movements do acquire a cortical representation in M1. These representations are already present at 10 months after allograft for the left hand. Interestingly, mapping of the right intrinsic hand muscles was only possible with high stimulation values, and complete intrinsic hand representation was only achieved as late as 26 months after the graft (Fig. 19.1). Recording from intrinsic muscles of a unilateral hand transplanted patient, Lanzetta et al.[43] reported the first signs of voluntarily driven electromyographic (EMG) activity in one of the tested intrinsic hand muscles at 11 months post-transplant. One month later, a first motor unit train was also identified in two other intrinsic hand muscles, and after 24 months, in the first lumbrical muscles. A similar time course was found by

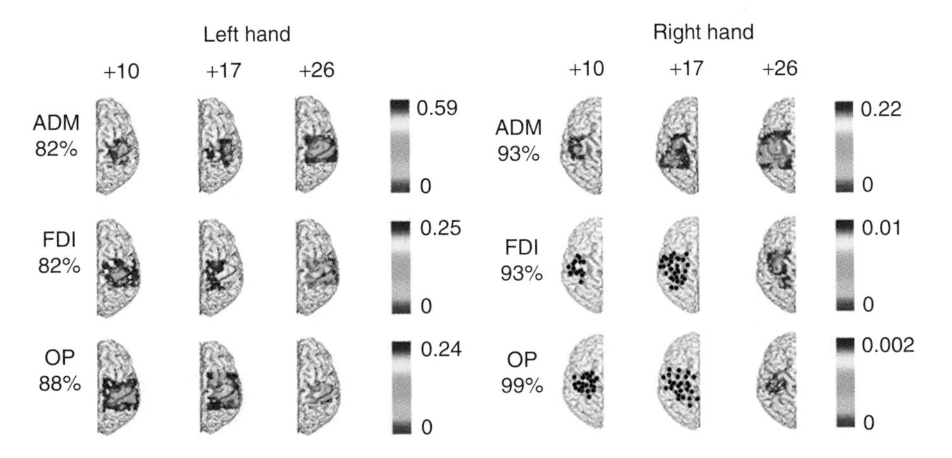

Fig. 19.1 Mean motor evoked potential (*MEP*) amplitudes recorded at each stimulated point and projected onto the 3-dimensional brain image of the bilateral allograft patient LB. Longitudinal progression of LB's left and right Abductor Digiti Minimi (*ADM*), First Dorsal Interosseous (*FDI*), Opponens Pollicis (*OP*) representations at 10, 17, and 26 months after graft. The amplitude of the recorded MEPs at each coil location is coded using a color map from *blue* (smaller MEP) to *red* (larger MEP). *Black dots* correspond to no MEP response at that stimulation intensity (Reprinted from Vargas et al.[19] With permission)

Schneeberger et al.[44] in a bilateral hand transplanted subject, with electromyographic signs of reinnervation first observed in the left hand 6 months after transplantation, followed by activation in the right hand 1 year after transplantation.[29]

Mercier et al 29 had previously shown by means of TMS that the central representation of digit movements is preserved in upper limb amputees. Recording from stump muscles, Reilly et al.[45] demonstrated that voluntary movements of the phantom hand trigger specific patterns of stump muscle activity, which differed from activity recorded in the same muscle groups during movements usually involving proximal limb musculature. Thus, our interpretation is that if central pathways survive deefferentation and deafferentation, the latent sensory-motor circuit might be functionally ready for the graft so that the intrinsic hand muscle representations could be reactivated in the recipient's brain as soon as some portion of the peripheral connections is reestablished.[45,46]

The relearning of finger movements is most likely another major factor influencing re-expansion and stabilization of the M1 hand representation. Finger movements that transplant recipients can perform immediately after graft are very different from the movements of an intact hand, and motor function gains are slow and require the subject to actively retrain fine hand movements. Hand-grafted patients are submitted to intense (twice a day) and varied rehabilitation training during the first year after the graft, continuing twice a week subsequently.[47] Thus, intensive physical rehabilitation for the grafted muscles probably influences the degree and range of reorganization found in M1, expanding the plastic possibilities of the hand allograft.

19.5 The Face Allograft Challenge

The idea that face allotransplantation could be used in reconstructive surgery was first supported by studies in animal models.[48] Human hand allograft had shown that the immunological obstacle of composite tissue transplantation could be overcome with usual immunosuppressive regimens.[49] Consequently, the remaining question in face transplantation was an ethical issue.[50] The French National Consultation Ethics Committee was the first worldwide committee to allow a partial functional allotransplantation to reconstruct the central part of a face, including the nose, both lips, and chin. On November 27, 2005, this surgery was performed on a young female patient in Amiens, France.[51] Since then, eight other face transplants have been performed in humans (review in[52]), and four of them have been extensively documented.[53-56] Although direct functional evidence on the brain reorganization following face lesion and allograft in humans is still poor, we will discuss below some basic tenets that might be of relevance.

19.5.1 What Hand Allograft Tells Us About Brain Plasticity Following Face Allograft

Although not yet directly demonstrated in humans, severe face injury might lead to plastic modifications in the corresponding face sensorimotor representations. For instance, one should expect that peripheral lesions in the face might correlate with an expansion of the hand sensorimotor representations over the original face territory. For instance, in rats, 2 h after a facial nerve transection, circumscribed regions of the forelimb representation expand medially into territory previously devoted to the vibrissae representation.[32] As stated before, these plastic changes have been extensively demonstrated in humans after hand amputation[14-16,23-25,28,29] and correlate with phantom limb pain.[30] Likewise, as for hand allograft,[18,20] face allograft might overturn the lesion-induced plasticity in the sensorimotor cortex.

Qualitative results gathered longitudinally from face allograft case reports[53-56] suggest that, as for hand allograft,[19,20,47] similar peripheral and central changes might be expected. For instance, as shown for hand allograft,[47] the return of sensory capacities in the grafted face has been shown to occur earlier in time than that of motor functions, with full sensory discrimination in the graft being identified from 3 to 6 months.[53-56]

Furthermore, as for hand allograft,[20] motor outcomes after face allograft seem to relate directly to the proper reconnection of peripheral nerves[54] with left hemibody motor functions progressing faster than those of the right side.[55] Likewise, as for the grafted hand,[20,47] physical therapy and rehabilitation protocols are of importance in regaining motor functions in the grafted face.[53,56] Interestingly, chronic pain associated to the missing face is reduced after allograft.[56] Pain in the missing arm has been classically correlated with the extent of plasticity in the sensorimotor cortex,[14,21-23,31] and its reduction could be taken as an evidence or plastic reorganization after face allograft. Taken together, these evidences point to similar plastic changes in the sensorimotor cortex after hand and face allograft.

As stated before, "the face organ"[57] is an indelible effector of sophisticate functions such as speech and emotional communication. Future studies might center on the brain mechanisms underlying the return of function after face allograft, focusing not only on the primary sensorimotor areas but also on brain plastic changes associated with complex cognitive functions such as self-image processing and the associated emotional states.

19.6 Conclusions

We conclude that, as for hand allograft, the process of motor cortical plasticity that follows face allograft might extend to the precise reconnection of sensory afferents as well as the recognition of newly transplanted muscles in order to allow novel motor synergies and functions restoration. Furthermore, unveiling the plastic changes that follow face allograft in the brain is of upmost importance due to the relevance of the "face organ" in the domain of social interaction.[58]

References

1. Hebb DO. *The Organization of Behavior*. New York: John Wiley & Sons, Inc.; 1949.
2. Kaas JH, Merzenich MM, Killackey HP. The reorganization of somatosensory cortex following peripheral nerve damage in adult and developing mammals. *Annu Rev Neurosci*. 1983;6:325-356.
3. Sanes JN, Suner S, Donoghue JP. Dynamic organization of primary motor cortex output to target muscles in adult rats. I. Long-term patterns of reorganization following motor or mixed peripheral nerve lesions. *Exp Brain Res*. 1990;79: 479-491.
4. Pons TP, Garraghty PE, Ommaya AK, et al. Massive cortical reorganization after sensory deafferentation in adult macaques. *Science*. 1991;252:1857-1860.
5. Recanzone GH, Merzenich MM, Jenkins WM, et al. Topographic reorganization of the hand representation in cortical area 3b owl monkeys trained in a frequency-discrimination task. *J Neurophysiol*. 1992;67:1031-1056.
6. Buonomano DV, Merzenich MM. Cortical plasticity: from synapses to maps. *Ann Rev Neurosci*. 1998;21:149-186.
7. Wall JT, Xu J, Wang X. Human brain plasticity: an emerging view of the multiple substrates and mechanisms that cause cortical changes and related sensory dysfunctions after injuries of sensory inputs from the body. *Brain Res Rev*. 2002; 39:181-215.
8. Florence SL, Taub HB, Kaas JH. Large-scale sprouting of cortical connections after peripheral injury in adult macaque monkeys. *Science*. 1998;282:1117-1121.
9. Merzenich MM, Kaas JH, Wall J, et al. Topographic reorganization of somatosensory cortical areas 3b and 1 in adult monkeys following restricted deafferentation. *Neuroscience*. 1983;8:33-55.
10. Wall JT, Kaas JH, Sur M, et al. Functional reorganization in somatosensory cortical areas 3b and 1 of adult monkeys after median nerve repair: possible relationships to sensory recovery in humans. *J Neurosci*. 1986;6:218-233.
11. Cohen LG, Bandinelli S, Findley TW, et al. Motor reorganization after upper limb amputation in man. A study with focal magnetic stimulation. *Brain*. 1991;114:615-627.
12. Kew JJ, Ridding MC, Rothwell JC, et al. Reorganization of cortical blood flow and transcranial magnetic stimulation maps in human subjects after upper limb amputation. *J Neurophysiol*. 1994;72:2517-2524.
13. Florence SL, Kaas JH. Large-scale reorganization at multiple levels of the somatosensory pathway follows therapeutic amputation of the hand in monkeys. *J Neurosci*. 1995;15: 8083-8095.
14. Flor H, Elbert T, Knecht S, et al. Phantom-limb pain as a perceptual correlate of cortical reorganization following arm amputation. *Nature*. 1995;8(375):482-484.
15. Flor H, Elbert T, Muhlnickel W, et al. Cortical reorganization and phantom phenomena in congenital and traumatic upper-extremity amputees. *Exp Brain Res*. 1998;119:205-212.
16. Roricht S, Meyer BU, Niehaus L, et al. Long-term reorganization of motor cortex outputs after arm amputation. *Neurology*. 1999;53:106-111.
17. Wu CW, Kaas JH. Reorganization in primary motor cortex of primates with long-standing therapeutic amputations. *J Neurosci*. 1999;19:7679-7697.
18. Lundborg G. Nerve injury and repair: a challenge to the plastic brain. *J Peripher Nerv Syst*. 2003;8:209-226.
19. Giraux P, Sirigu A, Schneider F, et al. Cortical reorganization in motor cortex after graft of both hands. *Nat Neurosci*. 2001;4:691-692.
20. Vargas CD, Aballéa A, Rodrigues EC, et al. Re-emergence of hand-muscle representations in human motor cortex after hand allograft. *Proc Natl Acad Sci USA*. 2009;28(106): 7197-7202.
21. Ramachandran VS, Hirstein W. The perception of phantom limbs. The D.O. Hebb lecture. *Brain*. 1998;121:1603-1630.
22. Flor H, Nikolajsen L, Staehelin Jensen T. Phantom limb pain: a case of maladaptive CNS plasticity? *Nat Rev Neurosci*. 2006;7:873-881.
23. Ramachandran VS. Behavioral and MEG correlates of neural plasticity in the adult human brain. *Proc Natl Acad Sci USA*. 1993;90:10413-10420.
24. Borsook D, Becerra L, Fishman S, et al. Acute plasticity in the human somatosensory cortex following amputation. *NeuroReport*. 1998;9:1013-1017.

25. Lotze M, Grodd W, Birbaumer N, et al. Does use of a myoelectric prosthesis prevent cortical reorganization and phantom limb pain? *Nat Neurosci.* 1999;2:501-502.
26. Ramachandran VS, Stewart M, Rogers-Ramachandran DC. Perceptual correlates of massive cortical reorganization. *NeuroReport.* 1992;3:583-586.
27. Dettmers C, Liepert J, Adler T, et al. Abnormal motor cortex organization contralateral to early upper limb amputation in humans. *Neurosci Lett.* 1999;263:41-46.
28. Ojemann JG, Silbergeld DL. Cortical stimulation mapping of phantom limb rolandic cortex. Case report. *J Neurosurg.* 1995;82:641-644.
29. Mercier C, Reilly KT, Vargas CD, et al. Mapping phantom movement representations in the motor cortex of amputees. *Brain.* 2006;129:2202-2210.
30. Karl A, Birbaumer N, Lutzenberger W, et al. Reorganization of motor and somatosensory cortex in upper extremity amputees with phantom limb pain. *J Neurosci.* 2001;21:3609-3618.
31. Lotze M, Flor H, Grodd W, et al. Phantom movements and pain. An fMRI study in upper limb amputees. *Brain.* 2001;124:2268-2277.
32. Huntley GW. Correlation between patterns of horizontal connectivity and the extent of short-term representational plasticity in rat motor cortex. *Cereb Cortex.* 1997;7:143-156.
33. Chen R, Corwell B, Yaseen Z, et al. Mechanisms of cortical reorganization in lower-limb amputees. *J Neurosci.* 1998;18:3443-3450.
34. Schwenkreis P, Witscher K, Janssen F, et al. Changes of cortical excitability in patients with upper limb amputation. *Neurosci Lett.* 2000;293:143-146.
35. Brenneis C, Loscher WN, Egger KE, et al. Cortical motor activation patterns following hand transplantation and replantation. *J Hand Surg Br.* 2005;30:530-533.
36. Roricht S, Machetanz J, Irlbacher K, et al. Reorganization of human motor cortex after hand replantation. *Ann Neurol.* 2001;50:240-249.
37. Neugroschl C, Denolin V, Schuind F, et al. Functional MRI activation of somatosensory and motor cortices in a hand-grafted patient with early clinical sensorimotor recovery. *Eur Radiol.* 2005;15:1806-1814.
38. Farne A, Roy AC, Giraux P, et al. Face or hand, not both: perceptual correlates of reafferentation in a former amputee. *Curr Biol.* 2002;12:1342-1346.
39. Kuiken TA, Marasco PD, Lock BA, et al. Redirection of cutaneous sensation from the hand to the chest skin of human amputees with targeted reinnervation. *Proc Natl Acad Sci USA.* 2007;104:20061-20066.
40. Wu CW, Kaas JH. Spinal cord atrophy and reorganization of motoneuron connections following long-standing limb loss in primates. *Neuron.* 2000;28:967-978.
41. Robinson GA, Madison RD. Motor neurons can preferentially reinnervate cutaneous pathways. *Exp Neurol.* 2004;190: 407-413.
42. Nguyen QT, Sanes JR, Lichtman JW. Pre-existing pathways promote precise projection patterns. *Nat Neurosci.* 2002;5: 861-867.
43. Lanzetta M, Pozzo M, Bottin A, et al. Reinnervation of motor units in intrinsic muscles of a transplanted hand. *Neurosci Lett.* 2005;373:138-143.
44. Schneeberger S, Ninkovic M, Piza-Katzer H, et al. Status 5 years after bilateral hand transplantation. *Am J Transplant.* 2006;6:834-841.
45. Reilly KT, Mercier C, Schieber MH, et al. Persistent hand motor commands in the amputees' brain. *Brain.* 2006;129: 2211-2223.
46. Reilly KT, Sirigu A. The motor cortex and its role in phantom limb phenomena. *Neuroscientist.* 2008;14:195-202.
47. Petruzzo P, Badet L, Gazarian A, et al. Bilateral hand transplantation: six years after the first case. *Am J Transplant.* 2006;6:1718-1724.
48. Siemionow M, Gozel-Ulusal B, Engin UA, et al. Functional tolerance following face transplantation in the rat. *Transplantation.* 2003;75:1607-1609.
49. Dubernard JM, Owen E, Herzberg G, et al. Human hand allograft: report on first 6 months. *Lancet.* 1999;353:1315-1320.
50. Agich GJ, Siemionow MJ. Until they have faces: the ethics of facial allograft transplantation. *Med Ethics.* 2005;31: 707-709.
51. Devauchelle B, Badet L, Lengelé B, et al. First human face allograft: early report. *Lancet.* 2006;368:203-209.
52. Siemionow M, Gordon CR. Overview of guidelines for establishing a face transplant program: a work in progress. *Am J Transplant.* 2010;10:1290-1296.
53. Dubernard JM, Lengelé B, Morelon E, et al. Outcomes 18 months after the first human partial face transplantation. *N Engl J Med.* 2007;357:2451-2460.
54. Guo S, Han Y, Zhang X, et al. Human facial allotransplantation: a 2-year follow-up study. *Lancet.* 2008;372:631-638.
55. Lantieri L, Meningaud JP, Grimbert P, et al. Repair of the lower and middle parts of the face by composite tissue allotransplantation in a patient with massive plexiform neurofibroma: a 1-year follow-up study. *Lancet.* 2008;372: 639-645.
56. Siemionow M, Papay F, Alam D, et al. Near-total human face transplantation for a severely disfigured patient in the USA. *Lancet.* 2009;374:203-209.
57. Siemionow M, Sonmez E. Face as an organ. *Ann Plast Surg.* 2008;61:345-352.
58. Frith C. Role of facial expressions in social interactions. *Philos Trans R Soc Lond B Biol Sci.* 2009;364:3453-3458.

Functional EEG Assessment of Face Transplantation

20

Vlodek Siemionow

Contents

V. Siemionow
Department of Plastic Surgery, Lerner Research Institute, Cleveland Clinic, Cleveland, OH, USA
e-mail: siemiov@ccf.org

Abstract Extensive traumatic loss of functional and composite structures of the face (skin, muscles, nerves, and bones) results in significant reorganization of the primary motor (M1) and somatosensory (S1) cortex. The first near-total US face transplant offers a unique opportunity to study the relearning process of integrating cortical representations of motor and sensory functions which were lost over a 5-year period following the patient's initial trauma. Using the functional EEG technique, we have found that trauma-induced cortical reorganization and associated loss of functions can gradually be reversed following face transplantation. The relearning of lost facial function governed by the somatosensory cortex confirms cortical plasticity and adaptation to the newly acquired functions. The restored functions in the transplant patient were found in the same areas of the motor cortex as in normal controls.

Abbreviations

ANOVA	Analysis of Variance
BESA	Brain Electromagnetic Source Analysis
CNS	Central Nervous System
EEG	Electroencephalography
EMG	Electromyography
ENG	Electroneurography
FFT	Fast Fourier Transform
fMRI	Functional Magnetic Resonance Imaging
IRB	Institutional Review Board
MEG	Magnetic Encephalography
MRCP	Motor-Related Cortical Potentials
M1	Primary Motor Cortex
MEP	Motor-Evoked Potentials

M.Z. Siemionow (ed.), *The Know-How of Face Transplantation*,
DOI: 10.1007/978-0-85729-253-7_20, © Springer-Verlag London Limited 2011

NP	Negative Potential
PET	Positron Emission Tomography
S1	Primary Somatosensory Cortex
SEF	Somatosensory-Evoked Magnetic Fields
SSEP	Somatosensory-Evoked Potentials
TMS	Transcranial Magnetic Stimulation

20.1 Background

Extensive traumatic loss of the functional and composite structures of the face (skin, muscles, nerves, and bones) creates significant reorganization in the primary motor (M1) and somatosensory (S1) cortex.[1] The first near-total face transplant in the USA offers the unique opportunity to study the relearning process, using EEG, of the motor and sensory functions which were lost over a 5-year period from the initial trauma to the time of face transplantation.

Changes that take place in the peripheral limbs of the body highly influence cortical organization of the adult human brain. Evidence of organizational changes in the human central nervous system (CNS) after successful transplantations of upper extremities was published recently.[2] After hand allograft, the primary sensory (S1) and primary motor (M1) cortical regions go through plastic modifications.[3] Cortical reorganization in the motor cortex was seen in the form of activation in the M1 pre- and post-graft of both hands.[4] Complex and sophisticated allograft transplants revealed positive but uneven motor and sensory recovery.[1] Functional magnetic resonance imaging (fMRI) had shown regained activity in previously inactive areas of the sensorimotor cortex after surgery.[4,5] While there are similarities to hand transplantation with both having sensory and motor components, the transplanted face needs to be treated as an analogous, but more complex organ.[6] To date, no electroencephalography (EEG) studies have been performed so far following human face transplantation. Here we present the results of an investigation into the functional reorganization of the somatosensory cortex in a patient following face transplantation using the well-established technique of EEG. The aim of this study is to determine, over time, the changes associated with regaining facial functions as assessed by physiological signals acquired from the sensorimotor cortex.

20.1.1 Animal Model Electrophysiological Studies

The first electrophysiological studies assessing the effects of face transplantation were performed using animal models. Successful transplants of full face, hemiface, and facial subunits were reported in several centers, but the functional recovery of this transplant remains unknown. A report of the functional aspects of the flap and its sensorimotor units was reported via an anatomical investigation of the whisker region following rat hemifacial transplantation.[7] This study was followed by an electrophysiological study consisting of electroneurography (ENG) and electromyography (EMG). Facial nerve conduction and voluntary motor activity of the transplanted hemiface were compared to potentials recorded in normal hemifaces. A qualitative observation of sensory and motor activity recovery after 6 weeks was also reported. The ENG and EMG results confirmed partial sensory recovery and moderate voluntary motor activity.[8,9]

The first brain recording of responses from a transplanted face in an animal model was done using a cortical sensory testing technique. Steel microelectrodes were used to record responses from the somatosensory cortex. The most extensive electrophysiological study in a rodent model was presented using both somatosensory-evoked potential testing (SSEP) and motor-evoked potential testing (MEP) for evaluation. Both modalities confirmed recovery of motor and sensory functions at 100 days post-transplant.[10]

In addition, functional recovery and cortical reintegration after hemifacial composite tissue allotransplantation was demonstrated through recording from the somatosensory cortex after whisker stimulation. The study examined animals transplanted with allogenic motor and sensory nerve appositions and showed progressing sensory adaptation following transplantation.[11]

20.1.2 Human Face Transplant Studies

The somatic motor and sensory representation in the cerebral cortex of humans was first described by Penfield and Boldrey in 1937.[12] Recently, facial representation in healthy human S1 was investigated using somatosensory-evoked magnetic fields (SEFs). Responses from tactile stimuli applied to different areas of the face (e.g., nose, cheek, and chin) were illustrated in details. However, there was a relatively large inter-individual variation contributing to a lack of statistical significance.[13]

Evidence of motor improvement in human, partial face transplantations is mostly limited to subjective physician observation. Investigators have found partial evidence that after rehabilitation training focused on the restoration of lip suspension and mouth occlusion, sensation recovered quickly.[14] But motor recovery was much slower and less robust in comparison. Facial movements were not visible up until 12 weeks after surgery. Even after 4 months, it was too early to evaluate true motor recovery except by progressive improvement in function.[15]

The electromyographic (EMG) evidence showing clear recovery of activity in the form of muscle contraction and response to facial nerve simulation was also reported. In the recording performed 3 months after surgery, an electroneuromyographic examination showed no evidence of reinnervation, with the exception of a minor motor response to facial nerve stimulation in the muscle. However in the next 6 months after surgery, EMG activity was observed and the testing after 1 year showed sensory reinnervation of the grafted skin.[16]

Functional mapping of the brain may provide insight into these cortical changes, and may indicate patterns of neuronal rehabilitation when applied to the transplanted face and associated brain plasticity. In the future, such information may have direct therapeutic consequences. It may help in the assessment of recovery progression and in planning proper training methods. Transcranial magnetic stimulation (TMS) was used in one patient who underwent bilateral hand transplantation 3 years after a traumatic amputation. The results of the study showed that the newly transplanted muscle could be integrated into the patient's motor cortex.[3]

In other cases of hand transplantation, functional magnetic resonance imaging (fMRI) was the method of choice because of its high-quality assessment of brain function changes.[4,5] But in this study, the patient was not approved for fMRI examination by the Institutional Review Board (IRB). Other evaluation methods can be fraught with problems. For example, positron emission tomography (PET) was not viable due to the risk of radiation exposure and increased cost. In the magnetic encephalography studies (MEG), additional activation was observed in the second somatosensory (S2) cortex, but the S2 responses are known to be less resilient to stimulus repetition than the S1 responses.[17]

Accordingly, the most feasible and realistic method for examining cortical reorganization in the brain is the well-established technique of electroencephalography (EEG) supported by computer-based assessment of quantitative power analysis, motor-related cortical potentials (MRCP), and analysis of source location. In addition to MRCP, the resting-state recording can provide information about brain activity that occurs when a person is not given a particular task to perform.[18] EEG methods are a low cost, easily accessible, and completely noninvasive approach to human investigation following face transplantation.

The aim of our research was to examine the changes associated with facial function recovery as assessed by physiological signals acquired from the sensorimotor cortex over time.

20.2 Subject and Methods

Three patients with extensive facial loss who had been screened as potential candidates for a face transplantation procedure and three age- and gender-matched healthy controls underwent electroencephalography (EEG) tests. One of the patients (a 48-year-old woman) would later be the recipient of the near-total face transplant 5 years after a traumatic gunshot blast to the face resulting in significant composite tissue loss. Cortical organizational changes in the sensorimotor cortex were recorded pre-transplant and 3, 6, 9, and 12 months post-transplant. Details of the surgical procedure have been published extensively elsewhere.[19] To obtain baseline, three age- and gender-matched healthy controls underwent testing under the same conditions and their results recorded.

20.2.1 Multi-Channel EEG Recording

Scalp EEG signals (referenced to the central electrode [Cz]) were recorded continuously from the scalp for the duration of rest periods and during performance motor tasks using a high-density, 128-channel EEG data acquisition system (Electrical Geodesics, Inc. Eugene, OR, USA). The 128 Ag-AgCl electrodes were arranged in a hat-like Geodesic Sensor Net and connected to each other by nylon strings. The distance between neighboring electrodes in the net was approximately 3 cm. The electrode net was applied to the head after it was soaked in an electrolyte solution consisting of 1-L distilled water, 1.5 teaspoons of potassium chloride, and a few drops of baby shampoo.[20] A small sponge in each electrode absorbed the liquid and served as a conducting media between the scalp and the electrode. An impedance map, based on impedance values of all the electrodes, was displayed on a computer monitor to inspect the quality of connection. If a particular electrode showed high impedance, adjustment (such as applying pressure or adding more water) was made to improve the connection. The EEG data recording did not begin until the impedance for all electrodes settled below 10,000 ohms. All channels were amplified (×75,000), filtered (0.01–100 Hz), digitized (250 sample/s), and recorded on the hard drive of a dedicated computer connected to the EEG acquisition hardware and software.

After obtaining informed consent, subjects were studied for approximately the same duration. They were placed in a sound-protected room and seated in an EEG investigation chair. During the entire session, subjects were instructed and observed by a technician. To avoid alpha rhythm occipital dominance during resting states and activation situations, subjects were instructed to keep their eyes open. Subjects were also carefully instructed to avoid any movements and relax to decreased body tension.

20.2.2 Experimental Procedures

Each subject underwent a brief training session prior to testing and in the middle of the experiment to ensure that the tasks were correctly performed (Fig. 20.1). A 2-min initial resting EEG (R1) was also recorded, during which the subject stayed in an undisturbed condition. Subjects were told not to move or tense the facial or neck muscles, as these activities would create signal disruption or "noise" in the EEG signals. In the next step, each subject was asked to repeat voluntary motor tasks of maximal facial animation including: smiling and pursing – the first task corresponding to mute *e* (like in word – *cheese*) and later *u* (as in word *wood*) vowels movements. The trigger was defined by low-volume external auditory cues generated by an S88 stimulator (Grass). Forty trials of each motor task were performed with enough break time to avoid motor fatigue (see Siemionow et al.[21]). While the patient could not physically perform the required tasks before and immediately after transplantation, she was asked to repeat them mentally (see Ranganthan et al.[22])

After completing the motor tasks, cortical responses to sensory tactile stimulation were recorded. All stimulations were performed by one experimenter who kept the stimulation parameters as uniform as possible between sessions; the stimulus intensity producing clear tactile sensation was tested before each recording. The subject was instructed to follow the stimulation and to keep alert. Repeatable tactile stimulations were applied to three chosen locations: upper lip, cheek, and nose (or places corresponding to these face parts – when applicable).

The final 2 min resting EEG (R2) was recorded immediately after completing the last motor task.

20.3 Data Processing and Analysis

Data were processed and sources modeled using the BESA program performed with the Brain Electromagnetic Source Analysis (BESA version 5.1 MEGIS Software GmbH, Graefelfing, Germany) system based on the FOCUS algorithm.[23] During offline processing, EEG signals were high-pass filtered at 3 Hz. All EEG data were also inspected visually. Recordings with artifacts caused by events such as eye blinks or head and body movements were excluded and the corresponding EMG signals discarded. The entire segment of the EEG

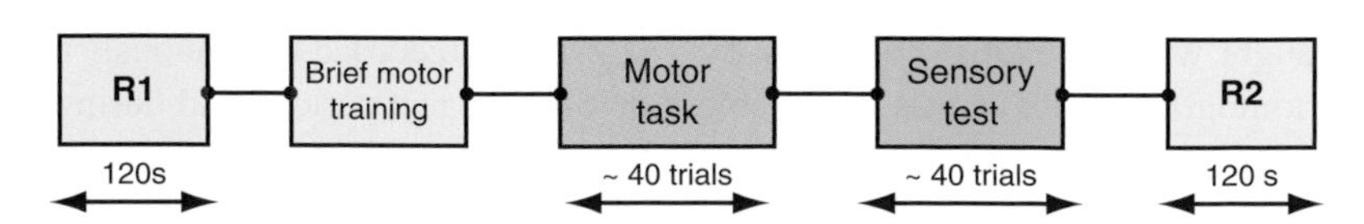

Fig. 20.1 Experimental paradigm. The first, initial 2-min rest period (*R*1); the battery of sensory tactile responses tests and the voluntary motor tasks of maximal facial animations: smiling and pursing followed by the second, final 2-min rest period (*R*2)

recordings was then marked and subsequently, the signals in each session were segmented into artifact-free epochs (256 samples in each epoch) without overlapping (mean of 40 clean epochs, ranging from 32 to 46 among the recording sessions).

Data were processed and sources modeled using the BESA program. The spatio-temporal dipole fit of the two sources during the motor task defines spatial vectors. The sources were localized and imaged in combination with model signal subspace of the measured data to separate the source activity of interest (sensorimotor) from other brain processes. Sequential brain source imaging (FOCUS and SBSI) methods were used to analyze the data.

20.3.1 EEG Amplitude Analysis

Each voluntary motor contraction of facial muscles was triggered in a defined time window of EEG (1,000 ms before and 3,000 ms after the trigger point). Time-locked averaging over all EEG windows corresponding to all trials was obtained (Fig. 20.2). Such an average is termed as a motor activity–related cortical potential (MRCP) because it is time-locked with each voluntary muscle contraction. This method eliminates random signals irrelevant to the motor task via averaging; only those signals directly related to the facial motor control (e.g., smiling and pursing) were retained. The MRCP amplitude was measured from the baseline (R1) to the peak of negative potential (NP). The baseline value was determined by taking the mean value of the R1 epoch (2 min) as a baseline. The NP (from baseline to peak) represents cortical activity directly related to the planning and execution of motor action.[24,25]

20.3.2 EEG Frequency Analysis

Power of the frequency bands as a function of time was derived to determine whether brain activity at a particular frequency at a given time changes with sensory and/or motor improvement.

Spectral analysis using a fast Fourier transform (FFT) was performed on raw EEG data associated with

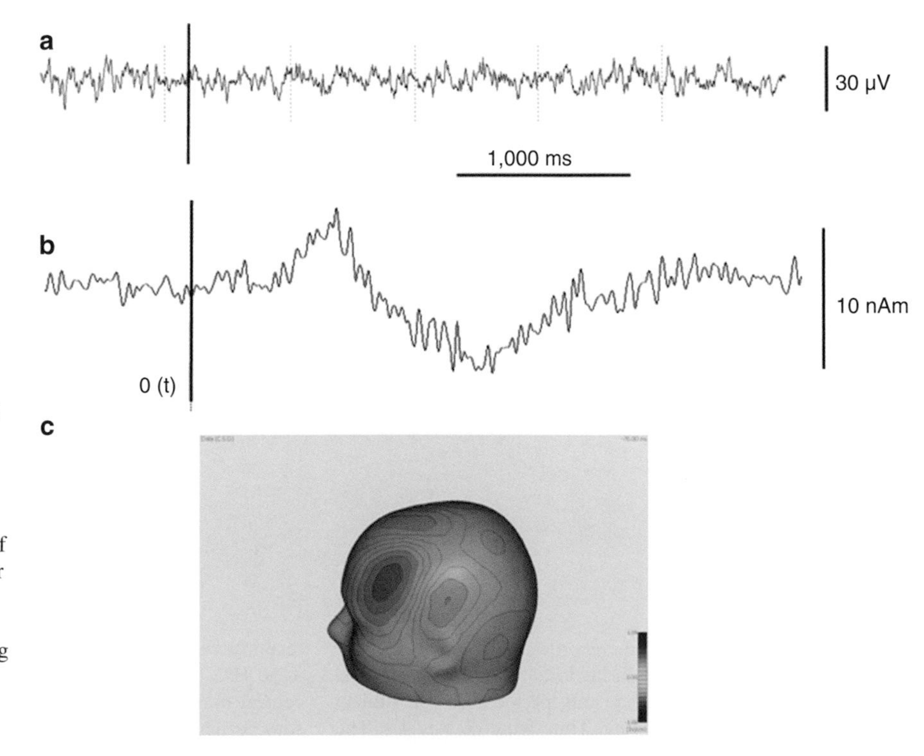

Fig. 20.2 Time-locked processing. The trigger time "0" (vertical line) was defined by an external auditory cue signal. (**a**) Raw EEG, (**b**) averaged brain activity, here presented in nAm as source strength magnitude, (**c**) map of brain activity during the motor task presented as current source density (C.S.D.) map (common reference). Resulting data can be further presented in time domain or frequency domain

R1 and R2 rest periods as well with tactile stimulation and performing the motor tasks. For each epoch, power for the following standard EEG frequency bands was derived: delta (0.5–4 Hz), theta (4–8 Hz), alpha (8–14 Hz), beta (14–35 Hz), and gamma (35–50 Hz). Subsequently, the mean relative power of each band across the number of epochs (e.g., 40 clean epochs for one task) was obtained.

20.3.3 Statistical Analysis

MRCP, power EEG frequency, and facial animation size were compared between tasks and between time points. Due to the repeated nature of the measurements, a two-way repeated measure analysis of variance (ANOVA) model was used. Separate analysis was performed for all dependent variables (MRCP, power, and size). A significance cut-off level of $p \leq 0.05$ was used.

20.4 Results

20.4.1 Resting EEG R1 and R2

The top view of the resting EEG recorded from one of the subjects is displayed in Fig. 20.3. Panel A shows data recorded during the first resting 2 min period R1 and on panel B data from the final resting period R2. Subjects kept eyes open during recording with a time window of 4 s. The responses have been filtered 3–50 Hz.

The resting EEG amplitude values were much higher than amplitudes recorded from healthy controls during an entire observation cycle (Fig. 20.4). In the beginning, the amplitude was sharply increased in the first 3 months after surgery (54%) and then dropped down toward control values, but remained consistently higher (29%). A similar propensity was observed in the post resting recording R2 with an even larger difference from controls.

20.4.2 Strength of Regional Sources

The frequency domain analysis demonstrated strength differences between hemispheres and a tendency in power frequency observed in regional sources (Fig. 20.5). The changes were most accentuated in the beta band (14–35 Hz) reflecting a complete healing process. After the first 3 months following surgery, the power was greater at 38.8 nAm²/Hz as compared to pre-transplantation (18.2 nAm²/Hz). After the next few months, the power value fell toward control values, but was still 8–15 nAm²/Hz above control levels.

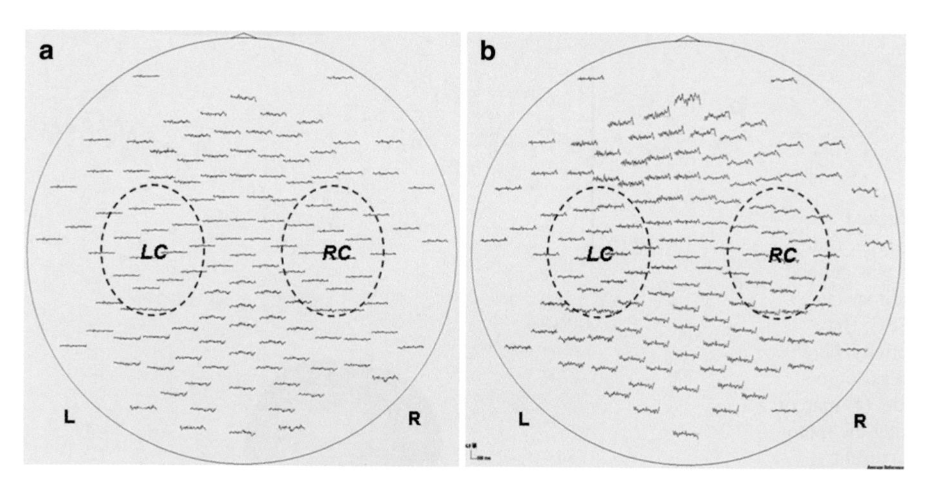

Fig. 20.3 Top view of the EEG recorded from one of the subjects. *Panel A*: Data recorded during initial resting 2 min period R1. *Panel B*: Data from final resting period R2. The subject kept eyes open during recording. Time window was 4 s. The responses collected from 128 channels and have been filtered 3–50 Hz. Two symmetrical oval lines mark the left and right central brain regions LC and RC associated with somatosensory cortices

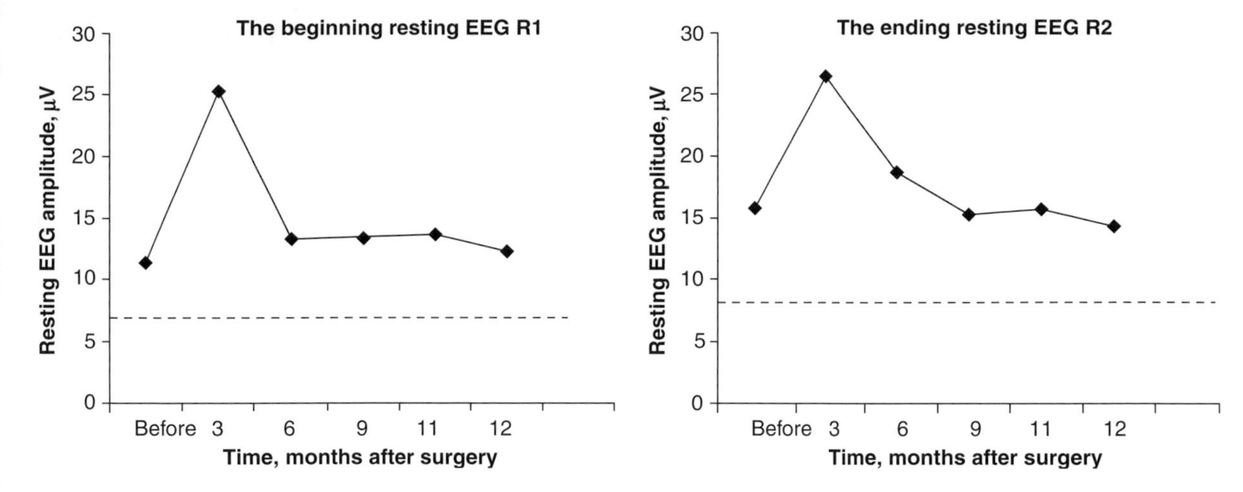

Fig. 20.4 The changes in the amplitude of resting EEG recorded during relearning process. *Left panel*: The first resting EEG R1 recording results and the *right panel* the second, final resting EEG R2. *Dashed lines* indicate data obtained from healthy controls

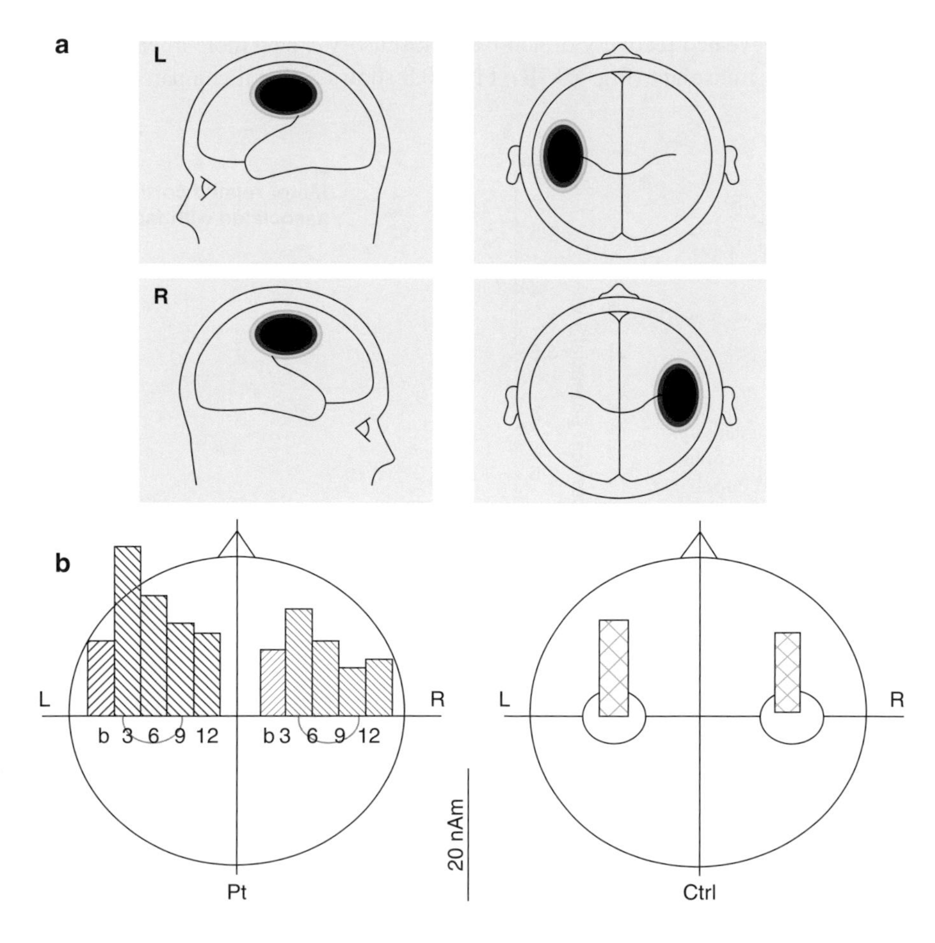

Fig. 20.5 The regional sources activities were recorded from the left (LC) and right (RC) central brain regions associated with contralateral and ipsilateral sensorimotor cortex (**a**). The signal magnitude reflects the estimated source activity if only one brain region is active. Source magnitude was greater in contralateral cortex (**b**). The spatial deconvolution FOCUS method was used to analyze the data

20.4.3 Motor-Related Cortical Potentials (MRCPs)

To quantify the MRCPs, two locations corresponding to the 10/20 international System were chosen: C3 and C4, which overlie the primary sensorimotor areas, contralateral and ipsilateral to the side performing the facial animations.[26] The chosen location showed prominent MRCP waveforms given the significant involvement of the underlying cortical fields of the primary sensorimotor cortex. An example of the obtained MRCPs is illustrated in Fig. 20.6 which shows the amplitude of the negative slope NP from the left and right brain locations in transplanted patients.

20.4.4 Responses to Tactile Stimulation

A top projection frequency map created from responses to tactile stimulation of the upper lip and of the nose is shown in Fig. 20.7 (A and B). Theta band changes in EEG power revealed recovery of non-existent sensory abilities after transplant (Fig. 20.7B). Fig. 20.8 shows subject response to stimuli applied to the upper lip before and after transplantation in all frequency bands. Power in the Alpha band for the first post-surgery sessions was higher by 37.4% (for 9 months time point). A similar change was observed in the Beta band when post-surgery power increased by 23.2%. Power in the Gamma band post-transplant decreased and remained low after 6 and 9 months.

20.4.5 Source Location Analysis

The spatio-temporal dipole fit of the two sources during the motor task defines the spatial vectors with a strong relationship to face and hand areas in the S1 (Fig. 20.9). This relationship consists with the homunculus drawn by Penfield and Rasmunssen[27] and other reports in humans. The current source location moved almost 11 mm toward normal location during time of retraining when compared with data recorded from healthy controls. The localization of the dominant brain side dipole was situated more inferior and lateral with regained lip localization if compared with value reported elsewhere.[4]

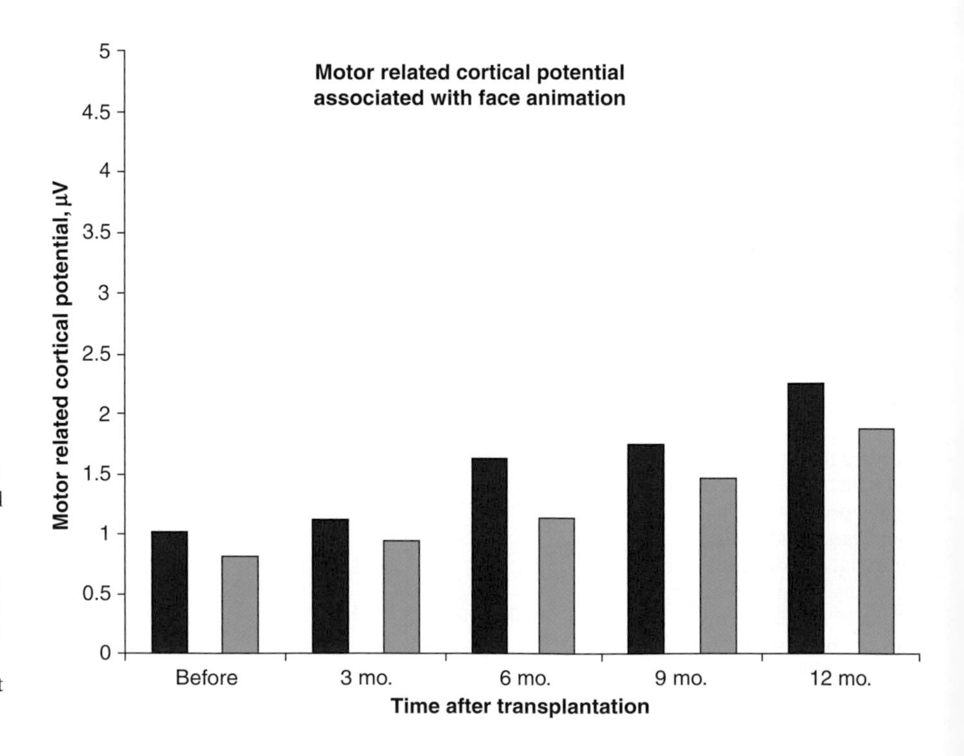

Fig. 20.6 Motor activity–related cortical potential (*MRCP*) derived during performing repeated voluntary motor tasks of facial animations: smiling task corresponding to mute *e* (like in word – *cheese*) vowel movements. Forty trials of each motor task were performed. While the patient could not physically perform the required tasks before and immediately after transplantation, he was asked to repeat them mentally

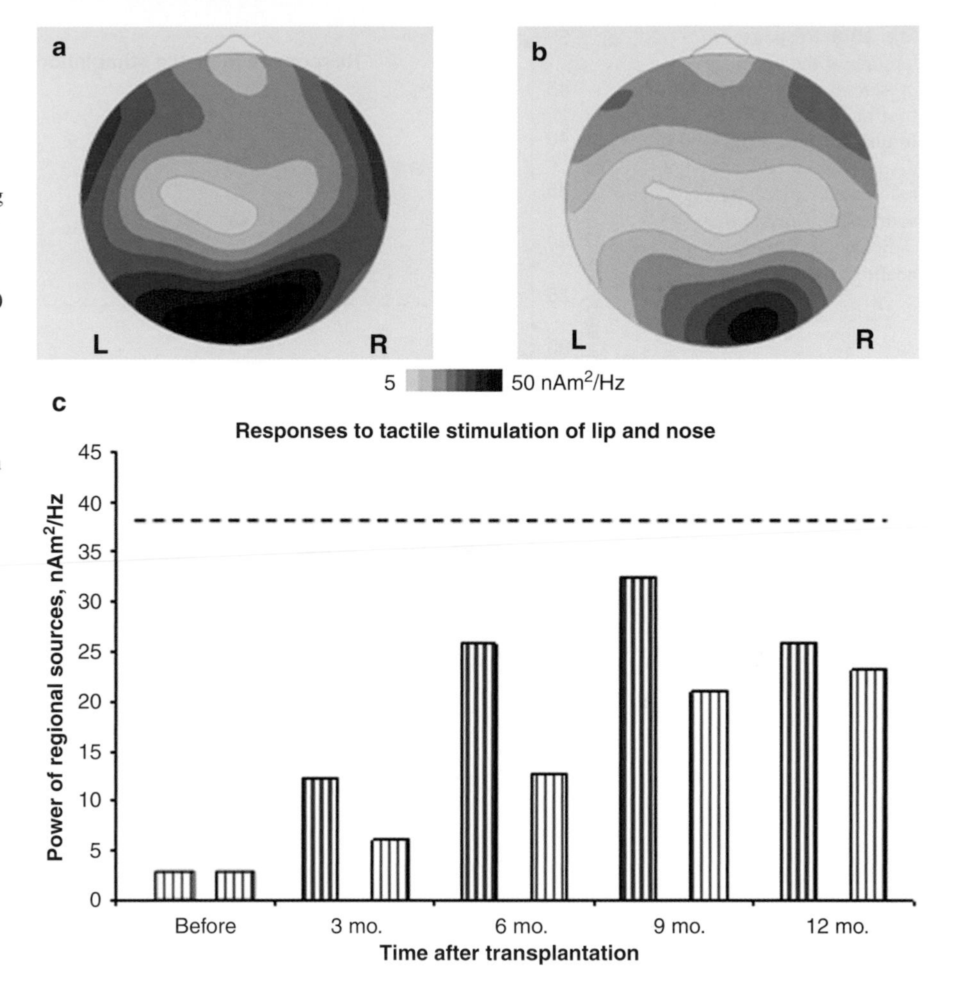

Fig. 20.7 *Panel A*: Responses to the tactile stimulation of the upper lip applied once every 10 s. *Panel B*: Responses of the nose once every 10 s. The subject kept eyes open during the recordings. The superimposed traces, measured with 128 channels EEG system, are each averages of about 40 responses. Power of frequency (beta band) maps expressed as relative power ($\mu V^2/Hz$). The maps show that the power is distributed more centrally in patient than in control. *Panel C*: Changes in power of frequency EEG signal recorded in nose tapping during healing time. *Dashed lines* indicate data obtained from healthy controls

20.5 Conclusion

This study examined EEG signals in a patient that underwent near-total face transplantation. The intensive surgical procedure resulted in clear changes of not only the amplitude or power of resting EEG but also a demonstrable recovery in sensorimotor abilities after transplantation. The spontaneous resting-state brain activity was influenced by various factors including tasks performed after (as R1) or before (R2) the recording. The activation areas were bilateral to lip stimuli, in full agreement with other studies using different types of somatosensory stimuli on intact subjects.[26]

This complex face transplant helped to restore the effectiveness of original connections and overtime, allowing for near-normal facial function and cortical reorganization. In the present study, we recorded remarkable changes in cortical reorganization following a near-total face transplant using a well-established EEG technique. We found that trauma-induced cortical reorganization was reversed following face transplantation. On the basis of above findings, we conclude that the process of cortical plasticity and its ability to reach the transplanted complex organ may occur following face transplantation up to 5 years following the initial trauma.

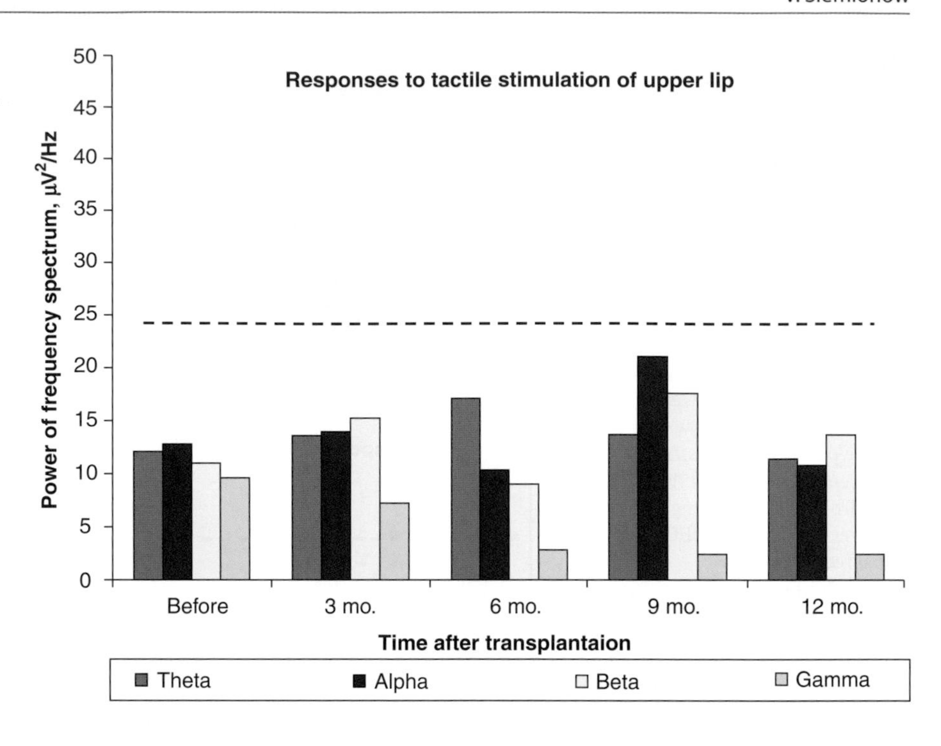

Fig. 20.8 Responses to tapping of the upper lip presented in form of power of four band. Power of frequency maps expressed in nAm^2/Hz as relative strength associated with regional source LC. *Dashed lines* indicate data obtained from healthy controls

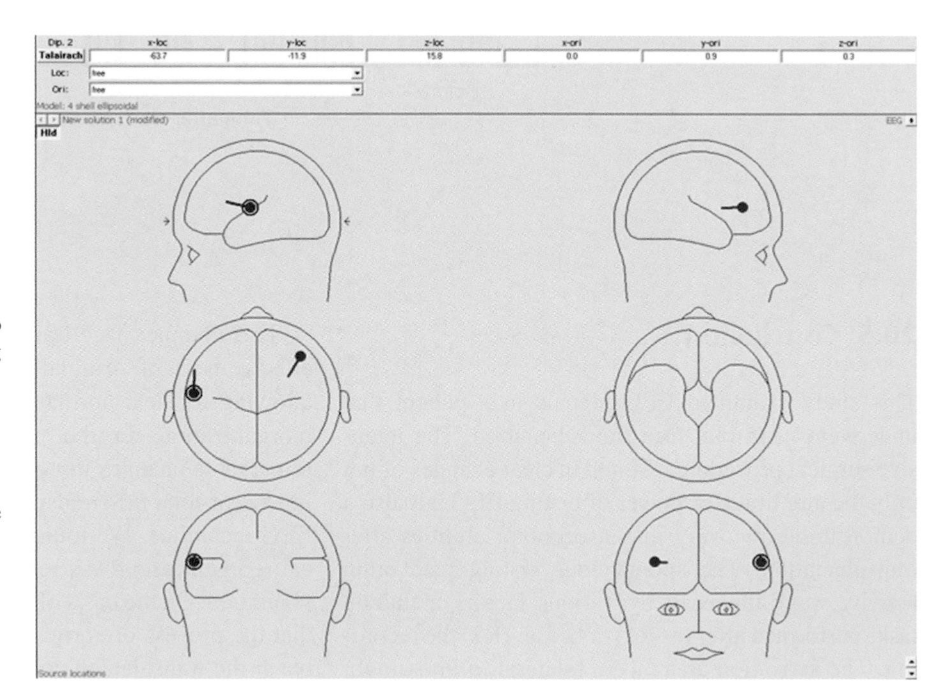

Fig. 20.9 The spatio-temporal dipole fit of the two sources (hemispheres) during the motor task defines spatial vectors. The sources were localized and imaged in combination with model signal subspace of the measured data to separate the source activity of interest (sensorimotor) from other brain processes. Sequential brain source imaging (SBSI) method was used to analyze the data

References

1. Gander B, Brown CS, Vasilic D, et al. A new frontier in transplant and reconstructive surgery. *Transpl Int*. 2006;19: 868-880.
2. Cavadas PC, Landin L, Ibañez J. Bilateral hand transplantation: result at 20 months. *J Hand Surg Eur*. 2009;34: 434-443.
3. Vargas CD, Aballe A, Rodrigues EC, et al. Re-emergence of hand-muscle representations in human motor cortex after hand allograft. *PNAS*. 2009;106(17):7197-7202.
4. Giraux P, Sirigu A, Fabien S, Dubernard J-M. Cortical reorganization in motor cortex after graft of both hands. *Nat Neurosci*. 2001;4:691-692.
5. Lanzetta M, Pozzo M, Bottin A, Marletti R, Farina D. Reinervation of motor units in intrinsic muscles of transplanted hand. *Neurosci Lett*. 2005;373:138-143.
6. Siemionow M. Sonmez E Face as an organ. *Ann Plast Surg*. 2008;61:345-352.
7. Landin L, Cadavas PC. Thy mystacial pad flap: a functional facial flap in rats. *Ann Plast Surg*. 2005;56:107-108.
8. Landin L, Cavadas PC, Gonzalez E, Rodriguez JC, Caballero A. Functional outcome after facial allograft transplantation in rats. *J Plast Reconstr Aesthet Surg*. 2008;61:1034-1043.
9. Landin L, Cavadas PC, Gonzalez E, Caballero-Hidalgo A, Rodriguez-Perez JC. Sensorimotor recovery after partial facial (mystacial pad) transplantation in rats. *Ann Plast Surg*. 2009;63:428-435.
10. Zor F, Bozkurt M, Nair D, Siemionow M. A new composite midface allotransplantation model with sensory and motor reinnervation. *Transpl Int*. 2009;23(6):649-656. Epub 2009 Dec 21.
11. Washington KM, Solari MG, Sacks JM, et al. A model for functional recovery and cortical reintegration after hemifacial composite tissue allotransplantation. *Plast Reconstr Surg*. 2009;123(2 Suppl):26S-33S.
12. Penfield W, Boldrey E. Somatic motor and sensory representation in the cerebral cortex of man as studied by electrical stimulation. *Brain*. 1937;60:389-433.
13. Nguyen BT, Tran TD, Hoshiyama M, Inui K, Kakigi R. Face representation in the human primary somatosensory cortex. *Neurosci Res*. 2004;50:227-232.
14. Dubernard J-M, Lengelé B, Morelon E, et al. Outcomes 18 months after the first human partial face transplantation. *N Engl J Med*. 2007;357(24):2451-2460.
15. Devauchelle B, Badet L, Lengelé B, et al. First human face allograft: early report. *Lancet*. 2006;368:203-209.
16. Lantieri L, Meningaud JP, Grimbert P, et al. Repair of the lower and middle parts of the face by composite tissue allotransplantation in a patient with massive plexiform neurofibroma: a 1-year follow-up study. *Lancet*. 2008;372:639-645.
17. Hari R, Karhu J, Hamalainen M, et al. Functional organization of the human first and secondsomatosensory cortices: a neuromagnetic study. *Eur J Neurosci*. 1993;5:724-734.
18. Kounios J, Fleck JI, Green DL, et al. The origins of insight in resting-state brain activity. *Neuropsychologia*. 2008;46: 281-291.
19. Siemionow M, Papay F, Daniel A, et al. Near-total human face transplantation for a severely disfigured patient in the USA. *Lancet*. 2009;374:203-209.
20. Electrical Geodesics Incorporated (EGI). *200 Technical Manual*. Eugene Oregon: Electrical Geodesics, Inc; 2006.
21. Siemionow V, Fang Y, Calabrese L, Sahgal V, Yue GH. Altered central nervous system signal during motor performance in chronic fatigue syndrome. *Clin Neurophysiol*. 2004;115:2372-2381.
22. Ranganathan VK, Siemionow V, Liu JZ, Sahgal V, Yue GH. From mental power to muscle power-gaining strength by using the mind. *Neuropsychologia*. 2004;42:944-956.
23. Berg P, Scherg M. A fast method for forward computation of multiple-shell spherical head models. *Electroencephalogr Clin Neurophysiol*. 1994;90:58-64.
24. Hallet M. Movement-related cortical potentials. *Electromyogr Cllin Neurophysiol*. 1994;34:5-13.
25. Siemionow V, Yue GH, Ranganathan VK, Liu JZ, Sahgal V. Relationship between motor activity-related cortical potential and voluntary muscle activation. *Exp Brain Res*. 2000;133:303-311.
26. Jasper HH. Report of the committee on methods of clinical examination in electroencephalography. Appendix: The ten twenty electrode system of the International Federation. *Electroencephalogr Clin Neurophysiol*. 1958;10:370-375.
27. Penfield W, Rasmussen T. *The Cerebral Cortex of Man A Clinical Study of Localization of Function*. New York: McMillan Press; 1950.

21 Assessment Methods of Sensory Recovery after Face Transplantation

Grzegorz Brzezicki and Maria Z. Siemionow

Contents

M.Z. Siemionow (✉)
Department of Plastic Surgery, Cleveland Clinic,
Cleveland, OH, USA
e-mail: siemiom@ccf.org

Abstract Sensory recovery is a prerequisite of successful functional rehabilitation after composite tissue face allograft transplantation including: speech, facial mimetics, swallowing, chewing, and drinking. Half of the patients out of four with reported outcomes received primary sensate grafts, while the other two had been transplanted with nonsensate facial allografts. Each transplant also significantly differed from each other in the area of the skin transplanted, tissues included, age of the recipient, and methods of sensory recovery assessment. While no direct comparisons can be made, some general conclusions can be drawn. All four patients achieved good recovery of light touch, punctate touch, and heat/cold sensation in follow-up times up to 2 years, starting as early as 2 weeks. Patients were not directly tested for pain sensation recovery; however, three of four needed regional anesthesia for routine graft biopsies 2–5 months posttransplantation. Primarily, nonsensate grafts were able to achieve comparable sensory results to innervated transplants. As the number of face transplant grows worldwide, more standardized approach to sensory testing should be initiated to allow easier comparisons, predict achievable recovery, and tailor rehabilitation to specific patient.

Abbreviations

DFNS	German Research Network on Neuropathic Pain
PSSD	Pressure-specified sensory device
PD	2 Point Discrimination
QST	Quantitative Sensory Testing

M.Z. Siemionow (ed.), *The Know-How of Face Transplantation*,
DOI: 10.1007/978-0-85729-253-7_21,

21.1 Introduction

The major goal of face transplantation is to attain good cosmetic result, which enables the patient to reintegrate with the society. However, without the restoration of motor and sensory function, the composite tissue allograft would act as an unanimated mask. Functional restoration directly impacts recipient's quality of life and accounts for the worthiness of such risky procedure with overall 18% death rate.[1] It was shown that, in other reconstructive procedures involving musculocutaneous or muscle flaps with split-thickness skin grafts, restoration of sensation is a prerequisite for full rehabilitation of the function. Sensation plays an important role in reconstructed areas like: hand (e.g., proper grip and protective sensation), oral cavity (e.g., protective sensation, fluent speech, and swallowing), and foot (e.g., effective gait and protective sensation).[2]

Out of 11 procedures performed only four patients had their functional outcomes reported so far[3-8] (Table 21.1). Both French patients received primary innervated flaps, at least partially, while the Chinese and American patients were transplanted with sensory noninnervated flaps. Facial nerve repair was attempted in all cases. Sensory recovery was assessed in different time-points with the longest follow-up of 2 years (Table 21.1)

Majority of the face is innervated by the trigeminal nerve with the exception of angle of the jaw, which is supplied by the anterior branch of the great auricular nerve originating from C2 to C3 nerve roots. Trigeminal nerve provides afferent innervation to the face by three major branches: the ophthalmic nerve (V_1), the maxillary nerve (V_2), and the mandibular nerve (V_3) (Fig. 21.1). Mandibular nerve also carries motor fibers to the mastication muscles, palate, and floor of the mouth muscles. The primary sensory receptors for touch/position and pain/temperature, types of nerve fibers, and assessment methods are presented in Table 21.2. The areas of cutaneous distribution (dermatomes) of the three branches of the trigeminal nerve have sharp borders with relatively little overlap compared to considerable overlap observed in dermatomes in the rest of the body.

Recent years brought major advancements in the field of sensory testing, predominantly as a function of increased interest in the diagnosis and treatment of peripheral neuropathies. Comprehensive sensory testing – Quantitative Sensory Testing (QST) – includes evaluation methods that test touch (light touch, punctate touch), pain, vibration, pressure, and temperature sensation.[9] The results of QST are expressed in thresholds for each modalities tested. Recently, German Research Network on Neuropathic Pain (DFNS) developed standardized protocol and reference values for testing upper and lower extremities, which enables reliable comparisons between patients and centers.[10,11] In the emerging field of face transplantation, no such standards for assessment of sensory recovery were set so far. Probably, the use of full QST battery for testing face sensation would be too cumbersome and time consuming; however, few standardized tests would certainly be beneficial for patients' monitoring.

21.2 Methods

All discussed face transplant recipients underwent assessment of sensitivity of the graft, specifically touch and temperature sensation at various time-points.

21.2.1 Discriminative Touch

Semmes–Weinstein Monofilament test is the most commonly used semiquantitative method of one-point discrimination (1PD) sensation assessment. It consists of series of nylon filaments embedded in plastic handle. Filaments are calibrated to provide a specified force measured in grams. Monofilaments are commercially available in sizes from 1.65 to 6.65 stratified by increasing predetermined pressure needed to bend the filament when applied to skin or mucosa.[12,13] However, the actual pressure delivered to the skin surface is the force divided by the surface area of the filament and may vary depending on the angle with the skin. Semmes–Weinstein test is relatively fast and easy technique of skin sensitivity evaluation but it also carries inherent interobserver variability, as one has to judge application of correct force by looking for bending of the filament.[14] The possibility of misjudgment is even greater in very sensitive (low sensory thresholds) regions of human body, face (lips, tip of the nose) being one of them.[15,16] Moreover, this method does not provide accurate static one-point discrimination (1PD) pressure thresholds (semiquantitative method) and is unsuitable for two-point discrimination (2PD) testing.

Table 21.1 Summary of the face transplant procedure with sensory outcomes reported

Team leader	Recipient	Follow-up (months)	Sensory nerves repaired	Motor nerves repaired	Sensory assessment methods	Touch sensation	Temperature	Pain
Dubernard	F/38	18	Bilateral V_2 and V_3	Mandibular branch of the facial nerve	Sensation (SW test), heat and cold sensation	Started at 10 weeks (skin and mucosa) 14 weeks almost all graft 6 months almost normal	Started at 2 weeks; almost all graft by month 4 All graft month 6	2 months
Zhang	M/30	24	No sensory nerves repaired	Right facial nerve (suboptimal)	Sensation (SW test), heat and cold sensation	Recovery at 3 months (skin and mucosa)	Full recovery at 8 months	N/A
Lantieri	M/29	12	Bilateral V_2	Bilateral facial nerves	Sensation (QST), heat and cold	Started at 3 months Ongoing improvement up to 12 months	Started at 3 months; ongoing improvement up to 12 months	4 months
Siemionow	F/49	9	No sensory nerves repaired	Bilateral facial nerves (donor's vagus and hypoglossal nerve cable grafts)	Sensation 1PD, 2 PD (PSSD), heat and cold, LT	At 3 months LT sensation present at the margins of the graft, 2PD 10–13 mm or absent At 9 months over almost whole graft, 2PD 6–8 mm	At 3 months temperature sensation at the margins of the graft, at 9 months over almost whole graft	5 months

1PD one point discrimination, *2PD* two point discrimination, *SW test* Semmes–Weinstein Monofilament test, *QST* Quantitative Sensory Testing, *LT* light touch

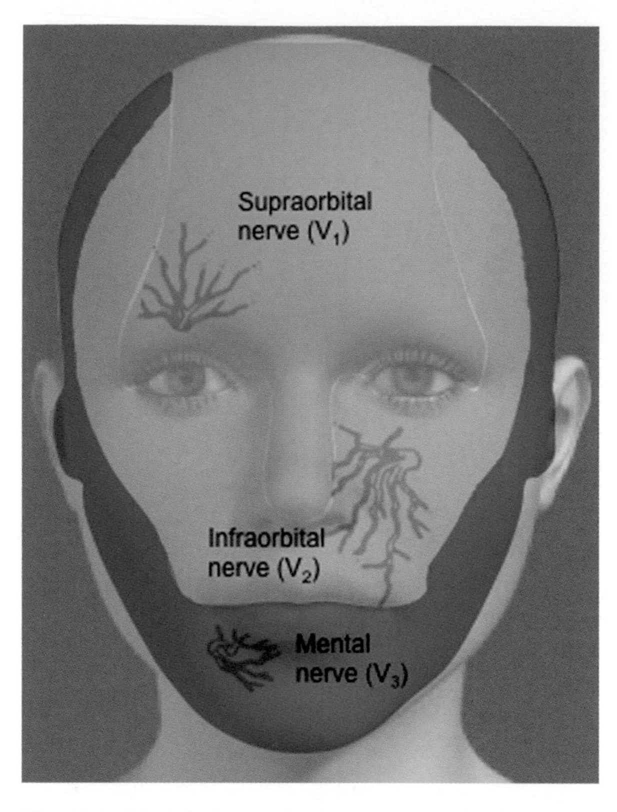

Fig. 21.1 Trigeminal nerve innervation of the face (*blue* ophthalmic nerve, *green* maxillary nerve, *red* mandibular nerve). Major sensory skin branches: supraorbital nerve, infraorbital nerve, and mental nerve are also shown

The 2PD test is a valuable diagnostic method in early stages of nerve dysfunction (entrapment) as it assesses innervation density in the analyzed area.[17] However, it is also commonly used in hand transplant patients as well as after nerve injury repair being a sensitive method of sensory recovery measurement.[18-21] Standard Weber test for 2PD, using Mackinnon–Dellon Disk-Criminator (Kom Kare Company, Middletown, Ohio) or other similar handheld devices is commonly incomparable between institutions or examiners since the prongs used for the assessment must be applied to skin in exactly the same manner to obtain repeatable results, thus the standardization is difficult.[21-23] Pressure-Specified Sensory Device™ (PSSD, Sensory Management Services LLC, Baltimore, MD) was introduced in 1992 by Dellon et al.[24] in an effort to eliminate the effect of examiner on the results. With this method, multiple authors were able to show reliable results in different anatomical parts of the body including: hand,[17,25-28] lower extremity,[17,28] and head and neck.[12,14] In few studies, PSSD was able to detect peripheral nerve injury (posttraumatic or entrapment), while electrodiagnostic tests were inconclusive.[17,29,30] PSSD is a handheld device with a probe consisting of two blunt prongs with adjustable interprong distance and sensitive transducers to measure and record the perception thresholds of pressure on the surface of the body in grams per square millimeter (g/mm^2) (Fig. 21.2). Each PSSD skin

Table 21.2 Sensory modalities transmitted by the trigeminal nerve and appropriate assessment methods

Type of sensation	Receptors involved	Transmitting nerve fibers	Assessment method
Light touch	Meissner's Pacinian, hair follicle	Aβ	Cotton swab, brush
Blunt (punctate)	Merkel, Ruffini	Aβ	Von Frey filaments, S-W monofilaments, PSSD
Sharp (punctate)	Unencapsulated	Aδ	Pin
Vibration	Pacinian	Aβ	Tuning fork
Deep pressure	Intramuscular afferents	Type III and IV	Pressure algometer
Cold	Unencapsulated	C	Ice-cold object
Heat	Unencapsulated	C, Aδ	Heated object

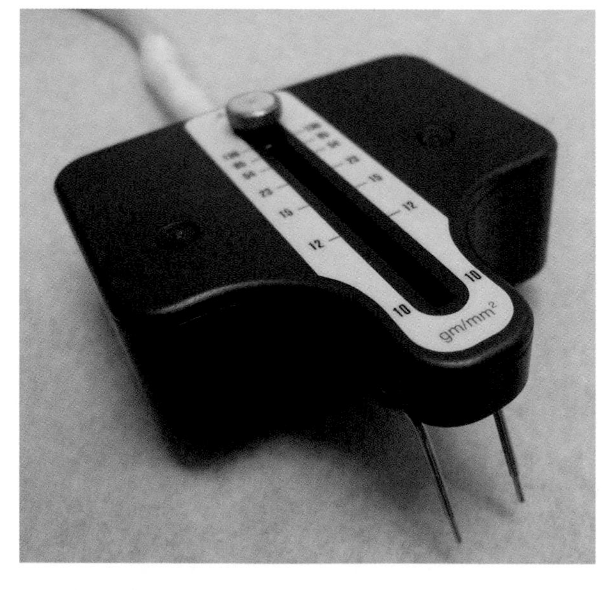

Fig. 21.2 PSSD device with two blunt prongs with adjustable interprong distance

test is repeated five times and the results are later averaged by computer software. Similarly, sensory threshold for one point (1PD) can be measured using the same contraption, but touching patient's skin with one prong only.

In three (Dubernard, Zhen, Lantieri) out of four reported sensory outcomes of face transplant recipients punctate blunt touch was assessed by the Semmes–Weinstein Monofilament test (Table 21.1). Normal pressure sensitivity thresholds for forehead and midface were recorded with this method by Weinstein.[31] Dubernard et al.[6] observed first signs of sensation restoration at 10 weeks. Four weeks later almost whole flap was sensate, while at 6 months sensitivity to touch was graded as almost normal. Chinese group reported sensory discrimination of the flap at 3 months.[8] In the second French face transplantation patient, first signs of touch sensation restoration were noted at 3 months and improved up to 1 year.[5] Siemionow's group assessed discriminative touch recovery by PSSD. In an effort to provide reliable results over time, based on area of the flap and previously reported techniques of facial sensitivity assessment,[13-15] same specific areas were tested symmetrically on both sides of the patient's face each time (Fig. 21.3). Using this methodology, Cleveland Clinic group was able to show ongoing punctate touch recovery by return of discriminative touch sensation with decreasing two-point distance[3,4] and decreasing sensory thresholds for 1PD and 2PD. Dellon et al.[32] published normative values for facial sensitivity, static and moving 1PD and 2PD (areas supplied by trigeminal nerve) obtained from 42 healthy subjects using PSSD. There was a significant difference in trigeminal nerve sensitivity in different age groups, defined as older and younger than 45 years. However, due to relatively small number of tested subjects these results cannot be treated as definite reference values. Moreover, assessment of branches of great auricular nerve would be valuable as the preauricular area might be reconstructed in full face transplantation.[3]

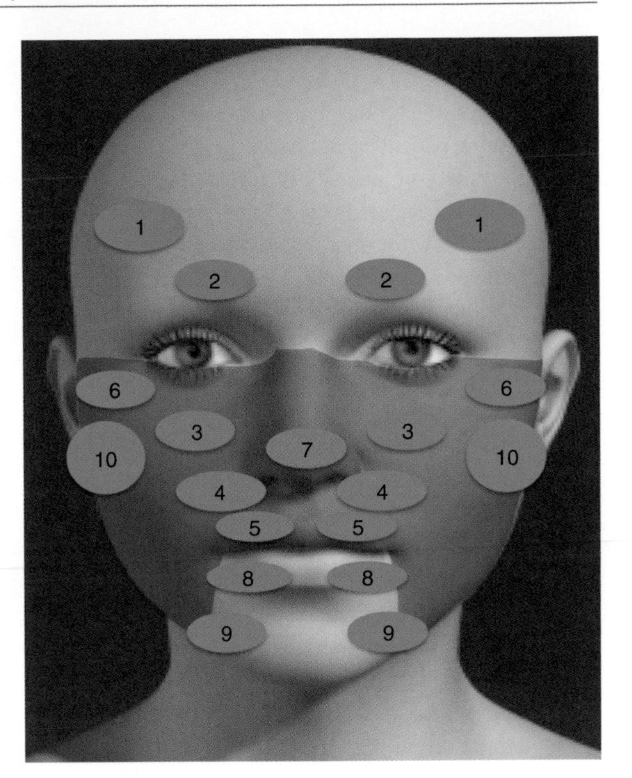

Fig. 21.3 Face allograft areas tested for 2PD recovery in Cleveland Clinic patient. Violet color represents the area of the composite tissue graft. *V1*: 1 – lateral forehead (3 cm lateral to the midpupillary line and 3 cm above eyebrows line) – supraorbital nerve, 2 – medial forehead (midpupillary line, 1 cm above eyebrows line) – supraorbital nerve; *V2*: 3 – maxilla (midpupillary line, 3 cm below the center of the pupil) – infraorbital nerve, 4 – nasolabial fold (in the middle of the fold) – infraorbital nerve, 5 – upper lip (1.5 cm lateral from the midline, 0.5 cm above the upper vermillion border) – infraorbital nerve, 6 – zygoma (3.5 cm lateral to the midpupillary line, 2 cm below the center of the pupil) – zygomaticofacial nerve, 7 – tip of the nose – external nasal nerve; *V3*: 8 – lower lip (1.5 cm lateral from the midline, 0.5 cm below the lower vermillion border) – mental nerve, 9 – mentum (medial epicanthal line, 2 cm below lower vermillion border) – mental nerve; *Great auricular nerve:* 10 – preauricular area (2 cm below the tip of the tragus, 2.5 cm medial to the tragus) – anterior branch of great auricular nerve

21.2.2 Light Touch

Light touch evaluation is performed by gently touching the skin over the tested area with a cotton swab or a special brush (Somedic brush – Horby, Sweden). The response may be measured as present/not present[3] or graded based on sensation intensity from 0 to –10.[33]

The only assessment of light touch recovery was reported by Cleveland Clinic group. Cotton swab test revealed ongoing reinnervation of the graft from the margins at 3 months, a typical pattern for the noninnervated free flaps. By 6 months almost all graft was innervated, leaving small areas on the cheeks and below the nose insensate.[1] In face and neck reconstruction with free uninnervated free flaps, light touch sensation recovery was the last to appear or was not achieved at all.[34,35]

21.2.3 Pain

Pain sensation can be simply tested using pinprick stimulus and noting patient's response as present/not present or grading it based on intensity of reported sensation similarly to light touch.[33] However, there are well-developed tests, which allow for quantitative testing for mechanical or temperature-related pain.[10,11,33] Mechanical pain can be tested as either pinch or pressure pain thresholds using different setups of pressure algometer. While it seems that pressure pain can be easily tested on the face, the pinch pain evaluation might be difficult to assess in different areas of the face as there is not sufficient amount of skin to be squeezed as in evaluation of fingers.[33] Cold and warm pain detection thresholds can be measured by different types of devices, which utilize thermodes to deliver heat or cold to the tested skin area in increments (e.g., ~5°C/s) up to designated cutoff temperature.[10,11]

None of the transplanted patients was tested specifically for pain sensation recovery. However, the restoration of pain sensation was noted when the patients were subjected to routine skin or mucosa biopsy. At 2 months posttransplant, Deavuchelle et al. reported the need for local anesthesia when taking biopsies. Lantieri and Siemionow reported similar findings at 4 and 5 months, respectively. Later pain sensation restoration in Cleveland Clinic patient was expected as the transplanted composite tissue allograft was not sensory innervated. However, it is impossible to draw any firm conclusions from these cases due to substantially different allografts and lack of knowledge where exactly those biopsies were taken in particular patients, as one, for example, can expect faster pain sensation recovery in noninnervated flaps at the periphery of the graft while the center of the flap will become sensate significantly later. In the battery of quantitative sensory testing (QST), there are pain-specific tests available for clinical use including pressure pain threshold and pinch pain threshold. Baad-Hansens et al. found it useful, as a part of complex sensory assessment in detection of sensory disturbances after orthognathic surgery.[33]

21.2.4 Temperature

Similarly to pain assessment, temperature sensation recovery can be evaluated in simple nonquantitative evaluation – present/absent, semiquantitative – graded 1–10 or using quantitative threshold tests from the QST armamentarium. When looking for temperature sensation threshold, typically lower temperature increments settings (1°C/s) compared to temperature pain sensation are used with the same thermode device.[10,11]

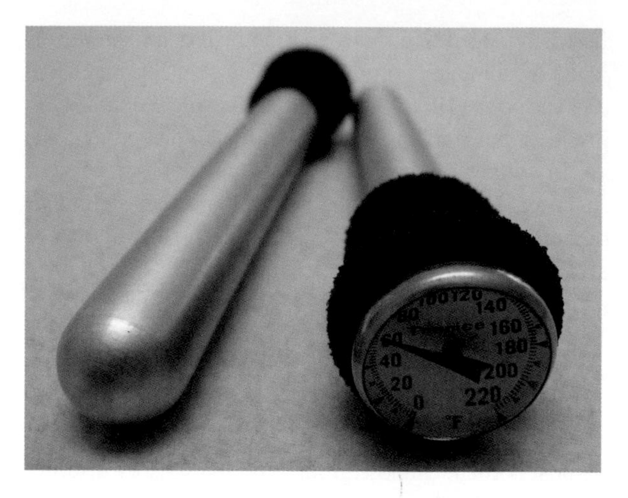

Fig. 21.4 Calorymetric metal cylinders used for evaluation of temperature sensation recovery

Temperature sensation was evaluated in all four patients with reported outcomes. In three cases (Dubernard, Zhen, Siemionow), authors used nonquantitative method of assessment, simply looking for absence or presence of heat/cold sensation after probe was applied to the tested area of the skin. Siemionow et al. used metal cylinder probes cooled to 5°C or heated to 50°C as previously used by others[36] (Fig. 21.4). First face allograft recipient had heat/cold sensation recovery 8 weeks before any sensation to touch was observed. In noninnervated facial composite tissue allograft temperature, light touch and punctate touch sensation started to recover simultaneously at 3 months, although the patient was not tested before 3 months when some degree of recovery might have already been observed.[3] Lantieri's group evaluated temperature sensation recovery by QST (method not exactly specified) and reported positive findings at 3 months (also first evaluation) and decreasing thresholds up to 1-year posttransplantation, very similar pattern to discriminative sensation recovery.[5]

21.3 Discussion

In summary, sensory recovery both to touch and heat/cold as well as pain was present in all four analyzed face allograft recipients. Very good results from two

French patients, who both received primary innervated flaps, at least partially, could be expected based on experience with hand allograft recipients, in whom protective sensation was achieved in all patients within 6–12 months and, as time progressed, 90% showed tactile and 72% of them discriminative sensibility. Moreover, further improvement should be expected up to 5 years after procedure with sufficient rehabilitation.[37] However, sensory outcomes observed in noninnervated facial composite tissue allografts are truly beyond what was expected.[3] Previous reports on sensory recovery in noninnervated flaps in face and neck reconstruction showed significantly worse results in all modalities tested when compared to the primary innervated flaps.[36,38,39] Generally some recovery occurs; however, it is often delayed or reduced, and also tends to be better in younger patients and smaller flaps.[40-42] Cleveland Clinic patient showed robust recovery as early as 3 months posttransplantation regardless of the graft size, covering 2/3 of the face and high complexity of the composite tissue transplant. Waris et al.[43] showed that primary noninnervated skin grafts are invaded by regenerating nerves from either graft bed or graft margins or both depending on the anatomy of the recipient site. Nerves tend to orientate toward the center of denervated graft area as a result of release of the chemotactic factors. Three weeks after skin grafting in human subjects, regenerated nerves were seen both at the subdermal level under the skin graft and at the margins of the graft. However, in myocutaneous flaps, the intact muscle–skin intersection suppresses nerve ingrowth into the skin even when the muscle was previously sensory-neurotized.[2] This effect can be reversed when the underlying muscle is trimmed enabling the outgrowth of the axons, which is not applicable in the face transplantation, as the whole composite tissue allograft is transplanted en bloc. Few other mechanisms can be found in literature elucidating excellent V2 sensory recovery in the first US face transplant patient, including spontaneous recovery by recruitment from surrounding intact nerves (greater auricular, greater and lesser occipital, and auriculotemporal nerves).[44] Development of cross-connections between facial and trigeminal nerves and between nerves within the graft (infraorbital, auriculotemporal, and branches of the greater auricular) with intact sensory nerves in the graft margins might also promote sensory restoration.[3,45] Sensory components within repaired facial nerves can provide sprouting and interconnections with the sensory nerves included into face allograft.[46] Moreover, it was shown that sympathetic nerve ingrowths within vascular pedicle could facilitate sensory recovery of the allograft.[47] Lastly, tacrolimus used as a mainstay of immunosuppressive regimen in face transplant patients is well known to promote faster nerve regeneration both in clinical and experimental setting.[48,49]

21.4 Conclusions

Face transplantation provides high grade of sensory restoration enabling further functional rehabilitation of graft recipient, better reintegration with the society, and, finally, improved quality of life. Recovery of sensation modalities as temperature, pain, and touch can be expected in patients receiving both sensory innervated and noninnervated flaps, even very large and complex ones. First signs of recovery might be observed as early as 2 weeks after the procedure in fully innervated grafts.[6]

Indications for this procedure will always be very limited to the most grievous facial disfigurements, which are often hardly comparable to each other due to complexity of the anatomical structure of the face. However, as the number of transplanted patients grows worldwide, it would be worthwhile to standardize timing and methods of assessment of functional recovery to predict patients' achievable outcomes and tailor rehabilitation therapy accordingly.

References

1. Gordon CR, Siemionow M, Papay F, et al. The world's experience with facial transplantation: what have we learned thus far? *Ann Plast Surg*. 2009;63:572-578.
2. Lu LI, Chuang DC. Sensory reinnervation of a musculocutaneous flap: an experimental rabbit study. *J Plast Reconstr Aesthet Surg*. 2006;59:291-298.
3. Siemionow M, Papay F, Alam D, et al. Near-total human face transplantation for a severely disfigured patient in the USA. *Lancet*. 2009;374:203-209.
4. Siemionow MZ, Papay F, Djohan R, Bernard S, Gordon CR, Alam D. First U.S. near-total human face transplantation: a paradigm shift for massive complex injuries. *Plast Reconstr Surg*. 2010;125:111-122.
5. Lantieri L, Meningaud JP, Grimbert P, et al. Repair of the lower and middle parts of the face by composite tissue allotransplantation in a patient with massive plexiform neurofibroma: a 1-year follow-up study. *Lancet*. 2008;372:639-645.

6. Dubernard JM, Lengele B, Morelon E, et al. Outcomes 18 months after the first human partial face transplantation. *N Engl J Med*. 2007;357:2451-2460.
7. Devauchelle B, Badet L, Lengele B, et al. First human face allograft: early report. *Lancet*. 2006;368:203-209.
8. Guo S, Han Y, Zhang X, et al. Human facial allotransplantation: a 2-year follow-up study. *Lancet*. 2008;372:631-638.
9. Walk D, Sehgal N, Moeller-Bertram T, et al. Quantitative sensory testing and mapping: a review of nonautomated quantitative methods for examination of the patient with neuropathic pain. *Clin J Pain*. 2009;25:632-640.
10. Rolke R, Baron R, Maier C, et al. Quantitative sensory testing in the German Research Network on Neuropathic Pain (DFNS): standardized protocol and reference values. *Pain*. 2006;123:231-243.
11. Rolke R, Magerl W, Campbell KA, et al. Quantitative sensory testing: a comprehensive protocol for clinical trials. *Eur J Pain*. 2006;10:77-88.
12. Zur KB, Genden EM, Urken ML. Sensory topography of the oral cavity and the impact of free flap reconstruction: a preliminary study. *Head Neck*. 2004;26:884-889.
13. Dellon AL. Testing for facial sensibility. *Plast Reconstr Surg*. 1991;87:1140-1141.
14. Grime PD. A pilot study to determine the potential application of the pressure specified sensory device in the maxillofacial region. *Br J Oral Maxillofac Surg*. 1996;34:500-503.
15. Posnick JC, Zimbler AG, Grossman JA. Normal cutaneous sensibility of the face. *Plast Reconstr Surg*. 1990;86:429-433. Discussion 34-35.
16. Weinstein S. Fifty years of somatosensory research: from the Semmes-Weinstein monofilaments to the Weinstein Enhanced Sensory Test. *J Hand Ther*. 1993;6:11-22. Discussion 50.
17. Siemionow M, Zielinski M, Sari A. Comparison of clinical evaluation and neurosensory testing in the early diagnosis of superimposed entrapment neuropathy in diabetic patients. *Ann Plast Surg*. 2006;57:41-49.
18. Karabekmez FE, Duymaz A, Moran SL. Early clinical outcomes with the use of decellularized nerve allograft for repair of sensory defects within the hand. *Hand NY*. 2009;4:245-249.
19. Bushnell BD, McWilliams AD, Whitener GB, Messer TM. Early clinical experience with collagen nerve tubes in digital nerve repair. *J Hand Surg Am*. 2008;33:1081-1087.
20. Keunen RW, Slooff AC. Sensibility testing after nerve grafting. *Clin Neurol Neurosurg*. 1983;85:93-99.
21. Rosen B. Recovery of sensory and motor function after nerve repair. A rationale for evaluation. *J Hand Ther*. 1996;9:315-327.
22. Bell-Krotoski JA, Buford WL Jr. The force/time relationship of clinically used sensory testing instruments. *J Hand Ther*. 1997;10:297-309.
23. Jerosch-Herold C. Should sensory function after median nerve injury and repair be quantified using two-point discrimination as the critical measure? *Scand J Plast Reconstr Surg Hand Surg*. 2000;34:339-343.
24. Dellon ES, Mourey R, Dellon AL. Human pressure perception values for constant and moving one- and two-point discrimination. *Plast Reconstr Surg*. 1992;90:112-117.
25. Dellon AL, Keller KM. Computer-assisted quantitative sensorimotor testing in patients with carpal and cubital tunnel syndromes. *Ann Plast Surg*. 1997;38:493-502.
26. Dellon ES, Keller KM, Moratz V, Dellon AL. Validation of cutaneous pressure threshold measurements for the evaluation of hand function. *Ann Plast Surg*. 1997;38:485-492.
27. Rosenberg D, Conolley J, Dellon AL. Thenar eminence quantitative sensory testing in the diagnosis of proximal median nerve compression. *J Hand Ther*. 2001;14:258-265.
28. Tassler PL, Dellon AL. Pressure perception in the normal lower extremity and in the tarsal tunnel syndrome. *Muscle Nerve*. 1996;19:285-289.
29. Coert JH, Dellon AL. Documenting neuropathy of the lateral femoral cutaneous nerve using the Pressure-Specified Sensory Testing device. *Ann Plast Surg*. 2003;50:373-377.
30. Coert JH, Meek MF, Gibeault D, Dellon AL. Documentation of posttraumatic nerve compression in patients with normal electrodiagnostic studies. *J Trauma*. 2004;56:339-344.
31. Weinstein S. Intensive and extensive aspects of tactile sensitivity as a function of body part, sex, and laterality. In: Kenshelo DR, ed. *The Skin Senses*. Springfield: Charles C Thomas; 1968:195-222.
32. Dellon AL, Andonian E, DeJesus RA. Measuring sensibility of the trigeminal nerve. *Plast Reconstr Surg*. 2007;120:1546-1550.
33. Baad-Hansen L, Arima T, Arendt-Nielsen L, Neumann-Jensen B, Svensson P. Quantitative sensory tests before and 1(1/2) years after orthognathic surgery: a cross-sectional study. *J Oral Rehabil.* [serial on the Internet]. Available from: http://www.ncbi.nlm.nih.gov/entrez/query.fcgi?cmd=Retrieve&db=PubMed&dopt=Citation&list_uids=20113390.
34. Shindo ML, Sinha UK, Rice DH. Sensory recovery in noninnervated free flaps for head and neck reconstruction. *Laryngoscope*. 1995;105(12 Pt 1):1290-1293.
35. Vriens JP, Acosta R, Soutar DS, Webster MH. Recovery of sensation in the radial forearm free flap in oral reconstruction. *Plast Reconstr Surg*. 1996;98:649-656.
36. Kim JH, Rho YS, Ahn HY, Chung CH. Comparison of sensory recovery and morphologic change between sensate and nonsensate flaps in oral cavity and oropharyngeal reconstruction. *Head Neck*. 2008;30:1099-1104.
37. Petruzzo P, Lanzetta M, Dubernard JM, et al. The international registry on hand and composite tissue transplantation. *Transplantation*. 2008;86:487-492.
38. Netscher D, Armenta AH, Meade RA, Alford EL. Sensory recovery of innervated and non-innervated radial forearm free flaps: functional implications. *J Reconstr Microsurg*. 2000;16:179-185.
39. Boyd B, Mulholland S, Gullane P, Irish J, Kelly L, Rotstein L. Reinnervated lateral antebrachial cutaneous neurosome flaps in oral reconstruction: are we making sense? *Plast Reconstr Surg*. 1994;93:1350-1359. discussion 60–62.
40. Aviv JE, Hecht C, Weinberg H, Dalton JF, Urken ML. Surface sensibility of the floor of the mouth and tongue in healthy controls and in radiated patients. *Otolaryngol Head Neck Surg*. 1992;107:418-423.
41. Cordeiro PG, Schwartz M, Neves RI, Tuma R. A comparison of donor and recipient site sensation in free tissue reconstruction of the oral cavity. *Ann Plast Surg*. 1997;39:461-468.
42. Petrosino L, Fucci D, Robey RR. Changes in lingual sensitivity as a function of age and stimulus exposure time. *Percept Mot Skills*. 1982;55(3 pt 2):1083-1090.
43. Waris T, Rechardt L, Kyosola K. Reinnervation of human skin grafts: a histochemical study. *Plast Reconstr Surg*. 1983;72:439-447.

44. Turkof E, Jurecka W, Sikos G, Piza-Katzer H. Sensory recovery in myocutaneous, noninnervated free flaps: a morphologic, immunohistochemical, and electron microscopic study. *Plast Reconstr Surg*. 1993;92:238-247.
45. Baumel JJ. Trigeminal-facial nerve communications. Their function in facial muscle innervation and reinnervation. *Arch Otolaryngol*. 1974;99:34-44.
46. Thomander L, Arvidsson J, Aldskogius H. Distribution of sensory ganglion cells innervating facial muscles in the cat. An anatomical study with the horseradish peroxidase technique. *Acta Otolaryngol*. 1982;94:81-92.
47. Lahteenmaki T. The regeneration of adrenergic nerves in a free microvascular groin flap in the rat. *Scand J Plast Reconstr Surg*. 1986;20:183-188.
48. Mackinnon SE, Doolabh VB, Novak CB, Trulock EP. Clinical outcome following nerve allograft transplantation. *Plast Reconstr Surg*. 2001;107:1419-1429.
49. Rustemeyer J, van de Wal R, Keipert C, Dicke U. Administration of low-dose FK 506 accelerates histomorphometric regeneration and functional outcomes after allograft nerve repair in a rat model. *J Craniomaxillofac Surg*. 2010;38:134-140.

22 Methods of Assessment of Cortical Plasticity in Patients Following Amputation, Replantation, and Composite Tissue Allograft Transplantation

Amanda Mendiola and Maria Z. Siemionow

Contents

Abstract The brain is constantly adapting to new inputs from the environment. New noninvasive techniques are available to scrupulously study cortical plasticity. Several studies have proven that changes in neural pathways occur due to denervation from injury such as amputation. Changes that occur rapidly are likely due to unmasking of established synapses that are latent, while changes that occur over long periods of time are more likely due to establishment of new neural connections. Cortical reorganization that occurs from traumatic amputation has been shown to be reversible with replantation and transplantation. With the new field of composite tissue transplantation, such as hand or face, it is critical to be aware of these changes in order to choose potential patients and to modify their rehabilitation based on our understanding of the cortical reorganization that occurs over time.

Abbreviations

BOLD	Blood oxygenation level dependent
fMRI	Functional Magnetic Resonance Imaging
GABA	Gamma-aminobutyric acid
Hb	Hemoglobin
HbO_2	Oxygenated hemoglobin
ICF	Intracortical facilitation
MEP	Motor-evoked potentials
MT	Motor threshold
NMDA	*N*-methyl-D-aspartate
PMC	Premotor cortex supplementary motor area (SMA)
SMA	Supplementary motor area
SICI	Short interval intracortical inhibition
TMS	Transcranial magnetic stimulation

A. Mendiola (✉)
Department of Surgery, Akron General Medical Center, Akron, OH, USA
e-mail: ammendio@neoucom.edu

M.Z. Siemionow (ed.), *The Know-How of Face Transplantation*,
DOI: 10.1007/978-0-85729-253-7_22, © Springer-Verlag London Limited 2011

22.1 Introduction

The field of composite tissue allograft (CTA) has significantly expanded in recent years. With advances in CTA, there is profound interest in functional recovery and cortical plasticity following hand and face transplant. Understanding cortical plasticity is critical to the surgeon to help his/her patient to undergo reconstruction followed by comprehensive rehabilitation.

It has been shown and published in the literature that the brain is a constantly changing structure. Injury to the upper extremity and subsequent reorganization can be assessed by several modern methods such as functional magnetic resonance imaging (fMRI) and transcranial magnetic stimulation (TMS). The establishment of quantitative methods of measuring functional loss and functional return following reimplantation and transplantation will add to our understanding of cortical plasticity after traumatic conditions.

First, one must understand how a normal brain is organized. The upper extremity and head compromise a large section of the sensorimotor cortex (Fig. 22.1). This is demonstrated in the homunculus by the disproportionate size of the face and hand versus a much smaller area for the lower extremity. Schematically, the face representation is located inferior and lateral, while the lower extremity's representation is superior and medially. Therefore, the hand and face areas are intimately associated and it has been demonstrated in the literature that tactile stimulation to the face after upper extremity amputation can induce sensation in the amputated limb (Fig. 22.2).[1]

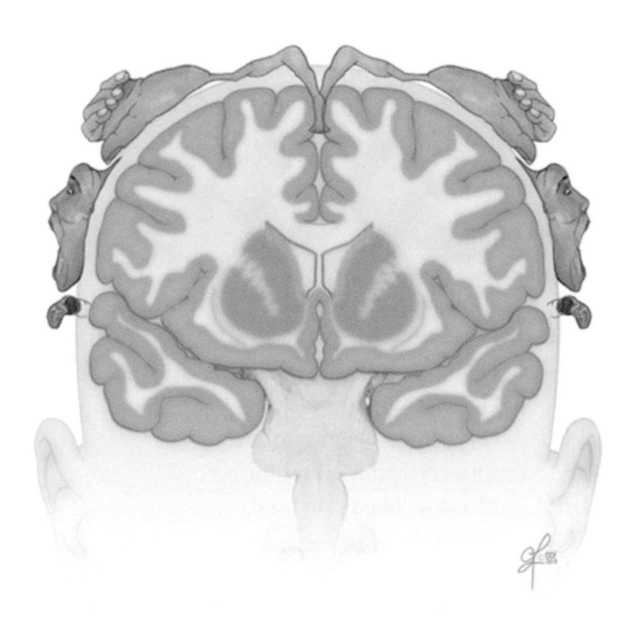

Fig. 22.1 The homunculus is a pictorial representation of the anatomic divisions of the primary somatosensory cortex. The hand and face are closely related and a large portion is dedicated to the thumb

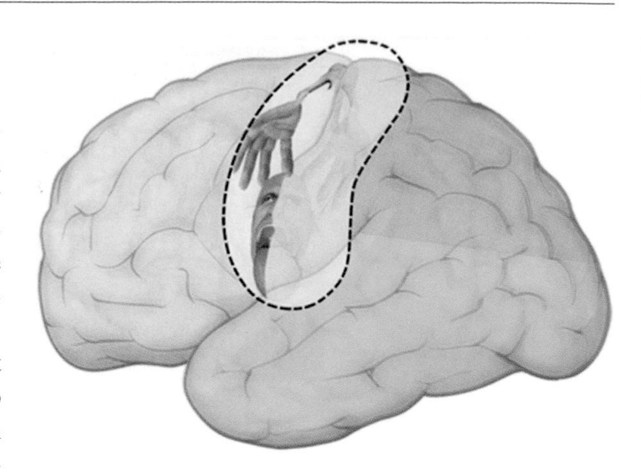

Fig. 22.2 An artistic interpretation of a lateral view of the homunculus demonstrating the intimate relationship between the hand and face

The purpose of this review is to summarize our knowledge thus far of cortical plasticity in patients after amputations, reimplantations, and transplantation procedures.

22.2 Methods of Assessing Cortical Plasticity

22.2.1 Functional Magnetic Resonance Imaging

Functional MRI is based on the principle that regional cerebral blood flow reflects neuronal activity.[2] Glucose and oxygen are consumed during neural activity and they are delivered to the neurons by the vascular system. Therefore, a change in blood delivery will reflect activity. Oxygen is transported in the blood by hemoglobin. The arterial system delivers the oxygen to capillary beds where the oxygen is extracted by the tissues and the venous system then carries the deoxygenated blood away. Hemoglobin that is carrying oxygen is referred to as oxygenated Hb (HbO_2) and once the oxygen has been extracted, it is called deoxygenated hemoglobin (Hb). Arterial blood contains mostly HbO_2 and blood in the capillary beds and venous system has a combination of HbO_2 and Hb. With an increase in neural activity, oxygen extraction rates in that location increase therefore increasing the concentration of deoxyhemoglobin.[2]

The most common method for brain mapping with fMRI is the BOLD (blood oxygenation level dependent) effect.[3] It is based on the principle that oxygenated blood has different magnetic properties than deoxygenated blood. Oxygenated blood is diamagnetic, while deoxygenated hemoglobin is paramagnetic. This creates magnetic distortion in and around capillary beds and venules. During a hemodynamic response, the ratio of oxygenated hemoglobin to deoxygenated hemoglobin increases thus creating a more homogeneous local magnetic field. Therefore, a change in ratio of HbO_2/Hb in a focal region changes the homogeneity of the magnetic field and acts as a marker of neural activity.[4]

Functional MRI provides both good spatial resolution as well as temporal resolution. Activation maps can be created while subject is performing tasks and while at rest.[5] It is an excellent tool to assess cortical plasticity due to the fact it is noninvasive.

22.2.2 Transcranial Magnetic Stimulation (TMS)

TMS is based on the principle that nerves and muscles can be stimulated by external electrical currents. It uses a brief, high-current pulse through a coil of wire to create magnetic field lines perpendicular to the coil. Then an electrical field is created perpendicular to the magnetic field (Fig. 22.3). These currents are capable of activating underlying structures.[6] TMS is mostly used to assess the primary motor cortex (M1). Using surface electromyography, motor evoked potentials (MEP) are easily obtained. The center of gravity (COG) is a measurement resulting from mapping studies and is the amplitude-weighted representative position of a motor map. Changes in COG show shifts in motor representations.[7]

Motor threshold (MT) is yet another measurement from TMS. MT is the lowest TMS intensity (expressed as a percentage of stimulator output) required to evoke

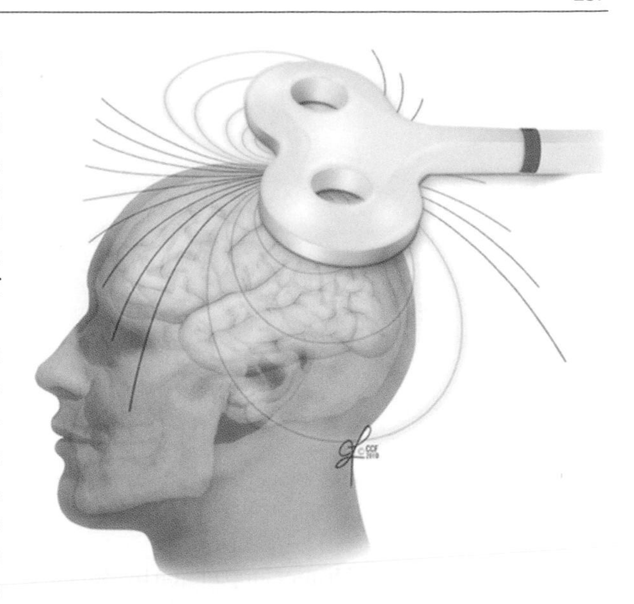

Fig. 22.3 Transcranial magnetic stimulation. The magnetic field is oriented perpendicular to the plane of the coil and produces currents in the induced electrical field lying parallel to the plane of the coil

an MEP from a certain muscle. MT represents neural membrane excitability. A lower MT indicates increased excitability.[8] MEP amplitudes are also another measurement of excitability. The ratio of the maximum MEP evoked by TMS to the maximum compound muscle action potential evoked by peripheral nerve stimulation is referred to as the percentage MMax (%MMax).[8] This ratio accounts for different muscle mass in individuals and different muscle groups.

Intracortical neural pathways can be evaluated by a conditioning-test TMS paradigm. A subthreshold conditioning stimulus is given followed by a suprathreshold test stimulus at intervals of 1–4 ms. The MEP is inhibited by the conditioning stimulus. This is known as short interval intracortical inhibition (SICI). Test stimuli given at 6–10 ms produce facilitation. This is referred to as intracortical facilitation (ICF). Changes in SICI and ICF are likely due to different mechanisms.[8]

22.3 Cortical Plasticity in Upper Extremity Amputees

There have been several studies demonstrating the cortical plasticity in motor and somatosensory cortex of upper extremity amputees. Cortical changes have been

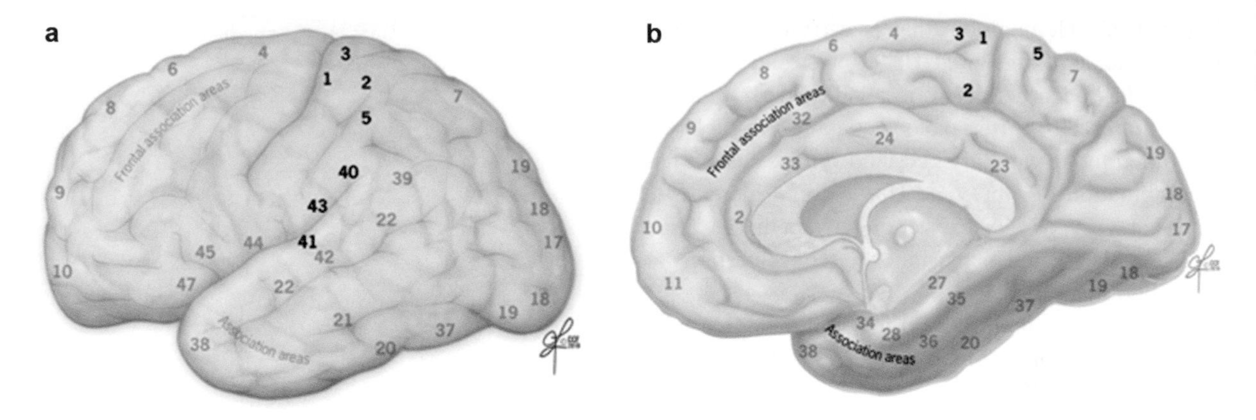

Fig. 22.4 (**a**) Lateral view of cerebrum with Brodmann's areas listed. Brodmann divided the areas based on cytoarchitectonics. (**b**) Medial view of cerebrum with Brodmann's areas listed

observed as quickly as 24 h after amputation by Borsook et al.[9] Their study consisted of one patient who was diagnosed with peripheral neuroectodermal tumor of his radius and forearm. He later underwent amputation of his left upper extremity at the distal humerus. Within 23 h of his surgery, stimulation to his face, tongue, chest, and neck caused sensation in the amputated hand. Functional magnetic resonance imaging (fMRI) was performed 1 month after surgery. He continued to have phantom sensations at this time point and fMRI revealed brush stimulation of corner of the mouth, lateral aspect of arm, and dorsum of foot on the affected side activated contralateral postcentral gyrus (Brodman area 7), which correlates with the Penfield map of the expected location of the hand in that area (Fig. 22.4). This study demonstrated that cortical changes occur very rapidly after amputation and are likely due to unmasking of existing connections and phantom perception after amputation activates distinct areas of the brain.

Functional magnetic resonance has the ability to map out cortical activation during motor tasks. Cruz et al.[10] studied seven patients, five of whom were amputees and two dysmelic patients. All patients underwent motor tasks, and fMRI was used to map out areas of cortical activation. All amputees showed activation of areas other than the contralateral primary motor cortex (SM1). There was prominent activation of ipsilateral SM1 and extension to accessory motor areas such as premotor cortex (PMC) and supplementary motor area (SMA). Patients who used a prosthesis on a daily basis had less activation of ipsilateral cortex as well as patients that were able to use their stump dexterously. Those who were an earlier age at amputation and longer duration since amputation had a more prominent contralateral activation than those who were not as well adapted. This article demonstrated that cortical reorganization is affected by the length of time since the amputation and the use of the affected limb.

Other studies have shown that changes occur in motor cortex excitability after upper limb amputation. Cohen et al.[11] used TMS to demonstrate that motor-evoked potentials (MEP) can be elicited at lower thresholds and are larger on amputated side. Also, TMS recruited a larger percentage of motor neuron pool on the side of the amputation. The excitability of motor system projecting to muscle immediately above amputation was increased.

Schwenkreis et al.[12] studied patients with traumatic upper limb amputation with phantom pain. Phantom limb pain is pain that is felt in a limb that is no longer present such as after an amputation. Transcranial stimulation was performed and found that cortico-cortical excitability was increased in the contralateral hemisphere to the amputated side. There was reduced cortical inhibition in patients mostly in forearm amputees and increased cortical facilitation in upper arm amputees. Separate mechanisms cause increased facilitation versus decreased inhibition. Gamma-aminobutyric acid (GABA) is a known inhibitory neurotransmitter and *N*-methyl-D-aspartate (NMDA) is an excitatory neurotransmitter. The results suggest a decrease in GABA-related motor cortical inhibition and an enhancement of NMDA-dependent excitatory pathways. The role of GABA and NMDA mechanisms of cortical plasticity were previously demonstrated in animal studies.[13,14]

22.4 Cortical Plasticity in Hand Replantations

Articles discussing cortical plasticity after amputation show that the brain undergoes reorganization once deafferented. When a patient undergoes replantation, the brain again undergoes reorganization to restore function. Bjorkman et al. studied a patient who had undergone traumatic amputation of their hand and immediately underwent replantation.[15] The patient underwent fMRI studies at 1, 2, 4, 8, and 12 months after replantation. Sensory stimulation of the replanted hand showed activation in the ipsilateral sensory cortex at 1 month. Throughout the different time points, the sensory activation became predominantly on the contralateral side showing initial recruitment on the ipsilateral side with normalization throughout the first year. The first motor stimulation was not achieved until 4 months. At this time, it showed activation at the contralateral area, demonstrating there was not much evidence of cortical reorganization. Likely this is because the muscles controlling the wrist and most of the hand are left intact when amputated at the wrist level and are therefore preserved and can be activated. Also, this was an immediate replant and thus not much cortical reorganization because of the short duration of deafferentation.

Long-term reorganization of the motor cortex was studied by Roricht et al.[16] where they looked at ten patients that had hand replantations after traumatic amputation with follow-up of up to 14 years. Using TMS they found the COG of biceps on replanted side shifted laterally toward the muscles of the replanted hand by 9.8 mm. The response amplitudes were larger and had lowered thresholds than the control side consistent with previous findings in the upper limb amputees.[11] The muscles of the replanted hand had normal COG and thresholds although the response amplitudes were enlarged. This demonstrated that there is different reorganization in the motor cortical areas supplying the muscles of the replanted hand and upper arm of the ipsilateral side. Likely the reorganization process is influenced by the difference in the extent of deafferentation and the muscles' different roles in hand motor control.

22.5 Cortical Plasticity in Hand Transplants

It has been shown that a significant amount of cortical reorganization happens in patients who were not candidates for immediate replantation. Piza-Katzer et al.[17] followed a patient who had a bilateral hand transplantation that occurred 6 years after the amputation. The patient was followed with fMRI and in the early postoperative period had strong activation of higher motor cortex area, weak activation of primary sensorimotor motor cortex, and no activation in primary somatosensory cortex. After 2 years, activation was seen in the primary somatosensory cortex. This article demonstrated that transplantation after long-term deafferentation resulted in cortical reorganization that continued for at least 2 years after the transplant.

Giraux et al.[18] were able to study a patient with fMRI before and after a bilateral hand transplant. Prior to surgery, they monitored flexion and extension of the missing fingers by palpating the corresponding extrinsic muscle contractions at the forearm. These movements showed activation in the most lateral part of the hand area in M1. Six months postoperatively, this activation expanded medially to occupy all of the hand region. Over 6 months, the center of gravity (COG) of hand activation shifted medially 10 mm for the right hand and 6 mm for the left hand. They also studied areas of activation with elbow movement pre- and postoperatively. Prior to the transplant, elbow movements showed activation in the contralateral central region of M1 that corresponded to the hand motor map. Six months postoperatively, these areas of activation shifted back toward the upper portion of the limb representation. The COG for elbow movements migrated from the central part of M1 to the superior part. It shifted 8 mm on the right and 7 mm on the left. This study demonstrated that new peripheral inputs were able to reverse the reorganization that occurred after the amputation.

22.6 Face and Hand Cortical Representation

As mentioned earlier, the areas of the somatosensory cortex of the hand and face are in close proximity. It is important for a surgeon in the field of composite tissue allograft transplantation to understand this relationship for both hand and face transplants. Currently, there are no published studies on cortical plasticity in facial transplantation. However, there have been studies that illustrate reorganization after deafferentation of the hand or face.

In adult rats, transection of the facial nerve resulted in enlargement of the forelimb region to occupy the area that represented the vibrissae.[19,20] Also, studies have been performed in nonhuman primates after deafferentiation of the upper limb, neck, and occiput.[21] These deafferented areas responded to tactile stimulation of the face.[21] It has also been observed that human patients have referred sensation in their amputated arm when the lower facial region is touched.[22]

Significant cortical reorganization can be observed in patients with facial palsy.[23] Using noninvasive imaging, it was observed that a larger portion of the cortex is activated while making fractionated finger movements than in normal volunteers. Most notably, there was a lateral extension of the activation laterally into the hand area in the contralateral primary sensorimotor cortex.

22.7 Discussion

It has been well established that neural pathways can be altered based on sensory information from the external environment. After amputation, there is increased cortical excitability and decreased inhibition mediated by GABA and NMDA neural pathways. These changes occur rapidly after loss of the limb suggesting these are latent pathways uncovered rather than new ones. Even after long-term deafferentation, the brain has the capabilities of restoring the original pathways and continues to undergo reorganization for up to 2 years after the transplant.

Functional MRI and TMS are critical tools in evaluating cortical plasticity. TMS allows for measuring inhibitory and excitatory pathways in the motor cortex, while fMRI has high spatial resolution. Both of the techniques are noninvasive and when used in conjunction provide extensive information regarding reorganization.

The recovery and rehabilitation of a patient after hand and upper extremity trauma is dependent on brain plasticity and cortical reorganization. While CTA is not lifesaving surgery, it is life changing for those patients who are recipients of a hand or face. There is much work needed to be done to understand changes in the nervous system when these reconstructive surgeries are performed to help aid in the selection of the patient and to predict the patient's recovery process. Changes that are beneficial as well as those that are harmful need to be identified so that recovery can be promoted. The study of cortical changes after amputation can lead to better understanding of the functional loss and better rehabilitation programs for amputees. Once the changes and their extent can be identified, they can help tailor rehabilitation programs to better assist amputees.

This chapter demonstrates the work that has been done thus far in researching cortical plasticity for CTA. Currently, reports about cortical plasticity following hand transplants, reimplants, and toe transfers are available. A whole new horizon awaits in the field of facial transplantation. To this date there has not been any published data on cortical changes after undergoing a face transplant. The principles of rehabilitation and sensory reeducation apply to face as well and much is needed to be done to understand this process to aid in the timing and tailoring of rehabilitation after transplant surgery.

References

1. Ramachandran VS, Stewart M, Rogers-Ramachandran DC. Perceptual correlates of massive cortical reorganization. *NeuroReport*. 1992;3:583-586.
2. Seong-Gi Kim, Bandettini Peter. Principles of functional MRI. In: Faro Scott, Mohammed Feroze, eds. *Functional MRI: Basic Principles and Clinical Applications*. New York: Springer; 2006:3-23.
3. Stippich Christopher. Introduction to presurgical functional MRI. In: Baert AL, Knauth M, Sator K, eds. *Medical Radiology Diagnostic Imaging: Clinical Functional MRI*. Berlin: Springer; 2007:1-7.
4. Ogawa S, Lee TM, Kay AS, Tank DW. Brain magnetic resonance imaging with contrast dependent on blood oxygenation. *Proc Natl Acad Sci USA*. 1990;87:9868-9872.
5. Cohen MS, Bookheimer SY. Localization of brain function using magnetic resonance imaging. *Trends Neurosci*. 1994;17:268-277.
6. Barker Anthony T. The history and basic principles of magnetic nerve stimulation. In: Pascual-Leone A, Davey NJ, Rothwell J, Wasserman EM, Puri BK, eds. *Handbook of Transcranial Magnetic Stimulation*. New York: Arnold; 2002:3-17.
7. Wassermann EM, McShane LM, Hallet M, Cohen LG. Noninvasive mapping of muscle representations in human motor cortex. *Electroencephalogr Clin Neurophysiol*. 1992;85:1-8.
8. Cohen LG, Mano Y. Neuroplasticity and transcranial magnetic stimulation. In: Pascual-Leone A, Davey NJ, Rothwell J,

Wasserman EM, Puri BK, eds. *Handbook of Transcranial Magnetic Stimulation*. New York: Arnold; 2002:346-357.
9. Borsook D, Becerra L, Fishman S, et al. Acute plasticity in human somatosensory cortex following amputation. *NeuroReport*. 1998;9:1013-1017.
10. Cruz VT, Nunes B, Reis AM, Pereira JR. Cortical remapping in amputees and dyslemic patients: a functional MRI study. *NeuroRehabilitation*. 2003;18:299-305.
11. Cohen LG, Bandinelli S, Findley TW, Hallet M. Motor reorganization after upper limb amputation in man. *Brain*. 1991;114:615-627. Pt 1B.
12. Schwenkreis P, Witscher K, Janssen F, et al. Changes of cortical excitability in patients with upper limb amputation. *Neurosci Lett*. 2000;293:143-146.
13. Jacobs KM, Donoghue JP. Reshaping the cortical motor map by unmasking latent intracortical connections. *Science*. 1991;251:944-947.
14. Garraghty PE, Muja N. NMDA receptors and plasticity in adult primate somatosensory cortex. *J Comp Neurol*. 1996; 367:319-336.
15. Bjorkman A, Waites A, Rosen B, Lundborg G, Larsson EM. Cortical sensory and motor response in a patient whose hand has been replanted: one-year follow up with functional magnetic resonance imaging. *Scand J Plast Reconstr Surg Hand Surg*. 2007;41:70-76.
16. Roricht S, Machetanz J, Niehaus L, Biemer E, Meyer BU. Reorganization of human motor cortex after hand replantation. *Ann Neurol*. 2001;50:240-249.
17. Piza-Katzer H, Brenneis C, Loscher WN, et al. Cortical motor activation patterns after hand transplant and replantation. *Acta Neurochir Suppl*. 2007;100:113-115.
18. Giraux P, Sirigu A, Schneider F, Dubernard JM. Cortical reorganization in motor cortex after graft of both hands. *Nat Neurosci*. 2001;4:691-692.
19. Sanes JN, Suner S, Donoghue JP. Dynamic organization of primary motor cortex output to target muscles in adult rats. I. Long term patterns of reorganization following motor or mixed peripheral nerve lesions. *Exp Brain Res*. 1990;79: 479-491.
20. Donoghue JP, Suner S, Sanes JN. Dynamic organization of primary motor cortex output to target muscles in adult rats. II. Rapid reorganization following motor nerve lesions. *Exp Brain Res*. 1990;79:492-503.
21. Pons TP, Garraghty PE, Ommaya AK, Kaas JH, Taub E, Mishkin M. Massive cortical reorganization after sensory deafferentation in adult macaques. *Science*. 1991;252: 1857-1860.
22. Yang TT, Gallen CC, Ramachandran VS, Cobb S, Schwartz BJ, Bloom FE. Noninvasive detection of cerebral plasticity in adult human somatosensory cortex. *NeuroReport*. 1994;5: 701-704.
23. Rijntjes M, Tegenthoff M, Liepert J, et al. Cortical reorganization in patients with facial palsy. *Ann Neurol*. 1997;41: 621-630.

Part IV

Approval Process of Face Transplantation

The Institutional Review Board Approval Process

23

Chad R. Gordon and Maria Z. Siemionow

Contents

M.Z. Siemionow (✉)
Department of Plastic Surgery, Cleveland Clinic,
Cleveland, OH, USA
e-mail: siemiom@ccf.org

Abstract Since 2005, ten face transplants have been performed in four countries: France, the USA, China, and Spain. These encouraging short-term outcomes, with the longest survivor approaching 5 years, have led to an increased interest in establishing face transplant programs worldwide. Therefore, the purpose of this chapter is to facilitate the dissemination of relevant details as per our experience in an effort to assist those medical centers interested in obtaining an Institutional Review Board (IRB)-approved protocol. In this chapter, we address the logistical challenges involved with face transplantation including: essential program requirements, pertinent protocol details, face transplant team assembly, project funding, the organ procurement organization (OPO), and the coroner. It must be emphasized that face transplantation is still *experimental* and its therapeutic value remains to be validated. All surgical teams pursuing this endeavor must dedicate an attention to detail and should accept a responsibility to publish their outcomes in a transparent manner in order to contribute to the international field. However, due to its inherent complexity, facial transplantation should only be performed by university-affiliated medical institutions capable of orchestrating a specialized multidisciplinary team with a long-term commitment to its success.

23.1 Background

On November 15, 2004, the Cleveland Clinic was granted the world's first IRB approval for human face transplantation.[1] Nearly 1 year later, the first successful face transplant was performed on November 27, 2005, in Amiens, France.[2] Currently,

M.Z. Siemionow (ed.), *The Know-How of Face Transplantation*,
DOI: 10.1007/978-0-85729-253-7_23,

Table 23.1 Recent timeline of the world's first eight face transplants

Transplant date	Indication	Place	Team
November 2005	Dog bite	France	Dubernard et al.
April 2006[a]	Bear attack	China	Zhang et al.
January 2007	Neurofibromatosis	France	Lantieri et al.
December 2008	Gunshot blast	Cleveland, Ohio	Siemionow et al.
March 2009	Gunshot blast	France	Lantieri et al.
April 2009	Burn	Boston, Massachusetts	Lantieri et al.
April 2009[b,c]	Fall/electrical injury	France	Pohamac et al.
August 2009[d]	Cancer resection	Spain	Cavadas et al.
March 2010	Gunshot blast	Spain	Barret et al.

[a] Expired June 2008, 2 months posttransplant
[b] Concomitant bilateral hand transplant
[c] Expired July 2009, 2 years posttransplant
[d] Concomitant tongue transplant

a total of ten face transplants have been performed in four countries: France, the USA, China, and Spain.[3-6] (Table 23.1)

There are undoubtedly many institutions worldwide planning to establish composite tissue allotransplantation (CTA) centers in the near future capable of performing facial transplantation.[7,8] Unfortunately, different countries require inconsistent protocols/approvals through an assortment of various government agencies. Therefore, providing a simple, generic recipe for program establishment program would be both misleading and simply impossible. In addition, none of the previous publications on facial transplantation effectively describe the optimal sequence, timing, and steps required for obtaining an Institutional Review Board (IRB)-approved face transplant protocol.[9,10]

The complexity of establishing of face transplant program/protocol goes far beyond performing the actual surgery. In this chapter, we describe the pretransplant requirements, such as the logistics of organizing a team, essential program requirements, IRB protocol details, project funding, the organ procurement organization (OPO), and the coroner. Although this overview may be more applicable to the US-based institutions, we hope that these blueprints for obtaining an IRB protocol will have worldwide application and be of significant interest to many plastic surgeons pioneering this new transplantation procedure in their respective countries.

23.2 Pretransplant Phase

23.2.1 Essential Program Requirements

It is intuitive that any medical institution assembling a facial allotransplant program should undoubtedly consider this project to be a long-term dedication of multiple decades and one in which needs significant time, money, and manpower prior to seeing fruition. It would be naive to think that hiring a single staff member familiar with microsurgery and/or interest in CTA simply translates into a successful, blossoming program. More importantly, in the best interest of the patients, a detailed plan should be in place given the unforeseen circumstance a team leader moves or retires.

In our experience, this project should be assessed by months and years, and not by hours and weeks. There are multiple checkpoints of success throughout this process prior to one entering the operating room, which begins with team assembly, IRB-protocol approval, patient evaluations, candidate selection, and numerous tailored mock cadaver transplants (Fig. 23.1).

We feel that one of the key requirements for success is that the transplant program be affiliated with a university hospital. This includes collaborating with an active, productive basic science laboratory which accelerates both the institution's facial CTA program as well as contributes to the overall advancement of the international field.[11-16]

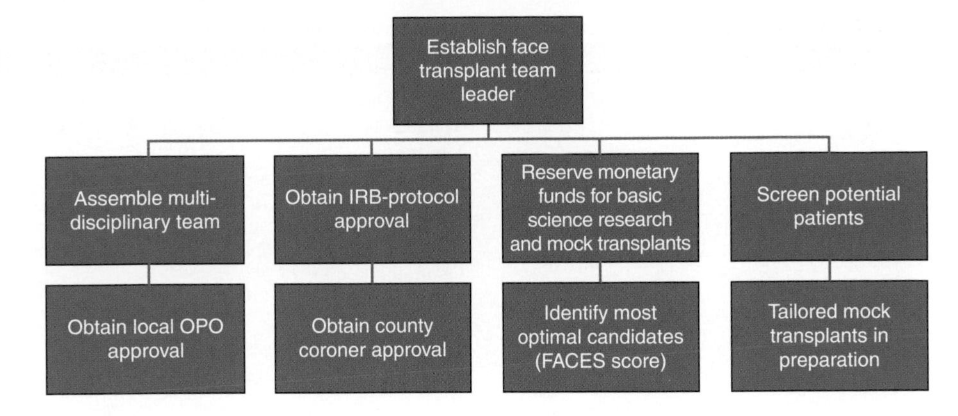

Fig. 23.1 Proposed checkpoints for establishing a face transplant program

Also, each hospital considering this endeavor must be capable of assembling a team available at *all* times, 7 days a week. Utilizing a broadly based cross-coverage schedule allows one to incorporate a talented network of reconstructive microsurgeons, craniomaxillo-facial surgeons, transplant surgeons, infectious disease specialists, transplant psychiatrists/psychologists, and immunologists. Each team member is *equally* valuable and all of those involved should mentally prepare for an extremely large time commitment, since postoperative management will be challenging and unprecedented.[17] The team leader should be well familiar with all of the technical, immunological, and legal aspects of CTA, along with this specialty's historical developments and future directions.[18,19]

23.2.2 Institutional Review Board (IRB) Protocol

Obtaining an IRB-approved face transplant protocol is a unique, complex process for which can easily and quickly become overwhelming and frustrating. It requires perseverance and a significant time commitment of 1–2 days/week. The principal investigator (PI) of the IRB protocol is most likely to become the FTT team leader and his/her collaborators will need to also reserve 1–2 days of nonclinical time each week to help work on establishing the protocol.

Besides having to detail the three critical phases of face transplantation (pre-, peri-, and posttransplant) from start to finish, he/she needs to provide the IRB with acceptable donor and recipient consent processes, which will also be equally challenging. Reason being is that any IRB in the USA needs to be in agreement of two important distinctions. The first is to make sure that the subject's overall risk is "reasonable" in relation to the anticipated benefit and that the important knowledge gained is "reasonably expected." Second, the IRB must be assured that adequate informed consent will be obtained. These are two separate, but tremendously important, requirements.[20]

The team leader should expect to attend a multitude of IRB-required meetings, respond timely to the committee members' questions, and provide scientific evidence to justify the protocol details. In the end, an estimation of 1–2 years seems to be a safe presumption for how long the start-to-finish time will entail. However, the sole exception being that if your institution already has a current hand or abdominal wall transplant IRB-approved protocol, then your approval process will be somewhat streamlined.

23.2.3 Assembling the Team

Once the IRB protocol has been finalized, the team leader should (1) act as an architect and design a project timeline with a tentative sequence of steps and (2) assemble accordingly an experienced team of experts capable of performing the task at hand. We recommend choosing a team leader established in the field of CTA and one that is devoted to a lifetime commitment, since he/she should possess a passion to succeed and the capability to lead and motivate.

As published recently, this innovative, experimental surgery is still in its infancy stage and is extremely complex in relative comparison to other CTA subtypes,

Table 23.2 Modified Gordon CTA classification system based on relative complexity

Type	Complexity	Allografts	Characteristics
I	Low	Flexor tendon Tongue Uterus Vascularized nerve	1. Absent skin 2. Reduced antigenicity
II	Moderate	Abdominal wall Facial subunit (ear) Genitalia (penis) Larynx Scalp Trachea Vascularized joint (knee)	1. Contain skin 2. Absent or less-challenging rehabilitation
III	High	Upper extremity (hand) Face	1. Requires multidisciplinary transplant team 2. Complex rehabilitation 3. Significant psychological obstacles 4. Complicated cortical reorganization
IV	Maximum	Concomitant CTA – Face/hand(s) – Tongue/mandible	1. High mortality risk 2. Extremely difficult rehabilitation

Source: Siemionow et al.[7]

such as abdominal wall, as defined by our recently modified classification system (see Table 23.2).[21] Therefore, facial transplantation should *not* be simply seen as another challenging reconstruction case. Plastic surgeons must envision this procedure analogous to an organ transplant with distinct indications and contraindications, and not as an additional rung on the reconstructive ladder.[22]

We suggest establishing a sizable, overlapping surgical team in the range of 6–10 staff surgeons, whose members are wholeheartedly devoted to the project. A complete combination of craniofacial and microsurgical trained plastic surgeons is ideal, with a potential need for adding an ENT/head and neck surgeon. In preparation for "The Big Day," all surgical team members should be required to participate in a series of mock, fresh-cadaver facial transplants (for our team, the early weekend mornings were most accommodating) and practice exercises to ensure all details are complete. These invaluable mock transplants should come with mandatory attendance, since they are crucial for both defining each surgeon's role and for perfecting the team's surgical chemistry in an effort to decrease potential delay and complication.[23-26] In the interim, your program's team leader and members should meet and present the protocol to various hospital ICU staff, anesthesiology staff, OR nurse manager(s), and surgical intensivist(s) as a means of perfecting timely execution and understanding of procedure complexity.

23.2.3.1 The Ancillary Staff Members

Unquestionably, a diverse team is needed for an optimal outcome. The world's finest reconstructive transplant surgeons could not perform successful face transplants if it were not for the right ancillary support. A large responsibility is delegated to a wide variety of surgical colleagues and nonsurgical staff, which includes a face transplant coordinator, transplant surgeon, transplant immunologist, transplant infectious disease expert, social worker, ethicist, and transplant psychiatrist/psychologist.[17]

23.2.3.2 Face Transplant Coordinator

The team leader needs to select and assign a knowledgeable face transplant coordinator (FTC). This person must be well informed as to the intricacies of face transplantation and should be either a physician, physician's assistant, or registered nurse. The true magnitude of this role cannot be underestimated. Their duties involve pre- and posttransplant coordination, aiding tasks such as candidate screening, coordinating all transplant-related activities, overseeing test results and prescription compliance, and helping to arrange follow-up care (main transplant hospital vs patient's local hospital).

For the first 3 posttransplant months, the FTC is on-call "24/7." A "back-up" FTC system may be needed in some cases. Responsibilities include monitoring/presenting daily drug levels and/or lab work since obvious abnormalities are of utmost importance and need to be confirmed by the FTC. Having a central figure in the middle of a large multidisciplinary team will theoretically increase communication efficiency and decrease the chance of miscommunication and/or wasteful duplication. Also, since most face transplant recipients will reside at a far distance from your university hospital, the FTC will arrange posttransplant monitoring via nearby hospital-subsidized housing for the first 2–4 months posttransplant. Scheduling and coordinating periodic follow-up visits are also his/her responsibility.

The FTC should prearrange a "satellite" medical team for all FT patients living at a significant distance from the hospital. This satellite team, which should obviously be in close proximity to the patient's primary residence, includes a surgeon (preferably a plastic, ENT, or transplant surgeon), an internist, and a physical therapist. This is of tremendous value to the patient if for some reason the FTT wants to request a tissue biopsy, medical exam, and/or alter any specific facial physical therapy.[19]

23.2.3.3 Medical Management

During the first year, between months 3 and 12, face transplant patients are to be followed closely by the team leader in line with each institution's approved treatment protocol. Postoperative care should include consultations to transplant immunology, infectious disease, and transplant psychiatry/psychology at the main hospital where the allotransplant was performed. During months 13–24, visits should be held quarterly and then every 6 months starting in the third year, unless there are intermittent signs of rejection and/or other transplant-related health problems. In addition, the PI will orchestrate all nonsurgical visits and medical exams as needed. Routine visits to the patient's primary physician are also encouraged and close relation should be developed with the PI of the IRB protocol to coordinate patient care.[9]

23.2.3.4 Social Worker

Transplant social workers are also assigned to the FTT. Their involvement is critical for the patient's social adjustment pre- and posttransplant. Their responsibilities include evaluating the patient's social/family support, current health insurance coverage, occupational status, potential for return to work after transplant, or need for job change/reeducation. In addition, the social worker may help to facilitate contact with a lawyer if any legal issues arise either before, during, or after surgery.[17]

23.2.3.5 Patient Advocate

It is strongly recommended that all potential candidate(s) assign either a family member or trustworthy friend/lawyer to act as their "patient advocate." Their role is similar to a "power of attorney," and involves deciding the patient's needs in certain instances during the entire process of facial transplantation. Of interest, the Royal College of Surgeons' guidelines also suggest that each candidate meet with other patients successfully managed by non-transplant, modern-day, facial reconstructive technique, which may be most relevant for pan-facial burn patients.[18]

23.2.3.6 Medical Ethics

An ethics committee is consulted for all related ethical questions, as they may pertain to pre- and post-facial transplantation. The role of the team bioethicist is to assess, identify, and investigate the patient's motivation and understanding of the procedure, discuss his/her perception of the risk versus benefit ratio of transplantation in exchange for life-long immunosuppression with its unavoidable side effects. The ethicist should also discuss the experimental aspect of the face transplantation and emphasize the fact that the final outcome cannot be fully predicted.[27]

23.2.3.7 Transplant Psychiatry/Psychology

A transplant psychiatrist/psychologist is assigned to the FTT. His/her responsibilities include performing pre- and postoperative assessments of the candidates and as required, oversee and provide treatments including psychopharmacological therapy, psychotherapy, and/or chemical dependency treatment. Beginning with their initial interview, the potential candidate undergoes emotional/cognitive evaluation for transplant potential, assessment of his/her decision-making capacity, and Thematic Apperception testing. Family support in combination with the candidate's socioeconomic status is investigated in order to identify their entire social support system, which may play a crucial role as to the transplant's success. Of significant concern is their medical compliance history, which includes degree of motivation, realistic expectation, potential for psychological regression, perceived body-image adaptation, and anticipated comfort with cadaveric facial allotransplant.[17,28]

Prophylactic social/family/marital interventions should be planned and an introductory transplant education should be provided. Psychological assessment of self-esteem, quality of life, and body image should be performed using standard "quality-of-life" measures. The inclusion of an experienced transplant psychiatrist/psychologist will minimize potential psychiatric morbidity throughout the entire process by aiding the recipient in reintegrating their "new" face both physically and psychologically. Psychosocial postoperative assessment is mandatory and should be conducted daily for the first 2–3 months, followed by a monthly rotation for the completion of the first posttransplant year. For the second year, the rotation may be decreased to a minimum of every 3 months and then biannually thereafter. The importance of diagnosing posttransplant depression cannot be understated, and should raise great cautionary measures since failure to comply with immunosuppression and/or rehabilitation will inevitably lead to failure, as witnessed with the world's first hand transplant and second face transplant patients.[3,17]

23.2.3.8 Physical Therapy and Rehabilitation

Physical therapy and speech therapy to perfect facial muscle reeducation are essential for obtaining optimal functional outcomes. A designated physical therapist and speech therapist should be heavily involved with the patient's cortical reeducation process from day 1 after transplant. It is essential that the patient be religiously compliant with his/her facial muscle exercises and speech therapy based on their individualized regimens. Access to a private gym, treadmill, and stationary bike will motivate the patient to continue physical therapy and may ultimately speed up the posttransplant recovery.[10]

23.2.3.9 Institutional Media

All interactions with the media are a collaborative effort between the FTT and a designated representative from the institution's Public Relations department. For patient safety and confidentiality, a media representative should meet with patient before and after transplant to discuss his/her level of willingness to disclose or conceal personal details during interactions with public media. In the early posttransplant period (<1 year), every effort should be made to conceal the patient's identity. For some programs, it may be prudent to admit face transplant candidates to the hospital using an alias, in order to allow full adherence to current privacy (i.e., HIPPA) regulations and to provide optimal protection from the press.

In our experience, an early meeting held between the FTT and the hospital's Public Relations (PR) office at the time of recipient identification is prudent for many reasons. This allows a team-designated PR individual (preferably one at the senior level with significant experience) to help schedule and attend all press conferences, personnel interviews, and any other public media sessions. It should be well explained to the patient as well as to the public media, that all photographs and videos relating to face transplantation are the property of the institution, and that they *must* receive written approval from both the FTT and the hospital prior to any public release (especially since FT will be confined to the ethics of your institution's IRB protocol). It should be also clear that *no* financial commitments with any outside agencies can be made for any transplant-related materials.[10]

23.2.3.10 Security

During the immediate posttransplant time period, all patients should be provided with private rooms isolated from the mainstream hospital access. In our experience, providing 24-h security at the front entrance of the patient's room provides an additional layer of privacy protection.

23.3 Funding

The overall cost attributed to face transplantation is dependent on a variety of factors, such as the geographic location for which the surgery is performed (i.e., county, state, or country). When compared to hand transplantation for example, its overall cost per patient is slightly greater, and may in fact range from $250,000 up to $1,500,000.[19] This gross estimation includes the complete cost of surgery (each surgeon's time and billing), an average stay of 2–3 weeks in the ICU, an entire hospital stay of 2–4 months, hotel room expenses thereafter (approximately 3–6 months), all related transplant medications, pertinent monitoring labs/biopsies, and rehabilitation therapy.

At this time, since this surgery is still considered "experimental," the inpatient bill will be, for the most part, *not* covered by the insurance company and therefore all costs are absorbed by the hospital. This financial deficit can, however, be offset by a combination of endowments, research grants, and/or departmental funds. Periodic and all unexpected posttransplant procedures during the first year posttransplant, such as skin biopsies, lab tests, and rejection therapies, are usually covered by insurance since they fall under "medical necessity." As for mandatory rehabilitation therapy, an adjusted cost schedule is provided to the recipient after 90–180 days depending on his/her medical insurance coverage and financial status. We recommend applying for, depending on your state's individual legislation, full medical coverage of the patient's posttransplant care since many states provide unlimited "transplant" benefits (i.e., kidney and liver transplant patients). Additionally, it may be prudent to contact the pharmaceutical companies to see if they will provide immunosuppression cost-free by way of an industry-sponsored grant.[19] More importantly, it would be unethical to perform such a procedure unless future provisions for postoperative rehabilitation and immunosuppression were allocated.

23.4 Organ Procurement Organization

The support of your institution's local organ procurement organization is essential for success; however, the entire process of obtaining OPO approval is lengthy and may be quite challenging. Under the federal system, OPOs are designated to a specific geographic region and must be 501C3 charitable nonprofit organizations. Each hospital is assigned to work with one particular OPO thereby limiting the options of a new face transplant program. Furthermore, donation/transplantation of "organs" involves a complex process overseen and coordinated by multiple organizations established through the direction of the US government including the US Department of Health and Human Services (HHS), the Organ Procurement and Transplantation Network (OPTN), the United Network of Organ Sharing (UNOS), the Health Resources Services Administration (HRSA), and the Centers for Medicare Services (CMS).[29]

The main role of the OPO is to oversee and coordinate the allotment of all donated organs, and therefore it is essential that any hospital entertaining face transplantation consult their OPO early on in the process as we did 2 years prior to IRB approval. Once the IRB protocol for face transplantation is approved by your hospital, it creates a basis for filling specific research protocol requests with your hospital-affiliated OPO. The PI or team leader is responsible for protocol presentation and if requested, an oral presentation at the OPO's Medical Board meeting may be quite valuable. Numerous meetings between the PI and the OPO's director/staff will be necessary for pertinent education about the logistical timeline during the day of surgery, as well as overall goals in identifying and recovering a facial allotransplant.[10]

Each OPO-employed transplant coordinator should use a CTA-tailored algorithm when approaching all potential donor families for facial organ donation. In summary, the vital organs such as the liver, kidneys, pancreas, heart, and lungs are discussed early on in the process so that facial tissue donation does not interfere with requesting lifesaving organs and tissues.[30]

Interestingly, in some particular instances, the living recipient is a human subject and falls under the federal IRB regulations, but the donor, however, is not. Therefore, the IRB may or may not be approving the consent form based on the legal requirements of the institution or country. This also raises important logistic steps since the donor may or may not be deceased at the hospital where the protocol is approved, and therefore donor transfer needs to be prearranged accordingly. This may limit a program's donor pool unless one has a large health system analogous to the Cleveland Clinic Health System (includes nine community hospitals).

Once supported by your local OPO, concise guidelines for the transplant coordinators working at affiliated hospitals should be established. Educational gatherings and presentations by various FTT members are then scheduled to facilitate full understanding of the complexity of the surgical procedure and for presentation of eligible candidates pursuing facial transplantation. We found it valuable to provide a short personal description (one paragraph) of the listed recipient (while at the same time limiting exact details so as to protect the recipient's identity) and how the donated facial organ would conceivably aid his/her reintegration into society. In retrospect, this seemed to greatly aid the transplant coordinator in his/her quest to obtain facial organ donor consent, given the large amount of public uncertainty related to facial transplantation.

23.5 Role of Coroner

The county coroner in your area (or medical examiner in some instances) is responsible for overseeing all deceased human bodies. If foul play is suspected as to the etiology of death or if the patient is a minor (<18 years old), the body is immediately placed into the possession of the county coroner until an official autopsy has been completed. Otherwise, an accelerated process is undertaken and the body is placed into the custody of either the funeral home and/or the hospital morgue.

In our experience, multiple meetings at the coroner's office were necessary for the establishment of a protocol for interhospital brain-dead donor transport. Each protocol will differ if perhaps donor transport is within the county or if donor consent is obtained at a community hospital within the same health system. Finally, different approvals are required when donor transport crosses a state border, which entails the local coroner office having to contact the coroner's office of the state where donor consent was originally signed. Full logistical understanding beforehand will expedite this complicated process.

Exact details should also be preestablished for medical transportation (i.e., ground vs airplane). An ICU physician should be preselected to serve as the accepting staff for the transfer of the beating-heart, brain-dead facial organ donor (preferably admitted to a neurosurgical ICU). We recommend using a different primary attending and separate surgical ICU for the recipient's direct admission, so that the two families (donor and recipient) are not coinciding prematurely. By having both patients in the same locale, concomitant recovery of the donor's facial organ along with the recipient's preparation for transplantation can be done efficiently in neighboring operating rooms. Obviously, this is of tremendous value, since many times preparation of the recipient's craniofacial defect overlaps with the time needed to recover the donor alloflap. This particular process can be quite intricate in detail and adjustments to your original surgical plan may be required.[23,24] Therefore, having the option of various surgeons walking between the two operating rooms for the purpose of observing each other's simultaneous progress is invaluable.

23.6 Conclusion

Face transplantation has progressed tremendously since the first partial allotransplant was performed by Dubernard et al. in 2005.[2] A total of nine patients have since followed and the results have been relatively successful. Early postoperative reports regarding aesthetic and functional outcomes are promising.[3] However, two recent face transplant-related deaths highlight the importance of patient selection and compliance with regard to immunosuppression, extreme psychological stability/social support, aggressive rehabilitation therapy, and constant patient motivation to succeed.[9,10,31] Complete evaluation of the current face transplant outcomes and further research pertaining to face transplantation, in areas such as bioethics and immunology, are undoubtedly warranted prior to anyone considering this surgical procedure as *non-experimental.*

References

1. Okie S. Facial transplantation: brave new face. *N Engl J Med.* 2006;354:889-894.
2. Dubernard JM, Lengele B, Morelon E, et al. Outcomes 18 months after the first human partial face transplantation. *N Engl J Med.* 2007;357:2451.
3. Gordon CR, Siemionow M, Papay F, et al. The world's experience with facial transplantation: what have we learned thus far? *Ann Plast Surg.* 2009;63:121-127.
4. Guo S, Han Y, Zhang X, Lu B, et al. Human facial allotransplantation: a 2-year follow-up study. *Lancet.* 2008;372: 631-638.
5. Lantieri L, Meningaud JP, Grimbert P, et al. Repair of the lower and middle parts of the face by composite tissue allotransplantation in a patient with massive plexiform neurofibroma: a 1-year follow-up study. *Lancet.* 2008;372: 639-645.
6. Associated Press. First total face transplant in Spain. MSNBC website. http://www.msnbc.msn. com/id/36731302/ ns/health-more_health_news/. Accessed May 14, 2010.
7. Siemionow M, Zor F, Gordon CR. Face, upper extremity, and concomitant transplantation: future challenges and potential concerns. *Plast Reconstr Surg.* 2010;126(1):308-315.
8. American Society of Reconstructive Transplantation. Website home page. http://www.a-s-r-t.com. Accessed November 29, 2009.
9. Siemionow M, Papay F, Alam D, et al. First U.S. near-total human face transplantation – a paradigm shift for massive facial injuries. *Lancet.* 2009;374:203-209.
10. Siemionow M, Papay F, Djohan R, et al. First U.S. near-total human face transplantation – a paradigm shift for massive facial injuries. *Plast Reconstr Surg.* 2010;125(1):111-122.
11. Gordon CR, Nazzal J, Lozano-Calderan SA, et al. From experimental rat hindlimb to clinical face composite tissue allotransplantation: historical background and current status. *Microsurgery.* 2006;26:566-572.
12. Siemionow M, Demir Y, Mukherjee A, Klimczak A. Development and maintenance of donor-specific chimerism in semi-allogenic and fully major histocompatibility complex mismatched facial allograft transplants. *Transplantation.* 2005;79:558-567.
13. Demir Y, Ozmen S, Klimczak A, Mukherjee AL, Siemionow M. Tolerance induction in composite facial allograft transplantation in the rat model. *Plast Reconstr Surg.* 2004;114: 1790-1801.
14. Siemionow M, Gozel-Ulusal B, Engin Ulusal A, Ozmen S, Izycki D, Zins JE. Functional tolerance following face transplantation in the rat. *Transplantation.* 2003;75:1607-1609.
15. Siemionow M, Ortak T, Izycki D, et al. Induction of tolerance in composite tissue allografts. *Transplantation.* 2002; 74:1211-1217.
16. Gozel-Ulusal B, Ulusal AE, Ozmen S, Zins JE, Siemionow MZ. A new composite facial and scalp transplantation model in rats. *Plast Reconstr Surg.* 2003;112:1302-1311.
17. Gordon CR, Siemionow M, Coffman K, et al. The Cleveland clinic FACES score: a preliminary assessment tool for identifying the optimal face transplant candidate. *J Craniofac Surg.* 2009;20(6):1969-1974.
18. Morris P, Bradley A, Doyal L, et al. Face transplantation: a review of the technical, immunological, psychological and clinical issues with recommendations for good practice. *Transplantation.* 2007;83:109-128.
19. Gordon CR, Siemionow M. Requirements for establishing a hand transplant program. *Ann Plast Surg.* 2009;63:262-273.
20. ASRM/ASPS. Facial Transplantation – ASMS/ASPS guiding principles. Available at: www.microsurg.org/ftGuide lines.pdf Accessed on June 13, 2010.
21. Gordon CR, Siemionow M, Zins J. Composite tissue allotransplantation: a proposed classification system based on relative complexity. *Transplant Proc.* 2009;41:481-484.
22. Siemionow M, Sonmez E. Face as an organ. *Ann Plast Surg.* 2008;61:345-352.
23. Siemionow M, Unal S, Agaoglu G, Sari A. A cadaver study in preparation for facial allograft transplantation in humans: part I. What are alternative sources for total facial defect coverage? *Plast Reconstr Surg.* 2006;117:864-872.
24. Siemionow M, Agaoglu G, Unal S. A cadaver study in preparation for facial allograft transplantation in humans: part II. Mock facial transplantation. *Plast Reconstr Surg.* 2006;117: 876-885.
25. Siemionow M, Papay F, Kulahci Y, et al. Coronal-posterior approach for face/scalp flap harvesting in preparation for face transplantation. *J Reconstr Microsurg.* 2006;22:399-405.
26. Siemionow M, Agaoglu G. The issue of "facial appearance and identity transfer" after mock transplantation: a cadaver study in preparation for facial allograft transplantation in humans. *J Reconstr Microsurg.* 2006;22:329-334.
27. Agich GJ, Siemionow M. Until they have faces: the ethics of facial allograft transplantation. *J Med Ethics.* 2005;12:707-709.
28. Brill SE, Clarke A, Veale DM, Butler PE. Psychological management and body image issues in facial transplantation. *Body Image.* 2006;3:1-15.
29. Glazier AK. Regulatory face-off: what agency should oversee face transplants? *Am J Transplant.* 2008;8:1393-1395.
30. Clarke A, Malloy J, White P, et al. A model for structuring the donor discussion in emergent transplant procedures. *Prog Transplant.* 2008;18:157-161.
31. Lengele BG. Current concepts and future challenges in facial transplantation. *Clin Plast Surg.* 2009;36:507-521.

Informed Consent for Facial Transplantation

24

Katrina A. Bramstedt

Contents

K.A. Bramstedt
California Transplant Donor Network, Oakland, CA, USA
e-mail: txbioethics@yahoo.com

Abstract Facial transplantation is a complex and innovative technique still in its infancy, thus it is critical that the informed consent process has integrity. The patient population is vulnerable due to their desire for esthetic change amid a society that places much value on personal appearance. The technology of transplant, itself, requires an alliance between the doctor and patient due to matters of behavioral and medication compliance. This said, there is much for the patient to know and understand about themselves and the technique, in a setting of many technological unknowns. This chapter describes key elements of the informed consent process for facial transplantation so as to optimize the aim of information transmission, comprehension, and voluntariness.

24.1 Introduction

Before addressing the informed consent process for facial transplantation, it is critical to understand the concept of informed consent. Specifically, a signed consent form does not necessarily equate to an informed, comprehending, and voluntary participant (patient or research subject). Informed consent requires the receiver of information to have received an adequate amount of information to make an informed choice. Also the person must be able to process the information in accordance with their level of cognitive functioning. Those who lack the functional ability to make decisions cannot give informed consent. Further, a person may have the functional ability to give consent, but if the information is too voluminous and complex, he/she may not have the ability to process it and therefore cannot truly analyze it for its relevance to their personal situation. Information might include a

M.Z. Siemionow (ed.), *The Know-How of Face Transplantation*,
DOI: 10.1007/978-0-85729-253-7_24,

description of the technology under discussion, how the technology interacts with the human body, the risks and benefits of using the technology, as well as the conceptual unknowns. Information can be descriptive text, verbal, or even pictorial images (e.g., video, photographs, drawings).

Voluntariness is also an important part of any informed consent process. If a person is pressured or coerced to consent to an intervention, there is concern that he/she may have not been able to give adequate reflection on the risks of participation. Similarly, the potential benefits of an intervention can cause a loss of objectivity during decision making if the person does not take care to reflect on benefits and risks (and alternatives) in tandem, or if the benefits posed are inflated. In some cases, it is not the information that is coercive, but rather the medical and contextual features of a patient's own, unique situation that can be coercive. An example of this is a patient with a chronic illness who lacks health insurance and seeks out research studies as a method of obtaining contact with physicians and attaining prescription medication that would otherwise be unobtainable. The patient's health status and financial status both are coercive. Both play roles in motivating the patient to consent to enroll in research studies that may or may not be in his/her best interest. In sum, understanding what motivates a patient to pursue a particular intervention is critical to understanding if he/she is being coerced or pressured along the way. In the text that follows, crucial elements of the facial transplant consent process will be presented. It should be noted that all such transplants to date have occurred using strict research protocols and countries have their own rules and regulations with regard to what elements must be included in research consent forms.[1]

24.2 Blazing a Trail

The path to creating the consent form for facial transplantation at the Cleveland Clinic began with the help of bioethicists. Once a draft form was created based on the US Federal regulations,[1] it was thoroughly reviewed and edited by a transplant bioethicist [KAB].[2] This was done by using literature on burns, trauma, facial/body disfigurement, and hand transplantation as a baseline. This group of literature contains bountiful information on issues in informed consent, as well as important psychosocial variables that are important to understanding the personal element of facial transplantation.[3-12] While there were few articles on ethics, consent, and facial tissue allografting, these were also studied.[13-16] Additionally, position statements and guidance documents from professional societies were studied for information that would optimize the consent form and the consent process.[17-20]

24.3 The Informed Consent Process

The informed consent process should ensure that the transplant candidate has decision-making capacity. This can be assessed by any physician, mental health worker, or trained bioethicist. Once this is assured, the candidate should be presented with the following information within a consent form in language that is easy to understand (e.g., grade 8 reading level). The information should be discussed with the candidate and all questions answered. Communication should be straightforward, sincere, and compassionate. It is also strongly suggested that supplemental information be given to the candidate, such as video or computer simulation images, educational reading material, and updates from peer-reviewed literature.[21,22] Candidates should also be given the opportunity to have contact with others who have undergone facial transplantation (with consent of all parties) in order to obtain first-hand information about the experience.

24.3.1 Facial Transplantation Is Research

The consent form and process should be clear to disclose that facial transplantation is still considered research, not standard clinical practice. Only a few of these procedures have been done around the world and the technique and associated medical care is still evolving. To this end, facial transplantation is still in its infancy and candidates need to understand that the knowledge base is small and growing. The proposed surgical procedure (including donor graft selection) and all associated medical care (e.g., immunosuppression, rehabilitation, and behavioral restrictions) should be fully disclosed. It should be made clear that immunosuppression is for a lifetime, and ongoing clinical follow-up will be needed, as well as compliance with medical regimens.

24.3.2 Reasonably Foreseeable Risks or Discomforts

Candidates for facial transplantation should be informed that there are risks that are currently unknown because the procedure is in its infancy. The reasonable foreseeable risks that should be disclosed can be divided into two categories: medical and psychosocial.

24.3.2.1 Medical Risks

Facial transplantation involves numerous medical risks that must be disclosed and discussed with candidates. As with any surgical intervention, infection is a risk. As with any transplant, acute rejection and chronic rejection leading to graft loss are also risks, as are the side effects of immunosuppression, including risks of diabetes and malignancy. In terms of graft loss (rejection, vascular insufficiency, and flap failure), this would require graft removal and restorative surgery. Other risks include delayed wound healing, surgical drain malfunction, hematoma, and bleeding or other vascular complications possibly requiring further surgery. Death is also a risk, possibly by way of infection leading to sepsis or surgical consequences involving vital structures in the head and neck region.[23] Additionally, there is a risk that even with physical therapy and rehabilitation the graft will not restore function to the face in terms of mouth or eyelid movement, or sensation, for example. Any risk statistics that are given to candidates should be discussed in simple terms that can be easily understood.

24.3.2.2 Psychosocial Risks

Because the procedure involves one's current disfigurement being altered with the transplantation of a donor facial graft, there is the risk of the recipient failing to accept the graft from a psychosocial perspective. For the transplant to be a success, the graft needs to be integrated into the recipient's body image, identity, and emotional responses in a positive way.[24] While the recipient will not look like the donor because the graft will integrate with the recipient's bone structure, the recipient may have anxiety about the identity of the donor, the circumstances that resulted in the donation, or other matters and these could cause stress for the recipient, impeding their emotional healing.[25,26] Additionally, although old marks and scars will be removed, new ones will appear. A new skin tone and/or texture might be evident. The recipient might not be pleased with the aesthetic or functional results of the graft and this could cause disappointment and distress. Also, the recipient might receive unexpected reactions from others to his/her new look. Adjusting to the new face and the responses of others may be easier for some and harder for others.

The recipient might be fearful about future failure of the graft, and if the graft fails and is removed, this might cause significant emotional trauma. Further, the fact that the recipient would have to undergo additional reconstructive surgery as part of graft removal and facial restoration, the new look (post-graft removal and reconstruction), could pose psychological challenges for the patient who would have already experienced three versions of appearance (pre-disfigurement, disfigurement, posttransplant).

24.3.3 Reasonably Foreseeable Benefits

The transplant team should be honest in its discussion of the potential benefits of facial transplantation, taking care not to inflate them. This is because this patient population likely has optimistic expectations.[24] Specifically, quality of life, esthetic appearance, and functional outcome are important variables in the expectations of disfigured patients with regard to facial transplantation.[27] While they may have coped fairly well with their disfigurement in the past, an outcome that does not meet their expectation could be met with poor adjustment.

24.3.4 Alternative Procedures or Courses of Treatment

The transplant team should be candid about informing the recipient candidate of other potential options instead of facial transplant. This includes taking no surgical action and waiting for future technologies to develop; taking no surgical action and exploring spiritual and psychological approaches to coping with disfigurement. The candidate should be counseled that facial transplantation is a voluntary procedure and there is no obligation to pursue it. Further, the candidate can decide to decline to participate, even on the day of surgery, after the consent process has been completed.

24.3.5 Confidentiality of Records

Many countries have policies with regard to patient privacy[28]; however, there remains the risk that the transplant recipient and family, and even the donor family could have their privacy invaded by media (e.g., newspaper, television) seeking the personal details of the parties involved. All parties should be counseled of the risks of interacting with the media. The medical records of the donor should be kept private and only information that is clinically critical to the donation (including donor age and gender) should be shared with the recipient. Details about the identity of the donor should remain confidential (as per any organ or tissue donation) unless next of kin chooses to release this information to the recipient.

It is likely that the donor family will become aware of the identity of the recipient. If the donor family wants to make contact with him/her, this should only be done with the consent of the donor and should be mediated by the regional tissue procurement organization.

24.3.6 What if the Procedure Fails?

The informed consent process must include a discussion of the possibility of transplant failure (e.g., rejection, flap failure), in terms of its causes, detection, and response by the team. The recipient's life should not be put at stake merely to save the graft and he/she should be informed of this before the procedure is initiated. The recipient should know under what conditions the graft would have to be removed, as well as the nature of the reconstructive rescue.

Because immunosuppressant noncompliance and tobacco smoking can be a cause of graft loss, transplant candidates need to be fully informed of their role in keeping the graft healthy. To this end, it is recommended that the candidate sign a behavior contract similar to that which is customarily signed by solid organ transplant recipients.[29]

24.3.7 Asking for Help

Pre- and posttransplant, the participant should know who to contact for help. There should be a list of identified personnel that the recipient can call for both medical and psychological assistance. The recipient should be encouraged to contact the transplant team if there are any immunosuppressant side effects rather than stopping or adjusting the medication regimen without counsel.

24.3.8 Refusal of Care After Transplantation

It is an accepted principle of medical ethics that patients with decision-making capacity have the right to refuse medical care, even lifesaving medical care. If a facial transplant recipient refuses care during recovery or after discharge from the hospital, this can be very serious, and the recipient needs to understand the consequences of his/her behavior. If refusal of care is occurring while the patient is hospitalized, the team should ascertain if the patient has decision-making capacity. If the patient lacks decision-making capacity, care can be given involuntarily. If the patient retains decision-making capacity, the reason for the patient's refusal should be extensively explored. The patient might benefit from further education or counseling or even a change of care provider. Diligent efforts should be undertaken to give the patient all medical care that provides benefit and minimizes harm.

24.3.9 Financial Matters

Recipients of facial transplantation need to be aware of the financial costs associated with this intensive technology. In countries like USA, where there is no universal health insurance, patients with economic hardship sometimes are unable to access medical care and medication. Patients need to understand the financial implications of the surgery, medical care, and lifelong immunosuppressant medication. If a patient were unable to access routine follow-up care and medication, graft success would be in jeopardy. From an ethics perspective, medical management should be accessible to all (organ and) facial graft recipients, regardless of their financial status.

24.3.10 Finalizing the Consent Process

It is suggested that the candidate, after signing the consent form, has a cooling-off period during which time he/she reflects on the decision made and has moments for additional contemplation. The consent form should be re-signed by the candidate after the cooling-off period has elapsed (e.g., 2–3 weeks). If new information about the risk-benefit profile of facial transplantation is received during the cooling-off period, this information must be given to the candidate and future consent forms should be updated to reflect this new knowledge.

24.4 The Advocate

Throughout the pre- and posttransplant experience, the transplant recipient should have access to an independent advocate to approach for advice, questions/concerns, and reflection. It is suggested that this person be a hospital bioethicist as this person has formal training and experience in matters such as privacy, confidentiality, consent, coercion, conflict of interest, medical decision making, and quality of life. Lacking a hospital bioethicist, a member of the hospital ethics committee could fill this role. At the University of Louisville, their hand transplant team allows the transplant recipient to select their own advocate.[13] Problems with this approach include selecting family members who might have a strong personal bias for the procedure or selecting those who lack strong grounding in bioethics making them unable to be suitable watchdogs for ethical problems. While the hospital bioethicist or ethics committee member is a hospital employee and could be argued to be biased to support decision making that favors transplantation (for statistics, media relations, or other heroics), the bioethicist/ethics committee member should be free from such conflicts of interest by the nature of their hospital role. In fact, because they know the inner workings of the hospital, they are better positioned as an advocate compared to others who lack the "inside track." In their role, the advocate should formally assess the candidate and should have veto power during the candidate selection process just as they frequently do in the situation of assessing living organ donor candidates.[30]

24.5 Conclusion

The transplant team and advocate must feel confident that the recipient candidate understands the technology and is motivated to proceed.[31] To achieve this, the recipient candidate must explain their rationale for seeking facial transplantation and must detail their goals (realistic and unrealistic) of participation. Through this process, the team and advocate will come to understand what motivates the patient to participate in such a novel experience and they will also learn the fears, myths, and misunderstandings the patient may hold. More education may be needed. Referrals to other providers may be needed (e.g., psychologist, clergy). If overt conflict of interest is identified (e.g., candidate is planning a movie or book about the experience), this should trigger formal discussions with the advocate and Institutional Review Board/Research Review Committee as such could impair the decision-making objectivity of the candidate. The team should not proceed with surgery until after the cooling-off period (24.3.10), with all the candidate's questions answered, and all medical, surgical, psychosocial, and ethics screenings satisfied.

References

1. US Code of Federal Regulations Title 45 Chapter 46 (Protection of Human Subjects), Section 116 (General Requirements for Informed Consent). http://www.hhs.gov/ohrp/humansubjects/guidance/45cfr46.htm#46.116. Accessed April 5, 2010.
2. Greenwald L. *Heroes with a Thousand Faces: True Stories of People with Facial Deformities & Their Quest for Acceptance*. Cleveland: Cleveland Clinic Press; 2007:252-253.
3. Ye M. Psychological morbidity in patients with facial and neck burns. *Burns*. 1998;24:646-648.

4. Balakrishnan C, Hashim M, Gao D. The effect of partial-thickness facial burns on social functioning. *J Burn Care Rehabil*. 1999;20:224-225.
5. Coull F. Personal story offers insight into living with facial disfigurement. *J Wound Care*. 2003;12:254-258.
6. Rumsey N, Clarke A, White P, Wyn-Williams M, Garlick W. Altered body image: appearance-related concerns of people with visible disfigurement. *J Adv Nurs*. 2004;48:443-453.
7. Bradbury E. Understanding the problems. In: Lansdown R, Rumsey N, Bradbury E, Carr T, Partridge J, eds. *Visibly Different: Coping with Disfigurement*. Oxford: Butterworth-Heinemann; 1997:180-193.
8. Mowlavi A, Bass MJ, Khurshid KA, Milner S, Zook EG. Psychological sequelae of failed scalp replantation. *Plast Reconstr Surg*. 2004;113:1573-1579.
9. Edgell SE, McCabe SJ, Breidenbach WC, Neace WP, LaJoie AS, Abell TD. Different reference frames can lead to different hand transplantation decisions by patients and physicians. *J Hand Surg Am*. 2001;26:196-200. Erratum in: *J Hand Surg* [*Am*]. 2001;26:565.
10. Kanitakis J, Jullien D, Petruzzo P, et al. Clinicopathologic features of graft rejection of the first human hand allograft. *Transplantation*. 2003;76:688-693.
11. Lanzetta M, Petruzzo P, Vitale G, et al. Human hand transplantation: what have we learned? *Transplant Proc*. 2004; 36:664-668.
12. Baumeister S, Kleist C, Döhler B, Bickert B, Germann G, Opelz G. Risks of allogeneic hand transplantation. *Microsurgery*. 2004;24:98-103.
13. Tobin GR, Breidenbach WC, Klapheke MM, Bentley FR, Pidwell DJ, Simmons PD. Ethical considerations in the early composite tissue allograft experience: a review of the Louisville ethics program. *Transplant Proc*. 2005;37: 1392-1395.
14. Siegler M. Ethical issues in innovative surgery: should we attempt a cadaveric hand transplantation in a human subject? *Transplant Proc*. 1998;30:2779-2782.
15. Clark PA. Face transplantation: part II – an ethical perspective. *Med Sci Monit*. 2005;11:RA41-RA47.
16. Wiggins OP. On the ethics of facial transplantation research. *Am J Bioeth*. 2004;4:1-12.
17. Royal College of Surgeons of England. Facial Transplantation: Working Party Report, 1st ed. November 2003.
18. Royal College of Surgeons of England. Facial Transplantation: Working Party Report, 2nd ed. November 2006.
19. National Consultative Ethics Committee for the Health and Life Sciences. Opinion No. 82: Composite Tissue Allotransplantation (CTA) of the Face (Full or Partial Face Transplant). February 6, 2004.
20. American Society of Plastic Surgeons. Facial Transplantation-ASRM/ASPS Guiding Principles. www.microsurg.org/ftGuidelines.pdf. Accessed 13.01.11.
21. Pomahac B, Aflaki P, Nelson C, Balas B. Evaluation of appearance transfer and persistence in central face transplantation: a computer simulation analysis. *J Plast Reconstr Aesthet Surg*. 2010;63:733-738.
22. Aflaki P, Nelson C, Balas B, Pomahac B. Simulated central face transplantation: age consideration in matching donors and recipients. *J Plast Reconstr Aesthet Surg*. 2010;63:e283-e285.
23. Barker JH, Brown CS, Cunningham M, et al. Ethical considerations in human facial tissue allotransplantation. *Ann Plast Surg*. 2008;60:103-109.
24. Rumsey N. Psychological aspects of face transplantation: read the small print carefully. *Am J Bioeth*. 2004;4:22-25.
25. Basch SH. The intrapsychic integration of a new organ: a clinical study of kidney transplantation. *Psychoanal Q*. 1973;42:364-384.
26. Castelnuovo-Tedesco P. Organ transplant, body image, psychosis. *Psychoanal Q*. 1973;42:349-363.
27. Barker JH, Furr LA, McGuire S, et al. Patient expectations in facial transplantation. *Ann Plast Surg*. 2008;61:68-72.
28. Health Insurance Portability and Accountability Act (HIPAA) of 1996 (P.L.104-191). 45 Code of Federal Regulations, Part 160 and 164.
29. Cupples SA, Steslow B. Use of behavioral contingency contracting with heart transplant candidates. *Prog Transplant*. 2001;11:137-144.
30. Bramstedt KA. Living donor transplantation between twins: guidance for donor advocate teams. *Clin Transplant*. 2007; 21:144-147.
31. Renshaw A, Clarke A, Diver AJ, Ashcroft RE, Butler PE. Informed consent for facial transplantation. *Transpl Int*. 2006;19:861-867.

25 Legal and Regulatory Aspects of Face Donation and Transplantation

Alexandra K. Glazier

Contents

A.K. Glazier
New England Organ Bank, Inc, Waltham, MA, USA
e-mail: alexandra_glazier@neob.org

Abstract Face transplantation, like many new experimental procedures, emerged into the field without clear regulatory oversight. New surgical techniques, squarely within the "practice of medicine," are generally not regulated in the USA unless the surgical procedure involves a device for which FDA approval is required. Transplantation of organs and/or tissues, however, is one of the most highly regulated fields within medicine. Multiple federal agencies, national standards, and state laws and regulations provide a tight system of oversight for the donation and transplantation of organs and tissues. This chapter examines how face transplantation may be adopted into this framework based on previous experience with solid organs in the USA.

Abbreviations

CMS	Centers for Medicare Services
FDA	Food and Drug Administration
HHS	Health and Human Services
HRSA	Health Resources Services Administration
IRB	Institutional Review Board
NOTA	National Organ Transplant Act
OHRP	Office of Human Research Protections
OPO	Organ Procurement Organization
OPTN	Organ Procurement and Transplantation Network
UAGA	Uniform Anatomical Gift Act

25.1 Introduction

Face transplantation, like many new experimental procedures, emerged into the field without clear regulatory oversight. New surgical techniques, squarely within the "practice of medicine," are generally not regulated in

M.Z. Siemionow (ed.), *The Know-How of Face Transplantation*,
DOI: 10.1007/978-0-85729-253-7_25, © Springer-Verlag London Limited 2011

the USA unless the surgical procedure involves a device for which FDA approval is required. Transplantation of organs and/or tissues, however, is one of the most highly regulated fields within medicine. Multiple federal agencies, national standards, and state laws and regulations provide a tight system of oversight for the donation and transplantation of organs and tissues. Given this level of regulation, it is curious that face transplantation seems to fall outside much of it.

25.2 Consent Considerations

25.2.1 Consent for Facial Graft Donation

In the USA, the donation of a facial graft, like the donation of any organ or tissue, is governed by state law through the Uniform Anatomical Gift Act (UAGA).[1] The UAGA, which follows the general law of gifts, provides the legal requirements for the voluntary and uncompensated transfer of an organ from a deceased donor to a recipient. Under the UAGA, an anatomical gift requires (1) donative intent expressed by the donor or donor's family; (2) recovery of the organ upon the donor's death; and (3) acceptance of the anatomical gift by a recipient.

There are some unique considerations for consent to facial graft donation including the potential for the donor and/or donor family's confidentiality to be compromised in light of media interest. Further, the UAGA provides that an adult can make a legally binding donation decision prior to death through a donor registry, donor card, or other signed document. However, given the newness of facial graft donation and the profound visual effect on the donor's appearance for burial, family consent for facial graft donation seems appropriate at this point (though not legally required) even if the potential donor was in the donor registry. These distinct considerations for facial graft donation can be addressed by using a specialized consent form and by developing consent procedures for facial graft donation.

25.2.2 Consent for Facial Graft Transplantation

Consent for transplantation is governed by the informed consent doctrine that is an entirely different legal standard than the UAGA establishes for consent to donation. The informed consent doctrine fundamentally requires the consenting party make a decision regarding a proposed healthcare treatment or procedure through a facilitated understanding of the attendant risks and benefits. The legal duty to obtain informed consent is born out of the doctor–patient relationship.[2]

In the case of transplantation, there are some additional regulatory requirements to ensure that any donor-specific risks are communicated and understood by the transplantation recipient.[3] Facial graft transplantation poses additional informed consent requirements. Because facial graft transplantation is experimental, it is likely to be under an IRB protocol as human subject research. As such, specific elements of informed consent applicable to human subject research are required including a description of the study and its purpose, the risks, benefits, alternatives, and costs of participation.[4] Also, institutional expectations regarding confidentiality must be carefully described and understood by a potential recipient as part of the informed consent process given the intense media attention the first several face transplants have garnered.

Overall, the consent for facial graft donation and informed consent by the recipient for the transplantation raise some unique considerations but also fits within existing legal and regulatory consent requirements.

25.3 Unclear Regulatory Oversight: Is a Facial Graft an Organ or a Tissue?

In contrast to the consent laws which provide a functional framework, the regulatory oversight of the operational process of facial graft donation and transplantation appears to fall outside of existing laws and jurisdictional authorities. This is due to the fact that there is a regulatory divide between the regulation of "organs" and "tissues" and facial grafts are not clearly in either category.

The regulatory schemes for organs and tissues are significantly different based on the distinctive clinical pathways that organs and tissues are donated and recovered for transplantation. One fundamental difference is that donated organs are transplanted almost immediately after recovery whereas donated tissues are quarantined and processed before use. Also, the medical necessity for biological compatibility combined with the fact that there over 110,000 Americans awaiting organ transplantation, makes fair and appropriate allocation a primary regulatory concern for organs.[5] In contrast, because there is no requirement for biological compatibility and no comparable waitlist exists for patients in need of tissue transplants, safety for transplantation is the sole regulatory concern for tissues.

25.3.1 Regulation of Organ Donation and Transplantation

The donation and transplantation of organs is overseen by the US Department of Health and Human Services (HHS). This authority was originally established by the National Organ Transplant Act (NOTA) and in subsequent regulations known as the "final rule,"[6] NOTA defines "organ" as heart, lung, liver, kidney, and pancreas and also grants authority to the Secretary of HHS to designate additional "organs" – small intestine and islet cells were subsequently added. NOTA also created the Organ Procurement and Transplantation Network (OPTN) to implement the donation and transplantation process.[7] The OPTN is currently operated by the United Network of Organ Sharing (UNOS) under contract with the Health Resources Services Administration (HRSA). The OPTN maintains the national organ transplant waitlist, regulates the allocation of organs, and implements applicable clinical policies.

The Centers for Medicare Services (CMS) also has authority over transplant centers and organ procurement organizations (OPOs) because the vast majority of kidney transplants are paid for by CMS under the End Stage Renal Disease program. This historical arrangement, first implemented when kidney transplants were the only kind available, has grown to encompass all aspects of organ donation and transplant. As a result, there are more than a combined 150 pages of CMS "conditions of participation" regulations that apply to transplant centers and to OPOs that address everything from initial federal designation to perform organ recoveries or transplants, to quality improvement, consent, and clinical standards.[8,9]

25.3.2 Regulation of Tissue Donation and Transplantation

The regulatory framework is much simpler for tissue. The Food and Drug Administration (FDA) has regulatory jurisdiction to ensure the safety and effectiveness of any "product" that is marketed for use in humans including tissues that are processed for transplant.[10] The FDA defines human tissue to include, for example, corneas, bone, heart valves, ligament, and skin and has promulgated comprehensive "good tissue practices" regulations that govern the donor screening, testing, recovery, and processing of human tissues distributed for transplantation.[11] The regulations specifically do not govern "vascularized human organs"[12] nor is it clear whether FDA's regulatory authority could appropriately extend to tissues that are not processed or, in FDA terms, are not "more than minimally manipulated."[13]

25.3.3 No Direct Regulation of Face Donation and Transplantation

Based on these regulatory schemes, it is unclear where face transplantation would fit. Composite grafts like those used in the face transplantation are a combination of skin, bone, muscle, and underlying vasculature. These grafts are not "organs" as currently designated by HHS under NOTA and the Final Rule and regulated by the OPTN. Nor are these grafts, due to their vasculature, "tissues" subject to FDA's regulations and the existing process for tissue transplantation.

As a result, it seems that none of the existing significant regulatory oversight for organ and tissue transplant directly applies. Because facial transplants are experimental, the protocols require IRB approval as human subject research and are thus regulated by the Office of Human Research Protections (OHRP) through the "Common Rule."[14] The Common Rule applies to human subject research conducted at institutions that have entered into an Assurance with OHRP, which is required if institutions are engaged in federally funded research. These regulations exact protections such as appropriate risk benefit ratio for protocol approval and other specific recipient informed consent and confidentiality requirements. Human subject regulations are not, however, designed to regulate the intricate clinical and allocation issues unique to transplant.

Table 25.1 Comparison of tissues, organs, and composite grafts

Tissues	Organs	Composite grafts
Corneas, skin, bone, heart valves, veins, tendons	Kidneys, liver, heart, lungs, pancreas, small intestines, islet cells	Facial grafts, hand grafts (bone, skin, muscle, and underlying vasculature)
Recovered from deceased asystolic donors	Recovered primarily from deceased (brain-dead) heart-beating donors	Will be recovered primarily from deceased (brain-dead) heart-beating donors
Nonvascular; no biocompatibility requirements	Vascularized; blood type and, for kidneys, tissue typing compatibility required	Vascularized; blood type and tissue typing compatibility likely required
Quarantined and processed before transplantation (2 days to 5 years range)	Transplanted almost immediately after recovery (<12 h on average)	Transplanted almost immediately after recovery (<12 h expected)
No waitlist	110,000 Americans waiting	Expectation of a future waitlist
Regulated by the FDA to ensure safety through donor screening, testing, recovery, and processing	Regulated by OPTN/UNOS and CMS to ensure appropriate consent, donor suitability, allocation, and transplantation	Unclear

Nor do they regulate any part of the process of deceased donation.[15]

The clinical aspects of face transplantation (and, for that matter, other composite grafts such as hand transplantation) are most like those of an organ transplant (Table 25.1). Composite grafts are vascularized and recovered from a brain-dead heart-beating donor (like organs) and are transplanted immediately (like organs) thereby requiring stat serology testing (like organs) and blood and tissue typing requirements for biological compatibility (like organs). Additionally, as a practical matter, the donation of composite facial tissue for transplant will necessarily be coordinated by OPOs (like organs) because these organizations are uniquely capable by virtue of their federal designation to identify, screen, and test potential donors and coordinate the recovery of composite grafts.

25.3.4 Is Regulation of Face Donation and Transplantation Necessary?

It seems likely that the answer will be yes. The donation of part of a person is at the core of all human transplantation, and face transplantation is no different. The public trust that makes such a gift possible demands that the highest clinical and ethical standards will be followed. Although face transplantation will start off with small case studies, there is every reason to expect it will continue to increase and eventually move from experimental to accepted medical practice. The appropriate practice of medicine may work to achieve a clinical standard of care for the surgical aspects of face transplantation, but issues of donor consent and suitability, allocation, and the like will be critical. Lessons learned, sometimes the hard way, from the organ donation and transplantation system should be heeded. Review of past advances in this field, such as living liver donation, highlight that the need to calibrate clinical and ethical considerations is best accomplished through an encompassing regulatory system.[16]

An assertion of oversight by one (or more) of the agencies involved in regulating either organ or tissue transplantation may be inevitable but in the case of face transplantation there need not be a delay between the cutting edge of medicine and the regulation of the same. The timing is particularly relevant because composite graft transplantation is in a very early and experimental stage. Regulatory precedent set now will serve future innovations in transplantation.

The FDA, with its proscribed jurisdictional scope to regulate the safety of biological products, is not well positioned to address the full range of concerns likely to be raised by face transplantation. The FDA could assure appropriate standards of donor suitability and disease testing. But, with clinical safety as a mandated priority, the FDA may lack a comprehensive ability to oversee other elements of face transplantation that require attention including considerations attendant to the donation and allocation of a vascularized composite graft. In particular, the allocation of composite grafts, even in the absence of a sufficient prevalence of patients to establish a true waitlist, does invoke considerations of utility and equity whenever more than one potential recipient is biologically eligible for a particular donor composite graft.

Likewise, relying on regulation of individual protocols through the rubric of human subject regulations would leave significant gaps. An IRB may be able to consider the intricacies of recipient informed consent and, perhaps in the best of circumstances, monitor the potential clinical concerns raised by face transplantation. But an IRB has no clear jurisdictional authority over the composite graft donation process. Even if the OPO responsible for the donation under a particular protocol voluntarily followed the IRB's instruction, disparity between IRB decisions would create variability that could undermine the regulatory goal of achieving appropriate and uniform clinical and ethical standards in this field.

The entity best suited to oversee composite graft transplantation in the USA is the OPTN, which should promptly request that HRSA undertake this responsibility by having HHS designate "vascularized composite grafts" as "organs" under the Final Rule and NOTA. In fact, HRSA held a hearing in the spring of 2008 to consider whether vascularized composite grafts "are more analogous to transplants of organs ... than to tissue transplantation" and "the potential benefits of subjecting such transplants to the oversight of the OPTN and HRSA."[17]

Vascularized composite graft transplantation should be regulated as an organ because the transplantation of these grafts raises the same regulatory concerns: (1) appropriate donor and recipient consent; (2) donor suitability; (3) maintenance of a waitlist; (4) creation of an allocation system for matching potential donors with potential recipients including confirmation of biological compatibility and considerations of utility and equity; and (5) oversight of the clinical procedures necessary for the transplantation to be performed almost immediately after recovery of a vascularized graft.

References

1. Uniform Anatomical Gift Act (2006) at www.nccusl.org.
2. See e.g. Cantebury v. Spence, 464 F.2d 772 (D.C. 1972).
3. 42 CFR §482.102(a).
4. 45 CFR §46.116.
5. United Network for Organ Sharing. www.unos.org.
6. 42 U.S.C. §274 et seq.; 42 C.F.R. Part 121.
7. 42 U.S.C. §274 et seq.
8. Medicare and Medicaid programs: Conditions for coverage for organ procurement organizations. *Fed Regist.* 2006;71(104): 30982-31054.
9. Medicare program; hospital conditions of participation; requirements for approval and re-approval of transplant centers to perform organ transplants; final rule. *Fed Regist.* 2007;72(61):15198-15280.
10. 21 U.S.C. 301 et seq.; 42 U.S.C. §201 et seq.
11. 21 C.F.R. Parts 16, 1270 and 1271.
12. 21 C.F.R. §1271(3)(d)(1).
13. 21 U.S.C. §301 et seq (The Food Drug and Cosmetic Act); 42 U.S.C. §201 et seq. (Public Health Services Act).
14. 45 C.F.R. Part 46.
15. 45 C.F.R. 46.102(f)(defining "human subject" as a "living individual").
16. Cronin D, Millis M, Siegler M. Transplantation of liver grafts from living donors into adults: too much, too soon. *N Engl J Med.* 2001;344(21):1633-1638.
17. HRSA. Organ procurement and transplantation network [notice]. *Fed Regist.* 2008;73(42):11429. Id. at 11422.

The Issue of Death and End of Life

26

Krzysztof Kusza, Jacek B. Cywinski, and Marc J. Popovich

Contents

J.B. Cywinski (✉)
Department of General Anesthesiology, Cleveland Clinic Foundation, Anesthesiology Institute, Cleveland, OH, USA
e-mail: cywinsj@ccf.org

Abstract This chapter will provide an overview of the concept of human death, highlighting sociological and philosophical controversies related to definition of the death. Currently acceptable medical diagnostic criteria of the brain death will be revived in the context of organ donation. Lastly the concept of donation after cardiac death will be reviewed with associated ethical controversies.

Death, brain death, human death

"He who learns but does not think is lost. He who thinks but does not learn is in great danger"

"Learn without thinking begets ignorance, think without learning is dangerous"

Confucius

26.1 General Considerations

26.1.1 What Death Actually Means?

Biological sciences can define most biological phenomena following thorough examination and proving their course and nature. We can then say we know the pathophysiology and causes of a disease. Biological events well known to the science, the nature of which is unfortunately irreversible, include human death. It may be concluded that its definition meets all requirements for certain diagnosis and the determination of the time when it occurred. So these days we consider death a dissociated event. It means that death occurs in body tissues and systems at various times. This leads to disintegration of the body as a functional whole, and

M.Z. Siemionow (ed.), *The Know-How of Face Transplantation*,
DOI: 10.1007/978-0-85729-253-7_26, © Springer-Verlag London Limited 2011

consecutive, permanent discontinuation of its individual functions at a varying time sequence. In fact, some functions of the body or their parts may still be maintained for some time, in dissociation from the others that disappeared before. The dissociated nature of this phenomenon can be seen in particular situations when brain death already occurred whereas blood circulation is still maintained. In these cases, the brain condition determines human life or death. In most clinical cases, brain edema resulting from cerebral damage increases from the side of the supratentorial space, and brain stem dies as its last part. In these situations, the factor qualifying brain death is irreversible lack of brain stem functions.[1] Permanent damage of the brain stem is determined on the basis of the lack of specific nerve reflexes and the lack of spontaneous respiratory function. This procedure, mainly based on clinical studies, is possible in the prevailing number of cases, and its result is certain. In particular circumstances, however, the examination of nerve reflexes is not fully feasible (e.g., in faciocranial injury), and its interpretation may be difficult (e.g., intoxication, pharmacotherapy). Moreover, in primarily infratentorial brain damage, brain death determination requires a special diagnostic procedure, as in this case, the clinical signs of permanent brain stem damage do not mean irreversible damage of the entire brain. In these cases, suspected brain death must be supported by instrumental examinations.[1-4]

Continuous improvement of medical knowledge and experience and the introduction of new methods and technologies are a constant process. This allows introducing even better and more secure diagnostic and therapeutic methods into the practice. The introduction of instrumental methods in the brain damage determination procedures is a valuable addition to clinical examinations, and in some cases it may be a conclusive procedure. Instrumental, electrophysiological or vascular examinations can be performed in brain damage: primary (e.g., direct injury) and secondary (e.g., hypoxia); they are also indispensable in particular cases of brain death diagnostics in children. Long-term clinical practice has clearly shown that in selected cases, replacing the concept of death of a human as a whole with brain death and death of a human as a whole is justified from the scientific and practical point of view. In the light of the advance in medicine and the dynamic development of intensive care, this position is apparently necessary and legitimate with every respect. Despite great opportunities for saving the human life and health, which are currently provided by modern medicine, they have their limits. One of them is brain death.[4-7]

26.1.2 Sociological and Philosophical Aspects of Human Death

Both human birth and death are always accompanied by emotions. In the case of birth, they are positive, expected, and raise hope for a better future in humans, as well as are positively disciplining. Death is the second, after birth, secret of life, which causes completely different, negative emotions. Death is not expected, does not raise hope for a better future, and is not positively disciplining. Death, not only in the common language, is often described with additional terms to give it a more profound dimension. Death can therefore be premature, unnecessary, stupid, nonsense, undeserved or deserved, sudden, inevitable, martyr's, and heroic. Depending on which culture and continent we live in, death and birth are always accompanied with specific ceremonials. It emphasizes the mysteriousness of these events and the human's powerlessness against death.[8,9] However, death has also its administrative dimension. When a doctor certifies death, the dead person is eliminated from the society, deprived of his/her civic rights, and his/her close relatives are obliged to bury him/her. All above events that accompany death are a very important part of human existence. The medical definition of death is deprived of all those elements that in us, humans, simply raise ordinary sorrow, often disease, inability to cope with life, and the awareness of the irreversibility of this event.[10,11]

According to the classical definition of death, it occurs at the time of confirming definitive termination of respiration and cardiac arrest, resulting in death of a human as a whole. In Hellenistic and Judaic-Christian cultures, it was believed that as a result of death, the soul leaves the human body at a vague time. The classical definition of death does not mean death of all cells of the body, when death of the whole human could be concluded. The first attempts to define death as a phenomenon lasting a certain period of time were made by Xavier Marie Bichat in 1880,[12] who defined life as persistent resistance to death. This definition is therefore based on the preservation of all physiological functions which are crucial for the existence of life. This form of phenomena defining is referred to as "ex negativo," that is, proving by negation. Further definitions by Schwarz, Singer, Engelhardt, and Fletcher are consistent in their nature with the form of determining the death phenomenon by the philosophy of negation, according to which the actual human life is determined by: consciousness and the ability of perception, ability of thinking, ability of justified action, ability to communicate, having a set of concepts, ability of self-control, understanding the past and the future, and the ability to make contacts. Therefore, the existence of a human person depends on the presence of the act of consciousness and the prevailing role of active cerebral cortex.[13-16] The Papal Academy of Sciences took a position on this matter, presenting it in the "Declaration about the artificial extension of the life and the determination of the moment of the death" of 1985,[17] according to which, "a person is dead when he/she has suffered an irreversible loss of all capacity of integrating and of coordinating the physical and mental functions of the body." Another similar definition was presented by John Paul II at the World Congress of Transplantologists in Rome in 2000, saying that "death is a phenomenon consisting in complete dissociation of a closely connected whole, that is, a human person."[18]

However, the above considerations do not refer to the medical criteria of death in any way. On the other hand, their construction is an open formula, which seems to await to be completed with the medical shape of definition of human death. In fact, the philosophical considerations indicate that a human person is considered a whole, but not in the whole. Therefore, philosophy separates the whole as a functioning unit from the whole which is the sum of functions of each organ, tissue, and cell. We already know that one can survive the death of one's own heart but not one's own brain. Therefore, philosophy clearly suggests that there is a single key structure in the human body, which allows the functioning of all other structures, systems, and organs. From the biological and medical point of view, we are sure that death is a process occurring in time, and it results from primary disintegration of vital processes; on the other hand, however, this disintegration accelerates and intensifies the dying process, secondarily to the primary event. The individual physiological functions of the organs are first disrupted in the form of partial failure, progressing to complete loss of function of a given organ.[15,16,19-21]

Considering the entropy phenomenon as a thermodynamic function of state, which determines the direction of spontaneously occurring processes in an independent thermodynamic system, and that it is a measure of disorder in this system, death occurs when the human body loses its ability to resist its entropy. This means that the thermodynamic system (human body) goes from one state of balance to another, without the interference of external factors, and its entropy always increases, which clearly defines the occurrence of death.[19]

26.1.3 Medical Considerations of Brain Death and Death of a Human as a Whole

In addition to the arguments resulting from the philosophical and physical sciences which determine brain death as human death, also medicine advocates this concept, basing on the physiological parameters of blood flow through the brain, routinely used during brain death certification, thus proving that the concept of isolated brain death, and hence death of a human as a whole, is correct (Table 26.1).

The second, "temporal" complex is represented by ranges of time intervals when specific elements of the brain are devoid of blood flow and its substrates, causing their irreversible damage. At 37°C, the following anoxia period determines limit time intervals allowing for "restitutio ad integrum," that is, a return to the baseline condition:

Table 26.1 Table number (26) show the relationship among appearing neurological symptoms versus cerebral blood flow (mL/100g/min)

Cerebral blood flow (mL/100 g/min)	Neurological symptoms and condition
≥150 mL/min	Convulsions, intracranial hypertension, brain edema
40–60 mL/min	Physiological blood flow (without pathological symptoms)
≤25 mL/min	Confusion, slow EEG tracing and its gradual disappearance
≤20 mL/min	No evoked potentials
≤15 mL/min	Coma ("penumbra") – limit of survival
≤8 mL/min	Irreversible damage – death of the neurons, brain death

- 3–8 min – cerebral cortex
- 5–10 min – most structures of the brain stem
- 15–30 min – neurons of the respiratory complex of the brain stem

The causes of brain damage may be classified into primary and secondary. Primary causes include: cranial and cerebral injuries, brain tumors, cerebral hemorrhage, cerebral infarct, neuro infections, and impaired flow of cerebrospinal fluid. Secondary causes include: impaired circulation (cardiac arrest, shock), intoxication, anoxia, metabolic disorders (hepatic coma, renal coma, and hypoglycemia). The course of brain damage differs depending on whether its cause is primary or secondary[22-24]

The brain death concept assumes that human death also occurs when brain functions are irreversibly lost, despite maintained blood circulation. To justify the correctness of the brain death concept, it is necessary to fulfil the following conditions[9]:

- Death is a process dissociated in time.
- The cerebral cortex is the structural basis for physiological processes crucial for the human nature.
- The brain stem has significant control functions ensuring the biological integrity of the body.
- There is a pathophysiological process in the course of which isolated brain death occurs.
- The criteria of this state can be determined.
- It is possible to develop a diagnostic procedure allowing to determine the brain death criteria using the dedicated clinical tests.

26.1.4 Brain Death Criteria and Diagnostic Tests: Problems Related to the Development of Diagnostic Protocols

Death criteria, due to its social, legal, religious, and medical consequences, are only a certain convention necessary for the normal functioning of the societies. They have therefore a significant role consisting of the regulation of the reality and putting it in order. The things are similar in the medical sciences. However, it is worth noting that in the medicine, the definitions and criteria usually have practical implication in the form of taking or not taking specific diagnostic or therapeutic actions, contrary to the other sciences in which they usually describe the factual state.[25] The brain death definition is not an exception, it is a real definition; it means that it unambiguously describes the defined object or phenomenon. It also has other properties, it is an operational definition, i.e., it is formulated for a concrete, useful purpose and it required empirical procedures aiming at the determination of the presence or absence of the specified criteria. In a certain sense, it is an arbitrary, practical definition, allowing to solve the actual problems. It does not define human death, but it is also the initial point for this definition, its forecast; it is the defining term in the definition of human death. The arbitrariness, which is a feature of the brain death definition, raises serious problems of nonmedical nature, as well as significant difficulties in the development of diagnostic protocols.[26,27]

The first of them results from the dissociated nature of death. At the 22nd Congress of Medical Associations in Sydney in 1968, it was concluded that the situation in which death, as a process taking some time, occurs extremely clearly, is isolated brain death. A consequence of the death nature, which is dissociated in time, is that the death criteria, including the brain death criteria, depending on at which moment of the body's disintegration process we want to use may differ rather significantly.[9,20]

The second problem is caused by the need to achieve a compromise between the maximum safety of the procedure (minimizing the risk of positive error) and avoiding a very complex diagnostic procedure. The requirement of the maximum safety of the diagnostic procedure assumes the use of a broad set of criteria; however, from the point of view of everyday clinical practice, it is justified to simplify the diagnostic procedure and to use only those clinical methods which are the least labor consuming and allow to determine the most representative signs of brain damage. The conflict between these tendencies affects most diagnostic protocols; the best example is the determination of the observation time. A long period of clinical observation reduces the risk of committing a diagnostic error; on the other hand, however, it leads to a significant worsening of the organ function, which is important from the transplantation point of view.[7]

The third problem results from the lack of ability to verify some of the assumptions of the diagnostic procedure (observation time in infants and newborns, procedure in anoxic brain damage) based on scientific studies of the highest degree of robustness (class one recommendation). Such studies cannot be designed for several reasons; first of all, the sample size would be low and the study groups would be scattered at various clinical sites; secondly, long duration of the observation has a negative effect on procurement of organs for transplantation, as well as generates additional costs of prolonged therapy of critically ill patients. Many elements of the diagnostic procedures are based on expert opinions and clinical reports (low level of recommendation), which makes it difficult to develop a single, commonly used diagnostic protocol.[27-29]

Finally, the fourth problem is related to the different pathologies of the events leading to brain damage, and the fifth one results from the coexistence of various concepts of brain damage: the concept of brain stem death, the concept of the entire brain death, and the death of the superior brain. All above factors contribute to the formation of various diagnostic codes; the situation is additionally complicated by the large number of testing methods and their varying availability. It means that the brain death criterion is sanctioned by a certain convention determining the time of human death. This fact can be used to criticize the brain death concept, because a patient with an identical picture of the lack of neurological function, depending on the criteria used, can be considered "still alive" in one country but "already dead" in another one or will be never dead if we consider religious point of view. This is an argument for the standardization of the diagnostic protocols.[30-32]

This argument, albeit justified, seems utopian. Taking into account the cultural and religious pluralism, as well as different ethical norms, a standardized protocol of certifying brain death or certifying an irreversible loss of the cardiac function seems impossible to implement at present time. Of course, we can say that death is the same for all people, regardless of their religion, ethnicity or geographical location; however, it should be pointed out that a religious believes or a lack of them very often determine what would happen to us after death. And this fact significantly affects the transplantation and certification procedures. It is therefore more important to proceed fairly and ethically during the certification procedures, and decide any doubt on the patient's favor. We often the lack of understanding the death mechanisms causes aggression in this social group, resulting from powerlessness and insufficient intellectual grounds during the construction of legislative acts, including those related to the protocols of certifying brain death or irreversible termination of the cardiac function.[30]

In fact, ethics is a science on morality, aiming to determine the binding norms of actions, and to describe and explain the actually existing morality; it includes the value system and norms binding in a given community or professional group in a given epoch. It is worth remembering, however, that ethics cannot be disposed of, that ethics cannot be introduced by law, although it may be taught. In fact, the intellect is the ability of humans to understand, infer, judge, enabling them to know the facts and to strive to determine them, and understanding means become aware of the connections between the things and facts; therefore, we deeply believe that the intellect must prevail for the benefit of our patients. As long as a ceremony of a repeated burial is practiced in the Indonesian islands, and the dead body is exhumed to reverse its bones and perform the final burial in the "famadihama" ritual in Madagascar, also cultivated by the North American Sioux, and the Hinduists consider the time of cranium rupture during dead body cremation as the time of death,[32] the standardized certification protocol is not possible, even in universal medicine. Should this happen, the above ceremonies require our respect, the willingness to understand them, and the right to fulfil them.

26.2 Definition of Death and Organ Donation

The demand for life-saving and quality-of-life-enhancing organ transplants has increased since the first successful organ transplants were performed. At the same time, societies realized the need for a uniform, medically and ethically acceptable definition of death, which would allow to procure suitable organs for transplants and prevent resource consuming futile life-sustaining efforts in terminally sick patients.

The definition of human death has always been a matter of ethical debate and great controversy. In 1968, a committee at Harvard Medical School published a landmark report to define irreversible coma.[33] Criteria described in that paper gradually gained acceptance in the medical community as well as in the societies as the definition of brain death. The concept of brain death was developed in part to allow patients with devastating neurologic injury to be declared dead before the occurrence of cardiopulmonary arrest to optimize organ donation and to provide justification for removal of the life support.[34]

The "Dead-donor rule," which requires that patients be declared dead before the removal of any life-sustaining organs has been used to justify procurement of the organs from brain-dead patients.[34] Advances in medicine and particularly in critical care made it possible to artificially sustain respiratory and circulatory functions making it impossible to apply the relatively straightforward old definition of death: The patient was dead when he/she ceased to have evidence of circulation, respiration, and neurologic functioning.[35] All three of these functions are interlinked and dependent on each other and lost over a very short period of time, with the loss of any one of them quickly leading to the loss of the other two.[34] However, with the development of mechanical ventilation and cardiac support measures, it became possible to have the continuation of respiration and circulation in the absence of any detectable neurologic functioning.[34] In 1981, a presidential commission articulated the Uniform Determination of Death Act, which states that "An individual who has sustained either (1) irreversible cessation of circulatory and respiratory functions, or (2) irreversible cessation of all functions of the entire brain, including the brain stem, is dead."[36,37] This rule has become the accepted standard for determining death and suitability of organ donation throughout the USA.[31] These regulations created a new medical diagnosis of "brain death" which is widely accepted in the USA and seems to be ethically sound based on the premise that an individual is dead when the brain is dead.

Clinical diagnosis of brain death allows organ donation or withdrawal of life support which will lead to cessation of circulatory and respirator functions. Declaration of brain death follows a certain set of examinations which determine irreversible loss of brain function: Brain death is declared when the brainstem reflexes, motor responses, and respiratory drive are absent in a normothermic, nonmedicated comatose patient with a known irreversible massive brain lesion and no contributing metabolic derangements.[38]

26.2.1 Diagnostic Criteria for Clinical Diagnosis of Brain Death

The American Academy of Neurology in a summary statement outlined the guidelines for determining of brain death[38]:

1. Prerequisites. Brain death is the absence of clinical brain function when the proximate cause is known and demonstrably irreversible.
   (a) Clinical or neuroimaging evidence of an acute CNS catastrophe that is compatible with the clinical diagnosis of brain death
   (b) Exclusion of complicating medical conditions that may confound clinical assessment (no severe electrolyte, acid-base, or endocrine disturbance)
   (c) No drug intoxication or poisoning
   (d) Core temperature ≥32°C (90°F)
2. The three cardinal findings in brain death are coma or unresponsiveness, absence of brainstem reflexes, and apnea.
   (a) Coma or unresponsiveness – no cerebral motor response to pain in all extremities (nail-bed pressure and supraorbital pressure)
   (b) Absence of brainstem reflexes
      - Pupils
        - No response to bright light
        - Size: midposition (4 mm) to dilated (9 mm)
      - Ocular movement
        - No oculocephalic reflex (testing only when no fracture or instability of the cervical spine is apparent)
        - No deviation of the eyes to irrigation in each ear with 50 mL of cold water (allow 1 min after injection and at least 5 min between testing on each side)
      - Facial sensation and facial motor response
        - No corneal reflex to touch with a throat swab
        - No jaw reflex
        - No grimacing to deep pressure on nail bed, supraorbital ridge, or temporomandibular joint
      - Pharyngeal and tracheal reflexes
        - No response after stimulation of the posterior pharynx with tongue blade
        - No cough response to bronchial suctioning

(c) Apnea – testing performed as follows:
- Prerequisites
  - Core temperature ≥36.5°C or 97°F
  - Systolic blood pressure ≥90 mmHg
  - Euvolemia. *Option*: positive fluid balance in the previous 6 h
  - Normal PCO_2. *Option:* arterial PCO_2 ≥ 40 mmHg
  - Normal PO_2 *Option:* preoxygenation to obtain arterial PO_2 ≥200 mmHg
- Connect a pulse oximeter and disconnect the ventilator.
- Deliver 100% O_2, 6 l/min, into the trachea. *Option*: Place a cannula at the level of the carina.
- Look closely for respiratory movements (abdominal or chest excursions that produce adequate tidal volumes).
- Measure arterial PO_2, PCO_2, and pH after approximately 8 min and reconnect the ventilator.
- If respiratory movements are absent and arterial PCO_2 is ≥60 mmHg (*option*: 20 mmHg increase in PCO_2 over a baseline normal PCO_2), the apnea test result is positive (i.e., it supports the diagnosis of brain death).
- If respiratory movements are observed, the apnea test result is negative (i.e., it does not support the clinical diagnosis of brain death), and the test should be repeated.
- Connect the ventilator if, during testing, the systolic blood pressure becomes ≤90 mmHg or the pulse oximeter indicates significant oxygen desaturation and cardiac arrhythmias are present; immediately draw an arterial blood sample and analyze arterial blood gas. If PCO_2 is ≥60 mmHg or PCO_2 increase is ≥20 mmHg over baseline normal PCO_2, the apnea test result is positive (it supports the clinical diagnosis of brain death); if PCO_2 is <60 mmHg or PCO_2 increase is <20 mmHg over baseline normal PCO_2, the result is positive (it supports the clinical diagnosis of brain death); if PCO_2 is <60 mmHg or PCO_2 increase is <20 mmHg over baseline normal PCO_2, the result is indeterminate, and an additional confirmatory test can be considered.

26.3 Care for the Brain-Dead Organ Donor

In order to increase the number of transplantable organs, United Network of Organs Sharing (UNOS) created the document called "The Critical Pathway for the Adult Organ Donor" which helps professionals in an organ donor's treatment plan. The Critical Pathway is a concise, one-page document, designed to help critical care staff and procurement coordinators understand and follow the steps required for effective donor management.[39] After brain death has been declared in potential organ donors and consent is given for donation, donors need to be medically managed to keep their organs viable until organ recovery can occur; this period of intensive care management is particularly important because it significantly affects the quality of the procured organs. The Critical Pathway describes optimal care for the organ donor and maps the process to improve the outcome for successful organ transplantation. The pathway encourages and promotes collaboration between organ procurement team members including coordinators and critical care unit personnel and delineates roles to prevent duplication of effort or confusion.[39] It has been shown that adherence to the Critical Pathway, which has been endorsed by four major transplantation associations, significantly increased the number of organs procured and transplanted from brain-dead donors. A study by Rosendale et al. demonstrated a 10.3% increase in organs recovered and an 11.3% increase in organs transplanted.[40] There is no sacrifice in the quality of the transplanted organs or an increase in donor management time.

Some brain-dead donors fail conventional resuscitation measures in the intensive care unit oriented to optimize cardiac output, perfusion pressures, and metabolic status. In these cases, three-drug hormonal resuscitation therapy has been shown to improve hemodynamic and metabolic parameters. Rosendale et al. demonstrated that administration of a methylprednisilone bolus, infusions of arginine vasopressin and triiodothyronine to 701 brain-dead donors resulted in a 22.5% increase in the number of organs transplanted per donor.[41]

26.4 Donation of Organs After Cardiac Death

Because the number of available organs available for transplantation continues to lag behind the number of patients awaiting transplantation, a small but growing proportion of organ donations are now being supplied after cessation of irreversible circulatory function, the so-called donation after cardiac death (DCD).[42] Declaration of cardiac death fulfils one of the two criteria defined in the Uniform Determination of Death Act, which is further permitted by the dead-donor rule – the latter being the fundamental principle of organ donation, simply stating that organs may only be harvested from those who are dead.[43] Unfortunately the DCD process in the USA remains without a definitive national standard, and is subject to local legal jurisdictions. Most hospitals that permit DCD use variations on protocols established by the University of Pittsburgh Medical Center (described below).[44]

Typically potential donors are those patients who have suffered severe, irreversible neurologic injuries but do not meet brain-death criteria.[42,43] Most frequently these patients are able to breathe spontaneously, which by definition precludes the diagnosis of brain death, but have no chance for sustained survival if removed from the ventilator.[42,45] However, other critically ill or injured patients have been considered candidates as well.[46] The process, more or less as originally described in the University of Pittsburgh protocol, is as follows[47]: After consent is obtained from relatives, the patient is taken to the operating room where he or she is placed on the operating room table and aseptically prepared and draped for organ procurement. The patient is then removed from all life supportive equipment and monitored for 60–90 min. If there is objective evidence of cardiac standstill, an additional period of time (conventionally 5 min, but the University of Pittsburgh protocol permits 2 min) is permitted to pass, after which the patient is declared dead. Only after this period may the process of organ harvesting begin.[46,47] Clearly, this technique of procurement is associated with difficulties and controversies.

26.4.1 Patient Selection

Organs obtained by DCD are obviously at risk for injury, related to inadequate blood flow that occurs while awaiting the dying process. As such, the selection of a potential donor becomes important because the time to death may be extremely variable.[42,46] The University of Wisconsin has published a tool[45] that permits evaluation of donors based on criteria such as age, presence of an endotracheal tube, presence and quality (i.e., frequency, tidal volume size) of spontaneous breathing, oxygenation difficulties, and the need for vasopressors. Points assigned with higher numbers indicated more severe dysfunction. A DCD tool score, which the sum of points, may then be used to determine the probability of expiration within 60 min after cessation of life support. Thus, potential donor with a low DCD tool score may not be further considered, on the grounds that the time to expiration may be prolonged.[45]

26.4.2 Is the Time-to-Death Declaration Sufficient?

There are examples in the medical literature of the so-called Lazarus syndrome, or auto-resuscitation after 10 or more minutes of cardiac standstill[46] While there are no reports of patients actually surviving meaningfully afterward, this phenomenon nevertheless has led to ethical controversies as to whether a 2- or 5-min wait is sufficient to declare death and to ensure compliance with the dead-donor rule. Currently, most jurisdictions in the USA that allow DCD permit these waiting times based on a preponderance of legal, medical, and ethical opinions.[47]

26.4.3 Conflict of Interest

Explicitly stated in most rules governing DCD practice is that the organ procurement team may in no way be involved in the care of the patient prior to declaration of death, except for the surgical preparation in the operating room.[42,44] The procurement team must be physically away from the patient until death is declared. Frequently the care and death declaration of the patient falls to an intensivist, but even this arrangement may result in a blurring of lines relative to conflict of interest because the intensivist may subsequently be involved in the care of the future organ recipient.[46]

26.4.4 Outcomes of Recipients with DCD Organs

Organs obtained via DCD are at risk for poor function and there is a body of outcomes literature which tends to confirm that hypothesis, both in kidney and liver transplantation. Morbidity and mortality have been reported to be higher in both groups.[46]

26.4.5 Special Consideration of Face Transplantation with a DCD Donor

So far an unprecedented consideration, it would seem likely that procuring a facial graft from a DCD donor would be extremely challenging given the intricate dissection involved, particularly relative to vascular patency. Although systemic heparin boluses may be administered antemortem to the potential donor[44] as long as explicit informed consent is obtained from the relatives, it is unknown whether this would help preserve graft vascular patency.

References

1. Wijdicks EF. Determining brain death in adults. *Neurology*. 1995;45:1003-1011.
2. Bohatyrewicz R, Nestorowicz A, Kusza K. Commentary on diagnosis procedures of brain death. *Anestezjol Intens Ter*. 2008;40:114-116.
3. The Quality Standards Subcommittee of the American Academy of Neurology. Practice parameters for determining brain death in adults (summary statement). *Neurology*. 1995;45:1012-1014.
4. Combes JC, Chomel A, Ricolfi F, d'Athis P, Freysz M. Reliability of computed tomographic angiography in the diagnosis of brain death. *Transplant Proc*. 2007;39:16-20.
5. Dupas B, Gayet-Delacroix M, Villers D, Antonioli D, Veccherini MF, Soulillou JP. Diagnosis of brain death using two-phase spiral CT. *AJNR Am J Neuroradiol*. 1998;19:641-647.
6. Okii Y, Akane A, Kawamoto K, Saito M. Analysis and classification of nasopharyngeal electroencephalogram in "brain death" patients. *Nihon Hoigaku Zasshi*. 1996;50:57-62.
7. Litscher G. New biomedical devices and documentation of brain death. *The Internet Journal of Advanced Nursing Practice*. 1999;3(2). http://www.ispub.com/ostia/index.php?xmlFilePath=journals/ijanp/vol3n2/brain.xml.
8. Jonas H. Gehirntod und menschliche Organbank: zur pragmatischen Umdefinierung des Todes. W: Tenże: Technik: Medizin und Ethik: zur Praxis des Prinzips Ver-antwortung. Frankfurt am Main, Insel. 1985:222.
9. Beecher HK. After the "definition of irreversible coma". *N Engl J Med*. 1969;281:1070-1071.
10. Komunikat w sprawie wytycznych Krajowych Zespołów Specjalistycznych w dziedzi-nach: anestezjologii i intensywnej terapii, neurologii i medycyny sądowej w sprawie kryteriów śmierci mózgu. Dz.Urz.MZOS, z 26 czerwca 1984 r., nr 6, poz.38.
11. Biesaga T. Kontrowersje wokół nowej definicji śmierci. *Med Praktyczna*. 2006;2:20-26.
12. Bichat X. Recherches physiologiques sur la vie et la mort. 1800.
13. Schwarz M. Biologische Grundphänomene von Lebenswesen. Wien 1995;109.
14. Singer P. *Praktische Ethic*. Stuttgart: Reclam; 1984.
15. Engelhardt HT. *The Foundation of Bioethics*. 2nd ed. Oxford, New York: Oxford University Press; 1966:chap 3, 108-111.
16. Fletcher J. Indicators of humanhood. A tentative profile of man. *Hastings Cent Rep*. 1972;2(5):223.
17. Papieska Akademia Nauk. Deklaracja o sztucznym przedłużaniu życia i dokładnym ustaleniu momentu śmierci z dnia 21.10. 1985.
18. Jan Pawel II. L'Osserv. Rom. 2000;228:11-12.
19. Korein J. The problem of brain death: development and history. *Ann NY Acad Sci*. 1978;315:19-38.
20. WMA Declaration of Sydney on the Determination of Death and the Recovery of Organs. In: 22nd World Medical Assembly; August 1968; Sydney, Australia.
21. Shewmon AD. The brain and somatic integration: insights into the standard biological rationale for equating "brain death" with death. *J Med Philos*. 2001;26:457-478.
22. Lassen N, Astrup J. Cerebral blood flow: normal regulation and ischemic thresholds. In: Wenstein PR, Faden AI, eds. *Protection of the Brain from Ischemia*. Baltimore: Williams and Wilkins; 1990.
23. Schmidt RF, Thews G. *Physiologie des Menschen*. Berlin: Springer; 1987.
24. Betz E. Cerebral blood flow: its measurement and regulation. *Physiol Rev*. 1972;52:595-630.
25. Spann W. Die Bestimmung des Todeszeitpunktes aus gerichtsärztlicher Sicht. In: W Krosl W. Scherzer E (editors): Die Bestimmung des Todeszeitpunktes. Wien 1973.
26. Kopania J. Śmierć organizmu, a śmierć człowieka. Wydawnictwo Uniwersytetu Śląskiego. 1999;51-60.
27. Young GB, Shemie SD, Doig CJ, Teitelbaum J. Brief review: the role of ancillary tests in the neurological determination of death. *Can J Anaesth*. 2006;53:620-627.
28. Report of the Study Group on Brain Death. Guidelines and criteria for diagnosis of brain death. *JMAJ*. 1985;94:1949-1972.
29. Report on the criteria for the determination of brain death in children. *JMAJ*. 45(8): 336–357, 2002.
30. Wennervirta J, Salmi T, Hynynen M, et al. Entropy is more resistant to artifacts than bispectral index in brain-dead organ donors. *Intensive Care Med*. 2007;33:133-136.
31. Wijdicks EF. Brain death worldwide: accepted fact but no global consensus in diagnostic criteria. *Neurology*. 2002;58:20-25.
32. Kerrigan M. *The History of Death: Burial Customs and Funeral Rites, from the Ancient World to Modern Times*. Guilford: Amber Books Ltd; 2007:7-23.

33. A definition of irreversible coma. Report of the Ad Hoc Committee of the Harvard Medical School to examine the definition of brain death. *JAMA*. 1968;205:337-340.
34. Truog RD, Robinson WM. Role of brain death and the dead-donor rule in the ethics of organ transplantation. *Crit Care Med*. 2003;31:2391-2396.
35. Pernick MS. Back from the grave: recurring controversies over defining and diagnosing death in history. In: Zaner RM, ed. *Death: Beyond Whole-Brain Criteria*. Boston: Kluwer Academic; 1988.
36. President's Commission for the Study of Ethical Problems in Medicine and Biomedical and Behavioral Research: Defining Death: A Report on the Medical, Legal, and Ethical Issues in the Determination of Death. Washington, DC, Government Printing Office, 1981.
37. Guidelines for the determination of death. Report of the medical consultants on the diagnosis of death to the President's Commission for the Study of Ethical Problems in Medicine and Biomedical and Behavioral Research. *JAMA*. 1981;246:2184-2186.
38. American Academy of Neurology. www.aan.com/practice/guideline/uploads/118.pdf.
39. United Network for Organ Sharing (UNOS) Resources: Donor Management. http://unos.org/resources/donerManagement.asp?index=2.Accessed May 24, 2010.
40. Rosendale JD, Chabalewski FL, McBride MA, et al. Increased transplanted organs from the use of a standardized donor management protocol. *Am J Transplant*. 2002;2: 761-768.
41. Rosendale JD, Kauffman HM, McBride MA, et al. Aggressive pharmacologic donor management results in more transplanted organs. *Transplantation*. 2003;75:482-487.
42. Steinbrook R. Organ donation after cardiac death. *N Engl J Med*. 2007;357:209-213.
43. Ethics Committee ACoCCM, Society of Critical Care Medicine. Recommendations for nonheartbeating organ donation. A position paper by the Ethics Committee, American College of Critical Care Medicine, Society of Critical Care Medicine. *Crit Care Med*. 2001;29:1826-1831.
44. Swoboda SM, McRann DA, Alexander CE, Hall A. Solid organ donation: a resource for practitioners caring for the potential organ donor. *Contemp Crit Care*. 2009;7:1-11.
45. Lewis J, Peltier J, Nelson H, et al. Development of the University of Wisconsin donation after cardiac death evaluation tool. *Prog Transpl*. 2003;13:265-273.
46. Rady MY, Verheijde JL, McGregor J. Organ procurement after cardiocirculatory death: a critical analysis. *J Intensive Care Med*. 2008;23:303-312.
47. Institute of Medicine. *Organ Donation: Opportunities for Action*. Washington, DC: National Academics Press; 2006.

27 Organ Procurement Organization Approval Process: OPO Requirements for CTA Retrieval

Gordon R. Bowen, Charles Heald, and Daniel J. Lebovitz

Contents

Abstract The near-total face transplant performed by a multidisciplinary team of doctors and surgeons at Cleveland Clinic could not have occurred without the help of another critical partner – Lifebanc, northeast Ohio's organ procurement organization (OPO). While the focus of this ground-breaking surgery centered on the recipient, the procedure ensued thanks to an organ and tissue donor and her generous family.

For four years prior to the transplant surgery, Dr. Siemionow and her colleagues worked alongside Lifebanc to define and refine the process leading up to and following surgery. The OPO wanted to ensure that the highest medical and ethical standards would be maintained – regarding both the recipient and donor – before it consented to partner with Cleveland Clinic. This involved meticulous planning and work by Lifebanc in areas such as due diligence, the donation consent procedure, recovery logistics, and donor privacy.

In 2007, Lifebanc agreed to collaborate with Dr. Siemionow. "If there's anything we can do to save or enhance the lives of other people, we should be doing it," says Gordon Bowen, Lifebanc Chief Executive Officer. It's a simple sound bite that summarizes Lifebanc's position, but reaching that point was complex. The following chapter hopes to illustrate this process.

Abbreviations

AOPO Association for Organ Procurement Organization
CTA Composite Tissue Allotransplantation
IRB Institutional Review Board
MAB Medical Advisory Board
OPO Organ Procurement Organization

G.R. Bowen (✉)
Lifebanc, Cleveland, OH, USA
e-mail: gordonb@lifebanc.org

M.Z. Siemionow (ed.), *The Know-How of Face Transplantation*,
DOI: 10.1007/978-0-85729-253-7_27,

27.1 An Introduction to Lifebanc

It is important to understand the primary role of an OPO like Lifebanc before delving into the part it played in the nation's first face transplant. For nearly 25 years, Lifebanc has saved and healed hundreds of thousands of lives as the nonprofit organ and tissue recovery organization for northeast Ohio. Lifebanc was one of the original seven independent OPOs in the USA. Today, there are 58 OPOs supporting organ and tissue donation nationwide.

As the region's federally designated OPO, Lifebanc serves more than 4 million people and works with 80 hospitals. It teams with two transplant centers in northeast Ohio: Cleveland Clinic and University Hospitals Case Medical Center. Each year, Lifebanc staff and volunteers educate nearly 100,000 students and adults to provide accurate information about organ donation, engage the public, and increase the number of registered organ and tissue donors.

Lifebanc's mission is:

- Lifebanc saves and heals lives through organ and tissue donation for transplantation.

"More than 105,000 men, women and children in the United States are waiting for transplants to continue living," says Bowen. "We are dedicated to helping them receive the gift of life that organ and tissue donation makes possible."

Lifebanc helps those on the organ waiting list through a series of steps: When brain death occurs in a patient, hospitals call the OPO. The organization evaluates the patient's suitability as an organ and tissue donor. Lifebanc checks the Ohio Donor Registry to see if the individual is a registered donor and consults with the patient's family. After donation status is confirmed, the OPO enters basic information, such as blood type and body size, into a national computer system run by the United Network for Organ Sharing (UNOS), which also holds similar data about patients awaiting transplants. UNOS matches donors to recipients, and Lifebanc works with the appropriate hospital staff to manage the donor, coordinate transplant teams, recover the organs and/or tissue, and deliver them to waiting transplant surgical teams and recipients. The OPO also provides memorial kits and offers grief support to donor families.

27.2 Organ Procurement Organizations and Research

When Lifebanc considered partnering with Dr. Siemionow on the face transplant, it had to weigh the possible ramifications that endorsing a life-enhancing procedure would have on the OPO's main purpose of providing life-saving organs. That concern would be voiced and debated by Lifebanc's staff, Board of Directors and Medical Advisory Board (MAB) many times. But one of the considerations that helped sway the OPO was its dedication to research.

"Part of our mission is to participate in new and innovative approaches to improve organ and tissue donation in the United States," says Daniel Lebovitz, M.D., Medical Director of Lifebanc, a Pediatric Critical Care specialist at the Cleveland Clinic Children's Hospital and an associate professor at the Cleveland Clinic Lerner College of Medicine. To that end, Lifebanc supports research that advances the transplantation field.

However, the OPO's commitment to research is not limited to transplantation. Researchers attempting to develop new therapies for a host of diseases, ranging from cancer to cystic fibrosis, rely on donated human organs and tissues. Lifebanc facilitates researchers in acquiring human organs and tissues needed to conduct studies, advance science, contribute to new treatments and save lives. For example, if a donor has an organ such as the pancreas that is unacceptable for transplantation and the family consents to research, this donor's pancreas may be used in diabetes research. Similarly, aortic tissue may be recovered and used to investigate specific aspects of heart disease.

Understanding the importance of research, Lifebanc was receptive to learning about Dr. Siemionow's studies on composite tissue transplantation and plans for the first face transplant.

27.3 Early Discussions Between Cleveland Clinic and Lifebanc

In the summer of 2004, Dr. Siemionow met with Bowen and Lifebanc's chief clinical officer. She presented more than 20 years of research on composite tissue transplantation and shared photos and stories of people with severely deformed faces who would benefit from the procedure. In turn, Lifebanc explained how it operates.

Intrigued by the concept, Bowen initiated conversations with Dr. Lebovitz and Lifebanc staff about whether face transplantation was part of the OPO's mission. "We had to decide if this was something we wanted to get involved with, because it had never been done before and at the time it seemed controversial," recalls Bowen.

That fall, Lifebanc received notice of approval for the surgery by Cleveland Clinic's Institutional Review Board (IRB). With a green light from Cleveland Clinic to move the research to reality, Lifebanc assembled its Board of Directors and Medical Advisory Board in November to consider its potential involvement in the procedure. "We all appreciated that this was going to be an innovative phenomenon," says Dr. Lebovitz. "But the question was how would it impact our primary mission and would it generate some backlash?"

Lifebanc's Medical Advisory Board created a Facial Transplant Subcommittee to perform due diligence and review the proposed protocol from Dr. Siemionow. The subcommittee consisted of an ethicist and two transplant surgeons – one from Cleveland Clinic and one from University Hospitals of Cleveland. Their task was to consider several issues, including whether this life-enhancing (rather than life-saving) procedure was consistent with Lifebanc's mission and what effect participation in this face transplantation procedure might have on organ donor registration.

"Any involvement would place Lifebanc front-and-center here and with the other organ procurement organizations across the United States," says Bowen. "We wanted to positively affect donations locally, regionally and nationally." The Medical Advisory Board decided it required more information from Dr. Siemionow, so the Facial Transplant Subcommittee drafted a list of more than 30 questions concerning clinical issues, tissue donation, ethics, the media and public perception. Here is a sampling of those questions:

- What steps will be put into place to ensure that other organs are not compromised?
- What is the maximum time frame from cardiac cessation to recovery?
- Where does the facial recovery process fall in the order of organ recovery?
- How do we handle operating room staff reactions?
- What are the social and psychological criteria for inclusion of the recipient, recipient's family and donor family?
- What is included in the consent, and who will perform the informed consent process?
- How does Cleveland Clinic plan to safeguard the core mission of Lifebanc to obtain life-saving organs for transplantation?
- Is the risk of death greater for the recipient than the benefit of the transplant?
- What happens if the recipient's immune system rejects the new face?
- What specialists are needed for the recovery and transplantation process?
- How many of these transplants do you project doing in the future?

"We needed to find out exactly what Dr. Siemionow and her team were going to do, how long the procedure was going to take and how involved Lifebanc would be," says Bowen.

27.4 Lifebanc Raises Ethical and Logistical Concerns

In August 2005, Lifebanc's Facial Transplant Subcommittee met to review responses from the Clinic's surgical team and determine if the OPO was prepared to participate. Although members of the subcommittee remained supportive of the concept of face transplantation, they still had a few reservations. Dr. Lebovitz wrote a letter declining participation for three main reasons: discomfort with the informed consent process, ambiguity surrounding Lifebanc's exact role in the process and the potential psychological impact of the surgery on the donor family, recipient and recipient family. Lifebanc offered to assist the facial transplantation team in addressing these issues.

Informed consent for donation is paramount to Lifebanc. "We tell donor families everything we are going to do and answer all their questions to the best of our ability," says Bowen. "In this case, we felt like we didn't know enough about face transplantation, so we could not oversee the informed consent process." Bowen and Lebovitz encouraged Cleveland Clinic to create its own consent form for the composite tissue transplantation. They shared Lifebanc's consent form for organ and tissue donations and provided input to representatives from the Clinic, including the hospital's in-house coordinator who was trained to obtain informed consent for the recovery of the face. In addition, Lifebanc and the Cleveland Clinic figured out how the consent process would transpire: A family support liaison from Lifebanc would handle consent for donation of standard organs and tissues first, such as the heart, lungs, kidneys, connective tissue and corneas. Afterward, the in-house coordinator from Cleveland Clinic would obtain separate consent for the necessary composite tissue to perform the facial transplantation.

To assuage Lifebanc's two other concerns with this proposal and gain more detailed information about Dr. Siemionow's plans, the OPO requested Cleveland Clinic complete the standard forms for inclusion of Lifebanc in any research study. Lifebanc Research Request forms are adapted from the national Association for Organ Procurement Organization's (AOPO) template outlining general guidelines for OPO research participation to help steer all 58 OPOs it serves. These forms ask researchers, in this case Dr. Siemionow's team, to delineate the goals of their project, explain how it could affect the timing of organ and tissue recovery, outline personnel and resources required of Lifebanc and so on.

27.5 OPO Seeks National Direction and Support

As talks between Lifebanc and Dr. Siemionow's team progressed and ideas on how to partner solidified, the OPO found itself in uncharted territory. Bowen and his staff were keenly aware that whatever steps Lifebanc took would be watched by AOPO and the other 57 OPOs. Therefore, Lifebanc sought the opinions of its peers on face transplantation.

Bowen presented an overview of the process and the status of Lifebanc's relationship with Dr. Siemionow's team at AOPO's 2006 annual meeting. They admitted interest in assisting the research team, predicted that the field of composite tissue allografts would likely continue to grow and advocated for Lifebanc and other OPOs to be active in the process. They reported some of the ambivalence of the Lifebanc Board of Directors was due to uncertainty of the effect on organ and tissue donation in northeast Ohio. Then Bowen made a bold request, asking AOPO and UNOS to survey their members on issues surrounding this topic. "We wanted to see what the perception was, because it was so new," says Bowen. "No one had a clue." AOPO and Lifebanc created a survey, which was completed by 25 of the association's 58 OPO members.

In January 2007, Bowen and Dr. Lebovitz presented a summary of the results to the AOPO Executive Committee. Some of the questions and responses are as follows:

Does your OPO have an official stance on whether a face transplant is an organ or tissue transplant?

Yes	12%
No	88%

Does your OPO have an official stance on participating with a researcher on a face transplant case?

Yes	12%
No	88%

Does face transplant fit into your OPO's mission statement?

Yes	76%
No	24%

Does the community need an official stance from one or more national organ donation/transplantation organizations?

Yes	56%
No	44%

Should UNOS be asked to address some of the issues associated with face transplants in its committee structure?

Yes	48%
No	52%

Do you consider this procedure research or transplant?

Research	48%
Transplant	52%

Do OPOs need a separate consent form in this case?

Yes	68%
No	32%

Do you believe that the face transplant issue might have a negative impact on organ and tissue donor rates locally and/or nationally?

Yes	36%
No	64%

Bowen subsequently made presentations to other national organizations to gain insight into a wider opinion. He and a representative from the Boston-area OPO were part of a roundtable discussion at the Health Resources and Services Administration's offices. Several organizations attended the event, including the American Association of Tissue Banks, the Centers for Medicare & Medicaid Services, the U.S. Food & Drug Administration and others. "Experts from the OPO and transplant community hoped to gain a better understanding of the potential number and types of CTA procedures and the role that government and regulatory agencies would have in these new procedures," says Bowen.

Meanwhile, Dr. Siemionow's team completed and returned the Research Request forms to Lifebanc in 2007. The MAB still had some questions, so it invited Dr. Siemionow to its next meeting. "Once she explained the procedure in person, we felt more comfortable," says Bowen. Now that Lifebanc understood and concurred with the logistics of the transplant surgery, it had one other main concern: How would they handle the onslaught of mainstream media attention?

27.6 Lifebanc Agrees to Partner with Cleveland Clinic

A couple years earlier, a news story was broadcast nationwide that identified the OPO as partnering with Cleveland Clinic on the first face transplantation. The story reinforced the importance of establishing a cohesive media and communication plan between Lifebanc and Cleveland Clinic. In January 2007, Lifebanc's MAB approved the facial transplantation protocol on the condition that the OPO and Cleveland Clinic create a joint media/communications plan.

The OPO and hospital strategized on how to present and control news about the facial transplant to internal staff, donor hospitals, transplant centers, local and national organizations and worldwide media outlets. They developed an extensive list of talking points for Bowen and Lifebanc's staff, as well as for Dr. Siemionow's team and Cleveland Clinic, to ensure a consistent, accurate message. The detailed media and communication plan proved invaluable leading up to the surgery and afterward. In particular, it helped protect the identity of the donor and donor family, an issue of utmost importance to Lifebanc.

Lifebanc's agreement to support Dr. Siemionow hinged on three additional stipulations:

- The donor needed to come from the Cleveland Clinic main campus only.
- Approval of this procedure would be for only one patient with reevaluation following.
- Lifebanc would screen the patient, collect and provide donor information to Dr. Siemionow, and be present with the donor family during the Cleveland Clinic in-house coordinator's informed consent process.

After nearly three years of discussions with Dr. Siemionow and her team, Lifebanc's ultimate decision to partner with Cleveland Clinic stemmed from its dedication to improving lives. "It was important that we as an OPO were willing to actively participate as a member of this team in this innovative facial transplant procedure that could enhance quality of life for an individual," says Dr. Lebovitz. "Lifebanc wanted to be a part of this."

With a partnership forged, there was still more work to be done. At Lifebanc's request, AOPO invited Dr. Siemionow to speak at its 2007 annual meeting on the clinical perspective of the facial transplantation process. (This was the same meeting where Bowen revealed survey results regarding the process from 25 of the 58 nationwide OPOs.)

In the meantime, Dr. Siemionow screened possible recipients and worked alongside Lifebanc to develop criteria for the appropriate donor. She required a donor between 18 and 70 years of age who would be the same gender as the recipient. Other criteria included skin tone matching and the nature of the injury (no face trauma). Lifebanc added to the list. "The key issue for us was to find an individual who previously affirmed his or her desire to be an organ donor on the state donor registry (driver's license) and a family that wanted to donate everything possible and consented to research," says Bowen. "In addition, it had to be a donor family that Lifebanc's family support liaisons felt was emotionally stable because of the predictions of high media involvement. They had to be prepared."

The narrow scope forced Lifebanc to reconsider its position about seeking a donor only from the hospital's main campus. In 2008, Lifebanc's Board of Directors agreed to expand its donor pool to Clinic affiliate hospitals. The Clinic administration notified all hospitals, which consented to transferring the identified donor to the main campus. Ultimately, Lifebanc agreed to consider donors from all hospitals within its service area. Conversations between Lifebanc and Dr. Siemionow continued as they waited for the right donor and recipient.

The two also educated one another: Just as Lifebanc needed to learn about composite tissue allografts, so too did Dr. Siemionow's team require a tutorial on the donation process. Focused on plastic and vascular surgery, the team had to comprehend the standard organ transplant and tissue portions of the recovery process. So Lifebanc arranged for the facial transplant team specialists to observe donor management in the intensive care unit and recovery in the operating room. "The facial transplant team witnessed how all these organ transplant teams work together – one team recovers the heart, another recovers the kidney and so on," says Dr. Lebovitz. "Then Dr. Siemionow could figure out how to bring her team into the same process as part of this whole symphony." Lifebanc assisted the plastic and vascular physicians and other medical staff in gaining an understanding of OPO regulations, differentiating between an organ and tissue donor, and learning about post-donation funeral and embalming practices.

Lifebanc was equally concerned with educating its own staff. A plastic surgeon on Dr. Siemionow's team presented information on face transplantation at a Lifebanc staff meeting. The OPO stressed the importance of donor confidentiality, reminding employees of the confidentiality form they sign annually agreeing to protect donor identities. Lifebanc understood that confidentiality in this case would be particularly hard to maintain, particularly for those working in the hospitals during recovery who might be hounded by the media.

27.7 Lifebanc's Role in the Face Transplantation

In December 2008, Lifebanc identified a potential facial composite donor, a woman listed on the Ohio Donor Registry who had consented to organ and tissue donation as well as research. The OPO moved swiftly into action, coordinating the well-orchestrated donation process.

Lifebanc's family support liaison approached the family asking if they would donate her heart, lungs, pancreas, kidneys, liver, small intestine, and tissue to save and heal other's lives. After they agreed, the family support liaison mentioned the face transplantation and asked if the family was willing to learn more. They were, so Cleveland Clinic's in-house coordinator was called upon to explain the procedure and receive separate consent. The family also granted permission to transport the donor's body to Cleveland Clinic's main campus and have the body cremated after recovery to avoid the potential for media sensationalism surrounding the funeral. Thanks to well-defined donor criteria and the sound judgment of the family support liaison, this special family was the first and only one asked to participate in the procedure.

Lifebanc representatives witnessed this second informed consent process, during which time the family was notified about unique aspects of this particular donation: The first near-total face transplant done in the USA would generate a media buzz, and they might get swept up in it. Once the recipient's name was released, the family would know who received their loved one's composite tissue. Lifebanc assured the donor family it would keep them updated throughout the entire process and offer the same support it provides all donor families.

Aside from the transfer to Cleveland Clinic, donor management proceeded as usual. Lifebanc's objective was to maintain homeostasis and organ perfusion until recovery could occur. The OPO undertook several steps that are critical to successful transplants, such as conducting a medical/social history of the donor, verifying blood and tissue type, and running serology tests.

The OPO contacted the organ transplant teams, who initiated recovery simultaneously with Dr. Siemionow's team. The in-house coordinator, along with Cleveland Clinic police, controlled access to the operating rooms where both the recovery and face transplantation occurred in yet another move to ensure confidentiality of the recipient and donor. After recovery, the donor body was escorted to the crematorium.

27.8 The Aftermath

Though the procedure was finished, Lifebanc's work was not complete. As part of the media plan carefully crafted by both the OPO and Cleveland Clinic, Bowen participated in a post-procedure news conference. Lifebanc also issued a statement reinforcing its position on the procedure and its overall mission. It read, in part, "While this case is a remarkable medical advancement, we must never forget the donor and donor family in this process. We remain focused on meeting the continuing need to increase organ and tissue donation for those individuals who are in need of life-giving transplants."

One of Lifebanc's main concerns throughout the process was that the number of registered donors might decrease when word spread about the face transplantation. Those apprehensions were alleviated by four years of scrupulous planning leading up to the transplantation. "We had a communication plan in place to show we did our due diligence," says Bowen. "We did not want this to negatively impact donation."

It did not. Each year since 2006, the number of new registrants on the Ohio Donor Registry in the 20 northeast Ohio counties served by Lifebanc has steadily increased. In December 2008, when the procedure was performed at Cleveland Clinic, 7,078 people in Ohio signed up as new donors. Six months later, after the recipient had come forward and shared her story with national media, the number of new registrants increased 44%.

There was no backlash against donation, partly because of two messages emphasized by Lifebanc. First, the OPO is committed to life-saving transplantations. Second, no one who is a registered organ or tissue donor is signed up as a face donor. This case involved a special consent process above and beyond what the other 80 million registered American donors have authorized.

"There was a lot more support for and less negativity about this procedure than we anticipated," says Dr. Lebovitz. "As an OPO, this experience makes us more willing to participate in high-profile research protocols in the future as long as we do our homework, prepare in advance and work with the right professional team." Lifebanc has agreed to team with Cleveland Clinic on three more composite tissue allografts. The OPO is prepared to be a leader in CTA, helping establish a roadmap for other organ procurement organizations in this pioneering field. "OPOs have a responsibility to assist people who have chosen to donate life-saving organs and tissues at the time of their deaths to help as many others as possible," says Dr. Lebovitz. "One of the ways we can do that is by participating in research in the area of organ donation. We need to be leaders in work that advances the science of organ donation."

Since 2008, Bowen has been a resource for others considering CTA. He continues to talk about Lifebanc's facial allograft transplant experience with OPOs in Baltimore, Boston, Chicago, New York and other cities. In those conversations, Bowen stresses the importance of performing due diligence, solidifying the consent process, and establishing a thought-out media plan. He also encourages OPOs to select the right partner. "We worked on this for years with Dr. Siemionow's team and held each other's hands," says Bowen. "We have a great relationship that we hope will continue to advance the science of composite tissue allograft transplantation to help more people with this need."

Part V

Societal, Financial, and Public Relations Issues in Face Transplantation

28 Addressing Religious and Cultural Differences in Views on Face Transplantation

Antonio Rampazzo, Bahar Bassiri Gharb, and Maria Z. Siemionow

Contents

Abstract Up to present more than 80 cases of composite tissue allotransplants (CTA) have been reported across the world. The geographical distribution of the reported cases is prominently in favor of Europe and North America. It is therefore questionable if the religious and cultural views could have a role in determining the diffusion of face transplantation practice in different countries. While Christianity and Islam encourage organ donation, Buddhism, Shinto, and conservative branches of Judaism have had a tormented process of acceptance of brain death concept and cadaveric organ donation. No religion is formally against transplantation from deceased donors. Specific religious bans against face transplantation do not exist; however, a wide gap is present between the religious stances and the popular beliefs. Education on indications, organ procurement procedures, and treatment of the donor is needed to clarify erroneous beliefs and address the fears, helping the diffusion of the practice.

Abbreviation

CTA Composite Tissue Allotransplantation

28.1 Introduction

According to the International Registry on Hand and Composite Tissue Transplantation, since 1998, when the first successful hand transplantation was performed, 79 composite tissue allografts have been transplanted across the world. With regard to the distribution of the cases, Europe leads the field with the highest number of the transplantations performed (30), followed by North America (18), Asia (16), South America (15), and Africa (1). Face transplantation was performed for the

M.Z. Siemionow (✉)
Department of Plastic Surgery,
Cleveland Clinic, Cleveland, OH, USA
e-mail: siemiom@ccf.org

M.Z. Siemionow (ed.), *The Know-How of Face Transplantation*,
DOI: 10.1007/978-0-85729-253-7_28,

first time in 2005 in Amiens (France). Since then, seven cases have been performed in Europe, two in USA and one case was performed in China. From the geographic distribution of the transplant cases, it is clear that the practice of composite tissue allotransplantation (CTA) is lagging behind in many countries (Figs. 28.1–28.4). Besides the technical and medical aspects, we wondered if religious and cultural factors could impose a barrier limiting the diffusion of CTA application, especially in face transplantation. We considered if there were specific religious or cultural bans against face transplantation and, how the conclusions or views of each faith and culture on organ transplantation, eventually could be applied to facilitate acceptance of face transplantation.

28.2 Religious Views on Transplantation

Reviewing studies on the influence of religions on transplantation in general and specifically with respect to views on CTA proved to be a very complex problem. Three deeply interconnected aspects which are of major consideration in transplantation are the criteria of declaration of death and the brain death concept, organ donation, and acceptance of the donated organ. This powerful philosophical triad has been thoroughly investigated in the literature pertinent to transplantation of human organs. Excluding ethical discussions, which are not addressed in this chapter, we briefly summarize the viewpoints of the major religions on the issue of transplantation.

28.2.1 *Christianity*

The Christian community has a long-standing tradition of involvement in the care of sick and dying. The care for the sick is to care for the Christ himself. Christians believe that God became human in Jesus Christ, and in the person of Jesus, God affirms the dignity of human body. The human body should be treated respectfully with scrupulous value for the wishes of the deceased or

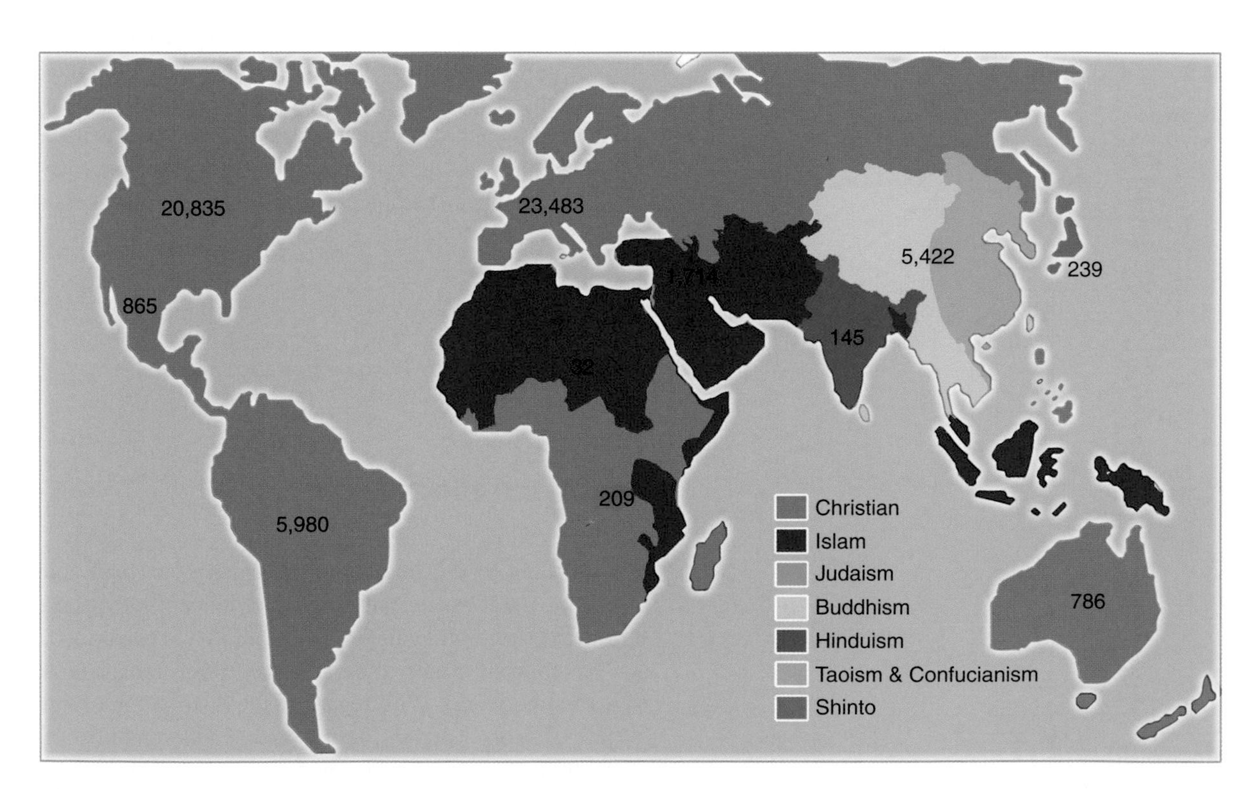

Fig. 28.1 The total number of solid organs transplanted in 2008 according to Global Observatory on Donation and Transplantation[49] is superimposed to world religious map. Europe and North America lead the field

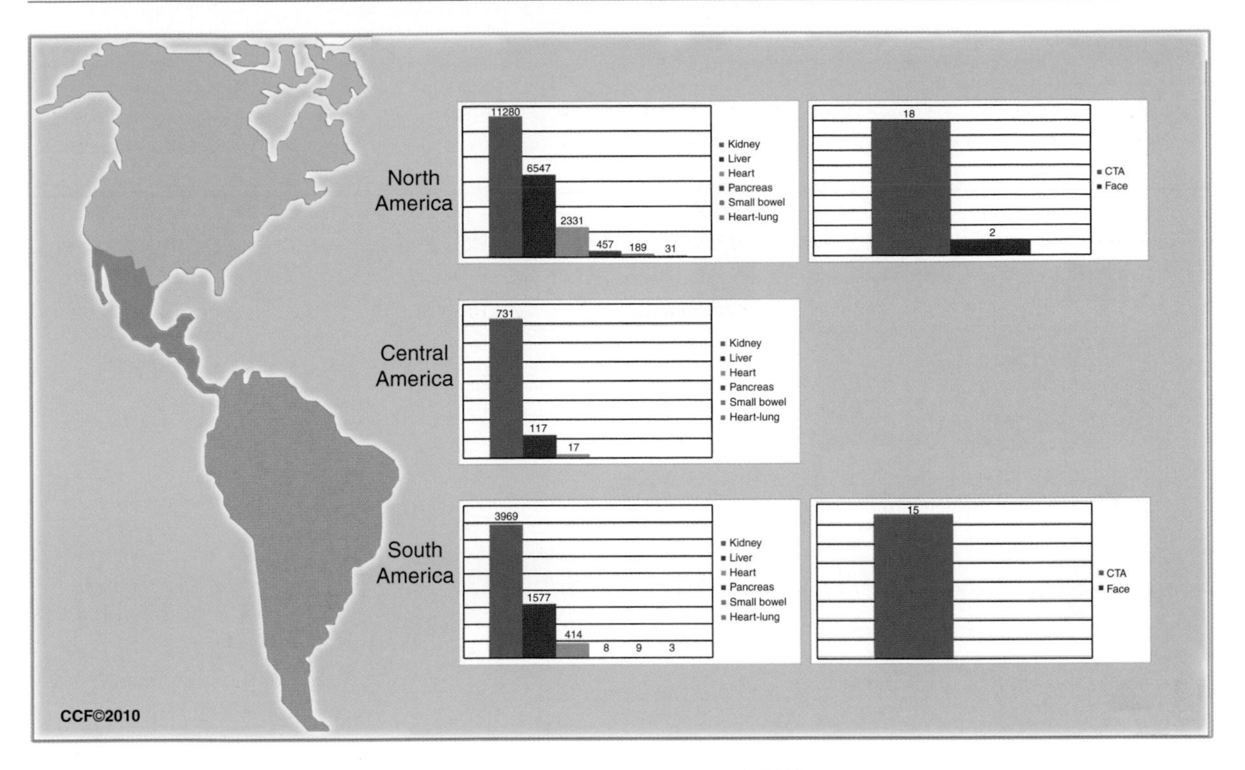

Fig. 28.2 The total number of solid organs and CTA transplanted in America in 2008

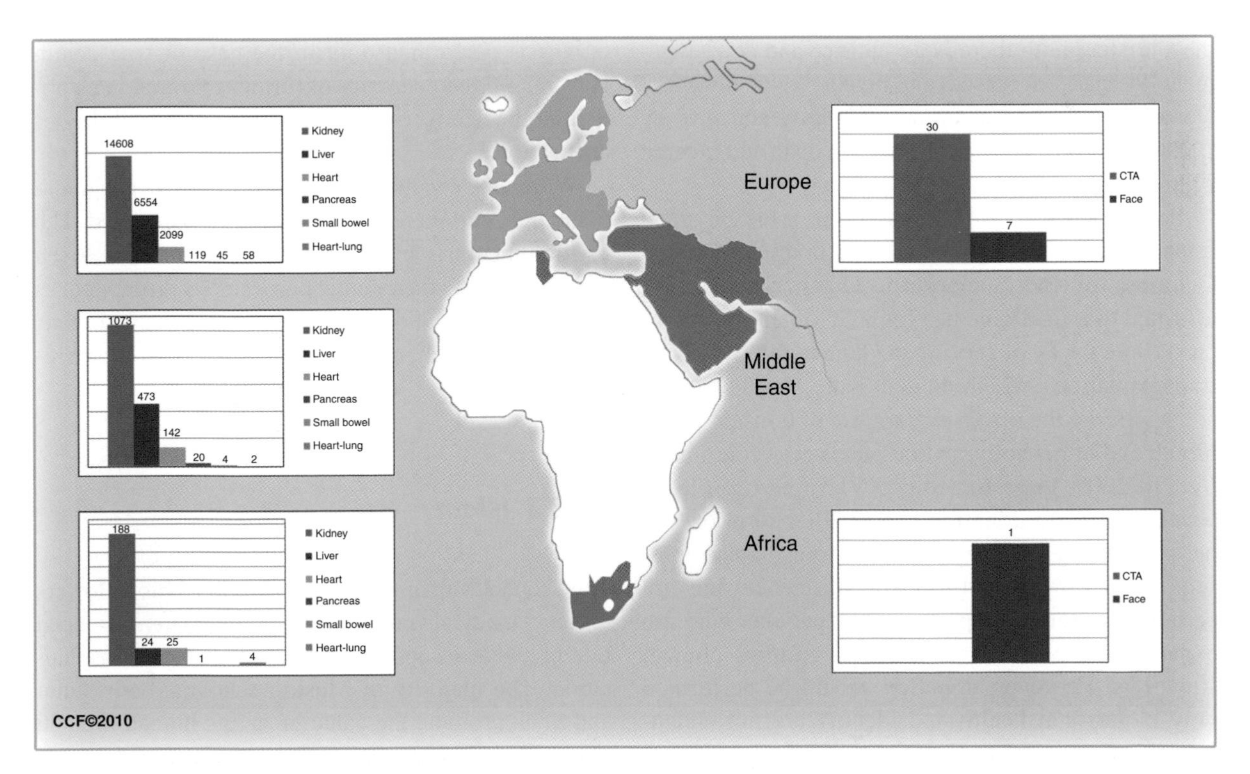

Fig. 28.3 The total number of solid organs and CTA transplanted in Europe, Middle East, and Africa in 2008

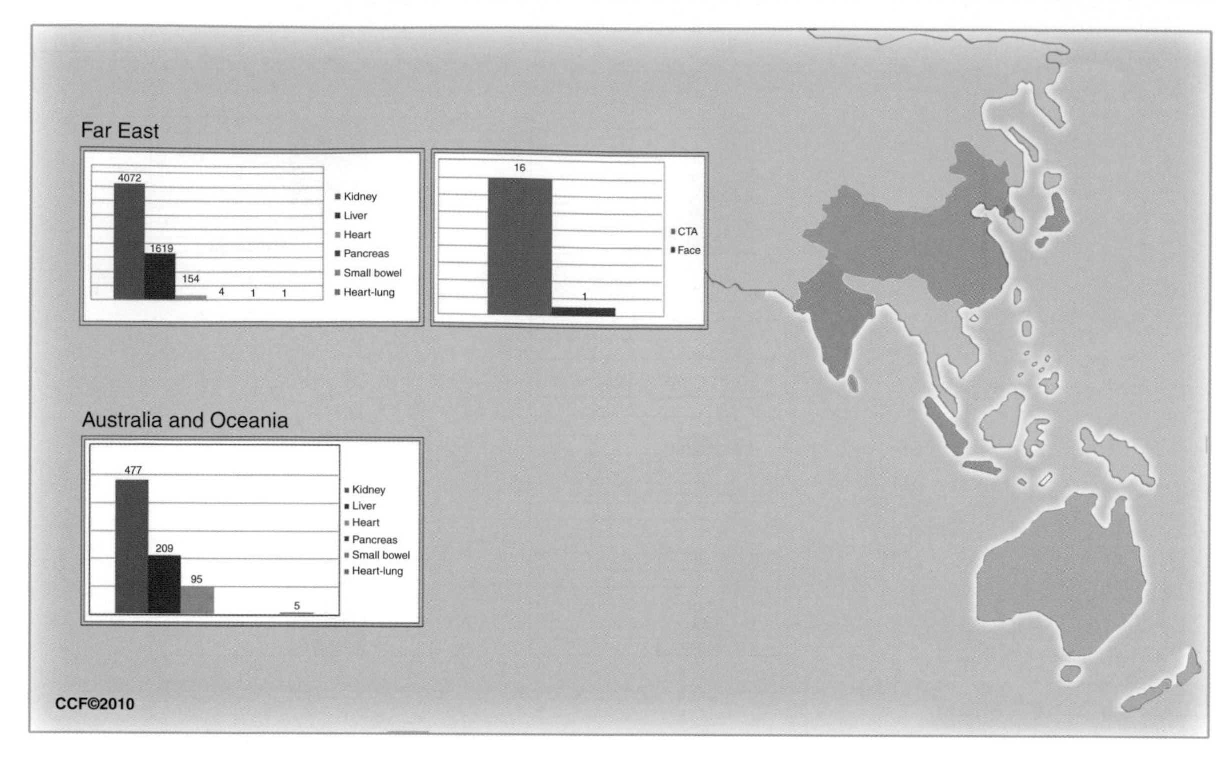

Fig. 28.4 The total number of solid organs and CTA transplanted in the Far East and Oceania in 2008

the next of kin.[1] With respect for the cadaver, within the Christian faith there is acceptance and encouragement to make deceased and living organ donation. Donation is viewed as an act of charity and love by Catholics, protestants, Baptists, Episcopal, and Luteran Churches.[2]

Pope John Paul II[3] addressed the issue of organ transplantation at the opening of the first International Congress of the Society for Organ Sharing. He affirmed that medicine has found in organ transplantation a new way of serving the human family. Organ transplantation, which began with blood transfusions, offered the man a way to give of himself, of his blood, and of his body, so that others may continue to live. In 2009, Pope Benedict XVI, in an open letter wrote that organ donation is a peculiar form of witness to charity. Tissue and organ transplants represent a great victory for medical science and are certainly a sign of hope for many patients who are experiencing grace and sometimes extreme clinical situations. However, donation should be performed only if personal health and identity are not endangered, and only for a morally valid and proportional reason. The principal criteria of respect for the life of the donor must always prevail, so the procurement of organs is performed only in the case of true death. He stressed the necessity of further research in establishing the death criteria to exclude any possibility of arbitration.[4]

Within the encouraging attitude toward organ transplantation, certain restrictions have been imposed. The use of human tissue "obtained by direct abortions even for research and therapeutic purposes" is prohibited, as is the transplantation of the brain or gonads, organs that are intimately connected to personal and procreative identity.[5]

28.2.2 Islam

The Prophet Mohammad has encouraged his followers to seek medical attention when ill.[6] This has been interpreted as a supportive reason for organ transplantation. The majority of Muslim scholars, both Sunni and Shia, promote the value of saving human life and hence allow organ transplantation as a necessary means to attain a noble end.[6,7]

The Islamic view of death is the departure of the soul from the body. However, it is the medical professionals who define death medically and clinically.[8] The Academy of Islamic Jurisprudence with members from several Islamic countries acknowledged the concept of brain death in 1986.[9] The Islamic jurists reached their ruling based on the major principles of Islam: Necessity makes forbidden things permissible; a need may be considered an extreme necessity, whether it be for an individual or general; prevention of evil takes precedence over acquisition of advantages; altruism; cooperation; seeking remedies, the value of human being; principle of doing no harm; the human body is the property of Allah (mutilation and suicide are forbidden). Islamic teachings and fatwas permit all types of organ transplantation if the required conditions are fulfilled.[8,10]

A few Muslim scholars have opposed cadaveric organ donation, because they viewed that the brain criteria of death are not the traditional Islamic view of death.[9,11] These scholars state that the traditional view of death in human beings consists in complete cessation of heartbeat, breathing, and whole brain function.[12] Furthermore, others consider removing any organ from a cadaver as an act of aggression against the human body: The human body is a divine trust; therefore, donation is not permissible as one cannot trade something of which one is not the true owner.[7]

Favorable opinions prevail[13] about donation to non-muslims, but some opposing opinions have been expressed by a few scholars.[2]

According to the Islamic rules, transplantation of tissues and skin is allowed as long as the expected results are good. Sex organs including organs responsible for fertilization (testicles and ovaries) and those used in satisfying the sexual desire (e.g., penis) should not be transferred from one human being to another. Organs inciting sexual desire must be covered and used only in marital relationship.[14]

28.2.3 Judaism

In traditional Jewish doctrine, salvaging organs from the deceased was akin to tampering with the divine image of God and viewed as blasphemy.[15] However, in Judaism, there is a general legal principle affirming "saving of a human life takes precedence over all other laws," including the prohibition from desecrating the body and delaying burial[16] and a basic tenet is not to stand by idly when your neighbor's life is in danger. In this aspect, Judaism mandates altruism.[17]

Of the three main branches of Judaism, namely, Orthodox, Conservative, and Reformist, the latter two not only permit but encourage organ donation. Among the Orthodox Jews, there is no consensus regarding the death criteria; however, some leaders have clearly come out in favor of brain death as acceptable criterion making the organ donation possible.[16,18]

28.2.4 Hinduism

The life after death and ongoing process of rebirth is a strong belief of Hindus. The important issue for a Hindu is that what sustains life should be accepted and promoted as Dharma (virtuous living). Hindus believe that every action has karmic implications and something as serious as replacing a major organ can carry some of the donor's karma to the recipient. Hindus also believe that the soul of the donor lives on in the inner world after death, and may influence the organ recipient. The fact that part of a deceased donor's physical body is still "alive" may interfere with his moving on to the next incarnation. These potentially could fade if the donor would finally move on to next incarnation, but some of the donor traits may have already been integrated into the recipient's personality.

All these religions and philosophical beliefs should be respected. Organ donation is an integral part of Hindu myths and writings which convey stories about body parts being used to benefit others, and thus can be paradigmatic for illustrating and encouraging altruism that enables organ donation.[19]

28.2.5 Buddhism

In Buddhism, the body is the source of attachment to worldly affairs, and is detrimental to the realization of nirvana, i.e., liberation from the cycle of suffering. Therefore, donating organs is a way to accumulate merits which will count toward realization of Liberation, with understanding that this is performed out of genuine desire to help others without regard to one's own self and body. Since the body and the person are strictly separated, donating the body or its parts would be in the same category as donating one's property. This is in accordance with the basic Buddhist teachings that emphasize selflessness and elimination of egoistic desires.[20]

As for the death criteria, Buddhism has adopted different positions according to the normative criterion of death in different countries.[21]

Some Buddhists, including those who follow Tibetan Buddhism, believe that consciousness may stay in the body for some time after breathing has stopped. Thus it is important that until the consciousness leaves the body it should remain undisturbed. For this reason, Tibetan Buddhists might be concerned that surgery performed soon after death will damage a person's consciousness and cause harm in their future lives.[22]

The debate on the time of death and its compatibility with brain death criteria is still active and uniform consensus has not been achieved, leaving the decision to the individual conscience.[23]

28.2.6 Chinese Popular Religion

Chinese cultures, including the Mainland, Taiwan, and Singapore, have been significantly shaped by Confucian ethics and Taoist tradition. According to *Confucian* philosophy, there is a great barrier to organ donation. A high regard is placed on filial piety: The body is a gift received from parents and ancestors. Thus, a person is not allowed to damage the body.[16]

According to the popular beliefs, the spirit takes about 8 h to separate from the body and for some period thereafter continues to linger near the corpse. The ghost of somebody subjected to organ donation might be expected to become angry and seek retaliation against whoever authorized the use of its body. The other issue is the nature of death: People dying through the accidents, homicides, suicides, or executions, which are most likely to produce a good organ donor, will become angry ghosts because they died before their time.[24]

Taoists believe that the body must be preserved to provide the soul with a resting place upon death to ensure immortality. To mutilate any of the body parts would be considered heretical[25] as it severs the natural process and should be avoided.[26] Modern Taoist scholars facing the challenge of medical technology argue, however, that Taoism sees human body only as a shelter for the soul that bears no substantial meaning. If physical body is simply a shelter, any attempt to change it or remove any part from it will not affect the essence of life.[26]

28.2.7 Shinto

In Shinto, the indigenous religion of Japan, injuring a dead body is considered a serious crime as the dead body is considered to be impure and dangerous, and thus quite powerful.[2] On the other hand, the body is a resting place for the soul, which explains a custom of attempting to call the soul back to the body of the dead, and allowing a waiting period after death to make sure that the soul will not return to the body. These religions and philosophical beliefs are the reasons for delaying official determination of someone's death as long as possible. The survivors remain attached to the body of the deceased and wish through the funeral ritual to launch the soul of deceased on a new journey.[27]

On these grounds, the law on "brain death," which declares that organ donation from diseased donors is allowed at any age groups whenever the family member approve, passed just recently in 2009 (www.TTS.org).

28.3 Cultural Views on Transplantation

The notion of replacing diseased organs with healthy ones dates back at least 3 millennia in the popular mythology. The legendary Chinese physician Pien Ch'iao, exchanged the heart of Kung He, who had a strong will and a weak spirit, with the heart of Ch'i Ying in whom the opposite prevailed.[28] Saints Cosmas and Damien was attributed the miracle of replacing the gangrenous leg of Deacon Justinian with the leg of a deceased Moor (AD 348).[29] In 1680, Pu Songling reported the story of the ghost Judge who replaced the ugly face of Zhu's wife with the beautiful face of a deceased girl.[28] Interestingly, all instances of the transplantation in the Indian sculptures have in common the use of animals as donors: God Siva beheaded the victim in a fit of anger and he performed the transplant as a penitence (the legend of Daksha and Ganesha).[28]

Racial and cultural differences do appear to play a role in the decision of organ donation. Explanation for donor concern include religious beliefs,[30] religious fears, other myths and superstitions,[31] racism,[31] family practices, and a general view on the world.[32]

Although after considerable debate, the western world has widely accepted the definition of brain death and as a consequence, the deceased donation process has evolved steadily in most of European countries and in the USA. However, low donation rates in certain regions of Europe, such as Eastern Europe, Greece and parts of Italy, highlight the presence of strong cultural, traditional, and religious beliefs that deter organ donation despite a favorable legal and religious framework.[33] Such variations are even more noticeable within multicultural societies. In Canada, lower donation rates have been reported among immigrant ethnic minorities who tend to uphold their traditional spiritual and cultural beliefs.[34,35] These differences are also visible in the multicultural environment of the UK and US societies, with lower donation rates among the native Asian, African, and Chinese minorities in the UK and African-American, American Indians, Hispanic, and Asian groups in the USA.[36-38] Variations in donation rates were also noted even in the context of a homogenous cultural background such as Switzerland. Here, population groups, based on their language background (German, French, Italian), showed substantial differences with regard to the patterns of knowledge, concerns, and motives underlying their willingness to donate.[39]

The Far Eastern countries have been slower to implement laws to support organ donation and recognizing brain death. The reluctance to donate has been attributed to: deeply engrained cultural superstitions regarding death; the fear of death; worrying that removing an organ after death violates the deceased; concerns about being "cut up" after death, suspicions that a potential donor's life will not be preserved; wanting to be buried or cremated whole; disliking that their organs be placed inside someone else and beliefs that cadaver donations are against their religion.[24,40] Cheung et al.[41] found that compared to Caucasians, Asians had more negative attitudes toward organ donation and cited the maintenance of body integrity after death as a significant impediment to organ donation. Countries such as Japan, China, and Korea have only recently acknowledged the concept of brain death.[25] Despite government legislative and advertising efforts, it remains difficult for the population to accept brain death as true death and therefore, these countries have a poor rate of deceased donor organs.[42]

Also in Islamic countries, the main reasons for refusal to donate the organs are inadequate knowledge about transplantation and misinformation regarding the organ donation process. For example, in Turkey, a survey showed that religious belief is a prime reason why people refuse to donate, although Islam is favorable to organ donation.[43] Unfortunately, the public is still largely ignorant about these religious rulings on transplantation in general and particularly about brain death.[44]

The same problem is faced by Judaism: Reconciling the removal of organs with concerns about the need for integrity of the body at the time of resurrection is difficult. Resurrection is cited as an incompatibility factor for organ donation.[45]

Cultural attitudes, through which religious beliefs are interpreted, often create a specific barrier to organ donation, despite the fact that religious rulings themselves do not prohibit organ donation. There is an ongoing process of society education increasing the awareness about organ donation and transplantation. Interestingly, the second- and third-generation immigrants are more open to donation than their parents and grand-parents.[16]

28.4 Face Transplantation

Although all religions stress the respect for the human cadaver and prohibit body mutilation, there are no specific religious bans against face transplantation.

Not many comments were made by religious leaders in response to the recent reports on human face transplantation.

One of these includes the comment of Dr Maher Hathout, Chairman of the Islamic Center of Southern California, on December 6, 2005, who on the permissibility of face transplantation said: "I do not think that the face is more important than the heart, the kidney, the liver, or the cornea of the eyes, transplantation of which was allowed by scholars. If the issue is a mere concern about the similarity in complexion and features, I would say that this similarity will not be completely identical; besides the medical technology now can make the transplanted face look as close as possible to the features of the patient before injury."[46] If, in the circumstances of necessity and when the donor, during his or her lifetime, decided to donate his or her organs after death to save or to improve the life of others, donation is an act of charity.[46]

In a study on cadaveric organ donation in Taiwan,[47] where families were questioned on donating the organs of their next of kin, it was found that religious beliefs regarding preservation of the ideal image of the deceased were the most important factors contributing to the decision to donate, especially the skin and the bone. The chances for the deceased to have a better afterlife were considered to be compromised if the appearance of the donor was damaged. The donation of skin and bones was related to bad death and had negative impacts on the donor's family.[47]

28.5 Conclusions

The systematic evaluation of religious positions represented by the major religions reveals that no single religion formally forbid the donation or receipt of the organs or are against transplantation from the deceased donors. While Middle Eastern religions and in particular Christianity and Islam strongly encourage organ donation, considering it as an act of charity and altruism, the same approach does not hold true for the Far Eastern religions. Here, the interpretation of the religious rulings has been more tormented. Eastern religions have been less open to the idea of organ donation for a variety of cultural and religious reasons, including different definitions of love, altruism, and personhood.[45] For some religious traditions, such as Buddhism, Shinto, and conservative branches of Judaism, the major challenge has been to determine whether the medical model of whole brain death has a basis in either the traditional doctrinal statements or in the religious sources. These three religious traditions view death as a gradual process and concern is expressed that there will be no tampering with the body until there is a cessation of respiration, circulation, and heart beat (Judaism), or until survivors have had a chance to launch the soul on the new journey (Shinto).[16] To retrieve an organ from a potential donor, before the dying process is completed and before the rituals enacted, is seen by some to violate the proper respect for the dying person, bringing misfortune to the surviving relatives.[16] Without distinction of faith, concerns raised by all religious groups regard respecting the freedom and autonomy of the donor; not making organ donation a means of economic gain; treating the body with dignity and respect of religious beliefs; determining the dying process and death and not hastening death in order to retrieve the organs.[16]

A close interrelation was discovered existing between religion, culture, and organ donation shown by the fact that many still cite faith as a reason why they cannot donate the organs after death. It is clear that the majority do not know that most religions view organ donation as an act of charity and a great gift to give.[2] In a study investigating the influence of knowledge and religiousness on the attitudes toward organ donation, the degree of a person's self-reported religiosity was negatively associated with favorable attitudes toward organ donation. The results indicated the impact of four variables: education regarding organ donation, knowledge of someone who had donated an organ after death, awareness of anyone who received a donated organ, and religious beliefs.[30] Feld et al.[48] observed that nearly half of the Jews in an Ontario community reported that they believed organ donation to be against Jewish law. In addition, among those unwilling to donate organs, 24% said that religion contributed to their decision. These studies suggest that religion can be used either as a facilitator or as a barrier to organ donation.

This review didn't find any specific religious bans on face transplantation, similar to specific prohibitions imposed on the transplantation of the sex organs by both Christianity and Islam. From the general positions assumed by the major religious authorities on organ transplantation and based on the criteria summarized previously, we can presume that face transplantation, as a necessary medical procedure for reconstruction of severely disfiguring and disabling facial defects, is allowed by all religious groups which have expressed themselves in favor of organ transplantation. This is proved by the favorable position of the Chairman of the Islamic Center of Southern California.[46]

However, because of the central role of human face in representing the individual identity, the mystical concerns, centering on the belief in resurrection of the dead in different cultures, could place an important weight on the mind of potential donors. These cultural and religious differences could be responsible of the different geographical distribution of face transplants, favoring prominently Europe where 8 out of 11 face transplants were performed and where the culture of organ transplantation and donation has a long history. This study showed that a wide gap exists between religious positions and popular beliefs: Deeply ingrained cultural values and beliefs often outweighed the favorable stance on organ donation of the religious groups to which the donor family belonged. Religious rulings can positively influence cultural opinions and facilitate acceptance of face donation at public level. Since religious beliefs have been used as a major argument against donation and donor handling, specific education on issues related to organ need, organ procurement procedures is necessary to clarify the erroneous beliefs and to address the fears. Distributing more accurate information to society and to leaders with religious influence will translate into the public knowledge, and should positively influence the acceptance of face donation and transplantation.

References

1. Teo B. Organ donation and transplantation: a Christian viewpoint. *Transplant Proc*. 1992;24:2114-2115.
2. Barlow BG. Religious views vary on organ donation. *Nephrol News Issues*. 2006;20:37-39.
3. John Paul II.Address of His Holiness John Paul II to participants of the First International Congress of the Society For Organ Sharing. Available at: http://www.vatican.va/holy_father/john_paul_ii/speeches/1991/june/documents/hf_jp-ii_spe_19910620_trapianti_en.html. Accessed May 28, 2010.
4. Pope Benedict XVI. A message from the holy father pope Benedict XVI. *Transplantation*. 2009;88:S96-S97.
5. United States Conference of Catholic Bishops. *Ethical and Religious Directives for Catholic Health Care Services*. 4th ed. Washington, DC: United States Conference of Catholic Bishops. Available at: http://www.usccb.org/bishops/directives.shtml#partfive. Accessed May 25, 2010.
6. UK's muslim law council approves organ transplants. *J Med Ethics*. 1996;22:99.
7. Goolam NM. Human organ transplantation – multicultural ethical perspectives. *Med Law*. 2002;21:541-548.
8. el Shahat YI. Islamic viewpoint of organ transplantation. *Transplant Proc*. 1999;31:3271-3274.
9. al Mousawi M, Hamed T, al Matouk H. Views of Muslim scholars on organ donation and brain death. *Transplant Proc*. 1997;29:3217.
10. Shaheen FA, Al-Jondeby M, Kurpad R, Al-Khader AA. Social and cultural issues in organ transplantation in Islamic countries. *Ann Transplant*. 2004;9:11-13.
11. Rady MY, Verheijde JL, Ali MS. Islam and end-of-life practices in organ donation for transplantation: new questions and serious sociocultural consequences. *HEC Forum*. 2009; 21:175-205.
12. Hedayat K. When the spirit leaves: childhood death, grieving, and bereavement in Islam. *J Palliat Med*. 2006;9:1282-1291.
13. Raza M, Hedayat KM. Some sociocultural aspects of cadaver organ donation: recent rulings from Iran. *Transplant Proc*. 2004;36:2888-2890.
14. Sellami MM. Islamic position on organ donation and transplantation. *Transplant Proc*. 1993;25:2307-2309.
15. Cohen KS. Choose life: Jewish tradition and organ transplantation. *Del Med J*. 1988;60:509-511.
16. Gillman J. Religious perspectives on organ donation. *Crit Care Nurs Q*. 1999;22:19-29.
17. Jotkowitz A. New models for increasing donor awareness: the role of religion. *Am J Bioeth*. 2004;4:41-42. discussion W35-W37.
18. Mayer SL. Thoughts on the Jewish perspective regarding organ transplantation. *J Transpl Coord*. 1997;7:67-71.
19. Muzammil Z. Hinduism and organ donation. Available at: http://raodtaac.com/pdf/Hindu_Dharma_and_organ_donation.pdf. Accessed May 29, 2010.
20. Tsomo KL. *Into the Jaws of Yama, Lord of Death: Buddhism, Bioethics, and Death*. 1st ed. Albany: State University of New York Press; 2006.
21. Hongladarom S. Organ transplantation and death criteria: Theravada Buddhist perspective and Thai cultural attitude. Available at: http://pioneer.chula.ac.th/~hsoraj/web/Organ%20 Transplantation-Buddh.pdf. Accessed May 29, 2010.
22. Lynch E. Faith in transplants. *Nurs Stand*. 2005;19:24-27.
23. Keown D. Buddhism, brain death and organ transplantation. *J Buddhist Ethics*. 2010;17:1.
24. Ikels C. Ethical issues in organ procurement in Chinese societies. *China J*. 1997;38:95-119.
25. McConnell JR 3rd. The ambiguity about death in Japan: an ethical implication for organ procurement. *J Med Ethics*. 1999;25:322-324.
26. Tai MC. An Asian perspective on organ transplantation. *Wien Med Wochenschr*. 2009;159:452-456.
27. Hardacre H. Response of Buddhism and Shinto to the issue of brain death and organ transplant. *Camb Q Healthc Ethics*. 1994;3:585-601.
28. Bhandari M, Tewari A. Is transplantation only 100 years old? *Br J Urol*. 1997;79:495-498.
29. Peltier LF. Patron saints of medicine. *Clin Orthop Relat Res*. 1997;334:374-379.
30. Rumsey S, Hurford DP, Cole AK. Influence of knowledge and religiousness on attitudes toward organ donation. *Transplant Proc*. 2003;35:2845-2850.
31. Callender CO, Washington AW. Organ/tissue donation the problem! education the solution: a review. *J Natl Med Assoc*. 1997;89:689-693.
32. Mays VM, Ponce NA, Washington DL, Cochran SD. Classification of race and ethnicity: implications for public health. *Annu Rev Public Health*. 2003;24:83-110.
33. Mavroforou A, Giannoukas A, Michalodimitrakis E. Organ and tissue transplantation in Greece: the law and an insight into the social context. *Med Law*. 2004;23:111-125.
34. Molzahn AE, Starzomski R, McDonald M, O'Loughlin C. Chinese Canadian beliefs toward organ donation. *Qual Health Res*. 2005;15:82-98.
35. Bowman KW, Richard SA. Cultural considerations for Canadians in the diagnosis of brain death. *Can J Anaesth*. 2004;51:273-275.

36. Fahrenwald NL, Stabnow W. Sociocultural perspective on organ and tissue donation among reservation-dwelling American Indian adults. *Ethn Health*. 2005;10:341-354.
37. Boulware LE, Ratner LE, Sosa JA, Cooper LA, LaVeist TA, Powe NR. Determinants of willingness to donate living related and cadaveric organs: identifying opportunities for intervention. *Transplantation*. 2002;73:1683-1691.
38. Morgan M, Hooper R, Mayblin M, Jones R. Attitudes to kidney donation and registering as a donor among ethnic groups in the UK. *J Public Health (Oxf)*. 2006;28:226-234.
39. Schulz PJ, Nakamoto K, Brinberg D, Haes J. More than nation and knowledge: cultural micro-diversity and organ donation in Switzerland. *Patient Educ Couns*. 2006;64:294-302.
40. Woo KT. Social and cultural aspects of organ donation in Asia. *Ann Acad Med Singapore*. 1992;21:421-427.
41. Cheung AH, Alden DL, Wheeler MS. Cultural attitudes of Asian-Americans toward death adversely impact organ donation. *Transplant Proc*. 1998;30:3609-3610.
42. Kim JR, Elliott D, Hyde C. The influence of sociocultural factors on organ donation and transplantation in Korea: findings from key informant interviews. *J Transcult Nurs*. 2004; 15:147-154.
43. Kececioglu N, Tuncer M, Yucetin L, Akaydin M, Yakupoglu G. Attitudes of religious people in Turkey regarding organ donation and transplantation. *Transplant Proc*. 2000;32:629-630.
44. Daar AS. The evolution of organ transplantation in the middle east. *Transplant Proc*. 1999;31:1070-1071.
45. LaFleur WR. From agape to organs: religious difference between Japan and America in judging the ethics of transplant. *Zygon*. 2002;37:623-642.
46. Hathout M. Face transplant surgery. Available at: http://www.islamonline.net/servlet/Satellite?cid=1133769952021&pagename=IslamOnline-English-Ask_Scholar%2FFatwaE%2FFatwaEAskTheScholar. Accessed May 29, 2010.
47. Shih FJ, Lai MK, Lin MH, et al. The dilemma of "to-be or not-to-be": needs and expectations of the Taiwanese cadaveric organ donor families during the pre-donation transition. *Soc Sci Med*. 2001;53:693-706.
48. Feld J, Sherbin P, Cole E. Barriers to organ donation in the Jewish community. *J Transpl Coord*. 1998;8:19-24.
49. WHO O. Global Observatory on Donation and Transplantation. Available at: http://www.transplant-observatory.org/Contents/World%20Transplant%20Information%20%E2%80%93/Data_Tables/Pages/default.aspx. Accessed July 9, 2010.

29 Comparative Cost Analysis of Conventional Reconstructions and the First US Face Transplantation

James R. Gatherwright, Frank Papay, Risal Djohan, Elliott H. Rose, Lawrence J. Gottlieb, and Maria Z. Siemionow

Contents

M.Z. Siemionow (✉)
Department of Plastic Surgery,
Cleveland Clinic, Cleveland, OH, USA
e-mail: siemiom@ccf.org

Abstract While the ethical and technical concerns regarding facial transplantation have been discussed throughout this book, there is a lack of information regarding its cost. Moreover, there is no information available in the literature on the financial implications of complete reconstructions in patients with severe craniomaxillofacial defects. We have had the unique opportunity to compare the cost of traditional reconstructive procedures and face transplantation in a single patient in our institution. In addition, we have also been able to compare the financial data of three other patients who underwent conventional reconstruction in other institutions: Mount Sinai Medical Center in New York and the University of Chicago. As the field of transplantation evolves, it will become increasingly important to disseminate this valuable information. Given the current economic climate in medicine, it will undoubtedly be a central issue to the face transplant debate. The following chapter illustrates these issues by comparing the cost of multistaged conventional reconstructive procedures in four patients with the cost of the first US face transplant.

Abbreviations

ATG	Anti-thymocyte Immunoglobulin
CMV	Cytomegalovirus
CPT	Current Procedural Terminology
CTA	Composite Tissue Allotransplantation
ESRD	End-Stage Renal Disease
G-CSF	Granulocyte Colony Stimulating Factor
MMF	Mycophenolate Mofetil
PCP	Pneumocystis Carinii Jerovici
PCV	Pneumococcal vaccine
PTSD	Post-Traumatic Stress Disorder

M.Z. Siemionow (ed.), *The Know-How of Face Transplantation*,
DOI: 10.1007/978-0-85729-253-7_29,

29.1 Background

There has been significant development in the field of facial transplantation with 12 occurring and discussion among those performing the procedure as to what the implications are post-transplant. While the technical and ethical concerns have been discussed at length,[1-3] the financial cost has yet to be addressed. Cost and ethics are separate issues, but are not mutually exclusive.

Whether or not facial transplantation should be pursued in the future will not only depend on the risk-benefit ratio to the patient but also on the cost-benefit ratio to society as a whole. However, unlike solid organ transplantation, the role of composite tissue allotransplantation (CTA) in the field of reconstructive surgery has yet to be defined. Most of the data and experience we have regarding transplants is with respect to solid organs and while hand transplantation may offer an acceptable comparison, it too is in its incipient stages with the longest follow-up limited to 10–12 years in only ten patients.[4] Many of the issues regarding CTA will continue to be debated including: transplantation of a nonessential organ, life-long immunosuppression, psychological implications, and societal acceptance. However, it may be that the largest impediment to facial transplantation will not be ethical or medical issues, but financial restrictions.[5]

There have been no reports thus far detailing the actual cost of such a procedure, nor comparing it to traditional reconstructive options. Undoubtedly the questions regarding face transplantation will turn to "who is responsible for paying and how?" Given that this question has been at the center of the US health care reform debate, it is important to start investigating the costs and potential benefits of face transplantation as a competitive alternative to conventional reconstructive procedures. Our experience with the US first face transplant patient, who had undergone multiple reconstructive procedures prior to transplant, provides a unique opportunity to directly compare the cost of conventional reconstructions and face transplantation in a single patient. In addition, two well-established reconstructive microsurgery authors (LG and ER) have provided us with important financial information regarding their experience with complex reconstructions of craniomaxillofacial deformities using conventional methods. In the following chapter, we present the comparative costs of these two approaches as well as identify potential areas of concern. We are confident that ensuing medical, ethical, and scientific debates will produce future studies and areas of additional interest on this subject.

29.2 The Patients

29.2.1 Transplant Patient

29.2.1.1 Pre-transplant Reconstructions

The patient is a 46-year-old Caucasian female who suffered significant facial trauma following a shotgun blast in September 2004. The injury resulted in multiple defects including loss of her nose, cheeks, lower eyelids, right globe, upper lip, upper jaw, maxillary alveolus, and teeth. She had also lost her ability to smell and her nutrition was maintained through a percutaneous endoscopic gastrotomy tube. Between 2004 and 2008, the patient underwent 23 separate, traditional reconstructive operations to restore the tridimensional anatomical defect and functional deficits. Despite exceptional surgical expertise, these attempts resulted in sub-optimal esthetic and functional outcomes.

29.2.1.2 Face Transplantation

In August of 2008, 4 years after the trauma, when all conventional surgical alternatives had been exhausted, the patient requested to be considered as a potential candidate for face transplantation as an alternative option with the understanding that it is an experimental procedure. Following a 22-h operation, the successful face transplant of a gender- and age-matched brain-dead donor was completed on December 10, 2008. Components of the face transplant included a Le Fort III composite graft (skin, bone, connective tissue), microvascular anastamoses, facial nerve repair, and reconstruction of the bilateral orbital floors and lower eyelids.[6,7] Functional units included the nose, lower eyelids, and upper lip. Immunosuppression and infectious prophylaxis regimens have been previously reported upon as well as graft monitoring and the treatment of acute rejections.[6,7] She has had marked improvement in esthetics, self-esteem, psychological well-being, pain, and functional recovery including the ability to eat and smell.[6-9]

29.2.2 Conventional Reconstruction Patients

The pertinent details regarding two patients who had undergone traditional facial reconstructions are provided by Dr. Elliott Rose and his experience. A detailed description of these two patients can be found in Sects. 10.6.1 and 10.6.2 of this book. Additionally, Dr. Lawrence Gottlieb provided information regarding his experience with one patient. Paraphrased summaries are provided below for ease of reading in *italics*.

29.2.2.1 Patient #1 – ER#1

The following is a paraphrased excerpt from Sect. 10.6.1 of this book:

Patient #1 was a 12 year old Irish boy who suffered a near total facial burn except for most of the mid-face and nose which were spared. He underwent eleven prior surgeries for eyelid ectropion repair, chin and lip correction but the results were sub-optimal and therefore he was referred to a unit at Mount Sinai Medical Center for more facial reconstruction. Conventional reconstruction procedures entailed insertion of bimalar fascia lata slings for lower lip suspension, patterned microvascular free radial forearm flap to the chin/ neck subunit, sequential patterned scapular flaps, placement of fascia lata slings from the malar arches to the lateral lip modioli for buttressing of the deep facial foundation and lateral lip support. Additional "refinement" procedures including debulking and contouring of the cheek and neck flaps, SAL, canthoplasty OU, insertion of Porex chin implant, multiple scar revisions, dermal placation of the nasolabial creases, and laser resurfacing of the facial scars. Facial contours were restored with sculpted soft tissue.

29.2.2.2 Patient #2 – ER#2

The following is a paraphrased excerpt from Sect. 10.6.2 of this book:

Patient #2 is a 10 year old girl who sustained 80% TBSA burns in a crib fire in her native Columbia who was subsequently adopted. Prior to her transfer to Mount Sinai for facial restoration, she underwent more than 10 prior reconstructive surgeries including Z-plasties and tissue expansion with little success. Conventional reconstruction entailed wide excision of keloid of the left hemi-face and scalp, insertion of a fascia lata sling for lateral lip suspension and support of the deep facial foundation, and resurfacing with a patterned, sculpted microvascular free scapular flap tailored to the defect, followed with a mirror image patterned, sculpted microvascular free scapular flap to the right hemi-face and placement of a fascia lata sling to the right lateral commissure. Peri-ocular reconstruction entailed re-allignment of the medial canthal ligament by transnasal wire fixation and re-suspension of the lateral canthal ligament by wire fixation to the lateral orbitial rim. Both upper and lower lids were resurfaced with a single sheet graft to the orbital subunit with a slit for the ciliary aperture. Nasal reconstruction included architectural modification of the nasal tip with conchal cartilage grafts and external resurfacing with a patterned, pedicled forehead flap. The divided base of the nasal pedicle was "piggy-backed" to the lower eyelid for ectropion repair prior to permanent inset. Nostril patency was re-established with full thickness skin grafts wrapped around a nasal stent. Additional procedures included debulking/ contouring of the nasal and cheek flaps, SAL, insertion Porex chin implant, levator advancement OS, dermal strip grafts for upper lip augmentation, nostril thinning and repositioning, multiple scar revisions, and laser resurfacing.

29.2.2.3 Patient #3 – LG#1

The following patient's reconstruction can be found in Dr. Gottlieb's Sect. 11.3:

> *This eight-year-old boy sustained a massive facial injury from a pit bull attack. His entire right cheek, right lower eyelid skin, nose, left cheek, 20% of left upper lip, 10% of left lower lip, entire left upper and lower eyelids, left forehead, and both ears were missing. During that initial procedure, the left eye was covered with conjunctival flaps, the right parotid duct was repaired (the left was not found), and avulsed small branches of the right facial nerve were inserted directly into muscle. The left facial nerve was not found. In addition, his left oral commissure was repaired and associated lacerations of surrounding skin were closed. A double pedicle deep inferior epigastric perforator (DIEP) lower abdominal free flap was transferred 12 days after the injury. The abdominal flap was suspended to bone using multiple bone-anchoring (Mitek™, DePuy Mitek, Inc.) sutures. Nasal lining was provided with a turned-in portion of the inferior portion of the abdominal flap and the left eye was totally covered with the flap. Temporary nasal projection was accomplished with cadaver cartilage grafts. The superior portion of the abdominal flap was thinned by dissecting the subscarpa's fat layer, which was then turned under the flap to provide more bulk in the malar area. As expected, the loss of both upper and lower left eyelids was the most significant challenge. The left eyelid conjunctival flaps had separated and a full thickness hard palate mucosal graft was used for conjunctival replacement. The thinned abdominal flap was used for eyelid skin and to revascularize the mucosal graft. One month later, autologous costochondral cartilage grafts were inserted between the mucosal graft and the outside skin. At a subsequent operation, a remnant of levator was found which was mobilized and secured to the neo-tarsus. Secondary nasal reconstruction was performed using a double paddle radial forearm flap that was able to provide nasal lining, nasal tip, and skin for the breakdown just medial to the left eye. Tissue expanders were used to reconstruct the left side of the forehead. Multiple small revisions of the eyelids, nose and lips were performed every three or four weeks for the first year, alternating sites to allow for swelling to settle.*

29.3 Data Collection

29.3.1 Face Transplant Patient

All data was collected between 11/18/2004 and 11/27/2009 with respect to this patient's treatment before and after transplantation at the Cleveland Clinic. Financial data for conventional reconstruction was collected from 11/8/2004 to 12/8/2008 (1,491 days), the date of face transplantation. Peri-transplant costs included the operation and subsequent hospitalization from 12/8/2008 until discharge on 2/5/2009 (58 days). Post-transplant costs were recorded from the patient's discharge to 11/27/2009 (295 days). The cost incurred during the first US face transplant was calculated using data collected by the Decision Support System, of the Cleveland Clinic.

29.3.2 Conventional Reconstruction Patients

All financial data were provided by ER's and LG's respective institutions based on Medicare reimbursement rates. The number of hospital admissions was also recorded to provide an estimate of the relative complexity.

29.4 Results

29.4.1 Cost Comparison

29.4.1.1 Total Costs

The face transplant patient underwent 23 conventional reconstructive procedures before undergoing face transplantation. The costs of conventional reconstruction and face transplantation were $353,480 and $349,959, respectively (Fig. 29.1). The net difference in cost between pre-transplant reconstruction and face transplantation was found to be $3,521. This is a

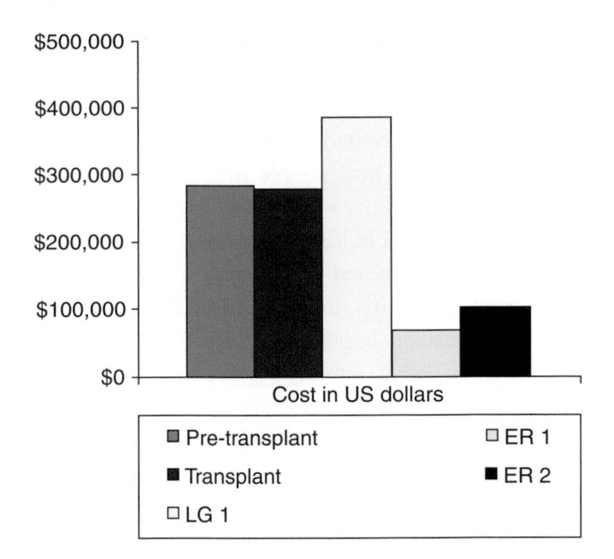

Fig. 29.1 Comparative direct costs of multiple conventional reconstructions compared to face transplantation. The comparative costs of a patient pre-transplant, the transplant itself, Dr. Gottlieb's conventional reconstruction patient (*LG 1*), and Dr. Rose's two conventional reconstruction patient's (*ER 1* and *ER 2*)

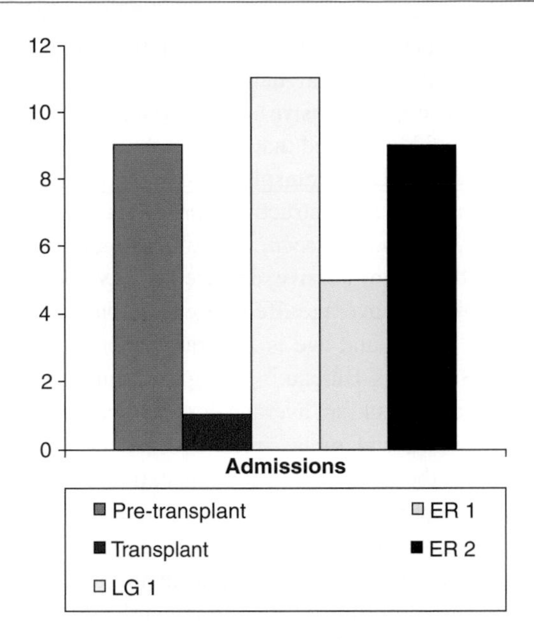

Fig. 29.2 Comparative number of admissions by patient. The comparative number admission of a patient pre-transplant, the transplant itself, Dr. Gottlieb's conventional reconstruction patient (*LG 1*), and Dr. Rose's two conventional reconstruction patient's (*ER 1* and *ER 2*)

calculated 1% relative decrease. ER #1's and ER #2's treatment costs totaled $84,517 and $111,046 respectively. The cost of treatment for LG's patient was calculated to be $484,391. A graphic summary of the cost comparison can be found in Fig. 29.1.

29.4.2 Inpatient Costs of Transplant Patient

The patient was admitted pre-transplant nine times for a total of 115 days and once for the transplantation encounter lasting 58 days. The average length of stay was 12.78 (1–46) days during the conventional reconstruction. The average cost per day was pre- $1,614.21, peri- $4,015.40, and $1,341.44 post-transplantation. Indications for post-transplant admissions included: tracheostomy removal, G-tube removal, CMV infection (twice), and neutropenia (4 admissions). LG's patient was admitted 11 times for a total of 66 days. ER #1 and ER #2 were admitted five and nine times, respectively. A summary of all the admissions are displayed in Fig. 29.2.

The average inpatient costs per day pre-transplant and peri-transplant were $3,073.74 and $6,033.78, respectively. Data was only available on LG's patient, totaling $7,339.27 per day of inpatient stay (Fig. 29.3).

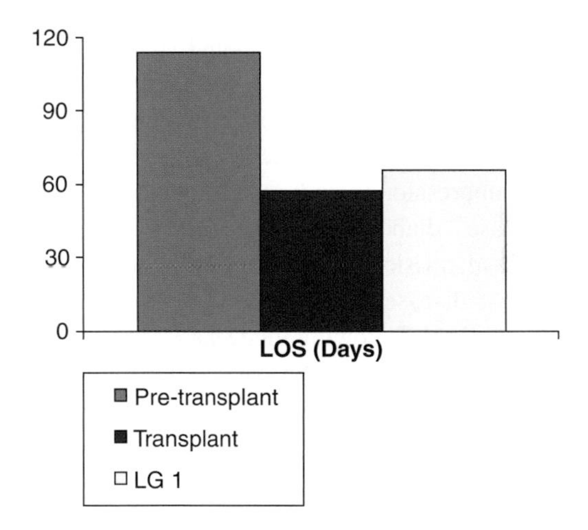

Fig. 29.3 Comparative Length of Stay (*LOS*) following face transplantation and conventional reconstructive procedures. The *LOS* of a patient pre-transplant, the transplant itself, Dr. Gottlieb's conventional reconstruction patient (*LG 1*), and Dr. Rose's two conventional reconstruction patient's (*ER 1* and *ER 2*)

29.5 Discussion

We have had the unique opportunity to directly compare the cost of conventional reconstructive options and face transplantation in one patient with the same anatomical defects. We have also had the good fortune to present the financial data on three patients who underwent comprehensive conventional reconstructions. It should be noted that the initial operative costs incurred during facial transplantation were comparable to conventional reconstructions in different patients. However, as can been seen, post-transplantation costs are significant, progressive, and their final values indeterminable. The average life expectancy based on our patient's gender and age is 37 years according to the 2010 US Census Bureau.[10] Using data published on kidney transplantation, average cost per year of immunosuppression can range from $10,000 to $14,000. Therefore, the overall cost for our patient's life-long immunosuppression can be projected to cost anywhere from $370,000 to $518,000. Therefore, while the initial surgeries may be comparable in initial cost, the total in the long-run will be significantly more. This highlights the need for continued investigation into alternative therapies that can obviate the need for immunosuppression.[11] This also does not include complication-related treatment costs secondary to the side effects of immunosuppressive therapies of the treatment of rejection episodes.

Both approaches have associated risks inherent to any reconstructive procedure: infection, bleeding, flap loss, and even death. Additional risks common to all transplant procedures include side effects related to immunosuppression including cardiovascular and kidney disease, diabetes, and secondary malignancy. There are also risks associated with rejection, grafts-versus-host disease, and medication noncompliance. The risk of death following face transplantation is significant as demonstrated by two face transplant related deaths.[4,12,13] Also face transplantation is unique in that if there is graft failure, either due to technical or rejection related issues, the salvage procedures are limited. As seen in our patient, transplant patients may have also undergone several reconstructive procedures prior to transplantation and therefore typical reconstructive options may have already been exhausted. A prior face transplant may preclude the use of necessary vascular territories as well as access to conventional reconstructive options. Additionally, scarring and modified anatomy can increase the difficulty of dissection leading to increased morbidity and suboptimal outcomes. Even if a patient's life is salvageable, graft loss may cause irreparable harm to the patient's physical and mental health in addition to increased costs.

Facial disfigurement can be one of the most socially devastating events that can happen to a person.[14] While some may argue that attractiveness is inconsequential, studies have demonstrated that this may not be true and may actually affect how people treat one another and themselves.[15-17] Patients with facial disfigurement also suffer numerous psychological problems including anxiety, low self-esteem and confidence, substance abuse, and marital problems, depression, PTSD, and suicide.[18-20] All previously published reports regarding face transplantation have subjectively reported improved self-esteem and reintegration, but for the first time, a recent article by Coffman et al. used objective measures to assess our patient's psychological well-being. These objective tests demonstrated a significant decrease in depression and verbal abuse and improved quality of life and social reintegration.[8] The role of the face in communication is invaluable[21] with up to two-thirds of our communication being nonverbal.[22] The functions of the face are usually not restored adequately by conventional means, while initial results regarding both sensory and motor recover are promising.[6,7,23,24] In the case of our patient, there was also a drastic reduction in chronic pain secondary to the removal of scarred and contracted tissues. She had originally rated her pain to be 8/10, and this dropped significantly to 1/10 following transplantation.[6,7] It can be argued that regardless of the method, be it conventional reconstruction or transplantation, it is impossible to quantify the value of these procedures using monetary means as they have immensely improved these patients' overall quality of life.

Another unique aspect specific to face transplantation is that conventional reconstructive techniques require composite, autogenous tissue and therefore has inherent morbidity. It begs the question whether these techniques and their attendant morbidity should be the standard choice for the severely disfigured patient. Common morbidities following the use of composite tissue flaps include: delayed wound healing,[25-28] nerve damage,[27,29,30] cold intolerance,[31] neuroma,[26,32] pathological fractures,[32,33] loss of function,[30,34-36] and loss of structural integrity.[35] More common morbidities may reach a complication rate as high as 50%.[25] In addition, the time required to complete this process may exceed 3–4 years. During this time, the patient is subject to the additional risk of multiple surgeries as well as the psychological impact of a delayed reconstruction. Therefore, given that the monetary cost of conventional reconstruction is comparable to face transplantation, the cost of a prolonged course and associated morbidity on the patient's quality of life and psyche may make face transplantation the first-line choice in a select group of severely disfigured patients.

It should also be high-lighted that the variability in cost between these patients is directly proportional to the complexity of the defects. Patients with more extensive defects or those involving important facial sub-units, such as the nose and mid-face, will incur a higher cost regardless of the method. The example of face transplantation used in this study included approximately 80% of the face and bone and most likely represents the upper end of complexity, resources, and cost.[37] Additional variables impact the cost and quality of life following either conventional reconstruction or face transplantation.[24,38-44] Therefore, these experiences may not be germane to other patients and institutions.

Finally, currently, society is financially responsible for patients with end-stage renal disease (ESRD) and pays for their transplant and immunosuppression, for up to 3 years.[45] The nonessential nature of facial transplantation will more than likely elicit questions regarding the true necessity of providing such a service. Although our patient's previous reconstructions were predominantly covered by Medicare, we believe that the final outcome was sub-optimal with comparable financial cost to transplantation. However, it may be argued that denying the disfigured a means of repair, either conventional or transplant, is unethical. While the face transplant debate will continue, objective comparisons of the financial cost and associated benefit between conventional reconstruction and face transplantation are necessary to make informed decisions that will impact patients and society alike.

29.6 Conclusion

To the best of our knowledge, this is the first report on the comparative cost between conventional reconstruction and face transplantation. We have successfully documented that the cost of the first face transplant in the USA is comparable to multistaged, conventional reconstructive procedures in four different patients. While many questions remain to be answered regarding the long-term financial and ethical costs, it is imperative to start addressing these issues. As face transplantation moves further away from "can or should we?" the question of fiscal responsibility will be pushed into the forefront and may become the central issue of the face transplant debate. While we have focused on the financial issues regarding facial transplantation, the alleviation of psychological and physiological suffering, the potential for functional recovery, and renewed hope given to our patient may be considered priceless.

References

1. Agich GJ, Siemionow M. Until they have faces: the ethics of facial allograft transplantation. *J Med Ethics*. 2005;31: 707-709.
2. Clark PA. Face transplantation: Part II-an ethical perspective. *Med Sci Monit*. 2005;11:RA41-RA47.
3. Alam DS, Papay F, Djohan R, et al. The technical and anatomical aspects of the World's first near-total human face and maxilla transplant. *Arch Facial Plast Surg*. 2009;11: 369-377.
4. The International Registry on Hand Composite Tissue Transplantation http://www.handregistry.com/. Accessed 7/1/2010, Monza, Italy and Lyon, France, 2009.
5. Concannon MJ. Discussion: an economic analysis of hand transplantation in the United States. *Plast Reconstr Surg*. 2010;125:599-600.
6. Siemionow M, Papay F, Alam D, et al. Near-total human face transplantation for a severely disfigured patient in the USA. *Lancet*. 2009;374:203-209.
7. Siemionow MZ, Papay F, Djohan R, et al. First U.S. near-total human face transplantation: a paradigm shift for massive complex injuries. *Plast Reconstr Surg*. 2010;125:111-122.
8. Coffman KL, Gordon C, Siemionow M. Psychological outcomes with face transplantation: overview and case report. *Curr Opin Organ Transplant*. 2010;15:236-240.
9. Gordon CR, Siemionow M, Papay F, et al. The world's experience with facial transplantation: what have we learned thus far? *Ann Plast Surg*. 2009;63:572-578.
10. The 2010 Statistical Abstract; Births, deaths, marriages, and divorces: Life expectancy Washington, D.C.: United States Census Bureau, 2010:Table 103.
11. Siemionow M, Klimczak A. Tolerance and future directions for composite tissue allograft transplants: part II. *Plast Reconstr Surg*. 2009;123:7e-17e.
12. Guo S, Han Y, Zhang X, et al. Human facial allotransplantation: a 2-year follow-up study. *Lancet*. 2008;372:631-638.
13. Davies L. First hand and face transplant patient dies. http://www.guardian.co.uk/world/2009/jun/15/hand-face-transplant-patient-dies. Accessed 7/1/2010. London, U.K. 2009.
14. Macgregor FC. Facial disfigurement: problems and management of social interaction and implications for mental health. *Aesthet Plast Surg*. 1990;14:249-257.
15. Maner JK, Kenrick DT, Becker DV, et al. Sexually selective cognition: beauty captures the mind of the beholder. *J Pers Soc Psychol*. 2003;85:1107-1120.

16. Stevenage SV, Spreadbury JH. Haven't we met before? The effect of facial familiarity on repetition priming. *Br J Psychol*. 2006;97(Pt 1):79-94.
17. Stevenage SV, Lee EA, Donnelly N. The role of familiarity in a face classification task using thatcherized faces. *Q J Exp Psychol A*. 2005;58:1103-1118.
18. Rumsey N, Clarke A, White P, Wyn-Williams M, Garlick W. Altered body image: appearance-related concerns of people with visible disfigurement. *J Adv Nurs*. 2004;48:443-453.
19. Levine E, Degutis L, Pruzinsky T, Shin J, Persing JA. Quality of life and facial trauma: psychological and body image effects. *Ann Plast Surg*. 2005;54:502-510.
20. Robinson E, Rumsey N, Partridge J. An evaluation of the impact of social interaction skills training for facially disfigured people. *Br J Plast Surg*. 1996;49:281-289.
21. Soni CV, Barker JH, Pushpakumar SB, et al. Psychosocial considerations in facial transplantation. *Burns*. 2010;36:959-964.
22. Morris PJ, Bradley JA, Doyal L, et al. Facial transplantation: a working party report from the Royal College of Surgeons of England. *Transplantation*. 2004;77:330-338.
23. Dubernard JM, Lengele B, Morelon E, et al. Outcomes 18 months after the first human partial face transplantation. *N Engl J Med*. 2007;357:2451-2460.
24. Lantieri L, Meningaud JP, Grimbert P, et al. Repair of the lower and middle parts of the face by composite tissue allotransplantation in a patient with massive plexiform neurofibroma: a 1-year follow-up study. *Lancet*. 2008;372:639-645.
25. Swanson E, Boyd JB, Manktelow RT. The radial forearm flap: reconstructive applications and donor-site defects in 35 consecutive patients. *Plast Reconstr Surg*. 1990;85:258-266.
26. Bardsley AF, Soutar DS, Elliott D, Batchelor AG. Reducing morbidity in the radial forearm flap donor site. *Plast Reconstr Surg*. 1990;86:287-292. discussion 93-94.
27. Anthony JP, Rawnsley JD, Benhaim P, Ritter EF, Sadowsky SH, Singer MI. Donor leg morbidity and function after fibula free flap mandible reconstruction. *Plast Reconstr Surg*. 1995; 96:146-152.
28. Zimmermann CE, Borner BI, Hasse A, Sieg P. Donor site morbidity after microvascular fibula transfer. *Clin Oral Investig*. 2001;5:214-219.
29. Goodacre TE, Walker CJ, Jawad AS, Jackson AM, Brough MD. Donor site morbidity following osteocutaneous free fibula transfer. *Br J Plast Surg*. 1990;43:410-412.
30. Boyd JB. *Deep Circumflex Iliac Groin Flaps*. New York: Churchill Livingstone; 1989.
31. Smith AA, Bowen CV, Rabczak T, Boyd JB. Donor site deficit of the osteocutaneous radial forearm flap. *Ann Plast Surg*. 1994;32:372-376.
32. Richardson D, Fisher SE, Vaughan ED, Brown JS. Radial forearm flap donor-site complications and morbidity: a prospective study. *Plast Reconstr Surg*. 1997;99:109-115.
33. Soutar DS, McGregor IA. The radial forearm flap in intraoral reconstruction: the experience of 60 consecutive cases. *Plast Reconstr Surg*. 1986;78:1-8.
34. Youdas JW, Wood MB, Cahalan TD, Chao EY. A quantitative analysis of donor site morbidity after vascularized fibula transfer. *J Orthop Res*. 1988;6:621-629.
35. Boyd JB, Rosen I, Rotstein L, et al. The iliac crest and the radial forearm flap in vascularized oromandibular reconstruction. *Am J Surg*. 1990;159:301-308.
36. Swartz WM, Banis JC, Newton ED, Ramasastry SS, Jones NF, Acland R. The osteocutaneous scapular flap for mandibular and maxillary reconstruction. *Plast Reconstr Surg*. 1986; 77:530-545.
37. Gordon CR, Siemionow M, Zins J. Composite tissue allotransplantation: a proposed classification system based on relative complexity. *Transplant Proc*. 2009;41:481-484.
38. Prihodova L, Nagyova I, Rosenberger J, Roland R, van Dijk JP, Groothoff JW. Impact of personality and psychological distress on health-related quality of life in kidney transplant recipients. *Transpl Int*. 2010;23:484-492.
39. Smith GC, Trauer T, Kerr PG, Chadban SJ. Prospective quality-of-life monitoring of simultaneous pancreas and kidney transplant recipients using the 36-item short form health survey. *Am J Kidney Dis*. 2010;55:698-707.
40. Knoll GA. Is kidney transplantation for everyone? The example of the older dialysis patient. *Clin J Am Soc Nephrol*. 2009;4:2040-2044.
41. Buchanan P, Dzebisashvili N, Lentine KL, Axelrod DA, Schnitzler MA, Salvalaggio PR. Liver transplantation cost in the model for end-stage liver disease era: looking beyond the transplant admission. *Liver Transplant*. 2009;15:1270-1277.
42. Simmons RG, Abress L, Anderson CR. Quality of life after kidney transplantation. A prospective, randomized comparison of cyclosporine and conventional immunosuppressive therapy. *Transplantation*. 1988;45:415-421.
43. Helderman JH, Kang N, Legorreta AP, Chen JY. Healthcare costs in renal transplant recipients using branded versus generic cyclosporin. *Appl Health Econ Health Policy*. 2010;8: 61-68.
44. Asgeirsdottir TL, Asmundsdottir G, Heimisdottir M, Jonsson E, Palsson R. Cost-effectiveness analysis of treatment for end-stage renal disease. *Laeknabladid*. 2009;95:747-753.
45. Kasiske BL, Cohen D, Lucey MR, Neylan JF. Payment for immunosuppression after organ transplantation. American Society of Transplantation. *JAMA*. 2000;283:2445-2450.

30 Media-Related Aspects From a Public Relations Perspective

Eileen M. Sheil, Angie Kiska, and Tracy Wheeler

Contents

T. Wheeler (✉)
Corporate Communications,
Cleveland Clinic, Cleveland, OH, USA
e-mail: wheelet2@ccf.org

Abstract On December 18, 2008, a *New York Times* story ran under the headline, "In an Extensive and Intricate Operation, a Face Is Remade." These stories – and hundreds of others like them – shared a common theme: they were overwhelmingly positive and they focused on the 22-h procedure, not the still-anonymous patient.

These stories did not happen by luck or by accident. They grew out of a long-term strategy within Cleveland Clinic's Corporate Communication's department to position Dr. Siemionow as a trusted go-to source on the topic of face transplants – no matter where in the world they may occur.

Public and media relations had a crucial role in how information was presented and released to the media. While the physicians are the unquestioned experts in the clinical aspects of facial transplantation, members of the public and media relations team are the experts in understanding how the media work and what it takes for the hospital to earn the most positive news coverage possible.

The following chapter examines the Cleveland Clinic's experience with public relation and media relations.

30.1 Introduction

On December 18, 2008, a *New York Times* story ran under the headline, "In an Extensive and Intricate Operation, a Face Is Remade."[1] The headline of the Associated Press story read, "Nation's first face transplant done in Cleveland."[2] And on CNN, Dr. Sanjay Gupta began his *House Call* program with "a groundbreaking surgery. Doctors at the Cleveland Clinic announcing for the first time a successful near total face transplant in the United

M.Z. Siemionow (ed.), *The Know-How of Face Transplantation*,
DOI: 10.1007/978-0-85729-253-7_30, © Springer-Verlag London Limited 2011

States. A woman getting 80 percent of her face replaced with skin from a cadaver. It's remarkable."[3]

These stories – and hundreds of others like them – shared a common theme: They were overwhelmingly positive and they focused on the 22-h procedure, not the still-anonymous patient.

These stories did not happen by luck or by accident. They grew out of a long-term strategy within Cleveland Clinic's Corporate Communication's department to position Dr. Siemionow as a trusted go-to source on the topic of face transplants – no matter where in the world they may occur. Ultimately, the fact that the media coverage played out the way it did is a testament to Dr. Siemionow and her team. They made all the right decisions, from the initial thoroughness of the protocol approved by the hospital's Institutional Review Board to choosing the right patient to their surgical mastery – all factors that influenced a positive outcome.

But public and media relations had a crucial role to play too. While the physicians are the unquestioned experts in the clinical aspects of facial transplantation, members of the public and media relations team are the experts in understanding how the media work and what it takes for the hospital to earn the most positive news coverage possible.

In what has become a widely quoted phrase, Dr. Maria Siemionow, upon announcing the surgery, said, "You need a face to face the world."[1] She was talking about her patient, Connie Culp. But Ms. Culp was not the one facing the world that day; she was still anonymous in December 2008. It was Dr. Siemionow and the nine other physicians on the surgical team who were facing the world that day, as was the Cleveland Clinic itself. What the hospital needed to face the world was a detailed, well-thought-out plan, which we had, every step of the way – from the first announcement of the IRB approval in 2004 to the announcement of the successful surgery in 2008 to the introduction of Connie Culp in 2009.

30.2 IRB Approval

After 10 months of debate on the medical, ethical, and psychological issues surrounding face transplantation, Cleveland Clinic's Institutional Review Board approved Dr. Siemionow's protocol on Oct. 15, 2004, making it the first hospital in the world to do so.[4]

We knew media interest would be immense. (In fact, the IRB process required a plan from Corporate Communications to show how the patients' best interests would be protected.) However, a press release or press conference was not the right approach. The procedure was too new and too complex. Trying to educate hundreds or thousands of media outlets all at once would be next to impossible. Instead, we offered exclusive access to the (Cleveland) Plain Dealer, Ohio's largest newspaper. Rather than educating hundreds or thousands of reporters, we focused on educating one.

The result was a straight-forward, factual 1,525-word front-page story on Sunday, Oct. 31, 2004, which announced to the world, "Clinic plans the first transplant of a human face."[4] The night before the story appeared, Eileen Sheil, the Cleveland Clinic's Executive Director of Corporate Communications, called Dr. Siemionow and told her: "Your life will never be the same after this story runs. Everything is going to change."

More than 3,000 phone calls and e-mails poured in over the next couple of weeks. National, international, and local news reporters; documentary film makers; tabloid TV and magazines; science publications; and popular consumer magazines all vied for the story. In many cases, multiple reporters called from the same outlet in hopes of not only scooping their competition but scooping their colleagues.

Initially, though, our official response was to let the Plain Dealer story – and the Associated Press report based on the Plain Dealer story[5] – to serve as the main source of information. Over the following weeks and months, we sorted through thousands of media requests and began reaching out to those reporters and outlets we could trust to present this procedure as a medical/science story, rather than as a sensationalistic science-fiction-has-become-reality story. Not knowing when a face transplant might take place in Cleveland, the goal was to establish Dr. Siemionow as a respected, go-to expert for the media.

About 5 months after announcing the IRB approval, we slowly inserted Dr. Siemionow back into the media, starting with National Public Radio's All Things Considered[6] in March 2005, followed by the Chicago Tribune[7] in June 2005, the *New York Times*[8] in July 2005, *Time*[9] magazine and the *Associated Press*[10] in September 2005, and a 6-min segment on NBC's Today show with Katie Couric[11] in November 2005.

Then on November 30, 2005, the floodgates opened. French surgeons announced that they had performed the world's first face transplant.[12] Our phone lines lit up and our e-mail in-boxes overflowed with media wanting Dr. Siemionow's opinion of the news out of

France. *The New York Times*,[13] *Newsweek*,[14] *New Scientist*,[15] *NPR*,[16] *AP*,[17] *CNN*,[18] *CBS*,[19] *ABC*,[20] and *NBC*[21] all wanted to know what Dr. Siemionow thought. She had become a key source to media throughout the world – even though she had yet to perform a human face transplant. In the media, the Cleveland Clinic and Dr. Siemionow were portrayed as conservative pioneers, while the French were labeled as reckless and unethical.

The New York Times, December 6, 2005, "Ethical Concerns on Face Transplant Grow":[22]

Dr. Maria Siemionow, director of plastic surgery research at the Cleveland Clinic, who has been preparing to perform a full face transplant, said that the way the transplant was conducted appeared to conflate two experimental protocols: the transplantation of facial tissue and the infusion of stem cells from the donor bone marrow into the patient in an attempt to prevent rejection of the new face.

The first procedure, although untried until now, has been well studied, and the microsurgical techniques involved are commonplace. But the second has been successful in human subjects only rarely and only recently. While pilot studies do suggest that an infusion of stem cells from the donor can help produce "chimerism" in humans, a state in which foreign tissue is tolerated by the body with comparatively little or no suppression of the immune system, it is far from standard practice in transplantation.

The French team's decision to perform two novel procedures simultaneously means that it may be difficult to determine the cause of success or failure of the transplant, Dr. Siemionow said.

"They should not be doing two experiments on the same patient," she added. "Ethics aside, it will make it difficult to get clean answers - if it works, why does it work, and if it goes wrong, was it the transplant or the stem cells?"

30.3 "We finally did it!"[23]

At 5:30 p.m. of December 9, 2008, Cleveland Clinic surgeons began the world's most extensive face transplant to date. They ended 22 h later, at 4:30 p.m., Dec. 10. However, it would be another week before anyone outside the operating room or the Corporate Communications office would know about this groundbreaking event.

Before going public, we wanted to see how the patient would respond to the surgery; mainly, would the transplant show signs of rejection? We knew that we could not wait long, though. News of the face transplant was sure to leak eventually, as employees secretly tell their families, who tell friends, who call the media.

We also needed time to prepare the announcement in the way that would be most effective. That meant preparing the doctors for every conceivable question that might come their way, in addition to writing detailed fact sheets and press releases, editing video and photos taken in the OR, creating graphics to illustrate the technical aspects of the operation, preparing a password-protected website, staging a press conference that would surely draw international interest, and establishing a toll-free call-in line for reporters unable to attend.

The process began in a lunch room next to Dr. Siemionow's office. The surgical team and the communications team gathered around a dining table. We set a tape recorder on the table and began asking questions: When did the surgery begin? When did it end? Can you explain the procedure, step by step? How many surgeons and support staff were involved? How is the patient doing? What if the patient begins to show signs of rejection? The questions and answers went on for over an hour.

As they answered, we coached them on their language: Don't say that the donor's face was "harvested"; say that it was "surgically removed" or "transferred to the recipient." Don't refer to the procedure as a "tissue allograft"; simply refer to it as a "transplant."

The coaching also included a mock press conference, in which we asked questions we expected to hear from the media, such as: Why did it take four years after IRB approval to perform the transplant? How is a face transplant different than a skin graft? What happens if the transplant is rejected? Will the patient look like the donor? Who is the patient? Is this ethical?

Essentially, we considered a few broad sets of questions: those about the procedure, those about the ethics, and those about the patient and donor. We addressed all of these head-on, as best we could. In the press release, fact sheet and graphic, we provided details of the surgery, addressed the ethics question, acknowledged the long-term risk of immunosuppression, and made it clear that the patient and the donor family wished to remain anonymous.

From the media alert: *Patient Privacy: For the protection of patient privacy, the patient and family wish to remain anonymous; they will not attend the news conference and will not be available for interviews.*[24]

From the press release: *For the privacy and protection of those involved, no information will be released on the patient, the donor or their families. (A written statement from the patient's sibling is available at www.clevelandclinic.org/face)*[25]

A fact sheet shared some details of the patient's condition, but not her identity.

The recipient, who wishes to remain anonymous, is a woman who had suffered severe facial trauma. She had no nose and no palate after her injury. She was unable to eat or breathe on her own, without a tracheostomy, and was missing bone support.

Clinic doctors have been treating her for several years, and she has undergone several reconstructive procedures; however, none of the available, conventional treatment options could restore her facial function.

After the transplant, the expectation is that the patient will be able to eat, speak, and breathe normally again.[26]

The strategy worked. The patient questions were kept to a minimum, placing the focus on the surgeons and their breakthrough. The Plain Dealer headlines read, "'We finally did it,' emotional doctor says of transplant"[23] and "Face transplanted at Clinic; First operation in U.S. is performed on woman."[27] On CNN, medical correspondent Elizabeth Cohen said, "Her doctors call it the first surgery of its kind, a near total face transplant was performed at the Cleveland Clinic. ... In a breakthrough 22-hour surgery, surgeons transplanted 80 percent of her face. From a cadaver came skin, facial muscles and nerves, lower eyelids, cheekbones, upper jaw, blood vessels, arteries."[3]

The message we had planned for came through loud and clear: The surgery is all about restoring function and quality of life to those who have been injured.

The accuracy and the tone of the coverage were heavily influenced by the press material we handed out. For instance, compare the CNN quote above to these two paragraphs from our fact sheet:

About the procedure:

In a 22-hour procedure, a team of eight surgeons replaced 80 percent of a trauma patient's face – essentially transplanting the full face except her upper eyelids, forehead, lower lip, and chin.

This is so far the largest and most complex face transplant in the world. The surgery integrated different functional components, such as nose and lower eyelids as well as different tissue types including, skin, muscles, bony structures, arteries, veins and nerves. Approximately 500 square centimeters of tissue were transplanted onto the recipient.[26]

All in all, the headline of our press release – CLEVELAND CLINIC PERFORMS NATION'S FIRST NEAR-TOTAL FACE TRANSPLANT; Team of Eight Surgeons Replaced 80Percent of a Trauma Patient's Face[25] – became the message heard around the world after the December 17, 2008, press conference.

Getting through the initial announcement and press conference was just the beginning. Calls and e-mails came in throughout the evening and the rest of the week. Fortunately, we had created a password-protected website where journalists could download everything that was available at the press conference. The doctors were not available for interviews at this point, but reporters were able to download photos, graphics, and information.

30.4 Introducing Connie

As winter 2008 turned to spring 2009 in Cleveland, the still-anonymous patient was progressing very well. She had experienced just one episode of mild rejection. Her facial functions were slowly returning. She could smell, eat solid food, and drink from a cup. She could feel the kiss of her grandson on her check. All of this, of course, was great news. It also meant she was almost ready to go home and live her life in public again. It was our job to help make the hospital-to-home transition as smooth as possible.

Hoping for a secret, quiet return home was not an option. Considering that the number of media requests now approached 5,000 since the IRB approval in 2004, media would place a premium on being the first to capture images of the patient who received the world's most extensive face transplant. Members of the media relations department met with Connie Culp to assess her hopes and concerns before building a strategy.

Ms. Culp was introduced on May 5, 2009.[28] Instead of a press conference, though, we invited a very select group of national and local reporters to a medical media briefing, in an attempt to focus the stories on the science behind the procedure, not just the appearance of Ms. Culp.

In an effort to avoid any media ambushes on the day of the briefing, the reporters, photographers, and videographers were escorted into the meeting room. Dr. Siemionow addressed the media, focusing on the medical advances made by the medical team and the immense impact the surgery has had on the patient's life and her ability to re-enter society (Fig. 30.1). The medical team gave the background on Ms. Culp's trauma, sharing before-and-after photos. Ms. Culp was then escorted into the briefing room by security (Fig. 30.2). After a brief but emotional statement, she was escorted out of the room by security without taking questions. Media were unable to follow her out of the room, though none attempted to. Surgeons, doctors, and nurses who cared for Ms. Culp were available for one-on-one interviews with the invited media after the briefing concluded (Fig. 30.3).[32]

That same day, Ms. Culp sat down for an interview with Diane Sawyer for ABC's "Good Morning America"[29] and "Nightline.[30]"

Fig. 30.1 Dr. Maria Siemionow addresses invited media and hospital staff at the May 5, 2009, medical briefing at which Connie Culp was publicly introduced. Large flat-screen TVs helped the surgical team illustrate the complexities of the surgery with before-and-after photos of Ms. Culp, as well as CT scans, artist renderings, and detailed animation of the surgical procedure

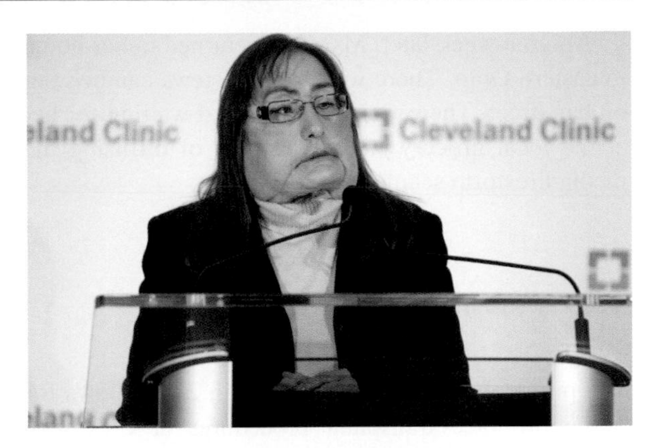

Fig. 30.2 Connie Culp decided to make a brief statement at the May 5, 2009, medical briefing. She was both funny and touching. "Well, I guess I'm the one you came to see today," she said with a chuckle. "While I know you all want to focus on me, I think it's more important you focus on the donor family that made it so I could have this Christmas present, I guess I should say"

Fig. 30.3 (**a**, **b**) Surgeons, doctors, and nurses who cared for Ms. Culp were available for one-on-one interviews with the invited media after the briefing concluded

About a week later, Ms. Culp returned to her home in eastern Ohio. There were no TV crews camped out on her street. There were no paparazzi waiting for her at the local grocery store. The goal of diffusing the media firestorm seemed to work.

30.5 Moving Forward

The protocol approved by the IRB in 2004 allows for two more face transplant procedures at Cleveland Clinic. Though it is impossible to predict when the next one may happen, it is likely that the media interest would not be as heavy as it was in 2008.

According to a Lexis search of "Cleveland Clinic and face transplant or facial transplant," 1,190 stories were written between 2004 and 2009:

2004	60 stories
2005	207 stories
2006	115 stories
2007	40 stories
2008	245 stories
2009	437 stories

Everyone expected Dr. Siemionow's work to generate worldwide media interest with each new announcement. So, while part of our job was to attract media attention, much of our efforts focused on managing and maximizing the news coverage.

We were able to do that at times by providing reporters with what they needed before they had even asked for it – detailed technical information, graphics, photos, video, and easy access to all of it. In making their jobs easier for them, we helped to ensure that our message was understandable, meaningful, and translatable to the general public. To do that, we provided – among other information – a timeline and a background sheet on Ms. Culp's injuries, surgery, complications, medications, and future.

Patient Care Timeline
Media Briefing
May 5, 2009

- *Surgery began at 5:30 p.m. Dec. 9, 2008, and ended at 4:30 p.m., Dec. 10.*
- *Stayed 12 days in the Intensive Care Unit.*
- *Stayed 45 days in the post-transplant unit.*
- *While in the post-transplant unit, the patient experienced one minor episode of rejection on the 47th day after surgery. There were no clinical signs of rejection, such as redness or swelling. However, a biopsy showed early signs of rejection in the mucosa, not the skin. The episode resolved in three days with high-dose immunosuppressants. There has been no sign of rejection since.*
- *Discharged Feb. 5; 58 days after leaving the operating room.*
- *Upon discharge, the patient resided in the Cleveland area for about three months, so that she could be close to the hospital for follow-up care, including blood tests, biopsies to check for rejection, physical therapy for her facial muscles, and meetings with social workers.*
- *Future: Maintaining proper immunosuppression is the most important aspect of the patient's on-going care. She is currently on a immune-suppressing regimen that is similar to that of a kidney transplant patient. Doctors will continue to monitor the growth of the facial nerve and her progression of functional recovery.*[31]

Media Briefing
May 5, 2009
RESTORING FUNCTION TO AN INJURED FACE

THE INJURY

In September 2004, Connie Culp sustained severe facial trauma when she was shot in the face. The 46-year-old lost most of her mid-face. She had no nose or lower eyelids. She had no upper jaw, no palate and no upper lip. She lost one eye and endured extensive damage to the other. The facial nerve – which controls the movement of all facial muscles – was completely missing from the left side of her face and partially missing on the right side.

With such severe injuries, she was unable to eat solid food, drink from a cup, smell, taste, or breathe on her own without a tracheostomy. Over time, her face became a wall of scar tissue and misaligned anatomy.

THE SURGERY

On Dec. 9, 2008, a team of eight Cleveland Clinic surgeons began a 22-hour operation to replace 80 percent of her face.

The surgeons' goal was to create a new, fully functional face for Connie, which meant freeing nerves encased in scar tissue and replacing missing muscle and bone. The procedure goes far beyond a skin graft, requiring microsurgical connections between the donor tissue and the patient's veins, arteries and nerves, in addition to rebuilding bony structures underlying it all.

For the new face to function, connection of the facial nerve is crucial, because without nerve signals, muscles won't move.

With Connie, surgeons traced the facial nerve back to its main trunk, near the ear – moving carefully – to find a scar-free area to which the transplanted nerve could be attached. On the left side of her face, the facial nerve was completely replaced, while the facial nerve on her right side was partially transplanted.

AFTER THE SURGERY

The key now is for the nerves to grow enough to reach the muscles. Growing at a pace of about an inch a month, the facial nerve should reach its full length about one year after the surgery which will improve Connie's facial function, giving her more facial movement.

Five months after the surgery, Connie can eat solid food, drink from a cup, wink, pucker her lips, smell and taste. Her follow up care includes future procedures, monitoring of immunosuppressant therapy and physical therapy.

"From the outset, this surgery was about restoring functionality, about making the patient feel presentable to society," Dr. Maria Siemionow said. "This is all about quality of life."[32]

Ultimately, public relations professionals cannot tell journalists what to write, but we can give reporters unexpectedly interesting information that they cannot ignore. If the facts are not clearly presented to reporters, they cannot report them.

References

1. Altman L. In an extensive and intricate operation, a face is remade. *New York Times*. December 18, 2008:A-18.
2. Marchione M. Nation's first face transplant done in Cleveland. *Associated Press*. December 17, 2008.
3. Gupta S. House call with Dr. Sanjay Gupta, CNN, December 20, 2008. Transcript available at: http://transcripts.cnn.com/TRANSCRIPTS/0812/20/hcsg.01.html
4. Spector H. Clinic plans the first transplant of a human face. *Plain Dealer*. October 31, 2004:A1.
5. Cleveland clinic plans first transplant of a human face. *Associated Press*. October 31, 2004.
6. Block M, Siegel R. Cleveland clinic clears first hurdle to perform face transplants. *National Public Radio. All Things Considered*. March 1, 2005.
7. Gorner P. Surgery's next step: Face transplants; the severely disfigured see hope for a normal life, but ethicists fear the risks may be too. *Chicago Tribune*. June 12, 2005.
8. Mason M. A new face, health & fitness. *New York Times 2005*. July 26, 2005:1.
9. Gorman C. Face transplant waits for the future. *Time*. September 20, 2005.
10. Marchione M. A tale of two faces: plans advancing for world's first face transplant. *Associated Press*. September 17, 2005.
11. Couric K. Dr. Maria Siemionow discusses preparations being made to perform the world's first face transplant. *NBC Today*. November 21, 2005.
12. French docs perform first face transplant. *UPI*. November 30, 2005.
13. Altman L. French, in first, use a transplant to repair a face. *New York Times*. December 1, 2005:A-1.
14. Springen K. The ultimate transplant; surgeons are ready to offer burn victims entire faces from donors, but there are risks. *Newsweek*. December 12, 2005:60.
15. Coghlan A. The face-transplant benefit that will speak for itself. *New Scientist*. December 10, 2005:10.
16. All Things Considered, Maria Siemionow discusses French breakthrough in facial transplants. *National Public Radio*. December 1, 2005.
17. Marchione M, Leicester J. Surgeons in first partial face transplant did not try normal procedures first, ethics questioned. *Associated Press*. December 1, 2005.
18. Cooper A. Facing the Future. *CNN, Anderson Cooper 360 Degrees*. December 2, 2005.
19. Ethical questions surrounding face transplants. *CBS Evening News*. December 2, 2005.
20. McKenzie J. Controversial surgery: Face transplant. *ABC World News Tonight*. December 2, 2005.
21. Bazell R. Risky face transplants could give patients better quality of life. *NBC Nightly News*. December 3, 2005.
22. Altman L, Mason M. Ethical concerns on face transplant grow. *New York Times*. December 6, 2005:A-12.
23. Spector H. "We finally did it," emotional doctor says of transplant. *Plain Dealer (Cleveland)*. December 18, 2008:A-1.
24. Cleveland Clinic Corporate Communications, Media Alert: Cleveland Clinic Performs Nation's First Near-Total Face Transplant; Team of Eight Surgeons Replaced 80 Percent of a Trauma Patient's Face. December 17, 2008.

25. Cleveland Clinic Corporate Communications, Press Release: Cleveland Clinic Surgeons Perform Nation's First Near-Total Face Transplant; Surgery Largest, Most Complex of Its Kind; 80 Percent of Trauma Patient's Face Transplanted. December 17, 2008.
26. Cleveland clinic corporate communications. *Fact Sheet*. December 17, 2008.
27. Spector H. Face transplanted at Clinic; First operation in U.S. is performed on woman. *Plain Dealer (Cleveland). 2008*. December 17, 2008:A-1.
28. Marchione M. Nation's first face transplant patient shows face. *Associated Press*. May 5, 2009.
29. Sawyer D. Exclusive Interview; Connie Culp: Face transplant. *ABC News, Good Morning America*. May 8, 2009.
30. Sawyer D. The Face of courage; a new beginning. *ABC News, Nightline*. May 8, 2009.
31. Cleveland clinic corporate communications. *Patient Care Timeline*. May 5, 2009.
32. Cleveland clinic corporate communications. *Media Briefing*. May 5, 2009.

Part VI

World Experience with Face Transplantation

31 Clinical Facial Composite Tissue Allotransplantation: A Review of the Global Experience and Future Implications

Helen G. Hui-Chou and Eduardo D. Rodriguez

Contents

H.G. Hui-Chou (✉)
Division of Plastic and Reconstructive Surgery,
The Johns Hopkins Hospital/University of Maryland
School of Medicine, Baltimore, MD, USA
e-mail: hhuichou@jhmi.edu

Abstract Since 2005, 11 facial composite tissue allotransplantations (CTA) have been performed in 8 different centers in 6 countries. Five teams have reported their work and outcomes in separate publications. We review in detail the first four global experiences and compare indications, donor/recipient matching criteria, CTA graft anatomy, immunosuppressive protocols, and postoperative course. A thorough review of the eight publications by five transplantation groups was conducted. Additional information gathered from official press releases were also included for review. Some details of the remaining transplants are also discussed to facilitate evaluation of indications, matching criteria and future directions for facial CTA.

Abbreviations

CMV	Cytomegalovirus
CTA	Composite Tissue Allotransplantation
EBV	Epstein–Barr Virus
EMG	Electromyelography
GVHD	Graft vs. Host disease
NCS	Nerve Conduction Studies
PSSD	Pressure-Specified Sensory Device

31.1 Introduction

In the 3-year period between November 2005 and December 2008, four facial composite tissue allotransplants (CTA) were completed. Since 2009, there are very early reports of seven additional facial CTA, but few details have been released. The first four facial CTA were performed in four different centers in three different countries. We examine the global experience

M.Z. Siemionow (ed.), *The Know-How of Face Transplantation*,
DOI: 10.1007/978-0-85729-253-7_31,

and compare indications, donor/recipient matching criteria, CTA graft anatomy, immunosuppressive protocols, and postoperative course. Many aspects of the transplants differed between the groups. Comparative analysis of their experiences may provide a better understanding of the outcomes and thus further elucidate future directions for clinical facial CTA.

Eleven total patients have received facial CTA for various facial disfigurements. There have been eight pivotal publications by the five transplantation groups. These publications were in *The Lancet*,[1-4] *The New England Journal of Medicine*,[5] and *Plastic and Reconstructive Surgery*.[6-8] Additional information was also drawn from another key article published in *Transplantation*,[9] as well as from various press releases. Data compiled from articles, press releases, and presentations are summarized in Table 31.1.[10] Over 20 different comparisons are included in the table. Several distinct comparisons are worth noting in further detail. Seven additional recent facial CTA are summarized in Table 31.2.

31.2 Current Factors in Facial Composite Tissue Allotransplantation

31.2.1 Patient Selection

Patient selection has proven to be a very important factor; poor patient selection has perhaps led to the first mortality in facial transplantation. The patient chosen for facial transplant by the Chinese group came from a remote village far from his transplant surgeons and hospital.[2] In addition, his lower socioeconomic and educational status perhaps led to the discontinuation of his maintenance immunosuppression regimen. Lack of compliance with the medication regimen and follow-up visits has been implicated as causes for his death; however, no autopsy was performed at the family's request.

The second facial transplant also demonstrated the importance of patient selection in terms of facial function restoration. The surgeon had not anticipated the degree of damage to the recipient's facial nerve and it was reported that the quality of the nerve repair was poor.[2] This led to an inanimate allograft that never provided any return of facial function or expression to the recipient. This CTA graft demonstrated no superiority to conventional reconstructive techniques that provide aesthetic subunit facial restoration, but often lack sensory or motor restoration.

31.2.2 Mucosal Component

The oral mucosal component of the facial CTA graft has proven to be a key element for early recognition of acute graft rejection in every case confirmed by the appearance of erythema and edema.[3,5,6,9] This mirrors what is observed with patients suffering from Graft-versus-Host Disease (GVHD) after bone marrow transplantation. In addition, not only does the mucosa herald clinical rejection episodes, but it also appears to be more antigenic than skin. As shown by the third and fourth facial transplant groups, the mucosa can show histologic evidence of mild rejection while skin biopsies and the appearance of the CTA graft remained clinically normal.[6]

31.2.3 Infectious Disease Factor

Viral serology and matching between donor and recipient also appears to be an important factor. The third facial transplant case developed severe cytomegalovirus (CMV) viremia due to CMV mismatch of the donor and recipient.[3] The recipient was CMV titer negative, but converted after his facial CTA. This seroconversion in addition to his immunosuppression regimen resulted in the viremia which required intravenous Foscarnet treatment. While treating the CMV viremia, the team also held one of the maintenance immunosuppression drugs, mycophenolate mofetil (MMF), until the infection was cleared. An infection of this severity jeopardizes not only the facial CTA graft but also the patient. This infection was not prevented with the use of standard CMV prophylaxis. Recently, the fourth facial CTA was revealed to have the same CMV mismatch with a CMV positive donor to a negative recipient. However, no adverse events have been reported to date related to this CMV mismatch.

Table 31.1 Facial composite tissue allograft transplant: first four cases (2005–2008)

Location	Amiens, France	Xi'an, China	Créteil, France	Cleveland, USA
Date of transplant	November 27, 2005	April 13, 2006	January 21, 2007	December 10, 2008
Surgeon/team	Jean-Michel Dubernard, M.D., Ph.D. and Bernard Devauchelle, M.D., DMD	Shuzhong Guo, M.D.	Laurent Lantieri, M.D.	Maria Siemionow, M.D., Ph.D., DSc
Hospital	Amiens University Hospital	Xi'an Hospital, Fourth Military Medical University	Henri Mondor Hospital	The Cleveland Clinic
Recipient	Isabelle Dinoire	Guoxing Li	Pascal Coler	Connie Culp
Age at transplant	38	30	29	45
Injury/defect and date	Dog mauling on May 28, 2005	Bear attack in October 2004	Massive plexiform neurofibroma	Close range shotgun trauma in September 2004
Blood type match (donor/recipient)	O+/O+	A/A	O+/O+	A/AB
HLA match				
Antigen	5	3	3	2
Donor	**A2 A3**, **B8 B44,** DR15 **DR3**	**A11** A9, **B38** B7, **DR10 DR15**	A2 **A24, B7** B72, **DR11** DR18	A11 A23, **B44** B64, Cw5 Cw8, **DR7** DR16, DRw51, DQ5 DQ9
Recipient	**A2 A3, B8 B44, DR3** DR7	**A11** A2, **B38** B52, DR4 **DR14**	**A23** A26, **B7** B51, **DR11** DR13	A1 A29, B8 **B44**, Cw7 Cw16, **DR7** DR17, DRw52, DQ2
Donor procurement criteria	Brain death, standard criteria	Deceased after cardiac death	Brain death, standard criteria	Brain death, standard criteria
Additional donor information (gender, age, cause of death)	Female, 46, severe irreversible cerebral ischemia	Male, 25, traffic accident	Male, 50	Female, 44
Infectious disease serology	None reported	None reported	CMV +/–, Syphilis +/–	CMV +/–
Donor restoration	Silicone face mask	None reported	Silicone face mask	Five organ donor, cremated
Face CTA components				
Facial subunit(s) – skin	• Perioral • Lips • Nasal	• Nose • Upper lip • Right cheek	• Perioral • Lips • Nose • Bilateral cheek	• Lower eyelids • Perioral • Upper lip • Nose • Bilateral cheek

(continued)

Table 31.1 (continued)

Soft tissue-mucosa, muscle, gland	• Oral mucosa • Nasal mucus • Orbicularis oris	• Right partoid gland • Partial buccal mucosa • Partial masseter	• Oral mucosa • Bilateral parotid gland	• Parotid gland • Lymph nodes • Muscles
Nerve(s)	• Infraorbital • Mental • Zygomatic • Buccal • Mandibular	• Right facial (not satisfactory)	• Infraorbital • Mental • Facial	• Bilateral facial (with donor vagus and hypoglossal n. interpositional graft)
Bone(s)	None	• Nasal bone • Lateral orbital and infraorbital wall • Front wall of maxillary sinus • Nasal septal cartilage • Right zygomatic arch	None	• Lefort III maxilla with teeth • Maxillary sinus • Zygoma
Vessel(s)	• Bilateral facial artery	• Right facial artery and vein	• Bilateral trunks of external carotid arteries • Thyrolinguofacial veins	• Bilateral common facial arteries • Bilateral external jugular veins • Left posterior facial vein
Sentinel graft/other	• Radial forearm flap	None	None	None
Transplant surgical procedure sequence	1. Tracheostomy 2. Vessels – bilateral facial artery and vein 3. Nerves 4. Sensory nerves: bilateral infraorbital and mental 5. Motor nerve: mandibular branch of left facial nerve 6. Sentinel graft: radial forearm to left submammary	1. Right facial artery and vein end-to-end 2. Bone reconstruction with titanium 3. Masseter muscle of CTA to obliterate recipient maxillary sinus 4. Right facial nerve not well coapted	1. Tracheostomy 2. Left then right external carotid end-to-end 3. Thyrolinguofacial trunks for end-to-end venous 4. Facial and infraorbital nerves were sutured and glued with fibrin glue 5. Mental nerves placed adjacent to foramen	1. *Had tracheostomy* 2. Bilateral facial vessels 3. Bone 4. Nerves 5. Muscles
Blood transfusion	None reported	Over 6 L of blood products transfused	Transfused 35 units of blood products	Intra-op: 500 mL Post-op: 2 units
Transplant time	Total time: 15 h Ischemia time: 230 min	Total time: 18 h	Total time: 15 h Ischemia: 2 h Procurement: 4 h	Total time: 22 h Procurement: 9 h Ischemia: 2 h 40 min Transplant: 12 h
Bone marrow	Infusions on POD 4 and POD 11	None	None	Vascularized bone marrow within maxillary bone segment

Induction therapy	• Thymoglobulin 1.25 mg/kg daily × 10 days	• Humanized Anti-IL-2 Receptor Monoclonal Ab (Daclizumab 50 mg) *Zenapax* • FK506 & Prednisone & MMF • X-ray irradiation (4 Gy) of CTA • Protein A immunoadsorption therapy	• Thymoglobulin 1.25 mg/kg daily × 10 days • FK506 (10-13 ng/mL) • MMF 2 g daily • Prednisone taper to 10 mg daily	• Thymoglobulin 1.2 mg/kg IV daily × 9 days • Steroids
Maintenance therapy	• FK506 (8–10 ng/mL) • MMF 2 g daily • Prednisone taper over 15 days to 25 mg daily	• FK506 (10–25 ng/mL) • MMF taper dose • Prednisone taper dose • Humanized IL-2 receptor monoclonal antibody @ POD 14	• FK506 10 mg daily (8–10 ng/mL) • MMF 500 mg daily • Prednisone 7.5 mg daily	• FK506 4 mg BID (12–15 ng/mL for first 3 months then, 10–12 ng/mL) • MMF 750 mg BID • Prednisone 10 mg daily
Rescue therapy	• Dose changes in Prednisone, FK506, and MMF • Clobetasol ointment, Protopic ointment, Prednisone mouthwash • Pulse steroids • Extracorporeal photopheresis (ECP) at 10 months	• Increase in FK506 • Pulse steroids	• Pulse steroids • Thymoglobulin for 7 days • ECP started at 3 months	• Solu-Medrol 1 g IV bolus × 1
Acute rejection episode(s)	2 episodes	3 episodes	2 episodes	1 episode (subclinical)
	POD 18 and POD 214	POD 90, 150, and 510	POD 28 and 64	POD 47
Histologic rejection	In mucosa and skin	In skin alone	In mucosa and skin	In mucosa biopsy alone (Banff grade III/IV rejection)
Prophylaxis regimen				
CMV	Ganciclovir then, Valanciclovir	Acyclovir	Valganciclovir	Ganciclovir (5 mg/kg BID) then, Valganciclovir (900 mg)
PCP	Trimethoprim sulfamethoxazole	Metronidazole	Trimethoprim sulfamethoxazole	Trimethoprim sulfamethoxazole (400 mg)
Other	None	Allicin	Penicillin V for syphilis	
Infectious complications	• Herpes Simplex Virus type 1 (HSV) on POD 185 • Molluscum contagiosum on POD 214	None reported	• Cytomegalovirus (CMV) on POD 64 requiring Foscarnet and temporary discontinuation of MMF	None reported

(continued)

Table 31.1 (continued)

Medical complications	Renal insufficiency – serum creatinine from 0.6 to 1.5 mg/dL Rx – switch to Rapamycin (Sirolimus)	Hyperglycemia/New Onset Diabetes – BS rise to 361 mg/dL Rx – insulin then, repaglinide and metformin	Steroid-induced confusion Rx – Thorazine	Leucopenia – First episode resolved with transient withdrawal of MMF and ganciclovir, Second episode required administration of leukocyte growth factor
Sensory recovery				
Semmes-Weinstein	6 months	3 months	4–12 months	PSSD at 3–6 months
Heat/Cold	6 months	8 months	4–12 months	5 months
Motor recovery	fMRI and Functional return by 18 months	No recovery	EMG return by 12 months	None reported to date
Chimerism	None: Microchimerism detected only once at level of 0.1% on POD 60, all other measures resulted in recipient cells	None reported	None detected on peripheral blood microchimerism PCR	None reported
Funding for protocol	University Hospitals of Amiens and Lyon	New Clinical Technique Foundation of Xijing Hospital	Hospital Clinical Research Program	Cleveland Clinic Foundation
Follow-up	• Published in *NEJM* on December 13, 2007 • 3 years posttransplant • Immunosuppression regimen: – Rapamycin target 8–12 ng/mL – MMF 2 g daily – Prednisone 10 mg daily • No further rejection episodes • Renal function stable with creatinine at 1.1 mg/dL	• Published in *Lancet* on August 23, 2008 • Immunosuppression regimen: – FK506 target 10–15 ng/mL – MMF at 0.25 g BID – Prednisone 15 mg daily • Returned to home village over 2 days travel away from Xi'an – Stopped immunosuppression – Last follow-up at 24 months • His death reported on December 21, 2008 – Died in July, 27 months after transplant – Autopsy refused by family	• Published in *Lancet* on August 23, 2008 • 2 years posttransplant • Immunosuppression regimen: – FK506 6 mg daily – MMF 0.5 g daily – Prednisone 7.5 mg daily • Social improvement – Full time employment 13 months after transplant – First relationship • No malignant transformations due to immunosuppression regimen • By 4/09: off prednisone	• 3 months posttransplant • Discharged from Cleveland clinic on February 5, 2009 • Immunosuppression regimen: – FK506 4 mg BID – MMF 0.5 g BID – Prednisone 10 mg daily • "She can eat pizza. And hamburgers. She can smell perfume, drink coffee from a cup and purse her lips as if to blow a kiss." • Still requires dental implants for maxilla
Final outcome		**Deceased: December 21, 2008**		

Adapted from Hui-Chou et al.[10]Details of the first four facial transplantation cases from 2006 to 2008

Table 31.2 Facial composite tissue allograft transplant – Recent cases (2009–2010)

Location	Créteil, France	Créteil, France	Boston, USA	Valencia, Spain	Seville, Spain	Créteil, France	Barcelona, Spain
Date of transplant	March 26, 2009	April 4, 2009	April 9, 2009	August 18, 2009	January 26, 2010	February 2010	March 20, 2010
Surgeon/team	Laurent Lantieri, M.D.	Laurent Lantieri, M.D.	Bohdan Pomahac, M.D.	Pedro C. Cavadas, M.D., Ph.D.	Tomas Gomez Cia, M.D., Ph.D.	Laurent Lantieri, M.D.	Joan Pere Barret
Hospital	Henri Mondor Hospital	Henri Mondor Hospital	Brigham and Women's Hospital	La Fe University Hospital	Hospital Virgen del Rocio	Henri Mondor Hospital	Vall d'Hebron University Hospital
Recipient	Andres Rodriguez	Male	James Perry Maki	Male	Rafael	Male	Male
Age at transplant	25	30	59	43	35	None reported	20–40
Injury/defect and date	Gunshot trauma	Burn in 2004: Bilateral hand and face	Burn/electrical injury after falling onto electrified rail tracks on June 30, 2005	Radiotherapy aggressive tumor	Neurofibromatosis Type I	None reported	Accidental self-inflicted shotgun farming in 2005
Blood type match (donor/recipient)	None reported	None reported	None reported	None reported	None reported	None reported	None reported
HLA match							
Antigen	None reported	None reported	None reported	None reported	None reported	None reported	None reported
Donor	None reported	None reported	None reported	None reported	None reported	None reported	None reported
Recipient	None reported	None reported	None reported	None reported	None reported	None reported	None reported
Donor procurement criteria	Brain Death, Standard Criteria in Lyon, France	Brain Death, Standard Criteria	James Helfgot, Brain Death, Standard Criteria	None reported	None reported	None reported	None reported
Additional donor information (gender, age, cause of death)	Male	Male	Male, 60, massive cerebral infarcts after heart txp; Heart, face, liver donor	35, traffic accident	30, traffic accident	None reported	None reported
Infectious disease serology	None reported	None reported	None reported	HIV –/+	None reported	None reported	None reported
Donor restoration	Silicone face mask	Silicone face mask and hands	Prosthetic mask	None reported	None reported	None reported	None reported

(continued)

Table 31.2 (continued)

Face CTA components							
Facial subunit(s) - skin	• Nose • Perioral	• Perioral, upper lip • Scalp • Forehead • Nose • Ears • Upper and lower eyelids	• Upper lip • Nose	• Lips • Nose • Cheek • Perioral • Periorbital	• Lips • Perioral • Lower 2/3 face	None reported	• Nose • Eyelids • Full facial • skin
Soft tissue-mucosa, muscle, gland	• Mucosa • Muscle	None reported	• Mucosa • Muscle	• Muscle • Tongue	• Tongue	None reported	• Muscles
Nerve(s)	• Bilateral Facial • Bilateral Trigeminal	None reported	• Bilateral infraorbital (V2) • Buccal (V3) • Buccal branch of CN VII	None reported	None reported	None reported	None reported
Bone(s)	• Nasal bone • Maxilla • Mandible symphysis	None reported	• Maxilla • Nasal bone	• Palate • Mandible with teeth	None reported	• Mandible symphysis • Maxilla	• Mandible with teeth • Maxilla
Vessel(s) – artery/ vein	• Bilateral external carotid arteries • Bilateral facial veins	None reported	• Right facial artery • Left external carotid • artery • Bilateral facial veins	None reported	None reported	None reported	None reported
Sentinel graft/other	None	• Bilateral Hands	• Radial forearm flap	None reported	None reported	None reported	None reported
Transplant surgical procedure sequence	1. *Had tracheostomy* 2. *Had PEG tube*	None reported	1. Tracheostomy 2. Vessel and nerve dissection 3. Removal of previous reconstruction flap 4. Vascular anastomosis on left, then right 5. Nerve repair 6. Bony fixation	None reported	None reported	None reported	None reported

Blood transfusion	None reported	None reported	None reported	None reported	None reported	None reported	None reported
Transplant time	Total time: 15 h Ischemia: 4 ½ hours	Total time: 30 h Hand: 7 h	Procurement: 7 h Total time: 17 h Ischemia: 1 h 15 min	15 h	30 h	None reported	24 h
Bone marrow	Vascularized bone marrow within mandibular segment	None reported	None reported	Vascularized bone marrow within mandible	None reported	Vascularized bone marrow within mandibu-lar segment	None reported
Induction therapy	None reported	None reported	None reported	None reported	None reported	None reported	None reported
Maintenance therapy	None reported	None reported	None reported	None reported	None reported	None reported	None reported
Rescue therapy	None reported	None reported	Steroids	None reported	None reported	None reported	None reported
Acute rejection episode(s)	None reported None reported	None reported None reported	2 episodes None reported	None reported None reported	None reported None reported	None reported None reported	None reported None reported
Histologic rejection	None reported	None reported	None reported	None reported	None reported	None reported	None reported
Prophylaxis regimen							
CMV	None reported	None reported	None reported	None reported	None reported	None reported	None reported
PCP	None reported	None reported	None reported	None reported	None reported	None reported	None reported
Other	None reported	None reported	None reported	None reported	None reported	None reported	None reported
Infectious complications	None reported	Pseudomonas infection	None reported	None reported	None reported	None reported	None reported
Medical complications	None reported	None reported	None reported	None reported	None reported	None reported	None reported
Sensory recovery							
Semmes-Weinstein	None reported	None reported	None reported	None reported	None reported	None reported	None reported
Heat/cold	None reported	None reported	None reported	None reported	None reported	None reported	None reported
Motor recovery	None reported	None reported	None reported	None reported	None reported	None reported	None reported
Chimerism	None reported	None reported	None reported	None reported	None reported	None reported	None reported
Funding for protocol	Hospital Clinical Research Program	Hospital Clinical Research Program	None reported	None reported	None reported	None reported	None reported
Follow-up	None reported	June 8, 2009 cardiac arrest after bacterial infection	None reported	Discharged September 12, 2010	None reported	None reported	None reported
Final outcome	None reported	**Deceased: June 8, 2009**	None reported	None reported	None reported	None reported	None reported

Details of the seven additional facial transplantation cases from 2009 to 2010

Infection risk is important, especially with patients placed on immunosuppression post-transplant. Patients can have active infections or be colonized with microorganisms resulting from their prior treatments. The first face CTA with double hand CTA transplant provided insight into this risk factor. A burn patient who was colonized with *Pseudomonas* developed infection of his CTA posttransplant which led to sepsis, and eventually death. The fourth face CTA transplant was noted to have negative cultures of oro- and nasal mucosa. The patient was further monitored postoperatively by interval cultures.[4,8] Considerations of active and past infections will be required to avoid reactivation after patients are placed on immunosuppression.

31.2.4 Blood Transfusions and Crossmatch

Another important consideration with facial CTA is the need for blood transfusions during and after the surgical procedure. The second and third facial transplant teams reported needing large amounts of blood products during the transplant.[2,3] For the second facial transplant, the blood loss was due to the use of a donor after cardiac death criteria and a rushed procurement which prevented hemostatic control of the CTA graft. For the third facial transplant, the blood loss was a result of resection of the highly vascular tumor. It is unclear what detrimental effects, if any, resulted from the blood transfusions with the addition of the circulating antibodies and antigens.

31.2.5 Acute Rejection

Each group reported two to three episodes of acute rejection requiring different combinations of rescue therapies to clear the clinical signs.[2,3,9] The advantage over solid organ transplants is that composite tissue allografts, such as hand and face CTA, allow for direct clinical monitoring of rejection. Solid organ transplants rely on secondary markers as initial signs of rejection. For facial CTA, the teams have been able to directly observe the mucosa and skin, as well as a sentinel graft placed by the first facial transplant team and the team from Boston, USA. The facial CTA graft is also easily accessible for biopsy to confirm clinical findings.

31.2.6 Chimerism and Tolerance

Lastly, a goal of achieving tolerance to a facial CTA graft would alleviate many concerns regarding this non-life-saving procedure and the associated risks of lifelong immunosuppression. This tolerance is measured by the existence of chimeric cells in the recipient after transplant. Unfortunately, there has been no documented evidence of durable chimerism in any of the transplant recipients.[3,5] The first facial CTA group utilized infusions of bone marrow harvested from the donor's iliac crest. This was performed on post-transplant days 4 and 11. They did not report sustained levels of chimerism, only one transient episode on day 60.

What may hold more promise is the inclusion of a bone segment within the facial CTA which would contain a source of vascularized bone marrow (VBM). Currently, four facial CTA have included a bony segment of either the maxilla or mandible. Animal studies have shown instances of sustained chimerism when VBM is included with the CTA.[11-13] Further studies of the four patients with a vascularized bone and VBM component are required to determine the clinical relevance.

31.3 Future Implications

The four facial CTA protocols reviewed were all slightly different. The indications for the procedure and outcomes for the patient were also quite varied. Photographs of the patients pre-transplant and different follow-up time points allow for evaluation of the different outcomes. (Figs. 31.1–31.4) Recipient selection has clearly proven to be a vital factor to achieve and maintain a successful facial CTA. It is important to select individuals who understand and commit to the rigorous immunosuppression regimen and post-transplant routine. This requires thorough psychological and social evaluations by a multidisciplinary team.

Another aspect of recipient selection is identifying patients who require a functional facial reconstruction in addition to an aesthetic restoration. This typically includes most facial deformities of the central and

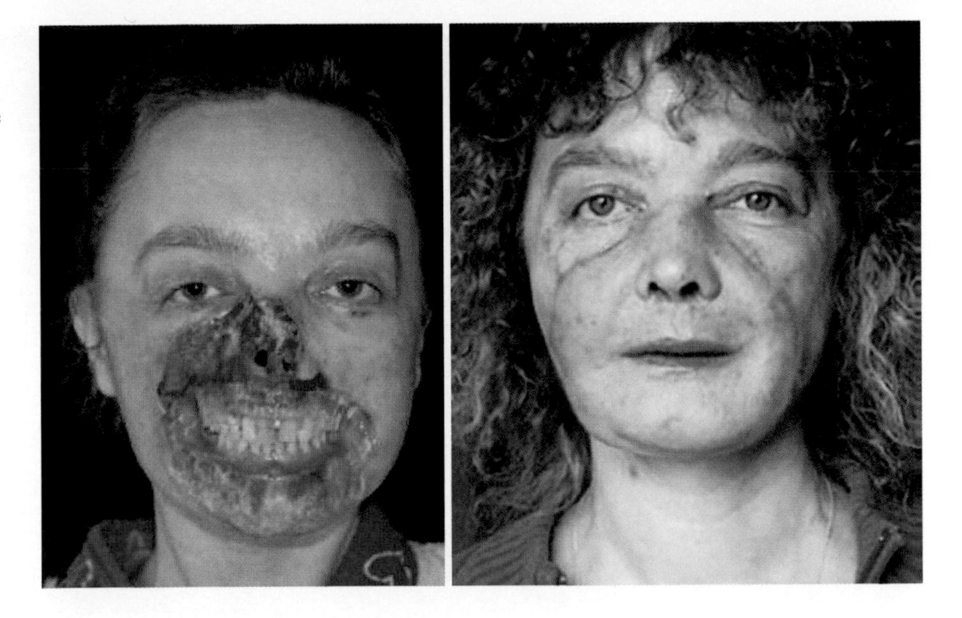

Fig. 31.1 Isabelle Dinoire before transplant and 18 months after transplant (Reprinted from Devauchelle et al.[1] With permission)

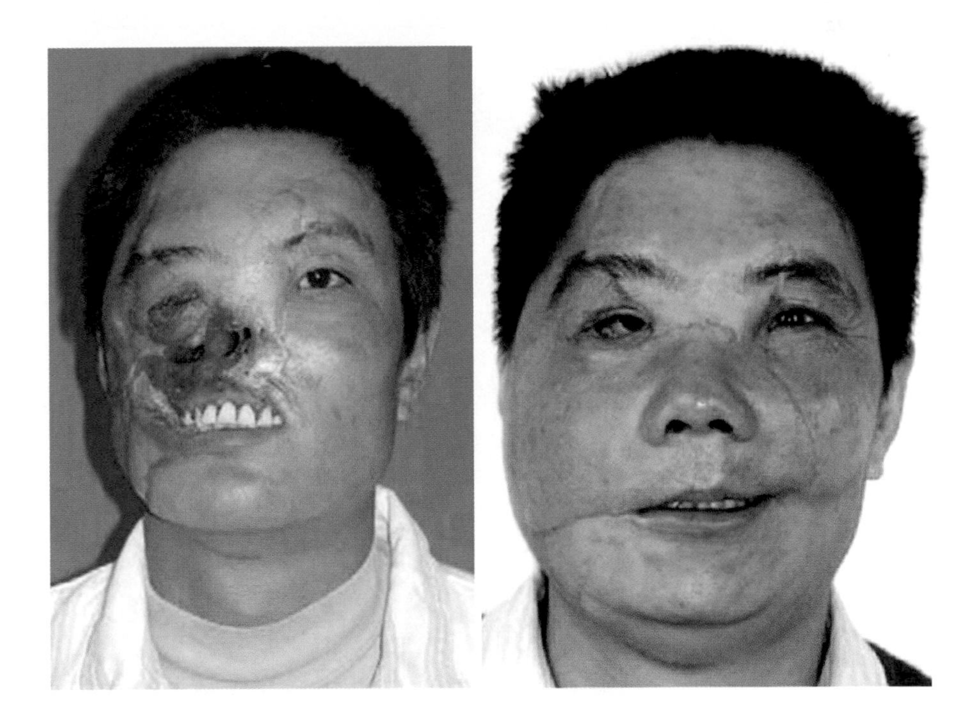

Fig. 31.2 Guoxing Li pre-transplant and at 24 months follow-up (Reprinted from Guo et al.[2] With permission)

midface, specifically the perioral and/or periorbital subunits. Complex facial components requiring eyelid and/or lip function are nearly impossible to restore with conventional techniques.[14,15] The first four transplant recipients all had defects involving the perioral subunit of the face. The transplants performed to date included three traumatic facial injuries and a benign tumor defect. The mechanism of the facial defect may be a secondary consideration if focus is shifted to replacement of these key facial subunits. Recent transplants have been more comprehensive including full face as well as one patient with bilateral hand CTA.

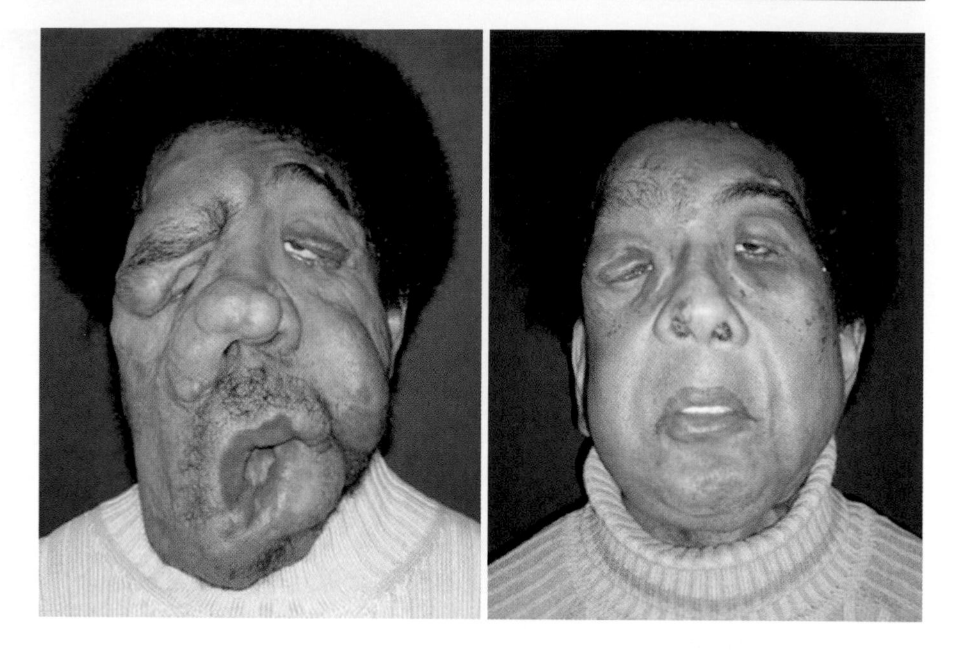

Fig. 31.3 Pascal Coler pre-transplant and at 12 months posttransplant (Reprinted from Lantieri et al.[3] With permission)

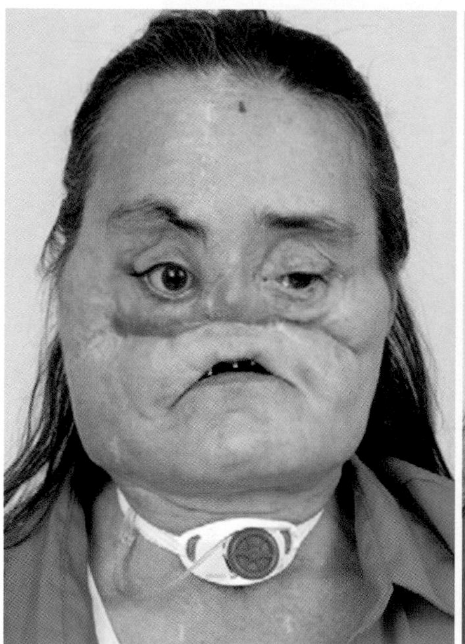

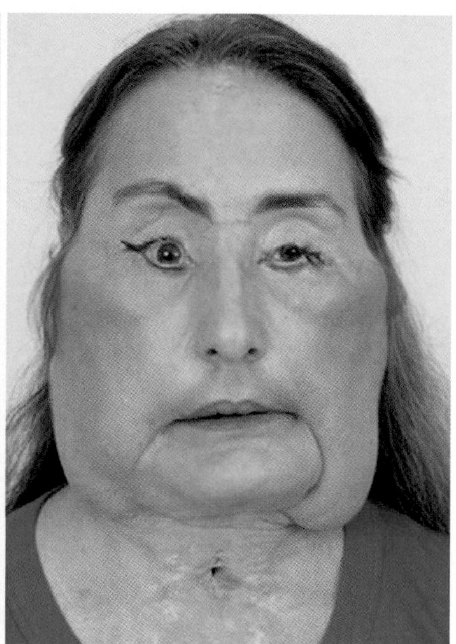

Fig. 31.4 Connie Culp pre-transplant and 6 months after transplant (Reprinted from Siemionow et al.[4] With permission)

Burn patients are ideal candidates for facial CTA as they often present with severe facial deformities involving the perioral and periorbital regions. However, consideration should be paid to the sensitization by tissue allografts in managing the acute burn patient. Other potential recipients are children born with severe congenital anomalies, such as mandibular agenesis. Pediatric patients may be ideal candidates since experiences in solid organ transplants have shown the development of durable chimerism and tolerance allowing for withdrawal of immunosuppression.[16,17] If issues with informed consent procedures could be addressed, pediatric patients could be the next group of potential recipients.

Another group of patients that warrant consideration are those afflicted with severe locally aggressive benign tumors including neurofibromatosis, vascular or lymphatic malformations. Lastly, patients devastated by infectious etiologies such as noma could be facial CTA recipients. This shift in focus from the mechanism of facial deformity to the selection of patients that can benefit from improved facial functioning by replacing damaged or missing perioral and periorbital subunits will expand transplantation indications. In addition, more rigorous selection of compliant and committed patients will ensure improved outcomes.

After recipient selection, donor selection and matching should be considered. As shown by the first four transplant donors, perhaps it is important to limit the donor pool to Brain Dead Standard Criteria donors to optimize procurement and minimize ischemia times. The transplant times of the first four cases ranged from 15 to 22 h.[2,5,6] Though all CTA grafts were re-perfused without any difficulty, longer ischemia times may make the graft more susceptible to rejection and failure.

Viral serology matching avoids exposing the recipient to new infections following their transplant while they are highly immunosuppressed. Important viral serology matching includes Hepatitis B, Hepatitis C, CMV, Epstein–Barr Virus (EBV), and Human Immunodeficiency Virus (HIV). Although a positive donor to positive recipient match in theory could be performed, as is often done with solid organ transplants, it is perhaps most prudent to transplant a recipient who has no other initial co-morbidities. However, there are early reports that a facial CTA recipient was HIV positive at the time of transplant.

Another critical donor/recipient factor is human leukocyte antigen (HLA) matching. Of the six major HLA antigens, the first four transplants had as few as two to as many as five antigens matched. In solid organ transplantation, perfect matches have become rare; more patients receive partially matched or even unmatched organs.[18-21] The implications of the HLA match are not completely understood in terms of facial transplantation; however, attempts must still be made at optimal donor/recipient HLA matches. The minimum number of matches will be debated and may depend on the outcome of the transplant performed at the Cleveland Clinic, which only had two antigens match.[4,8]

Additional goals for donor/recipient matching include age, with a margin of 10 years above and below the recipient, skin tone, gender, and race. Of these, skin tone and quality may be the final determinant as gender and race may be limited by HLA typing. However, if the HLA match is present, there should be no restrictions on donor/recipient gender or race, if the skin tone does offer the best match. This is due to the facial CTA response to the recipient's circulating hormones to determine hair production. In terms of racial matching, except for obvious disparities, some races have similar skin tone, and therefore may provide better matches outside of the recipient's own race demographic.

The immunosuppression protocol utilized by all groups thus far has paralleled the regimen for renal transplantation.[1-3,5] This triple drug standard maintenance regimen includes tacrolimus (FK506 or Prograf), mycophenolate mofetil (MMF or CellCept), and prednisone. As with renal transplant regimens, even the doses of prednisone used for facial CTA have been lower, which lessens some of the immunosuppressive side effects of these drugs. With low-dose and steroid wean protocols, facial CTA immunosuppression regimens may be dropped to two drug therapy. Currently, a few facial CTA transplants have been reported to be on only dual drug with the removal of steroids from their regimen.

Donor bone marrow can provide a source of hematopoietic stem cells which could partially populate a recipient's immune system and allow for tolerance of a foreign graft such as a face transplant. This tolerance is measured by chimerism, or the presence of two different cell lines within the recipient. The use of donor-derived bone marrow infusions after transplantation was described by the first transplant group in an attempt to achieve chimerism in their recipient.[5] However, no durable chimerism was ever documented. This could be due to the temporary nature of the bone marrow infusion and lack of engraftment of these donor cells through the peripheral blood system. However, if the donor bone marrow cells could be provided constantly at a low level, the cells may survive longer and possibly allow for engraftment within the recipient. An alternative which could provide a reliable and durable source of donor stem cells would be to include vascularized bone marrow.[18,22] Additionally, these cells are generated at a small level to avoid the transformation into GVHD.[18]

Using objective measures of facial function is essential to determine motor recovery. This should include functional magnetic resonance imaging (fMRI), electromyelography (EMG)/nerve conduction studies (NCS), and pressure-specified sensory device (PSSD).[23-25] These tests can objectively document the remaining motor and sensory functions of facial subunits. The fMRI and PSSD are noninvasive tests that could be performed preoperatively and postoperatively at frequent intervals to monitor for return of motor and sensory function. Though more invasive, the EMG/NCS allows for direct testing of motor junctions and remaining muscle. Postoperatively, an EMG/NCS will allow for monitoring at less frequent intervals.

Patients with different limitations resulting from their facial deformities may perceive varying levels of dysfunction and disability. Although facial deformities have been argued to be a non-life-threatening problem that is not worth subjecting patients to the risks of potentially life-threatening immunosuppression, some patients with severe limitations may argue that their deformities are indeed life limiting. The issue of improving quality of life is also echoed in renal dialysis patients who choose transplants over lifelong dialysis. In addition, some diabetics can be well managed with insulin therapy without any threat to their life. However, these same patients often choose a pancreas transplant to improve their quality of life. Hence, we may better appreciate a patient's perceived quality of life and justify why a facial deformity may result in sufficient disability to seek a facial transplant. Patient perceptions could be documented objectively with the use of several standard patient reported outcome measures such as Standard Form-36 and the Facial Disability Index.[26-29] These surveys could demonstrate and record each patient's own perception of disability and loss of function, as well as their progress following transplantation. These additional tools will further aid the evaluation and selection of recipients for facial transplantation.

The first facial transplant group used a radial forearm flap, placed in the submammary region with anastomosis to the axillary vessels, to monitor rejection. This was used for frequent and routine biopsies in lieu of direct biopsies of the facial CTA graft. This group reported biopsies that correlated with clinical signs of rejection both on the face CTA and sentinel graft.[9] The group from Boston also reported transplanting a radial forearm flap to reconstruct a burn contracture on the patient during his facial CTA transplant. However, local clinical monitoring and conservative biopsies of the facial CTA could provide more direct and accurate information, avoiding inaccurate or delayed interpretation from the sentinel graft.

In composite transplants, the mucosa could become the new focus of local monitoring for rejection.[30,31] In prior transplants, they noted earlier signs of rejection in the mucosa, which was followed several days later with diffuse skin erythema and edema. The third and fourth transplant groups showed the presence of subclinical rejection in the mucosa based on histologic criteria.[3,4,8] During these episodes of documented histologic rejection, the mucosa and skin appeared clinically normal and free of erythema and edema. The mucosa has proven to be more antigenic and an earlier predictor of acute rejection episodes than the skin. Therefore, the mucosa, which can also be biopsied without visible scarring, would be a better indicator of local tissue changes than a sentinel graft. However, concern is emerging over whether acute rejection is being overly treated in facial CTA grafts. Specifically, overtreatment of acute rejection may prevent the proliferation of regulatory T cells, which may be the key for graft tolerance.[18] In addition, frequent and aggressive treatment of rejection episodes places the patient at risk for the sequelae of the immunosuppressive regimens including renal dysfunction, avascular necrosis, and opportunistic infections.

Long-term considerations should be made regarding chronic graft rejection and the risks of lifelong immunosuppression. The risks of immunosuppression are well documented in solid organ and bone marrow transplantation including: metabolic disorders such as renal toxicity, diabetes, and avascular necrosis, opportunistic infections, and malignancies. However, what is not known is whether the composite tissue allograft which includes skin components could increase a recipient's risk of skin cancers or other malignant transformations. As shown by the third facial transplant group, even a patient with elevated malignant risks such as in neurofibromatosis could be a suitable facial transplant recipient. This team has been hypervigilant in surveillance for any increased malignant transformations.[3]

Lastly, although acute rejection has well-defined standards prescribed by the Banff Criteria,[32] chronic rejection is still undefined in facial CTA. It is yet unknown what mechanisms will contribute to chronic graft rejection. One concern is that overtreatment of acute rejection episodes may contribute to the development of chronic rejection.[18] Additionally, the time course for the onset of chronic rejection has been unpredictable in hand transplantation, and may be equally unpredictable in facial CTA grafts. However, if chronic rejection follows the typical presentation seen with solid organ transplantation, then fibrosis and scarring of the skin and vascular structures are likely to occur in facial CTA grafts as well. The effects of chronic rejection on facial CTA graft function may be less severe than as seen with solid organs; however, this remains to be seen.

A goal for future CTA is to clarify indications and matching criteria for facial CTA. Although CTA can be an exciting means for reconstruction of a severe facial deformity, there are still many unknown variables. Currently, there is an 18% (2/11) mortality rate for facial CTA. Long-term considerations are also warranted regarding concerns for chronic rejection and morbidity for patients on lifelong immunosuppression.

There is a need for an International Facial CTA registry to allow updated scientific evaluations, discussions, and innovations in immunosuppressive strategies. Two publications made attempts to objectively examine the clinical facial CTA which have provided insight for critical evaluation.[10,33] The International Registry on Hand and Composite Tissue Transplantation was created as a database, but information must be provided and updated more frequently by the groups performing these surgeries.

31.4 Conclusion

Joseph Murray performed the first kidney transplant between identical twin brothers in 1954 in hopes of discovering tolerance to non-self tissues. Fifty years later, plastic surgeons are building on Murray's initial contributions by expanding the field of transplantation and developing reconstructive transplantation as the next rung of the reconstructive ladder. It is through careful analysis of these groundbreaking cases that we will find the elusive keys to successful long-term tolerance of facial composite tissue allografts.

References

1. Devauchelle B, Badet L, Lengele B, et al. First human face allograft: early report. *Lancet*. 2006;368:203-209.
2. Guo S, Han Y, Zhang X, et al. Human facial allotransplantation: a 2-year follow-up study. *Lancet*. 2008;372:631-638.
3. Lantieri L, Meningaud JP, Grimbert P, et al. Repair of the lower and middle parts of the face by composite tissue allotransplantation in a patient with massive plexiform neurofibroma: a 1-year follow-up study. *Lancet*. 2008;372:639-645.
4. Siemionow M, Papay F, Alam D, et al. Near-total human face transplantation for a severely disfigured patient in the USA. *Lancet*. 2009;374:203-209.
5. Dubernard JM, Lengele B, Morelon E, et al. Outcomes 18 months after the first human partial face transplantation. *N Engl J Med*. 2007;357:2451-2460.
6. Meningaud JP, Paraskevas A, Ingallina F, et al. Face transplant graft procurement: a preclinical and clinical study. *Plast Reconstr Surg*. 2008;122:1383-1389.
7. Pomahac B, Lengele B, Ridgway EB, et al. Vascular considerations in composite midfacial allotransplantation. *Plast Reconstr Surg*. 2010;125:517-522.
8. Siemionow MZ, Papay F, Djohan R, et al. First U.S. near-total human face transplantation: a paradigm shift for massive complex injuries. *Plast Reconstr Surg*. 2010;125:111-122.
9. Kanitakis J, Badet L, Petruzzo P, et al. Clinicopathologic monitoring of the skin and oral mucosa of the first human face allograft: report on the first eight months. *Transplantation*. 2006;82:1610-1615.
10. Hui-Chou HG, Nam AJ, Rodriguez ED. Clinical facial composite tissue allotransplantation: a review of the first four global experiences and future implications. *Plast Reconstr Surg*. 2010;125:538-546.
11. Siemionow M, Ulusal BG, Ozmen S, et al. Composite vascularized skin/bone graft model: a viable source for vascularized bone marrow transplantation. *Microsurgery*. 2004;24: 200-206.
12. Siemionow M, Klimczak A. Advances in the development of experimental composite tissue transplantation models. *Transpl Int*. 2010;23:2-13.
13. Nasir S, Klimczak A, Sonmez E, et al. New composite tissue allograft model of vascularized bone marrow transplant: the iliac osteomyocutaneous flap. *Transpl Int*. 2009;23:90-100.
14. Langstein HN, Robb GL. Lip and perioral reconstruction. *Clin Plast Surg*. 2005;32:431-445. viii.
15. Spinelli HM, Jelks GW. Periocular reconstruction: a systematic approach. *Plast Reconstr Surg*. 1993;91:1017-1024. discussion 1025-1016.
16. Tzakis AG, Reyes J, Zeevi A, et al. Early tolerance in pediatric liver allograft recipients. *J Pediatr Surg*. 1994;29:754-756.
17. Mineo D, Ricordi C. Chimerism and liver transplant tolerance. *J Hepatol*. 2008;49:478-480.
18. Starzl TE. Acquired immunologic tolerance: with particular reference to transplantation. *Immunol Res*. 2007;38:6-41.
19. Takahashi K, Saito K, Takahara S, et al. Excellent long-term outcome of ABO-incompatible living donor kidney transplantation in Japan. *Am J Transplant*. 2004;4:1089-1096.
20. Tanabe K, Takahashi K, Sonda K, et al. ABO-incompatible living kidney donor transplantation: results and immunological aspects. *Transplant Proc*. 1995;27:1020-1023.

21. Tanabe K, Tokumoto T, Ishikawa N, et al. ABO-incompatible living donor kidney transplantation under tacrolimus immunosuppression. *Transplant Proc*. 2000;32:1711-1713.
22. Hewitt CW, Black KS, Henson LE, et al. Lymphocyte chimerism in a full allogeneic composite tissue (rat-limb) allograft model prolonged with cyclosporine. *Transplant Proc*. 1988;20:272-278.
23. Dellon ES, Mourey R, Dellon AL. Human pressure perception values for constant and moving one- and two-point discrimination. *Plast Reconstr Surg*. 1992;90:112-117.
24. Tassler PL, Dellon AL. Correlation of measurements of pressure perception using the pressure-specified sensory device with electrodiagnostic testing. *J Occup Environ Med*. 1995;37:862-866.
25. Coert JH, Meek MF, Gibeault D, et al. Documentation of posttraumatic nerve compression in patients with normal electrodiagnostic studies. *J Trauma*. 2004;56:339-344.
26. McHorney CA, Ware JE Jr, Lu JF, et al. The MOS 36-item Short-Form Health Survey (SF-36): III. Tests of data quality, scaling assumptions, and reliability across diverse patient groups. *Med Care*. 1994;32:40-66.
27. McHorney CA, Ware JE Jr, Raczek AE. The MOS 36-Item Short-Form Health Survey (SF-36): II. Psychometric and clinical tests of validity in measuring physical and mental health constructs. *Med Care*. 1993;31:247-263.
28. Ware JE Jr, Sherbourne CD. The MOS 36-item short-form health survey (SF-36). I. Conceptual framework and item selection. *Med Care*. 1992;30:473-483.
29. Van Swearingen JM, Brach JS. The Facial Disability Index: reliability and validity of a disability assessment instrument for disorders of the facial neuromuscular system. *Phys Ther*. 1996;76:1288-1298. Discussion 1298-1300.
30. Ruiz P, Garcia M, Pappas P, et al. Mucosal vascular alterations in isolated small-bowel allografts: relationship to humoral sensitization. *Am J Transplant*. 2003;3:43-49.
31. Ruiz P, Garcia M, Pappas P, et al. Mucosal vascular alterations in the early posttransplant period of small bowel allograft recipients may reflect humoral-based allograft rejection. *Transplant Proc*. 2002;34:869-871.
32. Cendales LC, Kanitakis J, Schneeberger S, et al. The Banff 2007 working classification of skin-containing composite tissue allograft pathology. *Am J Transplant*. 2008;8:1396-1400.
33. Gordon CR, Siemionow M, Papay F, et al. The world's experience with facial transplantation: what have we learned thus far? *Ann Plast Surg*. 2009;63:572-578.

32 Facial Allotransplantation in China

Robert F. Lohman

Contents

R.F. Lohman
Department of Orthopaedic and Plastic Surgery,
Cleveland Clinic, Cleveland, OH, USA
e-mail: Lohmanr@ccf.org

Abstract Between 2003 and 2006, there were two operations in China for partial face transplantation. Details about the operative technique, background research, immunosuppression, etc are summarized here. Although these operations were free of significant early complications, neither patient survived in the long run. Careful patient selection, and close, life-long follow up will probably be required to minimize late complications after face transplantation.

32.1 Introduction

Two instances of composite tissue allotransplantation have been performed in China and subsequently reported in English language journals. The first operation that can be considered a partial face transplant was described in a case report in *Plastic and Reconstructive Surgery* in 2005.[1] The operation took place in June of 2003, at the Jinling Hospital in Nanjing. The team was lead by Hui Jiang and the operation involved transplantation of skin, scalp, both ears, and lymph nodes. The recipient had a wound resulting from excision of a large melanoma. She died as a result of metastatic disease 6 months after the operation.

The second human face transplant was performed at the Xijing Hospital of the Fourth Military Medical University, in Shaanxi China.[2] Details of this operation are better documented. It was carried out in April of 2006, about 5 months after the partial face transplant that was performed in Amiens, France. The Xijing group was led by Shuzhong Guo and Yan Han. Their group transplanted the entire "nose, upper lip, parotid gland, front wall of the maxillary sinus, part of the infraorbital wall, and zygomatic bone." Their experience was reported in *The Lancet*, in August of 2008.

M.Z. Siemionow (ed.), *The Know-How of Face Transplantation*,
DOI: 10.1007/978-0-85729-253-7_32,

In December of 2009, the patient's death was reported. The precise circumstances surrounding his death are not known.

32.2 Background Research

Chinese research papers about questions related to the technique of facial transplantation began to regularly appear in English after the transplant of 2005. This research has primarily focused on three areas: (1) the choice of donor vessels to perfuse the transplanted tissue, (2) the use of University of Wisconsin (UW) Solution to preserve the donor tissue, and (3) the potential for recovery of neuromuscular function.[3]

Huiyong *et al* reported their experience in 2007 after practicing recovery of facial tissue in cadavers.[4] They were concerned that the dual external carotid model suggested by Siemionow and colleges was time consuming and that preparation of the donor tissue was difficult. Instead they proposed simpler method, using the superficial temporal artery on one side and the facial artery on the other of the donor. They compared these two methods in 12 fresh cadavers and found that recovery time was 232±6 min if the dual carotid artery method was used and 113±6 min if the superficial temporal and contralateral facial arteries were used. They thought donor tissue was easier to prepare using their method because it did not require dissection around the area of the carotid bulb and submental triangle, and "fewer vessels required ligation." They also noted that the length of available vessels and nerves was long enough to make suture repair at the recipient site feasible. Injection studies showed that the facial and superficial temporal arteries could perfuse the entire face.

A canine model for face transplantation was developed to investigate additional questions. The model involved recovery and transplantation of "scalps, ears, eyelids, conjunctivas, parotid glands, and mimetic muscles." Cyclosporine and corticosteroids were used for immunosuppression. Shengwu et al recovered allograft tissue, perfused it with UW, and stored the tissue at 4 °C for 12, 24, 36, or 48 h before transplanting it to another dog. The control group consisted of tissue preserved with normal saline and stored in similar fashion. A greater proportion of stored tissue preserved with UW solution survived than control tissue. However, there was no advantage to preservation with UW solution if the allograft tissue was transplanted immediately after recovery (100% survival in both groups).[5]

Dogs survived for up to 900 days after transplant. Reinnervation of the facial muscles occurred as shown by recovery of the blink reflex.[3]

This and other background work laid the foundation for the operative technique and plan used by Guo and Han for the more complex facial transplantation performed at Xijing Hospital. Additional cadaver studies by these authors, which were not published in English, led them to conclude that each facial artery would perfuse only the ipsilateral half of the human face. In order to successfully transplant portions of the face on both sides of midline, they thought it would be necessary to include arteries from both sides of the donor.[2]

32.3 Patient Selection

The patient who underwent transplantation at Jingling Hospital in 2003 was a 72-year-old woman who was initially diagnosed with melanoma in 2002.[1] It had not been possible to control her disease and despite two attempts at local excision, chemotherapy, and immunotherapy, her disease had continued to progress. The details of her metastatic evaluation were not described, but she was thought to have stage 3 C disease, with tumor deposits in four lymph nodes. In June of 2003, she underwent a more aggressive attempt at excision of the tumor, including a large portion of her scalp, both ears, cervicofacial skin, and adjacent lymph nodes. It was not possible to close the entire wound with skin grafts or free flaps. The remaining open wound involved about 3% of her body surface area. Therefore, on September 16, 2003 she underwent partial face transplantation. The stated goal of the operation was to close the wound, restore her appearance, and reconstruct the external ears (see Figs. 32.1 and 32.2).

"The transplant was designed and performed following the Declaration of Helsinki with the recipient's informed consent." The authors "made a comprehensive pretransplantation evaluation of the patient's general condition" and "informed her of the risks and benefits of transplantation surgery and chronic immunosuppression" including "all the risks

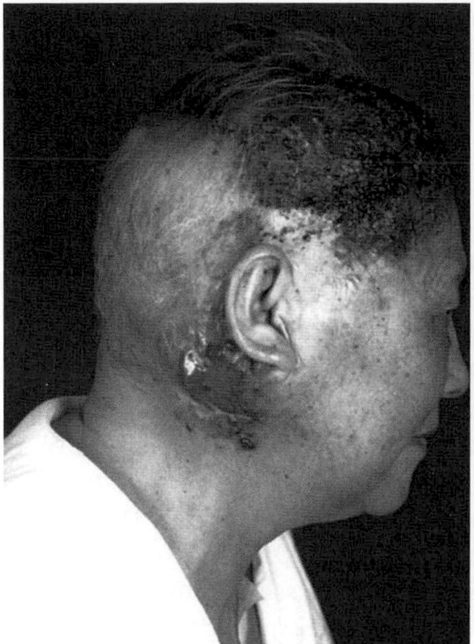
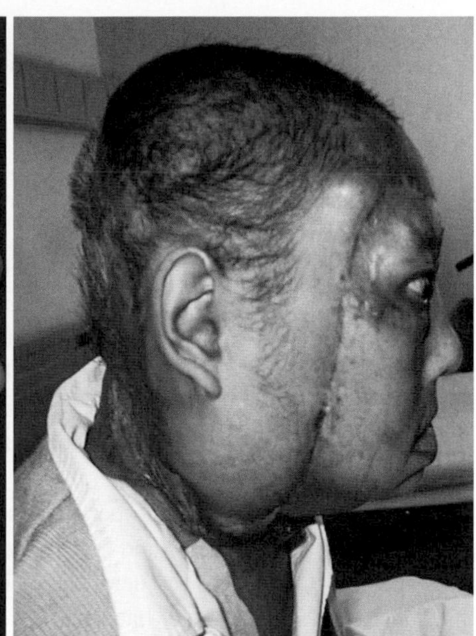

Fig. 32.1 Preoperative and 120 day postoperative appearance of the patient operated on at Jinling Hospital (Copyright Plastic and Reconstructive Surgery, used with permission[1])

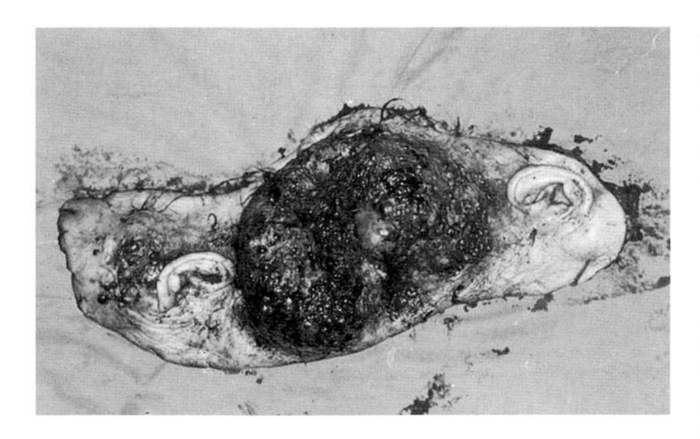

Fig. 32.2 The specimen, including scalp, ears, skin, and lymph nodes (Copyright Plastic and Reconstructive Surgery, used with permission[1])

of infection, rejection, and malignancy." It is not clear from the description published in English, when the evaluation and informed consent took place with respect to excision of the tumor in June 2003. Furthermore, there is no indication whether or not an Institutional Review Board, or the local equivalent, approved the protocol. The donor was a brain dead young male; it was not disclosed whether or not the donor's family was asked to consent to recovery of the necessary tissue.[1]

The donor and the recipient were matched for ABO blood type. The recipient was evaluated with a panel reactive assay (PRA) using a complement-dependant microlymphocytotoxicity test. The PRA was low (0.12% for HLA Class I and 0.61% for HLA Class II) indicating the recipient was not sensitized to alloantibodies. The HLA match between donor and recipient was not described; a mixed lymphocyte reaction test was "negative."[1]

The patient who underwent transplantation at Xijing Hospital was an otherwise healthy 30-year-old male, who lived in "a remote village of Yunnan province, China."[2] His face was injured when a bear attacked him in October of 2004. He was treated locally by debridement followed by wound closure with "a left forearm pedicle flap." However, his wounds failed to heal after these operations, and he subsequently came to Xi'an for additional treatment.

At the time of presentation, the patient had an open wound extending into the nasal cavity, right maxillary sinus, and oral cavity. The soft tissue defect included the entire nose and upper lip, as well as the skin of the right cheek, and portions of the right parotid gland. There was severe scar contracture involving the remaining skin of the right cheek, which also distorted both the upper and lower eyelids, and portions of the lower lip. The underlying boney skeleton, including the anterior wall of the right maxillary sinus, the orbital floor, and the zygoma were also absent. Because of the extensive injury involving specialized and anatomically unique tissue, and because of the deficient skeletal support, Gou and his team determined that conventional reconstructive operations would not be useful for treating the patient. They considered, and rejected, various options including tissue expansion, transfer of prefabricated flaps, grafts, *etc*. In light of the ongoing international discussion about facial transplantation, and the success of the team in France, they concluded that allotransplantation was "the first therapeutic option to reconstruct the face of the recipient." They did not undertake more conventional operations for facial reconstruction before patient a face transplant.[2]

As previously noted, the patient lived in a remote village, "without access to proper medical care." He was described as a farmer, and was socially isolated after the injury. Little other detail is available about his social and educational background. During the 18 months after his injury, the facial wounds failed to heal and the situation "seriously affected his appearance and function." The patient and his family were advised about the risks, benefits, and potential complications associated with the operation. The patient "strongly wanted surgery." Both the patient and his family provided written consent. All of this communication apparently took place in the time period between March 11, 2006 (when the patient initially presented to Xijing Hospital for treatment) and April 12, 2006 (when the transplant operation occurred).[2]

In addition to the recipient and his family's consent, the donor's family also provided consent. The donor was a 25-year-old male who died in a traffic accident. The local hospital ethics committee reviewed the proposed procedure and granted approval, as did the Shaanxi Provincial Health Department.

Before surgery, a PRA assay was performed. On two separate occasions, the PRA was found to be very high (99% and 98%). This indicated that the patient was highly sensitized to alloantibodies. The patient was thought to be the result of the high content of fungi in his diet, previous blood transfusions, the presence of a chronically infected, nonhealing wound, and/or other unknown factors. Therefore, he underwent immunoadsorption therapy and the PRA was reduced to <5% before transplantation. Highly sensitized patients are thought to be at risk for acute rejection. In addition, there is data suggesting that reduction of PRA values prior to transplant leads to improved graft function after solid organ transplant.

The donor and the recipient were both blood type A. Three of six HLA sites were matched: the recipient was A-11,12; B-38,52; DR-4,14 while the donor was A-11,9; B-38-7; DR-10,15. The mixed lymphocyte reaction of the recipient to donor antigens was <5%.[2]

32.4 Operative Technique

The operation at Jinling started with recovery of tissue from the donor. Initially the carotids were exposed and flushed with cold UW solution. Next the scalp, skin, ears, and vessels were removed and preserved in cold UW solution. The tissue was then irradiated (8 Gy) over 20 min. The warm ischemia time was 2 min and the cold ischemia time as 6 h. The recipient was prepared by exposing the vessels and "ablation of the residual melanoma focus." The donor external carotid arteries were anastomosed to the recipient's left external carotid artery and right superior thyroid artery. The donor external jugular veins were anastomosed to the recipient internal jugular veins. The external auditory canals were sutured together and then the skin was closed. The closure was supplemented with a negative pressure dressing.[1]

The operation at Xijing Hospital was more complex. After selection of a suitable donor, three-dimensional CT was used to evaluate the facial skeleton. Next the facial tissue was harvested by initially infusing 1,000 ml of UW solution at 4 °C into each of the donor's carotid arteries. Both the mandibular arteries and the facial veins were dissected to provide blood flow to the recovered tissue. The ipsilateral facial nerve was divided at its main trunk. Osteotomies were carried out along the floor of the maxillary sinus, the zygoma and orbit, including the lateral and inferior orbital rims, and nasal bones. Portions of the maxillia, zygoma, orbit, nasal bones, nasal septum, and cartilages were included. The entire right parotid gland, portions of the massiter and other facial muscles, all of the soft tissue of the nose, the upper lip, buccal mucosa, and the overlying skin were also included. The flap was recovered at the time of cardiac death, and no particular measures were taken to insure hemostasis. After recovery, the tissue was cooled on ice and irradiated with X-rays (4 Gy).[2]

The recipient was prepared by initially excising all of the infected and scarred soft tissue. The mucosa was removed from the right maxillary sinus. The infraorbital nerve was noted to be absent at the foramen, and the facial nerve was difficult to expose because of scar and trauma in the area of the stylomastoid foramen. The right facial vein and the right external maxillary artery were exposed and prepared for the vascular anastomoses. The vessels and nerves on the left side of the neck were not exposed.

The right facial veins, and then the right maxillary arteries were anastomosed to one another in an end-to-end fashion. Dextran-40 (500 ml) and papaverine (30 mg) were administered at the time the vascular anastomoses were carried out. Once blood flow to the transplanted tissue was established, it appeared to be well perfused. In fact, the authors report 5,000 ml of blood was lost from the edges of the tissue at this point. Approximately 2 h were then required for hemostatsis "and 6,250 ml of plasma and erythrocytes were used after the anastomsis." With the left facial vein occluded, there was no evidence of congestion, and arterial perfusion remained adequate. Therefore, the left-sided vessels, which had been recovered with the flap, were not anastomosed. However, they were preserved for later use in case vascular embarrassment occurred. During the 2 h needed for hemostasis, no evidence of hyperacute rejection, such as thrombosis or erythema, was noted.[2]

After hemostasis was established, the surgeons' attention was directed toward bony fixation. They noted some mismatch in the relative sizes of the orbit and nose. The bones were trimmed as necessary to insure apposition and then fixed with titanium microplates and screws. The donor masseter muscle was packed into the maxillary sinus; the nasal septum and other tissues around the piriform aperture were sutured to stabilize the nose and the nasal cavities were packed with iodoform gauze. The facial neurorrhaphy proved difficult because of the proximal location of the recipient's injury. The neurorrhaphy "was not satisfactory." Lastly, the skin was closed and entire operation took 18 h.

32.5 Immunosuppression

For the case at Jinling Hospital, a four-drug immunosuppressive protocol was used. Tacrolimus (3 mg) was administered orally 2 h before the operation and then continued (3 mg) orally twice a day after the operation. The Tacrolimus blood concentration levels were maintained between 20 and 25 ng/ml for the first 2 weeks after surgery, then reduced to 15–20 ng/ml for 2 weeks, and then reduced to 10 ng/ml after a month.[1]

Methylprednisolone (1 g) was administered intravenously during the operation, and continued at 20 mg a day for the first week after surgery. Prednisone (20 mg) was then administered orally for 3 months, and then reduced to a 15 mg daily maintenance dose.[1]

Mycophenolate mofetil (750 mg) was administered 2 h before surgery. Mycophenolate mofetil (1.5 g) was continued after surgery. Humanized monoclonal IL-2 antibody (50 mg) was administered during the operation, and then continued at the same dose twice a week.[1]

The group at Xijing Hospital used a similar protocol: Immunosuppressive therapy was also initiated prior to the operation. Three doses each of prednisone (25 mg) and mycophenolate mofetil (500 mg) were administered orally at 12-h intervals prior to the operation. As the operation started, tacrolimus (5 mg) was infused and during the operation, tacrolimus levels were maintained at 25 ng/ml by additional periodic infusions. Immediately before blood flow to the graft was reestablished, methylprednisolone (1 g) and humanized IL-2 receptor monoclonal antibody (50 mg) were administered. After the transplant, the four-drug immunosuppression protocol was continued.[2]

Tacrolimus blood levels were maintained at 20–25 ng/ml for the first 2 weeks after transplantation by administering an oral dose of 5–9 mg, twice a day. The dose was then reduced to 6 mg, twice a day, and again at 3 months to maintain blood levels at 20 ng/ml. The dose was then gradually reduced to 1 mg twice a day at 15 months, and then adjusted as necessary between 1 mg and 3 mg twice a day to maintain blood levels between 10 and 15 ng/ml.[2]

Mycophenolate mofetil was initially administered orally at a dose of 1.5 g twice a day for 6 months. The dose was then reduced to 1.0 g twice a day until 17 months after the transplantation. Between 17 months and 21 months after the transplantation, the dose was 0.5 g and 0.25 g, each given once a day. After 21 months, the dose was further reduced to 0.25 g taken twice a day.[2]

Methylprenisolone was administered for the first 5 days after the operation. On the first and second day after surgery, 0.5 g was administered intravenously; on the third, fourth, and fifth day after surgery, 0.25 g was administered. After day six, prednisone (80 mg) was administered orally for 13 months, and then tapered to 25 mg a day for 3 months, 20 mg a day for 3 months, and finally 10 mg a day for 3 months. Oral predisone was discontinued 22 months after the operation.[2]

Two doses of humanized IL-2 receptor monoclonal antibody (50 mg) were administered after surgery. One dose was administered at 2 weeks and the next was at 4 weeks.

32.6 Antimicrobial Therapy

The Jinling group administered broad-spectrum antibiotics to prevent infections, but specific details were not provided.[1]

At Xijing Hospital, Ceftizoxime (2.0 g) was administered intravenously at the start of the operation and then given every 8 h for 2 weeks as a prophylactic measure. Acyclovir, metronidazole, and allicin were also used for prophylaxis in the perioperative period. Bacterial and fungal cultures from the pharynx, nose, and sputum were also monitored after surgery. Enterobacter, *Enterococcus faecalis*, and *Staphylococcus epidermidis* were eventually isolated from the oropharynx and sputum, and vancomycin (1.0 g) was given every 12 h for 4 days.[2]

32.7 Outcome

In the first case, the patient was observed for clinical evidence of acute rejection and additionally, skin biopsies were performed at 7, 14, 30, and 120 days after the operation. No clinical or pathologic evidence of acute rejection was observed. The authors had planned to use topical Tacrolimus and methylprednisolone to manage episodes of acute rejection as needed. In addition, there was no evidence of graft versus host disease, and no abnormalities in blood glucose concentration or other laboratory abnormalities were noted. The tissue remained well perfused and hair growth was noted.

No other details about the patient's postoperative course are available. She was reported to have died as a consequence of metastatic disease 6 months after the operation. Further details are not available; it is has not been disclosed when the metastatic disease was diagnosed, whether or not the operation facilitated local control of her disease, or if the immunosuppressive therapy was thought to contribute to its progression.[1]

More detail is available from the group at Xijing Hospital. They reported that soon after surgery their patient began to tolerate an oral diet with no difficulties. Facial swelling was noticeably decreased after 1 week, and had largely resolved after 1 month. Wound healing progressed normally and the patient was discharged from the isolation unit after a month. Three days after surgery, the patient was noted to have an elevated blood glucose concentration. Glucose tolerance, in response to a 75 g glucose challenge, was impaired and insulin therapy was initiated. After 2 weeks, blood glucose concentrations had returned to normal, and the insulin therapy was discontinued. Blood glucose concentrations were again elevated 3 months after surgery. This was initially treated with insulin (70 units per day) but as the immunosuppressive therapy was tapered, the insulin requirements were also reduced. Insulin therapy was permanently discontinued 21 months after surgery and the blood glucose concentration was controlled with oral antiglycemics. An insulin function test suggested that islet cell function was impaired. Bone density and renal function remained normal at 1 year after surgery.[2]

During the first 3 months after surgery, there were no signs of acute rejection. However, at 3, 5, and 17 months after surgery, the transplanted tissue showed signs of acute rejection. The clinical evidence for acute rejection included swelling, erythema, and congestion of the skin. The first episode of acute rejection was treated by increasing the tacrolimus dose, and the second episode of acute rejection was treated by methylprednisolone for 5 days followed by oral prednosone. The patient appeared healthy, and returned to his village 14 months after surgery. However, he elected to discontinue immunosuppressive therapy 16 months after surgery, and began to ingest various unknown herbs for 3 weeks. At 17 months after surgery, another episode of acute rejection occurred, and the patient returned to the hospital. Conventional immunosuppressive therapy was reinstituted, including an increased dose of tacrolimus, and clinical signs of acute rejection resolved except for a minor degree of persistent swelling and congestion.[2]

A biopsy of the transplanted flap was performed at 1 month after surgery. The cuticular layer of the skin as well as the papillary dermis was thin. Hair follicles, sweat glands, arterioles, and venules were present deep to the epidermis. A scant mononuclear cell infiltrate was present affecting the glands and vessels. These changes were graded 0–1 according to the composite tissue allotransplantation rejection score system. A biopsy at 5 months showed a moderately dense mononuclear cell infiltrate, and was graded 1–2.

Guo and associates reported their patient's appearance was greatly improved by the operation, and this is obviously true as evidenced by the photographs they provided. His appearance was further refined by two additional operations, performed under local anesthesia in November of 2006 and April of 2007. Redundant skin was excised, autologous grafts were used to supplement the orbital floor, local flaps were used to correct ptosis of the lip, and other scars were revised. The authors reported that "the patient was able to eat, drink, and talk normally." However, recovery of facial nerve function was not ideal. The main trunk of the facial nerve had been avulsed at the time of the injury. The frontal branch recovered poorly and there was persistent ptosis of the upper eyelid. In addition, function of the buccal branches was also limited, and the patient was not able to "smile completely and symmetrically." In contrast to motor function, sensory function of the transplanted skin and mucosa recovered rapidly. Pressure sensation, determined by Semmes–Weinstein testing recovered at 3 months, and temperature discrimination recovered at 8 months.[2]

In July of 2008, the patient died in his home village. The cause of death remains unknown. It appears that the patient once again discontinued his immunosuppressive therapy, and began to ingest "local herbs." Some of the herbs may have been hepatotoxic, or caused an adverse reaction when combined with his other medications. Guo traveled to the patient's home to request an autopsy, but the family refused because he had already been buried.[6]

32.8 Comment

Jiang and associates carried out the first reported operation that can be thought of as a partial human face transplant. This operation was carried out to reconstruct a defect that resulted after excision of a locally advanced melanoma. The patient lived for 6 months after the operation and died as a result of metastatic disease. The operation was described in a brief case report. The authors note that allotransplantation can be useful in the short run for management of complicated defects that are not amenable to standard reconstructive options. This case represents a creative and unique solution to a very difficult problem.

However, this case highlights a number of potential problems related to novel techniques of composite tissue allotransplantation. There was no discussion about the IRB process; it is not known if an IRB reviewed and approved the protocol. By western standards, this operation is beyond the scope of innovative surgery, and was obviously experimental. Again by western standards, IRB approval would have been essential, and some comment about this would have been appropriate in the published text.

The patient selection in this case is also questionable. A 72-year-old with advanced cancer would not be a candidate for composite tissue allotransplantation in the west. No doubt her surgeons were faced with a difficult problem, and they were motivated by compassion and a desire to help her. However, it is not clear that she benefited from the operation. The surgeons knew she had locally advanced disease that had spread to at least four cervical and adjacent lymph nodes, and that she had a limited life expectancy. They noted that melanoma is typically resistant to available chemotherapy and immunotherapy regimes but that advanced melanoma could be treated by aggressive local excision. This patient was at high risk for distant metastatic disease; efforts to identify distant disease were not described. The authors also note that immunosuppressive therapy may have anticancer effects in some circumstances. However, the immunosuppressive protocol used here would reasonably be expected to accelerate the growth of occult metastatic disease.

The consent process in this case is also potentially problematic. One view of this operation is that it was a heroic effort to salvage a desperate situation. However the consent process may not have been optimal. It is not clear that she was offered or understood the potential for alternative methods of reconstruction. The authors felt that no method of reconstruction, other than composite tissue allotransplantation, would have been suitable. However, it seems likely that free flaps could have been used to cover this wound. If there were contraindications to such flaps, they were not mentioned. The result would have been less elegant, and the patient's appearance would have been inferior to what was achieved with transplantation, but immunosuppression would not have been required. Furthermore, it is not clear if the patient was offered composite tissue allotransplantation before or after the large tumor was excised from her head. If she were offered allotransplantation after tumor excision, and without a complete understanding of the alternatives, the consent would not have been proper by western standards. Commenting about the consent process for composite tissue allotransplantation, Yu implies that it is the surgeon rather than the patient who should make the decision about proceeding with a transplant "according to defined indications" because "patients usually lack professional knowledge to understand the side effects of immunosuppression." Although potentially true, these comments represent a more paternalistic attitude that is usually taken in western medicine.

The description of this case was published in a mainstream western medical journal, and the editors of *Plastic and Reconstructive Surgery* had a duty to understand the circumstances of the operation. Even if local Chinese standards did not require IRB approval or a more robust consent process, it would have been appropriate for the editors to speak to this at the time of publication.

Guo and coauthors also concluded that facial transplantation can be "successful in the short run" for treatment of severe facial deformities. They achieved a very excellent early outcome. The authors also note facial transplantation can be associated with complications such as acute rejection and new-onset diabetes. There four-drug regime was effective for immunosuppressive management and the diabetes was ultimately managed with oral agents. The three episodes of acute rejection experienced by their patent were successfully treated by adjustments to the regime.

This operation also may have been burdened by problems with patient selection and preparation. The patient presented to Xijing Hospital on March 11, 2006 and the operation was performed on April 13, 2006. During this relatively brief period, the necessary medical and psychological evaluation was carried out. It is possible that he lacked a sophisticated understanding of the proposed operation, its alternatives, and the complexity of the required aftercare. Given that he elected to discontinue his immunosuppressive therapy in favor of "local herbs," it seems possible that he did not fully understand the reasons for immunosuppression, and may not have understood the risks associated with forgoing this therapy. We can only speculate about his reasons for this decision. It is likely that failure to comply with the prescribed immunosuppressive regime contributed to his death.

The patient was a farmer from a village described as remote and rugged. He did not have access to "good medical care" before the operation. He was anxious to return to his village after surgery. However, travel back and forth between his village and Xijing Hospital was difficult. Furthermore, there apparently was a shortage of qualified local medical personnel to supervise his care. These factors combined to make follow-up difficult, and may also have contributed to the patient's death. The problems with compliance and follow-up that occurred after surgery illustrate the critical importance of careful patient selection for facial transplantation.

After their initial examination of the patient, Guo and his associates believed that reconstruction of the deficient tissues with conventional techniques "would not be possible. Allotransplantation was therefore chosen as the first therapeutic option to reconstruct the face of the recipient." This line of reasoning may be ahead of the curve of what most reconstructive surgeons currently believe. Facial transplantation remains an experimental and rare method of treatment. With experience, good results, and societal acceptance, facial transplant may one day become the first therapeutic option for certain patients with severe facial disfigurement. It probably is not the first therapeutic option today.

32.9 Summary

Two patients have undergone partial facial transplantation in China: a 72-year-old woman with advanced melanoma and a 30-year-old man who was injured by a bear. The short-term results for both of these patients were excellent; however, both of them died less than two and half years after transplantation. Both patients were managed using a four-drug immunosuppression protocol. One patient had episodes of acute rejection and diabetes that were successfully managed with drugs. These cases suggest that, to a large extent, surgeons have overcome many of the technical problems associated with facial transplantation; however, careful patient selection and follow-up are critically important for the long-term success of these operations.

References

1. Jiang HQ, Wang Y, Hu XB, et al. Composite tissue allograft transplantation of cephalocervical skin flap and two ears. *Plast Reconstr Surg*. 2005;115:31e-35e.
2. Guo S, Han Y, Zhang X, et al. Human facial allotransplantation: a 2-year follow-up study. *Lancet*. 2008;372:631-638.
3. Yu D, Li Q, Zheng S, et al. Some results of our research on composite facial allograft transplantation in dogs. *Transplant Proc*. 2010;42:1953-1955.
4. Huiyong W, Qingfeng L, Shengwu Z, et al. Cadaveric comparison of two facial flap-harvesting techniques for alloplastic facial transplantation. *J Plast Reconstr Aesthet Surg*. 2007;60:1175-1181.
5. Shengwu Z, Quinfeng L, Hao J, et al. Developing a canine model for composite facial/skin allograft transplantation. *Ann Plast Surg*. 2007;59:185-194.
6. AFP Newswire: China Face Transplant Patient Is Dead: Doctor. December 19, 2008.

The Cleveland Clinic Experience With the First US Face Transplantation

33

Maria Z. Siemionow, Risal Djohan, Steven Bernard, and Frank Papay

Contents

Abstract The ability to perform a complicated craniofacial reconstruction without the need for multiple surgical procedures is appealing. The putative application of composite face allograft transplantation in patients with complex, composite facial defects may be a viable alternative to conventional reconstructive options. In December 2008, the first near-total face transplantation in the USA was performed at the Cleveland Clinic. At that time, this was the largest and most complex face allograft reported in the world and included over 535 cm^2 of facial skin; full nose with nasal lining and bony skeleton; lower eyelids and upper lip; underlying muscles and bones, including orbital floor, zygoma, maxilla, alveolus with teeth, hard palate, and parotid glands; and pertinent nerves, arteries, and veins. Immunosuppressive treatment consisted of thymoglobulin, tacrolimus, mycophenolate mofetil, and prednisone. There have been no major complications and two rejection episodes on posttransplant days 47 and 452, which were effectively reversed by with corticosteroids and immunosuppression adjustment. The functional outcome has been excellent, with marked improvement in breathing, smell, taste, speech, drinking, and eating solid foods. We have demonstrated the feasibility using composite face allotransplantation in the treatment of the severely disfigured patient.

33.1 Introduction

The human face is unique in its structure and function, and in the specificity of its tissues and functional subunits, such as the nose, lips, and eyelids – none of which can be borrowed from another part of the body. To address the technical challenges of giving disfigured

M.Z. Siemionow (✉)
Department of Plastic Surgery,
Cleveland Clinic, Cleveland, OH, USA
e-mail: siemiom@ccf.org

M.Z. Siemionow (ed.), *The Know-How of Face Transplantation*,
DOI: 10.1007/978-0-85729-253-7_33,

patients normal-looking faces, we have investigated different composite tissue allograft models and tolerance-inducing protocols applicable to limb and face transplants during the past 20 years.[1-11] Based on our anatomical cadaver studies and numerous mock facial transplantations, we concluded that soft-tissue coverage of a full facial/scalp deficit would require over 1,200 cm^2 of autologous tissue obtained from a single flap.[12-15]

These observations served as the background for us seeking institutional review board approval. After nearly a year's effort, we received the nation's first institutional review board approval on November 15, 2004, and proceeded to implement the required multidisciplinary approach to facial transplantation.

After institutional review board approval, we began searching for approval from an organ procurement organization, at both the local and the national levels. In the meantime, three reports on partial face transplantation performed in France and China were published, with the first published in 2005.[16-21]

Our group at the Cleveland Clinic, Cleveland, Ohio, performed the first near-total face and maxilla transplant on December 9, 2008. This surgical procedure is the fourth in a series of facial transplant cases and the first in the USA.[22-24]

33.2 Methods

33.2.1 Institutional Review Board Protocol

In December of 2003, we had submitted to the Cleveland Clinic's Institutional Review Board the *Protocol for Composite Facial Allograft Transplant*. In this 40-page document, we outlined our study's aims and objectives, indicating the limited reconstructive options for patients with severe facial injuries, including aesthetic and functional units of the face such as nose, lips, eyelids, ears, and palate.

The protocol presented a multidisciplinary team approach of specialists including plastic surgeons, otolaryngologists, transplant surgeons, anesthesiologists, transplant psychiatrists, bioethicists, dentists, transplant infectious disease specialists, and immunologists. The study design was outlined in detail, including the selection process of the recipient, recipient inclusion and exclusion criteria, and donor inclusion and exclusion criteria. This was followed by a description of our planned immunosuppressive therapy and coexisting infection prophylaxis protocol. It also included descriptions of our surgical team members and respective roles, facial graft procurement and preservation, a coverage protocol for the residual donor defect, and preparation of the recipient. Finally, the issue of acute and chronic graft rejection was discussed, and rescue procedures were described in the event of graft failure and/or rejection, as were secondary reconstructive procedures that might be needed or purposely delayed for functional/aesthetic improvements.[23]

33.2.2 Patient Selection Process

According to our institutional review board-approved protocol, we were considering only candidates who had exhausted all conventional means of reconstruction and were severely disabled in performing basic facial functions. After review of several potential candidates who had approached us to be considered for face transplantation, we found the appropriate candidate who fit the protocol requirements and was motivated to enter the complex process of medical testing, screening, and a multidisciplinary team evaluation.

The patient was a 45-year-old woman who, in September 2004, sustained severe facial trauma to her midface from a close-range shotgun blast. She was taken to a nearby trauma center for stabilization and intensive care, and was eventually discharged to a nearby rehabilitation facility. At 2 months after injury, she was referred to our plastic surgery department (Cleveland Clinic, Cleveland, Ohio) for autologous craniofacial reconstruction.

In December of 2004 (3 months after the original injury), the patient underwent her first major operation, which included midfacial reconstruction by way of split-calvarial autografting, a free fibular microsurgical osteocutaneous flap, a temporoparietal myofasciocutaneous flap for palate reconstruction via a coronal approach, and a paramedian forehead flap for nasal soft-tissue coverage. Numerous operations followed thereafter, including an anterolateral thigh free flap for the purpose of right facial soft-tissue coverage and a radial forearm free flap reconstruction for reversal of midface collapse.

Nearly 4 years after her trauma, in August of 2008, a formal discussion was held by the face transplant team with the patient in regards to face transplantation due to her unsatisfactory results in terms of social acceptance and function (Fig. 33.1). By September of 2008, our patient had successfully passed all bioethical and psychiatric evaluations, had signed the informed consent, and was officially listed with a local organ procurement organization (LifeBanc, Cleveland, Ohio). We then improved our previously developed protocol with mock cadaver transplants tailored specifically to this patient's deformity and functional deficits, based on the patient's computed tomographic scan and stereolithic model (Fig. 33.2).[23]

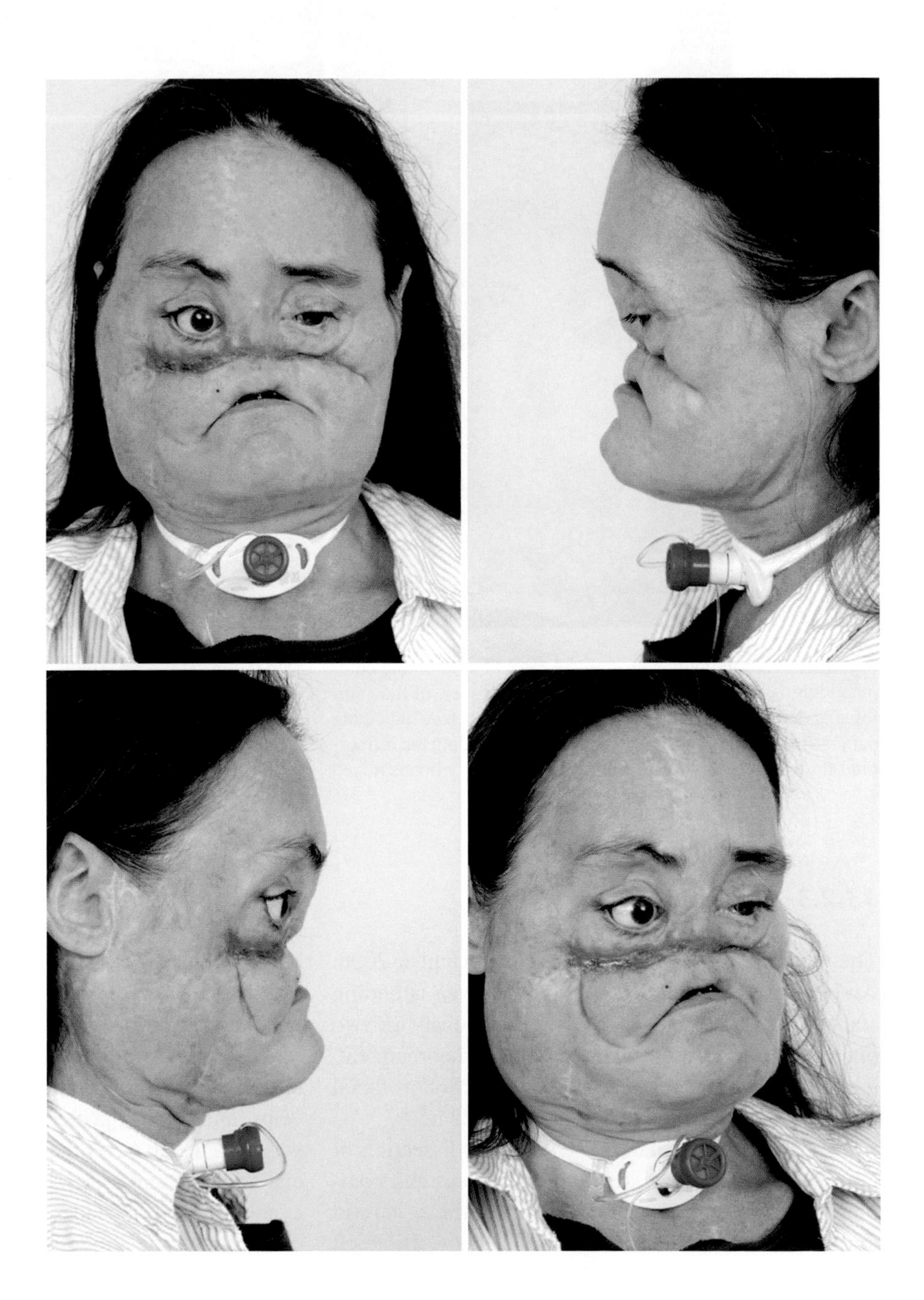

Fig. 33.1 Patient views before transplantation: *Frontal view*, *left profile*, and *right profile* indicate tridimensional craniofacial defect with missing nose, upper lip, and lower eyelids contracted by massive scar tissue

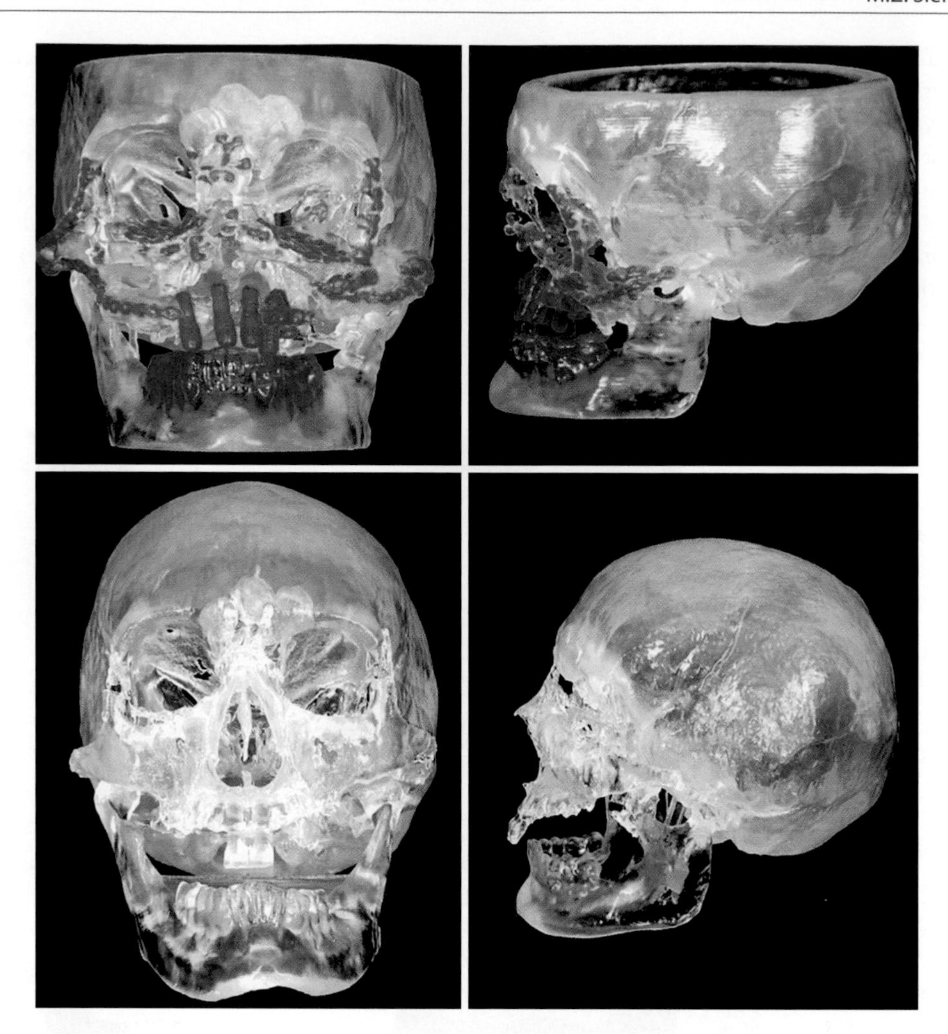

Fig. 33.2 The stereolithic anatomical model based on the computed tomographic scan of the patient. Frontal view of the craniofacial defect after gunshot injury to the patient's face indicating damage of the frontal and midface skeleton including the infraorbital floor and the nasal, zygomatic, and maxillary bones mixed with the metal pieces. The *left side* of the defect, with a significant tridimensional defect showing missing nose and nasal bones and upper jaw bony support. Three-dimensional reconstruction of the patient's preoperative bony skeleton: *frontal view* and *left*

33.2.3 Operative Procedure

The face transplant was performed in December 2008. An eight-surgeon team, consisting of seven attending staff and one fellow, worked simultaneously in two operating rooms. One team prepared the recipient, while the second team recovered the donor facial allograft. The entire procedure required 22 h.

In the recipient: bilateral neck vessel dissection of posterior facial veins, external jugular veins and common facial, external carotid, and common carotid arteries was performed to ensure patency. Bilateral superficial parotidectomy was performed in order to identify the patient's facial nerve anatomy. Soft tissue, bone, and hardware (plates and screws) were removed from previous reconstructions including calvarial bone grafts and an autogenous bone fibula flap (Fig. 33.3).

The composite tissue facial allograft was pedicled on bilateral common facial arteries, external jugular veins, and the left posterior facial vein. The allograft was designed to cover the recipient's anterior craniofacial skeleton, and it included about 80% of the surface area of the anterior face. It was based on a Le Fort III composite tissue allograft containing total nose, lower

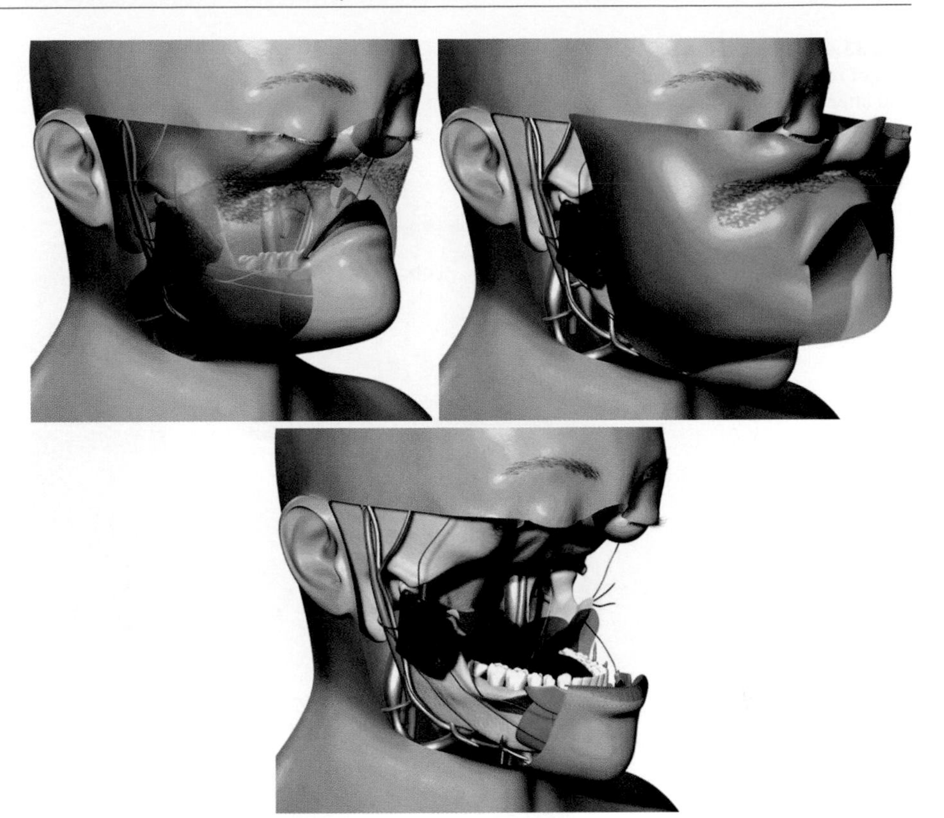

Fig. 33.3 Figures representing recipient defect before allograft *inset*, illustrating the need for a three-dimensional craniofacial reconstruction

eyelids, upper lip, total infraorbital floor, bilateral zygomas, and anterior maxilla with incisors, and included total alveolus, anterior hard palate, and bilateral parotid glands. The graft was then transferred to the recipient for the final inset (Fig. 33.4).

Once bone components of the facial allograft were secured and stable, we proceeded with bilateral microvascular anastomoses of both arteries and veins. On the left side, the facial common artery was anastomosed to the recipient facial common artery; the external jugular vein and the posterior facial vein were connected to the recipient external jugular vein and the posterior facial vein respectively. Total ischemia time was 2 h and 40 min. After the left side anastomoses, the allograft skin and mucosa pinked up, confirming the graft's viability based on a single anastomosis. On the right side, anastomoses were performed between the right common facial artery of the donor and right common facial artery of the recipient. The donor's right facial vein was attached to the right recipient facial vein.[22]

The bilateral facial nerves were repaired using standard epineural repair. On the right side, the donor's vagus nerve, used as an interpositional graft, was connected to the upper division of the trunk on the right side of the recipient's facial nerve. On the left side, the donor's hypoglossal nerve, used as an interpositional graft, was connected to the upper division of the trunk on the recipient's facial nerve (her lower division was previously injured). Both interposition grafts were connected to the main trunk of the donor's nerve.[23] Bilateral orbital floors were reconstructed using Medpor polyethylene facial implants (Porex Surgical, Inc., Newnan, GA). The bilateral eyelids were also reconstructed using the recipient's posterior lamella and the donor's anterior lamella. Skin closure completed the inset of composite facial allograft.

33.2.4 Medication and Physiotherapy

Immunosuppressive therapy and infection prophylaxis were administered according to our institutional review board protocol and based on induction with rabbit antithymocyte globulin (1–2 mg/kg IV for 9 days) and

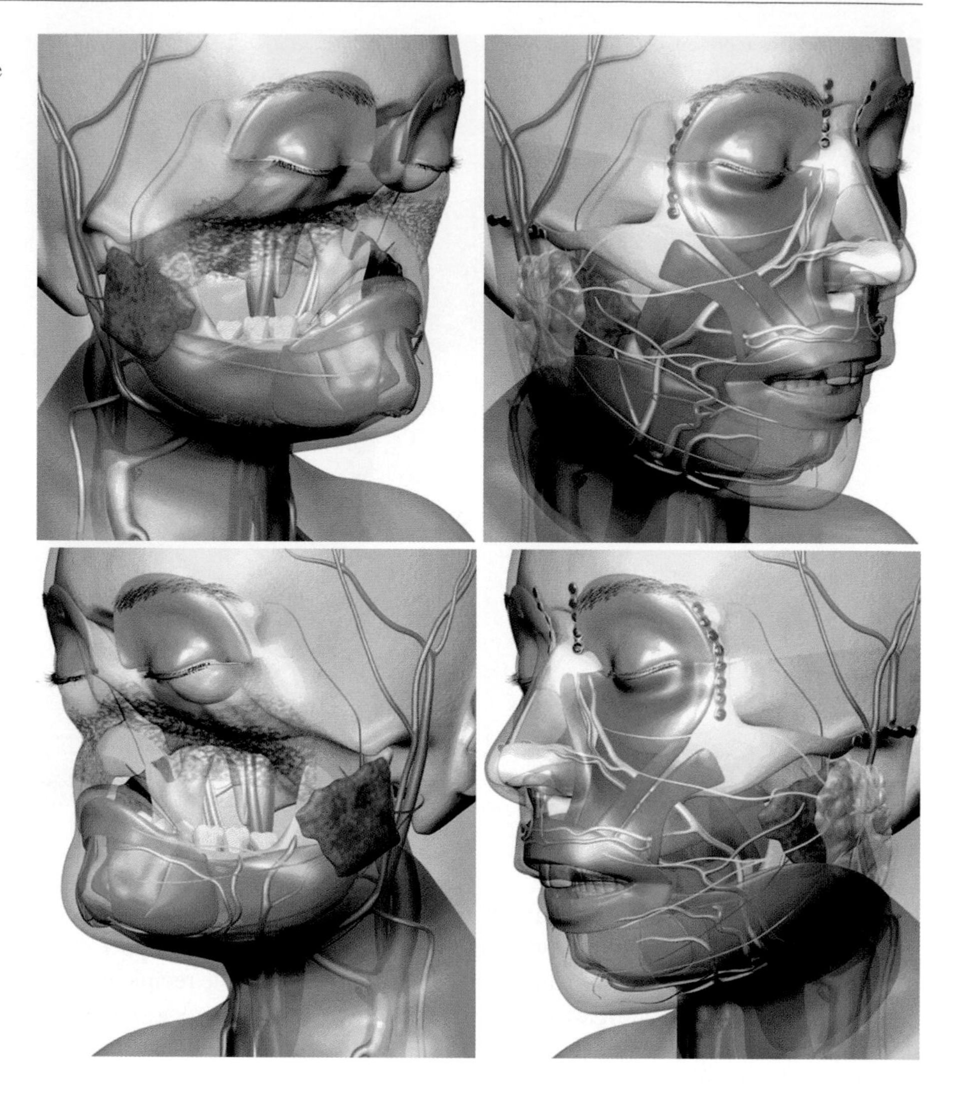

Fig. 33.4 Graphic illustration of recipient's face before and after transplantation confirming the complex multi-tissue construct required for facial reconstruction

standard triple therapy maintenance using corticosteroids (1,000 mg IV on the day of transplant and rapidly tapered thereafter), tacrolimus (dosed according to blood level (12–15 ng/ml)), and mycophenolate mofetil (discontinued at 6 months after transplantation). Prophylaxis for cytomegalovirus, consisting of ganciclovir followed by valganciclovir, was critical because there was a cytomegalovirus-positive donor/cytomegalovirus-negative recipient mismatch. As for *Pneumocystis jerovici* prophylaxis, our protocol included 400 mg of trimethoprim-sulfamethoxazole.

Rehabilitation, speech therapy, and sensory and facial acceptance reeducation were performed daily until the patient's discharge from the hospital. Quantitative neurosensory testing of two-point sensory discrimination of the facial skin was tested with the Pressure-Specified Sensory Device (Sensory Management Services, LLC, Baltimore, Md.) and by clinical evaluation using two-point sensory discrimination (DiskRiminator; Kom Kare Company, Middletown, Ohio). Motor recovery was evaluated by repose, pucker, smile, and performing vowel movements (a, e, i, o, and u).

The patient was evaluated with weekly mucosa and skin biopsies for the first 10 weeks, then bimonthly for 2 months, and monthly thereafter. In addition, urgent biopsies were planned for in the event of any clinical rejection.

33.3 Results

The initial postoperative course was uneventful. No microsurgical complications occurred, and there were no signs of ischemia and/or venous congestion in the facial allograft. Permanent sutures were removed at approximately 4 weeks. Only two units of blood were transfused during the entire postoperative period. She was transferred from the intensive care unit to the surgical floor at approximately 3 weeks postoperatively and was discharged from the hospital 2 months after transplantation.

Over the past 18 months following transplantation, there was one acute rejection episode on day 47 diagnosed when a routine mucosal biopsy demonstrated Banff grade III/IV rejection that resolved within 72 h after a single bolus of 1 g of Solu-Medrol (Pfizer, New York, NY).

At day 452 posttransplant, the first clinical episode of graft rejection occurred and presented as erythema, edema, and a maculopapular rash. To date, our patient has had no major or life-threatening postoperative complications. She developed several episodes of transient leukopenia at unacceptable levels. The first episode required the witholding of ganciclovir and mycophenolate mofetil for 48 h at posttransplant week. The second episode occurred at sixth months posttransplant and required a leukocyte growth factor (Neupogen; Amgen, Inc., 1,000 Oaks, Calif.) injection. She also presented with seven episodes of CMV viremia at days 247, 279, 284, 312, 348, 359, and 375 posttransplant, which were treated with adjusted dosages of ganciclovir and immunosuppressants.

The patient has regained most of her missing facial functions including nasal breathing, sense of smell, drinking from a cup, eating solid foods, and speaking intelligibly. Facial graft sensation recovery is nearly complete as confirmed by the presence of 7 mm two-point discrimination measured over the entire facial allograft. The use of local anesthesia was required during month 5 for routine skin biopsies of the allograft secondary to patient discomfort. The aesthetic outcome is improving with time. There was redundant skin on both sides of the facial graft which was removed in a revision procedure of excision, rotation, and lifting (Fig. 33.5).

Psychologically, the patient is doing quite well. Interestingly, she has become more "picky" regarding small aesthetic details as compared to her pretransplant expectations. She is very excited to reenter the public arena and has regained self-confidence entirely since the surgery. She now feels stronger emotionally, physically, and spiritually. We recently published the results of objective psychological tests demonstrating significant decreases in depression and verbal abuse in addition to improved quality of life and social reintegration.[25]

33.4 Discussion

Conventional reconstructive techniques should be the primary option when the surgical goal is skin coverage. However, we have shown that in extreme cases, face transplantation, although innovative, is a practicable alternative. While we received the Institutional Review Board approval to proceed with face transplantation in 2004, the preparation process for the first US case of near-total face transplantation took significantly longer than expected. There were several steps that needed approval beyond institutional review board acceptance. These included a lengthy and complex process of organ procurement organization approval; the process of fund-raising and financial approval within our own institution; and finally, the process of societal approval, including the media and medical community.

Here we report the successful functional outcome of the most complex three-dimensional reconstruction performed thus far of a craniofacial skeletal defect with a composite tissue allotransplant from a human donor. Therefore we have confirmed the feasibility of returning many essential functions – such as breathing through the nose, eating solid foods, and drinking from a cup – in one complex procedure of face transplantation.

Since our successful face transplant, there have been seven new cases of facial allotransplantation reported – three in France (Paris), one in the USA (Boston), three in Spain (Valencia, Seville, Barcelona) – a total of 11 cases worldwide within the past 4 years.[26-31] The recent "expansion" of face transplantation is encouraging and indicates that there is a paradigm shift in the surgical approach and the medicolegal, and ethical approval of this novel procedure. There is, however, a point of concern regarding the deaths of two face transplant patients, one in China at over 2 years after transplantation and one in France (April of 2009) who underwent concomitant bilateral hand and face transplantation.[29,32-35] This

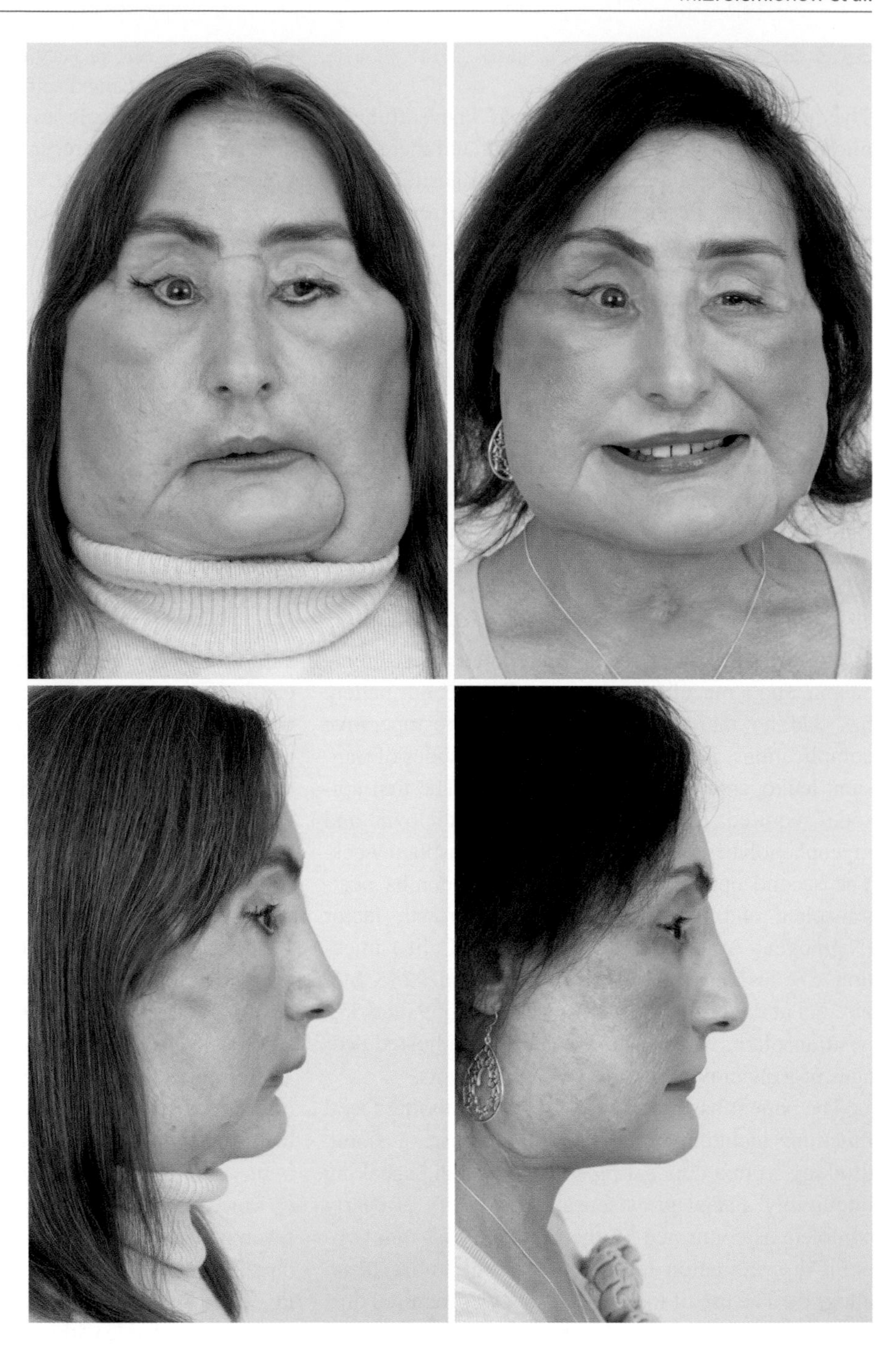

Fig. 33.5 Patient views after face transplantation at 5 months: frontal view (*above*, *left*), and profile view (*below*, *left*); patient views after face transplantation at 20 months: frontal view (*above*, *right*) and profile view (*below*, *right*) – demonstrating successful reconstruction of the facial defect

indicates an 18% failure rate, with the most significant and serious side effect – death. In contrast, hand transplant failures, where amputation of the transplanted hand is required because of rejection, there have been no reports of posttransplant death accompanying hand graft failure[36-38] except for the latest case of bilateral hand and face transplantation.[29]

We are pleased to report an excellent functional, psychological, and social outcome for our patient at 20 months following transplantation. We share our enthusiasm of supporting facial allograft transplantation, at early reconstructive stages, in a carefully selected group of severely disfigured patients with multiple functional deficits.

However, ethical challenges remain for patient selection, continuation of medical support, and appreciation of the moral, professional, and financial responsibilities to the patient. All these aspects should be taken into account before a novel, and still experimental, procedure is undertaken. The nature of these challenges will vary depending on the geographical, cultural, and economic conditions under which facial allotransplantation is performed. Nevertheless, in complex physically and functionally disabling cases, the patient's ethical right to make decisions should be respected, after being informed of the risks and benefits of the procedure and of the need for life-long immunosuppression.

References

1. Siemionow M, Gozel-Ulusal B, Engin Ulusal A, Ozmen S, Izycki D, Zins JE. Functional tolerance following face transplantation in the rat. *Transplantation*. 2003; 75:1607-1609.
2. Siemionow M, Ortak T, Izycki D, et al. Induction of tolerance in composite-tissue allografts. *Transplantation*. 2002;74: 1211-1217.
3. Gozel-Ulusal B, Ulusal AE, Ozmen S, Zins JE, Siemionow MZ. A new composite facial and scalp transplantation model in rats. *Plast Reconstr Surg*. 2003;112:1302-1311.
4. Demir Y, Ozmen S, Klimczak A, Mukherjee AL, Siemionow M. Tolerance induction in composite facial allograft transplantation in the rat model. *Plast Reconstr Surg*. 2004;114: 1790-1801.
5. Siemionow M, Demir Y, Mukherjee A, Klimczak A. Development and maintenance of donor-specific chimerism in semiallogenic and fully major histocompatibility complex mismatched facial allograft transplants. *Transplantation*. 2005;79:558-567.
6. Unal S, Agaoglu G, Zins J, Siemionow M. New surgical approach in facial transplantation extends survival of allograft recipients. *Ann Plast Surg*. 2005;55:297-303.
7. Yazici I, Unal S, Siemionow M. Composite hemiface/calvaria transplantation model in rats. *Plast Reconstr Surg*. 2006;118:1321-1327.
8. Yazici I, Carnevale K, Klimczak A, Siemionow M. A new rat model of maxilla allotransplantation. *Ann Plast Surg*. 2007;58:338-344.
9. Siemionow M, Klimczak A. Advances in the development of experimental composite tissue transplantation models. *Transpl Int*. 2010;23:2-13.
10. Kulahci Y, Siemionow M. A new composite hemiface/mandible/tongue transplantation model in rats. *Ann Plast Surg*. 2010;64:114-121.
11. Zor F, Bozkurt M, Nair D, Siemionow M. A new composite midface allotransplantation model with sensory and motor reinnervation. *Transpl Int*. 2010;23:649-656.
12. Siemionow M, Unal S, Agaoglu G, Sari AA. Cadaver study in preparation for facial allograft transplantation in humans: part I. what are alternative sources for total facial defect coverage? *Plast Reconstr Surg*. 2006;117:864-872. Discussion 873-875.
13. Siemionow M, Agaoglu G, Una Sl. A cadaver study in preparation for facial allograft transplantation in humans: part II. mock facial transplantation. *Plast Reconstr Surg*. 2006;117: 876-885. Discussion 886-888.
14. Siemionow M, Agaoglu G. The issue of "facial appearance and identity transfer" after mock transplantation: a cadaver study in preparation for facial allograft transplantation in humans. *J Reconstr Microsurg*. 2006;22:329-334.
15. Siemionow M, Papay F, Kulahci Y, et al. Coronal-posterior approach for face/scalp flap harvesting in preparation for face transplantation. *J Reconstr Microsurg*. 2006;22:399-405.
16. Devauchelle B, Badet L. Lengele' B, et al First human face allograft: early report. *Lancet*. 2006;368:203-209.
17. Guo S, Han Y, Zhang X, et al. Human facial allotransplantation: a 2-year follow-up study. *Lancet*. 2008;372:631-638.
18. Lantieri L, Meningaud JP, Grimbert P, et al. Repair of the lower and middle parts of the face by composite tissue allotransplantation in a patient with massive plexiform neurofibroma: a 1-year follow-up study. *Lancet*. 2008;372:639-645.
19. Dubernard JM, Lengele B, Morelon E, et al. Outcomes 18 months after the first human partial face transplantation. *N Engl J Med*. 2007;357:2451-2460.
20. Meningaud JP, Paraskevas A, Ingallina F, Bouhana E, Lantieri L. Face transplant graft procurement: a preclinical and clinical study. *Plast Reconstr Surg*. 2008;122:1383-1389.
21. Chenggang Y, Yan H, Xudong Z, et al. Some issues in facial transplantation. *Am J Transplant*. 2008;8:2169-2172.
22. Siemionow M, Papay F, Alam D. Near-total human face transplantation for a severely disfigured patient in the USA. *Lancet*. 2009;374:203-209.
23. Alam DS, Papay F, Djohan R, et al. The technical and anatomical aspects of the World's first near-total human face and maxilla transplant. *Arch Facial Plast Surg*. 2009;11:369-377.
24. Siemionow MZ, Papay F, Djohan R, et al. First U.S. near-total human face transplantation: a paradigm shift for massive complex injuries. *Plast Reconstr Surg*. 2010;125: 111-122.
25. Coffman KL, Gordon C, Siemionow M. Psychological outcomes with face transplantation: overview and case report. *Curr Opin Organ Transplant*. 2010;15:236-240.
26. World's fifth face transplant: man gets new nose, mouth and chin after shooting accident. Telegraph.co.uk Web site. Available at: http://www.telegraph.co.uk/scienceandtechnology/science/sciencenews/5063195/Worlds-fifth-face-transplant-Man-gets-new-nose-mouth-and-chin-after-shooting-accident.html. Accessed June 15, 2009.
27. Surgery News: World's 6th face transplant performed in Paris and 7th in Boston this week: Second face transplant in United States. MSNBC Web site. Available at: http://surgery.about.com/b/2009/04/12/surgery-news-worlds-6th-face-transplant-performed-second-face-transplant-inunited-states.htm. Accessed June 18, 2009.
28. Boston hospital performs face transplant. MSNBC Web site. Available at: http://www.msnbc.msn.com/id/30152143/. Accessed June 15, 2009.
29. Face-and-hands transplant patient dies. MSNBC Web site. Available at: http://www.msnbc.msn.com/id/31367511/. Accessed June 20, 2009.

30. Partial face transplant patient goes public in Spain. USA Today Web site. Available at: http://www.usatoday.com/news/health/2010-05-04-face-transplant_N.htm. Accessed June 11, 2010.
31. First Full-Face Transplant Claimed in Spain. CBSNEWS Web site: Available at: http://www.cbsnews.com/stories/2010/04/23/health/main6425492.shtml. Accessed June 9, 2010.
32. Chinese face transplant Li Guoxing dies. News.com.au Web site. Available at: http://www.news.com.au/story/ 0,27574,24829166-23109,00.html. Accessed June 19, 2009.
33. Infection kills face transplant patient. *Washington Times*. June 16, 2009. Available at: http://washingtontimes.com/news/2009/jun/16/infection-kills-face-transplant-patient/. Accessed June 17, 2009.
34. Lengele BG. Current concepts and future challenges in facial transplantation. *Clin Plast Surg*. 2009;36:507-521.
35. Morris P, Bradley A, Doyal L, et al. Face transplantation: a review of the technical, immunological, psychological and clinical issues with recommendations for good practice. *Transplantation*. 2007;83:109-128.
36. Gazarian A, Abrahamyan DO, Petruzzo P, et al. Hand allografts: experience from Lyon team. *Ann Chir Plast Esthét*. 2008;144:424-435.
37. Ravindra KV, Buell JF, Kaufman CL. Hand transplantation in the United States: experience with 3 patients. *Surgery*. 2008;144:638-643. Discussion 643-644.
38. Breidenbach WC, Gonzales NR, Kaufman CL, Klapheke M, Tobin GR, Gorantla VS. Outcomes of the first 2 American hand transplants at 8 and 6 years posttransplant. *J Hand Surg Am*. 2008;33:1037-1047.

The Spanish Experience With Face Transplantation

34

Pedro C. Cavadas, Luis Landin, Javier Ibañez, Alessandro Thione, Jose Rodrigo, Federico Castro, Marino Blanes, and Jose Maria Zarzalejos Andes

Contents

P.C. Cavadas (✉)
Department of Plastic and Reconstructive Surgery, Clínica Cavadas, Valencia, Spain
e-mail: pcavadas@telefonica.net

Abstract Herein, the first composite mandible–tongue lower face transplantation in humans is reported. The procedure was performed on an HIV-positive recipient who had suffered an epidermoid carcinoma of the tongue complicated by mandibular radionecrosis and multiple flap failure. We observed incipient return of function, as demonstrated by clinical exploration and electromyography. The early postoperative course was complicated by the presence of retromandibular abscesses that were controlled successfully. No episodes of rejection have been observed so far. It is still too early to report whether the procedure allowed the patient to return to a normal social life.

34.1 Introduction

Facial transplantation is extending the range of the reconstructive surgery armamentarium. Successful hand transplants performed in Lyon and elsewhere[1,2] encouraged to undertake a procedure of partial face transplantation in a collaboration between maxillofacial surgeon Devauchelle and transplant surgeon Dubernard.[3] To date, three cases of face transplantation have been described in detail,[3-7] and another five cases have been reported partially[8] or included in a systematic review.[9] The cases reported to date have included the transplantation of functional subunits, specifically the oral and orbital sphincters, in addition to esthetic units such as the cheeks and nose.

To the best of our knowledge, a composite mandible–tongue allograft transplantation procedure has never been performed before. In addition, composite tissue allograft (CTA) transplantation has never been performed in a recipient infected with human immunodeficiency virus (HIV). These factors made our case

M.Z. Siemionow (ed.), *The Know-How of Face Transplantation*,
DOI: 10.1007/978-0-85729-253-7_34,

unique, and prompted us to perform anatomical,[10] experimental,[11-14] and outcomes[15] research.

In this chapter, we report the reconstruction of the lower face of an HIV-positive recipient after an epidermoid carcinoma of the tongue, by use of a composite mandible–tongue allograft transplantation procedure.

34.1.1 Research Background

The most extensive research on facial transplantation has been done on rats. A full face transplant, followed by the hemiface transplant, hemiface/calvaria transplant, maxilla transplant, and composite maxilla and tongue transplant models, allowed the demonstration of peripheral blood chimerism and allograft survival after 350 days.[16-22] The model of maxilla allotransplantation showed graft acceptance and the growth of teeth, while no histopathological signs that suggested allograft rejection were noted for any component of the graft, including the teeth, mucosa, bone, muscle, cartilage, nerve, and vascular tissue.[23]

We developed a functional model of orthotopic hemifacial transplant to compensate for the limited information on functional recovery after face transplantation.[11] The mystacial pad (also known as the mystacial region, whiskers region, or vibrissal region) of rats was selected because of its high functional demands with respect to sensitivity and movement (Fig. 34.1) The movement of the whiskers is controlled by the motor branches of the facial nerve branches (namely, zygomatico-orbitalis, bucolabialis, and upper marginal mandibular), while the whiskers retrieve sensory information through the infraorbital branch of the trigeminal nerve.[12] We demonstrated statistically significant differences in nerve conduction studies and needle electrode examination, sensitivity, and histology of facial transplants after repair of the facial and trigeminal nerve branches, as compared with animals whose nerves were not repaired ($P<0.001$),[13,14] and these findings were later corroborated by others.[24]

Fig. 34.1 A facial transplant model was developed to evaluate the return of function following repair of the infraorbital and facial nerves after transplantation of the mystacial pad in rodents. We observed return of sensitivity to the whiskers of the rats when the nerves were repaired, as compared with transplants in which the nerves were not repaired

34.1.2 Clinical Background

To date, more that 52 hands have been transplanted in various countries, and the functional results have allowed to some recipients to return to work. In some instances, and in a number of our hand allografts (Fig. 34.2),[25,26] excellent intrinsic function has been recorded.[2]

Reconstructive surgery of postoncological patients using CTAs has been reported in three cases so far. A 56-year-old recipient of a kidney transplant received a latissimus dorsi allograft transplant after resection of a squamous cell cancer of the scalp (T3N0M0). Although the muscle allograft was covered with split thickness autologous skin grafts, several acute rejection episodes required that therapy with tacrolimus, mycophenolate mofetil (MMF), and prednisone be given to treat and prevent further rejection. Survival was reported to 259 days without recurrence of the tumor.[27] A 72-year-old female patient suffered a melanoma skin cancer (T4N3M0) that required total resection of the scalp and neck skin. A cephalocervical skin allograft was used to reconstruct the defect. The patient was given tacrolimus, MMF, and prednisone to prevent rejection.

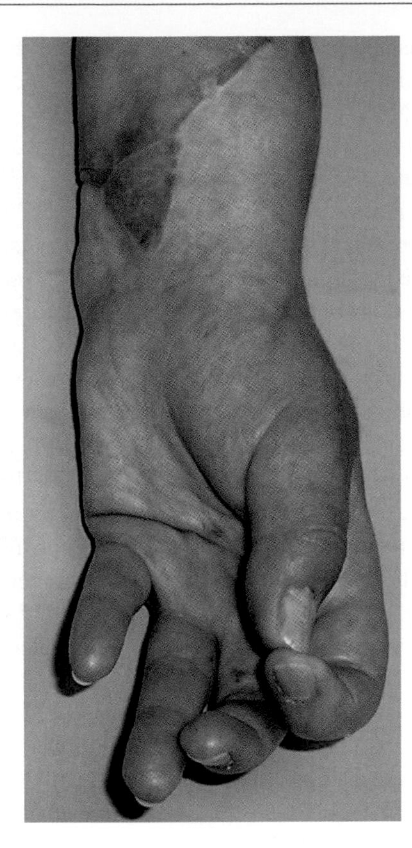

Fig. 34.2 An excellent degree of intrinsic function was reached in some of our hand allografts 2 years after transplantation

Survival was reported to 120 days without recurrence of the tumor.[28] Finally, a third patient suffered an epidermoid carcinoma of the tongue (T4N2bM0) that required excision of the floor of the mouth. The defect was reconstructed using a tongue allograft that was neurotized. The patient was given tacrolimus, MMF, and prednisone to prevent rejection. Survival was reported to 390 days without functional restoration of the allograft but with recurrence of the tumor.[29]

Patients who have been treated successfully for a pretransplantation malignancy can be considered suitable candidates for a transplant.[30] The inclusion criteria for kidney allograft transplantation in postoncological patients include at least 2 years free from recurrence or remission at the time of transplantation (excluding melanoma skin cancer).[31]

Patients infected with human immunodeficiency virus (HIV) have shown improved survival since the introduction of highly active antiretroviral therapy (HAART), but such patients are subject to end-stage renal disease (ESRD) as a result of HIV-associated nephropathy. In these patients, kidney allograft transplantation is a feasible treatment option for ESRD.[32] We performed a comprehensive review and meta-analysis of the outcomes of 12 series of kidney transplants in HIV-positive recipients under HAART.[15] Survival, rejection, and infections after transplantation were analyzed, among other variables. Among the 254 patients, 1-year survival rate was 0.93 (95% CI, 0.90–0.96) (Fig. 34.3), allograft rejection rate was 0.36 (95% CI, 0.25–0.49), and infection rate was 0.29 (95% CI, 0.17–0.43). These data were similar to those published in the 2006 OPTN/SRTR Annual report, in which survival of kidneys from non-expanded criteria

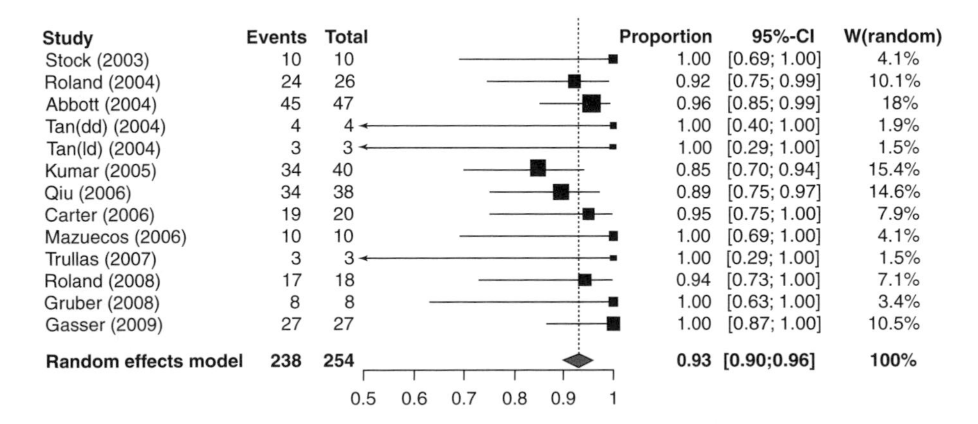

Fig. 34.3 Pooled estimated proportion of patients surviving the first year, analyzed using a random effects model. Of the 254 patients, 238 survived; this gave a pooled estimate of the survival rate in the first year of 0.93 (95% CI, 0.90–0.96). Analysis using the Q statistic revealed that there was no evidence of heterogeneity (Q_{12})=9.58; $P<0.653$; $I^2=0$; 95% CI, 0–0.45) (Reproduced from Landin et al.[15] With permission)

donors was 91% at 1 year.[33] Most authors observed a higher number of rejections among HIV-positive recipients than among HIV-negative recipients at their institutions.[34] In addition, the outcome in HIV-positive recipients under HAART was subject to three important factors. First, the use of protease inhibitors (PI), such as ritonavir, produced changes in concentration–time curves for cyclosporine when compared with patients under treatment with non-nucleoside retrotranscriptase inhibitors (NNRTI).[35] In addition, the requirement for calcineurin inhibitors (CNIs) was reduced dramatically when they were administered concomitantly with PI. Second, anti-lymphoproliferative drugs, such as MMF, showed a beneficial effect by decreasing HIV replication in infected lymphocytes, and favored a reduction in the dose and frequency of administration of the HAART.[36,37] Thirdly, the use of anti-thymocyte globulin affected the $CD4^+$ responses to Epstein–Barr virus (EBV) significantly, and increased the risk of reactivation of EBV.[38] Most of the studies concluded that HIV infection could no longer be considered a contraindication for kidney transplantation, although a high incidence of rejection should be expected.[15]

34.2 The Patient

The patient was a 42-year-old man who had been infected with HIV-1 virus 22 years before, and was in stage B-3 of the classification of the Centers for Disease Control and Prevention, due to $CD4^+$ T cell counts of 39/μL and the presence of oropharyngeal candidiasis. The virus showed mutations that conferred resistance to the antiretroviral drugs M41L, D67N, T69D, V75M, V118I, M184V, L210W, T215Y, K103N, Y181C, L10F, M46I, I54V, L63P, A71T, G73T, V82T, I84V, L90M, and I93L.

An epidermoid carcinoma of the tongue (T3N0M0) was diagnosed 10 years after HIV seroconversion, and was treated with chemotherapy and radiotherapy. Fifteen years after seroconversion, a severe infection of the mandible followed the removal of a tooth and was complicated by radionecrosis and the patient was referred elsewhere for reconstruction. Three attempts were made at microsurgical reconstruction of the mandible, including two iliac crest free flaps and one fibula free flap. However, total necrosis of the tongue occurred, and it was reconstructed with two radial forearm free flaps. Afterward, a latissimus dorsi musculocutaneous free flap was used for coverage; unfortunately, further necrosis required a pedicled musculocutaneous pectoral flap for definitive wound closure (Fig. 34.4).

At the time of patient evaluation, 20 years since seroconversion and 10 years after the diagnosis of tongue cancer, the patient was wearing a mask on the lower third of his face and there were two purulent fistulae on the left cheek. Swallowing was hindered by the absence of the tongue and a rigid pharyngolarynx. Furthermore, a paramedian palsy of the vocal cords impeded speech. The patient fed himself through a gastric percutaneous catheter. There was no sensitivity or movement in the remnants of the lower lip. The patient had never suffered anal intraepithelial neoplasia or condyloma, and his creatinine clearance (CrC) was 62 mL/min. The HAART consisted of five

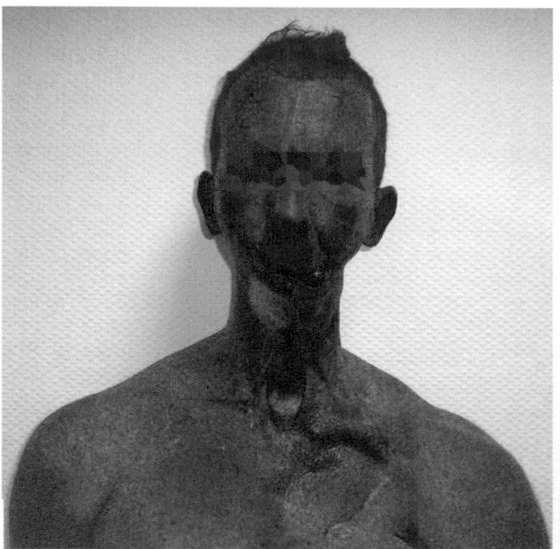
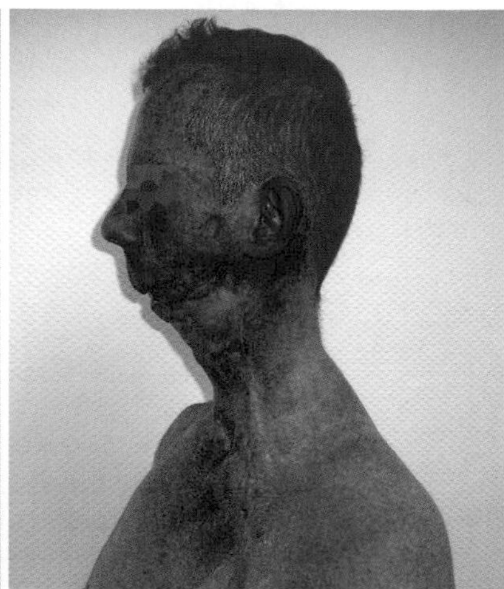

Fig. 34.4 Pretransplant appearance of the patient

drugs: darunavir coadministered with a booster dose of ritonavir, raltegravir, etravirine, and enfuvirtide. The patient's $CD4^+$ cell count at the time of our evaluation was 498 cells/μL, and HIV RNA was undetectable in plasma by polymerase chain reaction (<20 copies/mL). A panel for reactive antibodies was negative by flow cytometry (Labscreen, One Lambda, Canoga Park, CA, USA). The patient presented with bilateral avascular necrosis of the hips, as a consequence of steroid therapy during the microsurgical reconstruction procedures and the use of tenofovir, which had also produced renal tubular injury.[39]

In May 2009, the patient underwent an operation to identify the hypoglossal nerves, lingual nerves, inferior branches of the facial nerves, and inferior alveolar nerves. After a long and difficult dissection through irradiated and heavily scarred tissue, the nerve stumps were identified and marked. Fistulae and remnants of previous attempts at osseous reconstruction were debrided. The left condyle was found to be ankylosed, so it was debrided and burred and a conformed polymethylmethacrylate (PMMA) implant impregnated with gentamicin was put in place. The right condyle and ramus from the native mandible were preserved.

A CTA program for hand and face transplantation had been approved at our institution. The patient was informed of the current research knowledge, the results of previous facial transplants performed in human beings, the risks of transplantation after malignancy, the results of kidney allograft transplantation performed on HIV-positive recipients, and the novelty of a composite mandible–tongue allograft transplant. The recipient signed a statement of consent that included detailed information about the risks of malignancy after transplantation, metabolic complications and the risk of death associated with the procedure.

The entire process was supervised by the Ethics Committee at our institution and the National Organization for Organ Transplantation (Organización Nacional de Trasplantes, ONT). The approval certificates are available on request from the Editor.

34.3 The Procedure

The face transplant was performed on August 18, 2009. The preparation of the recipient was started simultaneously with the multi-organ donation. Pretransplant flow cytometry revealed a negative virtual cross-match. The donor and the recipient shared only one HLA antigen. First, a tracheostomy was performed under local anesthesia and sedation. A low pressure cuff tube with subglottic suction line (Tracoe Twist 306, Tracoe Medical, Frankfurt, Germany) was inserted and secured. Subsequently, two anesthesiologists who specialized in transplantation *and* replantation surgery performed general anesthesia. Due to the potential for interaction between antiretroviral therapy and some anesthetic drugs, anesthesia was give by inhalation using sevofluorane in combination with air or oxygen depending on the needs of the patient. Intraoperative analgesia was supplied using ramifentanile, because of its negligible effect on hemodynamics. Cisatracurium was used for intraoperative neuromuscular blockade. Propofol and midazolam were avoided because these are metabolized by cytochrome P450, which was inhibited by ritonavir.[40-48]

In the recipient, the skin was incised over the previous scars and the tagged nerves were identified easily. Afterward, the midline of the neck was incised and inverted T incisions were made above the clavicles. The external jugular veins (EJVs) were identified and preserved on both sides. On the right side, the sternocleidomastoid (SCM) was transected and retracted to gain full access to the right subclavian artery and vein, after cutting the anterior scalene muscle. The vessels were prepared for latero-terminal anastomoses. The right internal jugular vein (IJV) was dissected and prepared for termino-terminal anastomosis. The tissue of the lower third of the face was not removed until the allograft had been revascularized successfully.

The allograft was recovered from a non-heart-beating brain-dead multi-organ donor who was matched with the recipient on sex and blood group but not human leukocyte antigens (HLA). The donor had died because of head trauma, but the lower part of the face and mandible had not been injured. Sputum cultures were positive for *Pseudomonas aeruginosa*, but there was no bacteremia or fever, so the patient was considered to be a valid candidate for donation. A specific written agreement and consent form was signed by the family of the donor. During recovery of the heart and liver, the right and left common carotid arteries (CCA) were cannulated for perfusion and the right subclavian artery was ligated. After cross-clamping, 2 L of reconstituted Belzer–University of Wisconsin (UW) solution was used to perfuse the head.

The recovery of the face was begun by incising the neck at the suprasternal notch. The incision was extended bilaterally to the base of the neck at the posterior edge of the SCMs, which were cut transversely at their origin, with care to preserve both EJVs. The incision was continued laterally up to the ear lobes; a line was drawn to the labial commissure, coinciding with the line of the beard. Both EJVs were dissected cranially for 2 cm; then the platysma muscle was elevated. The digastric muscle was transected at its intermediate tendon. The anterior belly was included in the allograft. The common carotid arteries (CCAs) were dissected, along with the IJVs. Both the internal carotid arteries (ICAs) and the external carotid arteries (ECAs) were elevated cranially to the styloid process. The superior thyroid artery, the ascending pharyngeal artery, and the continuation of the ECAs were ligated while preserving the lingual and facial arteries on both sides. The hypoglossal nerves were dissected proximally at this point. The larynx was cut transversely just below the hyoid bone, preserving the attachments of the geniohyoid, hyoglossus, mylohyoid, and genioglossus muscles. The epiglottis was not included in the allograft. After the larynx had been cut, the lingual nerves were identified and dissected proximally; the cervicofacial branches of the facial nerves were referenced before entering behind the parotid gland, which was included in the allograft. The skin incisions were deepened to the oral mucosa by cutting the buccinator muscle, with care to preserve the Stensen's duct and preserving the maximum amount of mucosal lining. On the right side, the angle of the mandible was osteotomized using a cooled saw, with care taken to preserve the inferior alveolar nerve, while on the left side, the ramus and condyle were included in the allograft and separated from the temporal and masseter muscles. The pterygoid muscles, styloglossus, stylopharyngeus, stylohyoid, palatoglossus (posterior pillar of the fauces), and superior constrictor muscles were transected, as were the stylohyoid ligament and the pterygomandibular raphe. The palatine tonsil was included in the allograft. Posteriorly, the pharyngeal part of the tongue was preserved to the palatopharyngeal arch and the incision ended in the median glossoepiglottic fold. The submandibular glands were included in the allograft. After these maneuvers had been completed bilaterally, the allograft was raised (Fig. 34.5) and immersed in cold Belzer–UW preservation solution for transport.

Before revascularization, 20 mg of basiliximab was administered. Six hours after cross-clamping, the allograft was revascularized. First, the right IJV was anastomosed end-to-end using 8/0 nylon suture. The right CCA of the allograft was anastomosed end-to-side to the right subclavian artery 8/0. Then both EJVs were repaired end-to-end 8/0. Subsequently, an intra-allograft shunt was performed by end-to-end anastomosis of the right ICA to the left ICA 8/0, while the left CCA was ligated proximally (Fig. 34.6). After a successful revascularization, meticulous hemostasis was performed. Concomitantly, 250 mg of methylprednisolone was administered. The scar tissue was then removed from the lower third of the face. The hyoid of the allograft was sutured to the hyoid of the recipient; then the intraoral mucosa was sutured to render it watertight. The right ramus of the mandible was synthesized with a 2 mm locking plate (Synthes, Oberdorf, Switzerland) and the left condyle was inserted in its place and sutured to fibrous remnants after removal of the PMMA. The hypoglossal, lingual, inferior alveolar, and cervicofacial nerves were repaired primarily on the left side by 9/0 epineural neurorrhaphies. The lingual and inferior alveolar nerves in the right side were repaired primarily by 9/0 epineural neurorrhaphies. Since edema had increased in the allograft after revascularization, the right hypoglossal and cervicofacial nerves required nerve graft interposition for epineural 9/0 repair (donor: supraclavicular nerve). The skin was closed leaving several passive drains in both sides. At the end of the procedure, the patient had received 11 units of blood, six of fresh-frozen plasma, two of platelets, and one of fibrinogen.

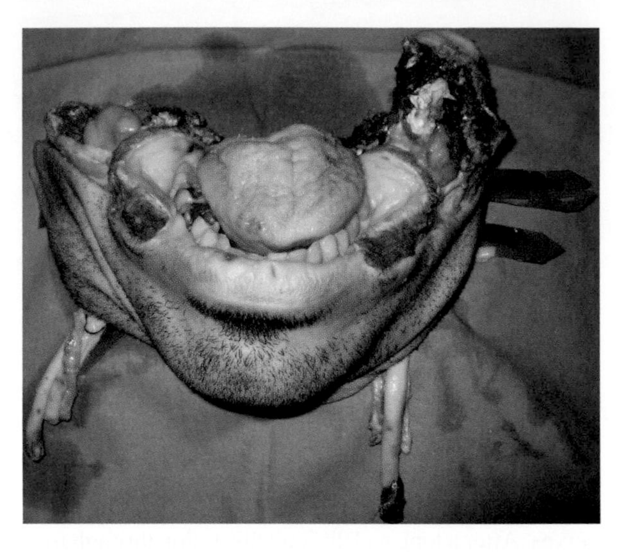

Fig. 34.5 Lower third composite mandible/tongue allograft

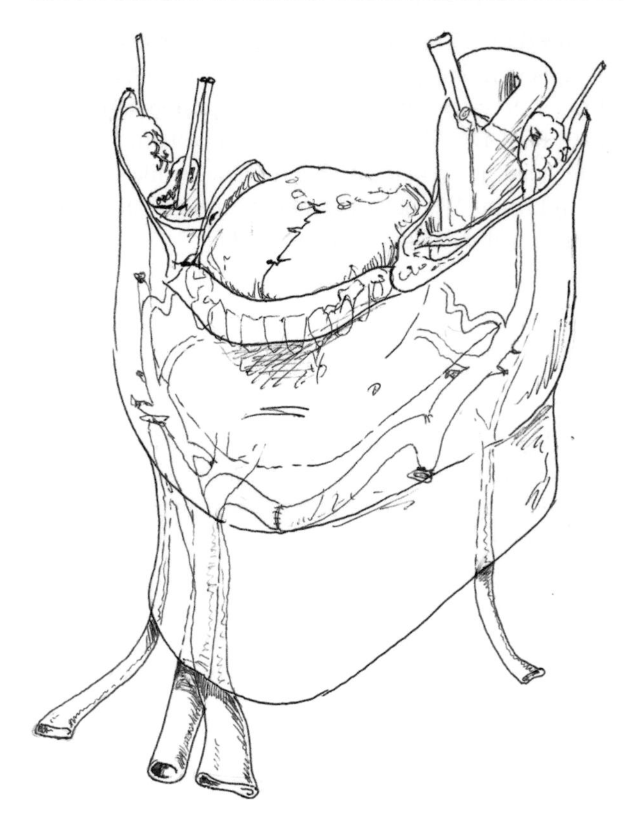

Fig. 34.6 A schematic drawing of the allograft shows intra-allograft vascular shunts and the vessels and nerves selected for repair. The lingual and facial arteries were preserved bilaterally. The left common carotid artery was ligated, as was the continuation of the superficial temporal vessels distally to the maxillary artery, which was only included on the left side

34.4 Follow-up and Outcomes

A complex postoperative course followed (Fig. 34.7). On postoperative day (POD) 1, CNI therapy was initiated with the introduction of tacrolimus 0.5 mg. Methylprednisolone was administered at 250 mg, 100 mg, 100 mg, and 100 mg and continued as shown in Fig. 34.8. Mycophenolate mofetil was introduced on POD 4 (1 g/day), and the second dose of basiliximab (20 mg) was also administered. HAART was resumed on POD 8; however, it was reduced to four drugs: ritonavir, darunavir, raltegravir, etravirine, because of the beneficial effect of MMF in preventing HIV replication.

On POD 15, the recovery of the patient was complicated by the presence of fever and severe edema of the face. X-ray examination revealed a right diaphragmatic paralysis, most probably because of inadvertent injury to the phrenic nerve. A computed tomography scan revealed the presence of purulent abscess in the medial side of the mandibular ramus bilaterally and at the stumps of the pterygoid muscles. However, the chest was clean. In addition, the mucosal lining was breached on the right side, exposing the mandible. Fluconazol and meropenem were introduced in the light of positive cultures for multiple bacteria and *Candida* from the wounds. The dose of immunosuppressive therapy was lowered and the abscesses were

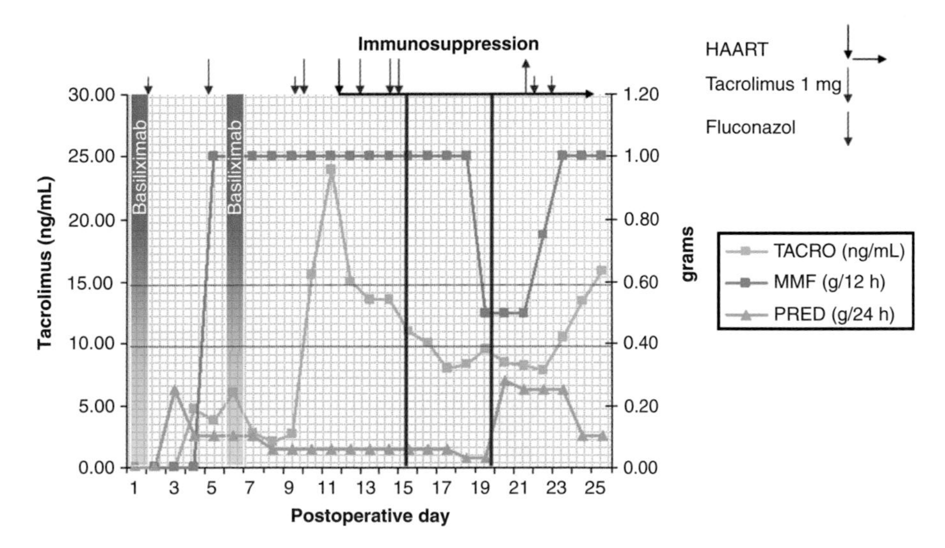

Fig. 34.7 The early postoperative course after face transplantation. Basiliximab was administered on days 0 and 4. Tacrolimus was introduced on day 1 at a dose of 0.5 mg. Subsequent doses of 1 mg and 0.5 mg were administered, and they had to be spaced after the introduction of HAART and fluconazol. Retromandibular abscesses were diagnosed on postoperative day 15 (*red vertical line*) and the immunosuppressive therapy was lowered. After 4 days (*blue vertical line*), extensive edema and the suspicion of acute rejection required adjustment of the triple therapy and steroid boluses

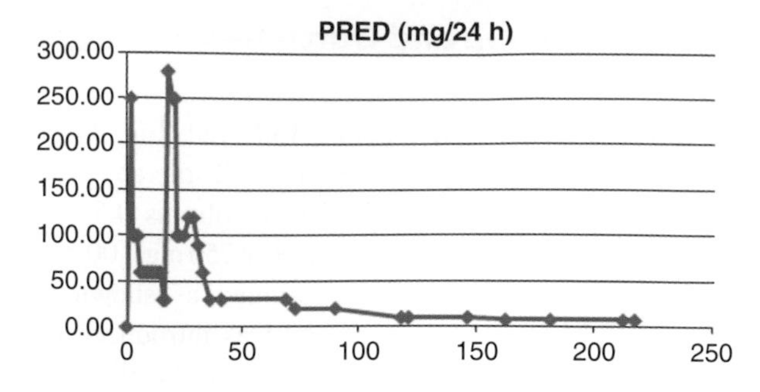

Fig. 34.8 Steroid tapering scheme

drained under general anesthesia. Consequently, the condition of the patient improved and immunosuppression was adjusted to be maintained at 12 ng/mL of tacrolimus, MMF 2 g/day, and prednisone 20 mg/day. Flexible videolaryngoscopy revealed an intact mucosal repair at the pharynx. The oral secretions were suctioned through the tracheostomy tube line every 2 h, and the patient was taught to check the cuff pressure and keep it between 20 and 30 mmH_2O to avoid tracheal pressure sores. The postoperative course was uneventful since then. The dose of prednisone was reduced to 7.5 mg/day after 8 months. Changes in the lymphocyte and $CD4^+$ T cell counts are summarized in Fig. 34.9. At the time of writing (postoperative day 246) the $CD4^+$ T cell count was 301 cells/μL, and the HIV load has remained undetectable, while CrC was 86 mL/min. We have not observed signs of clinical rejection at the skin or mucosa so far.

34.4.1 Rehabilitation Protocol

The day after the transplantation procedure, wiping of the mouth using nistatin followed by subglottic suctioning were started and the patient was allowed to shave 1 week after the transplant. Three months after transplantation, the intraoral wounds had healed secondarily and the patient started a rehabilitation protocol that consisted of lymphatic drainage and swallowing training for 2 h/day, in addition to mental compressive garments worn for 30 min every 4 h.

At the time of writing (POD 246), exploration has shown evidence of the presence of a Tinel sign at the emergence of the mental nerve; the tongue has

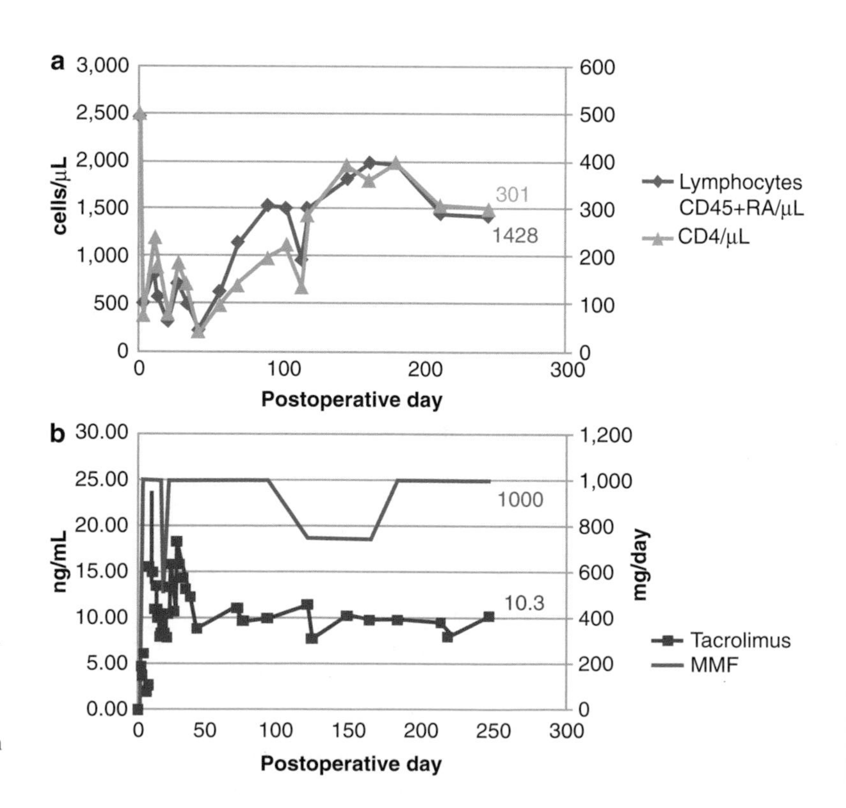

Fig. 34.9 (**a**) Changes in lymphocyte and $CD4^+$ counts after the transplant. The number of $CD4^+$/μL was over 200 on postoperative day 100. (**b**) Tacrolimus trough levels and the MMF administration scheme

shown movement on the left side with M2 power; needle electrode examination has shown the presence of voluntary motor unit potentials on both sides of the tongue, which were more powerful at the left side (Fig. 34.10). However, the orbicularis orii muscle still showed denervation activity. Functional magnetic resonance imaging examination was performed before the transplantation procedure, but it is still too early for posttransplant evaluation.

The patient has shown good psychological acceptance of the allograft. However, the impaired functional restoration of the tongue and the need for secondary surgery to improve the appearance have precluded the abandonment of the lower facial mask to this moment (Fig. 34.11).

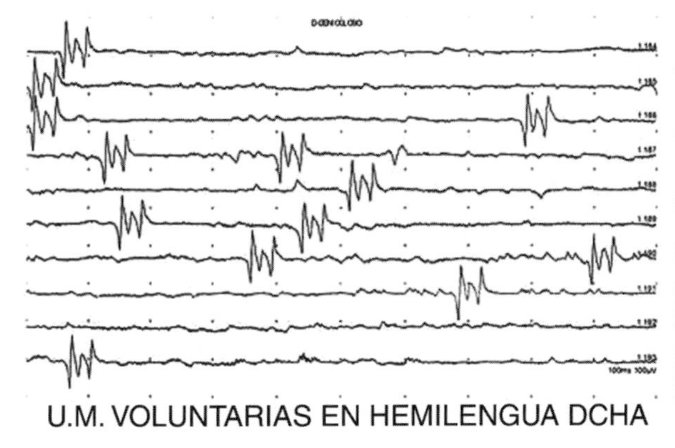

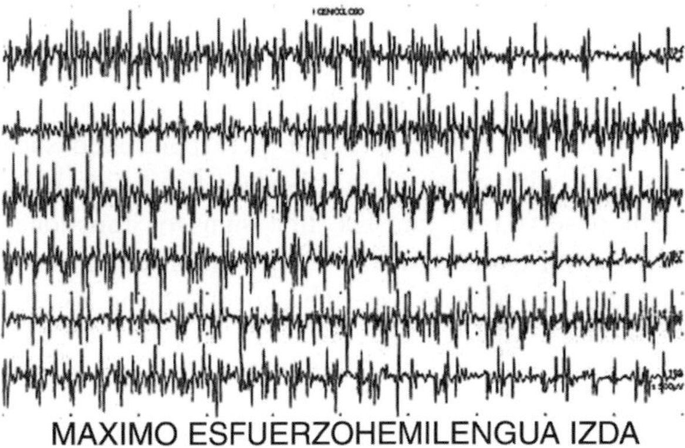

Fig. 34.10 Tongue electromyogram at the *left side* of the tongue (*right figure*) shows intense activity during voluntary contraction; at the *right side* of the tongue, where the nerve grafts were interpositioned, an incipient recovery with isolated potentials is observed (*left figure*)

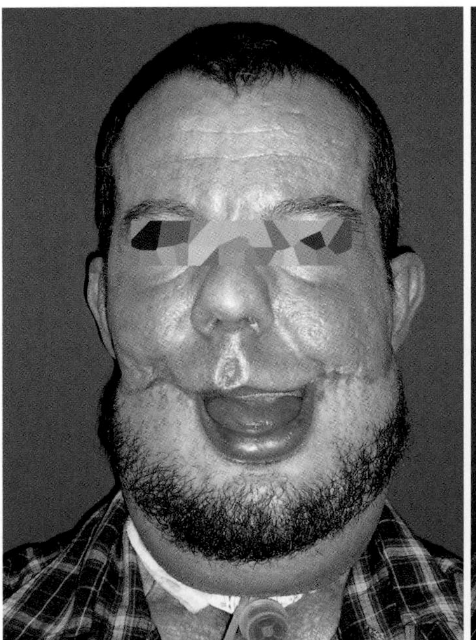

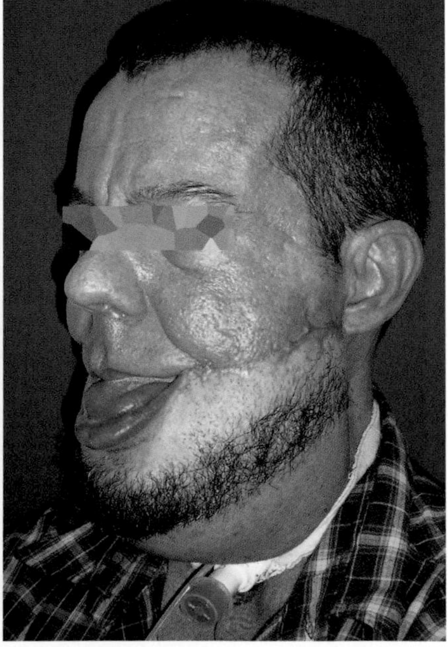

Fig. 34.11 Posttransplantation appearance on postoperative day 246

34.5 Conclusions

We have performed the first composite mandible–tongue lower face transplantation procedure in humans on an HIV-positive recipient who had suffered an epidermoid carcinoma of the tongue complicated by mandibular radionecrosis and multiple flap failure. Such a procedure required the development of a functional model of facial transplantation and the evaluation of the functional outcomes of kidney transplantation in HIV-positive recipients. We observed the incipient return of function as demonstrated by clinical exploration and electrodiagnostic tests. The early postoperative course was complicated by the presence of retromandibular abscesses that were controlled successfully with surgery. No episodes of rejection have been observed so far. The first signs of functional recovery of the allograft have been observed recently, and we hope that in time the patient will be able to have a normal social life.

Disclosure: There are no sources of support that require acknowledgment. The authors have no financial or personal relationships with other people or organizations that could influence this work inappropriately. The authors have no financial interest in any of the drugs mentioned in this work. Basiliximab, tacrolimus, sirolimus, and mycophenolate mofetil were used off-label to prevent face allograft rejection.

References

1. Dubernard JM, Owen E, Herzberg G, et al. Human hand allograft: report on first 6 months. *Lancet*. 1999;353:1315-1320.
2. Petruzzo P, Lanzetta M, Dubernard JM, et al. The international registry on hand and composite tissue transplantation. *Transplantation*. 2008;86:487-492.
3. Devauchelle B, Badet L, Lengelé B, et al. First human face allograft: early report. *Lancet*. 2006;368:203-209.
4. Dubernard JM, Lengelé B, Morelon E, et al. Outcomes 18 months after the first human partial face transplantation. *N Engl J Med*. 2007;357:2451-2460.
5. Siemionow M, Papay F, Alam D, et al. Near-total human face transplantation for a severely disfigured patient in the USA. *Lancet*. 2009;374:203-209.
6. Guo S, Han Y, Zhang X, et al. Human facial allotransplantation: a 2-year follow-up study. *Lancet*. 2008;372:631-638.
7. Lantieri L, Meningaud JP, Grimbert P, et al. Repair of the lower and middle parts of the face by composite tissue allotransplantation in a patient with massive plexiform neurofibroma: a 1-year follow-up study. *Lancet*. 2008;372:639-645.
8. Pomahac B, Lengele B, Ridgway EB, et al. Vascular considerations in composite midfacial allotransplantation. *Plast Reconstr Surg*. 2010;125:517-522.
9. Gordon CR, Siemionow M, Papay F, et al. The world's experience with facial transplantation: what have we learned thus far? *Ann Plast Surg*. 2009;63:572-578.
10. Landin L, Cavadas PC, Carrera A, et al. Human face/scalp alloflap harvesting technique. *Plast Reconstr Surg*. 2007;119:1114-1115.
11. Landin L, Gonzalez E, Cavadas PC. Functional recovery of the hemifacial allograft transplant in rats. *Am J Transplant*. 2006;6(Suppl 2):65-472.
12. Landin L, Cavadas PC. The mystacial pad flap: a functional facial flap in rats. *Ann Plast Surg*. 2006;56:107-108.
13. Landin L, Cavadas PC, Gonzalez E, et al. Functional outcome after facial allograft transplantation in rats. *J Plast Reconstr Aesthet Surg*. 2008;61:1034-1043.
14. Landin L, Cavadas PC, Gonzalez E, et al. Sensorimotor recovery after partial facial (mystacial pad) transplantation in rats. *Ann Plast Surg*. 2009;63:428-435.
15. Landin L, Rodriguez-Perez JC, Garcia-Bello MA, et al. Kidney transplants in HIV-positive recipients under HAART. A comprehensive review and meta-analysis of 12 series. *Nephrol Dial Transplant*. 2010;25:3106-3115.
16. Ulusal BG, Ulusal AE, Ozmen S, et al. A new composite facial and scalptransplantation model in rats. *Plast Reconstr Surg*. 2003;112:1302-1311.
17. Siemionow M, Ulusal BG, Ulusal AE, et al. Functional tolerance following face transplantation in the rat. *Transplantation*. 2003;75:1607-1609.
18. Unal S, Agaglou G, Zins J, et al. New surgical approach in facial transplantation extends survival of allograft recipients. *Ann Plast Surg*. 2005;55:297-303.
19. Demir Y, Ozmen S, Klimczak A, et al. Tolerance induction in composite facial allograft transplantation in the rat model. *Plast Reconstr Surg*. 2004;114:1790-1801.
20. Siemionow M, Demir Y, Mukherjee A, et al. Development and maintenance of donor-specific chimerism in semi-allogenic and fully major histocompatibility complex mismatched facial allograft transplants. *Transplantation*. 2005;79:558-567.
21. Yazici I, Unal S, Siemionow M. Composite hemiface/calvaria transplantation model in rats. *Plast Reconstr Surg*. 2006;118:1321-1327.
22. Kulahci Y, Siemionow M. A new composite hemiface/mandible/tongue transplantation model in rats. *Ann Plast Surg*. 2010;64:114-121.
23. Yazici I, Carnevale K, Klimczak A, et al. A new rat model of maxilla allotransplantation. *Ann Plast Surg*. 2007;58:338-344.
24. Washington KM, Solari MG, Sacks JM, et al. A model for functional recovery and cortical reintegration after hemifacial composite tissue allotransplantation. *Plast Reconstr Surg*. 2009;123:26S-33S.
25. Cavadas PC, Landin L, Ibañez J. Bilateral hand transplantation: result at 20 months. *J Hand Surg Eur*. 2009;34:434-443.
26. Cavadas PC. The valencia experience in hand transplantation. Presented at: The 9th International Symposium on Composite Tissue Allotransplantation; September 11, 2009; Valencia.
27. Jones TR, Humphrey PA, Brennan DC. Transplantation of vascularized allogenic skeletal muscle for scalp reconstruction

in a renal transplant patient. *Transplantation*. 1998;65: 1605-1610.
28. Jiang HQ, Wang Y, Hu XB, et al. Composite tissue allograft transplantation of cephalocervical skin flap and two ears. *Plast Reconstr Surg*. 2005;115:31e-35e.
29. Kermer C, Watzinger F, Oeckher M. Tongue transplantation: 10-month follow-up. *Transplantation*. 2008;85:654-655.
30. Bunnapradist S, Danovitch GM. Evaluation of adult kidney transplant candidates. In: Danovitch GM, ed. *Handbook of Kidney Transplantation*. 5th ed. Philadelphia: LWW; 2010.
31. Thompson JF, Mohacsi PJ. Cancer in dialysis and renal transplant patients. In: Morris PJ, Knechtle SJ, eds. *Kidney Transplantation*. Philadelphia: Saunders; 2008.
32. Frassetto LA, Tan-Tam C, Stock PG. Renal transplantation in patients with HIV. *Nat Rev Nephrol*. 2009;5:582-589.
33. From 2006 OPTN/SRTR Annual Report. Available at: http://optn.transplant.hrsa.gov/AR2006.exe.
34. Stock PG, Roland ME, Carlson L, et al. Kidney and liver transplantation in human immunodeficiency virus-infected patients: a pilot safety and efficacy study. *Transplantation*. 2003;76:370-375.
35. Frassetto LA, Browne M, Cheng A, et al. Immunosuppressant pharmacokinetics and dosing modifications in HIV-1 infected liver and kidney transplant recipients. *Am J Transplant*. 2007;7:2816-2820.
36. Margolis D, Heredia A, Gaywee J, et al. Abacavir and mycophenolic acid, an inhibitor of inosine monophosphate dehydrogenase, have profound and synergistic anti-HIV activity. *J Acquir Immune Defic Syndr*. 1999;21:362-370.
37. Kaur R, Bedimo R, Kvanli MB, et al. A placebo-controlled pilot study of intensification of antiretroviral therapy with mycophenolate mofetil. *AIDS Res Ther*. 2006;26: 16-20.
38. Gasser O, Bihl F, Sanghavi S, et al. Treatment-dependent loss of polyfunctional CD8+ T-cell responses in HIV-infected kidney transplant recipients is associated with herpesvirus reactivation. *Am J Transplant*. 2009;9:794-803.
39. Nelson MR, Katlama C, Montaner JS, et al. The safety of tenofovir disoproxil fumarate for the treatment of HIV infection in adults: the first 4 years. *AIDS*. 2007;21:1273-1281.
40. Adams J, Charlton P. Anesthesia for microvascular free tissue transfer. *Br J Anaesth (CEPD Rev)*. 2003;3:33-37.
41. Hagau N, Longrois D. Anesthesia for free vascularized tissue transfer. *Microsurgery*. 2009;29:161-167.
42. D'arminio M, Sabin CA, Phillips AN, et al. Cardio and cerebrovascular events in HIV-infected persons. *AIDS*. 2004;18: 1811-1817.
43. Bruegger D, Bauer A, Finsterer U, Bernasconi P, Kreimeier U, Christ F. Microvascular changes during anesthesia: sevoflurane compared with propofol. *Acta Anaesthesiol Scand*. 2002;46:481-487.
44. Lucchinetti E, Ambrosio S, Aguirre J, et al. Sevoflurane inhalation at sedative concentrations provides endothelial protection against ischemia-reperfusion injury in humans. *Anesthesiology*. 2007;106:262-268.
45. Egan TD. The clinical pharmacology of remifentanil: a brief review. *J Anesth*. 1998;12:195-204.
46. Baker MT, Chadam MV, Ronnenberg WC. Inhibitory effects of propofol on cytochrome P450 activities in rat hepatic microsomes. *Anesth Analg*. 1993;76:817-821.
47. Chen TL, Ueng TH, Chen SH, Lee PH, Fan SZ, Liu CC. Human cytochrome P450 mono-oxygenase system is suppressed by propofol. *Br J Anaesth*. 1995;74:558-562.
48. Suominen S, Svartling N, Silvasti M, Niemi T, Kuokkanen H, Asko-Seljavaara S. The effect of intravenous dopamine and dobutamine on blood circulation during a microvascular TRAM flap operation. *Ann Plast Surg*. 2004;53:425-431.

Microsurgical Aspects of Face Transplantation

35

Steven Bernard

Contents

S. Bernard
Department of Plastic Surgery, Cleveland Clinic,
Cleveland, OH, USA
e-mail: bernars2@ccf.org

Abstract Performing a face transplant requires expertise in both facial reconstruction and microvascular technique. The best way to approach the microsurgical aspects of face transplantation is to break them into smaller parts. Defining microsurgery loosely as procedures typically done under a microscope, this chapter will describe the details dealing with the arteries, veins, and nerves of face transplantation from anatomy to the technical aspects.

35.1 Introduction

The best way to approach the microsurgical aspects of face transplantation is to break them into smaller parts. Defining microsurgery loosely as procedures typically done under a microscope, this chapter will describe the aspects dealing with the arteries, veins, and nerves of face transplantation.

When performing the face transplant at the Cleveland Clinic, we found it helpful to break into surgical teams based on two subspecialties: microsurgery and maxillofacial surgery.[1-4] As we had a total of six experienced plastic surgeons all with fellowship training and one ENT surgeon, there was also a good deal of crossover between the two groups in terms of ability ultimately helping us with our goals. As described elsewhere in this book, the most critical factors for success were teamwork and practice.

Because of the large size of our team, we were able to further break our group into those working on the donor and the recipient. These two teams worked simultaneously in adjacent operating rooms. The ideal number of people working at the same time would be eight and would include a surgeon and an assistant on each side of the head for both the donor

M.Z. Siemionow (ed.), *The Know-How of Face Transplantation*,
DOI: 10.1007/978-0-85729-253-7_35,

and the recipient. Again the size of our team allowed us to maintain this number of surgical personnel throughout the more than 22 h case.

What this meant from a microsurgical standpoint was that typically two people were working on microsurgery at any given time during the case. From the standpoint of the donor, the blood vessels and nerves that would ultimately sustain the flap were carefully dissected in conjunction with the cutaneous incisions, soft-tissue dissection, and osteotomies as performed by the maxillofacial team. In the recipient room, the microvascular team needed to prepare adequate vessels to perfuse the flap while the maxillofacial team worked on defining the defect to be filled by the face transplant.

35.2 Anatomy

35.2.1 Introduction

Understanding of the anatomy of the blood vessels and nerves of the head and neck is critical to the success of a facial transplant. It is important to know both the macrovascular and microvascular anatomy to understand which artery will perfuse which part of the face transplant. On a macrovascular level, the course of the artery must be known so the dissection of the artery does not put the vessel itself in jeopardy. In addition, the course of the artery will, to an extent, determine which tissues can be carried with each vessel. The microvascular anatomy must be known to understand to what extent the two blood vessel will perfuse beyond the macrovascular angiosome.

Equally important one must also understand which vessels can be sacrificed in the recipient without jeopardizing the remaining tissues of the head and neck of the patient. Because of this, it is useful to view the microsurgical anatomy of facial transplant from the viewpoint of both the donor and the recipient.

35.2.2 Arteries

35.2.2.1 Common Carotid

A branch of the subclavian artery on the right and the aortic arch on the left, the common carotid arteries ultimately supply the entire face with blood. A dissection and harvest to this level could be used on the donor to gain extra length. If this is the level ultimately chosen for the donor vessels, the diameter of the common carotid needs to be taken into consideration. If the mismatch between donor and recipient vessel sizes is too great, the complexity of the operation increases as well as the potential for complication.

From the standpoint of the recipient, the common carotid is not a practical donor due to the risk of cerebral vascular accident or necrosis of remaining tissue (Fig. 35.1a, b).

35.2.2.2 Internal Carotid

The vertebral and internal carotid arteries play no role in facial transplantation. Due to rich collateral blood flow interconnection with the superficial vascular system, all conceivable transplants can be performed without counting on the terminal branches of the internal carotid artery: the supraorbital artery, supratrochlear, and dorsal nasal arteries.

35.2.2.3 External Carotid

The external carotid artery and its branches are the workhorses of facial transplantation. Taken as a whole, the entire face scalp maxilla and mandible will survive based on the external carotid arteries and its branches. The following is a description of the major branches and their contribution to face transplantation. The arteries are listed in order of branching caudal to cranial.

Superior Thyroid

The superior thyroid artery is the first branch of the external carotid artery. The course of the artery follows an acute turn in a caudal direction after its takeoff from the trunk before entering the superior part of the thyroid gland. Although this artery does not play a role in facial transplantation from a donor standpoint, this can be used as a recipient vessel.

Ascending Pharyngeal

This small branch parallels the internal carotid artery and supplies the pharynx and base of the skull. It has

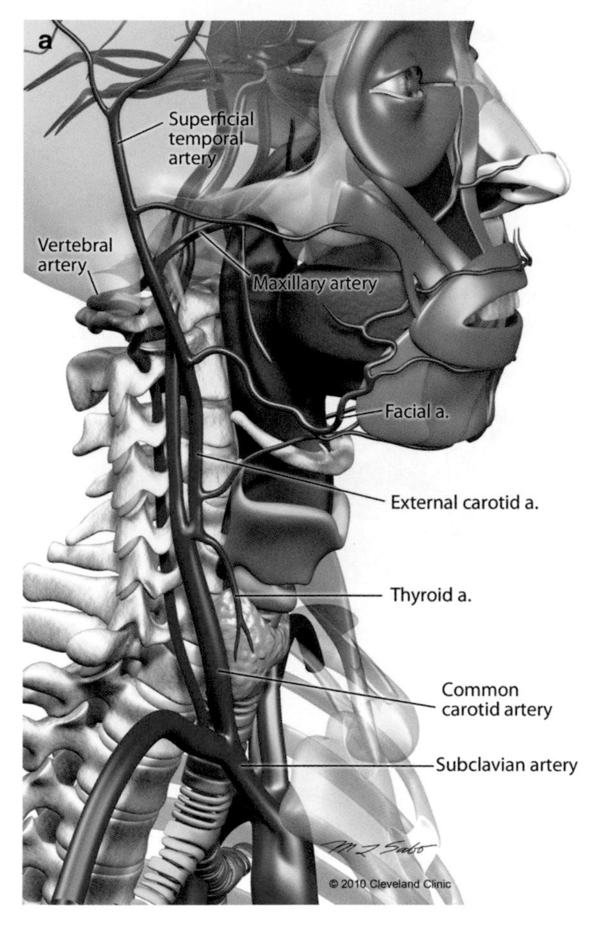

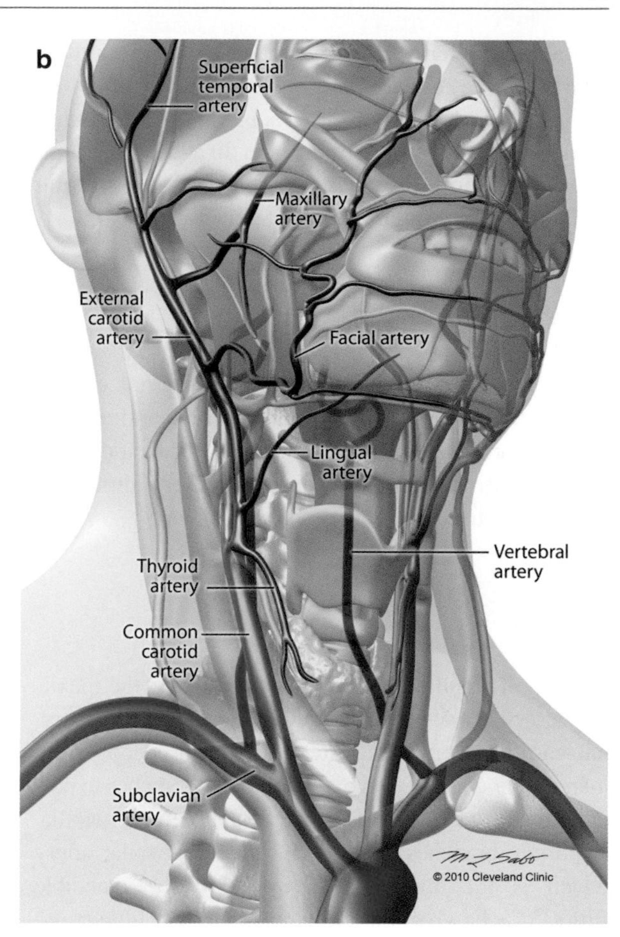

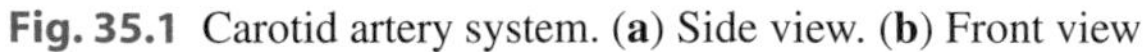
Fig. 35.1 Carotid artery system. (**a**) Side view. (**b**) Front view

no practical use in facial transplantation and all territories perfused by this artery are well perfused by others listed.

Lingual

The lingual artery is the direct blood supply to the tongue. The tongue itself is one of the structures that has poor cross perfusion between the left and right sides.[5] If the tongue is going to be included in the transplantation, the lingual artery should be included on both sides to ensure adequate perfusion.

Occipital

The next branch after the lingual artery is the occipital artery. This is the major branch of the external carotid artery with a posterior course. This artery terminates in the dorsal scalp and provides perfusion to both the ear and posterior scalp. Use of this vessel should be considered when these two structures are to be included in the transplant.

Facial

Also known as the external maxillary artery, the facial artery takes a tortuous course through the neck and into the face. From its takeoff, the artery runs deep to the margin of the mandible, traverses the submandibular gland, and emerges under the mandible at the mandibular notch to give off multiple important branches pertaining to facial transplantation. Lateral to the commissure, the facial artery gives off the submental artery and the superior and inferior labial arteries. These arteries form a circuit around the mouth and allow for excellent perfusion of the opposite side of the face due to the crossover. More cranially, the facial yields the lateral nasal artery (forming another circuit around the tip of the nose) and terminates in the angular artery on the sidewall of the nose. The facial artery is the main blood supply to the skin of the face. In addition, its rich collateral supply provides blood to both the maxilla and mandible through periosteal perforators.[5,6]

Posterior Auricular

A small branch directly off of the external carotid artery, the posterior auricular artery parallels the occipital artery and supplies the lateral scalp and external ear with blood. If possible, this artery should be included in transplants that contain the ear.

Maxillary

Also known as the internal maxillary artery, the maxillary artery is one of two terminal branches of the external carotid artery. The artery itself runs deep to the mandibular condyle and gives off many branches to portions of the face as well as the mandible and maxilla. Important branches include the inferior alveolar artery which is the main blood supply to the mandible (Fig. 35.2). Ex vivo studies demonstrate very little crossover between the left and right sides of the mandible and maxilla when considered from the standpoint of the inferior alveolar artery or maxillary artery.[6,7] This would be of clinical significance if the mandible or maxilla themselves were being used as a graft without the remaining face (skin and subcutaneous tissue) available for periosteal supply. The lack of crossover blood flow would make bilateral perfusion a recommended technique.

The second part of the maxillary artery yields branches to the muscles of mastication. These include branches to the pterygoids, masseter, temporalis, and buccinators muscles. Finally the artery itself terminates in branches to the palette including the greater palatine artery and posterior septal branches of the nose and pharynx.

Superficial Temporal

The other terminal branch of the external carotid artery is the superficial temporal artery. This artery begins in the substance of the parotid gland and heads in a cranial direction superficial to zygomatic arch to divide into frontal and parietal branches. This artery can supply blood to the scalp, forehead, ear, and lateral cheek.

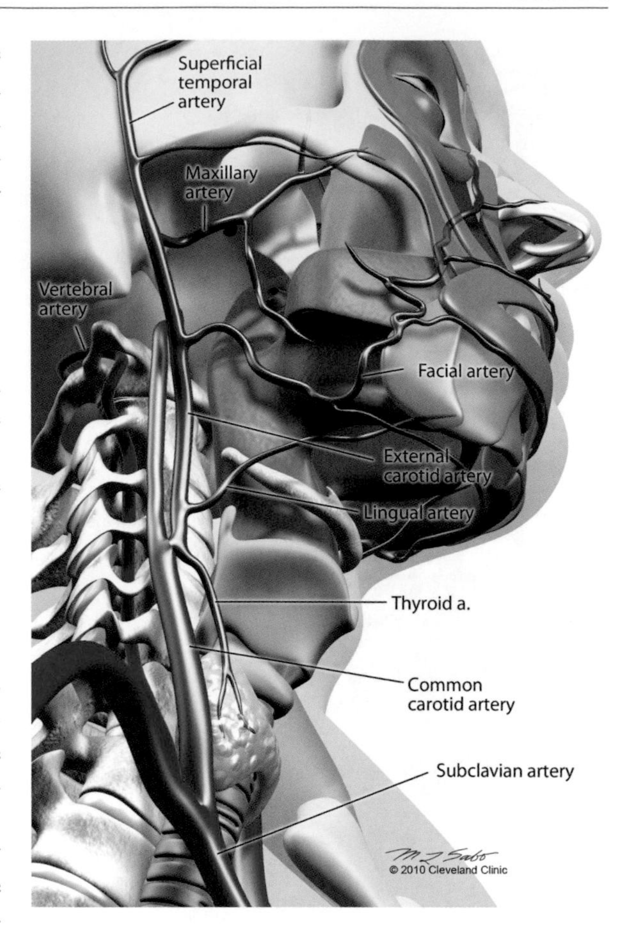

Fig. 35.2 Maxillary artery

35.2.3 Veins

35.2.3.1 Facial Vein

For the most part, the veins of the head and neck parallel the major arteries; however, there are important exceptions to this rule. One exception is the facial vein. This vein tends to run posterior to the artery and follows a much straighter course. This becomes important when dissecting out the facial artery near the margin of the mandible because dissecting too close to the artery itself may cause separation and damage to the facial vein. The facial vein itself is adequate to drain a facial transplant; however, it should be noted that in one clinical case of traumatic amputation and replantation, it was necessary to anastomose the contralateral facial vein.[8] In this case, one artery was sufficient to perfuse the entire face; however, post replantation, the contralateral face was noted to suffer venous congestion and it was necessary to anastomose a vein from the opposite side of the face to relieve that congestion (Fig. 35.3).

35.2.3.2 External Jugular Vein

Another important vein for facial transplantation is the external jugular vein. The external jugular is part of the superficial venous system of the face and its relationship to the internal jugular vein would be akin to the cephalic vein's relationship to the venae comitantes of the radial artery in a radial forearm free flap. It can help with venous drainage of the transplant on both the donor and recipient sides of the transplant.

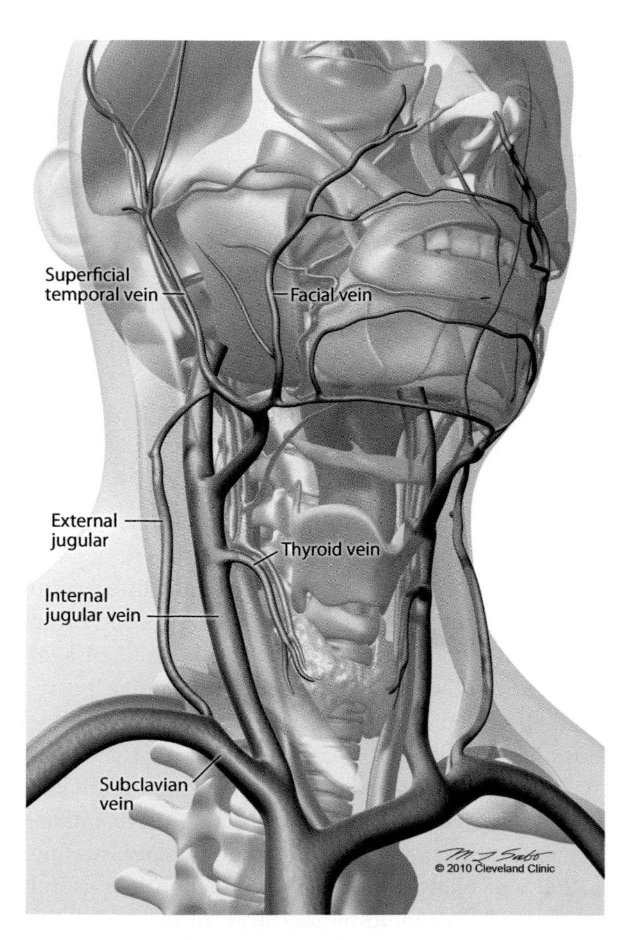

Fig. 35.3 Venous system of the head and neck

35.2.3.3 Internal Jugular Vein

Another option for facial transplant drainage is the internal jugular vein. This vein is readily available within the dissection and is often the only vein of significant and consistent size remaining. If it is to be used, the anastomosis is performed end to side to allow for both the inevitable size mismatch with the donor vein and to allow for adequate drainage in the remaining head and neck of the recipient. A side biting clamp can be used during the anastomosis to accomplish both goals.

35.2.3.4 Cephalic Vein

Although any vein of adequate size can be used for drainage, local veins are often missing from previous reconstruction attempts. If no local veins are available, work done for general facial reconstruction suggests the cephalic vein can be dissected from the deltopectoral groove to the level of antecubital fossa and transposed into the neck of the recipient for venous drainage.[9]

35.2.4 Nerves

Nerves for both motor and sensory function need to be considered in facial transplantation. Ideally both would be restored in all cases. The reality of facial transplantation is such that recipient nerves are often destroyed beyond the point of repair due to previous treatment of tumors or directly from trauma. Even when repair is possible, grafting will often be necessary as the lengths of both donor and recipient nerves are limited by the short distance from foramina to the substance of the transplant. From a motor standpoint, cranial nerve seven or the facial nerve is important for facial animation, while the muscles of mastication are controlled by cranial nerve five or the trigeminal nerve. The trigeminal nerve is of course the most important sensory nerve of the face as well.[1]

35.2.4.1 Trigeminal

Sensory

The trigeminal nerve is the largest cranial nerve and emerges from the base of the skull through multiple foramina including the superior orbital fissure for the ophthalmic sensory branch (V1), the foramen rotundum for the maxillary nerve (V2), and the foramen ovale for the mandibular branch (V3) (Fig. 35.4).

The ophthalmic branch proceeds along the superior part of the orbit to emerge into the subcutaneous tissue at the supraorbital foramen. This branch is responsible for sensation to the forehead and vertex of the scalp.

The maxillary nerve splits into multiple branches and is responsible for sensation to the midportion of the face including the lower eyelid, lower nose, upper lip, maxillary teeth, and gums. Although it emerges from multiple maxillary foramina, the major branch is the infraorbital nerve emerging below the orbit. It is this nerve that should be connected in the transplant to gain maximal sensation to the midface. Dissection through the floor of the orbit on the donor yields additional length for anastomosis to the remaining stump of the recipient.

The inferior-most branch of the trigeminal nerve is the mandibular branch giving sensation to parts of the tongue, lower face, and mouth. The inferior alveolar nerve follows the course of the inferior alveolar canal through the mandible itself to emerge from the mental foramen terminating in the mental nerve and yielding sensation to the chin. This nerve should be reconnected in a cutaneous transplant.

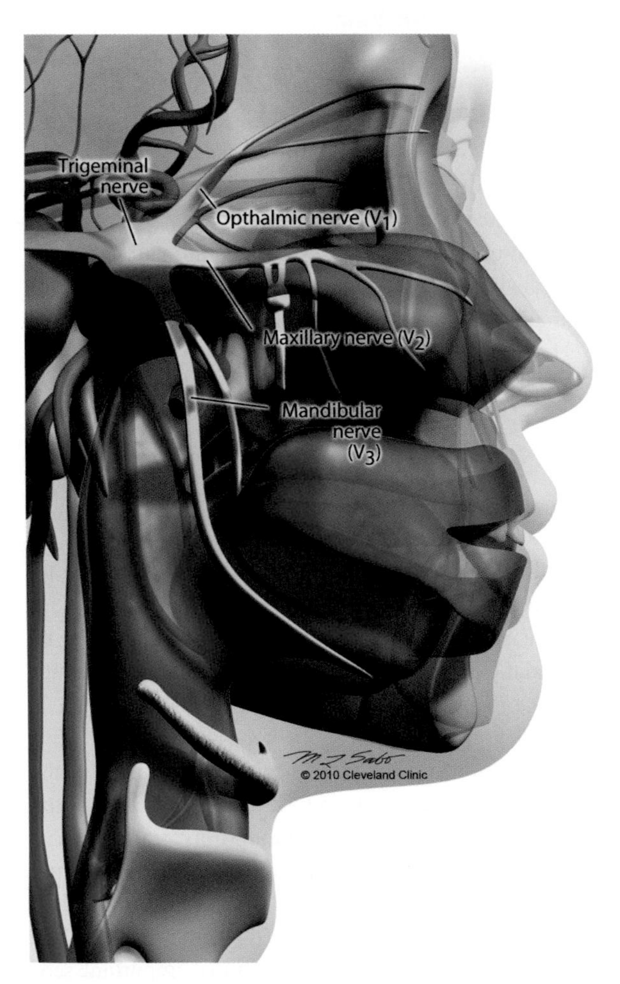

Fig. 35.4 Trigeminal nerve anatomy

Motor

Motor branches of the trigeminal nerve follow their own course to the muscles of mastication (masseter, temporalis, medial and lateral pterygoids) as well as several involved in speech and swallowing (tensor veli palatini, mylohyoid, anterior belly of the digastric, and the tensor tympani). This is an area yet to be defined by current techniques in facial transplantation and will present unique difficulties in total mandible replacement.

35.2.4.2 Facial

The facial nerve emerges from the skull through the stylomastoid foramen. The preauricular dissection technique used to find the nerve is no different in facial transplantation than for any other procedure and begins with identification of the trunk deep the tragal cartilage. On the donor side, as much length as possible is dissected in order to ease the eventual anastomosis. If possible, recipient branches that are still functioning should be left intact, but function of the transplant is ultimately more important than maintaining innervations to muscles that are already missing or that will likely be too deep to function in a useful manner. These should be sacrificed to maximize the function of the graft.

35.2.4.3 Innervation of the Tongue

Innervation of the tongue deserves special consideration. Full sensory and motor reanimation of the tongue would require coapting the lingual nerve (branch of the trigeminal nerve providing sensation to the anterior 2/3 of the tongue), chorda tympani (branch of cranial nerve seven supplying taste to the anterior 2/3 of the tongue), glossopharyngeal nerve (cranial nerve nine supplying both sensation and taste to the posterior 1/3 of the tongue), and finally the hypoglossal nerve (cranial nerve 12 providing most of the motor supply). From a practical standpoint, all of these would be difficult due to their deep location and short leash. It would be a great help to the patient if the hypoglossal and lingual nerves could be repaired providing some motor and sensation, but this awaits further study.

35.3 Facial Transplants Subunits

All reconstructions ultimately come down to defining what is missing both anatomically and functionally and coming up with the most accurate replacement. Facial transplantation finally gives us the ability to accurately and precisely replace the functions and esthetics of the face. To decide how to proceed from a microsurgical standpoint, we need to completely define the missing structures in the recipient's face. Once defined, we can decide what is necessary to take from the donor to replace these tissues. It is helpful to compartmentalize the face into functional subunits when assessing the blood vessel needs of the recipient. From a microsurgical standpoint, the subunits are defined by vascular territories. There has been a great deal of research devoted to the vascular territories of the face.[5-7,10,11] Initially, angiosomes were defined and assumptions made about what tissues might survive on each of the specific arteries outlined earlier in this chapter. The authors were able to make the determination that most of the soft tissues of the face could survive on the facial and superficial temporal arteries.[5] Concern was expressed for blood flow to the underlying structure such as the bone, however, as no direct connection could be found. Likewise, concern was expressed about the blood flow crossing the midline in various parts of the face, in particular of the forehead. Since the time of that early research, we have the experience of multiple facial transplantations that have all been based on one or both of the facial arteries. These have shown adequate if not excellent perfusion to all aspects of the transplants. In particular it should be noted that even though no direct branch from the facial artery to the maxillary bone exists, the posterior aspect of the palette and the maxilla included in some of the transplants demonstrate excellent perfusion with bleeding along the cut bony surfaces. The same has held true for the ends of the zygomatic arch, another area of concern from the standpoint of perfusion through the facial artery and ultimately the periosteal blood supply. Most recently, Cavadas et al. performed a transplantation of the lower face including the lower lip, tongue, and mandible.[12] This transplant was perfused by both the facial artery and lingual branches. The flap has shown excellent viability based on these vessels.

It may ultimately prove that the facial artery alone will perfuse all conceivable facial transplants; however, there are areas of perfusion beyond traditional angiosomes that remain unproven. The forehead and scalp may require the superficial temporal artery to maintain adequate blood supply while the ear and posterior scalp may require the occipital and postauricular arteries. To date the only experience with the forehead tissue is that of replantation in traumatic cases and these have been based on the superficial temporal artery.

35.3.1 Skin Only

With the subunit principle in mind the previous experience suggests that the facial artery will carry the entire face for soft tissue only reconstruction. This simplifies the microvascular dissection on both the donor and recipient. From the standpoint of the donor, the external carotid artery can be ligated above the level of the facial artery speeding up that dissection considerably. On the recipient's end, an anastomosis at the level of the facial artery allows the remaining external carotid artery to perfuse any tissues remaining in the recipient's face. Each higher level must be carefully assessed for maintenance of viability to the structures those arteries perfuse. As an example, sacrifice of the lingual artery on both sides may cause necrosis of the tongue. Sacrifice of the entire external carotid artery risks perfusion to the maxilla, mandible, and the face itself.

With this in mind, sacrificing both external carotid arteries as recipient vessels should be avoided if at all possible as collateral flow will have to come from the internal carotid system and vertebral arteries.

35.3.2 Mandible Only

Including the mandible in a facial transplantation is complicated because of the need to chew. The motor branch of the trigeminal nerve to the muscles of mastication is short and located medial to the mandible itself making access extremely difficult. In addition, if no skin is included, the blood supply to the mandible is via the maxillary artery which is deep and more difficult than the remaining arteries to dissect out. We will anxiously await the outcome of the Spanish experience to determine the ultimate function for mandibular reconstruction. The mandible is unlike the skin of the face and little crossover blood supply from left to right; therefore, both vascular systems should be transferred to maintain adequate blood supply. Cavadas et al. accomplished this by including the facial artery and performing a crossover anastomosis where the left external carotid artery was sutured end to side into the right external carotid artery.[12] The donor right internal carotid was ultimately connected to the subclavian artery of the recipient.

35.3.3 Maxilla at the Le Fort III Level

When skin is included with a Le Fort III bony subtotal facial transplant, it appears that there will be adequate perfusion through periosteal supply to all aspects of the bone and the facial artery alone is sufficient for this transplant.[13] If the bone itself is going to be transplanted without the skin, the maxillary artery will need to be included with the specimen to guarantee adequate blood flow.

35.3.4 Total Face

Total facial transplant at this time remains an unknown. Inclusion of the maxilla, the mandible, as well as the overlying facial skin and scalp may well be adequately perfused by the facial artery system alone; however, to be safe, inclusion of the internal maxillary arteries as well as the superficial temporal arteries will yield a less precarious result.

35.4 Influence of Previous Reconstructive Efforts

35.4.1 Primary Versus Secondary

35.4.1.1 Primary Defect

A primary recipient would be a recipient in which no previous attempts at reconstruction have been made and therefore untouched from a macrovascular standpoint. An example of this would be a patient with a benign disfiguring tumor such as neurofibromatosis who has not previously undergone reconstructive procedures. Given this parameter, all recipient vessels should be available for the reconstruction and there should be significant redundancy in the remaining vessels to allow perfusion of the recipient's remaining head and neck. This greatly simplifies access to the recipient blood vessels.

On the donor's side, inclusion criteria would mandate a pristine neck devoid of previous surgery as a preoperative workup with sophisticated radiologic studies cannot be performed in a timely fashion.

35.4.1.2 Secondary Defect

As a result of the morbidity of immunosuppression, current ethics of facial transplantation suggest that conventional efforts at reconstruction should be attempted prior to proceeding with transplantation. Given this fact, it is likely that most candidate patients will have had previous neck surgery in an attempt to reconstruct their defects. It is likely that several branches of the external carotid have been dissected out and used for previous free flaps. With that in mind, the preoperative workup of the candidate becomes even more critical. Preoperative imaging of the recipient's vascular network is necessary. The Cleveland Clinic experience suggests that 3D CT angiograms

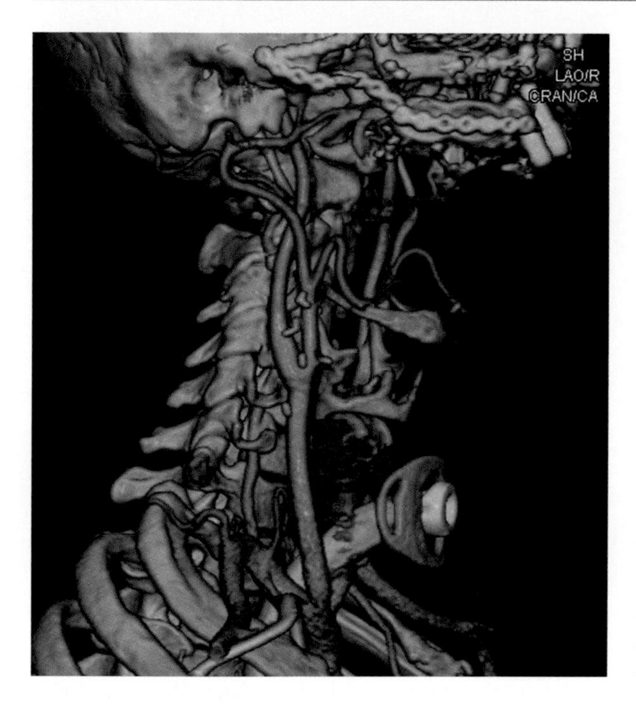

Fig. 35.5 3D CT Angiogram showing the arterial system in a face transplant candidate

have advanced enough to allow visualization of vessels down to 1–2 mm in size which is in fact adequate for visualization of recipient vessels that should be at least 2 mm in size (Fig. 35.5). The consideration of using both the left and right vessels will have to be made preoperatively as well.[3,4]

Taking into account the fact that structures remaining in the recipient patient must be perfused, the preoperative imaging will also suggest which vessels and at what level the vessels can be used for anastomosis to the face transplant. Tissues that have poor communication between the blood vessels from the left to the right side of the face or those that do not have a redundant supply from a separate arterial system are at risk if the feeding vessel is used as a donor vessel in the microvascular anastomosis. As previously stated, the tongue is an example where blood supply comes from the paired labial arteries and there is poor crossover from left to right. As a result, if there is damage to or loss of the labial artery, the ipsilateral side of the tongue is at risk for necrosis.

The imaging can also alert the transplant team to a situation where no viable recipient vessels exist in the neck and more creative solutions will be required. Ideally, the external carotid artery and facial vein of the donor would be anastomosed to the recipient's facial artery and external jugular vein. If the recipient does not have viable branches available on the external carotid, the next choice would be to perform the anastomosis to the external carotid itself either end to end or end to side. If the external carotid is not available, consideration could be given to an end-to-side anastomosis to the common carotid; however, cross clamping the carotid will add the risk of cerebral vascular accident if the Circle of Willis is inadequate. Finally, if all other possibilities are exhausted, the donor vessel can be taken at the level of the common carotid and anastomosed end to side to the subclavian artery.

On the venous side, the recipients include the external jugular vein, internal jugular vein (end to side), and the cephalic vein harvested through incisions in the arm. The team must also be prepared to use arterial and vein grafts if necessary for both anastomoses. The donor for the graft can either be the donor or recipient themselves.

35.5 Technique

The steps of harvesting the donor should be worked out prior to the actual procedure. From a microsurgical standpoint, adequate exposure of the blood vessels and nerves is required for transplant success. The initial maneuver would be to split the skin and the midline of the neck, elevate the lower neck skin just below the level of the platysma, and remove sternocleidomastoid muscles. This should yield good exposure to the carotid system (Fig. 35.6). All of the external planned incision should be made in the skin to allow as much exposure as possible. When gaining this exposure, as much periosteum as possible should be left intact to maintain the viability of the bone. The course of the facial artery and vein should be followed up to the level of the submandibular gland. At this point the dissection should go deep to the submandibular gland, therefore including the gland in the flap to protect the facial artery. From this point, the dissection of the facial artery should include a cuff of periosteum near the border of the mandible so as not to injure the vessels as they traverse the mandible itself. Appropriate mucosal incisions need to be made as well. If the remaining branches of the external carotid artery are deemed to be unnecessary, the artery itself can be ligated above

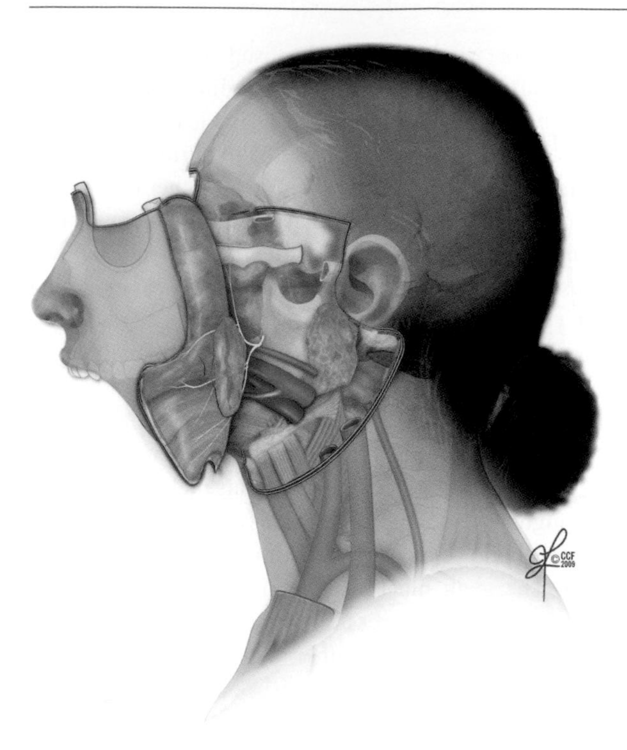

Fig. 35.6 Le Fort III osteotomies and mobilization of allograft

the level of the facial artery. As the dissection proceeds posterior to anterior, the facial nerve is identified and followed into the parotid gland. The superficial parotid gland is included with the transplant to protect the deep facial blood supply and facial nerve, while the recipient's parotid gland should be removed to gain access to the facial nerve. The parotid duct is ligated along its deep margin of the gland in the donor.

Sensory nerves are dissected down the level of the foramina with osteotomies performed in the donor to gain length if the bone itself is not needed for the transplant.

Once all the blood vessels and nerves have been identified and dissected as much as possible, the Le Fort osteotomies can be performed. The final step is to divide the donor vessel and harvest the face for transfer to the recipient. At this point, the flap should be stabilized through either fixation of the bone or suspension of the soft tissues to the bone and partial inset. It may be necessary to perform the nerve repairs or grafts at this time if they are deeper than the proposed transplant blood vessels. The arteries and veins should now be anastomosed. One side (artery and vein) should be completed and released to minimize ischemia time prior to completion of the opposite side. With revascularization complete, the final inset can be performed with trimming of excess donor tissue and repair of any remaining nerves.

35.5.1 Spare Donor Parts

Prior to leaving the operating room, consideration should be given to possible spare tissues required that can be harvested from the donor. These would include bone graft, skin grafts, cutaneous free flaps such as the radial forum, tendons to be used as static slings, nerves for nerve grafting.

35.6 Conclusion

Facial transplantation is in its infancy, but even with just a few cases to review, we can draw some conclusions about the microsurgical aspects of facial transplantation. The facial artery alone can perfuse the majority of conceivable facial transplants. We also know that coaptation of the facial nerve will result in useful function in the donor musculature. Likewise, coaptation of the branches of the trigeminal nerve results in sensation in the graft, but even without specific sensory nerve repair, sensation will return to portions of the face to a degree.

Other aspects of face transplantation remain unknown to be defined in the future. Previous research suggests that one vessel may be adequate to perfuse all soft-tissue portions of the face; however, when bone alone such as just the maxilla or mandible is going to be transplanted, bilateral perfusion is likely to be necessary. Another unknown is exactly how much of the recipient's external carotid system can be sacrificed before there is risk of the schema and loss of the recipient own tissues. These questions will all be answered as face transplantation becomes more common, but others will arise.

References

1. Alam DS, Papay F, Djohan R, et al. The technical and anatomical aspects of the world's first near-total human face and maxilla transplant. *Arch Facial Plast Surg*. 2009;11:369-377.

2. Siemionow M, Papay F, Bernard S, et al. Near-total human face transplantation for a severely disfigured patient in the USA. *Lancet*. 2009;374:203-209.
3. Siemionow M, Agaoglu G, Unal S. A cadaver study in preparation for facial allograft transplantation in humans: part II. Mock facial transplantation. *Plast Reconstr Surg*. 2006;117: 876-885.
4. Siemionow M, Unal S, Agaoglu G, Sari A. A cadaver study in preparation for facial allograft transplantation in humans: part I. What are alternative sources for total facial defect coverage? *Plast Reconstr Surg*. 2006;117:864-872.
5. Houseman ND, Taylor GI, Pan WR. The angiosomes of the head and neck: anatomic study and clinical applications. *Plast Reconstr Surg*. 2000;105:2287-2313.
6. Banks ND, Hui-Chou HG, Tripathi S, et al. An anatomical study of external carotid artery vascular territories in face and midface flaps for transplantation. *Plast Reconstr Surg*. 2009;123:1677-1687.
7. Pomahac B, Lengele B, Ridgway EB, et al. Vascular considerations in composite midfacial allotransplantation. *Plast Reconstr Surg*. 2010;125:517-522.
8. Wilhelmi BJ, Kang RH, Movassaghi K, Ganchi PA, Lee WP. First successful replantation of face and scalp with single-artery repair: model for face and scalp transplantation. *Ann Plast Surg*. 2003;50:535-540.
9. Takamatsu A, Harashina T, Inoue T. Selection of appropriate recipient vessels in difficult, microsurgical head and neck reconstruction. *J Reconstr Microsurg*. 1996;12:499-507. Discussion 508-13.
10. Mathes D. Discussion: vascular considerations in composite midfacial allotransplantation. *Plast Reconstr Surg*. 2010;125: 523-524.
11. Siebert JW, Angrigiani C, McCarthy JG, Longaker MT. Blood supply of the le fort I maxillary segment: an anatomic study. *Plast Reconstr Surg*. 1997;100:843-851.
12. Cavadas PC, et al. The Spanish experience with face transplantation. In: Siemionow M, ed. *The Know-How of Face Transplantation*. Springer; 2010.
13. Pomahac B, Aflaki P, Nelson C, Balas B. Evaluation of appearance transfer and persistence in central face transplantation: A computer simulation analysis. *J Plast Reconstr Aesthet Surg*. 2010;63:733-738.

36 The Sensory Recovery in Face Transplantation

Bahar Bassiri Gharb, Antonio Rampazzo, and Maria Z. Siemionow

Contents

M.Z. Siemionow (✉)
Department of Plastic Surgery, Cleveland Clinic, Cleveland, OH, USA
e-mail: siemiom@ccf.org

Abstract Recovery of normal function in face transplantation is fundamental to justify the necessity for lifelong immunosuppressive therapy. However, extensive soft tissue damage and scarring in face transplant patients has often hampered the repair of the sensory nerves. Nonetheless, it seems that near full return of sensation has been achieved in these patients. In this chapter we assessed the sensory outcome in face-transplanted patients and investigated the factors which could have impacted the final result. The results were compared to sensory return following replantation of face and scalp, repair of divided sensory nerves of the face, and in innervated and noninnervated vascularized free flaps used for head and neck reconstruction. Sensory recovery following face transplantation, even when the sensory nerves were not repaired, showed results comparable or superior to free autologus innervated tissue. Results were also comparable with the outcome of the microsurgical repair of the peripheral branches of the trigeminal nerve. Therefore, near normal sensory recovery can be expected following facial allotransplantation. Restoration of normal end organ receptors within the facial allograft, repair of the facial nerve, and immunosuppressive therapy with FK506 probably affect and accelerate the final outcome. We suggest a guideline on quantitative sensory testing and timing of the follow-up to allow comparison of results between different centers and improve our understanding of the mechanisms of sensory recovery in face transplantation.

36.1 Introduction

Composite tissue allotransplantation, in the recent years, has offered the possibility to restore optimally the lost facial anatomy and function in patients with

M.Z. Siemionow (ed.), *The Know-How of Face Transplantation*,
DOI: 10.1007/978-0-85729-253-7_36, © Springer-Verlag London Limited 2011

massive disfiguring defects. The ability to achieve the recovery of fine movements of facial expression is fundamental to obtain sphincterial control, avoiding ectropion and drooling, and to aid regular speech; while a normal sensation is essential to interact with the environment, to start defense reactions, to avoid drooling, and to draw pleasure and satisfaction from the external stimuli, which aid to perceive the face transplant as an integral part of the body rather than a foreign mask. The motor and sensory pathways of the face interact together closely and this interaction is fundamental to warrant a normal function. Although the recovery of the motor function of the face has been amply studied, the sensory return and its mechanisms have not been investigated with the same attention. The first near-total face transplantation was performed in Cleveland Clinic, in December 2008.[1,2] Due to extensive soft tissue loss and scarring following gunshot injury to the patient's mid-face, the infraorbital nerves, which represented the source of the main sensory input of the facial graft, were absent and thus the repair was not feasible. However, during the subsequent follow-up, the patient showed a steady and progressive recovery of sensation in all tested modalities, which was surprising and unexpected, as in our experience the recovery of sensation in non-innervated free flaps is slower, unpredictable, and full sensory return is not expected.

This chapter focuses on the sensory recovery in all of the reported cases of facial allotransplantation and comparing the results to three clinical situations which could help to obtain further insight about the mechanisms of sensory recovery in face transplantation: scalp and face replantation, repair of the severed peripheral branches of the trigeminal nerve, and free flaps commonly used for head and neck reconstruction. In order to compare the outcome of sensory recovery in the different clinical situations, the sensory return was graded using the Medical Research Council (MRC) Scale as modified by Mackinnon and Dellon[3] based on the evaluation of light touch, pain, and two-point discrimination (Table 36.1).

Table 36.1 Medical Research Council (MRC) Scale as modified by Mackinnon and Dellon[3]

Score	Interpretation
S0	No recovery
S1	Recovery of deep cutaneous pain
S2	Return of some superficial pain/tactile sensation
S2+	Return of some superficial pain/tactile sensation with overreaction
S3	Return of some superficial pain/tactile sensation without overreaction and the presence of static two-point discrimination (2pd) > 15 mm
S3+	As per S3 with good localization of stimulus, 2pd = 7–15 mm
S4	As per S3+, 2pd = 2–6 mm

36.2 Sensory Recovery in Facial Transplantation

Patients' data are summarized in detail in Table 36.2. It is evident that both mental and infraorbital nerves were repaired directly only in one patient.[4,5] In one patient repair of the sensory nerves was complicated by shortness of the recipient's nerve stumps: repair of the infraorbital nerves required fibrin glue and the mental nerves were not repaired.[6] In two patients repair of infraorbital nerves was not feasible.[1,7]

The first patient who underwent face transplantation[4,5] received early sensory reeducation and followed a cortical reintegration protocol (Fig. 36.1). The initial signs of sensory reinnervation appeared at 2 weeks for thermal stimuli. Sensory discrimination started at the lateral part of the upper lip and lateral area of the chin on both sides after 10 weeks. At 14 weeks, the sensation returned over the whole facial graft including the tip of the nose. At the same time the return of sensation to painful stimuli was registered. After 6 months, in the upper half of the flap the pressure thresholds were normal, while in the lower half the patient reported diminished light touch. Heat and cold sensation was nearly normal at 4 months and normal at 6 months over the entire graft.

In the Chinese face transplant patient[7] the sensory discrimination was restored at 3 months (Fig. 36.2). Heat and cold sensations over the whole graft returned at 8 months after transplantation.

The second French patient[6] showed return of thermal and mechanical sensation 3 months after surgery, with improvement of the sensory thresholds at 12 months (Fig. 36.3).

Our patient at Cleveland Clinic[1,2] recovered the discrimination of pain over a period of 5 months, and at 6 months the sensation returned to the entire transplanted face with an average two-point discrimination of 7 mm (S3+) (Fig. 36.4).

Table 36.2 Summarized clinical data of the 4 face transplant patients with documented long term follow-up

Recipient			Donor			Repaired structures			Immunosuppression	Sensory nerve recovery		
Age, sex	Tissue loss/ cause	Status previous to operation	Age, sex	CTA flap composition	HLA antigens/ rejections episodes	Vessels	Motor nerves	Sensory nerves	Drugs	Postoperative sensory reeducation	Follow-up, function	Methods utilized
38, woman[4,5]	Distal nose, full thickness upper and lower lips, chin, adjacent parts of right and left cheek/ dog bite	Preserved integrity of the proximal stumps of the zygomatic and levator anguli oris muscles; mouth opening 19 mm, surgical delay 6 months; no previous reconstruction	46, woman	skin, subcutaneous tissue, perioral muscles, mucosa of the oral and nasal vestibules, alar and triangular cartilages of the nose, anterior part of the septum	Five HLA sites matched/ day 24	Artery: right and left facial arteries end to end Vein: right facial veins end to end	Facial nerve: mandibular branch (left side) end to end; right side unrepaired because of scar	Mental nerves: right and left sides end to end Infraorbital nerve: right and left sides end to end	Thymoglobulin, Tacrolimus, Mycophenolate mophetil, Prednisone	Yes	2.5 months: lateral part upper lip, lateral mental area 3.5 months: whole skin surface including tip of nose 4 months: heat and cold sensation almost normal (normal at 6 months)	Pressure thresholds (Semmes-Weinstein test) Heat and cold sensation
30, man[7]	Extensive skin and soft tissue loss in the right buccal division, upper lip, total nose, front wall of the right maxillary sinus, lateral right orbital wall, infraorbital wall, right zygomatic bone and large portion of right parotid gland/ Bear attack	Severe cicatricial contracture deformity; First reconstruction radial forearm free flap, waited 17 months before definitive reconstruction	25, male	Parotid gland, partial buccal mucosa, partial masseter, partial zygomatic arch, the lateral orbital wall and infraorbital wall, the front wall of the maxillary sinus, total upper lip, total nose, nasal septal cartilage and nasal bone	Three HLA sites matched 3, 5, 17 months, eventually died due to not compliance to immuno-suppressive protocol	Artery: right external maxillary artery end to end Vein: right anterior facial vein end to end	Facial nerve: repair reported not satisfactory	Infraorbital nerve: not repaired	Tacrolimus, Mycophenolate mophetil, Prednisone, Humanized IL-2 receptor monoclonal antibody	ND	3 months: sensory discrimination, 8 months: heat and cold sensation	Pressure thresholds (Semmes-Weinstein test) Heat and cold sensation

(continued)

Table 36.2 (continued)

Recipient			Donor			Repaired structures			Immunosuppression	Sensory nerve recovery		
Age, sex	Tissue loss/ cause	Status previous to operation	Age, sex	CTA flap composition	HLA antigens/ rejections episodes	Vessels	Motor nerves	Sensory nerves	Drugs	Postoperative sensory reeducation	Follow-up, function	Methods utilized
29, male[6]	Massive plexiform neurofibroma middle and lower part of face	Complete facial paralysis (right side), partial paralysis (left side); Immediate recontruction	ND	Oral mucosa and skin middle and lower third of face, both parotid glands	Three HLA sites matched/ day 28, day 64, persistent mucosal grade 1 rejection	Artery: left and right external carotid artery end to end Vein: bilateral end to end to thyrolin-gualfacial trunk	Facial nerve: both nerves sutured and glued	Infraorbital nerve: both nerves sutured and glued Mental nerve: not repaired but just placed in front of the foramen	Thymoglobulin, Tacrolimus, Mycophenolate mophetil, Prednisone	ND	3 months: sensory reinnervation of skin for thermal, mechanical sensation 4 months: pain sensation 12 months: improvement in sensory thresholds	Not reported
45, female[1,2]	Absence of nose, nasal lining and underlying bone, contracted remnants of upper lip, loss of orbicularis oris and orbicularis oculi muscle functions, distorted and scarred lower eyelids Close-range shotgun blast	23 previous reconstructive procedures, Surgical delay to face transplanta-tion:51 months	ND	Le Fort III CTA containing total nose, lower eyelids, upper lip, total infraorbital floor, bilateral zygomatic bones, anterior maxilla with incisors, including total alveolus, anterior hard palate and bilateral parotid glands	Two HLA sites matched-day 47 (grade III/ IV mucosa)	Artery: common facial artery (right and left),end to end Vein: external jugular vein (right and left) end to end; left posterior facial vein end to end	Facial nerve: interposi-tional nerve graft from donor vagus nerve (right side); interposi-tional nerve graft from donor hypoglossal (left side),	Infraorbital nerve: not available for repair	Thymoglobulin, Tacrolimus, Mycophenolate mophetil, Prednisone	Yes	5 months: sensation to pinprick 6 months: sensory discrimination over entire flap	Pressure thresholds (Pressure-specified sensory device)/ Two-point discrimina-tion (Disk-Criminator)

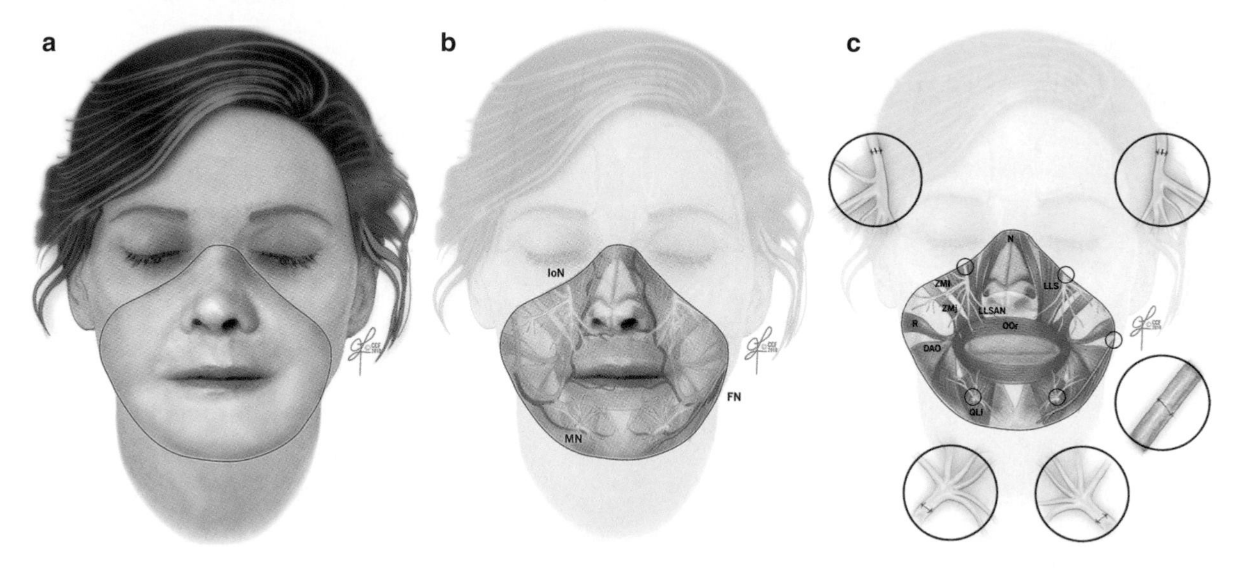

Fig. 36.1 (**a-c**) First face transplant performed in France in 2005.[4,5] (**a**) Outline of the facial allograft. (**b**) Infraorbital, mental and the left mandibular branch of the facial nerve were repaired; *IoN* infraorbital nerve, *MN* mental nerve, *FN* facial nerve. (**c**) In the *circles* the modality of nerve repair is shown. *DAO* depressor anguli oris, *LLS* levator labii superioris, *LLSAN* levator labii superioris alaeque nasi, *N* nasalis, *OOr* orbicularis oris, *QLI* quadratus labii inferioris, *R* risorius, *ZMI* zygomatic minor, *ZMj* zygomatic major

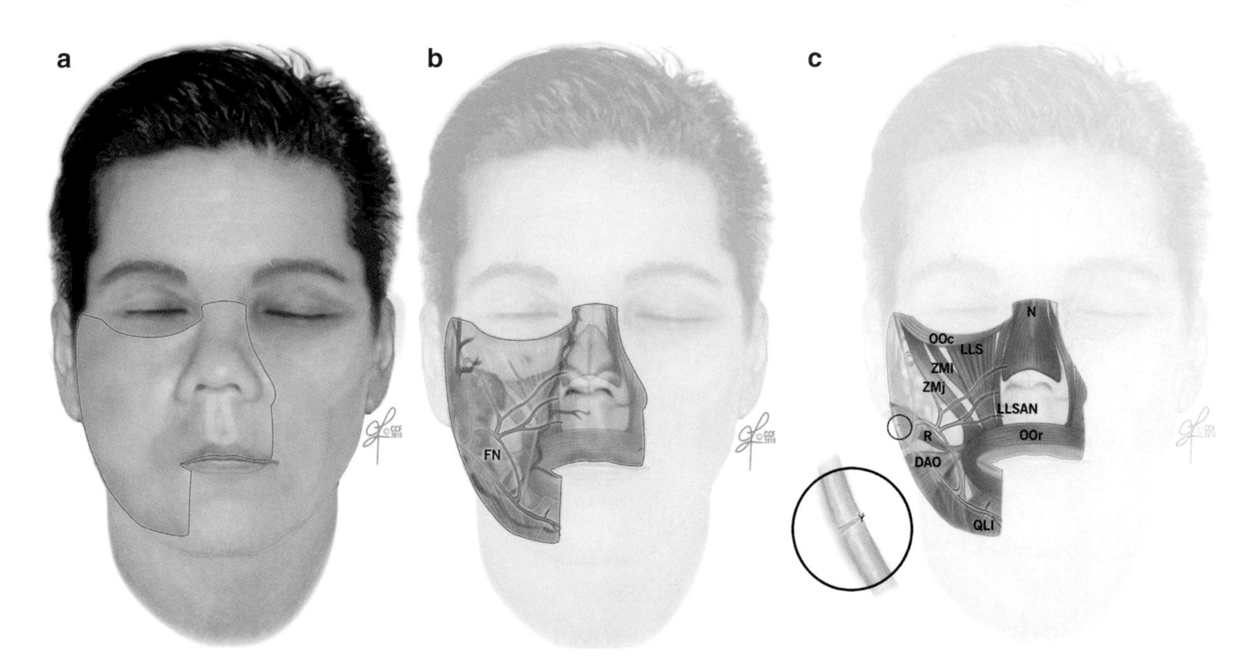

Fig. 36.2 (**a–c**) Second face transplant performed in China.[7] (**a**) Outline of the facial allograft. (**b**) Only the right facial nerve was repaired; *FN* facial nerve. (**c**) The repair of the facial nerve was reported as not satisfactory. Reinnervated muscles are showed; *DAO* depressor anguli oris, *LLS* levator labii superioris, *LLSAN* levator labii superioris alaeque nasi, *N* nasalis, *OOc* orbicularis oculi, *OOr*, orbicularis oris, *QLI* quadratus labii inferioris, *R* risorius, *ZMI* zygomatic minor, *ZMj* zygomatic major

36.3 Sensory Recovery After Scalp Replantation

We reviewed a total of 11 publications presenting results of sensory recovery following replantation of scalp and forehead in 34 patients (Table 36.3).[8–18] MRC scale could not be applied to any of these reports as the assessment methods were not specified or did not include all three determinants (touch, pain, and two-point discrimination) required by MRC.

Sensory nerves were repaired in 7 out of 34 patients and the mean two-point discrimination at an average

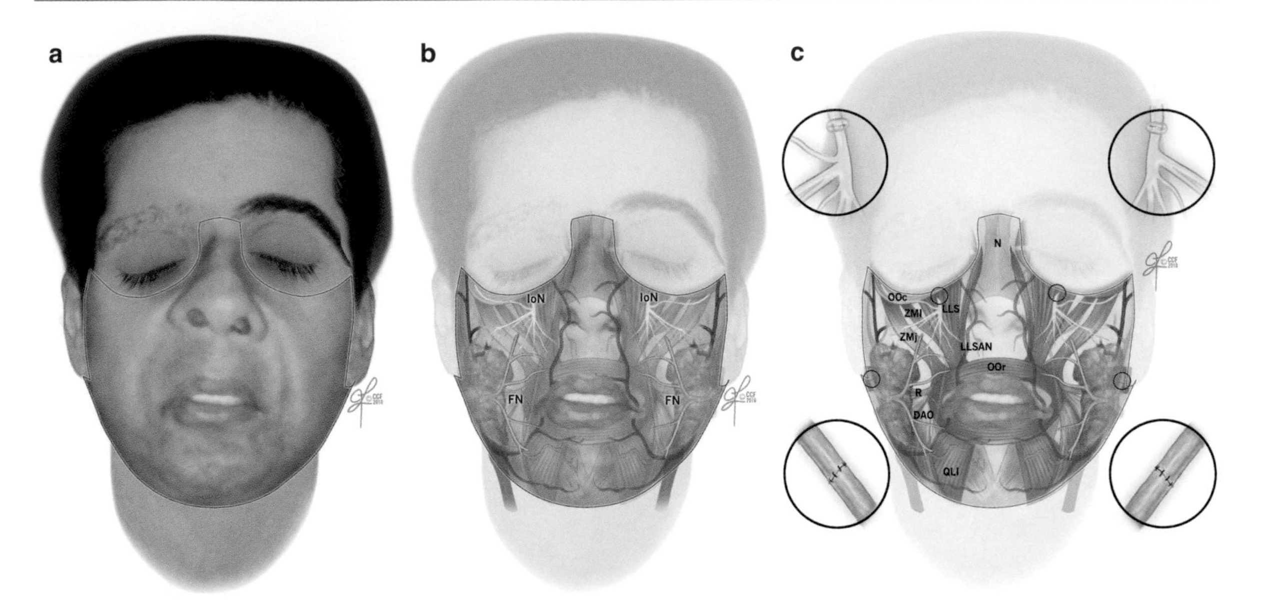

Fig. 36.3 (**a–c**) Third face transplant performed in France.[6] (**a**) Outline of the facial allograft. (**b**) Both facial and infraorbital nerves were repaired; *FN* facial nerve, *IoN* infraorbital nerve. (**c**) Facial nerves were repaired bilaterally while infraorbital nerves were sutured and glued on both sides. Mental nerves were not repaired microsurgically. *DAO* depressor anguli oris, *LLS* levator labii superioris, *LLSAN* levator labii superioris alaeque nasi, *N* nasalis, *OOc* orbicularis oculi, *OOr* orbicularis oris, *QLI* quadratus labii inferioris, *R* risorius, *ZMI* zygomatic minor, *ZMj* zygomatic major

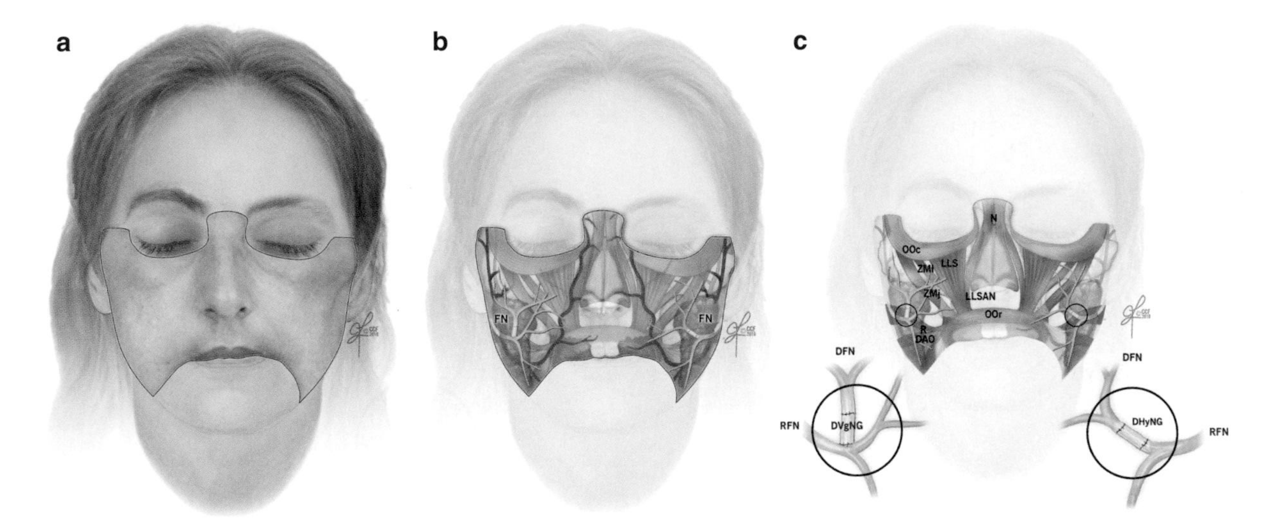

Fig. 36.4 (**a–c**) Fourth face transplant performed in USA.[1,2] (**a**) Outline of the facial allograft. (**b**) Because of the extension of the trauma only facial nerves were available for repair. *FN* facial nerve. (**c**) Facial nerve on the right side was repaired end to side to the recipient facial nerve with an interpositional nerve graft from the donor vagus nerve (DVgNG). On the left side the facial nerve was repaired end to end to the upper division of the recipient facial nerve with an interpositional Hypoglossus nerve graft from the donor (DHyNG). *DAO* depressor anguli oris, *LLS* levator labii superioris, *LLSAN* levator labii superioris alaeque nasi, *N* nasalis, *OOc* orbicularis oculi, *OOr* orbicularis oris, *R* risorius, *ZMI* zygomatic minor, *ZMj* zygomatic major

follow-up of 2 years was 15 mm. There was no information on which areas of scalp the measurements were taken from. In the remaining 27 patients, no nerve repair was performed and as result 4 patients had full or near full recovery of sensation, 7 patients recovered light touch, 7 patients reported protective sensation. "Acceptable" sensibility was reported for 6 patients. In the remaining three patients the two-point discrimination threshold was higher than normal (37.6 vs 23.2 mm for parietal scalp and 27.6 vs 22.3 mm for occipital scalp).

Table 36.3 Sensory recovery following scalp replantation

Author	Cases	Mean age (years)	Site	Nerve repair	Outcome measure	Outcome	Follow-up(months)
Cheng et al.[8]	7	26	Scalp, ear, forehead	7	2-PD	15 mm	24
Ueda et al.[9]	1	55	Right parietal and occipital scalp; right forehead skin; part of the right cheek and the right ear	No	Semmes-Weinstein test	2.44	36
Nahai et al.[10]	6	31	Scalp, ear, forehead	No	ND	Protective sensibility	ND
Yin et al.[11]	1	35	Scalp, forehead, right eyebrow	No	ND	Protective sensibility, No function frontalis	36
Cho et al.[12]	5	25	Scalp	No	2-PD	Supraorbital: 13.5 mm	42
						Forehead: 24 mm	
						Parietal Scalp: 39 mm	
						Occipital Scalp: 29.5 mm	
Chen et al.[13]	4	24	Scalp	No	ND	Light touch	6–12 (2 pts) ND (2 pts)
Chou et al.[14]	2	39	Scalp, forehead	No	ND	Light touch (forehead), deep pain (vertex)	6
Fogdestam et al.[15]	1	9	Scalp, forehead	No	ND	"Recovery of sensibility"	6
Oliva et al.[16]	1	33	Scalp, forehead, eyebrows	No	ND	Full recovery of sensation	6
Topalan[17]	1	15	Scalp	No	ND	Light touch	6
Sabapathy et al.[18]	5	24	Scalp, forehead	No	ND	"Acceptable recovery of sensation"	6–9

36.4 Sensory Recovery Following Repair of Peripheral Branches of the Trigeminal Nerve

Thirteen publications which assessed the neurosensory outcome following repair of the peripheral branches of the trigeminal nerve were assessed. These reports addressed the outcomes after severance of inferior alveolar nerve, lingual, infraorbital and mental nerves consequent to orthognathic surgery or facial trauma. In most of the studies the patient population was not uniform for the type of repaired nerve and the surgical procedure which included external neurolysis, internal neurolysis, nerve repair with either direct neurorraphy or interpositional nerve grafting. The presented outcomes were not differentiated on the basis of the method of the nerve repair. Although the quantitative sensory testing was employed as a method of assessment in all reports, the outcomes were mainly expressed as a degree of improvement in the global sensation. Four studies detailing results of objective neurosensory tests were available following repair of a completely transected nerve (Table 36.4).[19–22] MRC scale could be applied to three reported patients' series. In these reports sensory recovery ranged between S3+ and S4.

36.5 Sensory Recovery of Free Flaps

Twenty studies (13 on non-innervated flaps, including radial forearm flap, lateral thigh flap, anterolateral thigh flap, latissimus dorsi flap, trapezius flap, rectus abdominis musculocutaneous flap, fibula

Table 36.4 Sensory recovery following repair of the sensory nerves of the face

Author	Cases	Age (years)	Injured nerve	Procedure	Surgical delay (months)	MRC	Follow-up (months)
Robinson et al.[19]	13	31	Lingual	Direct suture	16	S3+	17
Robinson et al.[20]	53	30	Lingual	Direct suture	15	S3+	12
Hillerup et al.[21]	67	30	Lingual	Direct suture	8.5	NA	13
Tay et al.[22]	3	27	Inferior alveolar	Direct suture	0	S4	12

Table 36.5 Sensory recovery following reconstruction of defects in head and neck region with non-innervated free flaps

Author	Free flap	Cases	Age range (average years)	Indication, site	Radio therapy	MRC	Follow-up (average months)
Lahteenmaki et al.[25]	Dorsalis pedis flap	1	31	ND, face	ND	S3	36
	Latissimus dorsi flap	2	40	ND, face	ND	S2	39
	Trapezius flap	1	25	ND, face	ND	S0	28
Boyd et al.[24]	Radial forearm flap	10	56	Cancer, oral cavity	10	S2	14
Katou, f.[26]	Radial forearm flap	9	62	Cancer, oral cavity	ND	NA	25
Close et al.[27]	Radial forearm flap	4	62	Cancer, oral cavity	3	S3+	18
	Lateral thigh flap	4	53	Cancer, oral cavity	3	S3+	7
Shindo et al.[28]	Radial forearm flap	9	ND	Cancer, orofacial	8	S3	10
	Fibula osteocutaneous flap	9	ND	Cancer, orofacial	ND	S2	13
Vriens et al.[29]	Radial forearm flap	40	60	Cancer, oral cavity	28	S3+	38
Kimata et al.[30]	Anterolateral thigh flap	6	58	Cancer, oral cavity	ND	S1	12
	Rectus abdominis musculocutaneous flap	10	57	Cancer, oral cavity	ND	S1	27
Yu et al.[31]	Anterolateral thigh flap	5	60	Cancer, tongue	5	S1	15
Avery et al.[32]	Subfascial radial forearm flap	20	68	Cancer, oral cavity	ND	S3	≥6
	Suprafascial radial forearm flap	20	58	Cancer, oral cavity	ND	S3	≥6
Kerawala et al.[23]	Osteofascial radial forearm flap	12	65	ND, mandible	7	S2	38
	Radial forearm flap	38		ND, oral cavity	17	S3	
Shibahara et al.[33]	Radial forearm flap	30	60	Cancer, oral cavity	0%	NA	50
Kim et al.[34]	Radial forearm flap	12	55	Cancer, oral and oro-pharyngeal	ND	S2	6
Sabesan et al.[35]	Radial forearm flap	24	57	Cancer, oral cavity	28	S3	12
	Jejunal flap	10		Cancer, lateral pharyngeal wall		S3	
	Gastromental flap	6		Cancer, tongue base		S2	

osteocutaneous flap, jejunal flap, gastromental flap (Table 36.5)[23–35] and 7 on innervated free flaps including radial forearm flap, anterolateral thigh flap and rectus abdominis musculocutaneous flap (Table 36.6)[24,26,30,31,34,36,37]) were scored. MRC scale was applicable to 11 series of patients in the non-innervated flap group and to 6 series of patients in the innervated flap group. The median sensory recovery was graded S2+ in the non-innervated flaps and S3+ in the innervated flaps.

Table 36.6 Sensory recovery following reconstruction of defects in head and neck region with innervated free flaps

Author	Free flap	Nerves repaired	Cases	Age (average years)	Indication, site	Radio therapy	MRC	Followup (months)
Boyd et al.[24]	Radial forearm flap	Lateral antebrachial cutaneous nerve-lingual nerve	8	55	Cancer, oral cavity	8	S4	11
Katou, [26]	Radial forearm flap	Lateral antebrachial cutaneous nerve-lingual nerve	4	34	Cancer, oral cavity	ND	NA	13
Santamaria et al.[36]	Radial forearm flap	Lateral antebrachial cutaneous nerve-lingual (16), inferior alveolar (6), posterior auricular (3), cervical plexus (2), hypoglossal nerves (1)	28	45	Cancer, tongue	9	S4	18
Kimata et al.[30]	Anterolateral thigh flap	Lateral cutaneous nerve of the thigh-lingual nerve	8	63	Cancer, oral cavity	ND	S3+	18
	Rectus abdominis musculocutaneous flap	2 anterior cutaneous branches intercostal nerve-lingual nerve	5	60	Cancer, oral cavity	ND	S3+	14
Kuriakose et al.[37]	Radial forearm flap	Antebrachial cutaneous nerve-lingual nerve	17	51	Cancer, oral cavity	8	S3+	23
Yu et al.[31]	Anterolateral thigh flap	Lateral femoral cutaneous nerve-lingual nerve	6	62	Cancer, tongue	6	S4	16
Kim et al.[34]	Radial forearm flap	Antebrachial cutaneous nerve-lingual nerve (14), cervical plexus branch (1)	15	55	Cancer, oral and oro-pharyngeal	ND	S3+	6

36.6 Conclusions

Face allotransplantation is a new experimental surgical procedure and it is still reserved to patients with severe disfigurement who have exhausted other conventional reconstructive options. Therefore the first reported cases have been performed in the setting of destruction and distortion of the normal anatomical structures and functional facial units. The functional outcome of these first cases will have an important impact on the future development of this complex procedure and its potential application as a treatment modality for cases with different range of deformities. Review of the functional outcomes of these 4 cases published in the literature allows to conclude that reestablishment of motor and sensory function is essential for the success of face transplantation. In three out of four disclosed cases, with long term follow-up, repair of the sensory nerves was less than optimal or could not be performed. It was our aim to understand how the recovery will progress under these circumstances and how this will impact the final outcome; the next question was how would the normal restoration of the sensation be affected when the face transplantation surgery is performed long time after the original injury and what outcome could be expected when repair of the sensory nerves is not feasible. All these issues should be addressed in order to understand the mechanisms of sensory recovery following face transplantation.

During the follow-up of our patient[1,2] a comparison with the other reported cases was necessary to understand if the functional recovery could be considered as "normal" and what was the expected final sensory outcome. We reviewed the literature to understand the range of normal sensory thresholds of the human face and summarized the outcomes of the other reported face transplant cases. Since only a limited number of outcomes were available, we have revised other clinical conditions which would provide useful information about sensory recovery applicable to face transplantation. We evaluated the sensory outcomes following scalp and face replantation, the results of the microsurgical repair of the peripheral branches of the trigeminal nerve and reports on restoration of sensation in free flaps used for reconstruction of the head and neck defects both with and without nerve coaptation.

Most of the reviewed studies were case series. The neurosensory exams were poorly documented in some reports such as scalp replantation. In others, several different methods were used to assess the same variable, e.g., the pressure thresholds were evaluated with Semmes-Weinstein monofilaments, Von Frey's filaments, a cotton swab, wooden end, and cotton wrapped end of a cotton tip applicator. The results were expressed with different scoring systems applied for different variables, e.g., for Semmes-Weinstein test: logarithmic scale of applied force (the handle marking), the target force (g/mm^2), descriptive thresholds (normal, diminished light touch, diminished protective sensation, loss of protective sensation, deep pressure sensation only), or assessment of variable personal scoring scales. These factors made comparison of the results extremely difficult. We found that the only objective way to compare the outcomes of these studies was the application of the Medical Research Council (MRC) scale as modified by Mackinnon and Dellon,[3] whenever information on three variables including light touch, pain, and two-point discrimination was available. The advantage of using the MRC scale was to introduce an objective criterion in order to classify results from different case series and the ability to score them using available data.[38] For peripheral nerve injuries, an MRC score of S3 or higher is defined as useful sensory recovery,[39] therefore it allows to divide the outcomes into functional versus nonfunctional sensory recovery.

The sensory recovery in our patient was rated S3+, comparable to the outcome of microsurgical repair of the peripheral branches of the trigeminal nerve and to reinnervated free flaps used for the reconstruction within the head and neck region. The pattern of the sensory return in the allograft was similar to non-innervated free flaps as the sensory recovery started from the periphery and progressed toward the central part of the graft. The progress of the patient was rapid, and at 1 year follow-up she displayed 7 mm of two-point discrimination which is in the range of values reported for innervated free flaps or values reported after the repair of the trigeminal nerve, while non-innervated free flaps and scalps had no detectable two-point discrimination or very high thresholds.

What are therefore the factors which influence the sensory recovery in a face transplant in absence of microsurgical repair of the sensory nerves? The clinical studies have shown that several factors affect the sensory recovery of the free flaps: scarring of the recipient bed[40]; composition of the flap (presence of a skin component in the flap improves the sensory return, while the muscle or bone components can behave as a barrier to neurotization from the recipient bed)[41]; and finally recipient site, where the sensory recovery appears better in orofacial reconstruction as opposed to trunk and lower extremity reconstruction.[23,42,43] Thickness of the flap may impair the sensory return even in the absence of muscular or bony components, however there is no evidence in the literature to support this. Remarkable spontaneous return of the sensation has been reported in non-innervated radial forearm flaps used for orofacial reconstruction,[23] although a relevant difference remains when compared to the innervated flaps.[24] The innervation density of the transferred tissue has been reputed one of the factors impacting the sensory outcome when transferring flaps from the donor areas with lower innervation density to the areas with a higher concentration of terminal sensory endings.[44] Eventually, several studies have shown that sensory upgrading (improved 2-point discrimination compared to the donor site) can occur when flaps are transferred to the orofacial region. This has been explained with the wider cortical representation of the human face.[24]

The face transplant flaps probably share common patterns of reinnervation with the free tissue transfers, represented by regeneration of the nerves from the recipient bed and margins along the neural sheaths of the transferred tissue.[5,45] However, for such large flaps containing several components such as mimetic muscles and bone, used for reconstruction of full thickness defects, this mechanism does not appear sufficient to justify near full recovery of sensation.

In the reported face transplant flaps the near normal results of the neurosensory tests may indicate that the afferents are transmitted along already established neural pathways. In particular, the role of the facial nerve in conveying sensory signals should be considered, since this motor nerve has been always repaired. Sensory inputs could be transmitted by afferent fibers contained in the main trunk of the facial nerve,[46] by trigeminofacial communicating rami,[47] or by nervi nervorum of the facial nerve.[48] Non-innervated free flaps usually recover sensation from the periphery to the center,[24,49–51] while in innervated flaps the reverse occurs.[52] In the face allograft the observed direction of sensory return from the periphery to the central portion of the allograft parallels the direction of the facial nerve regeneration and thus may support this hypothesis.

The facial nerve normally contains somatic fibers, collecting the sensation of the external auditory meatus and posterior surface of the ear. Distal to the level where these components have left the nerve trunk, the sensory fibers are sparse. However, the presence of a sensory component in the facial nerve, mediating deep facial sensation (pressure, pressure pain and muscle sense) has been sustained. The presence of this component explained the preservation of deep facial sensation after trigeminal neurectomy.[53,54] This was confirmed also in physiological studies in cats.[55] The existence of purely sensory fibers has been shown in three major peripheral facial nerve branches of cats.[46] Afferent fibers found in the communicating rami appear to be concerned with deep sensibility or proprioception of the face.[47] The impulses conveying cutaneous sensation, which travel with the trigeminal nerve, contribute to appreciation of the facial movements,[56] therefore there is a close relationship between the two systems, and in absence of impulses from the trigeminal nerve, the afferents of the facial nerve could contribute to return of the sense of position and deep pressure as well as playing directly a role in transmission of the superficial touch and pain sensations.

The third neural pathway which could assume importance in these circumstances is represented by nervi nervorum. All cranial nerves are richly innervated by their own nerves called nervi nervorum, derived from fibers in the nerve trunk itself, which have nociceptive function.[57] Stimulation of nervi nervorum of the facial nerve trunk can be transmitted to trigeminocervical complex.[48] If these nerves can mediate somatic pain as well as referred pain is still to be discovered.

In considering different factors responsible for an improved sensory return in composite tissue allografts, such as face transplant, we should not forget that the immunosuppressive therapy with FK506 (Tacrolimus) confers an advantage for nerve regeneration which has been shown in limb allografting. FK 506 has been shown to increase the rate of axon regeneration, in a dose-dependent manner,[58] and to influence collateral sprouting of peripheral nerve fibers.[59] It was confirmed that FK506 doubles the number of axons that regenerate following a nerve injury, increases the number of myelinated axons by 40%, and significantly augments myelin thickness.[60] In addition FK506 reduces by half the time needed for neurological recovery after repair of nerve lesions.[61,62]

The drug regime of the first successful hand transplantation included FK 506, which allowed the allografted major nerves to regenerate and promoted sensory and motor function 14 years after the recipient's median and ulnar nerves had been severed.[63] The rate of nerve regeneration as indicated by advancing Tinel's sign was faster than expected. It was estimated at approximately 2–3 mm/day.[64]

In conclusion, near normal sensation can be restored following face transplantation. The pathways of sensory recovery are very complex. Other routes, besides the well known mechanisms of neurotization from the recipient bed and margins, should be investigated as potentially important mechanisms supporting functional sensory return in the absence of sensory nerve reconstruction. Afferent fibers contained in the facial nerve and communicating rami with the trigeminal nerve, as well as nervi nervorum of the facial nerve sheath, all may have an important impact on the final sensory outcome. The immunosuppressive therapy with FK506 can accelerate the existing mechanisms for nerve regeneration and contribute to an improved outcome when compared to traditional reconstructive techniques. Further insight into discovery of these complex mechanisms should be offered with documentation of sensory recovery in the reported cases of facial transplantation. To be able to compare the outcomes in patients operated in different centers across the world, we propose that at least pressure thresholds,[65,66] pain thresholds, and two-point discrimination should be evaluated in every patient. The follow-up visits for quantitative sensory testing should be scheduled at least once a month for the first 6 months, then 3 monthly during the first year, and 6 monthly thereafter until a plateau in sensory recovery is reached. Finally, inclusion of a sensory retraining program in the postoperative period has proved to affect significantly the outcomes following orthognatic surgery[67] and toe to hand transfers,[68] therefore sensory rehabilitation protocols should be included in the postoperative management of face transplant patients and the details of the outcomes should be disclosed to facilitate comparison of the results for this unique cases of face transplantation.

References

1. Siemionow M, Papay F, Alam D, et al. Near-total human face transplantation for a severely disfigured patient in the USA. *Lancet*. 2009;374:203-209.
2. Siemionow MZ, Papay F, Djohan R, et al. First U.S. near-total human face transplantation: a paradigm shift for massive complex injuries. *Plast Reconstr Surg*. 2010;125:111-122.
3. Mackinnon SE, Dellon AL. *Surgery of the Peripheral Nerve*. 1st ed. New York: Thieme Medical Publishers; 1988.
4. Devauchelle B, Badet L, Lengele B, et al. First human face allograft: early report. *Lancet*. 2006;368:203-209.
5. Dubernard JM, Lengele B, Morelon E, et al. Outcomes 18 months after the first human partial face transplantation. *N Engl J Med*. 2007;357:2451-2460.
6. Lantieri L, Meningaud JP, Grimbert P, et al. Repair of the lower and middle parts of the face by composite tissue allotransplantation in a patient with massive plexiform neurofibroma: a 1-year follow-up study. *Lancet*. 2008;372:639-645.
7. Guo S, Han Y, Zhang X, et al. Human facial allotransplantation: a 2-year follow-up study. *Lancet*. 2008;372:631-638.
8. Cheng K, Zhou S, Jiang K, et al. Microsurgical replantation of the avulsed scalp: report of 20 cases. *Plast Reconstr Surg*. 1996;97:1099-1106. Discussion 1107-1108.
9. Ueda K, Nomatsi T, Omiya Y, Tajima S. Replanted scalp recovers normal sensation without nerve anastomosis. *Plast Reconstr Surg*. 2000;106:1651-1652.
10. Nahai F, Hester TR, Jurkiewicz MJ. Microsurgical replantation of the scalp. *J Trauma*. 1985;25:897-902.
11. Yin JW, Matsuo JM, Hsieh CH, Yeh MC, Liao WC, Jeng SF. Replantation of total avulsed scalp with microsurgery: experience of eight cases and literature review. *J Trauma*. 2008; 64:796-802.
12. Cho BC, Lee DH, Park JW, Byun JS, Baik BS. Replantation of avulsed scalps and secondary aesthetic correction. *Ann Plast Surg*. 2000;44:361-366.
13. Chen IC, Wan HL. Microsurgical replantation of avulsed scalps. *J Reconstr Microsurg*. 1996;12:105-112.
14. Chou CK, Lin SD, Yang CC, Lai CS, Lin GT. Microsurgical replantation of avulsed scalp – two cases report. *Gaoxiong Yi Xue Ke Xue Za Zhi*. 1992;8:285-289.
15. Fogdestam I, Lilja J. Microsurgical replantation of a total scalp avulsion. Case report. *Scand J Plast Reconstr Surg*. 1986;20:319-322.
16. Zhou S, Chang TS, Guan WX, et al. Microsurgical replantation of the avulsed scalp: report of six cases. *J Reconstr Microsurg*. 1993;9:121-125. Discussion 125-129.
17. Topalan M, Ermis I. Replantation and triple expansion of a three-piece total scalp avulsion: six-year follow-up. *Ann Plast Surg*. 2001;46:167-169.
18. Sabapathy SR, Venkatramani H, Bharathi RR, D'Silva J. Technical considerations in replantation of total scalp avulsions. *J Plast Reconstr Aesthet Surg*. 2006;59:2-10.
19. Robinson PP, Smith KG. A study on the efficacy of late lingual nerve repair. *Br J Oral Maxillofac Surg*. 1996;34:96-103.
20. Robinson PP, Loescher AR, Smith KG. A prospective, quantitative study on the clinical outcome of lingual nerve repair. *Br J Oral Maxillofac Surg*. 2000;38:255-263.
21. Hillerup S, Stoltze K. Lingual nerve injury II. observations on sensory recovery after micro-neurosurgical reconstruction. *Int J Oral Maxillofac Surg*. 2007;36:1139-1145.
22. Tay AB, Poon CY, Teh LY. Immediate repair of transected inferior alveolar nerves in sagittal split osteotomies. *J Oral Maxillofac Surg*. 2008;66:2476-2481.
23. Kerawala CJ, Newlands C, Martin I. Spontaneous sensory recovery in non-innervated radial forearm flaps used for head and neck reconstruction. *Int J Oral Maxillofac Surg*. 2006;35:714-717.
24. Boyd B, Mulholland S, Gullane P, et al. Reinnervated lateral antebrachial cutaneous neurosome flaps in oral reconstruction: are we making sense? *Plast Reconstr Surg*. 1994;93:1350-1359. Discussion 1360-1362.
25. Lahteenmaki T, Waris T, Asko-Seljavaara S, Sundell B. Recovery of sensation in free flaps. *Scand J Plast Reconstr Surg Hand Surg*. 1989;23:217-222.
26. Katou F, Shirai N, Kamakura S, et al. Intraoral reconstruction with innervated forearm flap: a comparison of sensibility and reinnervation in innervated versus noninnervated forearm flap. *Oral Surg Oral Med Oral Pathol Oral Radiol Endod*. 1995;80:638-644.
27. Close LG, Truelson JM, Milledge RA, Schweitzer C. Sensory recovery in noninnervated flaps used for oral cavity and oropharyngeal reconstruction. *Arch Otolaryngol Head Neck Surg*. 1995;121:967-972.
28. Shindo ML, Sinha UK, Rice DH. Sensory recovery in noninnervated free flaps for head and neck reconstruction. *Laryngoscope*. 1995;105:1290-1293.
29. Vriens JP, Acosta R, Soutar DS, Webster MH. Recovery of sensation in the radial forearm free flap in oral reconstruction. *Plast Reconstr Surg*. 1996;98:649-656.
30. Kimata Y, Uchiyama K, Ebihara S, et al. Comparison of innervated and noninnervated free flaps in oral reconstruction. *Plast Reconstr Surg*. 1999;104:1307-1313.
31. Yu P. Reinnervated anterolateral thigh flap for tongue reconstruction. *Head Neck*. 2004;26:1038-1044.
32. Avery CM, Iqbal M, Hayter JP. Recovery of sensation in the skin of non-innervated radial flaps after subfascial and suprafascial dissection. *Br J Oral Maxillofac Surg*. 2006;44: 213-216.
33. Shibahara T, Mohammed AF, Katakura A, Nomura T. Long-term results of free radial forearm flap used for oral reconstruction: functional and histological evaluation. *J Oral Maxillofac Surg*. 2006;64:1255-1260.
34. Kim JH, Rho YS, Ahn HY, Chung CH. Comparison of sensory recovery and morphologic change between sensate and

nonsensate flaps in oral cavity and oropharyngeal reconstruction. *Head Neck*. 2008;30:1099-1104.
35. Sabesan T, Ramchandani PL, Ilankovan V. Sensory recovery of noninnervated free flap in oral and oropharyngeal reconstruction. *Int J Oral Maxillofac Surg*. 2008;37:819-823.
36. Santamaria E, Wei FC, Chen IH, Chuang DC. Sensation recovery on innervated radial forearm flap for hemiglossectomy reconstruction by using different recipient nerves. *Plast Reconstr Surg*. 1999;103:450-457.
37. Kuriakose MA, Loree TR, Spies A, Meyers S, Hicks WL Jr. Sensate radial forearm free flaps in tongue reconstruction. *Arch Otolaryngol Head Neck Surg*. 2001;127:1463-1466.
38. Dodson TB, Kaban LB. Recommendations for management of trigeminal nerve defects based on a critical appraisal of the literature. *J Oral Maxillofac Surg*. 1997;55:1380-1386. Discussion 1387.
39. Wyrick JD, Stern PJ. Secondary nerve reconstruction. *Hand Clin*. 1992;8:587-598.
40. Hermanson A, Dalsgaard CJ, Arnander C, Lindblom U. Sensibility and cutaneous reinnervation in free flaps. *Plast Reconstr Surg*. 1987;79:422-427.
41. Sonmez A, Bayramicli M, Sonmez B, Numanoglu A. Reconstruction of the weight-bearing surface of the foot with nonneurosensory free flaps. *Plast Reconstr Surg*. 2003; 111:2230-2236.
42. Lahteenmaki T, Waris T, Asko-Seljavaara S, Astrand K, Sundell B, Jarvilehto T. The return of sensitivity to cold, warmth and pain from excessive heat in free microvascular flaps. *Scand J Plast Reconstr Surg Hand Surg*. 1991;25:143-150.
43. Santanelli F, Tenna S, Pace A, Scuderi N. Free flap reconstruction of the sole of the foot with or without sensory nerve coaptation. *Plast Reconstr Surg*. 2002;109:2314-2322. discussion 2323–4.
44. Brown CJ, Mackinnon SE, Dellon AL, Bain JR. The sensory potential of free flap donor sites. *Ann Plast Surg*. 1989; 23:135-140.
45. Turkof E, Jurecka W, Sikos G, Piza-Katzer H. Sensory recovery in myocutaneous, noninnervated free flaps: a morphologic, immunohistochemical, and electron microscopic study. *Plast Reconstr Surg*. 1993;92:238-247.
46. Thomander L, Arvidsson J, Aldskogius H. Distribution of sensory ganglion cells innervating facial muscles in the cat. an anatomical study with the horseradish peroxidase technique. *Acta Otolaryngol*. 1982;94:81-92.
47. Baumel JJ. Trigeminal-facial nerve communications. their function in facial muscle innervation and reinnervation. *Arch Otolaryngol*. 1974;99:34-44.
48. Han DG. Pain around the ear in bell's palsy is referred pain of facial nerve origin: the role of nervi nervorum. *Med Hypotheses*. 2010;74:235-236.
49. Tindholdt TT, Tonseth KA. Spontaneous reinnervation of deep inferior epigastric artery perforator flaps after secondary breast reconstruction. *Scand J Plast Reconstr Surg Hand Surg*. 2008;42:28-31.
50. Place MJ, Song T, Hardesty RA, Hendricks DL. Sensory reinnervation of autologous tissue TRAM flaps after breast reconstruction. *Ann Plast Surg*. 1997;38:19-22.
51. Blondeel PN, Demuynck M, Mete D, et al. Sensory nerve repair in perforator flaps for autologous breast reconstruction: sensational or senseless? *Br J Plast Surg*. 1999;52: 37-44.
52. Schultes G, Gaggl A, Karcher H. Neuronal anastomosis of the cutaneous ramus of the intercostal nerve to achieve sensibility in the latissimus dorsi transplant. *J Oral Maxillofac Surg*. 2000;58:36-39.
53. Carmichael EA, Woolard HH. Some observations on the fifth and seventh cranial nerves. *Brain*. 1933;56: 109-125.
54. Ley A, Guitart JM. Clinical observations on sensory effects of trigeminal dorsal root section. *J Neurol Neurosurg Psychiatry*. 1971;34:260-264.
55. Davis LE. The deep sensibility of the face. *Arch Neurol Psychiatry*. 1923;9:283.
56. Trulsson M, Johansson RS. Orofacial mechanoreceptors in humans: encoding characteristics and responses during natural orofacial behaviors. *Behav Brain Res*. 2002;135: 27-33.
57. Thomas PK. The anatomical substratum of pain: evidence derived from morphometric studies on peripheral nerve. *Can J Neurol Sci*. 2004;31:398-403.
58. Wang MS, Zeleny-Pooley M, Gold BG. Comparative dose-dependence study of FK506 and cyclosporin A on the rate of axonal regeneration in the rat sciatic nerve. *J Pharmacol Exp Ther*. 1997;282:1084-1093.
59. Udina E, Voda J, Gold BG, Navarro X. Comparative dose-dependence study of FK506 on transected mouse sciatic nerve repaired by allograft or xenograft. *J Peripher Nerv Syst*. 2003;8:145-154.
60. Sulaiman OA, Voda J, Gold BG, Gordon T. FK506 increases peripheral nerve regeneration after chronic axotomy but not after chronic Schwann cell denervation. *Exp Neurol*. 2002; 175:127-137.
61. Sosa I, Reyes O, Kuffler DP. Immunosuppressants: neuroprotection and promoting neurological recovery following peripheral nerve and spinal cord lesions. *Exp Neurol*. 2005; 195:7-15.
62. Gold BG. FK506 and the role of the immunophilin FKBP-52 in nerve regeneration. *Drug Metab Rev*. 1999;31:649-663.
63. Dubernard JM, Owen E, Herzberg G, et al. Human hand allograft: report on first 6 months. *Lancet*. 1999;353: 1315-1320.
64. Bain JR. Peripheral nerve and neuromuscular allotransplantation: current status. *Microsurgery*. 2000;20:384-388.
65. Bell-Krotoski J, Tomancik E. The repeatability of testing with Semmes-Weinstein monofilaments. *J Hand Surg Am*. 1987;12:155-161.
66. Dellon AL, Andonian E, DeJesus RA. Measuring sensibility of the trigeminal nerve. *Plast Reconstr Surg*. 2007;120: 1546-1550.
67. Essick GK, Phillips C, Kim SH, Zuniga J. Sensory retraining following orthognathic surgery: effect on threshold measures of sensory function. *J Oral Rehabil*. 2009;36:415-426.
68. Graham B, Dellon AL. Sensory recovery in innervated free-tissue transfers. *J Reconstr Microsurg*. 1995;11:157-166.

Infectious Issues in Face Transplantation

37

Robin Avery, Chad R. Gordon, and Maria Z. Siemionow

Contents

Abstract Guidelines for infection prevention and management in solid organ transplant recipients have been published by the American Society of Transplantation. The classic principles of infection after solid organ transplantation, articulated by Rubin many years ago, continue to provide a framework for understanding the infection risk of the individual transplant recipient. Infectious risks that apply to all solid organ transplant candidates are considerations for face transplant recipients as well. In addition to these general principles, each kind of organ transplant has its own particular anatomic and functional considerations that have an impact on infection risk. Some risks can be reduced by careful pre-transplant screening, updating of immunizations, and patient education. However, the infection complication risk following face transplantation is still in its incipient stage. Future research is needed to define the optimal prevention strategy and duration of prophylaxis following face transplantation.

Abbreviations

AST ID	American Society of Transplantation Infectious Diseases
CAP	Community-Acquired Pneumonia
CDAD	*Clostridium difficile*–Associated Diarrhea
CMV	Cytomegalovirus
CNS	Central Nervous System
EBV	Epstein–Barr Virus
HAP	Hospital-Acquired Pneumonia
HBV	Hepatitis B Virus
HCV	Hepatitis C Virus
HHV	Human Herpes Virus
HSV	Herpes Simplex Virus
LTBI	Latent Tuberculosis Infection
MRSA	Methicillin-Resistant *Staphylococcus Aureus*

R. Avery (✉)
Department of Infectious Disease, Medicine Institute, Cleveland Clinic, Cleveland, OH, USA
e-mail: averyr@ccf.org

M.Z. Siemionow (ed.), *The Know-How of Face Transplantation*,
DOI: 10.1007/978-0-85729-253-7_37, © Springer-Verlag London Limited 2011

PICC Peripherally Inserted Central Catheter
PTLD Post-Transplant Proliferative Disorder
VAP Ventilator-Associated Pneumonia
VRE Vacomycin-Resistant *Enterococcous*
VZV Varicella-Zoster Virus

37.1 General Considerations Regarding Infection in Solid Organ Transplant Recipients

Guidelines for infection prevention and management in solid organ transplant recipients have been published by the American Society of Transplantation.[1] The reader is referred to this comprehensive work for more detailed discussions of donor/recipient screening, individual pathogens and infections, immunizations, and strategies for safer living.

The classic principles of infection after solid organ transplantation, articulated by Rubin many years ago, continue to provide a framework for understanding the infection risk of the individual transplant recipient. A combination of factors determines the transplant recipient's risk for infection, including the time post-transplant, antimicrobial prophylaxis, environmental exposures, and the "net state of immunosuppression."[2] The latter term is a composite of the effects of exogenously administered immunosuppressive medications and the underlying disease and co-morbidities including age, metabolic and renal factors, neutropenia, and disruption of mucosal barriers. Chapter 17 in this book provides definitions and guidelines for monitoring of immune function after face transplantation which can provide quantitative correlates for this concept. Hypogammaglobulinemia is another potentially correctable immune defect after transplantation which is associated with higher risk for a variety of infections, and some transplant centers monitor and replace IgG when below a threshold level (e.g., 400 mg/dl), or in patients with recurrent or severe infections and milder hypogammaglobulinemia.[3]

According to the classic timetable of infections after solid organ transplantation, the first month is largely dominated by postsurgical infections that can be seen with any major surgical procedure involving that organ. These include catheter-related infections, urinary tract infections, pneumonia, and wound or deep surgical site infections. Risk factors for such infections include long ICU stay and technical complications such as development of non-anatomic fluid collections and hematomas. In the era of increasing antimicrobial resistance, these infections may be caused by such organisms as methicillin-resistant *Staphylococcus aureus* (MRSA), vancomycin-resistant *Entercococcus* (VRE), or multi-drug resistant Gram-negative organisms including *Klebsiella*, *Pseudomonas*, and *Acinetobacter*. Also common during the first month are oral candidiasis (thrush) and reactivation of herpes simplex type I or II, and most solid organ transplant recipients receive prophylaxis for these (nystatin or clotrimazole for thrush prevention, and acyclovir or valacyclovir to prevent HSV in patients who are not receiving ganciclovir or valganciclovir for CMV prophylaxis).

The second phase of the post-transplant infection timetable is from the second to the sixth month. During this time, opportunistic infections that are commonly associated with transplantation begin to appear, some of which are donor-derived and some from internal reactivation or external exposures. Prominent among these are CMV, EBV (including post-transplant lymphoproliferative disease); other herpes viruses; parvovirus; polyomaviruses; early reactivation of hepatitis B and C; fungal infections including aspergillosis, cryptococcosis, and endemic mycoses; *Pneumocystis jiroveci* (formerly *P. carinii*); nocardiosis; tuberculosis, and non-tuberculous mycobacterial infection; strongyloidiasis and many others.

In the third period, after the sixth month, transplant recipients fall into three groups. Those that have done well with their allograft function and have had immunosuppression tapered become more like the general population in terms of risk, but never completely so. They still are susceptible to influenza and other respiratory viruses, pneumococcal pneumonia, and urinary tract infections, and should receive yearly influenza vaccine and updated pneumococcal vaccination. Those who have had more difficulties with rejection and intensified immunosuppression remain susceptible to all of the opportunistic infections seen in the second time period (see above). A third group of patients in this late time period appear to have done well initially but then develop late effects of long-term, immunomodulatory viral infections including BK polyomavirus, late CMV, HBV, HCV, and human papillomavirus infection.[2]

An understanding of this timetable, coupled with modern techniques of immune monitoring, can provide a framework for evaluation of the individual patient and also for establishing protocols for infection prevention for particular types of organ transplant.

37.2 Specific Considerations Regarding Infection in Face Transplant Recipients

37.2.1 The Unique Situation of the Face Transplant Recipient

In addition to the general principles enumerated above, each kind of organ transplant has its own particular anatomic and functional considerations that have an impact on infection risk. In the case of lung and intestinal transplantation, unlike kidney, liver, and heart, the graft is directly exposed to the external environment, and the microbiota of the mucosal surfaces plays a potentially important role in infection risk. The latter is also true for face and limb composite tissue allotransplantation. For face transplant recipients, theoretical risks include infections due to organisms colonizing the oral mucosa (including streptococci, anaerobes, *Capnocytophaga* spp., *Candida* spp., and sometimes Gram-negative aerobic bacilli); infections due to fungal spores colonizing the sinuses, nasal passages, and airways (including *Aspergillus* and other filamentous fungi); and organisms residing on the external skin (including staphylococci, *Propionibacterium,* corynebacteria, and in some cases Gram-negative bacilli). These risks may be compounded by nosocomial exposures particularly in the setting of long ICU and hospital stays (Tables 37.1 and 37.2).

37.2.2 Bacterial Infections

Given the risk of continuous exposure to oral organisms, immediate post-transplant prophylaxis should encompass oral organisms such as streptococci,

Table 37.1 Antimicrobial prophylaxis in four reported face transplant recipients

Report	Bacterial	Fungal	Viral	Other
Devauchelle et al., Dubernard et al.[14,15]	Amoxicillin-clavulanate	Not stated	Ganciclovir IV × 5 day, then valganciclovir × 5 mos	TMP-SMX × 4 mos
Guo et al.[21]	Ceftizoxime, metronidazole	Not stated	Acyclovir	Probiotics; allicin; IVIg; surveillance bacterial cultures and pre-emptive therapy
Lantieri et al.[11]	Not stated	Not stated	Valganciclovir × 6 mos	TMP-SMX × 6 mos; phenoxymethyl-penicillin for donor syphilis
Siemionow et al.[22]	Vancomycin and pipera cillin-tazobactam, then amoxicillin-clavulanate	Voricona-zole	Ganciclovir IV then valganciclovir × 5 mos	TMP-SMX × months

Table 37.2 Infectious complications in four reported face transplant recipients

	Bacterial	Fungal	Viral	Comments
Devauchelle et al., Dubernard et al.[14,15]	None reported	*Candida* stomatitis	HSV on lips; molluscum contagiosum	
Guo et al.[21]	*Enterococcus, Staph epidermidis, Enterobacter* on surveillance cultures, treated pre-emptively	None reported	None reported	Stated no opportunistic infections in 2-year follow-up
Lantieri et al[11]	None reported	None reported	CMV (ganciclovir-resistant)	Required foscarnet x 8 week; associated with rejection
Siemionow et al.[22]	*Pseudomonas* and *Staph. epidermidis* catheter-related BSI; *C. diff* and *Aeromonas* diarrhea	None	CMV (relapsing)	Neutropenia from ganciclovir and valganciclovir

anaerobes, *Capnocytophaga* spp., and *Candida* spp. Ampicillin-sulbactam is an excellent choice for the antibacterial part of prophylaxis. If the face transplant candidate has had known infection or colonization with other bacterial organisms in the past, prophylaxis may be modified accordingly. If there is concern for risk for MRSA and Gram-negative infections, a combination of vancomycin and piperacillin-tazobactam or other similar combination is a reasonable choice, which represents broader coverage but still includes the oral organisms above. The Cleveland Clinic face transplant recipient received the latter antibiotic combination. Duration of such prophylaxis should depend on the time to healing of anastomoses and suture lines, as well as other factors such as length of ICU stay.

Bacterial infections of the sinuses are an ongoing risk, and may relate to the particular anatomy of the individual recipient. Close follow-up is important in this regard, with imaging of the sinuses when an infection is suspected, and cultures of sinus material for microbiologic diagnosis whenever possible. Similarly for any deep surgical site infection, cultures of any purulence, fluid collections, or debrided tissue would be important, and ideally should be sent for anaerobic, aerobic, fungal, and mycobacterial cultures with corresponding stains. The Cleveland Clinic face transplant recipient did not experience any deep surgical site infections.

Pneumonia is a risk particularly for transplant recipients who are immobilized or bed-bound for prolonged periods of time, or those with risk for aspiration. Such risk can be decreased by measures such as the semi-sitting position in ICU patients, active physical therapy and early mobilization, and careful assessment of readiness to switch from gastric tube to oral feedings. When pneumonia occurs, it may be healthcare-associated pneumonia (HAP), ventilator-associated pneumonia (VAP), or community-acquired pneumonia (CAP), and the potential pathogens vary according to these circumstances. Since the face transplant recipient is immunosuppressed, it is important to consider unusual and opportunistic pathogens as well. Radiographic findings may provide clues to the nature of the pathogen: Lobar consolidations suggest a bacterial etiology; diffuse bilateral infiltrates are more suggestive of viral infection or *Pneumocystis* (though some bacterial infections such as legionellosis can present in this manner); nodular infiltrates particularly with cavitations suggest fungal, mycobacterial, or nocardial infection, or post-transplant lymphoproliferative disease. Since therapies are entirely different depending on the etiology of pulmonary infiltrates, efforts should be made to establish a microbiologic diagnosis either by expectorated sputum, tracheal aspirate, bronchoalveolar lavage with transbronchial biopsy, or in some cases open lung biopsy. The Cleveland Clinic face transplant recipient did not have any episodes of pneumonia, but did have episodes of bronchitis with negative cultures that responded to azithromycin.

Catheter-related infections are a risk for patients with indwelling long-term catheters including tunneled and non-tunneled central venous access lines. The Cleveland Clinic face transplant recipient had infections of peripherally inserted central catheter (PICC) lines, including a *Pseudomonas* catheter-related bloodstream infection and a coagulase-negative staphylococcal infection (which is the most common infection of indwelling catheters). Catheter-related infections other than coagulase-negative staphylococcal infections often require removal of the catheter; and any tunnel infection, or persistent positive blood cultures regardless of organism, also requires catheter removal. The long-term need for IV antibiotics, hydration, and other medications can sometimes lead to IV access issues; in addition, some patients have very difficult peripheral IV access for blood drawing, but the risk of infection must always be balanced against the convenience of leaving the intravenous catheter in place for this purpose.

Urinary tract infections are a risk for all solid organ transplant recipients, particularly renal transplant recipients who are subject to transplant pyelonephritis. Non-renal transplant recipients, including face transplant recipients, could develop urinary tract infections with presentations ranging from dysuria and frequency to sepsis without focal signs or symptoms.

Clostridium difficile–associated diarrhea (CDAD) is a major risk, particularly since the advent of the epidemic nosocomial strain of *C. difficile* in the mid-2000s.[4] Face transplant recipients are likely to need multiple courses of antibiotics, either as prophylaxis or treatment for infections, and are consequently at risk for CDAD. While CDAD usually presents with copious diarrhea (up to 15–20 times per day), ominous signs are high fever, high white blood count, and development of increasing abdominal pain, distention, and ileus, which may herald megacolon, a condition frequently requiring colectomy. The Cleveland

Clinic face transplant patient presented with refractory diarrhea at 13 months post-transplant; an assay for *C. difficile* and also a culture for *Aeromonas* were positive (*Aeromonas* is a water-associated bacterial organism that may cause cellulitis, and occasionally causes diarrheal disease in humans). Although her diarrheal episode was protracted, fortunately she never developed severe abdominal distention or megacolon, and the episode ultimately resolved without recurrences.

Sepsis remains a risk in immunocompromised patients, particularly those with complex post-transplant courses or predisposing factors such as deep abscesses, neutropenia, or breaches in mucosal defenses. There has been one death in a concomitant face and hand transplant recipient who reportedly developed "overwhelming infection, requiring surgical revisions, and subsequent cardiac arrest leading to death."[5]

37.2.3 Viral Infections: CMV

CMV continues to be a pathogen of major importance, despite an extensive literature on CMV prophylaxis[6] and pre-emptive therapy.[7] CMV is a betaherpesvirus that remains latent lifelong in infected individuals; CMV seropositivity in adults generally ranges from 60% to 80% depending on the region. Post-transplant CMV infection can occur either from reactivation of the recipient's own past CMV strain under the influence of immunosuppression, or can be acquired from the donor. Transplant recipients who are donor-seropositive, recipient-seronegative (D+/R–) are at highest risk for severe CMV, since they acquire primary CMV infection at a time that they are immunosuppressed, and have no antecedent-specific anti-CMV immunity.

There are internationally accepted definitions of CMV viremia and symptomatic CMV disease.[8] Clinically, CMV can present in three broad categories: asymptomatic viremia, CMV syndrome, and tissue-invasive CMV. Asymptomatic viremia, as the name suggests, refers to the detection of CMV on a peripheral blood assay such as the PCR or pp65 antigenemia test, in the absence of clinical symptoms referable to CMV infection. "CMV syndrome" refers to a flu-like syndrome with fevers, chills, myalgias, leukopenia, and occasional thrombocytopenia, and elevation of liver function tests. Tissue-invasive disease refers to the situation in which CMV inclusions can be visualized in tissue (or detected on a CMV immunostain); this category encompasses CMV pneumonitis, hepatitis, esophagitis, gastritis, enteritis, colitis, retinitis, meningoencephalitis, and other less common tissue sites of localization. The last category is clinically the most severe and is frequently associated with high CMV viral loads in peripheral blood, although there are exceptions.

There are two major prevention strategies for CMV: prophylaxis and pre-emptive therapy. "Prophylaxis" refers to the administration of antiviral therapy to all patients in a group for a defined duration, frequently 3–6 months; benefits of this strategy have been demonstrated in a meta-analysis.[6] By contrast, "pre-emptive therapy" refers to administration of antiviral therapy only to those who develop CMV viremia during monitoring with a sensitive early detection test such as CMV PCR or pp65 antigenemia.[7] The Cleveland Clinic transplant programs employ a combination of these strategies in order to maximize the benefits of both. CMV is an immunomodulatory virus, and the direct infectious syndromes are often followed by indirect effects including opportunistic infections such as fungal, PTLD, or *Pneumocystis* infections, and in some cases allograft dysfunction.[9] Pre-emptive therapy offers the opportunity for early detection and treatment of "late CMV" after prophylaxis[7] which can otherwise be a devastating complication if the viral load is allowed to rise to significant levels. The most common agent currently used for prophylaxis is valganciclovir, and the duration of prophylaxis has frequently been for 3 months post-transplant in other solid organ recipients,[10] but there is increasing evidence that longer durations of prophylaxis may confer added benefit.[11] The Cleveland Clinic face transplant patient received IV ganciclovir initially, followed by oral valganciclovir; the intended duration was 6 months, but prophylaxis had to be discontinued after 5 months due to neutropenia.

Ganciclovir- and valganciclovir-associated neutropenia is the major side effect of these anti-CMV drugs; other potential side effects include other cytopenias, and occasionally renal dysfunction, gastrointestinal side effects, and mental status or CNS effects. Ganciclovir derivatives are potentially teratogenic, which should be kept in mind for female transplant recipients of childbearing age.

The potential severity of CMV in the face transplant recipient is illustrated by the development of ganciclovir-resistant CMV in the second French face transplant recipient, which was associated with an episode of clinical acute rejection and required 8 weeks of therapy with foscarnet.[12] Although renal dysfunction was not mentioned in the case report, this length of foscarnet therapy is frequently accompanied by significant renal dysfunction which may continue after the discontinuation of this antiviral and may even eventuate in need for dialysis. The impact of CMV in composite allotransplantation has previously been described with reference to a cohort of hand transplant recipients, in which five of nine patients at risk developed CMV infection, including two with high CMV viral loads, and several with refractory or relapsing courses requiring treatment with foscarnet and/or cidofovir.[13] The association between rejection and CMV infection has been debated in other organ transplant settings, but may be particularly strong in composite allotransplantation.[5,12,13] This relationship provides another reason for implementing a vigorous CMV prevention strategy in face transplant recipients.

The Cleveland Clinic face transplant recipient, who is CMV D+/R–, did not develop ganciclovir-resistant CMV, but did develop refractory recurrent CMV viremia which was complicated each time by neutropenia due to ganciclovir or valganciclovir, despite use of filgrastim to support the WBC count. Since neutropenia confers a high risk of infections, particularly fungal infections, this was a very concerning situation. Once ganciclovir derivatives are not usable, either due to virologic resistance or adverse effects, the licensed alternative anti-CMV drugs pose significant risks of toxicity. Foscarnet is highly nephrotoxic and can cause electrolyte depletion and urogenital ulceration; intravenous cidofovir is nephrotoxic and can cause cytopenias and ophthalmologic complications. After careful deliberation as to the risks and benefits of available therapies, she ultimately received the investigational drug CMX001 under emergency IND through the FDA.[14] Her CMV viremia cleared after 6 weeks, and CMX001 was ultimately discontinued in the setting of protracted diarrhea after *C. difficile* and *Aeromonas* infection 13 months post-transplant, when multiple medications were stopped for the possibility that some of these might be prolonging the diarrheal syndrome.

Thus, CMV has been a major complication in two of the first four face transplant recipients reported in the literature as well as five of nine hand transplant patients at risk, with complex syndromes including ganciclovir resistance and relapsing courses.[5,12,13] If possible to avoid the high-risk CMV D+/R– status that would be ideal; however in many regions the CMV seropositivity rate makes this impractical, given all of the other considerations in identifying a suitable donor for face transplantation. Although the seropositivity rate was only 35.3% in the hand transplant cohort,[13] most regions have considerably higher CMV seropositivity rates in available donors. A combination of prophylaxis and pre-emptive therapy appears to offer a reasonable prevention strategy, but clearly further work remains to be done to identify the optimal duration and agent for such therapy. Some authors have advocated combinations of CMV hyperimmune globulin and extended antiviral prophylaxis.[13] The future availability of orally bioavailable anti-CMV drugs that do not cause neutropenia may alter management in this regard.

37.2.4 Viral Infections: Other Viruses

All members of the herpesvirus family share the property of latency and can reactivate after transplantation or may be donor-acquired. Epstein–Barr virus (EBV) replication may eventuate in post-transplant lymphoproliferative disease (PTLD) due to loss of virus-specific cytotoxic T-lymphocytes in the setting of immunosuppression. PTLD consists of a spectrum of clinical and histopathologic manifestations of lymphoproliferation, ranging from a polyclonal process to a monoclonal B-cell lymphoma. In the latter case, many organs may be involved including the allograft. In some high-risk settings (e.g., EBV D+/R– transplant recipients and/or pediatric transplant recipients), quantitative EBV DNA monitoring affords the opportunity for early intervention by reduction of immunosuppression for prevention of progression to PTLD.

Herpes simplex may reactivate as a cutaneous, mucosal (oral, genital, esophageal), or occasionally invasive internal pathogen (hepatitis, pneumonitis, meningoencephalitis). Although most commonly reactivation occurs early, the use of antiviral prophylaxis has sometimes resulted in later onset, as in the first French face transplant recipient, who developed HSV of the lips on day 185 which was treated with oral valacyclovir and topical acyclovir cream.[15]

Varicella-zoster virus (VZV) is the cause of chickenpox and zoster (shingles). About 90% of adults are varicella-seropositive, but the remaining 10% are susceptible to primary varicella and ideally should be vaccinated prior to transplantation. For varicella-seropositive individuals, reactivation may occur post-transplant in the form of dermatomal or occasionally disseminated zoster, depending on their global immune function. The latter condition can be clinically severe and may involve multiple organs.

Human herpesvirus-6 (HHV-6) and 7 (HHV-7) are the roseoloviruses, the causes of roseola in infants. Seropositivity in the general adult population is high, and reactivation post-transplant may occur earlier than CMV and may cause pancytopenia, pneumonitis, hepatitis, and meningoencephalitis. HHV-8 is the agent of Kaposi's sarcoma which may occasionally cause post-transplant KS. Parvovirus B19 is an under-recognized cause of severe anemia post-transplant and may be treated with intravenous immunoglobulin.

Hepatitis B and C have been studied extensively in transplant recipients, in one of several scenarios: that of a liver transplant performed for cirrhosis due to one of these viruses; or preexisting infection in the recipient of a non-liver allograft; or donor-derived transmission. None of the face transplant recipients reported to date have had clinical issues with viral hepatitis, but in some hyperendemic areas, such issues may arise and may complicate donor selection. Effective vaccination against hepatitis B is extremely important in the pre-transplant candidate.

Respiratory viruses are a threat to the organ transplant recipient particularly during the winter months. Influenza, parainfluenza virus (the agent of croup), respiratory syncytial virus, and adenovirus all may result in severe lower tract infection requiring, in some cases, ICU admission and intubation. Every effort should be made to avoid transmission of these viruses, including restricting visitation of ill family members. Stringent hospital infection control should be practiced on transplant wards with regard to these viruses, and some centers prohibit access of health care workers with respiratory symptoms to transplant wards. Yearly (non-live) influenza vaccination is recommended for transplant recipients, and during the 2009–2010 H1N1 pandemic influenza episode, H1N1 immunization was also administered to transplant recipients and their families.

37.2.5 Fungal Infections

Fungal infections of importance after transplantation include candidiasis (which may be mucosal or invasive); infections due to endemic mycoses such as histoplasmosis and coccidioidomycosis; cryptococcosis; and mold infections due to filamentous fungi such as *Aspergillus*, *Mucor* and other zygomycetes, *Scedosporium* and others. Oral candidiasis is very common in the early post-transplant phase, and most solid organ transplant patients receive topical mucosal prophylaxis with nystatin or clotrimazole. The first French transplant patient developed a candidal stomatitis on day 18 which presented when "diffuse erythema and edema were observed on the grafted mucosa."[16] This is important as this type of clinical appearance may coexist with, or be confused with, rejection. Candidiasis may also occur in the form of catheter-related candidal bloodstream infections and deep surgical site infections (the latter more common in abdominal organ transplantation).

Histoplasmosis[17] and coccidioidomycosis are common in particular endemic areas of the USA, and may reactivate late post-transplant in recipients. Histoplasmosis, most common in the midwest USA, may be manifested as multiple pulmonary nodules and infiltrates, central nervous system manifestations, or other organ localization, but also may present as fever with pancytopenia and no localizing symptoms, in which case the diagnosis is often delayed. Cocciodioidomycosis may reactivate in the lung, central nervous system, and other organs in a patient who has resided or traveled extensively in the southwest USA.

Cryptococcosis may cause meningitis, pulmonary nodules, cellulitis, or other manifestations. Central nervous system infection may be accompanied by elevated intracranial pressure requiring neurosurgical management. An unusual immune reconstitution syndrome has been described.[18]

Some of the most feared post-transplant infections relate to molds (filamentous fungi) such as *Aspergillus* spp. Lung transplant recipients are at higher risk than other solid organ recipients, partly because of exposure of the allograft to the external environment and the opportunity for fungal spores to colonize the graft; given these considerations, face transplant recipients are also likely at high risk. Aspergillosis often presents in the lungs with nodules (which may be cavitating and may show the characteristic "halo sign"), or in the sino-orbital area or CNS, but can involve any organ. Until the advent of newer azoles, the prognosis was grim. Non-*Aspergillus* mycelial fungi are increasingly common and may portend higher mortality than aspergillosis.[19] Risk factors for such infections include neutropenia, exposure to construction sites, gardening, farming, and marijuana smoking. Strategies for prevention of such infections include avoidance of exposures and, in some cases, antifungal prophylaxis with a mold-active agent. The Cleveland Clinic face transplant patient received prophylaxis with voriconazole, which was later discontinued due to elevation in liver function tests and difficulty managing the tacrolimus level (all azole antifungal agents increase levels of calcineurin inhibitors and sirolimus, although this is usually managed with close monitoring and dose adjustment.) In addition, this patient has been monitored with urinary *Histoplasma* antigen determinations approximately every 3–4 months (MiraVista Diagnostics) because of exposure to chickens early in life; all of her testing has so far been negative for evidence of reactivation of histoplasmosis. Fortunately, and perhaps surprisingly, none of the first four face transplant recipients have been reported to have an invasive fungal infection,[5] but this potential is very real and should be a matter of vigilance for transplant clinicians caring for face transplant recipients in the future.

37.2.6 Other Infections

Other infections that may cause substantial morbidity in transplant recipients include tuberculosis, non-tuberculous mycobacterial infections, and parasitic infections. The latter include strongyloidiasis (which may be disseminated with a high mortality), Chagas' disease, and other infections depending on geographic origin and/or travel history of the recipient. Pre-transplant screening for latent tuberculosis infection with either a tuberculin skin test or interferon-gamma release assay is very important, and patients with evidence of LTBI may be offered prophylaxis with isoniazid. A positive *Strongyloides* IgG serology should prompt consideration of pre-transplant ivermectin therapy to prevent the devastating complication of *Strongyloides* dissemination post-transplant.

37.2.7 Immunizations for Vaccine-Preventable Infections

Ideally, immunization status should be updated during the pre-transplant screening evaluation. Pneumococcal vaccine should be administered if not given during the previous 5 years, and tetanus-diphtheria-acellular pertussis (Tdap) vaccine should be administered if the last tetanus immunization was more than 10 years ago. Hepatitis A, hepatitis B, or combined hepatitis A/B vaccine series should be administered to seronegative patients. Seasonal influenza vaccination (including H1N1 influenza in the 2009–2010 season) should be administered according to current guidelines. Post-transplant patients should not receive live vaccines and thus should receive the injected non-live influenza vaccine rather than the live attenuated nasal vaccine.

Varicella-seronegative recipients should receive the live attenuated varicella vaccine pre-transplant. The utility of the zoster vaccine (a higher concentration of live varicella vaccine) in preventing post-transplant zoster when administered pre-transplant, is still unknown; however, this vaccine may be given prior to transplantation according to current guidelines (over age 60, not on immunosuppression). Human papillomavirus vaccine should be administered to girls and women aged 9–26 who have not yet been vaccinated. The AST ID Guidelines[20] contain detailed recommendations for immunization of both adult and pediatric transplant candidates and recipients.

37.3 Strategies for Safer Living

Once the immediate post-surgical issues have been addressed and the face transplant recipient enters the recovery phase post-transplant, detailed teaching regarding environmental risks is very important. Although transplant recipients can and do enjoy many activities of daily life, there are certain precautions regarding food, animals, water, travel, and other similar issues that can significantly decrease their risk for infections related to external exposures. In the case of the Cleveland Clinic face transplant recipient, the presence of *Aeromonas* in her stool during a diarrheal illness drew attention to the issue of well water versus city water.

Pet exposures are also important; many transplant recipients are owners of dogs, cats, and other pets, and the emotional attachment to these pets can be very strong particularly in individuals who have had difficulties in human society owing to their underlying medical conditions. However, exposure to cat litter carries a risk for toxoplasmosis, and pet birds or bird feeders can transmit fungal infections through bird droppings; reptiles, amphibians, and baby chicks can transmit salmonellosis, and there are many other such risks. With careful attention and detailed patient education, such risks can be minimized. The AST ID Guidelines contain a detailed description of post-transplant strategies for safer living, which can be provided to both transplant clinicians and patients.[21] If international travel should be contemplated, it is very important to visit a travel clinic with expertise in immunocompromised patients, for there are many general and destination-specific prophylaxis measures that can greatly mitigate the risks.[22]

37.4 Conclusions

Infectious risks that apply to all solid organ transplant candidates are considerations for face transplant recipients as well. Some risks can be reduced by careful pre-transplant screening, updating of immunizations, and patient education. CMV is a particular source of concern due to the high reported incidence of complex CMV syndromes and association with rejection in the face and hand transplant recipients to date,[5,12,13,15,16,23,24] despite prophylaxis and pre-emptive therapy or combinations thereof. Future research is needed to define the optimal prevention strategy and duration of prophylaxis.

References

1. American Society of Transplantation Infectious Disease Community of Practice. Guidelines for the management and prevention of infectious complications after solid organ transplantation. *Am J Transplant*. 2009;9(suppl 4):S1-S257.
2. Fishman JA, Rubin RH. Infection in organ-transplant recipients. *N Engl J Med*. 1998;338:1741-1751.
3. Mawhorter SD, Yamani MH. Hypogammaglobulinemia and infection risk in solid organ transplant recipients. *Curr Opin Organ Transplant*. 2008;13:581-585.
4. Riddle DJ, Dubberke E. Clostridium difficile infection in solid organ transplant recipients. *Curr Opin Organ Transplant*. 2008;13:592-600.
5. Gordon C, Siemionow M, Papay F, et al. The world's experience with facial transplantation: what have we learned thus far? *Ann Plast Surg*. 2009;63:572-578.
6. Kalil A, Levitsky J, Lyden E, Stoner J, Freifeld AG. Meta-analysis: the efficacy of strategies to prevent organ disease by cytomegalovirus in solid organ transplant recipients. *Ann Intern Med*. 2005;143:870-880.
7. Singh N, Wannstedt C, Keyes L, et al. Efficacy of valganciclovir administered as preemptive therapy for cytomegalovirus disease in liver transplant recipients: impact on viral load and late-onset cytomegalovirus disease. *Transplantation*. 2005;79:85-90.
8. Ljungman P, Griffiths P, Paya C. Definitions of cytomegalovirus infection and disease in transplant recipients. *Clin Infect Dis*. 2002;34:1094-1097.
9. Rubin RH. The indirect effects of cytomegalovirus infection on the outcome of organ transplantation. *JAMA*. 1989;261:3607-3609.
10. Paya C, Humar A, Dominguez E, et al. Efficacy and safety of valganciclovir vs. oral ganciclovir for prevention of cytomegalovirus disease in solid organ transplant recipients. *Am J Transplant*. 2004;4:611-620.
11. Palmer SM, Limaye AP, Banks M, et al. Extended valganciclovir prophylaxis to prevent cytomegalovirus after lung transplantation: a randomized, controlled trial. *Ann Intern Med*. 2010;152:761-769.
12. Lantieri L, Meningaud J-P, Grimbert P, et al. Repair of the lower and middle parts of the face by composite tissue allotransplantation in a patient with massive plexiform neurofibroma: a 1-year follow-up study. *Lancet*. 2008;372:639-645.
13. Schneeberger S, Lucchina S, Lanzetta M, et al. Cytomegalovirus-related complications in human hand transplantation. *Transplantation*. 2005;80:441-447.
14. Painter W, Mossad S, Siemionow M, et al. First use of CMX001, a novel antiviral drug, for the suppression of cytomegalovirus infection in solid organ transplant patients: A case series. American Transplant Congress 2010, oral late-breaking abstract presentation; 2010; San Diego, CA.
15. Dubernard J-M, Lengele B, Morelon E, et al. Outcomes 18 months after the first human partial face transplantation. *N Engl J Med*. 2007;357:2451-2460.
16. Devauchelle B, Badet L, Lengele B, et al. First human face allograft: early report. *Lancet*. 2006;368:203-209.
17. Cuellar-Rodriguez J, Avery RK, Lard M, et al. Histoplasmosis in solid organ transplant recipients: 10 years of experience at a large transplant center in an endemic area. *Clin Infect Dis*. 2009;49:710-716.
18. Singh N, Lortholary O, Alexander BD, et al. Allograft loss in renal transplant recipients with Cryptococcus neoformans associated immune reconstitution syndrome. *Transplantation*. 2005;80:1131-1133.
19. Husain S, Alexander BD, Munoz P, et al. Opportunistic mycelial fungal infections in organ transplant recipients: emerging importance of non-Aspergillus mycelial fungi. *Clin Infect Dis*. 2003;37:221-229.
20. Danzinger-Isakov L, Kumar D. Guidelines for vaccination of solid organ transplant candidates and recipients. *Am J Transplant*. 2009;9(suppl 4):S258-S262.

21. Avery RK, Michaels MG. Strategies for safe living following solid organ transplantation. *Am J Transplant*. 2009;9 (suppl 4):S252-S257.
22. Kotton CN, Hibberd PL. Travel medicine and the solid organ transplant recipient. *Am J Transplant*. 2009;9(suppl 4): S273-S281.
23. Guo S, Han Y, Zhang X, et al. Human facial allotransplantation: a 2-year follow-up study. *Lancet*. 2008;372:631-638.
24. Siemionow MZ, Papay F, Djohan R, Bernard S, Gordon CR, et al. First U.S. near-total human face transplantation: a paradigm shift for massive complex injuries. *Plast Reconstr Surg*. 2010;125:111-122.

Part VII

Future Directions in Face Transplantation

38 The Military Relevance of Face Composite Tissue Allotransplantation and Regenerative Medicine Research

Robert G. Hale

Contents

Abstract Maxillofacial battle injuries created the need for plastic and maxillofacial reconstructive surgery during World War I (WWI). Devastating maxillofacial injuries challenged the surgical professions to provide form and function to the defects inflicted by penetrating trauma on the battlefield. Local, regional, and distant flap transfers were developed to treat maxillofacial battle injuries in WWI and while still valued procedures, multiple surgical steps, donor site morbidity and limited esthetic-functional outcomes are often the result. During the 1980s, the advent of microsurgery added an invaluable tool to the reconstructive surgeons' armamentarium. However, despite significant technical advances, the results following major traumatic war injuries are less than satisfactory, both esthetically and functionally. Facial allotransplantation may offer an invaluable alternative to traditional reconstructive options.

Abbreviations

CMV	Cytomegalovirus
ENT	Ear, Nose, and Throat
IMF	Inter-maxillary Fixation
ISR	Institute of Surgical Research
NPWT	Negative pressure wound therapy
OMS	Oral and Maxillofacial Surgeons
WWI	World War I

38.1 Introduction

Maxillofacial battle injuries created the need for plastic and maxillofacial reconstructive surgery during World War I (WWI). Devastating maxillofacial injuries in WWI

R.G. Hale
Craniomaxillofacial Research,
U.S. Army Institute of Surgical Research,
San Antonio, TX, USA
e-mail: rghale56@hotmail.com

M.Z. Siemionow (ed.), *The Know-How of Face Transplantation*,
DOI: 10.1007/978-0-85729-253-7_38,

challenged the surgical professions to provide form and function to defects inflicted by penetrating trauma on the battlefield. Local, regional, and distant flap transfers developed to treat maxillofacial battle injuries during WWI are still valued procedures but multiple surgical steps, donor site morbidity, and limited esthetic-functional outcomes are often the result. Improvement over these multi-staged procedures occurred in the 1980s when microvascular techniques provided options to transfer distant composite tissues in fewer steps to close wounds and provide support. Although microvascular transfer of distant autogenous tissues was an improvement, the bar has recently been raised by face allotransplants. Now, severe facial defects can be repaired with "like" tissue in one step through allotransplantation with favorable prospects of achieving exquisite facial form and function but at the cost of lifetime immunosuppression. This chapter will describe past and present treatment of maxillofacial battle injuries, characterize maxillofacial battle injuries experienced by US service members in Iraq and Afghanistan, and extrapolate into the foreseeable future with recent developments in reconstructive allotransplantation and regenerative medicine.

38.2 Historical Perspective

Techniques based on procedures developed in WWI have been used with some degree of success for the past 90 years. WWI was the first conflict major industrial powers used explosive devices on a large and destructive scale. To avoid certain death on the battlefield, the opposing armies dug trenches and fought with grenades, machine guns, and artillery, exposing only their faces and hands momentarily to engage the opposing side. In essence, the trench served as the soldier's body armor. Over ten million allied soldiers were injured in WWI, and while injury statistics were crude it is estimated the allies treated over 20,000 severe maxillofacial injuries at specialized units in France and Britain.[1] Sir Harold Gilles, a pioneer of plastic and maxillofacial surgery, led the most famous maxillofacial unit of WWI in Britain.

Gilles developed innovative surgical procedures to treat never before seen maxillofacial battle injuries. The most severe facial injuries were characterized by open fractures, avulsions, and burns. In that pre-antibiotic era, bone surgery was limited to closed techniques and aggressive debridement of comminuted, open fractures. Aviation and naval engagements added complexity to the situation with severe facial burns.

To treat the devastating maxillofacial battle injuries Gilles utilized local and regional flaps from the cheeks, forehead, neck, and scalp to replace missing facial features. The central features of the nose and lips were especially challenging and required multiple steps to achieve any degree of success. Gilles used autogenous transfers of distant skin by attaching a tube of chest skin to the hand and then subsequently transferring the skin flap to the face; the procedure was called the "jumping flap."[2] Chest flaps based on supraclavicular pedicles were also elevated and advanced to resurface scarred faces. Since these skin flaps were not necessarily based on vessels, healing was unpredictable.

38.3 Current Concepts

Many autogenous flap techniques developed during WWI remain workhorses of facial reconstruction for mild to moderate face soft tissue defects caused by injury or cancer resection. These flaps unavoidably cause donor site deformity, preferably to a lesser extent than the untreated primary defect. The central facial features remain difficult to reconstruct because these structures are characterized by subtle changes in skin thickness, delicate contours, and varying projections that define a facial subunit. Adding to the complexity of face reconstruction is the replacement of avulsed eyelids and lips; not only are these structures anatomically distinctive in shape and tissue type, reconstruction must provide neuromuscular function in order to achieve an adequate result. To date, total eyelid reconstruction has eluded the surgical professions.[3]

Lip flaps described by Abbe, Eslander, Gilles, and Karapandzic repair partial lip defects up to 75% lip loss by rearrangement of remaining lip structures to restore function at the expense of a normal stoma aperture.[4] Beyond 75% lip loss, up to 100% loss of a lip, cheek advancement flaps and ventral tongue flaps, techniques also dating back to WWI, are described to close the wound; providing hair-bearing cheek skin to reconstruct a missing lip is an obvious advantage to a male who can grow a beard to hide scars. Function after lip reconstruction that uses cheek advancement flaps is obviously compromised due to loss of obicularis oris continuity.

Many large facial defects were inadequately treated until microvascular tissue transfers became available in the 1980s. These microvascular techniques can close large composite defects in one operation with radial, scapula, and fibula flaps. These flaps, however, are limited to restoring basic anatomy for support. The transferred skin bears little resemblance to facial skin and the result often has color, texture, and hair mismatch, with no sensory or motor function. These limitations are perhaps acceptable to elderly cancer survivors but not to young service members severely deformed by battle injuries.

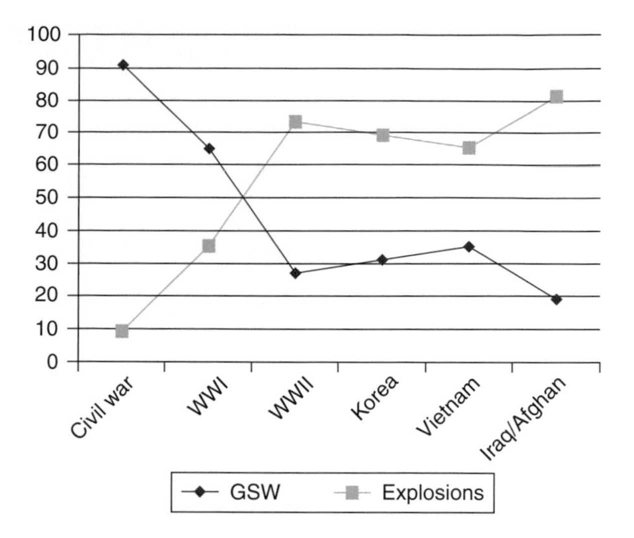

Fig. 38.1 Mechanism of injury, previous US wars

38.4 The US Armed Forces Experience

A recent study involving battle-injured US service members in Iraq and Afghanistan revealed 26% suffered wounds to the maxillofacial region.[5] A battle-injured service member received an average of 4.8 wounds to the body; the maxillofacial area averaged 2.4 wounds with a range of 1–8[5,6] (Table 38.1). The average age of the maxillofacial injured service member was 26 years old and 98% were male. Penetrating trauma is the mechanism of injury in 93% of maxillofacial battle injuries with gunshot wounds accounting for only 6% (Fig. 38.1). Explosive injuries are a major feature of the current conflict in Iraq and Afghanistan, accounting for 84% of maxillofacial battle injuries. Complicated penetrating maxillofacial soft tissue injuries and facial fractures occurred 14% and 27% of cases, respectively. Seventy-six percent of the facial fractures were compound in nature. Burns account for 5% of evacuated casualties; explosions were the primary cause of combat burns (86%), with the face involved 77% and hands 80%.[7]

Table 38.1 Comparison of wounds by body region

	Body area (%)	WWII (%)	Korea (%)	Vietnam (%)	Iraq/AFG (%)
H&N	12	21	21	16	29
Chest	16	14	10	13	6
Abdomen	11	8	8	9	11
Extremities	61	58	60	61	54

Adapted from Owens et al.[6]

Battlefield survivors of major face avulsions are characterized by loss of central facial features, notably portions of the jaws, lips, and nose. If the penetrating trauma is from a nearby explosion, second and third degree burns occasionally complicate the injury by burning skin adjacent to the area of tissue loss, which makes local flaps and tissue transfers difficult or impossible. Severe facial burns often lead to lid ectropian, microstomia, extra-articular ankylosis, and destruction of the cartilaginous portions of ears and noses. This combination of compound fractures, avulsions, and burns, conditions seldom seen in civilian trauma, create a challenge for face surgeons (Figs. 38.2–38.6).

US service members significantly injured in Iraq or Afghanistan are resuscitated and stabilized in the combat theater and then transferred to the Regional Army Medical Center in Germany. Serial and conservative debridement followed by facial fracture stabilization occurs throughout the evacuation process.[8] Within a week of injury in most cases, and within 24–48 h for burn cases, the injured service member is transferred to Walter Reed Army Medical Center, National Naval Medical Center or Brooke Army Medical Center. Polytrauma requires the coordination of specialists in several fields. Oral and maxillofacial surgeons (OMS) share the case load of maxillofacial trauma with otolaryngologists (ENT); in cases of severe facial injury, collaboration between OMS, ENT, and plastic surgery, and consultation between military and civilian institutions is the rule.

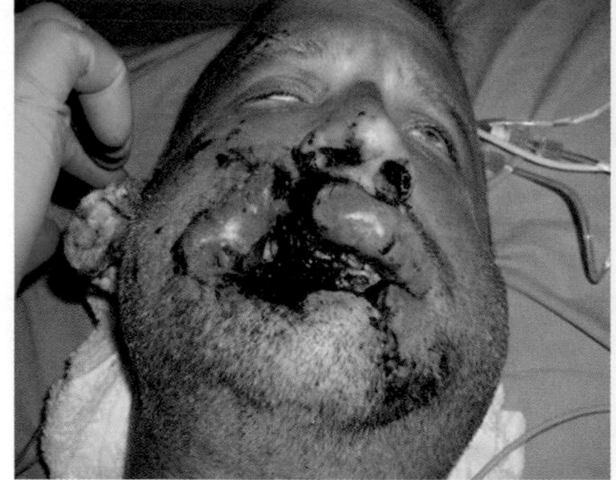
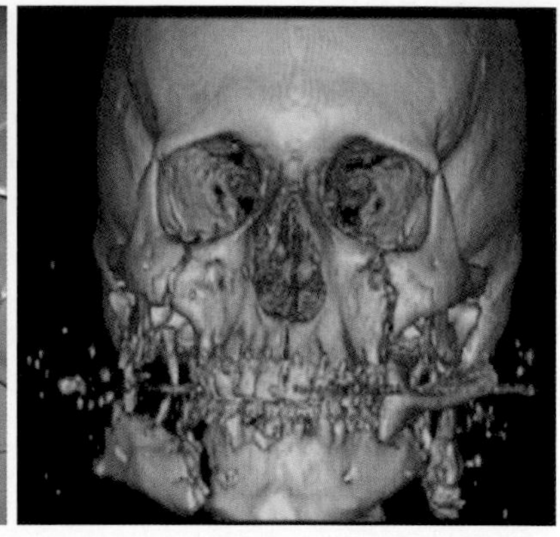

Fig. 38.2 Soldier injured in Iraq by exploding device caused avulsion of two-thirds of lips and open, comminuted mid and lower face fractures (Photo courtesy of COL Hale)

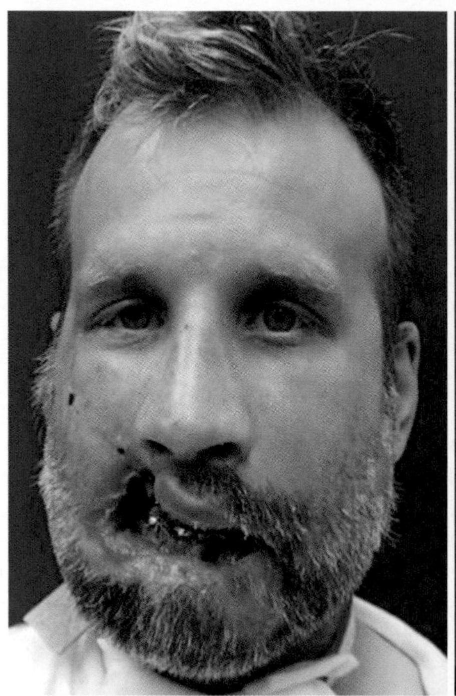
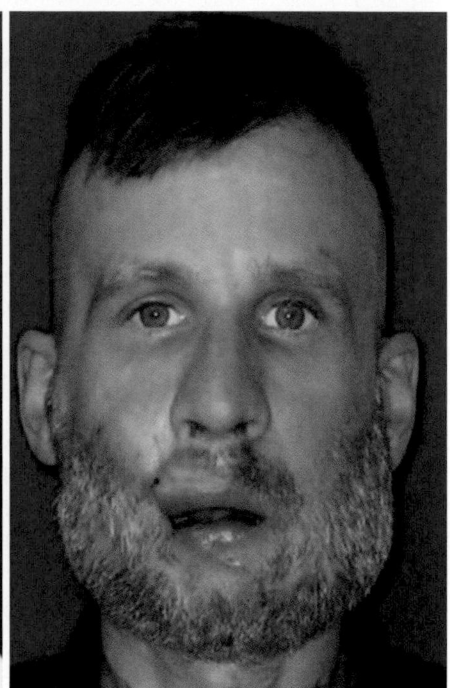

Fig. 38.3 Despite multiple soft tissue procedures to reconstruct the perioral structures, to include rhomboid flap, sliding cheek flaps, buccal mucosa advancement, bilateral rotational-advancement cervicofacial flaps, and ventral tongue flap, this soldier remains deformed and dysfunctional to a significant degree. His most severe disability is unintelligible speech due to lack of neuromuscular integration of reconstructed lip structures (Photo courtesy of COL Hale)

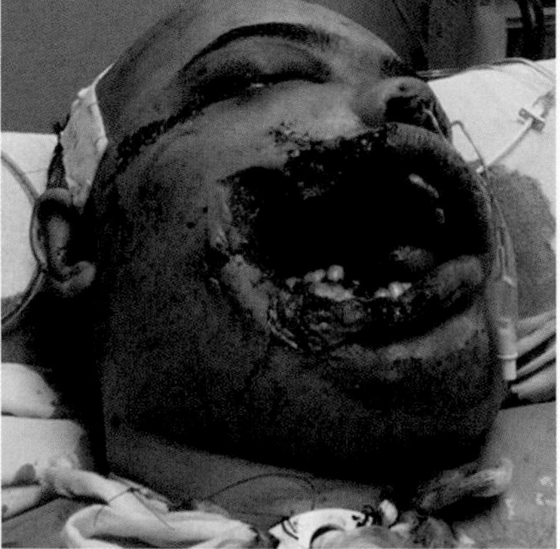
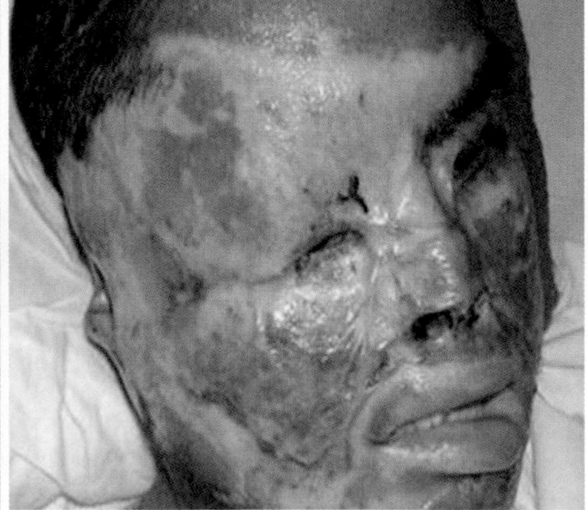

Fig. 38.4 Loss of the central features of the face, such as the lips, nose, and eyelids, by avulsion or burns, are the most difficult facial areas to reconstruct and the most important areas of function for the patient (Photos courtesy of COL Hale)

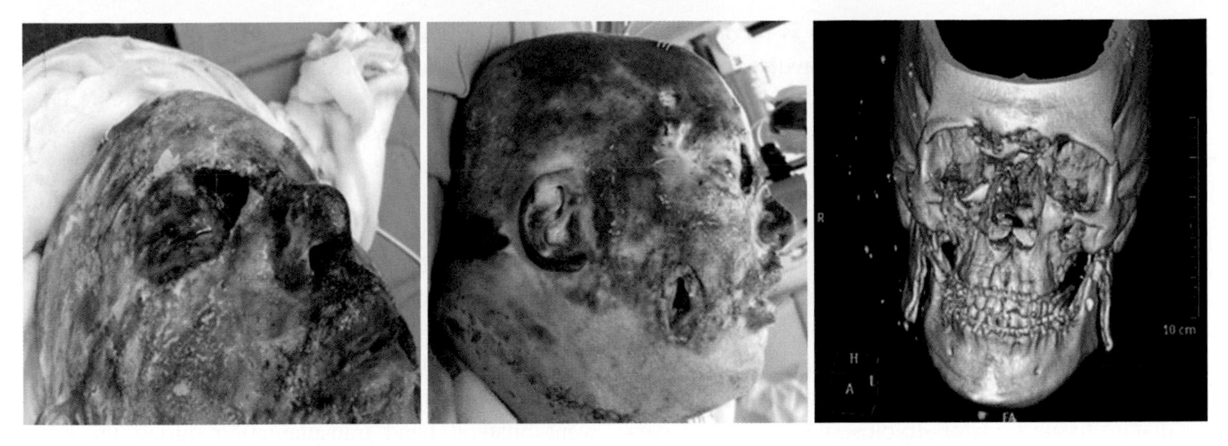

Fig. 38.5 Soldier sustained open, comminuted panfacial fractures and third and fourth degree burns from exploding device. Naso-orbital fractures inoperable due to condition of overlying skin. Despite over 20 surgeries, facial deformity persists (Photo courtesy of COL Hale)

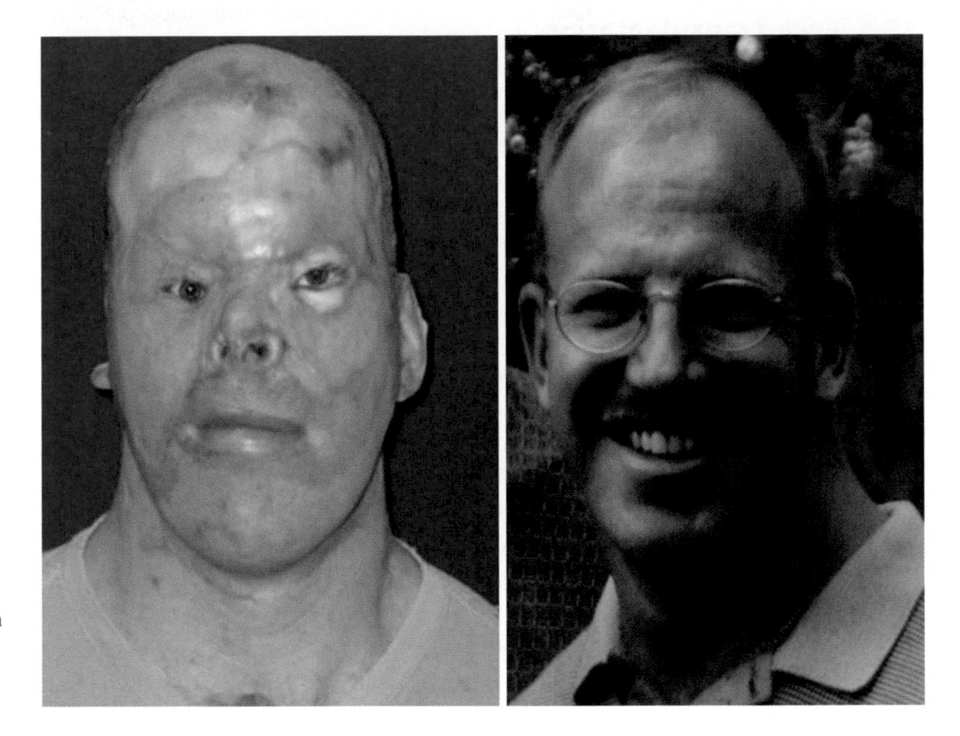

Fig. 38.6 Arguably, only a composite tissue allotransplant could reconstruct this soldier's face to normal form and function. Pre-injury photo seen on right (Photos courtesy of COL Hale)

Complex maxillofacial injuries caused by explosions are addressed by stabilizing the facial skeleton in a similar fashion as blunt trauma patients unless the overlying skin is burned or avulsed. In cases of severe soft tissue compromise, external fixation and inter-maxillary fixation is necessary until serial debridement, flaps and grafts can close the integument. Reestablishment of gross facial dimensions, occlusion, and facial projection guide treatment at this phase. Comminuted fractures deemed non-repairable are debrided and bone replaced with primary grafts in the upper face, midface, and mandibular condyle areas, provided soft tissue coverage is possible; primary bone grafts to repair continuity defects of the mandibular body are avoided until the zone of soft tissue injury is demarcated, debrided, and reconstructed with robust flaps.[9]

Once the existing facial skeleton is reconstructed and wounds closed, reevaluation of avulsed and damaged facial soft tissue features is performed. Treatment options to replace avulsed and damaged features are basically the same options used by reconstructive surgeons for decades: autogenous flaps and grafts with attendant donor site morbidity, and acceptance of multiple procedures and treatment limitations in cases of severe tissue loss or burns. Significant loss of lip structure creates a difficult deformity to reconstruct especially if there is significant involvement of the opposing lip. To avoid severe microstomia, regional or distant tissue transfers to close the wound are performed, but these reconstructions seldom provide acceptable appearance or function.[10]

38.5 Face Transplantation

In November of 2005, a team of surgeons in Amiens, France, led by Drs. Dubernard and Devauchelle performed the first face allotransplant to reconstruct a young woman's entire lower face, to include the nasal tip, lips, and chin.[11] The case was deemed successful from a reconstructive view but controversial due to patient selection criteria and lifetime use of immunosuppressants to prevent graft rejection. The face transplant successfully replaced the missing tissues with "like" tissue from a brain-dead, beating-heart donor. ABO blood type and major histocompatibility antigens were matched, as well as skin color, gender, and age. During the first 18 months after surgery, the patient had two acute rejection episodes requiring hospitalization and high doses of corticosteroids. Post-transplant cytomegalovirus (CMV) and fungal infections also required interventions. Five years later, the patient is reportedly stable with no signs of rejection; the allotransplant appears normal and well-integrated, and partial sensory/motor function has returned to the lips.

The favorable early outcome of the French team's face allotransplant encouraged further interest in the technology: In 2006, a rural Chinese farmer mauled by a bear with partial midface avulsion was reconstructed by Dr. Shuzhong Guo, of Xijing, China. A patient with severe facial neurofibromatosis was treated in 2007 by Dr. Lantieri, of Paris, France. A young woman with midface avulsion from a shotgun blast was treated at Cleveland Clinic in 2008 under the direction of Dr. Siemionow with replacement of the nose, lower eyelids, cheeks, upper lip, and all of the underlying bone supporting these facial features to include the maxilla with nine teeth (a close range shotgun blast is similar to a battle injury to the face in terms of composite tissue destruction). Two more patients were treated by Dr. Lantieri in 2009: a midface reconstruction after a shotgun blast and a severely burned patient who underwent resurfacing of nearly the entire face with a vascularized allogenetic flap that included the nose, ears, eyelids, forehead, and scalp; this burn patient also underwent bilateral hand transplantation during the same operation. Midface reconstruction after forth degree burns to the face with reconstructive allotransplantation was performed by Dr. Pomahac of Bingham and Women's Hospital in Boston. All of these patients previously had unacceptable results following extensive conventional treatment. All patients were treated with systemic immunosuppression to prevent rejection.

To date, 13 face allotransplantations have been performed worldwide but not without serious complications. The Chinese patient died after he returned to his rural home and discontinued immunosuppressant therapy. The burn patient with face and bilateral hand transplants developed multidrug resistant infection and died of sepsis. There are no reports of serious complications in the other face transplant patients; although follow-up is less than 24 months for the majority of patients.

Immunosuppression to prevent graft rejection downregulates the T-cells of the recipient through combinations of several classes of drugs: corticosteroids, antimetabolites, calcineurin blockers, and T-cell depleting antibodies. All of these drugs cause significant side effects beyond suppression of the immune system and increased risk of infections and malignancies.[12] Of the 45 hand transplants performed over the past 10 years, immunosuppression has led to cases of hyperglycemia, diabetes, nephrotoxicity, Cushing's syndrome, avascular hip necrosis, osteomyelitis, CMV infections, papilloma, herpes simplex, and cutaneous mycosis.[13] Besides rejection of the first-hand transplant due to medical noncompliance, one-hand transplant patient recently developed a condition suggestive of post-transplant lymphoproliferative disease and another rejected an allotransplant hand 6 months after transplantation due to accelerated atherosclerosis of the graft (chronic rejection) for unknown reasons (personal communication with Dr. Breidenbach).

Experience over the last 10 years with allotransplantation of hands and faces worldwide has proven the reconstructive technique is possible with current microvascular procedures and high levels of immunosuppression. Despite a composite tissue allotransplant engraftment success rate of over 90%, the long-term success and effects of lifelong immunosuppression is unknown in this group of patients. Additionally, acute flap failure due to vein thrombosis is expected to occur at a rate of 6–10%[14]; this complication would be catastrophic following face allotransplantation, therefore, "rescue" procedures are considered preoperatively. Currently, research in the field of reconstructive allotransplantation is focused on modulating the immune system utilizing fewer drugs, with a trend toward tacrolimus monotherapy for hand allotransplantation. Induction of immunotolerance, considered the "holy grail" by transplant specialists, would eliminate immunosuppressants entirely.[13]

Application of allotransplantation to reconstruct facial defects currently appears suitable for only the most severe cases of facial defects. As researchers develop predictable protocols to modulate the immune system safely, allotransplantation to repair composite facial defects or resurface facial burns will undoubtedly become more acceptable. The recent Cleveland Clinic face allotransplant case was a woman with a near total midface avulsion, dependent on a tracheotomy and feeding tube. This patient underwent multiple conventional surgeries, all predictably futile, before finally becoming a transplant candidate. Although not fully researched and appreciated, the burden of disease in patients with severe face defects must be significant, which is the most compelling argument in favor of face allotransplantation.[15,16] Dr. Maria Siemionow, head surgeon of Cleveland Clinic's face transplant team, said it well: "You need a face, to face the world."

38.6 The US Armed Forces Approach

The US Army Surgeon General (TSG) recognized the limitations of conventional treatment for severe facial battle injuries and established a Face Transplantation Advisory Board in June of 2009. The board is multidisciplinary and composed of an oral/maxillofacial surgeon, an otolaryngology/head and neck surgeon, a plastic surgeon, a maxillofacial prosthodontist, a nurse case manager, a social worker, a medical ethics representative, a psychologist or psychiatrist, an immunologist, Veterans Healthcare and Benefits Administration representatives, a pharmacist, a TRICARE representative, an Office of The Surgeon General representative, a Warrior Transition Command representative, and US Army Medical Research and Materiel Command representatives. The purpose of the board is: recommend face transplantation policy and procedures to TSG and oversee policy execution; be knowledgeable and stay current with the Institutional Review Board–approved face transplantation clinical trials that are established in the USA; when requested, meet in person or via video teleconferencing with soldiers, family members, and their physicians to present facts about face transplantation clinical trials, answer any questions, and provide assistance as needed; assist soldiers interested in pursuing face transplantation with visits to one or more approved clinical sites; and provide TSG with summaries of all meetings.

The Army's policy established the requirement for distribution of face transplantation clinical trial information to Army medical providers and assistance to potential soldier candidates with uncorrectable maxillofacial injuries or facial burns or unacceptable functional or esthetic results following recommended conventional surgery who are remaining on active duty (AD) in a non-deployable status or who are not remaining on AD. Further, the policy stated face transplantation is an option for those with severe maxillofacial injuries or facial burns deemed uncorrectable or for those with results unacceptable from a functional or esthetic standpoint following recommended conventional surgery. Candidates must be at least 18 years old, have severe maxillofacial injuries or face burns that result in unacceptable function or esthetics, and be eligible for TRICARE or Veterans Healthcare Administration (VHA)'s long-term care. Patient participation must be entirely voluntary after the soldier has been fully consented per Institutional Review Board (IRB) protocol.

The Army's policy declared face transplantation is not standard of care and it is only offered as a treatment in the context of a research protocol. If a research team with an approved IRB protocol selects the soldier for face or face/hand transplantation, the soldier will be placed on medical temporary duty at one of the transplant center locations in order to receive extensive preoperative and postoperative evaluations, treatment, and rehabilitation. Furthermore, since recipients will be subject to the risks/complications associated with immunosuppression, TRICARE and/or the VHA will need to bear the long-term costs for immunosuppressive medications at the end of the clinical trial.

The limited pool of brain-dead, beating-heart donors, issues with immunosuppression, and a lifetime expense of approximately $1,000,000 for each patient receiving face allotransplantation will spur science to regenerate the face. Already, regenerated bladders and a trachea "construct" have been successfully developed and transplanted into patients without the need for immunosuppression. As "constructs" of composite tissue are further developed, scientists will focus on the face, a highly vascularized and accessible body part of high value, as their target for reconstruction. Indeed, it is well within the realm of possibilities to regenerate the soft tissue envelope of the face in the next 5–10 years using growth factors, stem cells, scaffolds, engineered skin and, possibly, an in situ bioreactor (Biomask).

A Biomask based on negative pressure wound therapy (NPWT) principles is currently under development by collaborators of the US Army Institute of Surgical Research (ISR). Key to face regeneration is modulation of the inflammatory response responsible for scars and contractures seen during the healing process of an open wound; NPWT, stem cells and pharmaceuticals to modulate inflammation are currently under investigation at the ISR. It is hypothesized that skin grafts (autogenous or esthetically engineered skin) placed over regenerated dermis (neodermis), optimally vascularized by a Biomask, can be further enhanced toward normal skin by early transfer of fat grafts beneath the immature neodermis to favorably influence the tissue remodeling phase. Tissue engineered cartilaginous constructs can be inserted into this regenerated soft tissue envelope to form the projected facial features. Technology to regenerate the nerves and muscles of facial expression is hardly imaginable at this time but when achieved, a fully functional, regenerated face would be the result.

38.7 Conclusion

War once again has greased the wheels of innovation. The maxillofacial area is vulnerable in today's battlefield dominated by explosive devices. Severe maxillofacial battle injuries characterized by complicated lacerations and avulsions, open and comminuted fractures, and facial burns occur at a rate of approximately 10%.[5,7] Conventional treatments with autogenous flaps are futile in the most severe cases of facial injury; despite multiple surgical procedures by experienced surgeons, too often, surgical fatigue terminates reconstruction, not achievement of an adequate result.[8] Arguably, many service members are candidates for face composite tissue allotransplantation, especially severe burn casualties, but the risks of lifetime immunosuppression dampens enthusiasm for that approach. Composite tissue allotransplantation will remain the treatment of last resort until the immune system can be safely and predictably down-modulated or immunotolerance established in the recipients.

The importance of immunosuppression research has been recognized and a major congressional initiative is currently under consideration to fund basic, translational, and clinical research to specifically mitigate immunosuppression problems associated with composite tissue allotransplantation. Ultimately, regenerative medicine will provide "like" functional autogenous tissue to not only reconstruct maxillofacial battle defects but facial defects from all causes. For now, however, composite tissue allotransplantation is the bridge to correct severe maxillofacial battle injuries until regenerative medicine research offers a better solution.

Disclaimer The opinions or assertions contained herein are the private view of the author and should not be construed as official or reflecting the views of the Department of Defense or the US Government; the author is an employee of the US Government. All photographs courtesy of COL Robert G Hale.

References

1. Beebe GW, DeBakey ME. Location of hits and wounds. In: *Battle Casualties*. Springfield: Charles C. Thomas; 1952: 165-205.
2. Santoni-Ruigi P, Sykes PJ. *History of Plastic Surgery*. Heidelberg: Springer; 2007.
3. Vasilic D, Barker J, Blagg R, Whitaker I, Kon M, Gossman MD. Facial transplantation: an Anatomic and Surgical Analysis of the Periorbital Functional Unit. *Plast Reconstr Surg*. 2010;125(1):125-134.
4. Sikes JW, Ghali GE. Lip cancer. In: Miloro M, ed. *Peterson's Principles of Oral and Maxillofacial Surgery*. Vol. 1. Chapter 34. Hamilton, Ontario, Canada: BC Decker; 2004.
5. Lew TA, Walker JA, Wenke JC, Blackbourne LH, Hale RG. Characterization of maxillofacial battle injuries in U.S. service members in Operation Iraqi Freedom and Operation Enduring Freedom. *J Oral Maxillofac Surg*. 2010;68:3-7.
6. Owens BD, Kragh JF, Wenke JC, et al. Combat wounds in Operation Iraqi Freedom and Operation Enduring Freedom. *J Trauma*. 2008;64:295-299.

7. Kauvar DS, Wolf SE, Wade CE, et al. Burns sustained in combat explosions in Operation Iraqi and Enduring Freedom (OIF/OEF explosion burns). *Burns*. 2006;32:853-857.
8. Powers DB, Will MW, Bourgeois SL, et al. Maxillofacial treatment protocol. *J Oral Maxillfac Surg Clin North Am*. 2005;17:341-355.
9. Futran N. Maxillofacial trauma reconstruction. *Facial Plat Surg Cin North Am*. 2009;17:239-251.
10. Peleg K, Aharonson-Daniel L, Stein M, et al. Gunshot and explosion injuries: characteristics, outcomes, and implications for care of terror-related injuries in Israel. *Ann Surg*. 2004;239:311-318.
11. Dubernard JM, Lengele B, Morelon E, et al. Outcomes 18 months after the first human partial face transplantation. *N Engl J Med*. 2007;357:2451-2460.
12. Shuker ST. Maxillofacial blast injuries. *J Craniomaxillofac Surg*. 1995;23:91-98.
13. Siemionow M, Klimczak A. Basics of immune responses in transplantation in preparation for application of composite tissue allografts in plastic and reconstructive surgery: part I. *Plast Reconstr Surg*. 2008;121:4e-12e.
14. Jones NF, Johnson JT, Shestak KC, et al. Microsurgical reconstruction of the head and neck: interdisciplinary collaboration between head and neck surgeons and plastic surgeons in 305 cases. *Ann Plast Surg*. 1996;36:36-43.
15. Rumsey N. Psychological aspects of face transplantation: read the small print carefully. *Am J Bioeth*. 2004;4:22.
16. Morris PJ, Monaco AP. Facial transplantation, is the time right? *Transplantation*. 2004;77:329.

39 International Registry of Face Transplantation

Palmina Petruzzo, Marco Lanzetta, and Jean Michel Dubernard

Contents

P. Petruzzo (✉)
Department of Transplantation,
Edouard Herriot Hospital,
Lyon, France
e-mail: petruzzo@medicina.unica.it,
palmina.petruzzo@chu-lyon.fr

Abstract The International Registry on Hand and Composite Tissue Transplantation (IRHCTT) was founded in 2002 and its purpose was to collect detailed information from every case of hand and face transplantation providing a unique opportunity for the teams to share their experiences and to keep abreast of the latest developments. The registry of face transplantation presents some peculiar characteristics as face is considered a complex "organ" constituted of anatomic parts with different functions. In the Registry facial deficits are expressed as "aesthetic units," adding also the depth of the defect; and in the procedure section, it is not only important to report the transplanted aesthetic units but also to detail all the grafted tissues, vascular anastomoses, nerve repairs, and eventual additional surgical procedures. Immunosuppressive regimen, rejection episodes, side effects, or other complications are fully reported. The evaluation of outcomes is difficult as disfigurement involves different parts of face with loss of various functions. Consequently, it is important to evaluate both aesthetic and functional results, detailing the recovered functions, such as swallowing, eating and drinking, speaking, or opening and closing eyelids. Two items, such as "Psychological and social acceptance" and "Patient satisfaction and general well-being" are also reported as the goal of face transplantation is to allow the patient to have a social life and improved quality of life.

Abbreviations

CTA	Composite tissue allotransplantation
fMRI	functional magnetic resonance imaging
IRHCTT	International Registry on Hand and Composite Tissue Transplantation
SWMT	Semmes-Weinstein Monofilament Test
WEST	Weinstein Enhanced Sensory

M.Z. Siemionow (ed.), *The Know-How of Face Transplantation*,
DOI: 10.1007/978-0-85729-253-7_39, © Springer-Verlag London Limited 2011

39.1 Introduction

The International Registry on Hand and Composite Tissue Transplantation (IRHCTT) was founded in 2002, 4 years after the first hand allotransplantation performed in September 1998. The purpose of the Registry was to collect detailed information on voluntary basis from every case of hand transplantation providing a unique opportunity for centers performing hand transplantation to share their experiences and to keep abreast of the latest developments.[1-3]

The Registry is meant to help the scientific community to better understand what it should or should not do, providing data for discussion and critical analysis. In addition, this information might be also useful for the new teams wishing to start a composite tissue allotransplantation program. In these years the Registry included only uni- or bilateral hand transplantations;[1-3] however since 2010 the IRHCTT includes also cases of face transplantations therefore constituting two sections of the same Registry (www.handregistry.com).

A registration form, an annual update and two forms, which allow for the evaluation of functional recovery every year, have been created to collect data of hand transplantations. At the same time a registration form and an annual update, which allows to evaluate the recipient's general condition and recovery of facial functions, have been created to collect data of face transplantations.

39.2 Specifics of Face Transplant

The registry of face transplantation is very similar to the hand transplantation one as they both are concerned with composite tissue allotransplantations; however, it presents some peculiar characteristics as face is considered a complex "organ" constituted of anatomic parts with different functions.[4,5]

39.2.1 Anatomical Description

Firstly, recipients of facial transplantations present a severe disfiguration due to many causes and involving different anatomic parts of face with consequent loss of their function, while recipients of hand transplantation present uni- or bilateral amputation of the upper extremities at different levels with a consequent handicap which is easier to be evaluated. In the Registry, facial deficits are expressed as "aesthetic units," adding also the depth of the defect; it is also important to precise the number of previous surgical operations, which can influence the final result of transplantation (Table 39.1). Face transplantation is a very complex procedure, which may involve different parts of face, and for this reason it is not only important to report the transplanted aesthetic units but also to detail all the grafted tissues, vascular anastomoses, nerve repairs, and eventual additional surgical procedures (Table 39.2). At the end of the procedure the attachment of a picture might be useful.

39.2.2 Donor and Recipient Selection

Donor and recipient selection is always a crucial point in transplantation, although at present common criteria of selection do not exist for hand or face transplantation; in both sections of the Registry many data concerning donor and recipient are collected in order to create them in the near future.

39.2.3 Immunosuppression

Immunosuppressive regimen, rejection episodes, side effects, or other complications are fully reported in both sections of the Registry as they are essential elements in the evaluation of a "transplantation." For the first time in a CTA, bone marrow transplantation[6] and extracorporeal photopheresis[6,7] have been also used in face transplantation, thus remarking that it is not only a complex procedure of "reconstructive surgery," but also a "complex transplantation."

39.2.4 Infectious Complications and Prophylaxis

In hand as well as in face transplantation the prophylaxis is also reported, being important to avoid some

Table 39.1 Characteristics of disfiguration

Cause of disfiguration:		
Trauma ❍ yes ❍ no	Burn injury ❍ yes ❍ no	Congenital ❍ yes ❍ no
Date of disfiguration:		
Deficit (aesthetic units):		
Lateral		
Forehead (single side) ❍	Forehead (bilateral) ❍	
Brow (single side) ❍	Brow (bilateral) ❍	
Perioral (single side) ❍	Perioral (bilateral) ❍	
Cheek (single side) ❍	Cheek (bilateral) ❍	
Ear (single side) ❍	Ear (bilateral) ❍	
Central		
Nose ❍		
Upper lip ❍		
Lower lip ❍		
Chin ❍		
Tongue ❍		
Exposed tissue (depth):		
Subcutaneous: ❍ yes ❍ no		
Muscle: ❍ yes ❍ no		
Bone: ❍ yes ❍ no		
Blindness: ❍ yes ❍ no		
Speech: ❍ yes ❍ no		
Swallowing: ❍ yes ❍ no		
Surgery before: ❍ yes ❍ no		
Number of previous surgical operations: None ❍ Specify the number:		

infectious complications, such as cytomegalovirus infection. In addition, in face transplantation, oral mucosa can be included in the grafted tissues increasing the risk of infections.

39.2.5 Rejection Episodes

In the section concerning rejection episodes, the possibility to biopsy skin, mucosa, and muscle has been considered; the grade of severity has been scored on the basis of Banff score for CTA[8] used also for hand allotransplantation.

39.3 Cortical Re-organization and Rehabilitation

The patients undergoing hand allotransplantation have to follow a hard rehabilitation program, which includes a standard rehabilitation program for flexor and extensor tendons, sensory reeducation, and cortical reintegration. Indeed, after transplantation functional magnetic resonance imaging (fMRI) evidenced that hand representation shifts from lateral to medial region in the motor cortex and reoccupies the normal hand region.[9] fMRI should be also performed in recipients of face transplantation to investigate possible cortical alterations particularly when the disfiguration involves units

Table 39.2 Surgical procedure

Grafted aesthetic units
Scalp (full/partial/anterior/posterior/right/left) ❍
Forehead (full/right/left) ❍
Ear (right/left) ❍
Upper eyelid ❍
Lower eyelid ❍
Lacrimal glands/ducts:
Nose:
Cheeks (right/left):
Upper lip:
Lower lip:
Tongue:
Chin:
Neck:
Other:
Grafted tissues
Skin: ❍ yes ❍ no
Soft tissue: ❍ yes ❍ no
Bones: ❍ yes ❍ no
Cartilage: ❍ yes ❍ no
Salivary gland: ❍ yes ❍ no
Arterial anastomoses (specify):
Bilaterally: ❍ yes ❍ no
Vein anastomoses (specify):
Bilaterally: ❍ yes ❍ no
Intra-allograft vascular shunts:
Nerve repairs:
Facial nerve (specify branches):
Infraorbital nerve ❍
Mentonian nerve ❍
Alveolar inferior nerve ❍
Hypoglossal nerve ❍
Lingual nerve ❍
Zygomatico-orbital nerve: ❍
Supraorbital nerve ❍
Great auricular nerve ❍
Great occipital nerve ❍
Other:
Nerve transfers (specify):
Muscle transfers (specify):
Type of bone reconstruction
Topography:
Osteosynthesis:
Teeth involved: ❍ yes ❍ no
Blood transfusions: ❍ yes ❍ no
Number of ml:
Duration of procedure:

dedicated to functions such as speaking, and their modifications after transplantation.

In face transplantation as well as in hand transplantation, there are no standardized protocols for grafted patients' rehabilitation; therefore, the majority of teams apply the same protocols used after reconstructive procedures. For this reason in the Registry, in both sections, information concerning the rehabilitation procedures are collected in order to remark their importance and to develop these procedures.

39.3.1 Measuring Hand Outcomes

In hand transplantation, the evaluation of functional outcome was very difficult and largely discussed as the teams used different scores, which were created to evaluate hand/limb replantation or upper limb disabilities. At present in the Registry, all the recipients have been evaluated using the IRHCTT score[2] and the Dash score.[10] The Registry score was created to measure recipient ability and performances at the different time points of the follow-up, evaluating both cosmetic and functional results, including social behavior, patient satisfaction, and work status. The Disabilities of the Arm, Shoulder and Hand (DASH) Outcome Measure is a 30-item, self-report questionnaire designed to measure physical function and symptoms in patients with any or several musculoskeletal disorders of the upper limb. The Dash score is an international validated system of measure but it was not created to evaluate "grafted hands."

39.3.2 Measuring Face Outcomes

In face transplantation, the evaluation of outcomes is more difficult than in hand transplantation as the disfiguration involved different parts of face with consequent loss of various functions. It is important to evaluate both aesthetic and functional results (Table 39.3).

It is very difficult to score the aesthetic results; in the Registry color match, scar and volumetric evaluation are reported and scored from 0 to 5. The possibility to add a picture would be useful. Both sensibility and motion recovery are reported: recovery of sensibility was assessed using thermal test, the touch threshold test using monofilaments such as the Weinstein Enhanced Sensory

Table 39.3 Results

1. Aesthetic results – Color match – Scar – Volumetric evaluation
2. Functional results *Sensibility* – Heat and cold test – Two-point sensory discrimination test *Motility* – Muscular tonus – Active motion – Synkinesis
3. Daily activities – Open and close eyelid – Nose function – Chew – Swallow – Drink – Eat – Speak – Smile – Kiss – Blow

Test (WEST) or Semmes-Weinstein Monofilament Test (SWMT) and the recovery scored as absent, partial, or total. Motion recovery includes characteristics of the muscular tone, which can be normal, flaccid, hypotonic, little (slight), or very hypertonic; the grade of motion recovery of the muscles, which were involved in grafting; and presence of synkinesis. Moreover, the recovered functions after transplantation, such as swallowing, eating and drinking, speaking, or opening and closing eyelids, have also to be reported constituting the "daily activities" which modify the quality of life of these patients. Finally, two parts of the sections are dedicated to "Psychological and social acceptance" and to "Patient satisfaction and general well-being" in order to express how the patient feels after the transplantation as the goal of face transplantation is to allow the patient to have a social life and consequently to improve the quality of life. For this reason, in the near future the Rosenberg self-esteem questionnaire[11] will be evaluated by the teams involved in face transplantation and perhaps integrated in the Registry.

39.4 Conclusion

In conclusion, in the section "face transplantation" of the IRHCTT, we should be able to give the information which will allow the scientific community to know indications, limits, complications, and results of this type of transplantation and, consequently, what we would rather do or not do.

References

1. Lanzetta M, Petruzzo P, Margreiter R, Dubernard JM, et al. The International Registry on Hand and Composite Tissue Transplantation. *Transplantation*. 2005;15(79): 1210-1214.
2. Lanzetta M, Petruzzo P, Dubernard JM, et al. Second report (1998–2006) of the International Registry of Hand and Composite Tissue Transplantation. *Transpl Immunol*. 2007;18:1-6.
3. Petruzzo P, Lanzetta M, Dubernard JM, et al. The International Registry on Hand and Composite Tissue Transplantation. *Transplantation*. 2008;86:487-492.
4. Sieminonow M, Sonmez E. Face as an organ. *Ann Plast Surg*. 2008;61(3):345-352.
5. Gordon CR, Siemionow M, Coffman K, et al. The Cleveland clinic FACES score: a preliminary assessment tool for identifying the optimal face transplant candidate. *J Craniofac Surg*. 2009;20:1969-1974.
6. Dubernard JM, Lengele B, Morelon E, et al. Outcomes 18 months after the first human partial face transplantation. *N Engl J Med*. 2007;357:2451-2460.
7. Hivelin M, Siemionow M, Grimbert P, Lantieri L. Extracorporeal photopheresis: from solid organs to face transplantation. *Transpl Immunol*. 2009;21:117-128.
8. Cendales LC, Kanitakis J, Schneeberger S, et al. The Banff 2007 working classification of skin-containing composite tissue allograft pathology. *Am J Transplant*. 2008;8: 1396-1400.
9. Giraux P, Sirigu A, Schneider F, Dubernard JM. Cortical reorganization in motor cortex after graft of both hands. *Nat Neurosci*. 2001;4:1.
10. Gummesson C, Atroshi I, Ekdahl C. The disabilities of the arm, shoulder and hand (DASH) outcome questionnaire: longitudinal construct validity and measuring self-rated health change after surgery. *BMC Musculoskelet Disord*. 2003;16:11.
11. Rosenberg M. *Society and the Adolescent Self-image*. Princeton, NJ: Princeton University Press; 1965.

Concomitant Face and Upper Extremity Transplantation

40

Chad R. Gordon, Fatih Zor, and Maria Z. Siemionow

Contents

M.Z. Siemionow (✉)
Department of Plastic Surgery, Cleveland Clinic,
Cleveland, OH, USA
e-mail: siemiom@ccf.org

Abstract From its first origination involving successful rat hind limb allograft studies using cyclosporine, face and upper extremity transplantation (CTA) has since developed into an exciting and promising subset of reconstructive transplant surgery. Current surgical technique involving CTA has allowed optimal outcomes in patients with massive facial and/or upper extremity defects, however, with its coexisting immunological barrier; obligatory lifelong immunosuppression commits each patient to a daily risk of transplant-related complications with many unanswered questions. Since 1998, nearly 50 hands in 40 patients have been performed around the world at various levels ranging from wrist to shoulder. However, the risk-to-benefit ratio remains controversial in bilateral versus unilateral transplantation and has yet to be answered. A total of ten face transplants have been performed since 2005. Concomitant CTA, which involves a variable combination of allograft subtypes, has been performed in two of the nine face transplant patients. These have included simultaneous bilateral hand transplants and tongue with mandible. Future study is warranted to investigate the potential advantages and disadvantages of using this approach versus a staged manner for reconstruction.

Abbreviations

CTA	Composite tissue allotransplantation
CsA	Cyclosporine A
ECP	Extracorporeal phoresis
ICU	Intensive care unit
MMF	Mycophenolate mofetil

M.Z. Siemionow (ed.), *The Know-How of Face Transplantation*,
DOI: 10.1007/978-0-85729-253-7_40,

40.1 Background

From its first origination involving successful rat hind limb allograft studies using cyclosporine, face and upper extremity composite tissue allotransplantation (CTA) has since developed into an exciting and promising subset of reconstructive transplant surgery.[1] Numerous small and large animal models have since been developed for the investigation of applicability, anatomical study, and immunotolerance for achieving donor-specific tolerance. These models include those involving rats,[2,3] rabbits,[4] dogs,[5] nonhuman primates,[6,7] and miniature swine.[8] Valuable anatomical cadaver studies have also been insightful for many reasons such as providing roadmaps for donor allograft recovery and alloflap design,[9-11] demonstrating pertinent vascular territories,[12] and for identifying surface area estimations of potential antigenic tissue loads transplanted with face/scalp and upper extremity transplantation[10,13] (Fig. 40.1).

Current surgical technique has allowed optimal outcomes in patients with massive facial and/or upper extremity defects in the setting of CTA.[14] However, with its coexisting immunological barrier, obligatory lifelong immunosuppression commits each patient to a daily risk of transplant-related complications with many unanswered questions, analogous to solid organ transplantation.[15]

Since 1998, nearly 50 hand transplantations in 40 patients have been performed around the world.[16] Upper extremity transplantation, both unilateral and bilateral, has been shown to be safe and effective when performed in the appropriate recipient. Various levels of upper extremity transplantation have been performed ranging from wrist level to shoulder level. However, the risk-to-benefit ratio remains controversial in bilateral versus unilateral transplantation and has yet to be answered. From recent experience, the two most important determinants as to the success of each patient's upper extremity transplant are patient compliance and intense rehabilitation.[17]

In comparison, nine face transplants have been performed over the last 4 years. Overall long-term functional and aesthetic outcomes have been well received. In summary, multiple aesthetic subunits (i.e., nose, lips, eyelids) with or without underlying craniofacial skeletal defects (i.e., maxilla, mandible) have been successfully restored, thereby providing restoration of vital facial functions (i.e., smiling) in an unprecedented

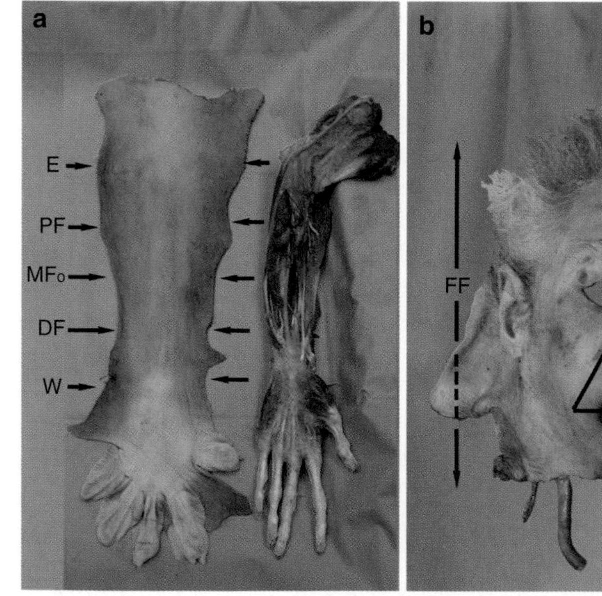

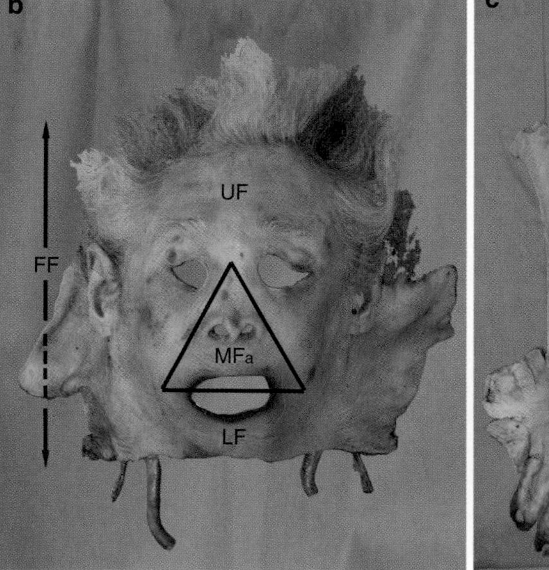

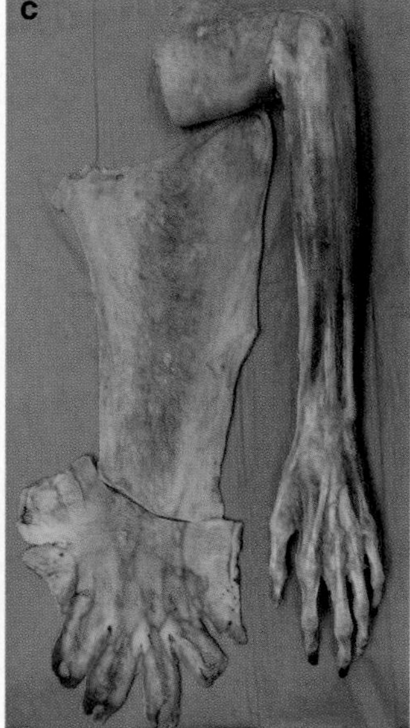

Fig. 40.1 (**a**) Right upper extremity transplant with various levels of accompanying skin component (E-elbow, PF-proximal forearm, MFo-middle forearm, DF-distal forearm, W-wrist) (**b**) Full face and scalp transplant with accompanying skin component (FF-full face, UF-upper face, MFa-middle face, LF-lower face) (**c**) Left arm with accompanying skin component

Table 40.1 Theoretical advantages and disadvantages of performing concomitant composite tissue allotransplantation

Advantages	Disadvantages	Remains unanswered
Single stage reconstruction	Increased perioperative mortality and morbidity	Rehabilitation outcomes
Cost	Increased overall anesthesia time	Patient satisfaction outcomes
Psychological acceptance	Need for a larger surgical team	Antigenicity of one donor versus two donors
Immunosuppression	Challenging cortical reintegration	
Single donor needed		
Single rehabilitation period		

manner.[18] Unfortunately, this type of surgery comes with an estimated 2 year mortality of 20% which is in contrast to hand transplantation where all hand transplant recipients are still alive today.

Concomitant CTA, which involves a variable combination of allograft subtypes, has been performed in two of the nine face transplant patients. These have included simultaneous bilateral hand transplants and tongue with mandible.[18-20] Future study is warranted to investigate the potential advantages and disadvantages of using this approach versus a staged manner for reconstruction (Table 40.1). Concomitant transplantation of non-vital organ structures, such as the face and upper extremity, for example, is still a challenging dilemma for many reconstructive surgeons. For this reason, we will explore the potential concerns and vast challenges that lie ahead.

40.2 Peri-transplant Challenges

40.2.1 Surgical and Medical Challenges

The first concomitant CTA transplantation involving face and bilateral hand transplants was performed in March 2009 by Lantieri and colleagues (Paris, France), but unfortunately the patient died due to severe infection with resulting septic shock at 2 months post transplant.[21] This outcome immediately opened a new discussion in the field of composite tissue allotransplantation, thereby changing the question from "Can we?" to "Should we?" perform simultaneous face/extremity transplantation. Numerous challenges exist and each one needs to be addressed appropriately prior to full acceptance of multi-CTA transplants by reconstructive transplant surgeons.

40.2.1.1 Surgical and Technical Challenges

Technically, simultaneous face and upper extremity transplantation is possible due to modern-day surgical expertise in reconstructive microsurgery and craniofacial principle. However, it is much more challenging with respect to single stage transplantation. In the case of both bilateral hand and concomitant face/upper extremity transplantation, a large multidisciplinary team (preferably 3–4 designated teams) is needed.[17,18]

Hand transplantation is immunologically challenging, rather than surgical challenging, since surgeons have mastered microsurgery and refined the technique of limb replantation through experience for nearly 40 years.[22] However, facial transplantation paints a rather different picture. The vascular anatomy of the face is well known but its surgical application to the harvesting of a free facial alloflap remains uncertain. The reconstructive need of each CTA patient is different and constitutes a new surgical technique and flap design. Thus, face transplantation is a true surgical challenge because of technical obstacles and uncertainty over the functional and anesthetic results[12]. Combining face transplantation with upper extremity transplantation would further extend complexity of the surgical procedure.[23]

40.2.1.2 Medical Challenges

Perioperative medical challenges can be related to anesthesia, maintenance of adequate hemodynamics, blood loss, and fluid/electrolyte balance. Performing simultaneous CTA significantly increases the operation time and prolonged exposure to anesthetic agents, which may induce potential life-threatening problems such as cardiac arrest, etc. Inhaled agents have a significant negative ionotropic effect with associated vasodilatation and decreased cardiac output. Anesthetic agents cause decreased systemic vascular resistance with subsequent reflex tachycardia. Additionally, the postoperative recovery time becomes longer with an increased risk of brief episodes of hypoxia and/or dyspnea. The intensive care unit (ICU) stay of the patient will most likely be longer with concomitant CTA. Postoperative pain management will also be more difficult in the combined procedure, so simultaneous face/upper extremity transplant patients are more likely to complain from severe postoperative pain and/or report perioperative anxiety.[24-26]

Blood loss during this type of surgery is an important factor that must be taken into account. During face transplantation, for example, a significant amount of surgical blood loss has been reported including up to 35 units by Lantieri and colleagues (world's third face transplant involving a patient with neurofibromatosis).[18] Simultaneous hand/upper extremity transplantation should not significantly increase the overall blood loss since this is performed under tourniquet control; however, the application of bilateral tourniquets for over 10–12 h may jeopardize the patient's hemodynamic status considering simultaneous blood loss from face transplantation. Thus, the recipient's requirement for blood transfusion, as previously reported in cases of face transplantation, brings additional hemodynamic and immunological risks.

Another challenge is the maintenance of fluid and electrolyte balance, which is more complicated with increased blood loss. Fluid and electrolyte disturbances have negative effects on intra-operative hemodynamics, so their cardiovascular status must be monitored intensively to prevent complications related to end-organ perfusion. Hemodynamic management of the CTA patient is based on maintaining normovolemia. Dextran and acute normovolemic hemodilution are often needed to decrease blood viscosity. Anesthesia management of transplant surgery requires choices targeted to the physiological variables that regulate microcirculatory flow variables such as vessel diameter, perfusion pressure, blood viscosity, coagulability, and blood volume.[27,28] Thus, maintaining normal hemodynamic parameters during combined upper extremity/face transplantation is much more difficult to manage. Finally, long tissue ischemia, blood loss, and hemodynamic instability may expose the patient to a higher risk of infection specifically in the context of immunosuppression.

40.2.2 Donor/Recipient Challenges

40.2.2.1 Donor-Related Challenges

Recovering concomitant tissues from a donor presents numerous concerns. These include dignified treatment of the deceased, donation decision making, and the psychological impact on the donor's family. Finding a proper donor for simultaneous hand and face transplantation would be of significant challenge.[29] This would require the family to choose either reconstruction of the created defects with multiple prosthetics or a decision for cremation.

40.2.2.2 Patient Selection Challenges

One of the most important predicting factors for a successful outcome in CTA is patient selection. The indications for bilateral hand transplantation are fast becoming widely accepted. Unfortunately, there are few articles addressing the screening indications and final patient selection of face and hand transplant candidates[30] (Table 40.2). In terms of concomitant face/upper extremity transplantation, the indications are not clear. The most likely scenario for patients suffering simultaneous facial and upper extremity disfigurement is in the case of severe burn injuries. However, there are many unanswered questions about whether or not to perform CTA in a severely burned patient.

Patients with severe burn injuries are often subjected to large amounts of transfused blood/products which may contribute to transient immunosuppression. In addition, some patients following debridement require cadaveric skin allograft coverage resulting in undesired immune-sensitization, which must be taken into account prior to performing CTA.[31] Therefore, the potential advantages/disadvantages of concomitant face/hand transplantation versus staged reconstruction in burn patients must be decided on a case-by-case basis.[32]

Table 40.2 A preliminary, screening tool used by the Cleveland Clinic for face transplant candidate selection

Category	Point system	
Functional status (SBSSS + KPS)	Straus–Bacon Social Stability Score	0–4
	Steady job for last 3 years = 1 pt	
	Same residence for the past 2 years = 1 pt	
	Married and lives with spouse/partner = 1 pt	
	Does not live alone = 1 pt	
	Karnovsky Performance Score	2–9
	Capable of normal activity, minor symptoms = 9 pts	
	Normal activity with effort = 8 pts	
	Cares for self, unable to carry on normal activity/work = 7 pts	
	Requires occasional assistance, can take care of most tasks = 6 pts	
	Requires considerable assistance, needs frequent medical care = 5 pts	
	Disabled, requires special care and assistance = 4 pts	
	Severely disabled, hospital admission indicated = 3pts	
	Very ill, urgently requiring admission = 2 pts	
Aesthetic deficit (i.e., aesthetic units)	Lateral	1–18
	Forehead (single side) = 1 pt	
	Brow (single side) = 1 pt	
	Periorbit (single side) = 1 pt	
	Cheek (single side) = 1 pt	
	Ear (single side) = 1 pt	
	Central	
	Nose = 2 pts	
	Upper lip (upper perioral area) = 2 pts	
	Lower lip (lower perioral area) = 2 pts	
	Chin = 2 pts	
Comorbidities	Cardiovascular status WNL = 1 pt	0–6
	Hematological status WNL = 1 pt	
	Hepatic status WNL = 1 pt	
	Nervous system status WNL = 1 pt	
	Pulmonary status WNL = 1 pt	
	Renal status WNL = 1 pt	
Exposed tissue (i.e., depth)	Subcutaneous tissue = 5 pts	5–15
	Muscle = 10 pts	
	Bone = 15 pts	
Surgical history (SH)/recipient vessel patency(RVP)	Extensive SH (>10 surgeries)/below-average RVP = 2 pts	2–8
	Moderate SH (5–10 surgeries)/average RVP = 4 pts	
	Minimal SH (<5 surgeries)/above-average RVP = 8 pts	

Source: Gordon et al. [30]
WNL within normal limits

40.2.3 Immunological Challenges

The most difficult challenge in transplantation of composite tissue allografts is the immunological aspect and the need for lifelong immunosuppression. The immunosuppressive protocols applied in cases of face and upper extremity transplantation prevent acute rejection and allow for long-term allograft survival. However, reports on hand and face transplantation confirm that there is currently no standard immunosuppressive protocol. Numerous centers employ varying induction and maintenance therapy protocols, as well as a different battery of antiviral, antifungal and antimicrobial medications. The most common regimen used for face and upper extremity transplantation is based on kidney transplant protocols and includes standard induction therapy followed by triple maintenance therapy.[33] However, deciding on a particular immunosuppressive regimen for concomitant face/upper extremity transplantation is more challenging considering the increased antigenic load of skin and other tissue components (Fig. 40.1). Of note, the face and bilateral hand transplant patient in France received induction with anti-lymphocyte serum, and then FK 506, prednisone, mycophenolate mofetil (MMF), and extracorporeal photopheresis (ECP).[21]

40.2.4 Rehabilitation and Recovery Challenges

The human brain plasticity allows for adaptation and cortical reorganization as proven with hand replantation.[34] It is not clear, however, what the true potential is for reversibility of the lost function of the hand or face when years have passed since the patient's original amputation or trauma. This often coincides with neuronal degeneration of the peripheral nervous system and subsequent regrowth of peripheral axons for which the reinnervation of new targets are expected. The patients, whose amputated body parts are replaced with allotransplantation, represent a unique example of neural plasticity.

Reorganization of sensory as well as cortico-motor representation is ubiquitous in the mature brain and occurs at the subcortical and cortical levels.[35] The mechanism underlying posttransplant cortical remapping remains unclear. The topography of the somatosensory maps of our body can be largely shaped by alterations of peripheral sensory inputs.[36]

Farne et al. investigated upper extremity tactile perception in transplanted patients using touch of the transplanted hand alone or in combination with another body part.[37] They found that newly acquired somatosensory awareness of the transplanted hand was hampered when the recipient's ipsilateral face was touched simultaneously. This study suggests that performing simultaneous face and upper extremity transplantation may in fact interfere with cortical reorientation and cortical representation of the face and hand. In this context, another challenge will be the choice of preference for organ specific rehabilitation – for example, face before hand or vice versa, as well as timing of rehabilitation.

40.3 Unresolved Challenges

40.3.1 Cost-Factor Challenges

It is obvious that single-stage operations are more cost effective than multistage surgeries. However, combining two or three major operations into one complex single-stage procedure may increase overall morbidity and mortality, thereby jeopardizing optimal outcomes. In order to truly determine the cost factor of concomitant CTA transplantation such as face and hand, all risks and benefits for each specific case must be clearly defined and discussed in detail with the patient.[15]

40.3.2 Ethical Challenges

Long-term clinical outcomes of upper extremity transplant patients and recent cases of face transplants suggests that in the properly selected patient population, face and hand transplants are ethically justified. There is, however, less data supporting unilateral hand transplantation.[38,39] It is clear that ethical challenges related to concomitant face and upper extremity transplantation will be even more difficult to predict considering overall risk-to-benefit ratio which will be specific to each patient. The fact that the only concomitant face and bilateral hand transplant patient died unexpectedly will make ethical justification even more challenging.[29]

40.3.3 Psychological Challenges

Psychologically, face and upper extremity transplants are well accepted in the immediate postoperative period. Rapid integration of the facial or extremity allograft into the patient's "new" body image is greatly assisted through the quick recovery of skin sensation. Overall, the psychological problems related to simultaneous hand and face transplantation seem to be manageable without complication.[15] However, in the context of complicated rehabilitation and its unknown effect on simultaneous cortical reorientation, the recipient's psychological acceptance following concomitant transplantation may be suboptimal leading to incompliance and frustration.

40.3.4 New Challenges

There are new emerging issues pertaining to CTA. The most significant are the unexpected death of the world's second face transplant recipient,[40] the death of the world's first concomitant CTA patient (face and bilateral hand),[18] the amputation of the world's first hand transplant recipient,[17] and the recent hand amputation of a Louisville patient.[41]

Many questions remain unanswered, such as "What is chronic rejection?"[42] "Should we be performing concomitant CTA?" and "Should we use a staged CTA algorithm?" Due to our field's limited experience and the overall extreme complexity of concomitant CTA,[23,43] we suggest performing the upper extremity transplant(s) at least 1 year prior to performing face transplantation in an effort to allow effective upper extremity rehabilitation and cortical reorganization.[35,36] Also, we must find answers to "What effect does concomitant CTA have in cortical reorganization in a staged versus concomitant approach?" and "Is a prolonged operative time (>30 h) involving simultaneous CTAs safe and ethically justified?[37]"

Another challenging dilemma in regards to candidate selection for face and/or upper extremity transplantation is patient blindness, and whether or not this should be considered a contraindication. Obviously, patients with significant craniofacial defects pursuing face transplantation are preselected to have some sort of mid-face injury jeopardizing vision. Prior research gathered in hand replantation demonstrates that vision is extremely important for rehabilitation, cortical reorganization and functional outcomes.[35] Therefore, complete blindness should be considered a contraindication for both face and upper extremity transplantation.[30] Of similar concern is advanced age of the CTA candidate when it comes to the long-term potential for cortical reeducation and the slow reintegration period following hand and/or face rehabilitation. It has been reported that hand function in replant patients over the age of 55 is worse when compared to younger patients, and therefore should be considered in choosing the optimal hand transplant recipient.[44]

40.4 Classification

Finally challenging is the lack of stable CTA classification for which is constantly changing based on the new cases of concomitant face transplantation performed recently. We introduce here a new modification to our classification scheme based on the relative complexity of the concomitant CTA and its distinct challenges[23] (Table 40.3).

40.5 Conclusion

Face and upper extremity transplantation has quickly become a clinical reality. The overall estimated mortality for face transplantation (within 2 years) remains around 20%, while no deaths have been reported in relation to upper extremity transplantation. Numerous factors have been identified as keys to achieving an optimal outcome. These include (1) the formation of a dedicated, university-hospital based, multidisciplinary team; (2) detailed screening with diligent donor/recipient selection criteria; (3) early, aggressive rehabilitation (within 48 h) of face and/or upper extremity allografts; and (4) superb recipient compliance. Future research is warranted to identify the exact clinical role of face/upper extremity transplantation, in order to decide whether it is justifiable as an early, primary method of reconstruction in the modern-day reconstructive algorithm.

Table 40.3 A new, modified Gordon CTA classification system based on relative complexity

Type	Complexity	Allografts	Characteristics
I	Low	• Flexor tendon • Tongue • Uterus • Vascularized nerve	1. Absent skin 2. Reduced antigenicity
II	Moderate	• Abdominal wall • Facial subunit – Ear • Genitalia – Penis • Larynx • Scalp • Trachea • Vascularized joint – Knee	1. Contain skin 2. Absent or less challenging rehabilitation
III	High	• Upper extremity – Hand • Face	1. Requires multidisciplinary transplant team 2. Complex rehabilitation 3. Significant psychological obstacles 4. Complicated cortical reorganization
IV	Maximum	• Concomitant CTA – Face/hand(s) – Tongue/mandible	1. High mortality risk 2. Extremely difficult rehabilitation

References

1. Gordon CR, Nazzal J, Lee WPA, Siemionow M, Matthews MS, Hewitt CW. Experimental rat hindlimb to clinical face composite tissue allotransplantation: Historical background and current status. *Microsurgery*. 2006;26:566-572.
2. Siemionow M, Gozel-Ulusal B, Engin Ulusal A, Ozmen S, Izycki D, Zins JE. Functional tolerance following face transplantation in the rat. *Transplantation*. 2003;75:1607-1609.
3. Gozel-Ulusal B, Ulusal AE, Ozmen S, Zins JE, Siemionow MZ. A new composite facial and scalp transplantation model in rats. *Plast Reconstr Surg*. 2003;112:1302-1311.
4. Nie C, Yang D, Li N, Liu G, Guo T. Establishing a new orthotopic composite hemiface/calvaria transplantation model in rabbits. *Plast Reconstr Surg*. 2008;122:410-418.
5. Shengwu Z, Qingfeng L, Hao J, et al. Developing a canine model of composite facial/scalp allograft transplantation. *Ann Plast Surg*. 2007;59:185-194.
6. Barth RN, Bluebond-Langner R, Nam A, et al. Facial subunit composite tissue allografts in nonhuman primates: I. Technical and immunosuppressive requirements for prolonged graft survival. *Plast Reconstr Surg*. 2009;123:493-501.
7. Cendales LC, Xu H, Bacher J, Eckhaus MA, Kleiner DE, Kirk AD. Composite tissue allotransplantation: Development of a preclinical model in nonhuman primates. *Transplantation*. 2005;80:1447-1454.
8. Kuo YR, Goto S, Shih HS, et al. Mesenchymal stem cells prolong composite tissue allotransplant survival in a swine model. *Transplantation*. 2009;87:1769-1777.
9. Baccarani A, Follmar KE, Baumeister SP, Marcus JR, Erdmann D, Levin LS. Technical and anatomical considerations of face harvest in face transplantation. *Ann Plast Surg*. 2006;57:483-488.
10. Siemionow M, Unal S, Agaoglu G, Sari A. A cadaver study in preparation for facial allograft transplantation in humans: Part I. What are alternative sources for total facial defect coverage? *Plast Reconstr Surg*. 2006;117:864-872. Discussion 873-875.
11. Siemionow M, Agaoglu G, Unal S. A cadaver study in preparation for facial allograft transplantation in humans: Part II. Mock facial transplantation. *Plast Reconstr Surg*. 2006;117: 876-885. Discussion 886-888.
12. Banks ND, Hui-Chou HG, Tripathi S, et al. An anatomical study of external carotid artery vascular territories in face and midface flaps for transplantation. *Plast Reconstr Surg*. 2009;123:1677-1687.
13. Gordon CR, Zor F, Siemionow M. Skin area quantification in preparation for concomitant upper extremity and face transplantation: A cadaver study and literature review. Transplantation 2011 [IN PRESS - Accepted January 15, 2011].
14. Siemionow M, Papay F, Alam D, et al. First U.S. near-total human face transplantation – a paradigm shift for massive facial injuries. *Lancet*. 2009;374:203-209.
15. Lengele BG. Current concepts and future challenges in facial transplantation. *Clin Plast Surg*. 2009;36:507-521.
16. International Registry of Hand and Composite Tissue Transplantation. Hand Registry home website page. http://www.handregistry.com. Accessed November 28, 2009.
17. Gordon CR, Siemionow M. Requirements for establishing a hand transplant program. *Ann Plast Surg*. 2009;63:262-273.
18. Gordon CR, Siemionow M, Papay F, et al. The world's experience with facial transplantation: What have we learned thus far? *Ann Plast Surg*. 2009;63:121-127.

19. Ravindra KV, Wu S, McKinney M, Xu H, Ildstad ST. Composite tissue allotransplantation: Current challenges. *Transplant Proc*. 2009;41:3519-3528.
20. American Society of Reconstructive Transplantation. Website home page. http://www.a-s-r-t.com. Accessed November 20, 2009.
21. Daily Telegraph webpage. Website home page. http://www.dailytelegraph.com.au/news/world/worlds-first-face-and-and-double-hand-transplant-patient-dies-in-france/story-e6frev00-1225736077918. Accessed November 26, 2009.
22. Hollenbeck ST, Erdmann D, Levin LS. Current indications for hand and face allotransplantation. *Transplant Proc*. 2009;41:495-498.
23. Gordon CR, Siemionow M, Zins J. Composite tissue allotransplantation: A proposed classification system based on relative complexity. *Transplant Proc*. 2009;41:481-484.
24. Patel SS, Goa KL. Sevoflurane. A review of its pharmacodynamic and pharmacokinetic properties and its clinical use in general anesthesia. *Drugs*. 1996;51:658-700.
25. McDonald DJF. Anesthesia for microvascular surgery: A physiological approach. *Br J Anaesth*. 1985;57:904-912.
26. Sigurdsson GH, Banic A, Wheatley AM, Mettler D. Effects of halothane and isoflurane anesthesia on microcirculatory blood flow in musculocutaneous flaps. *Br J Anaesth*. 1994;73:826-832.
27. Hagau N, Longrois D. Anesthesia for free vascularized tissue transfer. *Microsurgery*. 2009;29:161-167.
28. Hynynen M, Eklund P, Rosenberg PH. Anaesthesia for patients undergoing prolonged reconstructive and microvascular plastic surgery. *Scand J Plast Reconstr Surg*. 1982;16:201-206.
29. Agich GJ, Siemionow M. Facing the ethical questions in facial transplantation. *Am J Bioeth*. 2004;4:25-27.
30. Gordon CR, Siemionow M, Coffman K, et al. The Cleveland Clinic FACES score: A preliminary assessment tool for identifying the optimal face transplant candidate. *J Craniofac Surg*. 2009;20:1969-1974.
31. Murray JE. "The establishment of composite tissue allotransplantation as a clinical reality". In: Hewitt CW, Lee WPA, Gordon CR, eds. *Transplantation of Composite Tissue Allografts*. New York: Springer; 2008:3-12.
32. Siemionow M, Papay F, Djohan R, et al. First U.S. near-total human face transplantation – a paradigm shift for massive complex injuries. *Plast Reconstr Surg*. 2010;125:111-122.
33. Gander B, Brown CS, Vasilic D, et al. Composite tissue allotransplantation of the hand and face: A new frontier in transplant and reconstructive surgery. *Transpl Int*. 2006;19:868-880.
34. Giraux P, Sirigu A, Schneider F, Dubernard JM. Cortical reorganization in motor cortex after graft of both hands. *Nat Neurosci*. 2001;4:1-2.
35. Faggin BM, Nguyen KT, Nicoelis MA. Immediate and simultaneous sensory reorganization at cortical and subcortical levels of the somatosensory system. *Proc Natl Acad Sci*. 1997;94:9428-9433.
36. Pleger B, Dinse HR, Ragert P, et al. Shifts in cortical representations predict human discrimination improvement. *Proc Natl Acad Sci*. 2001;98:12255-12260.
37. Farnè A, Roy AC, Giraux P, Dubernard JM, Sirigu A. Face or hand, not both: perceptual correlates of reafferentation in a former amputee. *Curr Biol*. 2002;12:1342-1346.
38. Mathes DW, Kumar N, Ploplys E. A survey of North American burn and plastic surgeons on their current attitudes toward facial transplantation. *J Am Coll Surg*. 2009;208:1051-1058.
39. Mathes DW, Schlenker R, Ploplys E, Vedder N. A survey of North American hand surgeons on their current attitudes toward hand transplantation. *J Hand Surg Am*. 2009;34:808-814.
40. Chinese face transplant Li Guoxing dies." http://www.news.com.au/story/0,27574,24829166-23109,00.html. Website News.com.au. Accessed June 19, 2009.
41. Breidenbach W, Kadiyala R, Buell J, Blair B, Kaufman C. Update on the American Hand Transplant Experience: Five Hand Transplants Performed in Louisville. Oral presentation, American Transplant Congress; June 2, 2009; Boston, MA.
42. Unadkat JV, Horibe EK, Schneeberger S, Solari M, McLean K, Gorantla V, Lee WPA. Chronic rejection in experimental composite tissue allotransplantation: First evidence. American Society of Reconstructive Microsurgery, Annual Meeting; January 2008; Beverly Hills, CA.
43. Alam DS, Papay F, Djohan R, et al. The technical and anatomical aspects of the world's first near-total human face and maxilla transplant. *Arch Facial Plast Surg*. 2009;11:369-377.
44. Schneeberger S, Ninkovic M, Margreiter R. Hand transplantation: the innsbruck experience. In: Hewitt CW, Lee WPA, Gordon CR, eds. *Transplantation of Composite Tissue Allografts*. New York: Springer; 2008:234-250.

Immunosuppression in Composite Tissue Allotransplantation

41

Bijan Eghtesad and John J. Fung

Contents

B. Eghtesad (✉)
Department of HPB/Liver Transplant Surgery,
Cleveland Clinic, Cleveland, OH, USA
e-mail: eghtesb@ccf.org

Abstract Composite tissue allotransplantation (CTA) is emerging as a new modality in reconstructive surgery as potential treatment for complex tissue, anatomic and functional defects. While the technical aspects of such procedures are constantly undergoing refinement, the practical issues due to the need and as a result, lifelong risks of exposure to the immunosuppressive drugs, remains as a major drawback. The implementation of induction and maintenance therapy protocols used in solid organ transplantation (SOT) has resulted in excellent patient and graft survival in CTA and has minimized the risk of graft loss due to uncontrolled rejection. Further understanding of the mechanisms and tempo of graft acceptance and rejection in CTA may lead to protocols that minimize the need for maintenance immunosuppressive therapy and as a result reduce the risk for long-term side effects.

Abbreviations

AZA	Azathioprine
ATG	Antithymocyte globulin
CNI	Calcineurin inhibitors
CTA	Composite tissue allotransplantation
CsA	Cyclosporine
HSC	Hematopoietic stem cell
IL	Interleukin
MMF	Mycophenolate mofetil
MPA	Mycophenolic acid
RAPA	Rapamune
SOT	Solid organ transplantation
Tac	Tacrolimus
TDM	Therapeutic drug monitoring

M.Z. Siemionow (ed.), *The Know-How of Face Transplantation*,
DOI: 10.1007/978-0-85729-253-7_41, © Springer-Verlag London Limited 2011

41.1 Introduction

Composite tissue allotransplantation brings the hope for reconstruction of complex soft tissue and bony defects of the face, extremities, and other body parts, as a result of trauma, congenital anomalies, or variety of disease processes. The technical advances and surgical expertise in the science of reimplantation and reconstructive surgery in 1960s prompted the first hand transplant procedure by Gilbert in 1964.[1,2] While technically successful, the graft was lost to acute rejection 3 weeks after transplantation and required removal. The transplant was performed under the standard immunosuppressive regimen at the time, namely, azathioprine (AZA) and prednisolone. This failure and disappointing results of CTA in experimental animals put a halt on further attempts at CTA, mainly due to the perception that strong antigenicity of the skin and other tissues were not controllable. Thus, further attempts in CTA awaited the development of more effective immunosuppressive agents.

The introduction of cyclosporine A (CsA) in the late 1970s revolutionized the field of SOT by lowering the rate and severity of acute rejection, resulting in increased survival in renal allografts followed by liver and pancreas transplantation.[3-5] These encouraging outcomes in SOT resulted in renewed attempts in CTA in small and large animal models under CsA. Although these experiences showed a reduction in acute rejection, the incomplete control of rejection of various components of the CTA led to extension of the moratorium in clinical CTA transplantation.[6]

Advances in understanding and science of immunosuppression along with the availability of newer immunosuppressive drugs in the 1990s allowed rapid growth and success of SOT, increased graft survival, and almost complete eradication of graft loss resulting from acute rejection. These encouraging results were followed by successful outcomes in limb allotransplantation in animals.[6,7] As a result, clinical efforts at transplanting hand, larynx, and knee were initiated. The lessons learned clinically and in the laboratory during the 30 years from the first attempted hand transplant to the first successful one[8] resulted in significant achievement in understanding of immune response to allograft and generated enthusiasm in the CTA community.

With development of regimens based on antibody induction, calcineurin inhibitors (CNI), mycophenolate acid (MPA), and corticosteroids, short-term graft survival rates close to 90% have been achieved in over 100 reports of various CTAs, encompassing hand, knee, abdominal wall, face, trachea, larynx, scalp, tendon, penis, uterus, and tongue.

41.2 Transplantation of Face

The first successful face transplant was done in France in 2005.[9] In the period between this transplant and April of 2010, ten facial composite transplantations from different parts of the world have been reported. As of May 2010, 8/10 (80%) of the recipients are alive. There have been two deaths in this group. The first mortality occurred in China nearly 2 years after a successful transplantation due to noncompliance of the patient with immunosuppressive regimen. The second death happened in a patient from France who received concomitant face and bilateral hand transplants. He died of the complications of overwhelming infection 2 months after transplantation.

41.3 Overview of Immunosuppressive Agents

Currently available immunosuppression applied in clinical transplantation can be categorized into two types: pharmacologic and biologic.[10] Pharmacologic immunosuppression consists of: *glucocorticosteroids* – hydrocortisone, methylprednisolone, dexamethasone; *calcineurin inhibitors* (CNIs) which suppress cytokine production – tacrolimus (Tac) and cyclosporine (CsA); antiproliferative agents including *cell cycle inhibitors* – mycophenolic acid (MPA) and its derivative, mycophenolate mofetil (MMF), and azathioprine (AZA); and the class of mTOR inhibitors, rapamune (RAPA) and everolimus (RAD). Biologic immunosuppression includes the *monoclonal antibodies* such as anti-CD3 (OKT3), anti-CD20 (rituximab), anti-CD52 (alemtuzumab), and anti-CD25 (anti-IL-2 receptor antibodies (basiliximab and dacluzimab); the *polyclonal anti-lymphocyte preparations* – rabbit antithymocyte globulins (r-ATG) and equine antilymphocyte globulins (ATGAM). New classes of immunosuppressive agents include the JAK-2 inhibitors and costimulatory molecule inhibitors, which are in clinical trials in SOT and autoimmune diseases.

These agents can be used as induction agents; agents for maintenance therapy; or agents for rescue therapy in cases of rejection. The ultimate goal of using these agents is to effectively suppress rejection and at the same time to minimize their toxic side effects in the transplant recipient. In clinical practice this is achieved through the use of balanced dosages of multiple drugs that interfere with the immune response at various sites of the immune cascade.

Induction therapy: The main goal for induction therapy is to achieve rapid and intense immunosuppression for the first few weeks after the transplant to reduce the likelihood of early acute rejection. Generally, induction therapy includes the use of an intravenously administered monoclonal or polyclonal antibody preparation, which tends to decrease the severity of the first rejection and delays the time to first rejection. During this period of antibody induction, there is time to achieve target immunosuppressive levels of those agents that constitute the maintenance phase.

Maintenance therapy: The goal of maintenance therapy is to suppress the recipient immune system from recognizing and mounting an immune response to the allotransplanted organ or tissue. The immunosuppressive regimen is tailored to the individual patient to provide lifelong suppression of the immune system with minimal toxicity. In principle, the accommodation of the allograft and the acclimation of the recipient immune system allows for tapering levels of immunosuppression over time, while the use of multiple agents with differing targets of immunosuppression, allow synergistic or additive immunosuppression while minimizing individual immunosuppressive agent toxicity.

Treatment as rescue therapy: In both SOT and CTA, the first line of treatment in acute rejection episodes is increased dosing of glucocorticosteroids while optimizing the baseline immunosuppression. However, when this approach is unsuccessful, antibody-based therapy may be needed. Topical steroids and topical Tac have also been used in hand and facial tissue allotransplantation with good response, when the rejection is limited to the skin.

41.3.1 Calcineurin Inhibitors (CNI)

Both CsA and Tac are CNIs by virtue of their shared property of binding to their specific immunophilins, which leads to inhibition of calcineurin activity and as a result inhibition of transcription of several genes, including IL-2, IL-3, and IL-4. This ultimately interferes with T-cell signaling through cytokine production.[11] The routine application of CNIs in SOT has: (1) dramatically reduced the frequency and severity of rejection, (2) decreased the morbidity associated with treatment of rejection and graft loss, and (3) as a result improved patient survival. Because of clinical superiority of Tac, this drug has become the mainstay of immunosuppression in SOT and is also the mainstay in immunosuppressive regimens in CTA.[12-15] The principal side effects of CNIs include: nephrotoxicity, neurotoxicity, diabetogenicity, increased susceptibility to opportunistic infections, and increased risk of de novo virally associated malignancies.[16-18] The need for careful monitoring of drug levels is warranted for CNIs, due to the narrow therapeutic window between toxicity and efficacy. In general, therapeutic drug monitoring (TDM) is done frequently in the early posttransplant period, as well as following any substantial changes in CNI dosing. Target levels are higher in the early posttransplant period and following treatment of rejection, while lower levels are indicated in long-term stable patients or those patients with toxicity.

Although a large number of non-CNI immunosuppressive agents are available, their use in SOT and CTA has been primarily adjunctive to the use of CNIs. However, most immunosuppression protocols have attempted to reduce the overall intensity of CNI exposure and have included CNI-minimization by supplementing low levels of CNIs with less nephrotoxic agents such as MMF or RAPA.

41.3.2 Corticosteroids

By far the most used non-CNI agents in SOT are corticosteroids. Corticosteroids have been shown to prolong skin graft survival in rabbits.[19] Starzl et al.[20] and Murray et al.[21] in 1963 independently showed the benefit of combining corticosteroids with another immunosuppressive agent, AZA, to obtain meaningful survival after allogeneic renal transplantation in humans. Steroids continue to be used to control acute episodes of rejection and for prophylaxis in prevention of rejection both in SOT and CTA.[12,14,15]

Unfortunately, acute and chronic dosing of corticosteroids is associated with side effects, including hypertension, hyperglycemia, delayed wound healing, osteoporosis, glaucoma, suppressed growth, hyperlipidemia, increased risk of gastrointestinal ulceration, risk of fungal infections, and suppression of the pituitary-adrenal axis. Thus attempts to reduce or eliminate corticosteroid use have required the use of other non-CNI immunosuppressive agents.

41.3.3 Mycophenolic Acid (MPA)

Mycophenolic acid (MPA) and its pro-drug MMF, a semisynthetic derivative of MPA, are immunosuppressive agents that inhibit the de novo purine nucleotide synthesis and as a result, block DNA replication in T and B lymphocytes. MMF is rapidly absorbed after oral administration and is hydrolyzed to MPA. The bioavailability of MPA is approximately 90%. The drug is glucuronidated in the liver to the inactive MPA glucuronide and is excreted primarily by the kidneys. The incidence of adverse effects (nausea, vomiting, gastritis, abdominal pain, diarrhea, neutropenia) requiring dose reduction or withdrawal is high, ranging from 24% to 57%.[22-24] An enteric-coated preparation of MPA has been reported to reduce the GI side effects without compromising efficacy.

When MMF is used in combination with Tac and steroids, the dose of Tac required is often lowered, thus minimizing renal toxicity. This combination has shown to be effective in reducing acute cellular rejection in SOT with preservation of renal function. Most of the CTA immunosuppressive protocols have incorporated the combination of MMF and Tac.[14,15] The usual dose of MMF in adults is 1.0–1.5 g twice daily, however, the dosage should be reduced or discontinued in patients suffering major side effects. In particular, close attention to the leukocyte and platelet counts is indicated because of the bone marrow suppressive potential of the drug. The use of TDM for MPA levels is generally of little benefit, as there is poor correlation between MPA levels and efficacy/toxicity.

41.3.4 RAPA

RAPA and its analog, everolimus, are macrolide antibiotics structurally related to Tac. They bind to the same immunophilin, FKBP12, but do not inhibit cytokine gene transcription in T cells. Rather, RAPA and everolimus block the signals transduced from a variety of growth factor receptors to the nucleus by acting on phosphatidyl inositol kinases called mammalian targets of rapamycin (mTOR). Both agents inhibit fibroblasts migration and proliferation and synthesis of collagen and thus inhibit wound repair, although the pharmacokinetics and dosing of everolimus is associated with less inhibition on fibroblast activity. This observation has been reported in liver,[25] kidney,[26] and lung[27] transplantation. In addition, both drugs share similar side effects such as thrombocytopenia, leucopenia, anemia, elevated serum cholesterol and triglyceride levels and delayed wound healing. With RAPA, the development of oral ulcerations as a side effect makes this drug less attractive in face transplantation.

41.3.5 Antibody Induction

Antibody induction therapy has been limited to the perioperative period as a means to reduce early exposure to CNIs or to obviate the need for large doses of perioperative corticosteroids and CNIs. Antibody therapy can be depleting (OKT3, alemtuzumab), receptor modulating (IL-2 receptor Ab), or both (ATG). With the use of depleting antibody preparations, a phenomenon known as "first dose effect," related to intravascular release of cytokines by lymphocytes, can occur. The symptoms, including fever, chills, tachycardia, gastrointestinal disturbances, bronchospasm, and fluctuation of blood pressure, can be blocked by pretreatment with corticosteroids, diphenhydramine hydrochloride, and acetaminophen.

Antithymocyte antibody preparations have been widely used in CTA as induction agents. Their target is directed to the major T-cell surface molecules (CD2, CD3, CD4, CD8, CD28, as well as other T-cell receptors) as well as other leukocyte specificities (CD20 and CD40 on B-cells; CD16 on natural killer cells and macrophages).[28,29]

41.3.6 Rabbit Antithymocyte Globulin

Rabbit antithymocyte globulin (r-ATG) (Genzyme, Boston MA) has been the main induction therapy in reported cases of CTA and especially in cases of face transplantation. The usual dosing has been 1.5 mg/kg/day for 10 days in addition to corticosteroids, in most of the cases, with gradual institution of maintenance therapy with Tac and MMF. Side effects like leucopenia, thrombocytopenia, and fever are common but usually self limited.

41.3.7 Alemtuzumab

Alemtuzumab, a humanized, recombinant anti-CD52 monoclonal antibody, is an effective depleting agent which has been used increasingly in SOT trials[30,31] with the goal to reduce the overall amount of maintenance immunosuppression and as a result reduction of their side effects, and as part of "tolerogenic" protocols. Recently, the use of alemtuzumab as part of the induction protocol with the goal to minimize need for maintenance immunosuppression in hand transplantation was reported by Schneerberger et al.[32] Long-term follow-up is necessary to determine whether this strategy is effective, as the long-term effect of alemtuzumab on lymphocyte suppression may simply lead to postponing the onset of rejection for months.

41.3.8 IL-2 Receptor (IL-2R) Antibodies

IL-2 receptor (IL-2R) antibodies have been in use in SOT for more than a decade. Both daclizumab and basiliximab are humanized and chimeric IgG1 monoclonal antibodies, respectively, and both are directed against the IL-2 receptor alpha-chain (referred to as CD25 or T-cell activation antigen), which is upregulated on the surface of activated T lymphocytes. Binding of these antibodies results in internalization of the IL-2R-chain, rendering the IL-2 receptor unable to bind to IL-2. As an induction therapy, both agents have been used to delay or reduce the dose of CNIs in SOT recipients.[33] The use of IL-2R antibody in CTA has been reported and it has been used in one case of face transplantation.[34]

41.3.9 Topical Agents

Topical preparation of steroids and Tac have been used in CTA in combination with systemic immunosuppressive agents. Recently, with more experience in treatment of rejection in CTA, topical agents have been used for treatment of mild rejection episodes in the skin (grade I or II) without increase in intensity of systemic immunosuppression.

41.4 Immunosuppressive Protocols in Face Transplantation

Many in the transplant community believed that rejection in CTA would be hard to control and would require higher levels of immunosuppression, in part due to the large amount of skin, felt to be highly antigenic. However, from the initial clinical experiences, this has not been proven to be the case. Most of the patients have been maintained on levels of immunosuppression, similar to that used in SOT, consisting of Tac and MMF, or RAPA, and steroids. More recently Campath-1H lymphodepletion induction and steroid sparing maintenance with Tac and MMF have been successfully used in CTA.[32]

Immunosuppressive protocols for facial allotransplantation have been similar to those used in other CTA (Table 41.1) [9,34-36]; the most common protocols have included:

1. Induction therapy with:
   (a) rz-ATG, 1.5 mg/kg/day for 9–10 days.
   (b) IL-2R antibody induction depending on the type of IL-2R Ab used.
   (c) As mentioned earlier, alemtuzumab has been recently reported in hand transplantation but not yet in facial transplantation.
2. Maintenance therapy:
   (a) Tacrolimus: The route of administration depends on the patient's ability to take meds by mouth or feeding tube, otherwise intravenous administration may be needed. The target goals for TDM are:
      - Tac blood levels between 12 and 15 ng/ml in the first 2 months.
      - Tac blood levels between 8 and 12 ng/ml during the rest of the first year.
      - Tac blood levels between 5 and 8 ng/ml after the first year.
      - Tac blood levels should be modified according to changes in renal function or other events, such as level of sensitization (e.g., panel reactive antibody levels) as a result of previous transfusions or pregnancy, development of rejection, or toxicity.

Table 41.1 Clinical review of the first four facial allotransplantation

	Dubernard, France[9]	Guo, China[34]	Lantieri, France[35]	Siemionow, USA[36]
Induction	r-ATG 1.25 mg/kg 10 days	Daclizumab 50 mg total	r-ATG 1.25 mg/kg 10 days	r-ATG 1.25 mg/kg 9 days
Graft modulation	*Radiation to CTA*			
Maintenance	Tac 8–10 ng/ml MMF Prednisone	Tac 10–25 ng/ml MMF Prednisone	Tac 10–13 ng/ml MMF Prednisone	Tac 12–15 ng/ml MMF Prednisone
Acute rejection	Yes	Yes	Yes	Yes
Occurrence (day)	18, 214	90, 150, 510	28, 64	40, 450
Rescue therapy	Pulse steroid Increased Tac Clobetasole	Pulse steroid Increased Tac	Pulse steroid Increased Tac r-ATG X 7 days	Pulse steroid Increased Tac Protopic
Histology of ACR	Mucosa Skin	Skin	Mucosa Skin	Mucosa Skin
Prophylaxis	Gancyclovir Valacyclovir Trimethoprim-Sulfa	Acyclovir Metronidazole	Gancyclovir Valgancyclovir Trimethoprim-Sulfa	Gancyclovir Valgancyclovir Trimethoprim-Sulfa
Infection	Herpes simplex	None reported	CMV	CMV
Outcome	Alive	Dead	Alive	Alive

(b) Methylprednisolone: 1,000 mg IV in the operating room and a taper cycle as follow:
- 50 mg IV Q6 h × 4 doses
- 40 mg IV Q6 h × 4 doses
- 30 mg IV Q6 h × 4 doses
- 20 mg IV Q6 h × 4 doses
- 30 mg daily oral prednisone for 2 weeks and gradual taper by 2.5 mg reduction in daily dose every week until 10 mg daily. Further dose reduction will be based on events, such as episodes of rejection. The goal is to discontinue the drug in 3–6 months after CTA.

(c) Mycophenolate Mofetil: 1 g IV or PO bid starting on day one. In case of high PRA or positive cross-match, to be started in the operating room as IV until the patient is able to get it orally or through the GI access. Dosage should be modified with possible side effects like GI intolerance or drop in white blood cell count. In the instance of GI intolerance, the use of the enteric-coated MPA may assist in reduction of side effects. The goal is to stop the drug between 6 months and 1 year after transplantation.

Since many immunosuppressive agents are metabolized in the liver and their metabolites excreted in the urine, a full understanding and appreciation of the complex polypharmaceutical interactions will help avoid inadvertent toxicity or loss of efficacy.

Treatment for rejection: Treatment for rejection is based on the severity of the rejection determined by histopathological criteria (Table 41.2), including degree and nature of cellular infiltrates. With grade I and II skin rejection, the clinical management includes an increase in the Tac dose and topical steroid and/or Tac. In the event of higher grades of rejection, intravenous high-dose steroid (methylprednisolone) as single dose with or without taper cycle (as above) would be indicated. In those cases of steroid-resistant rejection, the use of anti-lymphocyte antibodies would be indicated.

Table 41.2 Side effects of immunosuppressive drugs

Agents	Key adverse effects
Cyclosporine	Nephrotoxicity, neurotoxicity, hyperglycemia, hyperlipidemia, gingival hyperplasia, hirsutism, hypertension, malignancy
Tacrolimus	Nephrotoxicity, neurotoxicity, hyperglycemia, gastrointestinal disturbances, hypertension, malignancy, alopecia
Corticosteroids	Hypertension, hyperglycemia, osteoporosis, growth retardation, cataract, gastrointestinal ulceration, poor wound healing
Azathioprine	Myelosuppression, hepatotoxicity
Mycophenolic acid	Myelosuppression, gastrointestinal disturbances, malignancy
Sirolimus	Hyperlipidemia, thrombocytopenia, anemia, poor wound healing, pneumonitis, mucosal ulceration
Anti-lymphocyte Ab	Cytokine release phenomenon, viral activation, immune complex syndrome

41.5 Immunological Monitoring in Patients After Face Transplantation

There are currently no objective means for evaluating the overall state of immunosuppression either SOT or CTA. As a result, clinical manifestations of under-immunosuppression and over-immunosuppression with presentation of rejection or infection and malignancies appears to be an indirect indicator of the degree to which the immune system is suppressed. The following monitoring may provide an aid to assess the state of immunosuppression in the CTA recipients:

1. Complete blood count with WBC differential daily paying attention to eosinophil counts, which have been associated with rejection in SOT, although not reported in rejection with CTA.
2. Weekly assessment of lymphocyte function by ATP level (Cylex ImmunKnow, IBT Lab, Lenexa, Kansas, KS, USA). While this assay has not been assessed in CTA, in SOT, it is important to perform serial determinations, although levels lower than 200 are usually a sign of over-immunosuppression, it is more important to look at the trend rather than absolute levels.
3. Serial donor specific antibody levels with or without virtual cross-match to mismatched donor antigens.
4. Short Tandem Repeat assay (STR) (AmpFLSTR Profile Plus, Applied Biosystem, Foster City, CA, USA) to look for evidence of donor cell chimerism, as a manifestation of graft-versus-host disease.

41.5.1 Monitoring for Infection

A complete guide to monitoring and treatment for infections in the transplant recipient is beyond the scope of this chapter and is reviewed elsewhere.[37] It is important to look clinically for evidence of any infections, both of the graft and systemically in the recipient. Clinical symptoms including GI symptoms (nausea and diarrhea), fever, cough, purulent drainage, or pustules. Opportunistic infections are potentiated by the use of immunosuppressive agents and a high level of suspicion should be raised in the early posttransplant period and after treatment for rejections. Invasive fungal infections may require biopsy and culture. Weekly CMV viral load determination and bimonthly EBV viral load measurement are essential for early detection and treatment of these fairly common viral infections.

41.6 Complications Related to Immunosuppression

Toxicity of immunosuppression with an increased risk of cancer, opportunistic infection, and organ failure in CTA is no different than in SOT (Table 41.3). Like the recipients of SOT, these patients need to be under continuous surveillance to predict and prevent, or recognize the potential complications and take the necessary steps to control them. Means of addressing toxicity and complications of immunosuppression depends on the clinical setting and should be considered individually. However, reduction of the dosage of one or more immunosuppressive agents or permanently discontinuing the drug with replacement to another immunosuppressant are options. However, many of the effects of immunosuppression can be long-lasting or even permanent;

Table 41.3 Rejection in composite tissue allotransplantation

Grading	Histological changes
Grade 0	Nonspecific change No or only mild lymphocytic infiltration without involvement of the superficial dermal structures or epidermis
Grade I (mild rejection)	Superficial perivascular inflammation with involvement of superficial vessels and without involvement of overlying epidermis
Grade II (moderate rejection)	Features of grade I with involvement of the epithelium and adnexal structures
Grade III (severe rejection)	Band-like superficial dermal infiltrates with more continuous involvement of the epidermis and middle and deep perivascular infiltrates
Grade IV (necrotizing rejection)	Features of grade III along with frank necrosis of the epidermis and other tissues

therefore, it is important to weigh the risks and benefits of significant changes in immunosuppression with the survival of the allograft.

In cases of bacterial infection, MMF and steroid should be lowered or stopped and in the presence of fungal or viral infections or the development of significant renal dysfunction, CNIs should be lowered or discontinued with replacement to RAPA or other less nephrotoxic agents.

41.7 Rejection in CTA

Reversible episodes of rejection have been experienced in most CTA recipients. These reports indicate that the skin is the prime target of rejection but that components other than skin may be affected by rejection too. The experimental data from early 1960s suggested that the level of immunogenicity was variable from one tissue to another but that skin was thought to carry the highest immunogenic load. This theory was introduced by Joseph Murray, who ranked the skin with the highest score of immunogenicity, comparing to other tissues tested.[38] This hierarchy of immunogenicity of tissues was widely accepted despite the findings by Lee et al.[39] that no single tissue was dominant in primarily vascularized limb allografts. In their experiments, it was demonstrated that the whole limb allograft was rejected more slowly than did allografts consisting of the individual components. However, this was not the case in further studies done by Goldwyn et al.[40] and Mathes et al.[41] In these animal studies, it was noticed that despite indefinite survival of the musculoskeletal components of the allografts, the skin component was rejected by the recipient.

There is limited information on involvement of other components of CTA than skin in rejection. Biopsy findings obtained from the allograft loss from China indicated that the skin was the principal target of rejection with the most severe changes while milder inflammation was found throughout the muscular and neural structures, while tendons, bones, and joints were spared.[42]

Pathological changes seen during graft rejection in face transplant recipients[43] are similar to the ones observed in rejection of other skin-contained CTAs such as hand[44,45] and abdominal wall,[46] suggesting that rejection happens in the skin with uniform characteristics. These skin changes were assessed by Cendales et al.[47] in reviewing 29 specimens from transplanted human limbs and abdominal walls recovered at various time points (1 month to 3 years) after transplantation. Based on their survey, a classification of clinical acute skin rejection was proposed (Table 41.2). The changes of rejection in facial transplantation show erythema of the skin and appearance of red macules. These findings have also been seen in the oral mucosa, which has an overall similar structure as the skin, except for not having skin appendages such as hair follicles, sebaceous, and sweat glands. A rich vascular supply of oral mucosa could be the possible explanation of denser inflammatory process seen on oral mucosal biopsies during rejection and also other inflammatory processes.[44] Clinical findings in skin and mucosa could present themselves before the histological changes observed in rejection. In contrast, these histological changes of rejection could be also present before clinical signs and symptoms appear. Nevertheless, the use of biopsies is warranted before treatment of rejection, in order to rule out other causes of allograft changes, e.g., drug reaction.

One major difference between SOT and CTA is that in SOT, frequent or severe acute rejection is often associated with high incidence of chronic rejection, due to incomplete control of the overall recipient immune system. However, this has not been the experience in CTA. Most acute rejection in these patients has been successfully controlled with conventional immunosuppressive regimens without any graft losses, except for two cases of graft loss in hand transplants due to chronic rejection secondary to noncompliance. The explanation for this may be the close follow-up these patients undergo with any change in the appearance of the skin or mucosa prompting rapid investigation. Although rejection episodes principally occur during the first year after transplantation, most patients are clinically stable on lower doses of immunosuppression in the long term. Nevertheless, in spite of these promising observations, it is important to look for signs of chronic rejection, including contractures, neuropathy, fibrotic changes to the skin, and diminished blood flow to the CTA. In fact, the hallmark of chronic rejection is the presence of arteriopathy, leading to graft ischemia.[48]

41.8 Composite Tissue Allografts, Chimerism, and Tolerance

The success of CTA is currently based on the chronic use of immunosuppressive drugs. The "ideal" immunosuppressive agents with no toxicity and selective efficacy has not yet been developed. A protocol for tolerance induction to CTA that would eliminate the need for chronic immunosuppressive therapy would result in application of CTA as a widespread treatment modality in reconstruction for large tissue and functional defects.

Tolerance is defined as a state of donor-specific hyporesponsiveness in the recipient in the absence of immunosuppression. This level of tolerance is achievable in hematopoietic stem cell (HSC) chimerism. Macrochimerism is usually the result of transplantation of the pluripotent HSC from the donor and engraftment of these cells in the recipient, and as a result production of all the donor origin cell lineages in the recipient. This will result in a bidirectional tolerance between the donor and the recipient. In SOT and CTA, as low as 1% donor chimerism is sufficient to induce a strong state of tolerance to the donor tissues.[49]

Microchimerism occurs when less than 1% of the donor-specific hematopoietic lineages are detectable. It happens as a result of migration of passenger leukocytes from the transplanted allograft into an unconditioned recipient. Specific persistent lineages from the donor origin then interact with the recipient immune system and participate in a clonal exhaustion/deletion process resulting in donor-specific tolerance,[50-52] or lead to generation of immunomodulatory cells, such as regulatory T cells, which have been reported in a CTA recipient.[53]

Mixed chimerism is when the donor and recipient hematopoietic cells coexist.[49] This has been shown to be effective in induction of donor-specific tolerance without a need for myeloablative preconditioning.[54] To achieve this goal, bone marrow infusion from the same donor has been used in SOT for induction of mixed chimerism as part of the tolerogenic protocols with different results. This method was used as part of the induction protocol in the first face transplant done in France.[9] In this case, some level of microchimerism was detected early on after the procedure but it disappeared in further evaluation. This transient microchimerism has been shown in cases of hand transplantation without donor bone marrow infusion but likely, as the result of vascularized bone marrow transplantation as part of the allograft.[55] Development of effective mixed chimerism through simultaneous donor bone marrow stem cell infusion or vascularized bone marrow transplantation with help of cell-based therapies could be the future in more effective induction of tolerance and less need for immunosuppressive drugs.

41.9 Conclusions

Composite tissue allotransplantation has become a reality in the field of soft tissue reconstruction and transplantation. Despite the excellent outcomes and progress in technical aspect of the field, the risk of the long-term immunosuppression for a procedure which is mostly considered to improve the quality of life rather than a lifesaving one is still present. Innovative immunomodulatory protocols designed to induce tolerance or to minimize the need for immunosuppression are much needed to promote the CTA as a novel reconstructive treatment for complex and functionally significant tissue injuries and limb defects in clinical practice.

References

1. Gilbert R. Transplant is successful with a cadaver forearm. *Med Trib Med News*. 1964;5:20-22.
2. Gilbert R. Hand transplanted from cadaver is re-amputated. *Med Trib Med News*. 1964;5:23.
3. A randomized clinical trial of cyclosporine in cadaveric renal transplantation. *N Engl J Med*. 1983;309:809-815.
4. Calne RY, Rolles K, White DJ, et al. Cyclosporine A initially as only immunosuppressant in 34 recipients of cadaveric organs: 32 kidneys, 2 pancreases, and 2 livers. *Lancet*. 1979;2: 1033-1036.
5. Iwatsuki S, Starzl TE, Todo S, et al. Experience in 1000 liver transplants under cyclosporine-steroid therapy: a survival report. *Transplant Proc*. 1988;20(1 suppl 1):498-504.
6. Press BH, Sibley RK, Shons AR, et al. Limb allotransplantation in rat: extended survival and return of nerve function with continuous cyclosporine/prednisone immunosuppression. *Ann Plast Surg*. 1986;16:313-321.
7. Benhain P, Anthony JP, Ferreira L, et al. Use of combination of low-dose cyclosporine and RS-61443 in a rat hindlimb model of composite tissue allotransplantation. *Transplantation*. 1996;61:527-532.
8. Dubernard JM, Owen E, Herzberg G, et al. Human hand allograft: report on first 6 months. *Lancet*. 1999;353:1315-1320.
9. Dubernard JM, Lengele B, Morelon E, et al. Outcomes 18 months after the first human partial face transplantation. *N Engl J Med*. 2007;357:2451-2460.
10. Fung J, Kelly D, Kadry Z, Patel-Tom K, Eghtesad B. Immunosuppression in liver transplantation: beyond calcineurin inhibitors. *Liver Transplant*. 2005;11:267-280.
11. European FK. 506 multicenter liver study group. Randomized trial comparing tacrolimus and cyclosporine in prevention of liver allograft rejection. *Lancet*. 1994;344: 423-428.
12. Whitaker IS, Duggan EM, Alloway RR, et al. Composite tissue allotransplantation: a review of relevant immunological issues for plastic surgeons. *J Plast Reconstr Aesthet Surg*. 2008;61:481-492.
13. Hadani H, Hettiaratchy S, Clarke A, et al. Immunosuppression in an emerging field of plastic and reconstructive surgery: composite tissue allotransplantation. *J Plast Reconstr Aesthet Surg*. 2008;61:245-249.
14. Gordon C, Siemionow M, Papay F, et al. The world's experience with facial transplantation: what have we learned thus far? *Ann Plast Surg*. 2009;63:572-578.
15. Ravindra K, Buell J, Kaufman C, et al. Hand transplantation in the United States: experience with 3 patients. *Surgery*. 2008;144:638-644.
16. Sigal NH, Dumont F, Durette P, et al. Is cyclophilin involved in the immunosuppressive and nephrotoxic mechanism of action of cyclosporin A? *J Exp Med*. 1991;173:619-628.
17. Rabkin JM, Corless CL, Rosen HR, et al. Immunosuppression impact on long-term cardiovascular complications after liver transplantation. *Am J Surg*. 2002;183:595-599.
18. Fung JJ, Jain A, Kwak EJ, et al. De novo malignancies following liver transplantation – A major cause of death. *Liver Transplant*. 2001;7(Supp 1):S109-S118.
19. Morgan JA. The influence of cortisone on the survival of homografts skin in the rabbit. *Surgery*. 1951;30:506-515.
20. Starzl TE, Marchioro TL, Waddell WR. The reversal of rejection in human renal homografts with subsequent development of homograft tolerance. *Surg Gynecol Obstet*. 1963;117:385-395.
21. Murray JE, Merrill JP, Harrison JH, et al. Prolonged survival of human kidney homografts by immunosuppressive drug therapy. *N Engl J Med*. 1963;268:1315-1323.
22. Jain A, Venkataramanan R, Hamad IS, et al. Pharmacokinetics of mycophenolic acid after mycophenolate mofetile administration in liver transplant patients treated with tacrolimus. *J Clin Pharmacol*. 2001;41:268-276.
23. Barkmann A, Nashan B, Schmidt HH, et al. Improvement of acute and chronic renal dysfunction in liver transplant patients after substitution of calcineurin inhibitors by mycophenolate mofetile. *Transplantation*. 2000;69:1886-1890.
24. Jain AB, Hamad I, Rakela J, et al. A prospective randomized trial of tacrolimus and prednisone versus tacrolimus, prednisone, and mycophenolate mofetilein primary adult liver transplant recipients: an interim report. *Transplantation*. 1998;66:1395-1398.
25. Trotter JF. Sirolimus in liver transplantation. *Transplant Proc*. 2003;35(suppl 3A):193S-299S.
26. Groth CG, Backman L, Morales JM, et al. Sirolimus (rapamycin)-based therapy in human renal transplantation: similar efficacy and different toxicity compared with cyclosporine. Sirolimus European Renal transplant Study Group. *Transplantation*. 1999;67:1036-1042.
27. King-Biggs MB, Cunitz JM, Park SJ, et al. Airway anastomotic dehiscence associated with the use of sirolimus immediately after lung transplantation. *Transplantation*. 2003;75: 1437-1443.
28. Rebellato LM, Gross U, Verbanac KM, et al. A comprehensive definition of the major antibody specificities in polyclonal rabbit antithymocyte globulins. *Transplantation*. 2003;75:657-662.
29. Michallet MC, Preville X, Flacher M, et al. Functional antibodies to leukocyte adhesion molecules in antithymocyte globulins. *Transplantation*. 2003;75:657 662.
30. Calne RY, Friend P, Moffatt S, et al. Prope tolerance, perioperative Campath 1h, and low-dose cyclosporine monotherapy in renal allograft recipients. *Lancet*. 1998;351:1701-1702.
31. Kirk AD, Hale DA, Mannon RB, et al. Results from a human renal allograft tolerance trial evaluating the humanized CD52-specific monoclonal antibody alemtuzumab (campath 1H). *Transplantation*. 2003;76:120-129.
32. Schneeberger S, Landin L, Kaufmann C, et al. Alemtuzumab: key for minimization of maintenance immunosuppression in reconstructive transplantation. *Transplant Proc*. 2009;41: 499-502.
33. Emre S, Gondolesi G, Polat K, et al. Use of daclizumab as initial immunosuppression in liver transplant recipients with impaired renal; function. *Liver Transplant*. 2001;7:220-225.
34. Guo S, Han Y, Zhang X, et al. Human facial allotransplantation: a 2-year follow-up study. *Lancet*. 2008;372:631-638.
35. Lantieri L, Meningaud JP, Grimbert P, et al. Repair of the lower and middle part of the face by composite tissue allotransplantation in a patient with massive plexiform neurofibromatosis: a 1-year follow-up study. *Lancet*. 2008;372:639-645.
36. Siemionow M, Papay F, Alam D, et al. First US near-total human face transplantation – a paradigm shift for massive facial injuries. *Lancet*. 2009;374:203-209.
37. Allen U, Avery RK, Blumberg E, Burroughs M, et al. Guidelines for the prevention and management of infectious

complications of solid organ transplantation. *Am J Transplant*. 2004;4(Suppl 10):5-166.
38. Murray JE. Organ transplantation (skin, kidney, heart) and the plastic surgeon. *Plast Reconstr Surg*. 1971;47:425-431.
39. Lee WP, Yaremchuk MJ, Pan YC, et al. Relative antigenicity of components of a vascularized limb allograft. *Plast Reconstr Surg*. 1991;87:401-411.
40. Goldwyn RM, Beach PM, Felman D, et al. Canine limb homotransplantations. *Plast Reconstr Surg*. 1966;37:184-195.
41. Mathes DW, Randolph MA, Solari MG, et al. Split tolerance to a composite tissue allogaft in swine model. *Transplantation*. 2003;75:25-31.
42. Wang HJ, Ding YQ, Pei GX, et al. A preliminary pathological study on human allotransplantation. *Chin J Traumatol*. 2003;6:284-287.
43. Kanitakis J, Badet L, Petruzzo P, et al. Clinicopathologic monitoring of the skin and oral mucosa of the first human face allograft: report on the first eight months. *Transplantation*. 2006;82:1610-1615.
44. Kanitakis J, Jullien D, Petruzzo P, et al. Clinicopathological features of graft rejection in the first human hand allograft. *Transplantation*. 2003;76:688-693.
45. Kanitakis J. Skin rejection in human hand allografts: histologic findings and grading system. In: Lanzetta M, Dubernard JM, eds. *Hand Transplantation*. Milan, Italia: Springer-Verlag; 2007:249.
46. Levi D, Tzakis A, Kato T, et al. Transplantation of the abdominal wall. *Lancet*. 2003;361:2173-2176.
47. Cendales L, Kirk A, Moresi M, et al. Composite tissue allotransplantation: classification of clinical acute skin rejection. *Transplantation*. 2005;80:1676-1680.
48. Demetris AJ, Murase N, Ye Q, et al. Analysis of chronic rejection and obliterative arteriopathy: possible contributions of donor antigen presenting cells and lymphatic disruption. *Am J Pathol*. 1997;150:563-578.
49. Ildstad ST, Sachs DH. Reconstitution with syngeneic plus allogeneic or xenogeneic bome marrow leads to specific acceptance of allografts or xenografts. *Nature*. 1984;307:168-170.
50. Burlingham WJ. Chimerism after organ transplantation: is there any clinical significance? *Clin Transplant*. 1996;10:110-117.
51. Starzl TE. Acquired immunogenic tolerance: with particular reference to transplantation. *Immunol Res*. 2007;38:6-41.
52. Starzl TE, Zinkernagel RM. Transplantation tolerance from a historic perspective. *Nat Rev Immunol*. 2001;1:233-239.
53. El Jaafari A, Badet L, Kanitakis J, et al. Isolation of regulatory T cells in the skin of a human hand allograft, up to six years posttransplantation. *Transplantation*. 2006;82:1764-1768.
54. Trivedi HL, Vanikar AV, Modi PR, et al. Allogeneic hematopoietic stem-cell transplantation, mixec chimerism, and tolerance in living related donor renal allograft recipients. *Transplant Proc*. 2005;37:737-742.
55. Granger DK, Briedenbach WC, Pidwell DJ, et al. Lack of donor hyporesponsiveness and donor chimerism after clinical transplantation of the hand. *Transplantation*. 2002;74:1624-1630.

Cellular Therapies in Face Transplantation 42

Maria Z. Siemionow, Maria Madajka, and Joanna Cwykiel

Contents

M.Z. Siemionow (✉)
Department of Plastic Surgery, Cleveland Clinic, Cleveland, OH, USA
e-mail: siemiom@ccf.org

Abstract The main purpose of cellular therapy application in face transplantation is the continuous need to develop new strategies that would eliminate use of toxic immunosuppressive protocols. Cellular therapy in transplantology can significantly benefit allograft survival and shorten healing time. Cells utilized for therapeutic purpose are isolated mostly from bone marrow (BM) and adipose tissues. They have the ability to proliferate and differentiate in the transplanted tissue, and have immunomodulatory activity. Most of the cellular therapies such as regulatory T-cells, dendritic and chimeric cells are still in the experimental stage. Molecular characterization of these cells as well as the mechanism of their participation in allograft acceptance and rejection is not well established and will contribute to the future of modern transplantology.

Abbreviations

AC	Adipocyte cell
APC	Antigen-presenting cell
ASCs	Adipose stem cells
BM	Bone marrow
BMDC	Bone marrow derived cell
BMSC	Bone marrow stromal cell
CsA	Cyclosporine A
CTA	Composite tissue allotransplantation
DC	Dendritic cell
FC	Fused cell
GVHD	Graft versus host disease
HGF	Hepatocyte growth factor
IGF-1	Insulin growth factor-1
Il	Interleukin
IFN	Interferon
mAb	monoclonal antibody

M.Z. Siemionow (ed.), *The Know-How of Face Transplantation*,
DOI: 10.1007/978-0-85729-253-7_42,

MAPK	Mitogen-activated protein kinase
MHC	Major histocompatibility complex
MSC	Mesenchymal stem cell
PEG	Polyethylene glycol
TCR	T-cell receptor
TNF	Tumor necrosis factor
VEGF	Vascular endothelial growth factor

42.1 Introduction

Cellular therapies in face transplantation are a relatively new approach. By definition, cellular therapy utilizes different populations of cells to support survival of the transplant and to provide a healing effect, ideally without or under minimal dosage of immunosuppressants. Successful cellular treatment requires careful experimental design starting from the source of the potential cells and the clinical evaluation of the therapy. Depending on the case and medical history of the patient, cellular treatment should be individualized. Primary decisions prior to therapy induction include the choice of cell origin (autologous or allogeneic?), optimal dosage, and delivery routes. Also essential is the viability of the cells in vivo, as well as ability to differentiate the cells after delivery. Maximal cellular effectiveness may also be dependent on in vitro manipulations including testing of different types of media enriched with growth factors or other cell growth modulators.

In addition to strictly biological and medical issues inherent to cellular therapy, there is also an issue of cell delivery route. The questions that arise include: the use of appropriate carriers in order to prevent cellular activity and also how to keep the delivered cells in the area of designated action. Construction of special delivery devices and development of new technologies is an important future dialogue which has to be initiated between bioengineers and medical doctors.

There is no strict classification of cellular therapies used in face transplantation. Each therapy may require different quantities of cells and delivery routes and is dependent on the individual needs of the patient. In general, currently used therapies can be divided based on the cell origin, which include cell population from bone marrow (BM), fat, and skin (Fig. 42.1); and also

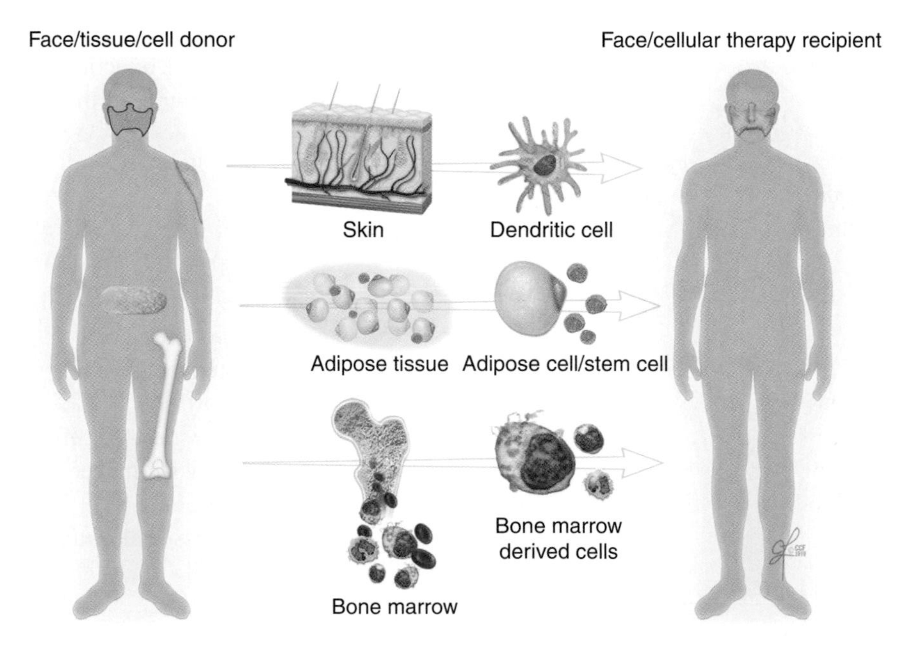

Fig. 42.1 Examples of diversity of therapeutic cells that can be considered to treat patients undergoing face transplantation. The advantage of cellular therapies includes procurement of facial graft and derived cells from the same donor. This approach can reduce the risk of transplant rejection. Potentially, donor tissue can serve as a reservoir of various cells isolated from skin, adipose tissue, or bone marrow. Each tissue is specifically processed ex vivo to isolate pure and functional population of cells. Cells of therapeutic properties such as dendritic cells originating from skin epithelium, adipose cells, and adipose stem cells as well as bone marrow stromal cells can be further introduced as supportive therapy following a face transplant

on the protocol of cells preparation for a particular clinical application, where isolated cells can be either processed in a standard way, genetically modified in vitro, or obtained by ex vivo cell fusion. In this chapter, based on the origin of cells we present a simple classification of cellular therapies with potential clinical applications.

42.2 Bone Marrow Derived Cells (BMDCs) Therapy as a Promising Approach in Face Transplantation

In the 1960s, Friedenstein and Petrakova isolated and identified a population of cells from the bone marrow (BM) which, when transplanted to the animal, have the ability to form bone, cartilage, marrow, fat cells, and stroma.[1,2] Their pioneering work introduced promising research based on therapeutic function of bone marrow stromal cells (BMSCs), called interchangeably mesenchymal stem cells (MSCs). The Greek word "stroma" means something upon which one rests or lies, the physical substrate,[3] and describes well the BMSCs responsible for the functional framework and architecture of cells in the bone marrow (BM) niche. Transplantation of BMSCs has advantages over the transplantation of the whole BM cell population. A large number of BMSCs can be obtained from a small volume of autologous BM, which does not require multiple biopsies/surgeries and does not impact the multipotential capability of collected cells in vitro.[3] The application of isolated BMSCs facilitates bone formation and regeneration, which are both extremely important in the clinical settings of facial reconstruction or transplantation. The choice of appropriate bone type as a source of cellular therapy is critical. In 2006, Akintoye investigated skeletal site-specific phenotypic and functional differences between orofacial (maxilla) and axial (iliac crest) BMSCc in vitro and in vivo.[4] Compared with iliac crest cells, orofacial bone marrow derived cells (BMDCs) proliferated more rapidly, expressed higher levels of alkaline phosphatase, and demonstrated more calcium accumulation in vitro. Orofacial BMDCs formed more bone in vivo, while iliac crest BMDCs formed more compacted bone and were more responsive in vitro and in vivo to osteogenic and adipogenic inductions. These data demonstrated that BMDCs originating from the same individual are unique cell populations, which may have a crucial impact on bone regeneration.[4]

Two years later, Steinhardt[4] reported that MSC isolated from maxillofacial BM are easily genetically modified to express osteogenic growth factor, which indicates their new potential in future applications. Extensive research performed on animal models using BMSCs confirmed the feasibility of their potential use in clinical practice in humans.[5,6]

The first face allograft transplant was performed by Dubernard and Devauchelle in 2006[7] and was supported by bone marrow cellular therapy from the same donor.[8] Bone marrow cells were collected from the donor directly before face dissection and BM infusion was performed on days 4 and 11 after transplantation under standard immunosuppression protocol.[7-9] The chimerism level defined as the presence of the donor cells in the blood and BM of the recipient was measured at different time points by PCR, and was recorded between 0.1% and 1% 2 months after transplant.[7] The low incidence of chimerism could be due to the poor quality of hematopoietic cells engraftment or depletion of T-cells population in BM compartment under immunosuppressive therapy. The mechanism of tolerance induction and its correlation with the level of chimerism is unclear and remains open for discussion and the investigation as to whether chimerism is a consequence or side effect of cellular therapy.

Based on the solid organ transplantation experience[10,11], microchimerism was undetectable in majority of the patients, and did not prevent rejection. However, in cases of composite tissue transplantation, multiple studies suggest that BM is immunomodulatory and may facilitate allograft acceptance. For example, vascularized bone marrow allotransplantation (VBMT) supported with the donor BM cells showed prolonged survival of donor skin graft and resulted in stable chimerism.[12] Also, induction of donor-specific chimerism was confirmed after bilateral VBMT.[13] Despite the questionable role of chimerism in tolerance maintenance, BMDCs therapy in face transplantation is still in a developmental stage. BMDCs are currently in trials for graft versus host disease (GVHD)[14] and it is unknown if the properties of these cells change during differentiation process. The future of BMDCs will depend on molecular biology and modern genetic studies and their ability to characterize these cells and create banks for cell storage and distribution for therapeutic applications.

42.3 Adipose Stem Cells (ASCs) and Adipose Cells (ACs) Therapy in Face Transplantation

Adipose tissue has been a subject of investigation for over the past 100 years and is used in reconstructive and esthetic surgery. Human and animal adipose tissues are derived from the embryonic mesoderm and contain heterogenous stromal cell population[9,15-17] carrying markers of monocytes, macrophages, and endothelial cells.[18]The multilineage capacity of ASCs offers a great regenerative potential. The advantage of the use of ASCs over MSC includes greater efficacy of cell differentiation,[19] lower effect on T-cell activation,[20] as well as a minimally invasive and yield-effective way of cell isolation.

Autologous transplantation of ASCs has been used in breast reconstructive surgery and closed fistulas associated with Crohn's disease.[21,22] In animal models, ASCs were used for myocardial regeneration.[23,24] Currently, clinical trials with ASCs are applied to patients after myocardial infarction and patients with lipodystrophy.[25] ASCs are interesting candidates for use in face transplantation due to their immunomodulatory function. Some reports confirmed that ASCs can prevent graft versus host disease (GVHD)[26] and sepsis.[27] In 2008, Lu et al. published that ASCs have the ability to enhance blood supply in skin flaps.[28] Transplanted fat cells can also positively influence the surrounding facial tissue and improve the healing process.[29] The mechanism of action of ASCs in transplantation surgery is not well established. It is known that adipose progenitor cells can mediate protective effects through enhanced angiogenesis by their ability to direct differentiation into endothelial cells and cytokine stimulated production of vascular endothelial growth factor (VEGF).[28] In 2006, Wang et al. reported[30] that human ASCs are able to produce not only VEGF, but also hepatocyte growth factor (HGF) and insulin growth factor-1 (IGF-1) in the response to tumor necrosis factor (TNF) via a p38 MAP kinase-dependent mechanism. Production of both VEGF and HGF stimulates cell growth, which is beneficial for tissue neovascularization, remodeling, and is important in wound healing. Additionally, ASCs can increase the level of anti-apoptotic factors acting against tissue ischemia[31] and can control tissue inflammation by their ability to respond to TNF, which leads to the increased production of VEGF. This is the mechanism by which adipose cells can regulate cell growth and angiogenesis and can promote tissue and wound healing, which would have beneficial effect following face transplantation.

42.4 Regulatory T-Cells (T-Reg) and Dendritic Cells (DCs) as Promising Candidates for Cellular Therapy in Face Transplantation

The establishment of clinically applicable cellular therapies represents an example of new strategies for tolerance induction with hope of eliminating life-long immunosuppression. During the past 20 years of experience with composite tissue allotransplantation (CTA), Siemionow's laboratory used various immunosuppressive protocols tested in animal models.[12,32-38] This resulted in more than 3,000 experiments performed in different composite tissue transplantations, including over 1,200 face transplantations. A well-tested protocol of selective targeting of T-cell receptor (TCR) by application of anti-TCR monoclonal antibody (mAb) combined with cyclosporine A (CsA) therapy, supported with BM transplantation, resulted in face allograft survival over 495 days posttransplant. This immunosuppressive protocol combined with cellular therapy eliminated the need for chronic immunosuppression. Long-term survival of animals was associated with over 25% of donor–origin chimerism in the peripheral blood of face transplant recipients. The same immunosuppressive protocols are currently used in combination with cellular therapy utilizing T-reg and/or DCs populations. The population of T-reg cells belongs to the family of T cells present in the thymus, but the process of their maturation is still not clear. They express CD4 and CD25 antigens, and promote tolerance to self and foreign antigens.[39,40] Adoptive therapy with CD4/CD25 T-reg was applied in allograft rejection,[41] as well as in organ transplantation.[42]

The use of T-reg in combination with DCs could be used as an optimal cellular therapy in face transplantation (Fig. 42.2). DCs are a resident population of T cells in the skin and are different from lymphoid $\gamma\delta$ T cells and $\alpha\beta$ T cells in terms of oncogeny, tissue tropism, and antigen receptor diversity.[43]

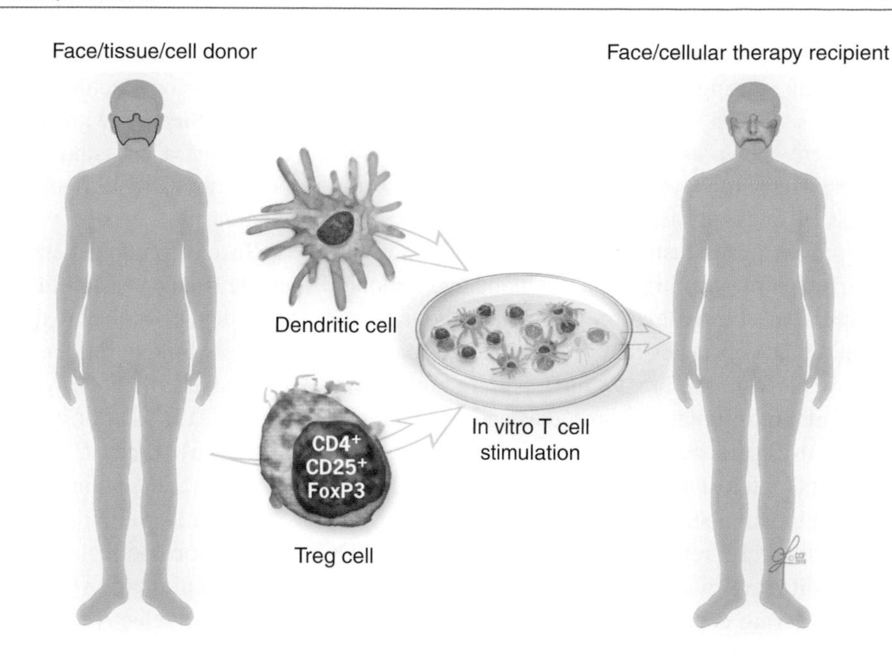

Fig. 42.2 Diagram representing application of T-reg cells in combination with DCs in face transplantation. T-reg represent a subpopulation of T cells that act to suppress the immune system and maintain immune homeostasis. T-reg activation is dependent upon the interactions with CD80, CD86, and CD40 cell surface receptors present on the DCs. Isolation of T-reg and DCs from the donor and their in vitro activation is the ultimate goal to generate the required number of T-regs in order to support face transplant acceptance

DCs respond to skin injury by attaching to keratinocytes via tight-junction proteins with epithelium. DCs are active immune cells that capture antigens, possess migratory capacity, and act as antigen-presenting cells (APC).[44] DCs provide the instructive role for T cell antigen recognition by establishing the immunological synapses with naïve T cells, which stimulate T cells and their commitment to proliferate.[44] As proposed by Kedl, there is a competition of T cells for binding in vivo to DCs, which can be compromised by an increased number of DCs.[45] During the first steps of an immunological response, the frequency of naïve T cells for a given antigen is low, and when T cells proliferate, competition for TCR receptor stimulation increases, especially among the cells occupying the same niche. This leads to functional diversification, where T cells achieving sustained stimulation differentiate to the effector cells.[44] Due to such interclonal competition and the low number of DCs, their potential use in face transplantation should be further developed. Following DCs injection, successful therapy must contain a specified number of active cells, defined time of reinjection if required, and the most efficient delivery route. Due to these facts and the biological nature of T-reg and DCs, their application is currently being tested in experimental models of composite tissue allografts.[46-48]

42.5 Ex Vivo Created Fused Chimeric Cells: A Step into the Future

The idea of chimeric cell research originates from the observation of spontaneous cell fusion occurring during normal embryogenesis and morphogenesis. It is involved in a variety of biological processes such as development of trophoblasts, placenta, muscles, and bones, as well as during immune responses and tumorogenesis. The earliest observation of spontaneous fusion in vitro between two different mammalian cells was reported in 1961 by Barski.[49] Interest in cell fusion was revived in 2002 by the discovery of spontaneous fusion in vitro of pluripotent embryonic stem cells with mouse BM cells[50] and brain progenitor cells.[51] These studies revealed that fused cells (FCs) can express characteristics of undifferentiated cells or properties of both types of fused cells. Further studies in in vivo models demonstrated that BMDC can fuse spontaneously with skeletal muscle cells, cardiac muscle cells, liver and intestine cells, and Purkinje neurons, resulting in formation of stable multinucleated heterokaryons.[52-54] In these experiments, fused cells were able to adopt phenotype and function of the recipient cells and contribute to the regeneration process. It has been shown that in vivo cell fusion could be triggered by cytokine microenvironmental changes that occur during injury. Several proinflammatory cytokines such as IL-4, IL-13, IFNγ, TNFα, IL-1, and IL-3 were shown as potential participants in a fusion process of monocytes, macrophages, and osteoblasts.[55,56] In experimental and clinical models, it was also proven that infusion of BMDC and the presence of mixed chimerism in peripheral blood and lymphoid organs after transplantation facilitated allograft survival.[57,58] The obstacles that diminished the success of cell-to-cell fusion in in vivo models were highly cytotoxic immunosuppressive protocols and a low number of cells that underwent fusion. These factors decreased scientists' interest in the in vivo fusion process and affected the number of published reports.

Bonde et al.[59] reported that fusion between BMDC of two different mouse strains occurred spontaneously in vitro during coculturing and in vivo after allogenic and syngeneic transplantation. This study showed that fused cells (FC) expressed both the donor and recipient major histocompatibility complex (MHC) antigens on their surfaces. Siemionow's team performed experiments with cell fusion of BMDC (Cwykiel J PSRC supplement 2010), proving the feasibility of creating in vivo donor–recipient FC. These cells, used as supportive therapy, facilitated face allograft survival. In order to improve engraftment of donor–recipient FC, a short immunomodulatory protocol of αβ- T cell receptor monoclonal antibody and CsA was used. This supportive therapy with donor–recipient chimeric cells represents a new and potentially breakthrough therapeutic modality for solid organ and composite tissue allotransplantation.

Success of the generation of in vivo cell fusion may create many possibilities for the use of BMDC as a specific tool applied to the development of novel therapies. Although further research is required to fully understand the underlying mechanisms of cell fusion process, the idea of creating tolerance-inducing cells ex vivo by direct in vitro cell-to-cell fusion would be of great value in clinical transplantation.

Recently, Siemionow's group created an innovative animal protocol for tolerance induction in face allotransplantation model. The protocol combines short-term selective use of αβ-T cell receptor mAb and CsA supported by cellular therapy. Cellular therapy was created by ex vivo fusion of donor and recipient BMDC using polyethylene glycol (PEG) technique and cell labeling as depicted in Fig. 42.3. The expression of MHC characteristic for the donor and recipient was detected on the surface of chimeric fused cells and was confirmed by the karyotype assay. This pioneering work opens new treatment options supporting long-term survival of face allograft transplants.

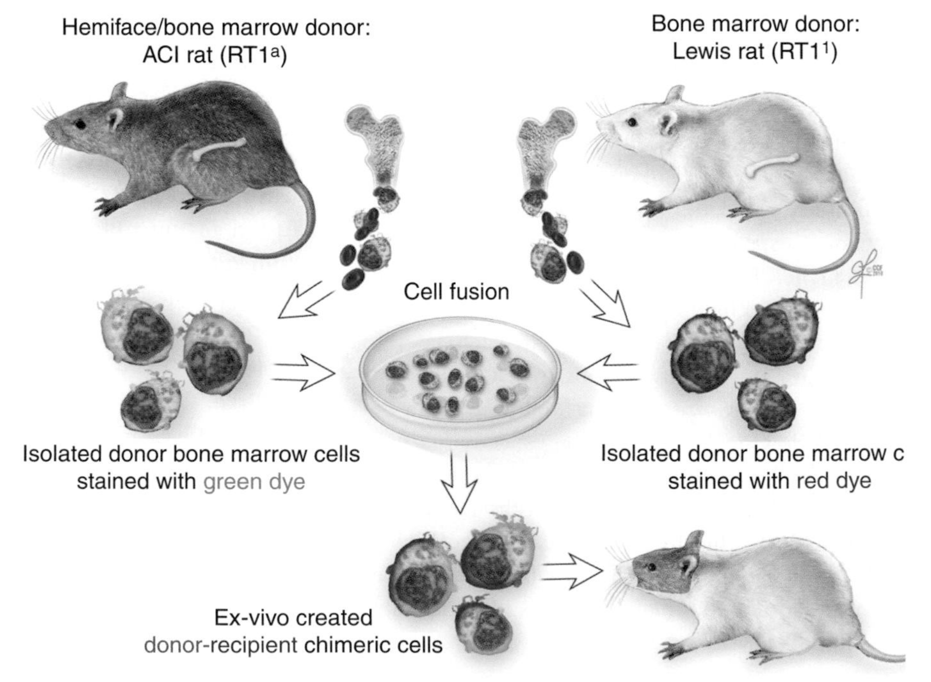

Fig. 42.3 Ex vivo creation of chimeric cells. This novel experimental approach toward tolerance induction is currently tested on an animal model and utilizes bone marrow–derived cells. During the first step, cells are isolated from the hemiface donor as well as from the future hemiface recipient. Cells are stained with two different fluorescent dyes to identify the donor cell origin (*red*) and recipient cell (*green*). Next, ex vivo fusion is performed using standard polyethylene glycol (PEG) method. Based on the fluorescence fused chimeric cells are separated and collected for injection as a supportive cellular therapy to the hemiface recipient rat

The future of cellular therapies will be based on individualized, custom-made protocols, with, e.g., chimeric human cells created ex vivo and utilized for development of donor(s)-specific transferable tolerance. In the future, this novel approach could serve as a supportive therapy for patients receiving composite tissue allografts including face transplants.

References

1. Friedenstein AJ, Petrakova KV, Kurolesova AI, Frolova GP. Heterotopic of bone marrow. Analysis of precursor cells for osteogenic and hematopoietic tissues. *Transplantation.* 1968;6:230-247.
2. Owen M, Friedenstein AJ. Stromal stem cells: Marrow-derived osteogenic precursors. *Ciba Found Symp.* 1988;136: 42-60.
3. Krebsbach PH, Kuznetsov SA, Bianco P, Robey PG. Bone marrow stromal cells: Characterization and clinical application. *Crit Rev Oral Biol Med.* 1999;10:165-181.
4. Steinhardt Y, Aslan H, Regev E, et al. Maxillofacial-derived stem cells regenerate critical mandibular bone defect. *Tissue Eng A.* 2008;14:1763-1773.
5. Mankani MH, Kuznetsov SA, Wolfe RM, Marshall GW, Robey PG. In vivo bone formation by human bone marrow stromal cells: Reconstruction of the mouse calvarium and mandible. *Stem Cells.* 2006;24:2140-2149.
6. Mankani MH, Kuznetsov SA, Shannon B, et al. Canine cranial reconstruction using autologous bone marrow stromal cells. *Am J Pathol.* 2006;168:542-550.
7. Devauchelle B, Badet L, Lengele B, et al. First human face allograft: Early report. *Lancet.* 2006;368:203-209.
8. Hequet O, Morelon E, Bourgeot JP, et al. Allogeneic donor bone marrow cells recovery and infusion after allogeneic face transplantation from the same donor. *Bone Marrow Transplant.* 2008;41:1059-1061.
9. Hausman GJ, Campion DR, McNamara JP, Richardson RL, Martin RJ. Adipose tissue development in the fetal pig after decapitation. *J Anim Sci.* 1981;53:1634-1644.
10. Caillat-Zucman S, Legendre C, Suberbielle C, et al. Microchimerism frequency two to thirty years after cadaveric kidney transplantation. *Hum Immunol.* 1994;41: 91-95.
11. Suberbielle C, Caillat-Zucman S, Legendre C, et al. Peripheral microchimerism in long-term cadaveric-kidney allograft recipients. *Lancet.* 1994;343:1468-1469.
12. Arslan E, Klimczak A, Siemionow M. Chimerism induction in vascularized bone marrow transplants augmented with bone marrow cells. *Microsurgery.* 2007;27:190-199.
13. Klimczak A, Agaoglu G, Carnevale KA, Siemionow M. Applications of bilateral vascularized femoral bone marrow transplantation for chimerism induction across the major histocompatibility (MHC) barrier: part II. *Ann Plast Surg.* 2006;57:422-430.
14. Robey PG, Bianco P. The use of adult stem cells in rebuilding the human face. *J Am Dent Assoc.* 2006;137: 961-972.
15. Hausman GJ. Adipocyte development in subcutaneous tissues of the young rat. *Acta Anat (Basel).* 1982;112: 185-196.
16. Loffler G, Hauner H. Adipose tissue development: the role of precursor cells and adipogenic factors. Part II: The regulation of the adipogenic conversion by hormones and serum factors. *Klin Wochenschr.* 1987;65:812-817.
17. Hauner H, Loffler G. Adipose tissue development: The role of precursor cells and adipogenic factors. Part I: Adipose tissue development and the role of precursor cells. *Klin Wochenschr.* 1987;65:803-811.
18. Planat-Benard V, Silvestre JS, Cousin B, et al. Plasticity of human adipose lineage cells toward endothelial cells: Physiological and therapeutic perspectives. *Circulation.* 2004;109:656-663.
19. Izadpanah R, Trygg C, Patel B, et al. Biologic properties of mesenchymal stem cells derived from bone marrow and adipose tissue. *J Cell Biochem.* 2006;99:1285-1297.
20. Keyser KA, Beagles KE, Kiem HP. Comparison of mesenchymal stem cells from different tissues to suppress T-cell activation. *Cell Transplant.* 2007;16:555-562.
21. Yoshimura K, Sato K, Aoi N, Kurita M, Hirohi T, Harii K. Cell-assisted lipotransfer for cosmetic breast augmentation: Supportive use of adipose-derived stem/stromal cells. *Aesthetic Plast Surg.* 2008;32:48-55. discussion 56-57.
22. Garcia-Olmo D, Herreros D, Pascual I, et al. Expanded adipose-derived stem cells for the treatment of complex perianal fistula: A phase II clinical trial. *Dis Colon Rectum.* 2009;52:79-86.
23. Madonna R, De Caterina R. Adipose tissue: A new source for cardiovascular repair. *J Cardiovasc Med (Hagerstown).* 2010;11:71-80.
24. Jumabay M, Matsumoto T, Yokoyama S, et al. Dedifferentiated fat cells convert to cardiomyocyte phenotype and repair infarcted cardiac tissue in rats. *J Mol Cell Cardiol.* 2009;47:565-575.
25. Tran TT, Kahn CR. Transplantation of adipose tissue and stem cells: role in metabolism and disease. *Nat Rev Endocrinol.* 2010;6:195-213.
26. Yanez R, Lamana ML, Garcia-Castro J, Colmenero I, Ramirez M, Bueren JA. Adipose tissue-derived mesenchymal stem cells have in vivo immunosuppressive properties applicable for the control of the graft-versus-host disease. *Stem Cells.* 2006;24:2582-2591.
27. Gonzalez MA, Gonzalez-Rey E, Rico L, Buscher D, Delgado M. Adipose-derived mesenchymal stem cells alleviate experimental colitis by inhibiting inflammatory and autoimmune responses. *Gastroenterology.* 2009;136:978-989.
28. Lu F, Mizuno H, Uysal CA, Cai X, Ogawa R, Hyakusoku H. Improved viability of random pattern skin flaps through the use of adipose-derived stem cells. *Plast Reconstr Surg.* 2008;121:50-58.
29. Coleman SR. Facial augmentation with structural fat grafting. *Clin Plast Surg.* 2006;33:567-577.
30. Wang M, Crisostomo PR, Herring C, Meldrum KK, Meldrum DR. Human progenitor cells from bone marrow or adipose tissue produce VEGF, HGF, and IGF-I in response to TNF by a p38 MAPK-dependent mechanism. *Am J Physiol Regul Integr Comp Physiol.* 2006;291: R880-R884.

31. Togel F, Hu Z, Weiss K, Isaac J, Lange C, Westenfelder C. Administered mesenchymal stem cells protect against ischemic acute renal failure through differentiation-independent mechanisms. *Am J Physiol Renal Physiol.* 2005;289:F31-F42.
32. Agaoglu G, Carnevale KA, Zins JE, Siemionow M. Bilateral vascularized femoral bone transplant: A new model of vascularized bone marrow transplantation in rats, part I. *Ann Plast Surg.* 2006;56:658-664.
33. Demir Y, Ozmen S, Klimczak A, Mukherjee AL, Siemionow M. Tolerance induction in composite facial allograft transplantation in the rat model. *Plast Reconstr Surg.* 2004;114:1790-1801.
34. Ozmen S, Ulusal BG, Ulusal AE, Izycki D, Siemionow M. Composite vascularized skin/bone transplantation models for bone marrow-based tolerance studies. *Ann Plast Surg.* 2006;56:295-300.
35. Siemionow M, Ulusal BG, Ozmen S, Ulusal AE, Ozer K. Composite vascularized skin/bone graft model: A viable source for vascularized bone marrow transplantation. *Microsurgery.* 2004;24:200-206.
36. Ozer K, Gurunluoglu R, Zielinski M, Izycki D, Unsal M, Siemionow M. Extension of composite tissue allograft survival across major histocompatibility barrier under short course of anti-lymphocyte serum and cyclosporine a therapy. *J Reconstr Microsurg.* 2003;19:249-256.
37. Ozer K, Oke R, Gurunluoglu R, et al. Induction of tolerance to hind limb allografts in rats receiving cyclosporine A and antilymphocyte serum: Effect of duration of the treatment. *Transplantation.* 2003;75:31-36.
38. Siemionow MZ, Izycki DM, Zielinski M. Donor-specific tolerance in fully major histocompatibility major histocompatibility complex-mismatched limb allograft transplants under an anti-alphabeta T-cell receptor monoclonal antibody and cyclosporine A protocol. *Transplantation.* 2003;76:1662-1668.
39. Shevach EM. Suppressor T cells: Rebirth, function and homeostasis. *Curr Biol.* 2000;10:R572-R575.
40. Shevach EM. Regulatory T cells in autoimmmunity*. *Annu Rev Immunol.* 2000;18:423-449.
41. Dorsch S, Roser R. Recirculating, suppressor T cells in transplantation tolerance. *J Exp Med.* 1977;145:1144-1157.
42. Zheng SG, Meng L, Wang JH, et al. Transfer of regulatory T cells generated ex vivo modifies graft rejection through induction of tolerogenic CD4+CD25+ cells in the recipient. *Int Immunol.* 2006;18:279-289.
43. Jameson J, Havran WL. Skin gammadelta T-cell functions in homeostasis and wound healing. *Immunol Rev.* 2007;215: 114-122.
44. Langenkamp A, Casorati G, Garavaglia C, Dellabona P, Lanzavecchia A, Sallusto F. T cell priming by dendritic cells: thresholds for proliferation, differentiation and death and intraclonal functional diversification. *Eur J Immunol.* 2002;32:2046-2054.
45. Kedl RM, Rees WA, Hildeman DA, et al. T cells compete for access to antigen-bearing antigen-presenting cells. *J Exp Med.* 2000;192:1105-1113.
46. Du JF, Li SY, Bai X. T(reg)-based therapy and mixed chimerism in small intestinal transplantation: does T(reg)+BMT equal intestine allograft tolerance?. *Med Hypotheses* 2011; 76(1):77-78.
47. Maury S, Lemoine FM, Hicheri Y, et al. CD4+CD25+ regulatory T cell depletion improves the graft-versus-tumor effect of donor lymphocytes after allogeneic hematopoietic stem cell transplantation. *Sci Transl Med.* 2010;2:41-52.
48. Fandrich F. Cell therapy approaches aiming at minimization of immunosuppression in solid organ transplantation. *Curr Opin Organ Transplant.* 2010;15:703-708.
49. Barski G, Sorieul S, Cornefert F. "Hybrid" type cells in combined cultures of two different mammalian cell strains. *J Natl Cancer Inst.* 1961;26:1269-1291.
50. Terada N, Hamazaki T, Oka M, et al. Bone marrow cells adopt the phenotype of other cells by spontaneous cell fusion. *Nature.* 2002;416:542-545.
51. Ying QL, Nichols J, Evans EP, Smith AG. Changing potency by spontaneous fusion. *Nature.* 2002;416:545-548.
52. LaBarge MA, Blau HM. Biological progression from adult bone marrow to mononucleate muscle stem cell to multinucleate muscle fiber in response to injury. *Cell.* 2002;111:589-601.
53. Alvarez-Dolado M, Pardal R, Garcia-Verdugo JM, et al. Fusion of bone-marrow-derived cells with Purkinje neurons, cardiomyocytes and hepatocytes. *Nature.* 2003;425:968-973.
54. Wang X, Willenbring H, Akkari Y, et al. Cell fusion is the principal source of bone-marrow-derived hepatocytes. *Nature.* 2003;422:897-901.
55. Anderson JM. Multinucleated giant cells. *Curr Opin Hematol.* 2000;7:40-47.
56. Merkel KD, Erdmann JM, McHugh KP, Abu-Amer Y, Ross FP, Teitelbaum SL. Tumor necrosis factor-alpha mediates orthopedic implant osteolysis. *Am J Pathol.* 1999;154: 203-210.
57. Siemionow MZ, Demir Y, Sari A, Klimczak A. Facial tissue allograft transplantation. *Transplant Proc.* 2005;37:201-204.
58. Rahhal DN, Xu H, Huang WC, et al. Dissociation between peripheral blood chimerism and tolerance to hindlimb composite tissue transplants: Preferential localization of chimerism in donor bone. *Transplantation.* 2009;88:773-781.
59. Bonde S, Pedram M, Stultz R, Zavazava N. Cell fusion of bone marrow cells and somatic cell reprogramming by embryonic stem cells. *FASEB J.* 2010;24:364-373.

Tissue Engineering for Facial Reconstruction

43

Tsung-Lin Yang, James J. Yoo, Maria Z. Siemionow, and Anthony Atala

Contents

A. Atala (✉)
Director, Wake Forest Institute for Regenerative Medicine, Wake Forest University School of Medicine, Medical Center Blvd., Winston-Salem, NC, USA
e-mail: aatala@wfubmc.edu

Abstract Craniofacial structures are essential for many physiological functions, including vision, olfaction, hearing, and food intake. In addition, facial features are critical for the development of personal identity, communication, and social interaction. Thus, damage to the face resulting from traumatic injury or disease can be particularly devastating to a patient's quality of life, and the development of methods to restore normal craniofacial structures is essential. In recent years, facial transplantation using microsurgical techniques has become a reality, but this technique is limited by a shortage of donor tissue and the need for chronic administration of immunosuppressive drugs to prevent graft rejection. Recent advances in tissue engineering and regenerative medicine provide opportunities to create biological substitutes that can be used in reconstructive surgery. This field applies the principles of cell transplantation, material science, and bioengineering to develop tissues and organs in the laboratory that can then be implanted into a patient to replace damaged or missing structures. In this chapter, we will discuss these techniques in detail, and we will illustrate how they can be used to revolutionize the concepts of facial reconstruction.

Abbreviations

AFPS	Amniotic-fluid and placental-derived stem
BSM	Bladder submucosa
EC	Endothelial cells
ECM	Extracellular matrix
ES	Embryonic stem
FDA	Food and Drug Administration
iPS	induced Pluripotent State
MEFs	Mouse embryonic fibroblasts
PGA	Polyglycolic acid

M.Z. Siemionow (ed.), *The Know-How of Face Transplantation*,
DOI: 10.1007/978-0-85729-253-7_43,

PLA Polylactic acid
PLGA Poly(lactic-co-glycolic acid)
SIS Small-intestinal submucosa
VEGF Vascular endothelial growth factor

43.1 Introduction

The face is the most prominent part of the body, and it plays many important roles. Physiologically, it is the entry point of both the respiratory and digestive systems, and contains structures that are essential in respiration and food intake. In addition, the sensory organs for vision, olfaction, hearing, and taste are housed in the craniofacial area. Psychologically, the face is essential for development and maintenance of personal identity, which is critical for social communication, expression of emotion, and mutual interaction. Therefore, when the facial area is damaged or disfigured due to disease processes or injury, a patient's quality of life is severely decreased and the ego is frequently disturbed.

Conditions involving birth defects, injuries, diseases, or certain therapeutic modalities such as surgery or radiation therapy often cause facial disfigurement, which result in severe physiological and psychological trauma.[1,2] Facial reconstructive surgery aims to restore the function and esthetics of each subunit of the face using tissue substitutes. As such, the reconstruction is usually customized for each individual patient. Various surgical methods have been used in facial reconstruction, including grafts, vascularized tissue flaps, microvascular free flaps, or combinations of these.[3,4] When larger facial defects are encountered, extensive reconstruction might be the treatment of choice, and this often requires multiple surgeries and different reconstructive approaches. Despite the recent progress in reconstructive surgery, the results of large-scale facial reconstruction remain unsatisfactory.[5] The application of free tissue transfers, expanders, and tissue prefabrication allows for facial defect coverage; however, functional deficits are not restored to the normal state. Moreover, regardless of the methods used, the structure and composition of the tissues that make up the face are specific, and it is not easy to transfer tissues from other parts of the body. In addition, tissues in the facial area are composite structures consisting of multiple tissue types that require coordinated function.

In order to achieve replacement of the tissues, according to Gille's rule of replacing "like tissues with alike," transplantation of the face from human donors is currently a novel approach for reconstruction of severe defects in patients after trauma, burn injuries, cancer, or congenital malformations. Since the first human face transplant was performed, several additional cases have been reported worldwide.[6,7] Despite the current success in face transplantation, several challenges must be addressed before the technique can be applied in mainstream clinical medicine. These include the availability of a suitable donor, complex surgical techniques, use of immunosuppressive medication, and ethical and psychosocial issues.

In recent years, advances in tissue engineering and regenerative medicine have provided various opportunities in medicine. The science of tissue engineering aims to generate tissues that would replace the structure and function of failing organs.[8-10] Using techniques from cell biology, material science, and transplantation, investigators have been able to construct tissue substitutes in the laboratory that can restore the normal structure and function of missing or diseased organs. This has been accomplished by combining cells, biomaterial "scaffolds" on which the cells can attach and grow, and appropriate signaling molecules in bioreactors. Although this field is still in the early stages of development, some successful approaches have already been applied to the human body, which suggests that this type of treatment is promising and may be used in the future.[11-13]

The use of tissue engineering techniques for the craniofacial area has several advantages. First, replacement tissues could be custom-designed for different individuals depending on their needs. For example, the use of an appropriate scaffold to guide tissue growth might simplify the reconstruction of a variety of facial contours and shapes, and this could lead to reconstructed tissues that are more similar to a patient's natural facial structure. New interactive biomaterial scaffolds are now being investigated in order to accomplish this goal.[14] In addition, if autologous cells were used to engineer replacement facial tissues, the use of immunosuppressant medications after tissue transplantation could be reduced or eliminated.[15] However, the generation of complex tissues such as those present in the craniofacial area remains a challenge. Normal tissue is made up of numerous cell types that are arranged in a specific and well-organized manner, and this structural complexity is often difficult to replicate in the laboratory. Within the living body, cell–cell and tissue–tissue interactions are dynamic and topographically oriented, and thus in order to engineer a functional tissue, it is necessary to establish an environment that is capable of providing temporal and spatial cues appropriate for tissue formation. In addition, in order for a tissue to be functional, it must be integrated to the vascular and nervous systems within the body. Tissue engineering provides an encouraging and exciting basis toward reconstructive options in facial reconstruction. This chapter will introduce the basic principles of tissue engineering and outline the current advances in this field. In addition, evolving methods for using tissue engineering technologies to reconstruct craniofacial tissues and organs will be discussed.

43.2 The Basics of Tissue Engineering

Tissue engineering employs aspects of cell biology and transplantation, materials science, and engineering to develop biological substitutes that can restore and maintain the normal function of damaged tissues and organs. It includes techniques such as the injection of functional cells into a nonfunctional body site to stimulate regeneration and the use of biocompatible materials to create new tissues and organs. These biomaterials can be natural or synthetic matrices, often termed scaffolds, which encourage the body's natural ability to repair itself and assist in determination of the orientation and direction of new tissue growth. Often, tissue engineering uses a combination of both of these techniques. For example, biomaterial matrices seeded with cells can be implanted into the body to encourage the growth or regeneration of functional tissue.

43.2.1 Biomaterials Used in Tissue Engineering

The design and selection of a biomaterial for use in regenerative medicine is critical for the proper development of engineered tissues. The selected biomaterial must be capable of controlling the structure and function of the engineered tissue in a predesigned manner by interacting with transplanted cells and/or host cells. In addition, it should be biocompatible, able to promote cellular interaction and tissue development, and it should possess the proper mechanical and physical properties required for tissue support and function in the body site of interest.

Appropriate biomaterials should be biodegradable and bioresorbable to support the reconstruction of a completely normal tissue without inflammation. Thus, the degradation rate and the concentration of degradation products in the tissues surrounding the implant must be maintained at a tolerable level.[16] Such behavior avoids the risk of inflammatory or foreign-body responses that are often associated with the permanent presence of a foreign material in the body. In addition, the biomaterial should provide appropriate regulation of cell behavior (e.g., adhesion, proliferation, migration, differentiation) in order to promote the development of functional new tissue. Cell behavior in engineered tissues is regulated by multiple interactions with the microenvironment, including interactions with cell-adhesion ligands[17] and with soluble growth factors.[18] Cell-adhesion promoting factors (e.g., Arg-Gly-Asp [RGD]) can be presented by the biomaterial itself or incorporated into the biomaterial in order to control cell behavior through ligand-induced cell receptor signaling processes.[19,20] In vivo, the biomaterials must provide temporary mechanical support sufficient to withstand forces exerted by the surrounding tissue and maintain a potential space for tissue development. The mechanical support of the biomaterials should be maintained until the engineered tissue has formed sufficient structural integrity to support itself.[21] This can be achieved by an appropriate choice of mechanical and degradation properties of the biomaterials.[22]

Finally, the chosen biomaterial must have properties that allow it to be processed into specific configurations. A large ratio of surface area to volume is often desirable to allow the delivery of a high density of cells. A high porosity, interconnected pore structure with specific pore sizes promotes tissue ingrowth from the surrounding host tissue. Several techniques, such as electrospinning, have been developed, and they allow precise control of porosity, pore size, and pore structure.[23-28]

Various biomaterials have been used in tissue engineering and regenerative medicine. These include naturally derived materials, such as collagen and alginate; acellular tissue matrices, such as bladder submucosa (BSM) and small-intestinal submucosa (SIS); and synthetic polymers, such as polyglycolic acid (PGA), polylactic acid (PLA), and poly(lactic-*co*-glycolic acid) (PLGA). Naturally derived materials and acellular tissue matrices have the potential advantage of biologic recognition. However, synthetic polymers can be produced quickly and reproducibly on a large scale with controlled properties of strength, degradation rate, and microstructure.

Collagen is the most abundant and ubiquitous structural protein in the body, and it may be readily purified from both animal and human tissues with an enzyme treatment and salt/acid extraction.[29] Collagen has long been known to exhibit minimal inflammatory and antigenic responses,[30] and it has been approved by the US Food and Drug Administration (FDA) for many types of medical applications, including wound dressings and artificial skin.[31] Collagen contains cell-adhesion domain sequences (e.g., RGD) that exhibit specific cellular interactions. This may help to retain the phenotype and activity of many types of cells, including fibroblasts[32] and chondrocytes.[33]

Alginate, a polysaccharide isolated from seaweed, has been used as an injectable cell delivery vehicle[34] and a cell immobilization matrix[35] owing to its gentle gelling properties in the presence of divalent ions such as calcium. Alginate is a family of copolymers of D-mannuronate and L-guluronate. The physical and mechanical properties of alginate gel are strongly correlated with the proportion and length of the polyguluronate block in the alginate chains.[34] Efforts have been made to synthesize biodegradable alginate hydrogels with mechanical properties that are controllable in a wide range by intermolecular covalent cross-linking and with cell-adhesion peptides coupled to their backbones.[36]

Acellular tissue matrices are collagen-rich matrices prepared by removing cellular components from tissues. The most common tissue that has been used for this purpose has been bladder tissue. The matrices are prepared by removing the cellular material from a segment of bladder tissue using mechanical and chemical processes.[37-40] The resulting matrix can be used alone or seeded with cells. The matrices slowly degrade after implantation and are replaced and remodeled by extracellular matrix (ECM) proteins synthesized and secreted by transplanted or ingrowing cells. Acellular tissue matrices support cell ingrowth and regeneration of many tissue types with no evidence of immunogenic rejection.[40,41] Because the structures of the proteins (e.g., collagen, elastin) in acellular matrices are well conserved and normally arranged, the mechanical properties of the acellular matrices are not significantly different from those of native bladder submucosa.[37]

Polyesters of naturally occurring α-hydroxy acids, including PGA, PLA, and PLGA, are widely used in regenerative medicine. These polymers have gained FDA approval for human use in a variety of applications, including sutures.[42] The degradation products of PGA, PLA, and PLGA are nontoxic, natural metabolites that are eventually eliminated from the body in the form of carbon dioxide and water.[42] Because these polymers are thermoplastics, they can easily be formed into a three-dimensional scaffold with a desired microstructure, gross shape, and dimension by various techniques, including molding, extrusion,[43] solvent casting,[44] phase-separation techniques, and gas-foaming techniques.[45] More recently, techniques such as electrospinning have been used to quickly create highly porous scaffolds in various conformations.[25-27,46]

Many applications require a scaffold with high porosity and a high ratio of surface area to volume. This need has been addressed by processing biomaterials into configurations of fiber meshes and porous sponges using the techniques described previously. A drawback of the synthetic polymers is lack of biologic recognition. As an approach toward incorporating cell recognition domains into these materials, copolymers with amino acids have been synthesized.[19,20,47] Other biodegradable synthetic polymers, including poly(anhydrides) and poly(ortho esters), can also be used to fabricate scaffolds with controlled properties.[48]

43.2.2 Cells Used in Tissue Engineering Applications

When cells are used for tissue engineering, donor tissue is removed and dissociated into individual cells, which are then implanted directly into the host or expanded in culture, attached to a support matrix, and then implanted as a cell-scaffold construct. The donor tissue can be heterologous, allogeneic, or autologous.

Autologous cells are the ideal choice, as their use circumvents many of the inflammatory and rejection issues associated with a nonself donor. In the past, one of the limitations of applying cell-based regenerative medicine techniques to organ replacement was the inherent difficulty of growing certain human cell types in large quantities. However, the discovery of native targeted progenitor cells in virtually every organ of the body has led to improved culture techniques that have overcome this problem for a number of cell types. Native targeted progenitor cells are tissue-specific unipotent cells derived from most organs. By noting the location of the progenitor cells, as well as by exploring the conditions that promote differentiation and/or self-renewal, it has been possible to overcome some of the obstacles that limit cell expansion in vitro. For example, urothelial cell culture has been improved in this way. Urothelial cells could be grown in the laboratory setting in the past, but only with limited success. It was believed that urothelial cells had a natural senescence that was hard to overcome. Several protocols have been developed over the last 2 decades that have improved urothelial growth and expansion.[49-52] Using these methods of cell culture, it is possible to expand a urothelial strain from a single specimen that initially covers a surface area of 1 cm^2 to one covering a surface area of 4,202 m^2 (the equivalent area of one football field) within 8 weeks.[49]

An advantage of native targeted progenitor cells is that they are already programmed to become the cell type needed, and no in vitro differentiation steps are necessary for their use in the organ of origin. An additional advantage in using native cells is that they can be obtained from the specific organ to be regenerated, expanded, and used in the same patient without rejection, in an autologous manner.[39,49,53-68] However, a major concern has been that, in cases where cells must be expanded from a diseased organ, there may no longer be enough normal cells present in that organ to begin the expansion process. Recent research suggests that this may not be the case, however. For example, one study has shown that cultured neuropathic bladder smooth muscle cells possess and maintain different characteristics than normal smooth muscle cells in vitro, as demonstrated by growth assays, contractility, and adherence tests in vitro.[69] Despite these differences, when neuropathic smooth muscle cells were cultured in vitro, and then seeded onto matrices and implanted in vivo, the tissue-engineered constructs showed the same properties as the constructs engineered with normal cells.[70] It is now known that genetically normal progenitor cells, which are the reservoirs for new cell formation, are present even in diseased tissue. These normal progenitors are programmed to give rise to normal tissue, regardless of whether they reside in a normal or diseased environment. Therefore, the stem cell niche and its role in normal tissue regeneration remains a fertile area of ongoing investigation.

Most current strategies for tissue engineering depend upon a sample of autologous cells from the diseased organ of the host. In some instances, primary autologous human cells cannot be expanded from a particular organ, such as the pancreas, or there is not enough normal tissue remaining in the diseased organ to use for the procedures described above. In addition, the use of autologous cells from tissues containing malignancies is not recommended, as abnormal cells could be harvested and would grow within the newly generated organ as well. In these situations, pluripotent human stem cells are envisioned to be an ideal source of cells, as they can differentiate into nearly any replacement tissue in the body.

Embryonic stem (ES) cells exhibit two remarkable properties: the ability to proliferate in an undifferentiated, but still pluripotent state (self-renewal), and the ability to differentiate into a large number of specialized cell types.[71] They can be isolated from the inner cell mass of the embryo during the blastocyst stage, which occurs 5 days postfertilization. Many protocols for differentiation of ES cells into specific cell types in culture have been published. However, many uses of these cells are currently banned in a number of countries due to the ethical dilemmas that are associated with the manipulation of embryos in culture.

Adult stem cells, especially hematopoietic stem cells, are the best understood cell type in stem cell biology.[72] Despite this, adult stem cell research remains an area of intense study, as their potential for therapy may be applicable to a myriad of degenerative disorders. Within the past decade, adult stem cell populations have been found in many adult tissues other than the bone marrow and the gastrointestinal tract, including the brain,[73,74] skin,[75] and muscle.[76] Many other types of adult stem cells have been identified in organs all over the body and are thought to serve as the primary repair entities for their corresponding organs.[77] The discovery of such tissue-specific progenitors has opened up new avenues for research.

A notable exception to the tissue-specificity of adult stem cells is the mesenchymal stem cell, also known as the multipotent adult progenitor cell. This cell type is derived from bone marrow stroma.[78,79] Such cells can differentiate in vitro into numerous tissue types[80,81] and can also differentiate developmentally if injected into a blastocyst. Multipotent adult progenitor cells can develop into a variety of tissues including neuronal,[82] adipose,[76] muscle,[76,83] liver,[84,85] lungs,[86] spleen,[87] and gut tissue,[79] but notably not bone marrow or gonads.

Research into adult stem cells has, however, progressed slowly, mainly because investigators have had great difficulty in maintaining adult non-mesenchymal stem cells in culture. Some cells, such as those of the liver, pancreas, and nerve, have very low proliferative capacity in vitro, and the functionality of some cell types is reduced after the cells are cultivated. Isolation of cells has also been problematic, because stem cells are present in extremely low numbers in adult tissue.[84,88] While the clinical utility of adult stem cells is currently limited, great potential exists for future use of such cells in tissue-specific regenerative therapies. The advantage of adult stem cells is that they can be used in autologous therapies, thus avoiding any complications associated with immune rejection.

The isolation of multipotent human and mouse amniotic-fluid and placental-derived stem (AFPS) cells that are capable of extensive self-renewal and give rise to cells from all three germ layers was reported in 2007.[89] Undifferentiated AFPS cells expand extensively without a feeder cell layer and double every 36 h. Unlike human embryonic stem cells, AFPS cells do not form tumors in vivo. Lines maintained for over 250 population doublings retained long telomeres and a normal complement of chromosomes. AFPS cell lines can be induced to differentiate into cells representing each embryonic germ layer, including cells of the adipogenic, osteogenic, myogenic, endothelial, neural-like, and hepatic lineages. Since the discovery of the AFPS cells, other groups have published on the potential of the cells to differentiate to other lineages, such as cartilage,[90] kidney,[91] and lung.[92] Muscle differentiated AFPS cells were also noted to prevent compensatory bladder hypertrophy in a cryo-injured rodent bladder model.[93]

Recently, exciting reports of the successful transformation of adult cells into pluripotent stem cells through a type of genetic "reprogramming" have been published. Reprogramming is a technique that involves dedifferentiation of adult somatic cells to produce patient-specific pluripotent stem cells, without the use of embryos. Cells generated by reprogramming would be genetically identical to the somatic cells (and thus, the patient who donated these cells) and would not be rejected. Yamanaka was the first to discover that mouse embryonic fibroblasts (MEFs) and adult mouse fibroblasts could be reprogrammed into an "induced pluripotent state (iPS)."[94] iPS cells in this study possessed the immortal growth characteristics of self-renewing ES cells, expressed genes specific for ES cells, and generated embryoid bodies in vitro and teratomas in vivo. Another study shows that teratomas induced by these cells contained differentiated cell types representing all three embryonic germ layers. More importantly, the reprogrammed cells from this experiment were able to form viable chimeras and contribute to the germ line like ES cells, suggesting that these iPS cells were completely reprogrammed.[95] It has recently been shown that reprogramming of human cells is possible.[96,97] However, despite these advances, a number of questions must be answered before iPS cells can be used in human therapies. One concern is that these cells contain three to six retroviral integrations, which may increase the risk of eventual tumorigenesis. Although this is an exciting phenomenon, our understanding of the mechanisms involved in reprogramming is still limited.

43.2.3 Generation of Tissue-Engineered Constructs

The basic strategy for engineering a tissue or organ involves seeding a biomaterial scaffold with appropriate cell types. This construct is then incubated for a period of time to allow the cells to attach to the scaffold and begin to grow. However, simply placing the cells onto a scaffold in a culture dish may not be the most efficient or effective method for growing tissue in vitro. For example, it has been shown that a number of cell types, such as muscle, may require exposure to mechanical forces in order to mature and develop the proper cellular orientation required for a functional tissue. In addition, oxygen and nutrient exchange is limited in a culture dish, and this may hamper the development of normal tissue. Finally, in order to engineer more complex tissues and organs that are made up of a number of different cell types, it is necessary to place each cell type in a very specific spatial orientation within a construct, and the simple culture dish method of in vitro tissue culture does not provide this capability. Therefore, over the last few years, a number of new technologies for creating a tissue-engineered construct have been developed.

Bioreactors for tissue engineering applications are designed to provide mechanical stimulation and mimic physiological conditions in vitro, as exposure to stimuli such as pulsatile flow and pressure changes has been shown to enhance tissue formation, organization, and function.[98-100] These components work in concert to provide an environment that allows preconditioning of cells on scaffolds in vitro and promotes the enhancement of cell–matrix interaction, cellular proliferation, and organization. Various tissues have been grown using bioreactors. For example, Lee and colleagues have shown that engineered heart valves become more completely endothelialized when they are preconditioned in a bioreactor system that mimics physiological blood flow,[101] and similarly, several studies have shown that a more complete endothelial layer forms in engineered blood vessels when they are preconditioned in a bioreactor.[26,102] In addition, bioreactors have been shown to improve the function of engineered muscle tissue, including bladder muscle.[103-105] Ladd and colleagues have shown that human skin can be expanded through a gentle stretching process in a bioreactor system to generate significantly larger pieces of skin for grafting.[9] Further development of bioreactors for both skin and muscle is particularly important in facial reconstruction using tissue engineering, as these are integral components of most craniofacial structures.

Another concern in tissue engineering is appropriate nutrient and gas exchange for the growing tissue, both during culture and immediately after implantation but before the new tissue becomes fully vascularized. While bioreactors can address this concern in vitro, it has been difficult to resolve this issue in vivo, and the size of most tissue constructs has been limited by the diffusion distance of oxygen within them. While the biological approach using vascular endothelial growth factor (VEGF) and endothelial cells is able to stimulate and promote neovascularization, it is unable to provide vascular supply to a large tissue mass within a short period of time. Recently, Oh et al. have developed a novel scaffold material that generates a sustained release of oxygen over a period of time, and their research indicates that use of this material may allow for prolonged cell survival and growth in the period after implantation but before adequate vasculature has developed (Fig. 43.1).[106]

It is known that the cells within a tissue have a very specific spatial organization, and this organization is required for appropriate tissue function. Recently, a bioprinting technique was developed to deliver cells and biomaterials to target locations to achieve spatial orientation of tissue constructs (Fig. 43.2). Natural materials such as alginate and collagen have been used as "bioinks" in this technique, which is based on inkjet technology.[107,108] Using this technology, these scaffold materials can be "printed" into a desired scaffold shape using a modified inkjet printer. In addition, several groups have shown that living cells can also be printed using this technology.[109-111] This exciting technique can be modified so that a three-dimensional construct containing a

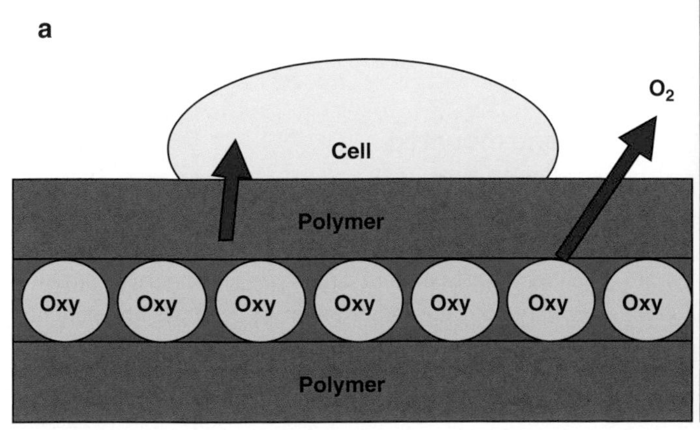

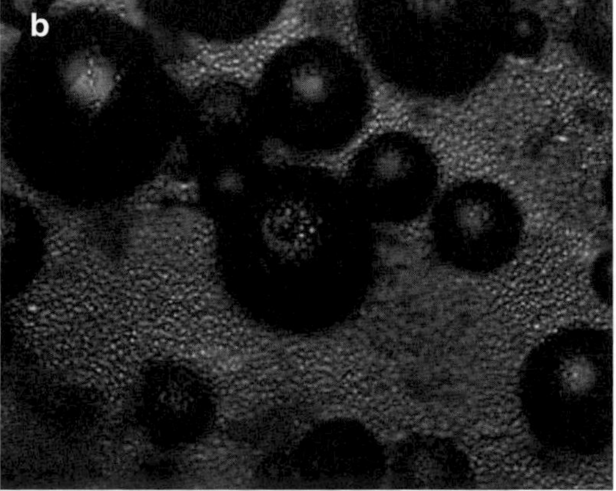

Fig. 43.1 (**a**) Oxygen-generating particles are incorporated into a polymeric biomaterial. This material is designed to generate a sustained release of oxygen over time. (**b**) Oxygen bubbles are released from a polymeric material containing oxygen-generating particles

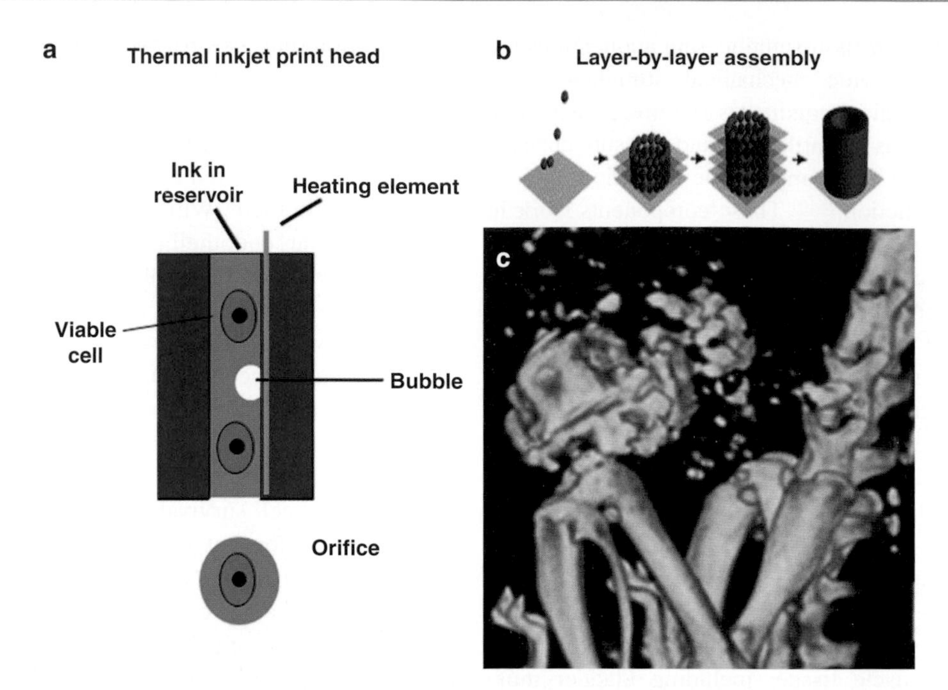

Fig. 43.2 (**a**) A schematic drawing of cell delivery system. Individual cells are pushed through the nozzle (orifice) by the bubbles generated by the heating element of the thermal inkjet print head. (**b**) Single cells are delivered to target locations layer by layer to form a three-dimensional structure. (**c**) Micro CT scan of a mouse 18 weeks after implantation of printed constructs consisting of amniotic-fluid stem cell-derived bone cells

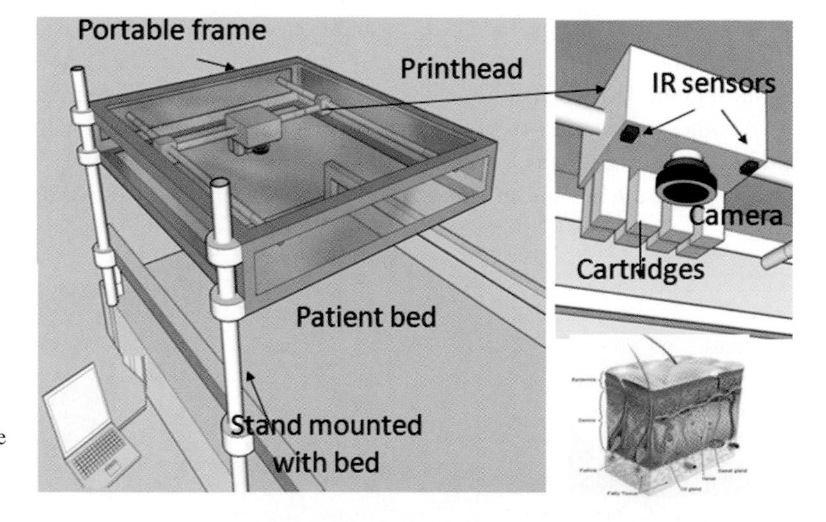

Fig. 43.3 A schematic drawing of a portable skin bioprinter. A multi-nozzle printhead, a digital camera, and infrared sensors are integrated into a portable printing operation frame. The printhead consists of multiple cartridges loaded with various skin printing materials. The portable operation frame has a computer-driven moveable module

precise arrangement of cells, growth factors, and extracellular matrix material can be printed.[111-113] Such constructs may eventually be implanted into a host to serve as the backbone for a new tissue or organ. Our group is currently developing a bioprinting system that will allow for on-site, in situ repair of burn injuries using tissue-engineered skin grafts produced with a portable skin printing system (Fig. 43.3). This printing system will deliver several dermal cell types and matrices simultaneously onto the injured skin to generate anatomically and functionally adequate dermal tissues. The amount and ratio of cells and matrices, as well as the thickness of the skin layers to be printed, can be precisely controlled using the inkjet bioprinter. The delivery of major skin tissue elements onto the injured site will allow for a rapid restoration of the skin and may minimize scarring and enhance cosmetic recovery. Such a system could eventually be modified to print skin grafts for other applications, including restoration of skin during facial reconstruction.

43.3 Reconstruction of Specific Craniofacial Structures Using Tissue Engineering

43.3.1 Skin

The skin, along with the mucosa, is the outermost layer of tissue in the facial area. The skin and the mucosa serve as barriers against exogenous pathogens and irritants, and they prevent the loss of body fluid and other components. In clinical situations, the loss of this covering layer, as in burn wounds, can cause severe metabolic disturbances that lead to increased morbidity and mortality. These issues are encountered in facial transplantation as well. If allogeneic skin grafts are used, the risk of immunologic rejection and pathogen transmission are always present. In addition, the shortage of donor skin and mucosa limits the clinical application of this technique. Tissue engineering techniques could potentially resolve some of these issues. Currently, many research groups are attempting to construct composite skin equivalents.[114-117] By culturing epithelial cells and dermal tissues, it is possible to generate a functional skin or mucosa equivalent that could be used clinically.[118,119] Skin equivalents produced by these approaches seem to have intact dermal–epithelial junctions, and have been studied both in experimental and clinical settings. The epithelium of these skin equivalents usually contains several stratified cell layers resembling normal skin.[120] Moreover, skin equivalents that contain both epidermis and dermis could be grafted using a simple procedure. Similar progress has been reported in the development of mucosal equivalents as well.[121-123] Engineered-mucosal equivalents used for intraoral reconstructions are able to promote vascular network formation. Histologically, the differentiation pattern of the epithelium mimics that of the original tissue.[124] However, the current engineered-mucosal tissue seems to be thinner than normal tissue and lacks prominent epithelial ridges.[125] Methods for improving microstructure formation within the tissue equivalent are under investigation.

43.3.2 Soft Tissue and Bone

In addition to the skin coverage, underlying soft tissue and bone play important roles in maintaining the facial contours and function. Application of tissue engineering methods in regenerating musculoskeletal tissue has demonstrated that both bone and cartilage can be generated both in vitro and in vivo.[13,126,127] In addition, in order to engineer tissue with appropriate physiological and mechanical properties for use in the facial area, a significant effort must be made to integrate different tissues into the reconstructive process. The esthetics of the face and the ability to express emotion are mainly governed by the underlying musculoskeletal function. To ensure that muscle and skeletal tissue work coordinately, soft tissue, cartilage, and bone must be integrated properly so that they will function together. In order to achieve this using regenerative medicine and tissue engineering, numerous distinct cell types must be combined with scaffolds composed of different materials and structural designs so that tissue heterogeneity that exists in native organs can be developed in a controlled manner. In such cases, tissue interface engineering techniques could be applied.[128,129] Using these techniques, scaffolds are created to minimize stress on the implant by effectively balancing the weight and strength loads between various tissues.

It is well known that the wound healing process in the body employs different strategies than the regenerative process. In wound healing, a fibrovascular tissue layer, or scar tissue, is generated in the defective region rather than normal tissue. The formation of scar tissue results in inadequate fixation at the junction between soft tissues and bone, which leads to restricted movement. Similarly, in engineered tissue, to provide a layer where the muscle and tendon can directly connect with cartilage and bone could promote tissue integration. In addition, the approach might be ideal for cell–cell interactions that mediate interface regeneration.[130,131] Moreover, the interaction between osteoblasts and fibroblasts is associated with recruitment and differentiation of progenitor cells into fibrochondrocytes, which could facilitate the formation of tissue structure.

On the other hand, in order to successfully engineer an integrated tissue composed of both soft tissue and bone, the structural features and material characteristics of the implanted scaffold must be well identified. The development of a biomimetic scaffold serves as a critical benchmark for the outcome of engineered tissue.[132-134] An optimally designed supporting scaffold should be able to provide structural and mechanical support as well as an appropriate environment for facilitating cell growth and differentiation.[135,136] Triphasic scaffolds composed of three different regions designed for soft tissue, fibrocartilage, and bone, respectively, might be a way to accomplish this.[137] The feasibility of the triphasic scaffold has been demonstrated both in vitro and in vivo. It is suggested that the application of integrative scaffolds might play a decisive role in functional tissue engineering.[138,139] Finally, in order for engineered tissue to be used in clinic, the issue of scale-up challenges must be addressed. Tissue engineering approaches used to repair small defects may not necessarily be ideal for use in repairing larger defects, which are frequently encountered in the clinic. Larger defects require tissue grafts with greater dimensions, but the nutrient diffusion through the immature vascular system in these large grafts might be limited. The ability to ensure that the engineered tissue is supplied with sufficient metabolic exchange is the first step to animate the transplanted tissue graft. In addition, the differences between various animals and humans must be considered since most successful tissue engineering approaches are first demonstrated in animal models. Moreover, for the purpose of facial reconstruction, the assembly of different tissue-engineered products designed for specific parts of the face might be another challenge. Although some attempts have been made toward this direction, few successful reports have been noted. The ability to regenerate a complex tissue on a large scale is an extraordinary achievement, and will revolutionize the next generation of facial reconstruction.

43.3.3 Vascularization

Most structures in the face are highly vascularized. In particular, the craniofacial area contains predominantly skeletal muscle, which is vascularized by numerous branching vascular networks nearby. To maintain the function of the tissues around the craniofacial area, adequate blood supply is required to meet the metabolic demands of the structures. Thus, the creation of highly vascularized tissue is required to allow engineered grafts to remain viable in vivo.[140] Currently, it has been shown that smaller engineered tissues are able to recruit vascular support from the host to maintain their physiological demands.[11,141,142] However, for larger implants, vascular structures derived from host cells cannot develop quickly enough to support the entire implant. The development of a new approach to the vascularization of tissue-engineered implants is required. However, the growth of a new microvascular system has been one of the major limitations to the successful introduction of tissue engineering products to clinical practice.

Numerous efforts have been made to overcome this limitation and attempts to enhance angiogenesis within the host tissue have been pursued using several approaches. These include the delivery of growth factors and cytokines that play central regulatory roles in the process of angiogenesis, which is thought to induce ingrowth of capillaries and blood vessels into an engineered implant, thus diminishing hypoxia-related cell damage. The delivery of such angiogenic factors has been achieved either by incorporating the desired factors into the scaffold material to be used or by genetic modification of the cells to be used in the engineering process, which forces the cells to express factors such as vascular endothelial growth factor (VEGF). VEGF is one of the most potent angiogenic factors.[140] A recent study by Mooney's group evaluated controlled release of VEGF by incorporating VEGF directly into PLGA scaffolds or by incorporating VEGF encapsulated in PLGA microspheres into scaffolds.[140] VEGF incorporated into scaffolds resulted in rapid release of the cytokine, whereas the pre-encapsulated group showed a delayed release. These studies demonstrated the delivery of VEGF in a controlled and localized fashion in vivo. This angiogenic factor delivery system was applied to bone regeneration and its potent ability to enhance angiogenesis within implanted scaffolds was followed by enhanced bone regeneration. This outlines a novel approach for engineering tissues in hypovascular environments.[140] In another study, VEGF delivery was tested in a study in which human vascular endothelial cells (EC) and skeletal myoblasts transfected with adenovirus encoding the gene for VEGF were injected subcutaneously in athymic mice.[140] The transfected cells formed a vascularized muscle tissue mass, while the non-transfected cells resulted in less angiogenesis and led to the growth of a significantly smaller tissue mass. This study demonstrates that the use of cells producing biological factors can be another powerful tool in tissue engineering. Another approach involved the development of a method to form and stabilize endothelial vessel networks in vitro in engineered skeletal muscle tissue.[143,144] This study used a 3D multiculture system consisting of myoblasts, embryonic fibroblasts, and EC co-seeded on highly porous biodegradable polymer scaffolds. These results showed that prevascularization of the implants improved angiogenesis and cell survival within the scaffolds. Moreover, they emphasize that cocultures with EC and muscle cells may also be important for inducing differentiation of engineered tissues. A breakthrough in engineering vasculature could provide a solution to regenerate bulky skeletal muscle, which increases the potential for clinical application to facial reconstruction.

43.3.4 Innervation

In the craniofacial area, proper innervation of muscle is critical. For example, the ability to form facial expressions, as well as many physiological functions such as mastication, requires functional coordination between many different muscles. Under normal circumstances, most of the muscles in the face are innervated by the facial nerve. If the facial nerve is damaged, these muscles would atrophy and lose the capacity for conscious movement. Therefore, when engineering components for facial reconstruction, the establishment of proper nerve connections between the muscular tissues of the host and the implanted tissues is required.

In some cases, the peripheral nervous system has been shown to regenerate and achieve functional recovery when injury occurs. Nonetheless, nerve recovery often takes a long time and functional recovery might be incomplete. Using tissue engineering, it may be possible to restore proper innervation of the muscle. However, there are several important issues to be addressed. The first issue is the regeneration and elongation of the motor axon, and the second is the regeneration of the neuromuscular junction. For successful regeneration, the neuron itself must survive and be able to restart the axonal growth process after injury. If the axon fails to regrow, then the connection between the central nervous system and the musculature will not be reestablished, and control over muscular function will not be restored. For this to happen, the growing axon must receive adequate nutritional and trophic support from the distal nerve stump. Next, the regenerated axon must be able to reinnervate the target muscle by forming a neuromuscular junction. Once this occurs and signaling is restored, the muscle must regenerate from atrophy caused by denervation.[145] The process is complicated by the fact that even if the axon is successfully regenerated, misdirected axonal guidance might cause a muscle to become reinnervated by an inappropriate axon.[146] This usually results if an axon is misrouted along the improper fascicle or if a muscle is simultaneously reinnervated by several motor neurons.[147] Thus, a number of situations might lead to dysfunctional innervations of muscle and make complete recovery impossible.

Currently, microsurgical treatment with nerve grafts is sometimes effective in repairing nerve damage.[148] Autologous nerve grafts are regarded as the treatment of choice in grafting procedures. Nonetheless, even with this treatment, residual disability is often encountered. Furthermore, there are several problems with this approach, including functional deficit and functional impairment at the donor site created by the graft harvest, and the frequent shortage of suitable graft nerve tissue.

In light of these disadvantages, the development of a tissue-engineered nerve conduit that serves as an alternative to the autologous nerve graft is the subject of intense interest. In order to prepare an artificial nerve guide that is suitable for nerve regeneration, several concepts should be considered. For example, the nerve guidance conduit must provide an appropriate scaffold for axon regeneration. Based on numerous clinical experiences, it is well known that physical support is vital for axon regeneration. In addition, because trophic support from the distal stump is important, the engineered nerve conduit must be permeable to these critical factors. Researchers are currently studying methods of controlling the interaction between the surrounding environment and the growing axons in the conduit. It has been shown that the application of permeable scaffolds for engineering nerve conduits might facilitate nerve regeneration. In this design, metabolic exchange and diffusion of growth and trophic factors could be achieved. Moreover, after a functional axon has regenerated, the scaffold must degrade in a controlled manner. The scaffold should provide a stable conduit for support and directional guidance during axon regeneration, so that the nerve is able to grow and reorganize its connections. If the degradation rate of this scaffold is too fast, the regenerated nerve might undergo biological, mechanical, or chemical damage. Therefore, the appropriate material for a nerve conduit must fit the above criteria to be suitable for clinical application.

Numerous studies have investigated the application of synthetic materials in nerve grafting. Scaffolds made of silicone were the first synthetic material employed in this manner. The silicone scaffold was developed for nerve reconstruction 2 decades ago[149] and it was shown to be useful in several studies. However, silicone is not a biodegradable material, and a foreign-body response to it may occur after implantation. Nerve grafts made of biodegradable materials are a more promising alternative for this reason. A variety of biodegradable materials have been tested in nerve regeneration. For example, conduits made of collagen, PGA, PLLA, and PLCL have been examined.[150] In addition, acellular matrices obtained from the decellularization of donor nerve tissue have also been successfully applied in nerve regeneration.[151] Successful results using these conduits have been reported clinically.[152] These results suggest that biodegradable conduits are a promising treatment for promoting the repair of motor nerve defects. More importantly, the results demonstrated that the axon regeneration assisted by biodegradable materials is similar to that achieved by autologous grafts. Many of these materials are now approved by the US Food and Drug Administration for human application, and they are frequently used in other clinical treatments.

Another key to reestablishing the innervation of a muscle is the formation of a neuromuscular junction between the motor neuron and the target muscle fiber. Experimentally, it has been shown that in coculture of myotubes and neural cells, neuromuscular-like junctions could be generated.[153] This indicated that the regeneration of neuromuscular junctions might be possible. Nonetheless, until now, the development of neruomuscluar junctions using tissue engineering strategies is still under investigation. It is well known that the formation of the neuromuscular junction is affected by numerous factors, such as chemotropic and electrical stimulation,[154] but it is not yet known how these stimulatory factors should be delivered. It is critical that these factors are delivered in a proper temporal and spatial manner for the successful establishment of a functional neuromuscular junction. Elucidation of this process is the next step in investigating the factors involved in complete regeneration of muscle innervation. It is likely that this process will require application of appropriate developmental and trophic factors, as well as the application of electrical, chemotactic, and mechanical stimulation.

43.4 Conclusions

Tissue engineering techniques have the potential to revolutionize reconstructive surgery, including facial reconstruction. The ability to generate new tissue structures that are genetically matched to each individual patient would render the current concerns in organ transplantation, such as donor shortages and the need for immunosuppressive therapy, obsolete. Some engineered tissues and organs, such as skin substitutes and urinary bladders, have already been introduced to the clinic, and the design of new tissue engineering approaches that one day may restore the original architecture and function of other, more complex tissues is still underway. However, although many advances have been made in the field to date, there are still numerous challenges that need to be addressed before the use of engineered tissue can be made a reality in facial reconstruction. In order to engineer fully functional facial structures, further research into the fundamental mechanisms of cellular interaction within facial tissues, including skin, muscle, cartilage, and bone, must be performed. In addition, the developmental biology and the intricate interactions between cells, scaffolds, and growth factors must be defined in order to generate the complex composite tissue structures required for facial reconstruction. In addition, adequate oxygen and nutrients to a newly implanted engineered tissue construct is critical, and this might be accomplished either by designing novel oxygen-generating biomaterials or by accelerating angiogenesis through the use of angiogenic growth factors and cytokines. Importantly, the mechanisms governing the establishment of new nerve signaling pathways between the host tissues and the implant must be studied in more detail, as proper innervations are required for the facial structures to work together to provide natural movement and facial expression.

Acknowledgments The authors wish to thank Dr. Jennifer L. Olson for editorial assistance with this manuscript.

References

1. Sarwer D, Bartlett S, Whitaker L, Paige K, Pertschuk M, Wadden T. Adult psychological functioning of individuals born with craniofacial anomalies. *Plast Reconstr Surg*. 1999;103:412-418.
2. Dropkin M. Body image and quality of life after head and neck cancer surgery. *Cancer Pract*. 1999;7:309-313.
3. Wallace CG, Wei FC. The current status, evolution and future of facial reconstruction. *Chang Gung Med J*. 2008; 31:441-449.
4. Menick F. Artistry in aesthetic surgery aesthetic perception and the subunit principle. *Clin Plast Surg*. 1987;14:723-735.
5. Birgfeld C, Low D. Total face reconstruction using a pre-expanded, bilateral, extended, parascapular free flap. *Ann Plast Surg*. 2006;56:565-568.
6. Devauchelle B, Badet L, Lengele B, et al. First human face allograft: Early report. *Lancet*. 2006;368:203-209.
7. Siemionow M, Papay F, Alam D, et al. Near-total human face transplantation for a severely disfigured patient in the USA. *Lancet*. 2009;374:203-209.
8. Langer R, Vacanti J, Vacanti C, Atala A, Freed L, Vunjak-Novakovic G. Tissue engineering: Biomedical applications. *Tissue Eng*. 1995;1:151-161.
9. Ladd MR, Lee SJ, Atala A, Yoo JJ. Bioreactor maintained living skin matrix. *Tissue Eng A*. 2009;15:861-868.
10. Atala A. Tissue engineering of artificial organs. *J Endourol*. 2000;14:49-57.
11. Atala A, Bauer SB, Soker S, Yoo JJ, Retik AB. Tissue-engineered autologous bladders for patients needing cystoplasty. *Lancet*. 2006;367:1241-1246.
12. Macchiarini P, Jungebluth P, Go T, et al. Clinical transplantation of a tissue-engineered airway. *Lancet*. 2008;372: 2023-2030.
13. Langer R, Vacanti JP. Tissue engineering. *Science*. 1993;260: 920-926.
14. Zaky SH, Cancedda R. Engineering craniofacial structures: Facing the challenge. *J Dent Res*. 2009;88:1077-1091.
15. Buxton PG, Cobourne MT. Regenerative approaches in the craniofacial region: Manipulating cellular progenitors for oro-facial repair. *Oral Dis*. 2007;13:452-460.
16. Bergsma JE, Rozema FR, Bos RR, Boering G, de Bruijn WC, Pennings AJ. In vivo degradation and biocompatibility study of in vitro pre-degraded as-polymerized polyactide particles [see comment]. *Biomaterials*. 1995;16:267-274.
17. Hynes RO. Integrins: Versatility, modulation, and signaling in cell adhesion. *Cell*. 1992;69:11-25.
18. Deuel TF. Growth factors. In: Lanza R, Langer R, Chick WL, eds. *Principles of Tissue Engineering*. New York: Academic; 1997:133-149.
19. Barrera DA, Zylstra E, Lansbury PT, Langer R. Synthesis and RGD peptide modification of a new biodegradable copolymer poly (lactic acid-*co*-lysine). *J Am Chem Soc*. 1993;115:11010-11011.
20. Cook AD, Hrkach JS, Gao NN, et al. Characterization and development of RGD-peptide-modified poly(lactic acid-*co*-lysine) as an interactive, resorbable biomaterial. *J Biomed Mater Res*. 1997;35:513-523.
21. Atala A. Engineering tissues, organs and cells. *J Tissue Eng Regen Med*. 2007;1:83-96.
22. Kim BS, Mooney DJ. Development of biocompatible synthetic extracellular matrices for tissue engineering. *Trends Biotechnol*. 1998;16:224-230.
23. Yoo JJ, Lee JE, Kim HJ, et al. Comparative in vitro and in vivo studies using a bioactive poly(epsilon-caprolactone)-organosiloxane nanohybrid containing calcium salt. *J Biomed Mater Res B Appl Biomater*. 2007;83:189-198.
24. Lee SJ, Van Dyke M, Atala A, Yoo JJ. Host cell mobilization for in situ tissue regeneration. *Rejuvenation Res*. 2008;11:747-756.
25. Choi JS, Lee SJ, Christ GJ, Atala A, Yoo JJ. The influence of electrospun aligned poly(epsilon-caprolactone)/collagen nanofiber meshes on the formation of self-aligned skeletal muscle myotubes. *Biomaterials*. 2008;29: 2899-2906.
26. Lee SJ, Liu J, Oh SH, Soker S, Atala A, Yoo JJ. Development of a composite vascular scaffolding system that withstands physiological vascular conditions. *Biomaterials*. 2008;29: 2891-2898.
27. Lee SJ, Oh SH, Liu J, Soker S, Atala A, Yoo JJ. The use of thermal treatments to enhance the mechanical properties of electrospun poly(epsilon-caprolactone) scaffolds. *Biomaterials*. 2008;29:1422-1430.
28. Lee SJ, Yoo JJ, Lim GJ, Atala A, Stitzel J. In vitro evaluation of electrospun nanofiber scaffolds for vascular graft application. *J Biomed Mater Res A*. 2007;83: 999-1008.
29. Li ST. Biologic biomaterials: Tissue derived biomaterials (collagen). In: JD B, ed. *The Biomedical Engineering Handbook*. Boca Raton, FL: CRS Press; 1995:627-647.
30. Furthmayr H, Timpl R. Immunochemistry of collagens and procollagens. *Int Rev Connect Tissue Res*. 1976;7:61-99.
31. Cen L, Liu W, Cui L, Zhang W, Cao Y. Collagen tissue engineering: Development of novel biomaterials and applications. *Pediatr Res*. 2008;63:492-496.
32. Silver FH, Pins G. Cell growth on collagen: A review of tissue engineering using scaffolds containing extracellular matrix. *J Long Term Effects Med Implants*. 1992;2: 67-80.
33. Sams AE, Nixon AJ. Chondrocyte-laden collagen scaffolds for resurfacing extensive articular cartilage defects. *Osteoarthritis Cartilage*. 1995;3:47-59.
34. Smidsrod O, Skjak-Braek G. Alginate as immobilization matrix for cells. *Trends Biotechnol*. 1990;8:71-78.
35. Lim F, Sun AM. Microencapsulated islets as bioartificial endocrine pancreas. *Science*. 1980; 2010: 908–910.
36. Rowley JA, Madlambayan G, Mooney DJ. Alginate hydrogels as synthetic extracellular matrix materials. *Biomaterials*. 1999;20:45-53.
37. Dahms SE, Piechota HJ, Dahiya R, Lue TF, Tanagho EA. Composition and biomechanical properties of the bladder

acellular matrix graft: Comparative analysis in rat, pig and human. *Br J Urol*. 1998;82:411-419.
38. Piechota HJ, Dahms SE, Nunes LS, Dahiya R, Lue TF, Tanagho EA. In vitro functional properties of the rat bladder regenerated by the bladder acellular matrix graft. *J Urol*. 1998;159:1717-1724.
39. Yoo JJ, Meng J, Oberpenning F, Atala A. Bladder augmentation using allogenic bladder submucosa seeded with cells. *Urology*. 1998;51:221-225.
40. Chen F, Yoo JJ, Atala A. Acellular collagen matrix as a possible "off the shelf" biomaterial for urethral repair. *Urology*. 1999;54:407-410.
41. Probst M, Dahiya R, Carrier S, Tanagho EA. Reproduction of functional smooth muscle tissue and partial bladder replacement. *Br J Urol*. 1997;79:505-515.
42. Gilding D. Biodegradable polymers. In: Williams D, ed. *Biocompatibility of Clinical Implant Materials*. Boca Raton, FL: CRC Press; 1981:209-232.
43. Freed LE, Vunjak-Novakovic G, Biron RJ, et al. Biodegradable polymer scaffolds for tissue engineering. *Biotechnology (NY)*. 1994;12:689-693.
44. Mikos AG, Lyman MD, Freed LE, Langer R. Wetting of poly(L-lactic acid) and poly(DL-lactic-co-glycolic acid) foams for tissue culture. *Biomaterials*. 1994;15:55-58.
45. Harris LD, Kim BS, Mooney DJ. Open pore biodegradable matrices formed with gas foaming. *J Biomed Mater Res*. 1998;42:396-402.
46. Han D, Gouma PI. Electrospun bioscaffolds that mimic the topology of extracellular matrix. *Nanomedicine*. 2006;2: 37-41.
47. Intveld PJA, Shen ZR, Takens GAJ. Glycine glycolic acid based copolymers. *J Polym Sci Polym Chem*. 1994;32: 1063-1069.
48. Peppas NA, Langer R. New challenges in biomaterials [see comment]. *Science*. 1994;263:1715-1720.
49. Cilento BG, Freeman MR, Schneck FX, Retik AB, Atala A. Phenotypic and cytogenetic characterization of human bladder urothelia expanded in vitro. *J Urol*. 1994;152 (2 Pt 2):665-670.
50. Scriven SD, Booth C, Thomas DF, Trejdosiewicz LK, Southgate J. Reconstitution of human urothelium from monolayer cultures. *J Urol*. 1997;158(3 Pt 2):1147-1152.
51. Liebert M, Hubbel A, Chung M, et al. Expression of mal is associated with urothelial differentiation in vitro: Identification by differential display reverse-transcriptase polymerase chain reaction. *Differentiation*. 1997;61:177-185.
52. Puthenveettil JA, Burger MS, Reznikoff CA. Replicative senescence in human uroepithelial cells. *Adv Exp Med Biol*. 1999;462:83-91.
53. Atala A, Vacanti JP, Peters CA, Mandell J, Retik AB, Freeman MR. Formation of urothelial structures in vivo from dissociated cells attached to biodegradable polymer scaffolds in vitro. *J Urol*. 1992;148(2 Pt 2):658-662.
54. Atala A, Cima LG, Kim W, et al. Injectable alginate seeded with chondrocytes as a potential treatment for vesicoureteral reflux. *J Urol*. 1993;150(2 Pt 2):745-747.
55. Atala A, Freeman MR, Vacanti JP, Shepard J, Retik AB. Implantation in vivo and retrieval of artificial structures consisting of rabbit and human urothelium and human bladder muscle. *J Urol*. 1993;150(2 pt 2):608-612.
56. Atala A, Kim W, Paige KT, Vacanti CA, Retik AB. Endoscopic treatment of vesicoureteral reflux with a chondrocyte-alginate suspension. *J Urol*. 1994;152(2 pt 2): 641-643. Discussion 4.
57. Atala A, Schlussel RN, Retik AB. Renal cell growth in vivo after attachment to biodegradable polymer scaffolds. *J Urol*. 1995;153:4.
58. Atala A, Guzman L, Retik AB. A novel inert collagen matrix for hypospadias repair. *J Urol*. 1999;162(3 Pt 2):1148-1151.
59. Atala A. Tissue engineering in the genitourinary system. In: Atala A, Mooney DJ, eds. *Tissue engineering*. Boston, MA: Birkhauser Press; 1997:149.
60. Atala A. Autologous cell transplantation for urologic reconstruction. *J Urol*. 1998;159:2-3.
61. Yoo JJ, Atala A. A novel gene delivery system using urothelial tissue engineered neo-organs. *J Urol*. 1997;158 (3 Pt 2):1066-1070.
62. Fauza DO, Fishman SJ, Mehegan K, Atala A. Videofetoscopically assisted fetal tissue engineering: Skin replacement. *J Pediatr Surg*. 1998;33:357-361.
63. Fauza DO, Fishman SJ, Mehegan K, Atala A. Videofetoscopically assisted fetal tissue engineering: Bladder augmentation. *J Pediatr Surg*. 1998;33:7-12.
64. Machluf M, Atala A. Emerging concepts for tissue and organ transplantation. *Graft*. 1998;1:31-37.
65. Amiel GE, Atala A. Current and future modalities for functional renal replacement. *Urol Clin North Am*. 1999;26: 235-246.
66. Kershen RT, Atala A. New advances in injectable therapies for the treatment of incontinence and vesicoureteral reflux. *Urol Clin North Am*. 1999;26:81-94.
67. Oberpenning F, Meng J, Yoo JJ, Atala A. De novo reconstitution of a functional mammalian urinary bladder by tissue engineering [see comment]. *Nat Biotechnol*. 1999;17: 149-155.
68. Park HJ, Yoo JJ, Kershen RT, Moreland R, Atala A. Reconstitution of human corporal smooth muscle and endothelial cells in vivo. *J Urol*. 1999;162(3 Pt 2):1106-1109.
69. Lin HK, Cowan R, Moore P, et al. Characterization of neuropathic bladder smooth muscle cells in culture. *J Urol*. 2004;171:1348-1352.
70. Lai JY, Yoon CY, Yoo JJ, Wulf T, Atala A. Phenotypic and functional characterization of in vivo tissue engineered smooth muscle from normal and pathological bladders. *J Urol*. 2002;168(4 Pt 2):1853-1857. Discussion 8.
71. Brivanlou AH, Gage FH, Jaenisch R, Jessell T, Melton D, Rossant J. Stem cells. Setting standards for human embryonic stem cells [see comment]. *Science*. 2003;300: 913-916.
72. Ballas CB, Zielske SP, Gerson SL. Adult bone marrow stem cells for cell and gene therapies: Implications for greater use. *J Cell Biochem Suppl*. 2002;38:20-28.
73. Jiao J, Chen DF. Induction of neurogenesis in nonconventional neurogenic regions of the adult central nervous system by niche astrocyte-produced signals. *Stem Cells*. 2008;26:1221-1230.
74. Taupin P. Therapeutic potential of adult neural stem cells. *Recent Pat CNS Drug Discov*. 2006;1:299-303.
75. Jensen UB, Yan X, Triel C, Woo SH, Christensen R, Owens DM. A distinct population of clonogenic and multipotent murine follicular keratinocytes residing in the upper isthmus. *J Cell Sci*. 2008;121(Pt 5):609-617.
76. Crisan M, Casteilla L, Lehr L, et al. A reservoir of brown adipocyte progenitors in human skeletal muscle. *Stem Cells*. 2008;26:2425-2433.
77. Weiner LP. Definitions and criteria for stem cells. *Meth Mol Biol*. 2008;438:3-8.

78. Devine SM. Mesenchymal stem cells: Will they have a role in the clinic? *J Cell Biochem Suppl*. 2002;38:73-79.
79. Jiang Y, Jahagirdar BN, Reinhardt RL, Schwartz RE, Keene CD, Ortiz-Gonzalez XR. Pluripotency of mesenchymal stem cells derived from adult marrow [see comment][erratum appears in Nature. 2007;447:879–880]. *Nature*. 2002;418:41-49.
80. Caplan AI. Adult mesenchymal stem cells for tissue engineering versus regenerative medicine. *J Cell Physiol*. 2007;213:341-347.
81. da Silva Meirelles L, Caplan AI, Nardi NB. In search of the in vivo identity of mesenchymal stem cells. *Stem Cells*. 2008;26:2287-2299.
82. Duan X, Chang JH, Ge S, Faulkner RL, Kim JY, Kitabatake Y. Disrupted-in-schizophrenia 1 regulates integration of newly generated neurons in the adult brain [see comment]. *Cell*. 2007;130:1146-1158.
83. Luttun A, Ross JJ, Verfaillie C, Aranguren X, Prosper F. Unit 22 F.9: Differentiation of multipotent adult progenitor cells into functional endothelial and smooth muscle cells. *Current Protocols in Immunology*. Hoboken, NJ: Wiley; 2006.
84. Mimeault M, Batra SK. Recent progress on tissue-resident adult stem cell biology and their therapeutic implications. *Stem Cell Rev*. 2008;4:27-49.
85. Ikeda E, Yagi K, Kojima M, et al. Multipotent cells from the human third molar: Feasibility of cell-based therapy for liver disease. *Differentiation*. 2008;76:495-505.
86. Nolen-Walston RD, Kim CF, Mazan MR, et al. Cellular kinetics and modeling of bronchioalveolar stem cell response during lung regeneration. *Am J Physiol Lung Cell Mol Physiol*. 2008;294:L1158-L1165.
87. in't Anker P, Noort WA, Scherjon SA, Kleijburg-van der Keur C, Kruisselbrink AB, van Bezooijen RL. Mesenchymal stem cells in human second-trimester bone marrow, liver, lung, and spleen exhibit a similar immunophenotype but a heterogeneous multilineage differentiation potentia. *Haematologica*. 2003;88:845-852.
88. Hristov M, Zernecke A, Schober A, Weber C. Adult progenitor cells in vascular remodeling during atherosclerosis. *Biol Chem*. 2008;389:837-844.
89. De Coppi P, Bartsch G Jr, Siddiqui MM, Xu T, Santos CC, Perin L. Isolation of amniotic stem cell lines with potential for therapy [see comment]. *Nat Biotechnol*. 2007;25: 100-106.
90. Kolambkar YM, Peister A, Soker S, Atala A, Guldberg RE. Chondrogenic differentiation of amniotic fluid-derived stem cells. *J Mol Histol*. 2007;38:405-413.
91. Perin L, Giuliani S, Jin D, et al. Renal differentiation of amniotic fluid stem cells. *Cell Prolif*. 2007;40:936-948.
92. Warburton D, Perin L, Defilippo R, Bellusci S, Shi W, Driscoll B. Stem/progenitor cells in lung development, injury repair, and regeneration. *Proc Am Thorac Soc*. 2008;5:703-706.
93. De Coppi P, Callegari A, Chiavegato A, et al. Amniotic fluid and bone marrow derived mesenchymal stem cells can be converted to smooth muscle cells in the cryo-injured rat bladder and prevent compensatory hypertrophy of surviving smooth muscle cells. *J Urol*. 2007;177:369-376.
94. Takahashi K, Yamanaka S. Induction of pluripotent stem cells from mouse embryonic and adult fibroblast cultures by defined factors. *Cell*. 2006;126:663-676.
95. Wernig M, Meissner A, Foreman R, et al. In vitro reprogramming of fibroblasts into a pluripotent ES-cell-like state. *Nature*. 2007;448:318-324.
96. Takahashi K, Tanabe K, Ohnuki M, et al. Induction of pluripotent stem cells from adult human fibroblasts by defined factors. *Cell*. 2007;131:861-872.
97. Yu J, Vodyanik MA, Smuga-Otto K, et al. Induced pluripotent stem cell lines derived from human somatic cells. *Science*. 2007;318:1917-1920.
98. Engelmayr GC Jr, Hildebrand DK, Sutherland FW, Mayer JE Jr, Sacks MS. A novel bioreactor for the dynamic flexural stimulation of tissue engineered heart valve biomaterials. *Biomaterials*. 2003;24:2523-2532.
99. Hoerstrup SP, Sodian R, Daebritz S, Wang J, Bacha EA, Martin DP. Functional living trileaflet heart valves grown in vitro. *Circulation*. 2000;102(19 Suppl 3):III44-III49.
100. Hoerstrup SP, Sodian R, Sperling JS, Vacanti JP, Mayer JE Jr. New pulsatile bioreactor for in vitro formation of tissue engineered heart valves. *Tissue Eng*. 2000;6:75-79.
101. Lee DJ, Steen J, Jordan JE, et al. Endothelialization of heart valve matrix using a computer-assisted pulsatile bioreactor. *Tissue Eng A*. 2009;15:807-814.
102. Tillman BW, Yazdani SK, Lee SJ, Geary RL, Atala A, Yoo JJ. The in vivo stability of electrospun polycaprolactone-collagen scaffolds in vascular reconstruction. *Biomaterials*. 2009;30:583-588.
103. Farhat WA, Chen J, Haig J, et al. Porcine bladder acellular matrix (ACM): Protein expression, mechanical properties. *Biomed Mater*. 2008;3:025015.
104. Wallis MC, Lorenzo AJ, Farhat WA, Bagli DJ, Khoury AE, Pippi Salle JL. Risk assessment of incidentally detected complex renal cysts in children: Potential role for a modification of the Bosniak classification. *J Urol*. 2008;180: 317-321.
105. Donnelly K, Khodabukus A, Philp A, Deldicque L, Dennis RG, Baar K. A novel bioreactor for stimulating skeletal muscle in vitro. *Tissue Eng C Meth*. 2010;16: 711-718.
106. Oh SH, Ward CL, Atala A, Yoo JJ, Harrison BS. Oxygen generating scaffolds for enhancing engineered tissue survival. *Biomaterials*. 2009;30:757-762.
107. Campbell PG, Weiss LE. Tissue engineering with the aid of inkjet printers. *Expert Opin Biol Ther*. 2007;7: 1123-1127.
108. Boland T, Xu T, Damon B, Cui X. Application of inkjet printing to tissue engineering. *Biotechnol J*. 2006;1:910-917.
109. Nakamura M, Kobayashi A, Takagi F, et al. Biocompatible inkjet printing technique for designed seeding of individual living cells. *Tissue Eng*. 2005;11:1658-1666.
110. Laflamme MA, Gold J, Xu C, et al. Formation of human myocardium in the rat heart from human embryonic stem cells. *Am J Pathol*. 2005;167:663-671.
111. Xu T, Rohozinski J, Zhao W, Moorefield EC, Atala A, Yoo JJ. Inkjet-mediated gene transfection into living cells combined with targeted delivery. *Tissue Eng A*. 2009; 15:95-101.
112. Ilkhanizadeh S, Teixeira AI, Hermanson O. Inkjet printing of macromolecules on hydrogels to steer neural stem cell differentiation. *Biomaterials*. 2007;28:3936-3943.
113. Roth EA, Xu T, Das M, Gregory C, Hickman JJ, Boland T. Inkjet printing for high-throughput cell patterning. *Biomaterials*. 2004;25:3707-3715.
114. Dai N, Williamson M, Khammo N, Adams E, Coombes A. Composite cell support membranes based on collagen and polycaprolactone for tissue engineering of skin. *Biomaterials*. 2004;25:4263-4271.

115. Yang E, Seo Y, Youn H, Lee D, Park S, Park J. Tissue engineered artificial skin composed of dermis and epidermis. *Artif Organs*. 2000;24:7-17.
116. El-Ghalbzouri A, Gibbs S, Lamme E, Van Blitterswijk C, Ponec M. Effect of fibroblasts on epidermal regeneration. *Br J Dermatol*. 2002;147:230-243.
117. Izumi K, Tobita T, Feinberg S. Isolation of human oral keratinocyte progenitor / stem cells. *J Dent Res*. 2007;86: 341-346.
118. Mansbridge J. Tissue-engineered skin substitutes in regenerative medicine. *Curr Opin Biotechnol*. 2009;20:563-567.
119. Mansbridge J. Skin tissue engineering. *J Biomater Sci Polym Ed*. 2008;19:955-968.
120. Gibbs S, van den Hoogenband H, Kirtschig G, et al. Autologous full-thickness skin substitute for healing chronic wounds. *Br J Dermatol*. 2006;155:267-274.
121. Scheller E, Krebsbach P, Kohn D. Tissue engineering: State of the art in oral rehabilitation. *J Oral Rehabil*. 2009;36: 368-389.
122. Sauerbier S, Gutwald R, Wiedmann-Al-Ahmad M, Lauer G, Schmelzeisen R. Clin application tissue engineered transplants I mucosa. *Clin Oral Implants Res*. 2006;17:625-632.
123. Song J, Izumi K, Lanigan T, Feinberg S. Development and characterization of a canine oral mucosa equivalent in a serum free environment. *J Biomed Mater Res A*. 2004; 71:143-153.
124. Izumi K, Feinberg S, Iida A, Yoshizawa M. Intraoral grafting of an ex vivo produced oral mucosa equivalent: A preliminary report. *Int J Oral Maxillofac Surg*. 2003;32:188-197.
125. Liu J, Lamme E, Steegers-Theunissen R, et al. Cleft palate cells can regenerate a palatal mucosa in vitro. *J Dent Res*. 2008;87:788-792.
126. Crane G, Ishaug S, Mikos A. Bone tissue engineering. *Nat Med*. 1995;1:1322-1324.
127. Freed L, Marquis J, Nohria A, Emmanual J, Mikos A, Langer R. Neocartilage formation in vitro and in vivo using cells cultured on synthetic biodegradable polymers. *J Biomed Mater Res*. 1993;27:11-23.
128. Thomopoulos S, Williams G, Gimbel J, Favata M, Soslowsky L. Variation of biomechanical, structural, and compositional properties along the tendon to bone insertion site. *J Orthop Res*. 2003;21:413-419.
129. Woo S, Gomez M, Seguchi Y, Endo C, Akeson W. Measurement of mechanical properties of ligament substance from a bone-ligament-bone preparation. *J Orthop Res*. 1983;1:22-29.
130. Kobayashi M, Watanabe N, Oshima Y, Kajikawa Y, Kawata M, Kubo T. The fate of host and graft cells in early healing of bone tunnel after tendon graft. *Am J Sports Med*. 2005;33:1892-1897.
131. Wang I, Shan J, Choi R, et al. Role of osteoblast-fibroblast interactions in the formation of the ligament-to-bone interface. *J Orthop Res*. 2007;25:1609-1620.
132. Geckil H, Xu F, Zhang X, Moon S, Demirci U. Engineering hydrogels as extracellular matrix mimics. *Nanomedicine (Lond)*. 2010;5:469-484.
133. Ma P. Biomimetic materials for tissue engineering. *Adv Drug Deliv Rev*. 2008;60:184-198.
134. von der Mark K, Park J, Bauer S, Schmuki P. Nanoscale engineering of biomimetic surfaces: Cues from the extracellular matrix. *Cell Tissue Res*. 2010;339:131-153.
135. Kumbar S, James R, Nukavarapu S, Laurencin C. Biomed Mater Electrospun nanofiber scaffolds engineering soft tissues. *Biomed Mater*. 2008;3:034002.
136. Zhang YZ, Su B, Venugopal J, Ramakrishna S, Lim CT. Biomimetic and bioactive nanofibrous scaffolds from electrospun composite nanofibers. *Int J Nanomedicine*. 2007;2: 623-638.
137. Spalazzi J, Doty S, Moffat K, Levine W, Lu H. Development of controlled matrix heterogeneity on a triphasic scaffold for orthopedic interface tissue engineering. *Tissue Eng B Rev*. 2006;12:3497-3508.
138. Spalazzi J, Dionisio K, Jiang J, Lu H. Osteoblast and chondrocyte interactions during coculture on scaffolds. *IEEE Eng Med Biol Mag*. 2003;22:27-34.
139. Jiang J, Tang A, Ateshian G, Guo X, Hung C, Lu H. Bioactive stratified polymer ceramic-hydrogel scaffold for integrative osteochondral repair. *Ann Biomed Eng*. 2010;38: 2183-2196.
140. Phelps E, Landázuri N, Thulé P, Taylor W, García A. Bioartificial matrices for therapeutic vascularization. *Proc Natl Acad Sci USA*. 2010;107:3323-3328.
141. Nomi M, Atala A, Coppi P, Soker S. Principals of neovascularization for tissue engineering. *Mol Aspects Med*. 2002; 23:463-483.
142. Ohashi K, Yokoyama T, Yamato M, et al. Engineering functional two- and three-dimensional liver systems in vivo using hepatic tissue sheets. *Nat Med*. 2007;13:880-885.
143. Levenberg S, Rouwkema J, Macdonald M, et al. Engineering vascularized skeletal muscle tissue. *Nat Biotechnol*. 2005; 23:879-884.
144. Amiel G, Komura M, Shapira O, et al. Engineering of blood vessels from acellular collagen matrices coated with human endothelial cells. *Tissue Eng*. 2006;12:2355-2365.
145. Fu S, Gordon T. The cellular and molecular basis of peripheral nerve regeneration. *Mol Neurobiol*. 1997;14:67-116.
146. Brushart T, Seiler W. Selective reinnervation of distal motor stumps by peripheral motor axons. *Exp Neurol*. 1987;97: 289-300.
147. Ito M, Kudo M. Reinnervation by axon collaterals from single facial motoneurons to multiple targets following axotomy in the adult guinea pig. *Acta Anat*. 1994;151:124-130.
148. Kretschmer T, Antoniadis G, Braun V, Rath S, Richter H. Evaluation of iatrogenic lesions in 722 surgically treated cases of peripheral nerve trauma. *J Neurosurg*. 2001;94: 905-912.
149. Lundborg G, Longo F, Varon S. Nerve regeneration model and trophic factors in vivo. *Brain Res*. 1982;232:157-161.
150. Gibson K, Remson L, Smith A, Satterlee N, Strain G, Daniloff J. Comparison of nerve regeneration through different types of neural prostheses. *Microsurgery*. 1991;12: 80-85.
151. Kim B, Yoo J, Atala A. Peripheral nerve regeneration using acellular nerve grafts. *J Biomed Mater Res A*. 2004;68: 201-209.
152. Inada Y, Hosoi H, Yamashita A, et al. Regeneration of peripheral motor nerve gaps with a polyglycolic acid-collagen tube: Technical case report. *Neurosurgery*. 2007;61: E1105-E1107.
153. Wagner S, Dorchies O, Stoeckel H. Functional maturation of nicotinic acetylcholine receptors as an indicator of, murine muscular differentiation in a new nerve–muscle coculture, system. *Pflugers Arch*. 2003;447:14-22.
154. Pedrotty D, Koh J, Davis B. Engineering skeletal myoblasts: Roles of three-dimensional culture and electrical stimulation. *Am J Physiol Heart Circ Physiol*. 2005;288: H1620-H1626.

Part VIII

Current Status of Face Transplantation

An Update on Face Transplants Performed Between 2005 and 2010

44

Maria Z. Siemionow, Can Ozturk, and Selman Altuntas

Contents

M.Z. Siemionow (✉)
Department of Plastic Surgery,
Cleveland Clinic, Cleveland, OH, USA
e-mail: siemiom@ccf.org

Abstract Background: Since 2005, 13 facial allotransplantation cases were performed throughout the world. The major indications for facial allotransplantations were neurofibromatosis and trauma injuries including animal bites, burns, falls, and shotgun blasts. The ratio of male to female was 11:2. Two male patients died at 2 months and 2 years after transplantation due to transplant-related problems. The composite tissue allotransplantations included cutaneous, myocutaneous, and osteomyocutaneous components and functional units of nose, eyelids and lips. Most of these were partial, a few near-total, and two were announced as total face transplantations. The analysis of the anatomical details, microsurgical techniques, and the functional outcomes of all cases with follow-up information from literature, meeting presentations and media reports is presented in this chapter.

44.1 Introduction

Tissue transplantation became the basis of modern plastic and reconstructive surgery.[1] Until 1962, it was impossible to replant an amputated tissue or to perform composite tissue transfers.[2] Advances in microsurgical techniques and instrumentation opened the field of replantation surgery as well as reconstructive surgery, making it applicable to the free autologous tissue transfers. Later, with the investigation of immunosuppressive agents, composite tissue allotransplantation became feasible. In 1998, Dubernard performed the first human hand allotransplantation.[3] Despite the controversy, to date more

M.Z. Siemionow (ed.), *The Know-How of Face Transplantation*,
DOI: 10.1007/978-0-85729-253-7_44,

than 50 hand allotransplants have been performed all over the world.[4] Larynx and knee allotransplantation followed hand allotransplantation. Finally, the first world Institutional Review Board (IRB) approval for face transplantation in humans was granted to Siemionow in 2004, and in 2005 facial allotransplantation was performed by Devauchelle and Dubernard.[5] Until this time, facial transplantation was a subject of magic or science fiction for most people. Although there was a great interest worldwide, many controversies and debates were taking place around the world regarding facial transplantation similar to the reactions following first heart transplantation. Despite these debates, currently 13 face transplants have been accomplished in France, China, USA, and Spain.[5-25]

There are several differences between transplantation of solid organs and composite tissue allograft, such as anatomical complexity and importance of functional recovery. The deficiencies of extremities and solid organs such as the kidney can be compensated; however, the patients with complex deficiencies of face have greater problems to maintain vital activities and social relationships. The most problematic and difficult cases for plastic surgeons are those presenting with severe facial disfigurement – traumatic, inborn or acquired. Our primary goal during reconstruction of such deficits is to replace the "like with like." However, most of the time it is impossible to reconstruct the face with "like tissue" because we do not have any extra components in our body that represent the missing parts of face such as lips, eyelids, or nose. The facial composite tissue allograft (CTA) would be a perfect solution for patients who have previously undergone multiple reconstructive procedures which lead to less-than-optimal outcomes both for the patient and the surgeon.

The side effects of lifelong immunosuppressive therapy, which is required after face transplantation, present the major disadvantage of CTA. Moreover, we do not have enough scientific data available yet about long-term results in this patient population. On the other hand, advantages of performing CTA include reconstruction of "like with like" tissue without creation of donor site morbidity. The conversation between a patient and a surgeon has also changed from the statement "there is nothing more we can do to improve your face" to the statement "there are some new options to make you look and function better." We have learned from past experiences that the choice of patient who will be compliant before and after surgery is as important as performing face transplantation surgery.[26]

Science is under constant evolution, and new techniques and technologies are becoming available every day. Once utopic, facial allotransplantation has been successfully managed technically, but the question arises: How can we yield the best outcomes for the patients? This chapter summarizes available data, in literature and media, on world experience with facial transplantation, emphasizing microsurgical techniques and difference between cases in terms of vascular and nerve repairs and relevant functional outcomes.

44.2 Background Data

The first successful face transplantation was a partial transplant and the last one that was performed in France, in June 2010, was a total face transplantation. Totally, 13 face transplantations were performed between November 2005 and July 2010. Most of these face transplants were partial, a few near-total, and two were announced as total face transplantations. The indications for transplantation were traumatic facial injury, neurofibromatosis, and deformity after tumor treatment. Traumatic injuries included animal bites, shotgun blast injury, burns, and falls. The male to female ratio was 11:2. Two male recipients died due to transplant-related problems. Details on patients and types of microsurgical repair are shown (Tables 44.1 and 44.2). Sensory and motor recovery results of reported patients are summarized (Table 44.3).

Table 44.1 Details of patients and microsurgical repair

Patient details & Team leader	Type of transplant	Artery repair	Venous repair	Sensory nerve repair	Motor nerve repair
38 y, F France, 2005 Alive **Dubernard**	Partial myocutaneous	Bilateral facial a. (end-end)	Bilateral facial v. (end-to-end)	Bilateral infraorbital and mental n.	Left mandibular branch (end-end)
30 y, M China, 2006 Died **Guo**	Partial osteomyocutaneous	Right external maxillary a. (end-end)	Right anterior facial v. (end-end)	None repaired	Right facial nerve not well coapted
29 y, M France, 2007 Alive **Lantieri**	Partial myocutaneous	Bilateral external carotid a. (end-end)	Bilateral thyrolinguofacial trunks (end-end)	Bilateral infraorbital n. Mental n. placed near foramen	Bilateral facial nerves (suture and fibrin glue)
45 y, F USA, 2008 Alive **Siemionow**	Nearly total osteomyocutaneous	Bilateral facial common a. (end-end)	Left external jugular, posterior facial, and right facial v(end-end)	None repaired	Bilateral facial nerves
28 y, M France, 2009 Alive **Lantieri**	Partial osteomyocutaneous	Bilateral external carotid a.	External jugular vein and thyrolinguofacial trunk	None repaired	Bilateral facial nerves
37 y, M France, 2009 Died **Lantieri**	Partial myocutaneous	Bilateral external carotid a.	Bilateral thyrolinguofacial trunk	Bilateral supraorbital, and infraorbital n.	Bilateral facial nerves
59 y, M USA, 2009 Alive **Pomohac**	Partial osteomyocutaneous	Left external carotid, right facial a. (end-end)	Bilateral facial v.	Bilateral infraorbital and buccal sensory nerves	Bilateral facial nerve branches
33 y, M France,2009 Alive **Lantieri**	Partial osteomyocutaneous	Bilateral external carotid a.	External jugular vein and thyrolinguofacial trunk	None repaired	Bilateral facial nerves
43 y, M Spain, 2009 Alive **Cavadas**	Partial osteomyocutaneous	Left common carotid and right internal carotid a.	Bilateral external jugular and right internal jugular v.	Bilateral inferior alveolar and cervicofacial nerve	Bilateral lingual and hypoglossal nerves
27 y, M France, 2009 Alive **Dubernard**	Partial osteomyocutaneous	Bilateral facial a.	Bilateral facial v.	Bilateral infraorbital and mental nerves	Bilateral facial nerves
34 y, M Spain, 2010 Alive **Montero**	Partial myocutaneous	Bilateral external carotid a.	Bilateral internal jugular v.	Bilateral infraorbital and mental nerves	Bilateral facial nerves
30 y, M Spain, 2010 Alive **Barret**	Total osteomyocutaneous	Bilateral external carotid a.	Bilateral external jugular, unilateral anterior jugular and retromandibular v.	Bilateral supraorbital, infraorbital and mental nerves	Bilateral facial nerves
35 y, M France, 2010 Alive **Lantieri**	Total myocutaneous	Bilateral external carotid a.	Bilateral external jugular veins and thyrolinguofacial trunks	Bilateral supraorbital, infraorbital and mental nerves	Bilateral facial nerves

a artery, *v* vein, *n* nerve, *y* years, *M* male, *F* female

Table 44.2 Anatomical details of CTA cases

Cases	Bone	Soft tissue	Skin and oral					
			Scalp forehead and ear	Eyelid and cheek	Nose	Lip and chin	Tongue, palate, and mucosa	Total amount of CTA (%)
Case 1 38 y, F France 2005	–	–	–	Partial cheek	Partial nose	Chin, upper and lower lip	Partial mucosa	25–30%
Case 2 30 y, M China 2006	Unilateral infraorbital, maxilla, nasal, and zygoma	Parotid gl., Septal cart	–	Partial cheek	Total nose	Upper lip	Partial mucosa	30–35%
Case 3 29 y, M France 2007	–	Bilateral parotid gl. and ducts, septal cart	–	Bilateral cheeks	Total nose	Chin, upper and lower lip	Partial mucosa	70–75%
Case 4 45 y, F USA 2008	Bilateral orbital floor, zygoma, nasal, maxilla with teeth and hard palate	Bilateral parotid gl. and ducts, septal cart	–	Bilateral cheeks and lower eyelids	Total nose	Upper lip	Partial mucosa	75–80% 535 cm^2
Case 5 28 y, M France 2009	Bilateral anterior maxilla (Premaxilla)	Bilateral parotid gl. and ducts, septal cart	–	Bilateral cheeks	Total nose	Chin, upper and lower lip	Partial mucosa	70–75%
Case 6 37 y, M France 2009	Nasal bone	Bilateral parotid gl. and ducts, septal and ear cart	Fore-head, scalp, and ear	Bilateral cheek, lower and upper eyelids	Total nose	–	Partial mucosa	70–75% + scalp
Case 7 59 y, M USA 2009	Bilateral maxilla with teeth and palate	Bilateral parotid ducts, septal cart	–	Bilateral cheek	Total nose	Upper lip	–	45–50%

Case 8 33y, M France 2009	Maxilla and mandible	Partial mid and lower face	–	Partial bilateral cheek	Partial nose	Chin, upper and lower lip	Partial mucosa	60–65%
Case 9 43 y, M Spain 2009	Mandible	–	–	Partial bilateral cheek	–	Chin and lower lip	Tongue and partial mucosa	30–35%
Case 10 27 y, M France 2009	Mandible	–	–	–	Partial nose	Chin, upper and lower lips	Partial mucosa	30–35%
Case 11 34 y, M Spain 2010	–	Bilateral parotid g. and ducts	–	Bilateral cheek	Partial nose	Chin, upper and lower lip	Partial mucosa	60–65%
Case 12 30 y, M Spain 2010	Bilateral orbital floor, zygoma, nasal, maxilla with teeth, palate, and mandibula	Bilateral parotid g. and ducts, septal cart	Fore-head,	Bilateral cheek, lower and upper eyelids	Total nose	Chin, upper and lower lip	Partial mucosa	95–100%
Case 13 35 y, M France 2010	–	Bil. parotid gl. and ducts, septal cart, tear gl. and lacrimal canal	Fore-head, ears	Bilateral cheek, lower and upper eyelids with gl.	Total nose	Chin, upper and lower lip	Partial mucosa	95–100%

M male, *F* female, *y* years, *gl* gland, *cart* cartilage, *bil* bilateral

Table 44.3 The details of sensory and motor recovery of reported patients

	Dubernard	Guo	Lantieri	Siemionow
Sensation	Semmes-Weinstein Light touch (10 week) Temperature (6 mo.)	Semmes-Weinstein Light touch (3 mo.) Temperature (8 mo.)	Quantitative sensory testing Light touch (3 mo.) Temperature (3 mo.)	Quantitative sensory testing Smell (2 day) Light touch (5 mo.) Temperature (5 mo.)
Motor recovery	Eat and drink (1 week) Upper lip (12 week) Lower lip (4 mo.) Labial contact (6 mo.) Chin and nose pyramidal muscle motion (12 mo.) Smile (14–18 mo.)	Unable to smile completely and symmetrically Facial nerve not fully functional Eat, drink, and talk normally	EMG Eat and speak (10 day) Orbicularis oris and oculi voluntarily contract (6 mo.) Spontaneous mimicry (9 mo.) Trigeminal and facial motor functions (12 mo.)	Upper lip occlusion Facial mimicry Eats and drinks from a cup Speaks more clearly and intelligibly

mo. month, *EMG* electromyography

44.3 Review of Current Cases of Facial CTA Recipient

Figures 44.1a, b through 44.13a, b present diagrams of 13 face transplant recipients indicating major soft tissue, bone, and functional components included in face allotransplant.

Case 1

Date: November 2005
Place: Lyon, France
Team Leader: Jean-Michelle Dubernard
Recipient: 38-year-old woman
Donor: Brain-dead 46-year-old woman
Indication: Midface trauma after dog bite
CTA design: Partial face transplant including nose, lips, chin, partial cheeks, mucosa (Fig. 44.1a, b)
Nerve repair: Left mandibular branch of facial nerve, bilateral infraorbital and mental branches of trigeminal nerve.
Vascular repair: Bilateral facial arteries and veins.
Time of surgery: 15 h
Status: Alive

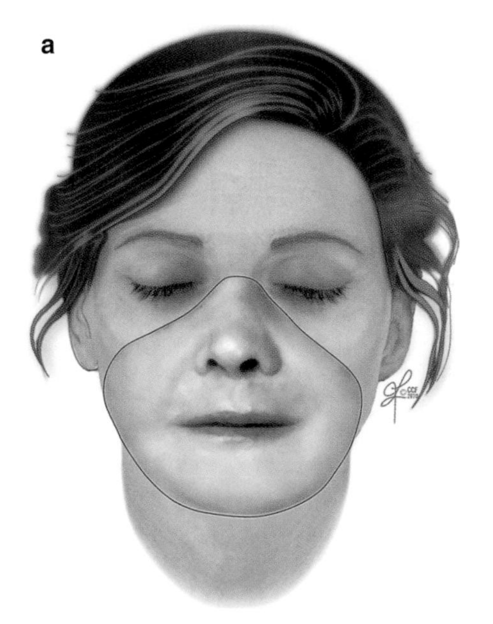

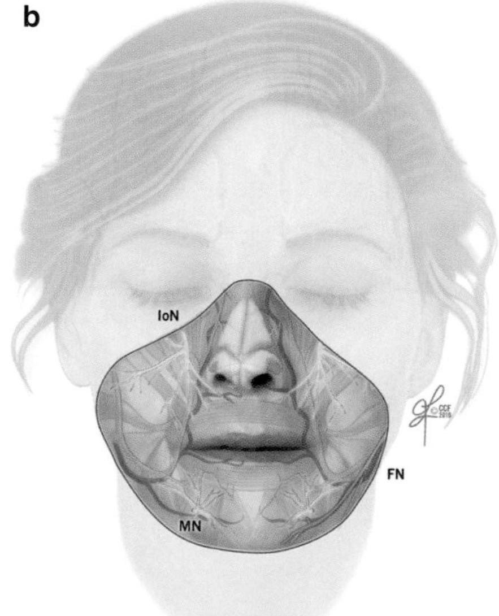

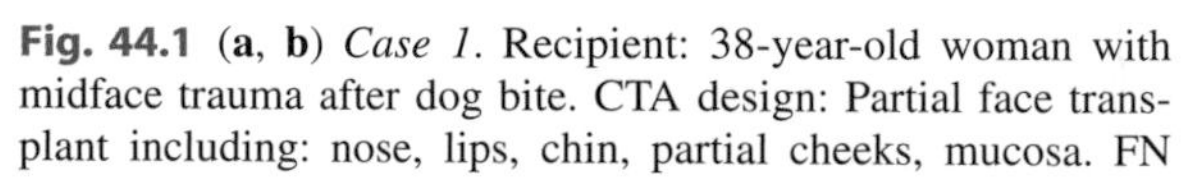

Fig. 44.1 (**a**, **b**) *Case 1*. Recipient: 38-year-old woman with midface trauma after dog bite. CTA design: Partial face transplant including: nose, lips, chin, partial cheeks, mucosa. FN facial nerve, IoN infraorbital nerve, MN mental nerve. Transplant date: November 2005, Lyon, France

Case 2

Date: April 2006
Place: Xian, China
Team Leader: Shuzhong Guo
Recipient: 30-year-old man
Donor: Brain-dead 25-year-old man
Indication: Midface trauma after bear bite

CTA design: Partial face transplant including overlying skin, soft tissue, upper lip, nose, the right anterior maxilla, sinus, right zygoma with lateral orbital wall, right parotid gland and partial masseter with intraoral mucosa (Fig. 44.2a, b)

Nerve repair: Right facial nerve (reported as "suboptimal")

Vascular repair: Right external maxillary artery and anterior facial vein

Time of surgery: 18 h

Status: Died due to complications around 2 years posttransplant

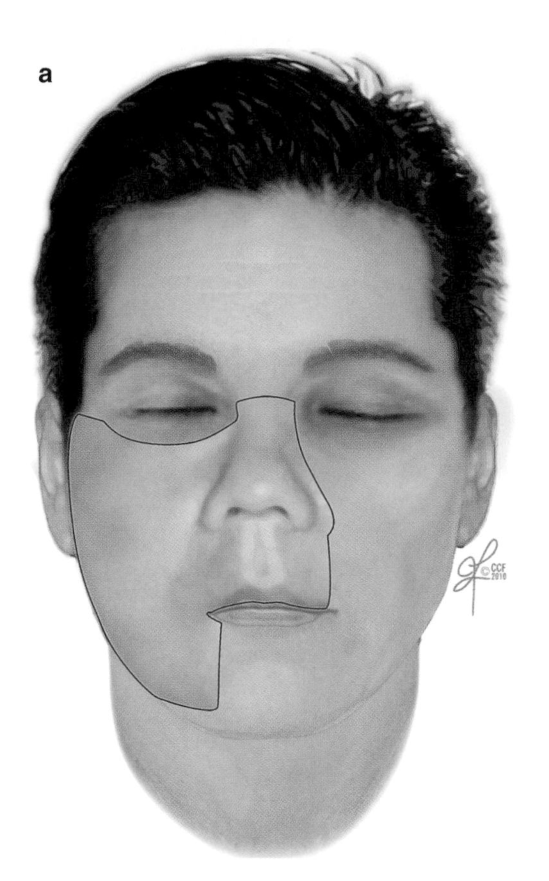

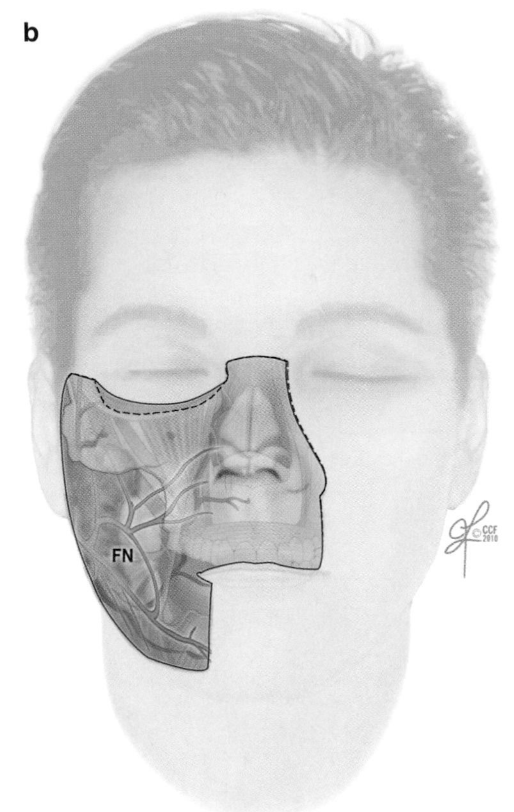

Fig. 44.2 (**a**, **b**) *Case 2*. Recipient: 30-year-old man with midface trauma after bear bite. CTA design: Partial face transplant including: overlying skin, soft tissue, upper lip, nose, the right anterior maxilla, sinus, right zygoma with lateral orbital wall, right parotid gland, and partial masseter with intraoral mucosa. FN facial nerve. Transplant date: April 2006, Xian, China

Case 3

Date: January 2007

Place: Paris, France

Team Leader: Laurent Lantieri

Recipient: 29-year-old man

Donor: Brain-dead male

Indication: Neurofibromatosis Type I

CTA design: Partial mid/lower face transplant including lower two-thirds of face skin, soft tissue, lips, chin, cheeks, nose, bilateral parotid glands, parotid ducts, and intraoral mucosa (Fig. 44.3a, b)

Nerve repair: Bilateral facial, infraorbital, and mental nerves. (reported as “mental nerves placed near foramen”)

Vascular repair: Bilateral external carotid arteries and bilateral thyrolingofacial trunks.

Time of surgery: 15 h

Status: Alive

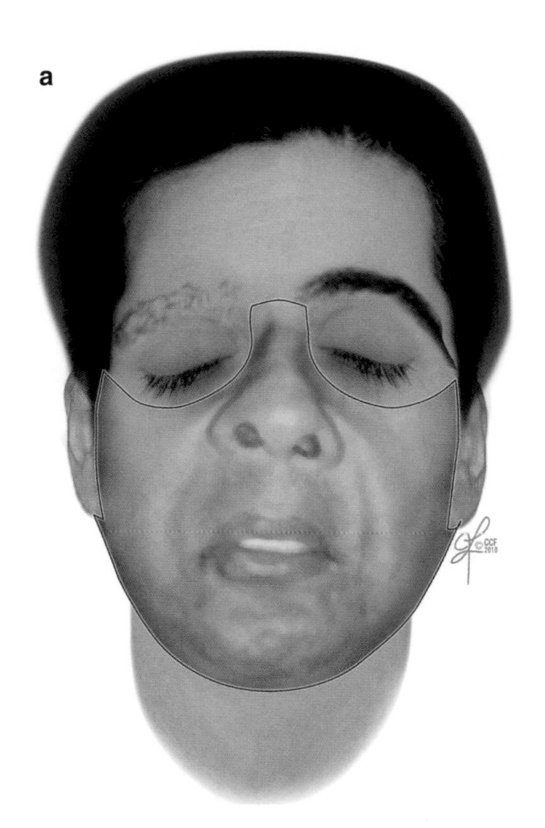

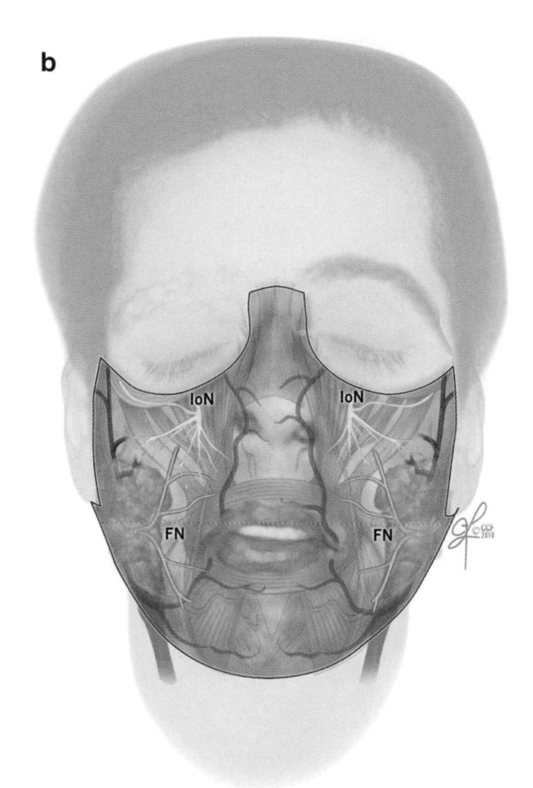

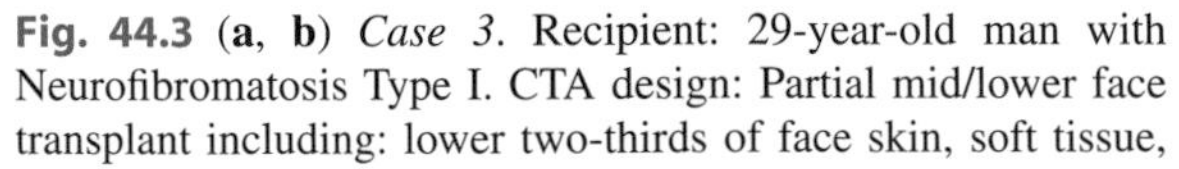

Fig. 44.3 (**a**, **b**) *Case 3*. Recipient: 29-year-old man with Neurofibromatosis Type I. CTA design: Partial mid/lower face transplant including: lower two-thirds of face skin, soft tissue, lips, chin, cheeks, nose, bilateral parotid glands, parotid ducts, and intraoral mucosa. FN facial nerve, IoN infraorbital nerve. Transplant date: January 2007, Paris, France

Case 4

Date: December 2008

Place: Cleveland, USA

Team Leader: Maria Z. Siemionow

Recipient: 45-year-old woman

Donor: Brain-dead 44-year-old woman

Indication: Shotgun injury

CTA design: Near-total face transplant including composite LeFort III midfacial skeleton, overlying skin, soft tissue, nose, lower eyelids, upper lip, total infraorbital floor, bilateral zygomas, bilateral parotid glands, anterior maxilla with central maxillary incisors, total alveolus, anterior hard palate, and intraoral mucosa (Fig. 44.4a, b)

Nerve repair: Bilateral facial nerves with grafts

Vascular repair: Bilateral common facial arteries; left external jugular, posterior facial, and right facial veins.

Time of surgery: 22 h

Status: Alive

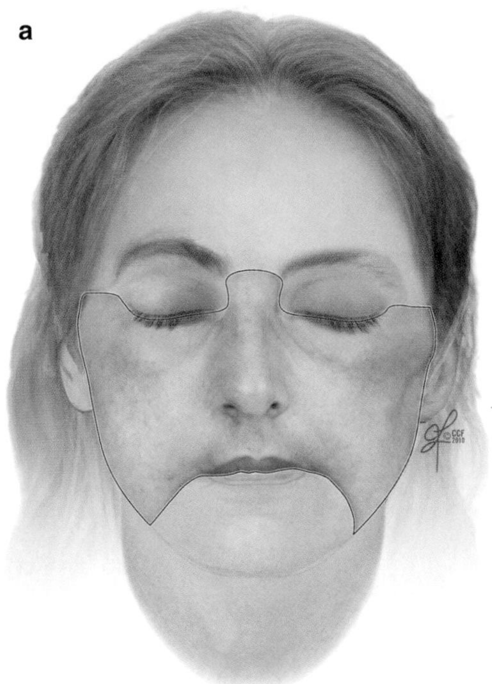

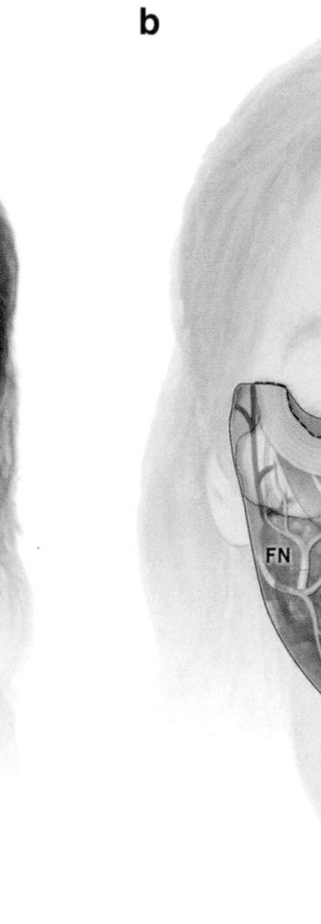

Fig. 44.4 (**a**, **b**) *Case 4*. Recipient: 45-year-old woman after shotgun injury to the face. CTA design: Near-total face transplant including: composite LeFort III midfacial skeleton, overlying skin, soft tissue, nose, lower eyelids, upper lip, total infraorbital floor, bilateral zygomas, bilateral parotid glands, anterior maxilla with central maxillary incisors, total alveolus, anterior hard palate, and intraoral mucosa. FN facial nerve. Transplant date: December 2008, Cleveland, USA

Case 5

Date: March 2009
Place: Paris, France
Team Leader: Laurent Lantieri
Recipient: 28-year-old man
Donor: Data not available
Indication: Shotgun injury

CTA design: Partial face transplant including premaxilla, chin, nose and overlying lower third of face skin, soft tissue, lips, cheeks, bilateral parotid glands, parotid ducts, and intraoral mucosa (Fig. 44.5a, b)

Nerve repair: Bilateral facial, nerves.

Vascular repair: Bilateral external carotid arteries, external jugular and thyrolinguofacial trunk.

Time of surgery: 15 h
Status: Alive

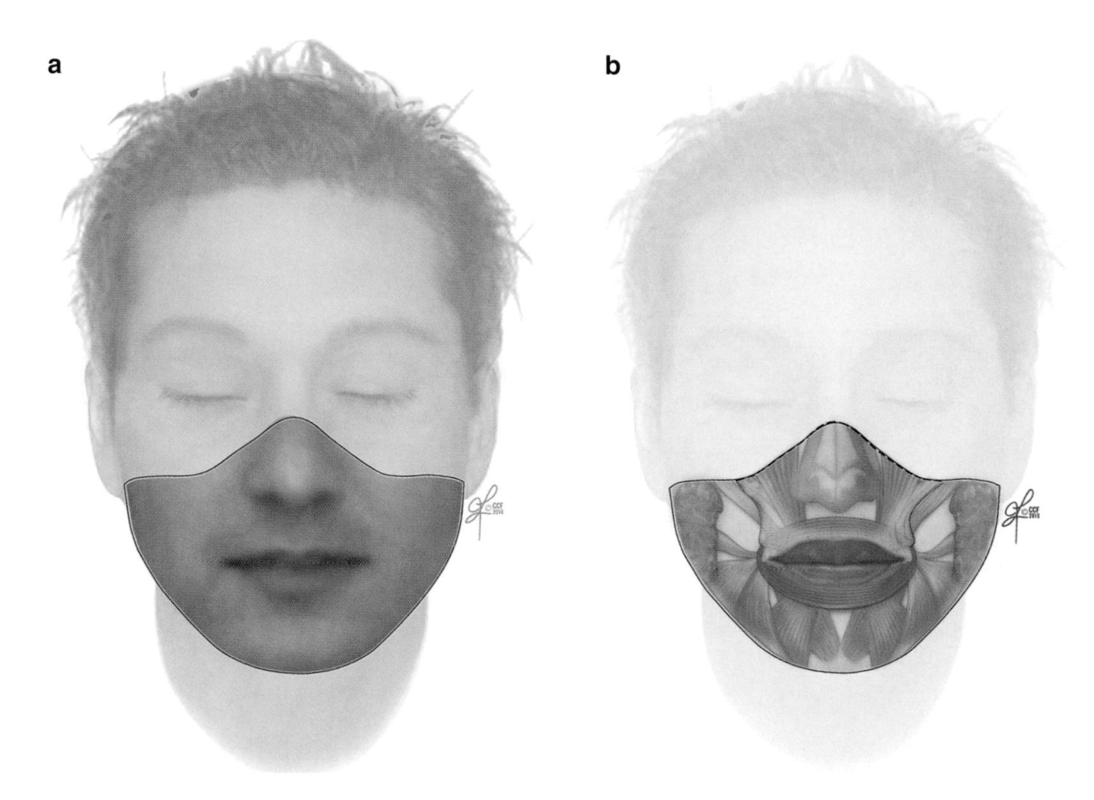

Fig. 44.5 (**a**, **b**) *Case 5* Recipient: 28-year-old man after shotgun injury to the face. CTA design: Partial face transplant including: premaxilla, chin, nose and overlying lower third of face skin, soft tissue, lips, cheeks, bilateral parotid glands, parotid ducts, and intraoral mucosa. Transplant date: March 2009, Paris, France

Case 6

Date: April 2009

Place: Paris, France

Team Leader: Laurent Lantieri

Recipient: 37-year-old man

Donor: Brain-dead man

Indication: Extensive burn sequela

CTA design: Partial upper two-third of face including entire scalp with bilateral ears, forehead, lower eyelids, nose, bilateral cheeks, bilateral parotid glands and parotid ducts, intraoral mucosa (Fig. 44.6a, b)

Concomitant bilateral hand transplantation

Nerve repair: Bilateral facial, supraorbital, and infraorbital, nerves

Vascular repair: Bilateral external carotid arteries, bilateral thyrolinguofacial trunk

Time of surgery: 30 h. (Including bilateral hand transplant)

Status: Died due to complications after 2 months posttransplant

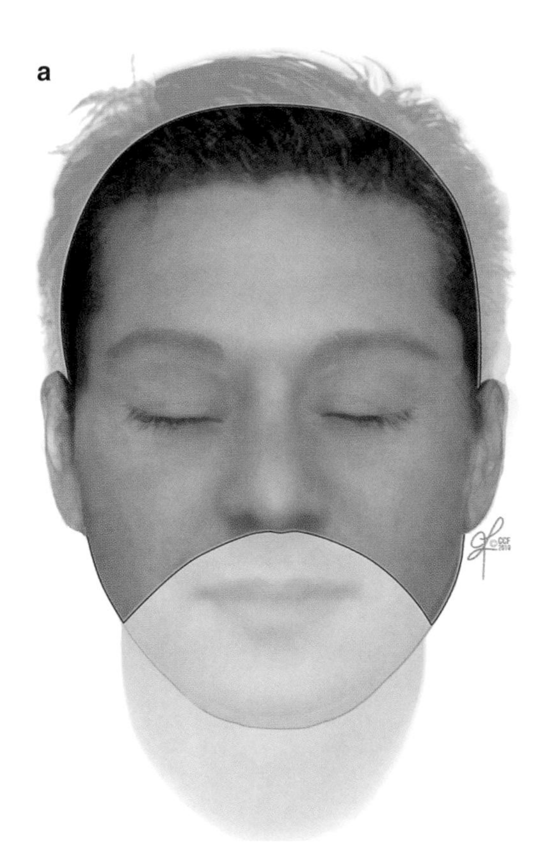

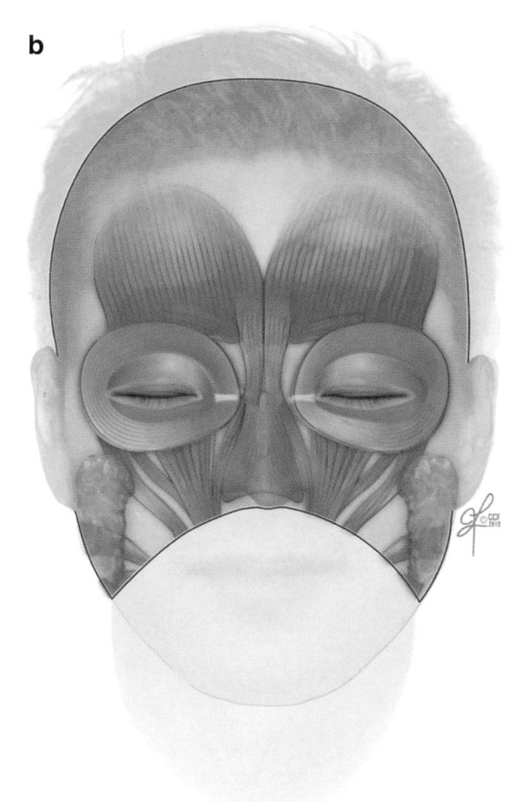

Fig. 44.6 (**a**, **b**) *Case 6*. Recipient: 37-year-old man with extensive burn sequela. CTA design: Partial upper two-third of face including: entire scalp with bilateral ears, forehead, lower eyelids, nose, bilateral cheeks, bilateral parotid glands and parotid ducts, intraoral mucosa. Concomitant bilateral hand transplantation. Transplant date: April 2009, Paris, France

Case 7

Date: April 2009

Place: Boston, USA

Team Leader: Bohdan Pomahac

Recipient: 59-year-old man

Donor: Brain-dead 60-year-old man

Indication: Fall/electrical injury

CTA design: Partial face transplant including midface tissue with overlying skin, soft tissue, nose, upper lip, bilateral cheek, bilateral parotid ducts, maxilla with teeth and palate (Fig. 44.7a, b)

Nerve repair: Bilateral facial and infraorbital nerves

Vascular repair: Left external carotid, right facial artery and bilateral facial veins

Time of surgery: 17 h

Status: Alive

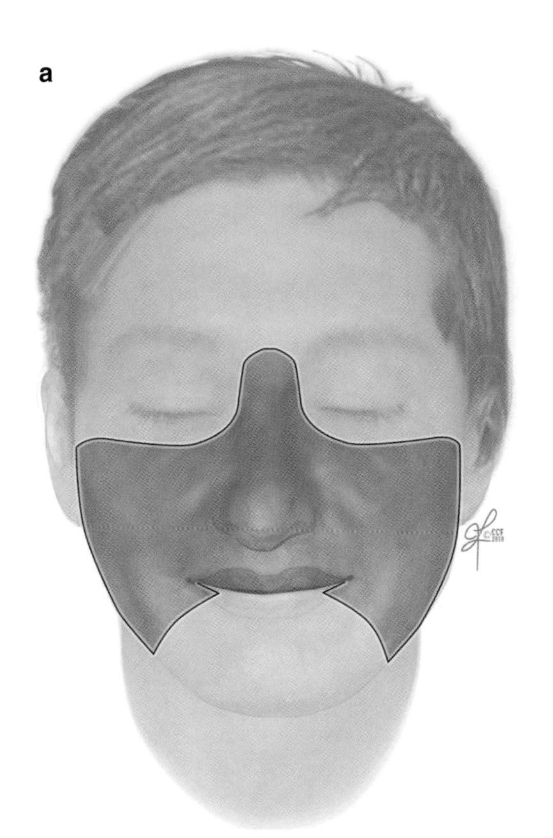

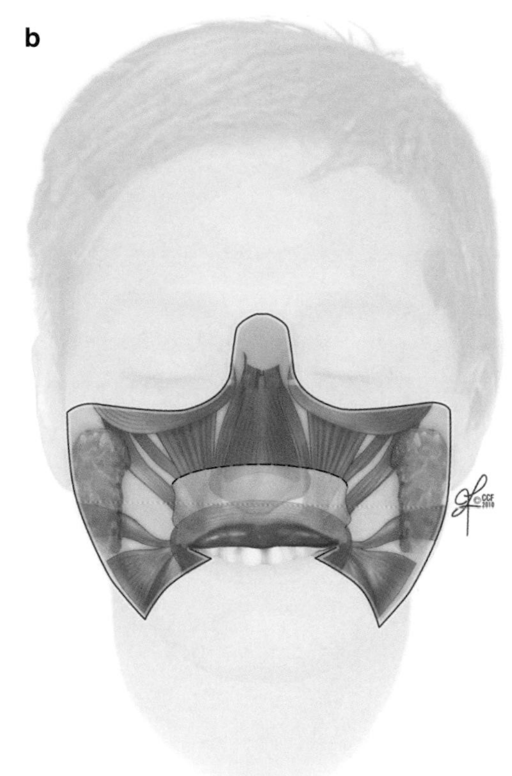

Fig. 44.7 (**a**, **b**) *Case 7*. Recipient: 59-year-old man after fall/electrical injury. CTA design: Partial face transplant including: midface tissue with overlying skin, soft tissue, nose, upper lip, bilateral cheek, bilateral parotid ducts, maxilla with teeth and palate. Transplant date: April 2009, Boston, USA

Case 8

Date: 2009

Place: Paris, France

Team Leader: Laurent Lantieri

Recipient: 33-year-old man

Donor: Data not available

Indication: Gunshot wound

CTA design: Partial mid/lower face transplant including maxilla, mandible, overlying skin, soft tissue, nose, bilateral cheek, lips and intraoral mucosa. (Fig. 44.8a, b)

Nerve repair: Bilateral facial nerves

Vascular repair: Bilateral external carotid arteries, external jugular veins and thyrolinguofacial trunks.

Time of surgery: Data not available

Status: Alive

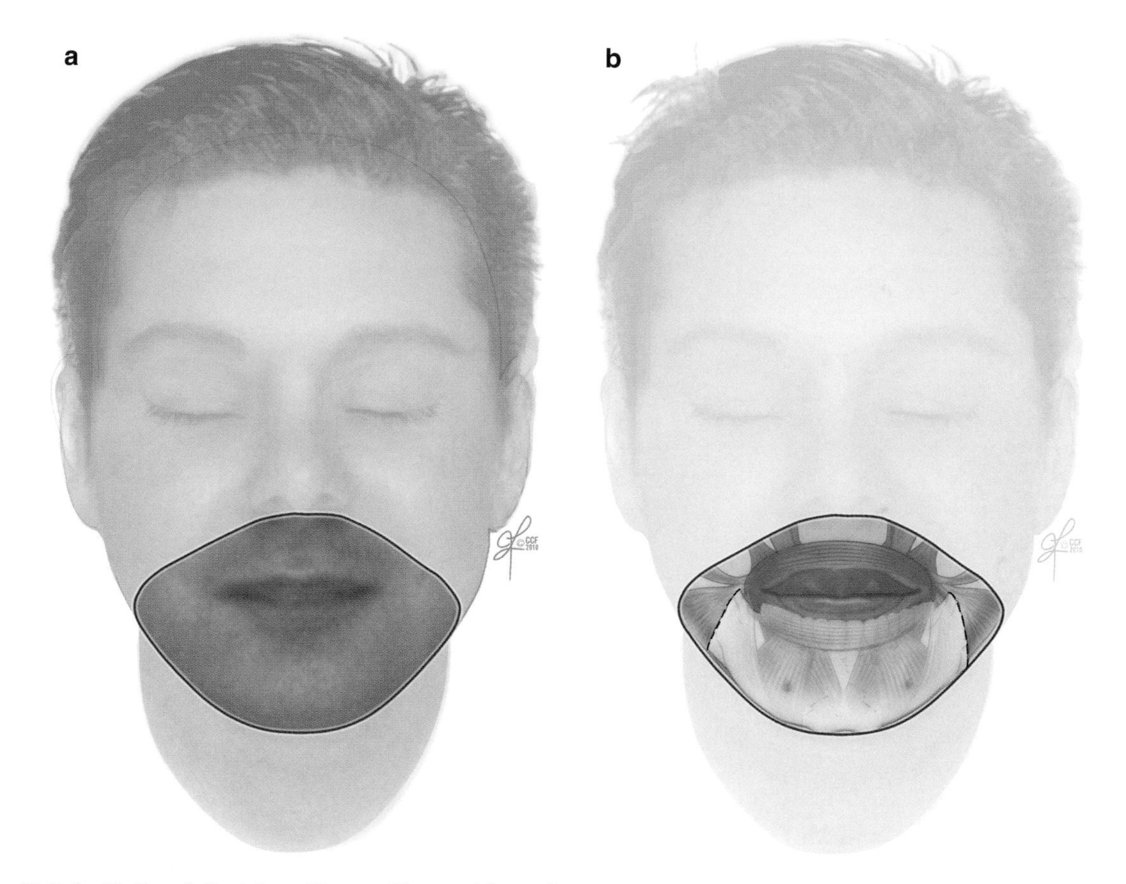

Fig. 44.8 (**a**, **b**) *Case 8*. Recipient: 33-year-old man with gunshot wound. CTA design: Partial face, transplant including: maxilla, mandible, overlying skin, soft tissue, nose, bilateral cheek, lips and intraoral mucosa. Transplant date: 2009, Paris, France

Case 9

Date: August 2009
Place: Valencia, Spain
Team Leader: J. Pedro Cavadas
Recipient: 43-year-old man
Donor: Brain-dead 35-year-old man
Indication: Cancer treatment sequela
CTA design: Partial face transplant including lower third of facial tissue with overlying skin, soft tissue, lower lip, partial cheek, chin, neck, mandible, tongue, and intraoral mucosa (Fig. 44.9a, b)
Nerve repair: Bilateral hypoglossal, inferior alveolar, lingual and cervicofacial nerves
Vascular repair: Left common carotid and right internal carotid arteries, bilateral external jugular and right internal jugular veins
Time of surgery: 15 h
Status: Alive

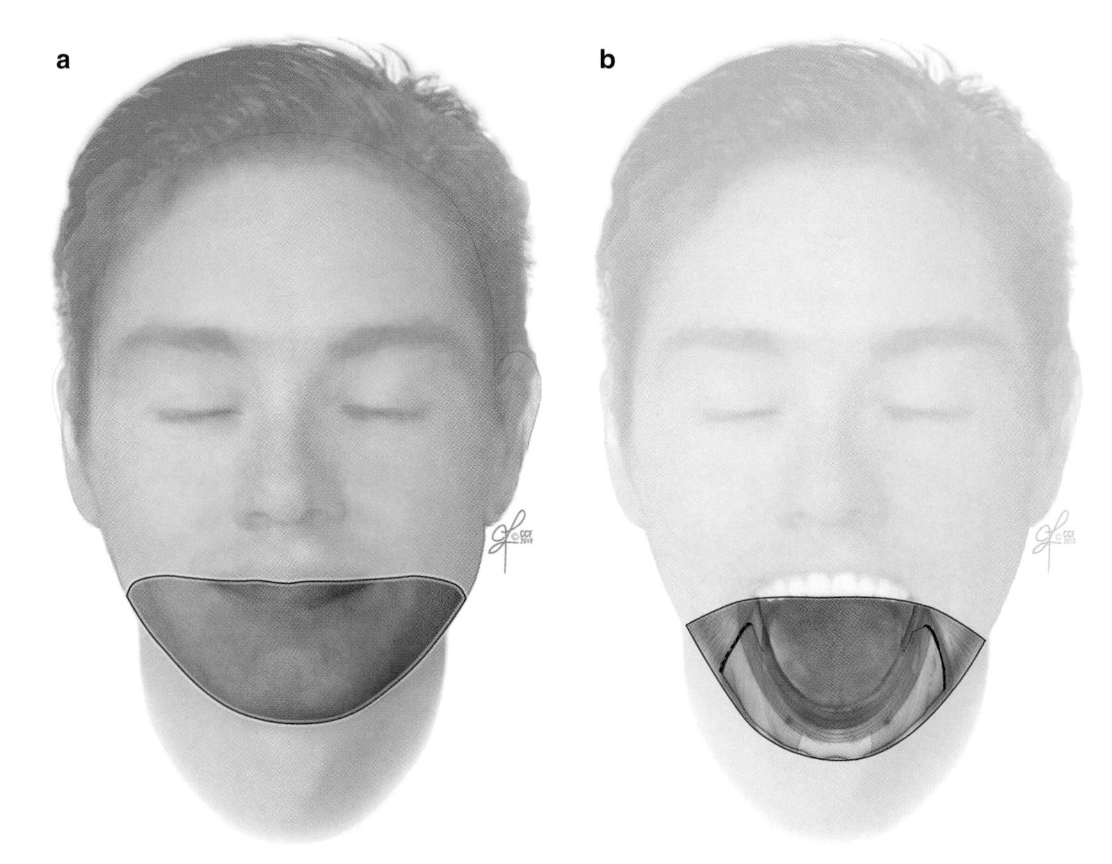

Fig. 44.9 (**a**, **b**) *Case 9*. Recipient: 43-year-old man with face deformity due to cancer treatment sequela. CTA design: Partial face transplant including: lower third of facial tissue with overlying skin, soft tissue, lower lip, partial cheek, chin, neck, mandible, tongue, and intraoral mucosa. Transplant date: August 2009, Valencia, Spain

Case 10

Date: November 2009
Place:Lyon, France
Team Leader: Jean-Michelle Dubernard
Recipient: 27-year-old man
Donor: Brain-dead man
Indication: Explosion trauma
CTA design: Partial face transplant including: lower third of facial tissue with overlying skin, soft tissue, nose, lips, chin, mandible and intraoral mucosa. (Fig. 44.10 a, b)
Nerve repair: Bilateral facial, infraorbital and mental nerves.
Vascular repair: Bilateral facial arteries and veins.
Time of surgery: Data not available.
Status: Alive

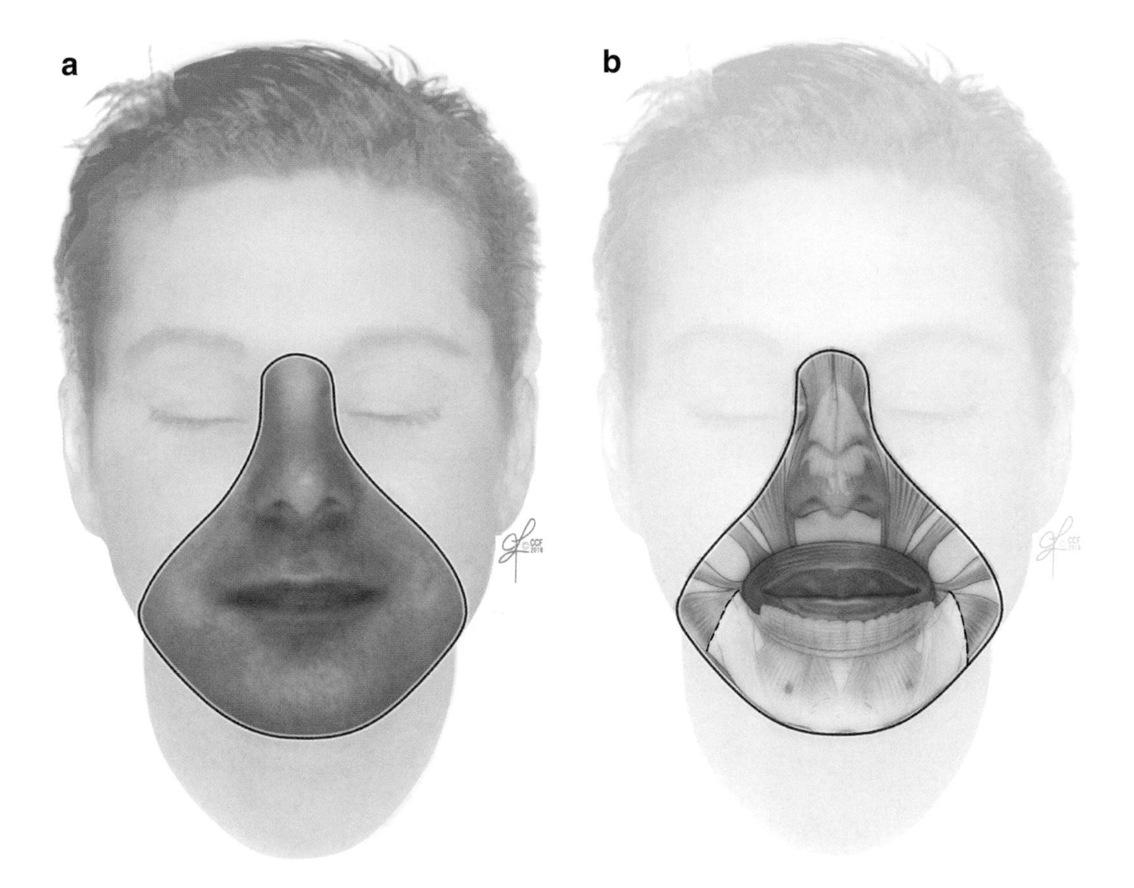

Fig. 44.10 (**a**, **b**) *Case 10*. Recipient: 27-year-old man with face deformity due to ballistic trauma.CTA design: Partial face transplant including: lower third offacial tissue with overlying skin, soft tissue, nose, lips, chin and mandible. Transplant date: November 2009, Amiens, France

Case 11

Date: January 2010
Place: Sevilla, Spain
Team Leader: Cia Montero
Recipient: 34-year-old man
Donor: Brain-dead 30-year-old man
Indication: Neurofibromatosis
CTA design: Partial face transplant including lower part of face with overlying skin, soft tissue, lips, chin, cheeks, bilateral parotid glands, parotid ducts, and intraoral mucosa (Fig. 44.11a, b)
Nerve repair: Bilateral facial, infraorbital and mental nerves.
Vascular repair: Bilateral external carotid arteries and internal jugular veins.
Time of surgery: 32 h
Status: Alive

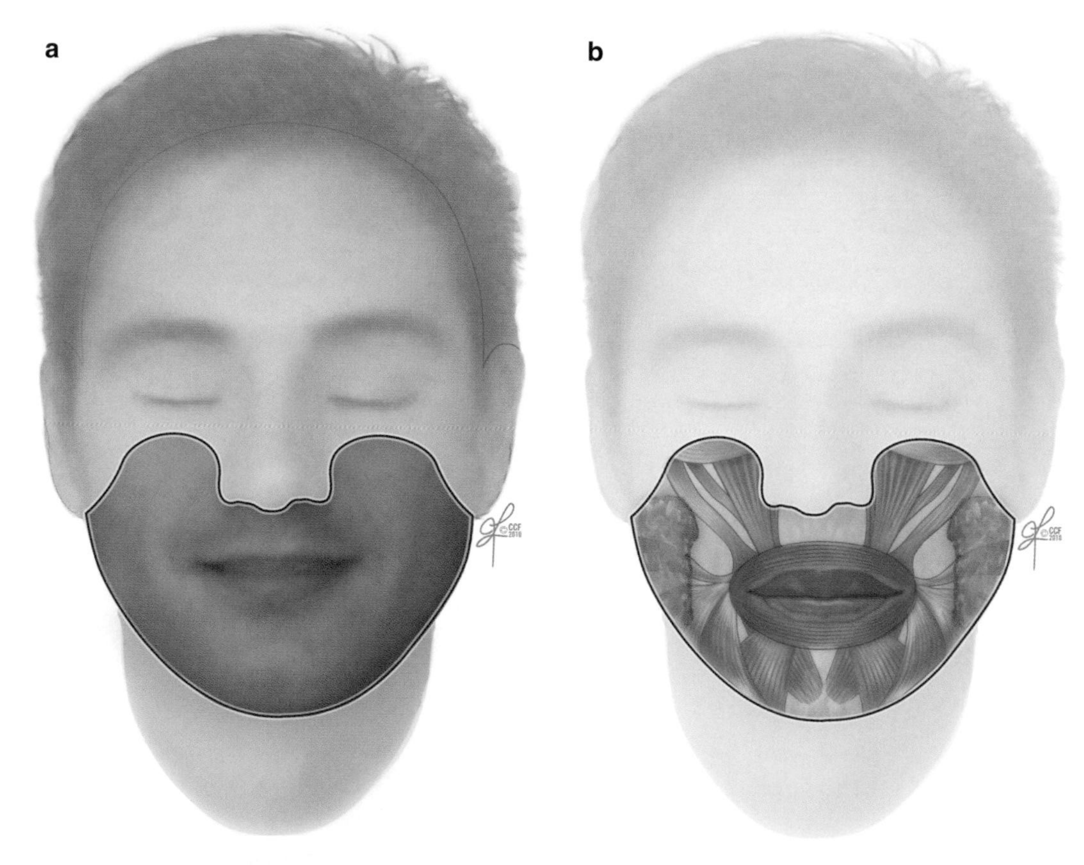

Fig. 44.11 (**a**, **b**) *Case 10*. Recipient: 34-year-old man with Neurofibromatosis. CTA design: Partial face transplant including: lower part of face with overlying skin, soft tissue, lips, chin, cheeks, bilateral parotid glands, parotid ducts, and intraoral mucosa. Transplant date : January 2010 Sevilla, Spain

Case 12

Date: March 2010
Place: Barcelona, Spain
Team Leader: Joan Pare Barret
Recipient: 30-year-old man
Donor: Data not available
Indication: Shotgun injury
CTA design: Total facial transplant including overlying skin, soft tissue, nose, lips, cheeks, eyelids, maxillary with teeth, palate, mandible and intraoral mucosa (Fig. 44.12a, b)
Nerve repair: Bilateral facial, supraorbital, infraorbital and mental nerves
Vascular repair: Bilateral external carotid arteries and external jugular veins, unilateral anterior jugular and retromandibular veins
Time of surgery: 22 h
Status: Alive

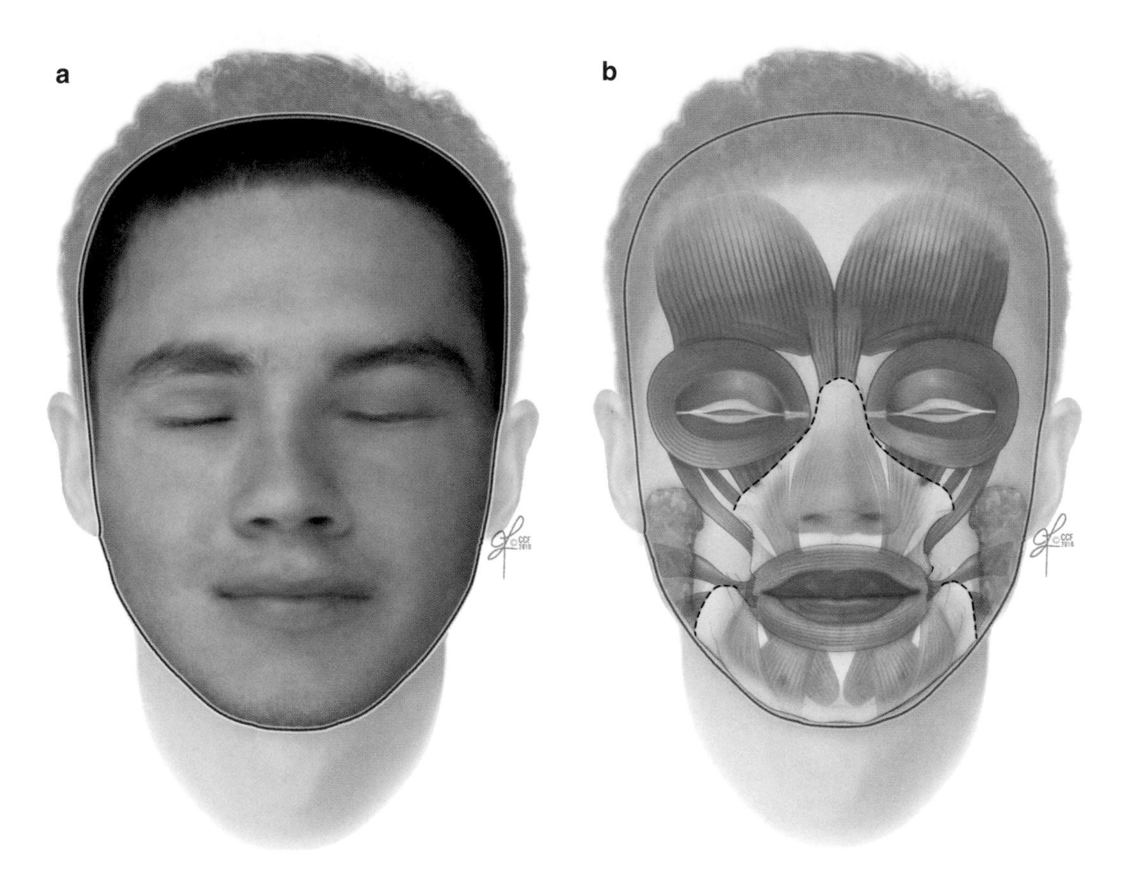

Fig. 44.12 (**a**, **b**) *Case 11*. Recipient: Man 31-year-old after shotgun injury to the face. CTA design: Total facial transplant including: overlying skin, soft tissue, nose, lips, cheeks, eyelids, maxillary with teeth, palate, mandible, and intraoral mucosa. Transplant date: March 2010, Barcelona, Spain

Case 13

Date: June 2010
Place: Paris, France
Team Leader: Laurent Lantieri
Recipient: 35-year-old man
Donor: Data not available
Indication: Neurofibromatosis
CTA design: Total face transplant including overlying skin, soft tissue, nose, lips, cheeks, eyelids, lacrimal glands and lacrimal ducts, and intraoral mucosa (Fig. 44.13a, b)
Nerve repair: Bilateral facial, supraorbital, infraorbital and mental nerves
Vascular repair: Bilateral external carotid arteries, external jugular veins and thyrolinguofacial trunks
Time of surgery: 12 h
Status: Alive

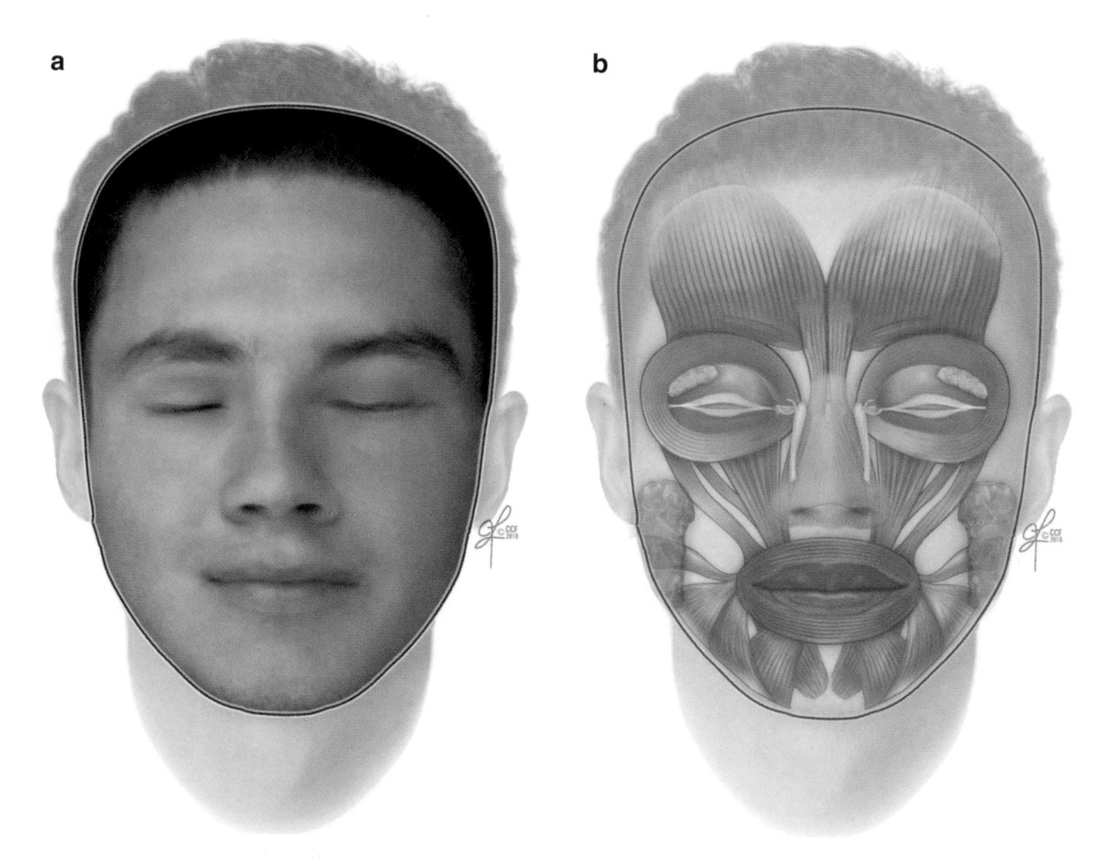

Fig. 44.13 (**a**, **b**) *Case 12*. Recipient: 35-year-old man with Neurofibromatosis. CTA design: Total face transplant including: overlying skin, soft tissue, nose, lips, cheeks, eyelids, lacrimal glands and lacrimal ducts, and intraoral mucosa. Transplant date: June 2010, Paris, France

44.4 Discussion

Since 2005, there are 13 partial, near-total, and total face transplantation cases performed around the world according to the published data and media reports. The technical details were available only for seven patients reported in the literature and data for the other six patients available based on meeting presentations and media reports.[5-25]

Revascularization of the face transplant by anastomosing between the donor and recipient vessels is the first priority for successful transplantation. According to facial transplantation reports, the preferred vessels for arterial anastomoses were external carotid arteries and their branches, such as facial and external maxillary arteries. The preferred vessels for venous anastomoses were facial, external jugular veins, and thyrolinguofacial trunks. Both arterial and venous anastomoses were performed using conventional microsurgical techniques. All of the arterial and venous repairs were done by standard end-to-end anastomosis; interestingly, there was not any report on end-to-side vessel anastomosis nor on vessel repair using vein grafts. This may be due to the fact that the calibers of vessels were nearly of equal size and adequate length and were available; thus, it was technically easy to perform end-to-end anastomosis. This differs from the experience in the free flap reconstruction within head and neck region, where vessel caliber discrepancy is a possibility and end-to-side anastomoses are often preferred in cases where sizes of recipient and donor vessels differ.

Takamatsu et al. described the selection of recipient vessels within head and neck region, and suggested the superficial temporal and facial vessels as the first choice vessels for midface and upper face reconstruction. For mandibular or lower face reconstruction, the external carotid artery and its branches are eligible.[27] Nearly half of the face transplantation cases included osteomyocutaneous components and most cases included the midface which is the most important and technically challenging part during facial reconstruction. The authors that performed osteomyocutaneous face transplants preferred to connect vessels at proximal levels, such as common facial artery and external carotid artery.[6-11] This approach may be important in order to maintain the viability of these composite allografts including bone components. However, Pomahac et al. reported that facial artery alone can perfuse both the soft tissues and bony elements of midfacial allografts that include maxilla and zygoma.[10] Their findings from cadaver studies elucidated that perfusion of bony components is supplied by vascular communications between facial artery and maxillary artery. The facial artery may nourish the midface anatomical structures; however, we have no data confirming that facial artery alone can perfuse total osteomyocutaneous face transplant. Lantieri et al. reported that in preclinical study, complete revascularization of a full facial and entire scalp flap was feasible when based on a single external carotid artery.[11] Furthermore, they observed sufficient revascularization of the CTA after external carotid artery end-to-end anastomosis in their clinical case; however, they carried out the contralateral external artery anastomosis to secure vascularity of the facial graft. In order to make recommendations on best design of vascular anastomosis, we need more experimental and clinical data confirming best arterial and venous supplies in case of near-total and total osteomyocutaneous face transplants.

During reconstructive microsurgery and replantation cases, the surgeons perform at least two venous anastomoses for one arterial repair. According to the facial transplant data, two arterial and two venous anastomoses were performed in general, except for one case where only single artery and single vein were connected and another case where two arteries and three veins were anastomosed. Due to the rich interconnecting vascular network within the face, single artery and venous repair are usually sufficient for adequate vascularization in most of the partial face transplant cases. There are reports describing successful replantation managing survival of large segments of face, scalp, and soft tissues based on a single vessel anastomosis.[28-29] However, we cannot afford failure of a facial transplantation solely due to technical problems with arterial or venous repair. Thus, to minimize the risk of transplant failure or other related problems, at least two arterial and two venous anastomoses should be performed. The cases discussed in this article had no early vascular complications such as transplant failure due to vascular thrombosis or vascular spasm and all transplants showed good perfusion without any problems after clamp release.

Another technically important issue when regarding vessel choice is the condition of recipient area. Most of transplant candidates present with scar tissues, anatomical variations, and disruptions of vascular territories due to traumatic injuries, tumors, or previous reconstructive procedures. It is sometimes quite difficult to identify the exact anatomic location of recipient vessels during surgery. Surgeons can easily damage those suitable for repair vessels during preparation of the recipient site. Thus, to prevent this problem, adequate information about vascular anatomy and vessel patency should be provided and should be supported by imaging studies such as angio CT, 3D MRI, or Doppler ultrasound monitoring.

The next step after vessel repair is the coaptation of sensory and motor nerves. The trigeminal nerve (CNV) is the main sensory nerve and the facial nerve is the main motor nerve repaired during face transplantation. Trigeminal nerve has three sensory branches: the ophthalmic branch (V1) conveys sensory information from the skin of the forehead, upper eyelids, and lateral aspects of the nose; the maxillary branch (V2) conveys sensory information from the lower eyelids, zygoma, and upper lip; and the mandibular branch (V3) conveys sensory information from the lateral scalp, skin anterior to the ears, lower cheeks, lower lips, and anterior aspect of the mandible. The main body of the facial nerve is somatomotor and controls the muscles of facial expressions through its five branches: temporal, zygomatic, buccal, mandibular, and cervical branches.

According to the facial allotransplantation literature reports, out of seven patients, the infraorbital nerves were coapted in four cases. In addition, mental, buccal, and supraorbital nerves were repaired in some of these four cases. In one case, instead of direct coaptation of mental nerves, donor stumps were placed near the mental foramen. In the remaining three cases, no sensory nerve repair but bilateral facial nerve repair was performed (Table 44.1). Three or six months after transplantation, all of the authors observed near-normal sensory recovery of the transplanted part of the face. Also, normal sensory recovery was reported for face transplant patients by news and during media interviews; however, the full data have not been published yet. In reported patients, authors analyzed sensory recovery by quantitative sensory tests, including Semmes-Weinstein, two-point discrimination, pressure-specified sensory devices, and heat/cold tolerance. Interestingly, there was almost total sensory recovery in two patients who had only facial nerves repaired but trigeminal branches were not repaired, due to the trauma. The question arises, how could this be feasible? As described in the literature, this could happen due to interconnections between facial and trigeminal nerve branches.[30-31] In a third patient, reported by Lantieri et al., bilateral mental nerves were not repaired since they were transected at the level of submental foramen.[7] Instead, the donor nerve was placed near the mental foramen. They reported that 3 months after surgery, quantitative sensory testing showed sensory reinnervation of the skin for both the thermal and mechanical sensation. They concluded that this was possible due to the regrowth of donor nerve into the recipient nerve or via direct sprouting of the donor nerve into this region of the face. In summary, sensory recovery of face CTA was reported as nearly perfect 3–8 months after surgery, in all patients. As expected, sensation to light touch at skin and oral mucosa recovered at about 3 months posttransplant; however, thermal sensation recovered at about 6–8 months. There is almost no difference between the results of sensory reinnervation after transplantation when compared to results reported for free tissue transfers and direct repair of autologous nerves.[32-33]

While the vascular perfusion based viability of allotransplant is important during acute period, the sensory and motor recovery is becoming more important at long-term follow-up. When assessing outcomes of extremity replantation as being successful, the viability of the replanted part was the most important feature in the first cases. But with development of techniques, it became clear that replantation success is equal to the degree of functional recovery. We believe that the degree of motor and functional recovery will be of utmost importance in evaluating long-term results of facial allotransplantation in the future. Motor recovery of CTAs was slower and often less optimal when compared to the sensory recovery outcomes. There are only few long-term follow-up results of motor recovery. In the first CTA case, Dubernard et al. reported that patient was able to move upper lip 3 months after transplant, there was a motion of lower lip at about 4 months, complete labial contact was present at 6 months, and normal smiling was recorded at 18 months. Endobuccal pressure increased progressively, contraction of the chin and nose muscles were present at the end of the first year.[5] In contrast, in the second patient, the facial nerve motor function was not satisfactory and did not improve over time.[6] The patient was unable to smile completely and symmetrically, and the functions of other muscles innervated by the facial nerve have not improved. According to the author Guo, this result was the result of poor coaptation of facial nerves during surgery. In the third patient, Lantieri et al. reported that EMG showed no electrical activity of reinnervation of facial nerve 3 months after operation, but after 1 year, EMG showed signs of motor reinnervation of both the trigeminal and facial nerves.[7] They also reported an unexpected result indicating recovery of involuntary reflex contraction of facial muscles in response to stimulation of the supraorbital branches of the trigeminal nerves. In the fourth patient, Siemionow et al. reported that at 6 month motor recovery including facial mimetics progressed at a slower but steady rate, as demonstrated by improved facial mimetics with symmetric smiling and upper lip occlusion.[8,9] The patient's upper lip and lower eyelid movements were imperfect. As expected, motor recovery was slowly progressing; however, it was almost fully recovered at 1 year posttransplant. We do not have any recent data available about motor recovery of facial nerve in eight out of 13 cases. The question arises how can we measure, quantitatively, the outcomes of motor nerve recovery? Several methods such as Carroll test have been used to determine functional recovery after hand transplantation and similar methods are required for facial transplants.[34] In a previous review article, Hui-chou et al. offered functional magnetic resonance imaging and EMG studies for determining the motor recovery.[26] The sensory reinnervation results are more satisfactory than the functional recovery of facial muscles. These findings are similar and comparable to hand transplant recipients. Hand transplant experience for 10 years showed that patients have shown 90% sensory recovery and satisfactory motor recovery within 1 year after transplant.[35]

Patients who are initially satisfied with early results will expect more improvement over time. Even patients with facial palsy without additional disfigurements are not satisfied with their appearance and are often ready to undergo extensive surgical reconstructive procedures to achieve only minimal improvements. It is unrealistic to expect better outcomes following allotransplantation, at least at current stage since complete functional recovery of motor nerves is not satisfactory even in acute cases with primarily nerve repair. To obtain optimal results, great care must be taken during preparation of recipient nerve ends and distal segments of facial nerve should be preferably used for coaptation.

This review confirmed that the functional outcomes were different in each patient, due to complexity of trauma and acquired deficits. The restoration of osteomyocutaneous defects is more challenging compared to partial myocutaneous defects. In general, all patients were satisfied with functional results. According to first four patient reports, all of them were able to eat, drink, and speak within 7–10 days after transplantation.[5-9] Siemionow et al. reported that functional recovery of three-dimensional facial defect was excellent with restoration of major functions which the patient had lacked before, including improvement of speech was improved after hard palate reconstruction with facial CTA and palatal obturator support.[8,9] According to the media reports, we acknowledge that Cavadas et al. performed the first tongue transfer as a part of their face CTA, while Lantieri transferred lacrimal gland and lacrimal ducts in their full face CTA.[12,16,25] However, at this point, we do not have scientific data confirming functional results of these recent facial transplants.

44.5 Conclusion

The human beings represent complex biological, psychological, and social entities. Illness easily occurs when one of these factors is missing. The facial allotransplantation candidates present with both the biological defect and illness, as well as with psychological and social problems, creating a challenging group of patients. The facial allotransplantation should not only reconstruct the anatomical defect but should also improve the functional and aesthetic deficits leading to improvement of the psychological condition and social problems which these patients encounter. Therefore, facial allotransplantation remains one of the most complex microsurgical reconstructions.

In fact, both for the surgeon and the patient, this operation is still "experimental" since we do not have sufficient data about long-term outcomes of the first 13 patients. We hope that this chapter will be of value to surgeons interested in facial allotransplantations and will help to plan the logistics of surgery and will give an idea about possible functional outcomes. We specially thank the authors who published and presented their results, other media and related resources which made this review possible.

References

1. Lee WP, Butler PE. Transplant biology and applications to plastic surgery. In: Aston SJ, Beasley RW, Thorne CHM, eds. *Grabb and Smith's Plastic Surgery*. Philadelphia: Lippincott-Raven; 1997.
2. Malt RA, McKhann C. Replantation of several arms. *JAMA*. 1964;189:716-722.
3. Dubernard JM, Owen E, Herzberg G, et al. Human hand allograft: report on first 6 months. *Lancet*. 1999;353: 1315-1320.
4. International registry on hand and composite tissue transplantation. http://www.handregistry.com/page.asp?page=4.
5. Dubernard JM, Lengele B, Morelon E, et al. Outcomes 18 months after the first human partial face transplantation. *N Engl J Med*. 2007;357:2451-2460.
6. Guo S, Han Y, Zhang X, et al. Human facial allotransplantation: a 2-year follow-up study. *Lancet*. 2008;372:631-638.
7. Lantieri L, Meningaud JP, Grimbert P, et al. Repair of the lower and middle parts of the face by composite tissue allotransplantation in a patient with massive plexiform neurofibroma: a 1-year follow-up study. *Lancet*. 2008;372:639-645.
8. Siemionow M, Papay F, Djohan R, et al. First U.S. near-total human face transplantation – a paradigm shift for massive facial injuries. *Plast Reconstr Surg*. 2010;125:111-122.
9. Siemionow M, Papay F, Alam D, et al. Near-total human face transplantation for a severely disfigured patient in the USA. *Lancet*. 2009;374:203-209.
10. Pomahac B, Lengele B, Ridgway EB, et al. Vascular considerations in composite midfacial allotransplantation. *Plast Reconstr Surg*. 2010;125:517-522.
11. Meningaud JP, Benjoar MD, Hivelin M et al. The procurement of total human face graft for allotransplantation: a preclinical study and the first clinical case. Plast Reconstr Surg 2010 June 15 [ePub ahead of print].
12. Spain's first face transplant patient smiling. Usa Today. http://www.usatoday.com/2009-08-22-spain-face-transplant_N.htm. Accessed August 22, 2009.
13. Spain's second face transplanted completed in Sevilla. Spain news. http://www.typicallyspanish.com/news/publish/article_24869.shtml. Accessed January 27, 2010.
14. World's First Full Face Transplant A Success Say Spanish Doctors. Medical news today.http://www.medicalnewstoday.com/articles/186612.php. Accessed April 26, 2010.
15. Gordon CR, Siemionow M, Papay F, et al. The world's experience with facial transplantation: what have we learned thus far? *Ann Plast Surg*. 2009;63:572-578.
16. Doctors perform face transplant with eyelids. Msnbc. http://www.msnbc.msn.com/id/38144667.
17. Devauchelle B. Microsurgical reconstruction of the face: Some outstanding indications. In *Proceedings of the 7th Congress of Polish Society of Oral and Craniofacial Surgery*, Jachranka, Poland. 2010;14-15.
18. Lantieri L, Grimbert P, Meningaud JP, Bellivier F. Face transplantation outcomes: Feasibility, reproducibility and efficacy. Paper presented at: 21st Annual Meeting of The European Association of Plastic Surgeons; Manchester, UK. 2010;27-30.
19. Eaton L. Spanish doctors carry out first transplantation of a full face. *BMJ*. 2010;340:c2303.
20. Eaton L. First patient to receive complete face transplant can leave hospital. *BMJ*. 2010;341:c4088.
21. Perez RH, Lopez P, Escoresca-Ortega A.M, et al. Severe rhabdomyolysis after allogeneic transplantation of facial structures: A case report. *Transplant Proc*. 2010;42(8): 3081-3082.
22. Barret J. Composite facial allotransplantation: The Barcelona full face transplant. In *Proceedings of the Advanced Educational Courses in Plastic Surgery Meeting of British Association of Plastic, Reconstructive and Aesthetic Surgeons*, Manchester, UK. 2010;1-2.
23. Hivelin M. Face transplantation outcomes: Report on the Mondor series. In *Proceedings of the Advanced Educational Courses in Plastic Surgery Meeting of British Association of Plastic, Reconstructive and Aesthetic Surgeons*, Manchester, UK. 2010;1-2.
24. Morelon E, Testelin S, Petruzzo P, et al. New partial face allograft transplantation: Report on first six months. Paper presented at: 2010 American Society for Reconstructive Transplantation 2nd Biennial Meeting; Chicago, Ill. 2010; 18-20.
25. Pedro Cavadas. Clinical Update & Update on Hand and Face Transplant Outcomes. In *Proceedings of the 2010 American Society for Reconstructive Transplantation 2nd Biennial Meeting*, Chicago, Ill. 2010;18-20.
26. Hui-Chou HG, Nam AJ, Rodriguez ED. Clinical facial composite tissue allotransplantation: a review of the first four global experiences and future implications (Review). *Plast Reconstr Surg*. 2010;125:538-546.
27. Takamatsu A, Harashina T, Inoue T. Selection of appropriate recipient vessels in difficult, microsurgical head and neck reconstruction. *J Reconstr Microsurg*. 1996;12:499-507; Discussion 508-513.
28. Wilhelmi BJ, Kang RH, Movassaghi K, Ganchi PA, et al. First successful replantation of face and scalp with single artery repair: model for face and scalp transplantation. *Ann Plast Surg*. 2003;50:535-540.
29. Yin JW, Matsuo JM, Hsieh CH, et al. Replantation of total avulsed scalp with microsurgery: experience of eight cases and literature review. *J Trauma*. 2008;64:796-802.
30. Tohma A, Mine K, Tamatsu Y, et al. Communication between the buccal nerve (V) and facial nerve (VII) in the human face. *Ann Anat*. 2004;186:173-178.
31. Baumel JJ. Trigeminal-facial nerve communications. Their function in facial muscle innervation and reinnervation. *Arch Otolaryngol*. 1974;99:34-44.
32. Kimata Y, Uchiyama K, Ebihara S, et al. Comparison of innervated and noninnervated free flaps in oral reconstruction. *Plast Reconstr Surg*. 1999;104:1307-1313.
33. Lahteenmaki T, Waris T, Asko-Seljavaara S, et al. Recovery of sensation in free flaps. *Scand J Plast Reconstr Surg Hand Surg*. 1989;23:217-222.
34. Carroll D. A quantitative test of upper extremity function. *J Chron Dis*. 1965;18:479-491.
35. Petruzzo P, Lanzetta M, Dubernard JM, et al. The international registry on hand and composite tissue transplantation. *Transplantation*. 2008;86:487-492.

Index

M.Z. Siemionow (ed.), *The Know-How of Face Transplantation*,
DOI: 10.1007/978-0-85729-253-7, © Springer-Verlag London Limited 2011